2ND EDITION
FOUNDATIONS
OF ELECTRONICS

RUSSELL L. MEADE

 DELMAR PUBLISHERS INC.™

I T P™

NOTICE TO THE READER

DEDICATION

To my beloved wife, Betty, for her patience, love, support, and encouragement throughout this challenging project.

Cover photo courtesy of Miller Electric Manufacturing Company of Appleton, Wisconsin
Background photo by W. Cody/Westlight
Cover design by Bob Clancy/Spiral Design

Delmar Staff:

Publisher: Michael McDermott	Senior Project Editor: Christopher Chien
Administrative Editor: Wendy Welch	Senior Production Supervisor: Larry Main
Developmental Editor: Mary Clyne	Art and Design Coordinator: Lisa Bower

For information address Delmar Publishers Inc.
3 Columbia Circle, Box 15-015
Albany, New York 12212-5015

Printed in the United States of America
Published simultaneously in Canada
by Nelson Canada,
a division of The Thomson Corporation

10 9 8 7 6 5 4 3 2   XX 00 99 98 97 96 95 94

Library of Congress Cataloging-in-Publication Data

Meade, Russell L.
  Foundations of electronics / Russell L. Meade.—2nd ed.
    p. cm.
  Includes index.
  ISBN 0-8273-5971-3
  1. Electronics. I. Title.
TK7816.M43 1994
621.381—dc20   93-38058
                            CIP

CONTENTS

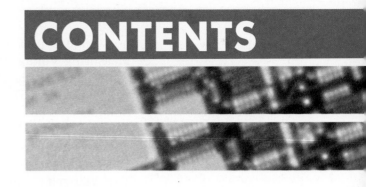

PREFACE   xvii
INTRODUCTION   xxvii

PART I   FOUNDATIONAL CONCEPTS

1   BASIC CONCEPTS OF ELECTRICITY   2

General Information Regarding the Field of Electronics   5
Definition, Physical and Chemical States of Matter   7
Composition of Matter   9
Structure of the Atom   10
Concept of Electron Theory   13
Ions   14
Energies That Change Electrical Balance   15
Conductors, Semiconductors, and Insulators   16
Sample of an Electrical System   17
Basic Principles of Static Electricity   17
Electrical Potential   21
Charges in Motion   23
Three Important Electrical Quantities   25
Basic Electrical Circuit   26
Summary   29
Formulas and Calculator Sequences Summary   31
Review Questions   32
Test Yourself   32

2   ELECTRICAL QUANTITIES AND COMPONENTS   34

Basic Electrical Units and Abbreviations   37
Using the Metric System to Help   38
Conductors and Their Characteristics   41
Resistors   47
Resistor Color Code   50

Measuring Voltage, Current, and Resistance with Meters   54
Diagrams Used for Electronic Shorthand   60
Safety Hints   67
Summary   68
Formulas and Calculator Sequences Summary   69
Review Questions   69
Challenge Questions   70
Test Yourself   71
Performance Projects Correlation Chart   71

3 OHM'S LAW   72

Ohm's Law and Relationships Between Electrical Quantities   74
Rearranging to Find the Unknown Quantity   79
Powers of Ten and Scientific Notation   81
Sample Application of Metric Prefixes and Powers of Ten   84
Some Basic Rules Relating to Powers of Ten   85
Direction of Current Flow   88
Polarity and Voltage   89
Comparison of Circuit Current Directions from DC and
AC Sources   90
Work, Energy, and Power   92
Measuring Electrical Energy Consumption   93
Commonly Used Versions of Ohm's Law and Power Formulas   96
Summary   100
Formulas and Calculator Sequences Summary   101
Review Questions   102
Challenge Questions   103
Test Yourself   103
Performance Projects Correlation Chart   104

PART II BASIC CIRCUIT ANALYSIS

4 SERIES CIRCUITS   106

Definition and Characteristics of a Series Circuit   109
Resistance in Series Circuits   109
Voltage in Series Circuits   112
Kirchhoff's Voltage Law   116
Using a Scientific Calculator   118
Power in Series Circuits   125
Effects of Opens and Troubleshooting Hints   127
Effects of Shorts and Troubleshooting Hints   128
Designing a Series Circuit to Specifications   133
Special Applications   134

Critical Thinking and the SIMPLER Troubleshooting
Sequence   144
Introduction to Troubleshooting Skills   144
The SIMPLER Sequence for Troubleshooting   144
Troubleshooting Levels   149
A Block-level Troubleshooting Example   150
A Component-level Troubleshooting Example   151
Summary   152
Formulas and Calculator Sequences Summary   153
Review Questions   154
Challenge Questions   156
Test Yourself   156
Performance Projects Correlation Chart   157
Chapter Troubleshooting Challenge   159

5 PARALLEL CIRCUITS   162

Definition and Characteristics of a Parallel Circuit   164
Voltage in Parallel Circuits   166
Current in Parallel Circuits   166
Kirchhoff's Current Law   167
Resistance in Parallel Circuits   170
Methods to Calculate Total Resistance (R_T)   170
Power in Parallel Circuits   180
Effects of Opens in Parallel Circuits and Troubleshooting
Hints   182
Effects of Shorts in Parallel Circuits and Troubleshooting
Hints   184
Similarities and Differences Between Series and Parallel
Circuits   186
Designing a Parallel Circuit to Specifications   186
Sources in Parallel   189
Current Dividers   189
Summary   192
Formulas and Calculator Sequences Summary   194
Review Questions   195
Challenge Questions   197
Test Yourself   198
Performance Projects Correlation Chart   198
Chapter Troubleshooting Challenge   199

6 SERIES-PARALLEL CIRCUITS   202

What is a Series-Parallel Circuit?   205
Approaches to Recognize and Analyze Series and Parallel Circuit
Portions   206

Total Resistance in Series-Parallel Circuits   209
Current in Series-Parallel Circuits   213
Voltage in Series-Parallel Circuits   217
Power in Series-Parallel Circuits   221
Effects of Opens in Series-Parallel Circuits and Troubleshooting Hints   223
Effects of Shorts in Series-Parallel Circuits and Troubleshooting Hints   225
Designing a Series-Parallel Circuit to Specifications   228
Loaded Voltage Dividers   228
The Wheatstone Bridge Circuit   238
Summary   243
Formulas and Calculator Sequences Summary   244
Review Questions   244
Challenge Questions   247
Test Yourself   248
Performance Projects Correlation Chart   249
Chapter Troubleshooting Challenge   251

7 BASIC NETWORK THEOREMS   254
Some Important Terms   257
Maximum Power Transfer Theorem   257
Superposition Theorem   263
Thevenin's Theorem   267
Norton's Theorem   273
Converting Norton and Thevenin Equivalent Parameters   277
Summary   279
Formulas and Calculator Sequences Summary   280
Review Questions   280
Challenge Questions   281
Test Yourself   282
Performance Projects Correlation Chart   282

8 NETWORK ANALYSIS TECHNIQUES   284
Assumed Loop or Mesh Current Analysis Approach   286
Nodal Analysis Approach   292
Conversions of Delta and Wye Networks   297
Summary   305
Formulas and Calculator Sequences Summary   306
Review Questions   307
Challenge Questions   307
Test Yourself   308
Performance Projects Correlation Chart   308

PART III PRODUCING & MEASURING ELECTRICAL QUANTITIES

9 CELLS AND BATTERIES   310

Chemical Action in a Cell   312
The Common Dry Cell   314
Operating Characteristics of Dry Cells   316
Rechargeable (Secondary) Cells and Batteries   317
Battery Capacity Ratings   318
Data on Recharging Rechargeable Batteries   320
Smart Chargers   320
Construction and Operational Characteristics of the Lead-acid
Battery   320
Battery Charging Methods   321
Examples of Other Types of Primary and Secondary Cells   323
Summary   326
Review Questions   327
Test Yourself   327
Chapter Troubleshooting Challenge   329

10 MAGNETISM AND ELECTROMAGNETISM   332

Background Information   335
Fundamental Laws, Rules, and Terms to Describe Magnetism   335
Elemental Electromagnetism   340
Some Practical Applications of Magnetism   347
Important Magnetic Units, Terms, Symbols, and Formulas   351
Practical Considerations About Core Materials   355
The *B-H* Curve   356
The Hysteresis Loop   356
Induction and Related Effects   359
Faraday's Law   363
Lenz's Law and Reciprocal Effects of Motors and Generators   364
Other Topics Related to Magnetism   367
Summary   370
Formulas and Calculator Sequences Summary   372
Review Questions   373
Challenge Question   374
Test Yourself   374

11 DC MEASURING INSTRUMENTS   376

General Requirements of Basic Meter Movements   379
The D'Arsonval (Moving Coil) Movement   379
Current Meters   383
Current Meter Shunts   385

Voltmeters   390
Calculating the Multiplier Values   392
Ohmmeters   398
VOMs, DMMs, and Other Related Devices   401
The Digital Multimeter (DMM)   402
Troubleshooting Hints   406
Summary   412
Formulas and Calculator Sequences Summary   413
Review Questions   414
Test Yourself   415
Performance Projects Correlation Chart   415
Chapter Troubleshooting Challenge   417

12 BASIC AC QUANTITIES   420

Background Information   423
Generating an AC Voltage   426
Rate of Change   431
Introductory Information Summary Regarding AC Sine-wave
Signals   432
Some Basic Sine-wave Waveform Descriptors   433
Period and Frequency   434
Phase Relationships   437
Important AC Sine-wave Current and Voltage Values   439
The Purely Resistive AC Circuit   442
Other Periodic Waveforms   444
Summary   449
Formulas and Calculator Sequences Summary   450
Review Questions   451
Challenge Questions   452
Test Yourself   452

13 THE OSCILLOSCOPE   454

Background Information   456
Key Sections of the Scope   458
The CRT   458
Combining Horizontal and Vertical Signals to View a Waveform   465
Measuring Voltage and Current with the Scope   467
Using the Scope for Phase Comparisons   474
Measuring Frequency with the Scope   475
Summary   478
Review Questions   478
Challenge Questions   479
Test Yourself   480
Performance Projects Correlation Chart   480
Chapter Troubleshooting Challenge   481

PART IV REACTIVE COMPONENTS

14 INDUCTANCE 486

Background Information   488
Review of Faraday's and Lenz's Laws   491
Self-inductance   492
Factors That Determine Inductance   493
Inductors in Series and Parallel   495
Energy Stored in the Inductor's Magnetic Field   498
The L/R Time Constant   500
Using Epsilon (ϵ) to Help Calculate Circuit Parameters   504
Summary Comments About Inductors   510
Troubleshooting Hints   511
Summary   512
Formulas and Calculator Sequences Summary   513
Review Questions   515
Challenge Questions   516
Test Yourself   517
Performance Projects Correlation Chart   517

15 INDUCTIVE REACTANCE IN AC 518

V and I Relationships in a Purely Resistive AC Circuit   521
V and I Relationships in a Purely Inductive AC Circuit   521
Concept of Inductive Reactance   523
Relationship of X_L to Inductance Value   524
Relationship of X_L to Frequency of AC   525
Methods to Calculate X_L 527
Inductive Reactances in Series and Parallel   529
Final Comments About Inductors and Inductive Reactance   531
Inductor Q Factor   533
Summary   534
Formulas and Calculator Sequences Summary   535
Review Questions   536
Challenge Questions   537
Test Yourself   538
Performance Projects Correlation Chart   538
Chapter Troubleshooting Challenge   539

16 RL CIRCUITS IN AC 542

Special Notations Regarding Scientific Calculators   545
Review of Simple R and L Circuits   545
Using Vectors to Describe and Determine Magnitude and Direction   547
Introduction to Basic AC Circuit Analysis Techniques   550

Fundamental Analysis of Series *RL* Circuits   554
Fundamental Analysis of Parallel *RL* Circuits   565
Examples of Practical Applications for Inductors and
RL Circuits   570
Summary   572
Formulas and Calculator Sequences Summary   573
Review Questions   574
Challenge Questions   576
Test Yourself   576
Performance Projects Correlation Chart   577
Chapter Troubleshooting Challenge   579

17 BASIC TRANSFORMER CHARACTERISTICS   582

Background Information   584
Coefficient of Coupling   585
Mutual Inductance and Transformer Action   586
Mutual Inductance Between Coils Other Than Transformers   588
Important Transformer Ratios   590
Transformer Losses   598
Characteristics of Selected Transformers   599
Troubleshooting Hints   602
Summary   606
Formulas and Calculator Sequences Summary   608
Review Questions   609
Challenge Questions   610
Test Yourself   611
Performance Projects Correlation Chart   611

18 CAPACITANCE   612

Definition and Description of a Capacitor   614
The Electrostatic Field   615
Charging and Discharging Action   616
The Unit of Capacitance   619
Energy Stored in Capacitor's Electrostatic Field   621
Factors Affecting Capacitance Value   621
Capacitance Formulas   624
Total Capacitance in Series and Parallel   626
Finding Voltage When Three or More Capacitors are in Series   629
The *RC* Time Constant   631
Types of Capacitors   636
Typical Color Codes   641
Typical Problems and Troubleshooting Techniques   641
Summary   644
Formulas and Calculator Sequences Summary   646

Review Questions   647
Challenge Questions   648
Test Yourself   648
Performance Projects Correlation Chart   649
Chapter Troubleshooting Challenge   651

19 CAPACITIVE REACTANCE IN AC   654

V and I Relationships in a Purely Resistive AC Circuit   657
V and I Relationships in a Purely Capacitive AC Circuit   657
Concept of Capacitive Reactance   659
Relationship of X_C to Capacitance Value   659
Relationship of X_C to Frequency of AC   659
Methods to Calculate X_C 660
Capacitive Reactances in Series and Parallel   663
Voltages, Currents, and Capacitive Reactances   667
Final Comments About Capacitors and Capacitive Reactance   669
Summary   669
Formulas and Calculator Sequences Summary   670
Review Questions   670
Challenge Questions   672
Test Yourself   672
Performance Projects Correlation Chart   672
Chapter Troubleshooting Challenge   673

20 RC CIRCUITS IN AC   676

Review of Simple R and C Circuits   679
Series RC Circuit Analysis   679
Phase-shift Application of Series RC Network   684
Parallel RC Circuit Analysis   690
Similarities and Differences Between RC and RL Circuits   695
Linear Waveshaping and Nonsinusoidal Waveforms   696
Other Applications of RC Circuits   697
Troubleshooting Hints and Considerations for RC Circuits   700
Safety Hints   701
Summary   701
Formulas and Calculator Sequences Summary   702
Review Questions   703
Challenge Questions   704
Test Yourself   705
Performance Projects Correlation Chart   705

21 RLC CIRCUIT ANALYSIS   706

Basic RLC Circuit Analysis and Power in AC Circuits   709
The Series RLC Circuit   714

The Parallel *RLC* Circuit   720
Power in AC Circuits   723
Rectangular and Polar Vector Analysis   729
Algebraic Operations   738
Application of Rectangular and Polar Analysis   741
Summary   745
Formulas and Calculator Sequences Summary   746
Review Questions   748
Challenge Questions   749
Test Yourself   750
Chapter Troubleshooting Challenge   751

22 SERIES AND PARALLEL RESONANCE   754

X_L, X_C, and Frequency   756
Series Resonance Characteristics   757
The Resonant Frequency Formula   759
Some Resonance Curves   763
Q and Resonant Rise of Voltage   764
Parallel Resonance Characteristics   767
Parallel Resonance Formulas   768
Effect of a Coupled Load on the Tuned Circuit   769
Q and the Resonant Rise of Impedance   770
Selectivity, Bandwidth, and Bandpass   770
Measurements Related to Resonant Circuits   773
Safety Hints   777
Special Notations About Single Components   778
Filter applications of Nonresonant and Resonant *RLC* Circuits   778
Summary   787
Formulas and Calculator Sequences Summary   789
Review Questions   790
Challenge Questions   791
Test Yourself   792
Performance Projects Correlation Chart   794
Chapter Troubleshooting Challenge   795

APPENDIXES

A COLOR CODES   802

**B ANSWERS TO IN-PROCESS LEARNING CHECKS AND PRACTICE PROBLEMS
ANSWERS TO ODD-NUMBERED REVIEW QUESTIONS, CHALLENGE QUESTIONS, AND TEST YOURSELF   804**

C CHAPTER CHALLENGE TEST RESULTS   837

D COPPER-WIRE TABLE   838

E USEFUL CONVERSION FACTORS 839

F SCHEMATIC SYMBOLS 840

G GLOSSARY 841

INDEX 853

PREFACE

The Book's Aim

The purpose of this book is to present electrical and electronics fundamentals in an interesting and easy-to-understand style. Knowledge of fundamentals is *critical* to success in any facet of the electrical or electronics professions. *Foundations of Electronics* is designed to communicate basic facts, concepts, and principles in a "reader-friendly" manner. Furthermore, the text fosters your ability to apply this knowledge to "real-life" situations. This text stresses the development of logical thinking patterns and practical applications of knowledge, not just the presentation of cold theory.

The Book's Approach

Several important strategies are used in the book's presentation. Some of these include:

- Helping you understand rather than just memorize important formulas.

- Providing calculator sequences for every formula.

- Giving meaningful overviews to provide a context for content details.

- Moving from the known to the unknown.

- Providing immediate application of knowledge gained.

- Offering clear and frequent examples.

- Making liberal use of realistic pictorial and diagrammatic information.

- Using color to improve the pedagogical value of the illustrations.

- Formatting that allows either group or individualized instruction approaches.

- Using frequent practice problems to reinforce concepts.

- Offering strategically located checkpoints.

- Focusing on troubleshooting, safety, and providing practical hints.

- Including useful summaries that distill the essence of what should be learned.

- Providing a unique troubleshooting technique, called the SIMPLER Method that sharpens critical thinking skills as it teaches troubleshooting.

- Supplying Chapter Challenge circuit problems to allow practice of troubleshooting with the SIMPLER technique.

The Book's Arrangement

The text begins with a brief overview of the wonders of today's electronic technology and the great opportunities afforded in this field. Next, crucial terms and elemental concepts are presented as foundations for the topics addressed throughout the remainder of the book. Concepts such as the electron theory, basic electrical units and symbols and how electronic circuitry can be illustrated with diagrams provide the stepping stones to what follows.

Essential circuit fundamentals are then presented so that students can develop the ability to analyze various circuit configurations. In the process of learning these fundamentals, several passive devices are studied; including resistors, inductors, and capacitors. Also, important test instruments and their uses are discussed. The book then applies the fundamental concepts, principles, and theories, first to DC circuits, and then to AC circuits.

It is our hope that as you move from the introductory concepts of electronics through the principles of the circuit analysis to practical circuit applications and on to the introduction of modern electronic devices, you will find the book interesting, informative, and very useful in your training.

Using the Book

The format of *Foundations of Electronics,* 2nd Edition is "user-friendly." Each chapter features key terms, an introductory outline, a preview of the chapter's contents, and learning objectives. The body of the chapter includes frequent step-by-step examples and practice problems, and In-Process Learning Checks to reinforce concepts. Formulas are highlighted throughout the text, and are immediately followed by calculator keystroke sequences. The formulas and sequences are summarized at the end of each chapter for easy reference. Each chapter closes with a detailed summary of important points as well as Review, "Challenge," and "Test Yourself" questions. (Answers to the practice problems, In-Process Learning Checks, and odd-numbered review, challenge, and test yourself questions may be found in Appendix B.)

A system of icons has been incorporated through the text to quickly identify the features described below:

 IN-PROCESS LEARNING CHECKS provide an opportunity for you to check your understanding of the subject each step of the way. Answers are provided in Appendix B.

PRACTICAL NOTES provide special hints, practical techniques, and information to the reader.

CALCULATOR SEQUENCES appear after each formula in the book. Formulas and sequences are summarized at the end of each chapter.

 SAFETY HINTS highlight requirements for safety, both while learning, and when applying electronic principles in "hands-on" situations.

TROUBLESHOOTING HINTS offer experienced-based information about circuit problems—what to look for and how to fix them.

 CHAPTER CHALLENGE troubleshooting problems appear throughout the text. Using the SIMPLER troubleshooting method, you will be able to guide yourself through each Chapter Challenge circuit and think critically about the problem you need to solve. One possible approach to solving each Chapter Challenge is provided in step-by-step illustrations following each challenge. The SIMPLER sequence and accompanying challenges are explained fully in Chapter 4.

The illustrations on pages xxii–xxiii show a sample Chapter Challenge and give instructions on how to complete the challenge. We hope that you and your instructor will have great success with the SIMPLER sequence technique and that these Chapter Challenges will make learning and teaching troubleshooting skills easier and more enjoyable.

Changes to the Second Edition

The development of this new edition began with an extensive review of the first edition. Twenty-two reviewers made detailed analyses of the first edition, and nine additional reviewers read the first draft. Their suggestions for improving content, pedagogy, and art presentation formed the core of our plan for developing the second edition.

CREATING TWO VERSIONS. The most important feedback that we received from our customers suggested that they wanted more thorough coverage from a devices text. To answer this request, we are offering two versions of this text. This version, *Foundations of Electronics, second edition,* provides enough information for a two-semester DC/AC circuits course. We have also created a second version to offer more detailed coverage of devices. *Foundations of Electronics: Circuits and Devices* offers all the DC/AC coverage, plus nine rewritten and expanded chapters on devices.

USING COLOR TO ADD LEARNING VALUE. As you use this book, you will notice that color has been used systematically to clarify the illustrations. For example, each resistor pictorial displays an actual color code, adding a practical, real-world dimension to these illustrations. In addition, voltage sources are always highlighted with yellow, resistance values are highlighted with brown, capacitive elements with blue, and so forth. We intend for this planned system to improve tangibly the value of the illustrations for students.

CALCULATOR SEQUENCES.   Calculator sequences have been added after every formula to demonstrate the keystrokes. They are summarized at the end of each chapter to make study and homework more convenient. The sequences are also used throughout the *Laboratory Projects Manual*.

EXPANDED CONTENT.   We asked our users which topics were most important to them. Based on their responses, we made several changes.

Ohm's Law is now covered in its own chapter to give students a detailed understanding and plenty of practice in this fundamental topic.

The *series and parallel circuits* chapters provide more examples, illustrations, and practice problems to guide students through this difficult area.

Network Analysis Techniques have been added as a new chapter. Chapter 8 covers mesh, loop, and nodal analysis, and conversions of wye and delta networks.

Troubleshooting With the SIMPLER Sequence

To become logical troubleshooters, technicians must develop good critical thinking skills. This book introduces a useful troubleshooting approach called the SIMPLER sequence. This successful troubleshooting method integrates critical thinking all along the way and encourages you to solve problems logically and efficiently. Instructors may use and teach other sequences, and the concepts outlined in the SIMPLER sequence are easily used with a wide variety of approaches.

CHAPTER CHALLENGE.   As you move through this book, you will encounter a series of troubleshooting problems called Chapter Challenge (for an example of this feature, see pages xxii–xxiii). Each Chapter Challenge includes a challenge circuit schematic and starting point symptoms information. Using the SIMPLER sequence, you will step through and solve the simulated troubleshooting problem.

SELECTING TESTS AND FINDING THE RESULTS.   To simulate making a parameter test on a particular portion of the challenge circuit, you may select a test from the test listing provided with the Chapter Challenge. You may then look up the results of the test you selected by looking in Appendix C. Look in the appendix for the identifier number assigned to the test on the Chapter Challenge page. Next to that number, you will see the parameter value or condition that would be present if you made that same test on the actual circuit simulated by the Chapter Challenge. *Please note:* The tolerances of components and of test instruments will create differences in the theoretical values and those actually shown in the answers. The theoretical values would be achieved only if all components were precisely rated and the test instruments had zero percent error.

The Learning Package

Foundations of Electronics has a well-correlated lab manual that can multiply the effectiveness of the learning process. The manual has 89 easy-to-use projects that provide hands-on experience with the concepts and principles you study in the theory text.

The format of the manual ensures that you will not miss the forest for the trees. You will not perform a whole series of cook-book steps, or the whole project, only to find out at the end you did not understand what you should have been learning in the project. Instead, you can learn important concepts each step of the way.

The projects are arranged as stand-alone, single-concept projects; however, they are grouped logically to allow maximum instruction flexibility. Each project can be performed individually or several related projects can be performed during one session.

Again, we wish you success as you have opportunity to use the two books as partners in your training. Students can also purchase *Quality Hand Soldering and Circuit Board Repair,* by H. Ted Smith. This practical handbook takes a quality-first approach to this critical and often overlooked skill. As an additional study aid, flashcards are available to help students practice key concepts from each chapter.

For the instructor, transparency masters, an instructor's guide for the text and lab manual, a testbank, and a computerized testbank are available from Delmar Publishers Inc. An instructor's resource manual has been developed for the second edition, and contains teaching outlines, hints, handouts, worksheets, demonstrations, and other resources. *PC Lab: Experiments in Basic Circuits* is a computer simulation program tailored to accompany this text. Students can use the program to practice experimenting with circuits before they move into the lab.

Scheduling Instructions

Ideally, your program can use all the material provided in the text and the correlated laboratory manual to maximize learning and practice opportunities. The following suggestions are provided as possible scenarios for those whose time is more limited:

- For a one-year (three-quarter or two-semester) program; Chapters 1–22 may generally be covered in their entirety.

- Time may be saved by deleting or minimizing reading assignments in the section of Chapter 3 covering powers of ten if you already have covered this topic elsewhere.

- Certain portions of the Network Theorems and Network Analysis Techniques in Chapters 7 and 8 may also receive light treatment, as appropriate to your course.

- Also, some programs prefer to cover cells and batteries in briefer fashion as simply a lecture, which does not include all the data in Chapter 9.

- Additionally, not all programs require the exhaustive treatment on analog meters provided in Chapter 11.

- If vector algebra and manipulation of complex numbers are not required in your program, the last half of Chapter 21 may be omitted.

1

Study the challenge circuit diagram and the starting point information.

2

Following the instructions, work through the SIMPLER sequence to find the problem with the circuit.

CHAPTER TROUBLESHOOTING CHALLENGE

Following the SIMPLER troubleshooting sequence, find the problem with this circuit. As you follow the sequence, record your responses to each step on a separate sheet of paper.

SYMPTOMS Gather, verify, and analyze symptom information. (Look at the "Starting Point Information.")

IDENTIFY Identify initial suspect area for the location of the trouble.

MAKE Make a decision about "What type of test to make" and "Where to make it." (To simulate making a decision about each test you want to make, select the desired test from the "TEST" column listing below.)

PERFORM Perform the test. (Look up the result of the test in Appendix C. Use the number in the "RESULTS" column to find the test result in the Appendix.)

LOCATE Locate and define a new "narrower" area in which to continue troubleshooting.

EXAMINE Examine available information and again determine "what type of test" and "where."

REPEAT Repeat the preceding analysis and testing steps until the trouble is found. What would you do to restore this circuit to normal operation? When you have solved the problem, compare your results with those shown on the following pages.

CHALLENGE CIRCUIT 4

STARTING POINT INFORMATION

1. Circuit Diagram
2. Cells are 1.5 V cells
3. With S_1 closed, $V_R = 5$ V

TEST	Results in Appendix C

(With S_1 Open)

V_{A-B}	.(18)
V_{B-C}	.(1)
V_{C-D}	.(51)
V_{D-E}	.(99)
V_{A-F}	.(77)

(With S_1 Closed)

V_{A-B}	.(7)
V_{B-C}	.(50)
V_{A-C}	.(81)
V_{C-D}	.(112)
V_{D-E}	.(73)
V_{C-E}	.(10)
V_{A-F}	.(105)

3

As you work through the SIMPLER sequence, choose tests (one at a time) from the list that you think might lead you to the problem solution.

4

Look at the number across from the test you chose in the Results column.

Chapter Challenge Test Results

Find the number listed next to the test you chose and record the result.

1. 1.4 V
2. signal normal
3. 2 kΩ
4. 2 V
5. 1.5 V
6. high
7. 1.5 V
8. 0 V
9. 0 V
10. 3 V
11. 0 V
12. 7 V
13. ∞ Ω
14. 0 Ω
15. ≈0 V
16. 1 kΩ
17. 400 Ω
18. 1.5 V
19. ∞ Ω
20. 1.5 V
21.
22. 10 V
23. no signal
24. circuit operates normally
25. 10 kΩ
26. slightly high
27. 0 V
28. 0 V
29. 0 Ω
30. greatly below max value

31. 5 V
32. 400 Ω
33. ∞ Ω
34. 6.4 V
35. noticeably high
36. 4 kΩ
37. 10 kΩ
38. 3.3 V
39. 14 V
40. 0 V
41. ≈1.5 mH
42. 3.0 V
43. 7 V
44. 1.5 V
45. high
46. 0 Ω
47. 0 V
48. 2.3 V
49. 14 V
50. 0.5 V
51. 1.5 V
52. 10 V
53. (14 VAC)
54. 10 kΩ
55. max value
56. 10 V
57. signal normal
58. ∞ Ω
59. 1.5 kΩ
60. 275 pF

61. 14 V
62. 1 kΩ
63. ∞ Ω
64. 1.5 V
65. ≈6 VDC
66. 0 V
67. 7.5 V
68. 10 kΩ
69. low
70. signal normal
71. no change in operation
72. 0 Ω
73. 1.5 V
74. ᘛ(120 VAC)
75. 10 V
76. slightly low
77. 5.9 V
78. 3.1 V
79. 7.5 V
80. 7 V
81. 2 V
82. 0 V
83. 10 V
84. 0 Ω
85. slightly below max value
86. 1.5 V
87. 5 Ω
88. 14 V
89. 2.3 V

90. ∞ Ω
91. 0 V
92. normal
93. 10 kΩ
94. ≈5 V
95. 10 V
96. ≈11 V
97. 0 V
98. low
99. 1.5 V
100. ∞ Ω
101. 0 V
102. 0 V
103. 0 V
104. ≈5 kΩ
105. 0 V
106. no signal
107. 50 pF
108. ∞ Ω
109. 12 kΩ
110. 0 V
111. no change in operation
112. 1.5 V
113. signal normal
114. 297 Ω
115. 0 V
116. ≈5 V
117. within normal range

Turn to Appendix C and find that number from the results column. Next to the number in the Appendix you will find the result this test would yield. As you locate the number of the test you chose, record the result listed next to it on a sheet of paper.

When you have completed the challenge and pinpointed the circuits problem, check your work against the data in the answer pages. This answer offers one possible sequence of steps that might have been used to get to the solution, along with step-by-step color illustrations.

- Further customizing of the program to your program's needs may be done by appropriate selection of the specific laboratory experiments provided in the *Foundations of Electronics* laboratory manual.

- For programs that must cover the dc and ac fundamentals in a two-quarter timespan, use the "trimming" ideas listed.

- Additionally, limit assignments and discussions in Magnetism and Electromagnetism (Chapter 10) to the "fundamental principles" portions, deleting the discussion of the many SI magnetic units and presentation of "motor" and "generator" effects.

ACKNOWLEDGMENTS

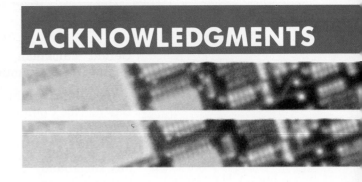

Many individuals have helped shape this textbook package. We wish to thank the following text reviewers for their time and valuable insights.

Richard Ackerman
ITT, Arlington

Robert Arndt
Belleville Area College

Patricia Aylesbury
Specs Howard School of Broadcast

Jack Babb
Solano Community College

Joseph Barron
Pikes Peak Community College

Edward Bowling
Johnston Community College

David Bryan
Alabama Aviation and Technical
College

John Carpenter
Sandhills Community College

John Cross
Sacramento, California

John Ephraim
East Tennessee State College

Joe Etminan
Rock Valley College

Arnie Garcia
Texas State Technical College

Thomas Giasomo
Bryant-Stratton

Carol Grant
Indiana Vocational Technical
College

John Hamilton
Spartan School of Aeronautics

John Hart
DeVry, Chicago

Mark Hughes
Orangeburg-Calhoun Community
College

Carl Jensen
DeVry, Woodbridge

Ted Johnson
Berkshire Community College

Dan Lookadoo
New River Community College

Robert Martin
Northern Virginia Community
College

Harbans Mathur
Kent State University

David McDonald
Lake Superior Technical University

Larry Parker
ITT, Indiana

Hal Sappington
Central Missouri State University

Gerald Schickman
Miami Dade Community College

Marshall Seawright
Bay de Noc Community College

Harlow Thielke
Wadena Technical College

Richard Williams
Duluth Technical College

The following instructors reviewed the laboratory projects manual:

Ray Engelland
Willmar Technical College

Marc Kalis
Winona Technical College

Raymond Litts
Piedmont Technical College

Ed Steinblock
St. Cloud Technical College

We also thank two students, Keith Enright and Tom Rhoades of NEI College of Technology, for their valuable comments on the lab manual.

INTRODUCTION

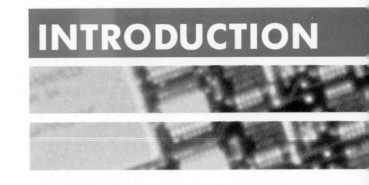

The Wonders of Electronics

Today, because of electronics, we have capabilities that could not even be dreamed about just a few years ago! In the field of *communications,* Samuel F. B. Morse invented the telegraph in the 1830s, Alexander Graham Bell invented the telephone in the 1870s, Heinrich Rudolf Hertz succeeded in transmitting a radio signal a short distance in the late 1880s, and Guglielmo Marconi invented the first practical "wireless" telegraph in the 1890s. As remarkable as their achievements were, none of these inventors could have imagined today's satellite communications systems, high-definition television, computers "talking" to each other via data transmission systems, or even virtual reality on a computer.

In *manufacuring,* who could have foreseen that engineering drawings would be drawn by moving a "mouse" or punching keys on a keyboard of a computer-aided drafting and design (CADD) system. This same CADD system may then be interfaced with a computer-aided manufacturing (CAM) system that automatically takes the computer-developed drawings and controls the systems that create the actual parts drawn. Automation in manufacturing involves everything from robots that make welds in automobiles to machines that automatically drill, punch, mill, shape, bend, and form metals, plastics and other materials into useful products. Computer-controlled "processing" plants develop and produce integrated circuits and semiconductor devices, chemicals, paints, and a myriad of other products.

Today's *service businesses,* such as banks, real estate, insurance, fast food, department stores, and hundreds of others could not work without electronic transfer of funds, computerized property listings, computerized billing of premiums and point of sale (POS). Computerized cash registers can keep a "running inventory," keep track of pricing of every item, print receipts, and even compute the change. Governmental agencies (such as the IRS) could not perform their daily duties without help from electronic systems.

Virtually all that we do in our everyday living is affected by electronics. Home appliances used to cook, wash, clean, and provide entertainment are electronic marvels.

Even our automobiles utilize great amounts of electronic wizardry. Imagine getting in your car, punching in your "user" number on a digital panel, and having the automobile's electronic systems then automatically set all mirrors, seat positions, and the car's suspension and steering systems to suit the user. Unrealistic? No, not really! Many of these types of amenities are found already in our newer automobiles.

What tomorrow may bring can hardly be imagined. At any rate we know electronics is here to stay. You have chosen a good profession.

The Importance of Learning Fundamentals

Becoming a competent technician in the field of electronics demands that you get a thorough grasp of fundamentals. In any field of endeavor, whether it be nursing, law, or a craft, there are fundamental terms, symbols, concepts and principles that must be learned. In electricity and electronics, it is vital that these basic principles be learned for they are the foundation upon which you build an understanding of all that follows in this profession.

Set your mind to work diligently at learning new terms, concepts, and the important foundational truths! You will reap large benefits later as you get further into this study.

Mastering the terms, symbols, formulas, and basic principles taught in this book will launch you into this stimulating field and will allow you to pursue it as far as your desire will carry you.

We wish you the best of success as you engage in the learning process . . . and beyond!

PART

FOUNDATIONAL
CONCEPTS

Basic Concepts
of Electricity

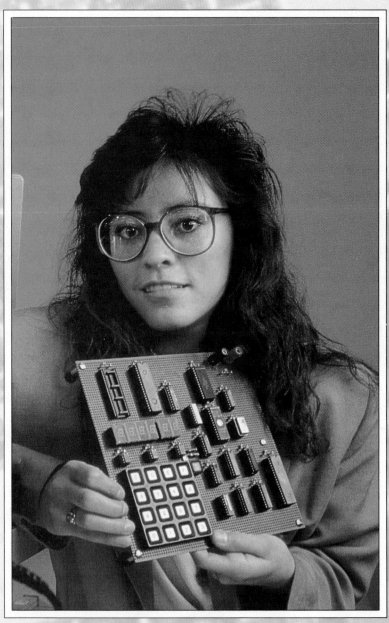

Courtesy of DeVry Institutes

1

KEY TERMS

Ampere
Atom
Circuit
Compound
Conductor
Coulomb
Current
Electrical charge
Electron
Element
Energy
Force
Free electrons
Insulator
Ion
Load
Matter
Mixture
Molecule
Neutron
Ohm
Polarity
Potential
Proton
Resistance
Semiconductor
Source
Valence electrons
Volt

OUTLINE

General information regarding the field of electronics

Definition, physical and chemical states of matter

Composition of matter

Structure of the atom

Concept of the electron theory

Ions

Energies that change electrical balance

Conductors, semiconductors, and insulators

Sample of an electrical system

Basic principles of static electricity

Electrical potential

Charges in motion

Three important electrical quantities

Basic electrical circuit

CHAPTER PREVIEW

This chapter will introduce you to concepts and theories proven useful as "thinking tools" when studying electricity and electronics. It also provides the foundation for understanding you will need as a technician when analyzing, installing, designing, and troubleshooting electronic circuitry.

OBJECTIVES

After studying this chapter, you should be able to:

• Describe the types of tasks performed by electronics technicians, technologists, and engineers

• Define the term **matter** and list its physical and chemical states

• Describe the difference between **elements** and **compounds**

• Discuss the characteristics and structure of an **atom, molecule,** and **ion**

• Define the electrical characteristics of an **electron, proton,** and **neutron**

• Explain the terms **valence electrons** and **free electrons**

• List the methods used to create electrical imbalances

• Describe the characteristics of **conductors, semiconductors,** and **insulators**

• State the law of electrical charges

• Discuss the terms **polarity** and **reference points**

• Define charge and its unit of measure the **coulomb**

• Define **potential** (emf) and give its unit of measure

• Define **current** and explain its unit of measure

• Calculate current when magnitude and rate of charge motion is known

• Define **resistance** and give its unit of measure

• List the typical elements of an electrical circuit

• Describe the difference between closed and open circuits

GENERAL INFORMATION REGARDING THE FIELD OF ELECTRONICS

You are about to start your preparation in a field of work that is both interesting and almost limitless in opportunity. There are three broad categories of work in this field. They are:

1. *Technicians* install, adjust, troubleshoot, repair and maintain a wide variety of electronic equipment and systems. See Figure 1–1.

2. *Technologists* assist in the design, development, and testing of electronic equipment and systems. See Figure 1–2.

3. *Engineers* design electronic equipment and systems. Engineers may also be involved as "customer engineers." These engineers serve in a technical service role for industry, providing expertise in making correct choices of equipment or systems for specific applications. See Figure 1–3.

Within the field of electronics, numerous specializations are possible. Computers, communications, consumer electronics, medical electronics, and industrial electronics are just a few. As you can see, you have chosen a great field of study!

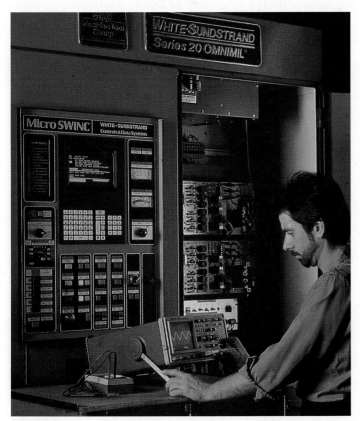

FIGURE 1–1   Electronics technician making tests on equipment (Courtesy of Cleveland Institute of Electronics, Inc.)

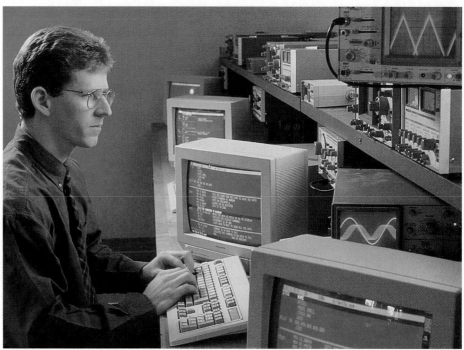

FIGURE 1–2   Electronics technologist performing computer calculations
related to equipment design (Courtesy of DeVry Institutes)

FIGURE 1–3   Electronics engineer working with industrial customer to
determine customer's system needs. (Courtesy of Cleveland Institute of
Electronics, Inc.)

FIGURE 1–4   A scene full of matter (Photo by Bruce Parker/Parker Productions)

DEFINITION, PHYSICAL AND CHEMICAL STATES OF MATTER

Definition of Matter

Everything that we see, touch, or smell represents some form of matter. In Figure 1–4 the workbench, the test equipment, and the air in the room represent **matter** in various forms. Although many definitions have been used, matter may be defined as anything that has weight and occupies space. We can also say that matter is what all things are made of and what our senses can perceive.

   The basic building block of all matter is the atom. Later in this chapter, atoms will be thoroughly discussed because of their importance to the understanding of electronics.

Physical States of Matter

Matter exists in three *physical* states. Matter is either a **solid,** such as the chair you are sitting on, a **liquid,** such as water, or a **gas,** such as oxygen.

   When matter exists in the liquid or gaseous state, its dimensions are determined by the container.

Chemical States of Matter

The chemical states of matter are **elements, compounds,** and **mixtures.** An element, Figure 1–5, is a substance that cannot be chemically broken into simpler substances. In fact, an element has only one kind of atom. Examples of chemical elements are gold, iron, copper, silicon, oxygen, and hydrogen.

FIGURE 1–5   Some familiar metallic elements (Photo by Michael A. Gallitelli)

**FIGURE 1–6   Some commonly used compounds (Photo by
Michael A. Gallitelli)**

A compound is formed by a chemical combination of two or more elements. In other words, compounds are built from two or more atoms that
combine into molecules. Also, a compound has a definite structure and characteristic (i.e., same weight and atomic structure). Figure 1–6 shows two
examples of compounds. Examples of compounds are water, which is the
chemical union of hydrogen and oxygen, and sugar, which contains carbon
(a black, tasteless solid) plus hydrogen and oxygen (two gases).

Mixtures are a combination of substances where the individual elements possess the same properties as when they are alone. There is no
chemical change resulting from the combination as there is with compounds. In Figure 1–7, mixing gold dust and sand would not yield a new,

**FIGURE 1–7   Mixtures are
combinations of substances in which the
individual elements possess the same
properties as when they are alone. A
mixture is different from a compound.**

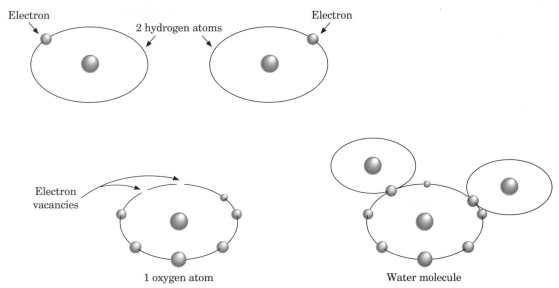

FIGURE 1–8   A molecule of water is formed from oxygen and hydrogen atoms.

chemically different entity or compound. The substances would simply be mixed or combined.

COMPOSITION OF MATTER

The building blocks of all matter are atoms. Atoms combine chemically to form molecules, and the new compound (matter) created is different than each element that went into the compound's makeup, Figure 1–8. Also, a specific substance's molecules are different than the molecules of other types of matter.

In summary, *all matter is composed of atoms and molecules.* The smallest particle into which a compound can be divided but retain its physical properties is the molecule. The smallest particle into which an element can be divided but retain its physical properties is the atom, Figure 1–9.

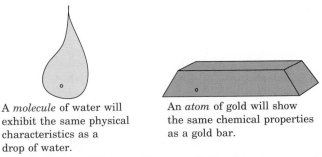

A *molecule* of water will exhibit the same physical characteristics as a drop of water.

An *atom* of gold will show the same chemical properties as a gold bar.

FIGURE 1–9   The smallest particle of a compound that has the same characteristics as the compound is a molecule. The smallest part of an element that will show the same characteristics as the element is an atom.

STRUCTURE OF THE ATOM

The Particles

The particles of the atom that interest electronics students are the **electron, proton,** and **neutron.** Although science has identified other particles (mesons, positrons, neutrinos), studying these particles is not necessary to understand the electron theory.

The Model

A Danish scientist, Niels Bohr, developed a model of atomic structure that explains the electron theory. In his model the atom consists of a nucleus in the center of the atom with electrons orbiting the nucleus. The nucleus has two particles: protons, which are electrically positive in charge, and neutrons, which are neutral in electrical charge. The orbiting electrons are negatively charged particles, Figure 1–10. The common analogy for this model is our solar system: Planets orbit the sun like electrons orbit the nucleus of the atom.

The net charge of the atom is neutral because the orbiting electrons' total negative charge strength equals the total positive charge strength of the protons in the nucleus. Also, the number of protons and electrons in an electrically balanced atom are equal, Figure 1–11.

Some interesting characteristics, Figure 1–12, of these three atomic particles are:

1. *The electron:*

   • has a negative electrical charge

   • has small mass or weight (9×10^{-28} grams)

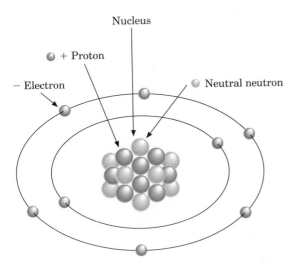

FIGURE 1–10   Bohr's model of the atom. The nucleus (center) contains protons and neutrons. Electrons orbit the nucleus.

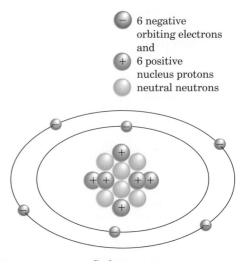

Carbon atom

FIGURE 1–11   Typically, atoms are electrically balanced.

ELECTRONS	PROTONS	NEUTRONS
Negative charge	Postive charge	No charge
Small mass/weight (about 9×10^{-28} grams)	1,836 times heavier than electron	Similar in mass/ weight to proton
Travel in orbits outside of nucleus	Located in the nucleus	Located in the nucleus
Rates of speed (trillions of orbits per sec.)	Equal in number to atom's electrons	

FIGURE 1–12   Some facts about atomic particles

- travels around the nucleus at unbelievable speed (trillions of times a second)

- helps determine the atom's chemical characteristics

2. *The proton:*

- has a positive electrical charge

- is located in the nucleus of the atom

- is approximately 1,800 times heavier than an electron

- is equal in number to the atom's electrons

3. *The neutron:*

- has no electrical charge

- is located in the nucleus of the atom

- is about the same mass or weight as a proton

- may vary in number for a given element to form different "isotopes" of the same element; for example, hydrogen has three isotopes: protium, deuterium, and tritium, Figure 1–13.

Atomic Number and Weight

Although it is not necessary to thoroughly understand atomic numbers and atomic weights, it is appropriate to mention them.

Deuterium
(heavy hydrogen)
nucleus =
1 proton
1 neutron

Tritium
(heavy, heavy hydrogen)
nucleus =
1 proton
2 neutrons

Protium
(light hydrogen)
nucleus =
1 proton

FIGURE 1–13   What atomic difference is there between these isotopes?

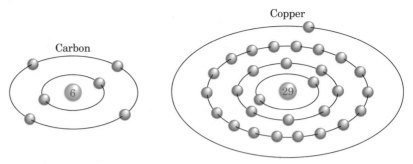

FIGURE 1–14   Carbon and copper atoms

The atomic number of an element is determined by the number of protons in each of its atoms. For example, the atomic number for copper is 29 and the atomic number for carbon is 6, Figure 1–14.

The atomic weight of an element is determined by comparing the weight of its atoms to an atom of carbon-12. For example, hydrogen, the simplest atom, has an atomic weight of about 1.007 "atomic mass units," and copper's atomic weight is 63.54.

✓ IN-PROCESS LEARNING CHECK I

It will be a good idea at this point to initiate you into a technique that will be used from time to time throughout the book: the "In-Process Learning Check." Rather than wait until the end of the chapter to find out whether you're missing any of the key points, these special "checkups" will provide you with a chance to assure that you're learning the key points. Fill in the blanks for the following statements. If you have trouble with any of them, simply move back to that topic and refresh your memory. Answers are in Appendix B.

1. Matter is anything that has _____ and occupies _____.

2. Three physical states of matter are _____, _____, and _____.

3. Three chemical states of matter are _____, _____, and _____.

4. The smallest particle that a compound can be divided into but retain its physical properties is the _____.

5. The smallest particle that an element can be divided into but retain its physical properties is the _____.

6. The three parts of an atom that interest electronics students are the _____, _____, and _____.

7. The atomic particle having a negative charge is the _____.

8. The atomic particle having a positive charge is the _____.

9. The atomic particle having a neutral charge is the _____.

10. The _____ and _____ are found in the atom's nucleus.

11. The particle that orbits the nucleus of the atom is the _____.

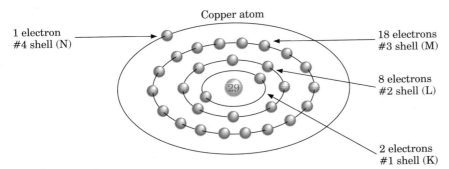

FIGURE 1–15   Electrons align at different distances from the nucleus.

Concept of Atomic Shells

As our illustrations indicate, all electrons traveling around the nucleus of the atom do not travel in the same path or at the same distance from the nucleus. Electrons align themselves in a structured manner, Figure 1–15.

Each ring, or shell, of orbiting electrons has a maximum number of electrons that can locate themselves within that shell if the atom is stable. The formula to use is $2n^2$, where n equals the number of the shell. For example, the innermost shell (closest to the nucleus) may have a maximum of 2 electrons; the second shell, 8 electrons; the third shell, 18 electrons; and the fourth shell, a maximum of 32 electrons. The outermost shell, whichever number shell that is for the given atom, can never contain more than 8 electrons.

CONCEPT OF ELECTRON THEORY

Purpose

The electron theory helps visualize atoms and electrons as they relate to electrical and electronic phenomena, and the previous descriptions of atomic structure are part of this electron theory. The next discussion of valence and free electrons illustrates some practical aspects of the electron theory.

Valence Electrons

Valence electrons are those electrons in the outermost shell of the atom. The number of valence electrons in the atom determines its stability or instability, both electrically and chemically, Figure 1–16.

For all atoms, this outermost shell is *full* when it has *8 electrons*. If there are 8 outermost ring electrons, then the material is stable and does not easily combine chemically with other atoms to form molecules. Also, these electrons are not easy to move from the atom. Examples of stable atoms are inert gases, such as neon and argon.

If the material's atoms have fewer than 8 outermost ring electrons, the material is chemically and electrically active. Often the material will chemically combine with other atoms to gain stability and to form molecules and/or atomic bonds. Electrically, these valence electrons can be easily

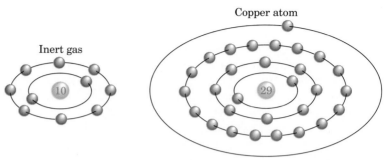

FIGURE 1–16   The number of valence electrons affects the chemical and/or electrical stability of an atom.

moved from their original home atom and are sometimes referred to as free electrons. Examples of such materials are copper, gold, and silver.

Also, there are materials that have 4 outermost ring electrons. They are halfway between being stable and very unstable. Germanium and silicon are two examples. These materials are used in most of today's solid-state, "semiconductor" devices, such as transistors and integrated circuits.

In summary, the concepts of the electron theory are based on atomic structure and help explain the various electrical phenomena discussed in this book.

IONS

When an electron leaves its original home atom because of chemical, light, heat, or other types of energy, it leaves behind an atom that is no longer electrically neutral. An **ion** is any atom that is not electrically balanced (or neutral) and that has gained or lost electrons. A positive ion is an atom having fewer electrons than protons (a deficiency of electrons). A negative ion is an atom having more electrons than protons (an excess of electrons), Figure 1–17.

When an electron is torn from a neutral atom, leaving a positive ion, or when an electron is added to a neutral atom, producing a negative ion, the process is called ionization. Later in your study of electronics, you will see how this process can be useful in various electronic devices.

11 protons in nucleus with only 10 orbiting electrons = "+ ion"

11 protons in nucleus with 12 orbiting electrons = "− ion"

FIGURE 1–17   Ions are atoms that have lost or gained electrons.

ENERGIES THAT CHANGE ELECTRICAL BALANCE

Why should anyone want to change the electrical balance of atoms, or control electron movement? As your study of electronics continues, you will see that the ability to control the movement of electrons, or electron flow, is the basis of electronics.

Figure 1–18 displays some common sources of energy causing electron movement and/or separation of charges. These sources are:

- Friction (static electricity)
- Chemical energy (batteries)
- Mechanical energy (a generator or alternator)
- Magnetic energy
- Light energy
- Heat energy

(a)

(b)

(c)

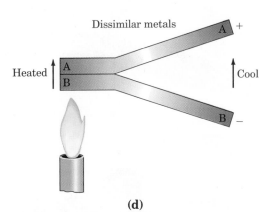

(d)

FIGURE 1–18 Some energy sources used to create useful energy: (a) A battery uses the chemical energy in its cell to create voltage. (b) A generator converts mechanical energy into electricity. (Courtesy of AC-Delco) (c) The sun is a source of light energy. (Photo courtesy of Brian Yacur.) (d) A thermocouple uses heat energy to produce an electric current.

By appropriately using one or more of these energies, sources of electrical energy may be developed. These sources develop electrical potential, or the potential to move electrons through whatever circuits are electrically connected to these sources. Electrical **circuits** are closed paths designed to carry, manipulate, or control electron flow for some purpose. Later in this chapter, electrical potential and movement of electrons are discussed to show how the concepts presented thus far are foundations for the study of electronics.

CONDUCTORS, SEMICONDUCTORS, AND INSULATORS

Conductors, such as gold, silver, and copper, have many free electrons. These materials conduct electron movement easily because their outermost ring electrons are loosely bound to the nucleus. In other words, their outermost ring contains 1, 2, or 3 electrons rather than the 8 electrons needed for atomic stability, Figure 1–19a.

Semiconductors, Figure 1–19b, are materials that are halfway between the conductors' characteristic of few outermost shell electrons and the stable, inert materials with 8 valence electrons. Semiconductor materials have 4 outermost ring electrons. Germanium and silicon are examples of semiconductor materials.

Insulator materials, Figure 1–19c, do not easily allow electron movement because their 5 to 8 outermost shell electrons are tightly bound to the atom. Insulators have few free electrons. Examples of insulator materials are glass and ceramic.

Electrical and electronic circuits are made up of a variety of components and interconnections consisting of conductors, insulators, and semiconductors.

(a)   (b)   (c)

FIGURE 1–19   (a) Conductors (Photo by Michael A. Gallitelli); (b) Semiconductors (Photos by Michael A. Gallitelli); and (c) Insulators (Courtesy of Cooper Industries/Bussmann Division)

FIGURE 1–20   Parts of a basic electrical system

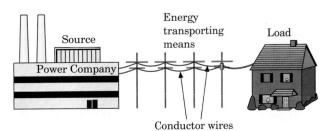

FIGURE 1–21   Components of a commonly used electrical system

SAMPLE OF AN ELECTRICAL SYSTEM

An electrical system typically has a source of electrical energy, a way to transport that electrical energy from one point to another, and an electrical load, Figure 1–20.

The source supplies energy through the transporting means to the load. The load may convert that electrical energy into another form of electrical energy or into another type of energy, such as heat, light, or motion.

Figure 1–21 illustrates an electrical lighting circuit. Observe that the method to transport the electrical energy to the light bulb (the load) is the conductor wires. The load then converts the electrical energy into another useful form, light.

The rest of this chapter will discuss basic electrical laws and principles to help you in understanding what is happening in this basic light circuit. You should seek answers to the following questions: How does the source cause electrical energy or electricity to move through the conductors to the load? In what manner and/or form does the electricity move from one point to another? What determines the quantity of electricity moved from the source to the load?

BASIC PRINCIPLES OF STATIC ELECTRICITY

Before studying the movement of electrons in electrical circuits, it will be helpful to learn some basic principles about charges and static electricity as a frame of reference.

What Is Static Electricity?

Everyone has been electrically shocked after walking on a thick pile rug and touching a door knob or after sliding across an automobile seat and touching the door handle. Where did that electricity come from? Sometimes this electricity is called static electricity. This term is somewhat misleading because static implies stationary and electrons are in constant motion around the nucleus of the atoms. Think of this type of electricity as being associated with nonconductors or insulator materials.

The Basic Law of Electrical Charges

In many science courses, the experiment of rubbing a rubber rod with fur or a glass rod with silk to develop charges is frequently performed, Figure

FIGURE 1–22   Static electricity is usually associated with nonconductive materials.

1–22. After doing this experiment, paper or other light materials are sometimes attracted to the charged body or object. What is meant by a charged body? This means that the object has more or less than its normal number of electrons. In the case of the rubber rod, it gained some electrons from the fur, and the fur lost those electrons to the rod. Thus, the net charge of the rubber rod is now negative, since it has an excess of electrons. Conversely, the fur is positively charged since it lost some of its electrons but retains the same number of protons in its nuclei.

With the silk and glass rod experiment, the glass rod loses electrons and becomes positively charged. Furthermore, another experiment would show there is some attraction between the charged rubber rod (negative) and the charged glass rod (positive). This leads to a conclusion that has been accepted as the basic electrical law: Unlike charges attract each other, and like charges repel each other. Remember this important electrical law, Figure 1–23.

Polarity and Reference Points

You might have noticed that a common way of showing the difference between charges is by identifying them with a minus sign for negative or a plus sign for positive. Using positive (+) or negative (−) signs frequently indicates electrical **polarity.** Polarity denotes the relative electrical charge of one point in an electrical circuit "with reference to" another point, Figure 1–24. The notion of something "with reference to" something else is not really mysterious. Everyone has heard statements such as "John is taller

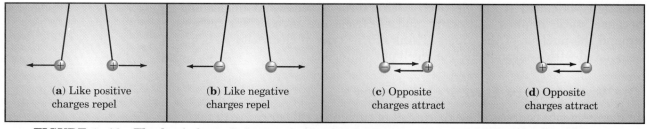

FIGURE 1–23   The basic law of charges is that like charges repel and unlike charges attract.

**FIGURE 1–24   In a flashlight cell, point A
is positive with respect to point B, and
point B is negative with respect to point A.**

than Bill," or "Bill is shorter than John." In the first statement, Bill is the
reference point. In the second statement, the reference point is John.

Another familiar example of polarity and reference points is that the
upper tip of the earth's axis is called the North Pole and the lower tip, the
South Pole. These terms refer to their geographic locations.

Review some key facts:

1. Electrons are negatively charged particles and protons are positively
charged particles.

2. Any substance or body that has excess electrons is negatively charged.

3. Any substance or body that has a deficiency of electrons is positively
charged.

4. Unlike charges attract and like charges repel.

Refer to Figures 1–25a, b, and c and answer the following questions:

1. In Figure 1–25a, which ball has a deficiency of electrons?

2. In Figure 1–25b, will the balls attract or repel each other?

3. In Figure 1–25c, if ball B has an equal number of electrons and protons, what is the polarity of ball A with respect to B?

(a)　　　　**(b)**　　　　**(c)**

FIGURE 1–25   Some arrangements of electrical charges

The answers are 1) **B;** 2) **repel;** and 3) **negative.** In number one, the positive polarity indication on ball B indicates a deficiency of electrons. In number two, both balls have the same polarity charge, so they repel each other. In number three, if ball B has an equal number of electrons and protons, it is electrically neutral. However, ball A remains negative with respect to ball B, since ball A is not electrically neutral. In this case, ball A is negatively charged with respect to anything that does not have an equal or greater negative charge.

Coulomb's Formula Relating to Electrical Charges

Examining the forces of attraction or repulsion (repelling force) between charged bodies, Charles Coulomb, a French physicist, determined that the amount of attraction or repulsion between two charged bodies depended on the amount of charge of each body and the distance between the charged bodies. He expressed this relationship with the formula:

$$\text{Force} = \frac{\text{Charge on body \#1} \times \text{Charge on body \#2}}{\text{Distance between them squared}}$$

| Formula 1–1 | $F = k\dfrac{Q_1 \times Q_2}{d^2}$ |

where F = force (newtons)
k = the constant 9×10^9 (for air or vacuum)
Q = charge in units of charge (coulombs)
d = distance between charged bodies (meters)

k value, $\boxed{\times}$, Q_1 value, $\boxed{\times}$, Q_2 value, $\boxed{=}$, $\boxed{\div}$, distance value, $\boxed{x^2}$, $\boxed{=}$

Coulomb's law of charges states that the force of attraction or repulsion between two charged bodies is directly related to the product of their charges and inversely related to the square of the distance between them.

Unit of Charge

In honor of Charles Coulomb, the unit of charge is designated the **"coulomb."** Many electrical units that define quantity or magnitude of electrical parameters are named after famous scientists who perform related experiments.

Electrical units of measure are powerful reference values used every day in electronics. Even as someone had to determine liquid quantities as a gallon, quart, pint, or cup, scientists define units of measure for electrical parameters so that each has its own "reference point" with respect to quantity or value.

In the case of electrical charge, the unit is the coulomb. This is the amount of electrical charge represented by 6.25×10^{18} electrons, or 6,250,000,000,000,000,000 electrons. Now you know the unit for charge used for the "Qs" in Formula 1–1.

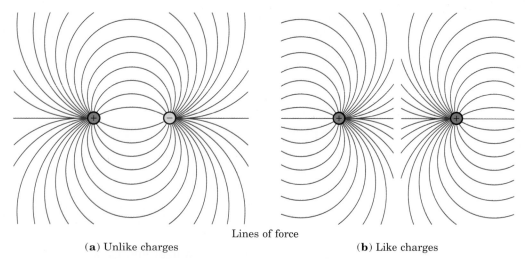

Lines of force

(**a**) Unlike charges   (**b**) Like charges

FIGURE 1–26   Electrostatic fields represented by lines of force

Fields of Force

You are familiar with several types of "fields," for example, gravitational fields and magnetic fields. Fields of force are represented by imaginary lines, which represent the field of influence for the force involved. Figure 1–26 shows electrostatic fields between unlike charges, Figure 1–26a, and between like charges, Figure 1–26b. This only shows a visualization of the nature of these fields, as represented by the lines. It does not attempt to indicate the strength of the charges, the strength of the field of force, or the distance between the charges. This representation helps us visualize the tangible forces between electrical charges.

ELECTRICAL POTENTIAL

What Is Electrical Potential?

The implication drawn from these tangible forces is that they can be harnessed to do work. In this case, the difference of charge levels at two points has the potential to move electrons from a point of excess electrons to a point of electron deficiency, if a suitable path is provided. (Remember, the protons are in the nucleus and are not free to move). This difference between two points having different charge levels is often termed a **potential** *difference*. This force, which moves electrons from one place to another, is called an **electromotive force** (emf). Remember these important terms because you will use them often.

The Unit of Electrical Potential or Electromotive Force

Scientists have established a measure for the unit of charge (the coulomb), and the unit of measure for this difference of potential. They have named this unit of potential difference the **volt** in honor of Alessandro Volta who invented the electric battery and the electric capacitor. Often, this potential difference between two points, which is measured in volts, is called voltage. An in-depth discussion will come later. For now mentally picture that charges and/or different points are positive or negative relative to each other (that is, have polarity); that there are different quantities of charge depending on the number of excess electrons or electron deficiency; and that there is a potential difference between such points, which provides an electromotive force capable of causing electron movement from one point to another point.

What Means Can Produce Electrical Potential?

Since electrical potential difference (electromotive force), or voltage, performs the electrical work of moving electrons from place to place, what establishes and maintains this voltage?

Earlier in this chapter, we discussed various types of energy that can cause "electrical imbalance," which you now know as electromotive force, or voltage. Static electricity, chemical energy, mechanical energy, magnetic energy, light energy, and heat energy were all mentioned. Look at Figure 1–27 to see some methods of using these various energies to establish and maintain differences of potential (voltage) between two points.

(a)

(b)

(c)

(d)

FIGURE 1–27   Several methods of producing voltage or emf: (a) A battery converts chemical energy into voltage. (Courtesy of Eveready Battery Company, Inc.) (b) A solar cell harnesses light energy. (Photo by Michael A. Gallitelli) (c) Heat energy allows the thermocouple to produce useful electric current. (Courtesy of CGS Thermodynamics) (d) Generators turn mechanical energy into electricity. (Courtesy of AC-Delco)

CHARGES IN MOTION

What happens if the two balls in Figure 1–28 touch each other? If you said that the excess electrons on ball A move to ball B and attempt to overcome the deficiency of electrons on ball B, you are right. In fact, electrons would continue to move until ball A and ball B had equal charges, or were neutral with respect to each other.

    If the balls do not touch each other but are connected with a copper conductor wire, Figure 1–29, what happens? If you surmise that some electrons move through the conductor wire until the charges on ball A and ball B are equal, you are right.

**FIGURE 1–28
Two different
charges**

**FIGURE 1–29   If
a conductor wire
is connected to
ball A and ball B,
what happens?**

Current

Are the electrons that move to ball B through the conductor the same electrons that left ball A? Probably not. As you have already studied, the conductor wire has many free electrons moving within it. Having a positive charge at one end of the conductor and a negative charge at the other end (a potential difference between its ends) causes the electrons to move from the negative end to the positive end. This movement of electrons is known as electrical **current flow,** or **current.**

An Analogy of Current Flow

Figure 1–30 illustrates the movement of electrons (current flow) from atom to atom within the conductor (from negative to positive ends). Bin A has a large quantity of rubber balls (representing electrons) and bin B has a few. The row of people between the two bins represent the conductor. They pass the balls from person to person, similar to the conductor material passing electrons from atom-to-atom when a current flows through the conductor. The progressive movement of balls from bin A to bin B simulates the movement of electrons from ball A through the conductor to ball B. In the con-

**FIGURE 1–30   An analogy relating to
electron movement**

**FIGURE 1–31   Another analogy relating
to electron movement**

ductor, as a free electron leaves its original atom and moves to an adjacent atom, it is replaced by another electron from another atom. This process is multiplied millions of times when there is current through a conductor. In Figure 1–29, as an electron leaves one end of the conductor (attracted by ball B's positive charge), it is replaced by an electron entering the conductor from ball A. Figure 1–31 also illustrates current flow.

Obviously, these analogies are oversimplifications of the physics involved in current flow, but they will aid your understanding as we discuss current flow and the specific electrical phenomena involved.

The Unit of Current

Recall that the unit of charge, the coulomb, was established in honor of Charles Coulomb, and the unit of potential difference, the volt, was named after Alessandro Volta. In like manner, the unit of measure for current was named in honor of the French mathematician and physicist, Andre Ampere.

An **ampere** of current is the quantity of electron movement represented by a flow rate of one coulomb of charge per second. Restated, *a flow of one coulomb per second = one ampere.*

A Formula That Relates Current to Charge Movement and Time

A common formula that relates current (in amperes) to coulombs of charge and time (in seconds) is:

$$\text{Formula 1–2} \quad \bigg| \quad I = \frac{Q}{T}$$

where I = current in amperes
Q = charge in coulombs
T = time in seconds

 charge in coulombs, $\boxed{\div}$, time in seconds, $\boxed{=}$

This formula allows you to calculate the current in amperes, if you know the charge in coulombs and the time in seconds. For example, if 10 coulombs of charge move from one point to another point in an electrical circuit over 2 seconds, current equals 5 amperes ($I = 10/2$). Also, this formula can be transposed to calculate the charge if the current and time are known. This transposition yields $Q = I \times T$. If at this point you do not understand how the formula was transposed, be encouraged by knowing that some techniques of transposing will be discussed again later in the book. Look at this formula and the examples, and use your knowledge about direct and inverse relationships to understand the formula. Is current *directly* or *inversely* related to the quantity of charge moved? Is current *directly* or *inversely* related to the duration of time taken to move the given charge? Your answers should have indicated a *direct* relationship between current and quantity of charge moved and an *inverse* relationship between current in amperes and the time it took to move the given charge. Recall that the direct relationship between quantities means that as one quantity increases, the other quantity increases. The inverse relationship indicates that as one quantity increases, the other quantity decreases, or vice versa.

✔ IN-PROCESS LEARNING CHECK II

Fill in the blanks as appropriate.

1. An electrical system (circuit) consists of a source, a way to transport the electrical energy, and a _____.

2. Static electricity is usually associated with _____ type materials.

3. The basic electrical law is that _____ charges attract each other and _____ charges repel each other.

4. A positive sign or a negative sign often shows electrical _____.

5. If two quantities are *directly* related, as one increases the other will _____.

6. If two quantities are *inversely* related, as one increases the other will _____.

7. The unit of charge is the _____.

8. The unit of current is the _____.

9. An ampere is an electron flow of one _____ per second.

10. If two points have different electrical charge levels, there is a difference of _____ between them.

11. The volt is the unit of _____ force, or _____ difference.

THREE IMPORTANT ELECTRICAL QUANTITIES

This chapter has discussed the magnitude of electrical charges (amount or value of charge); polarity (either negative or positive); and the difference of charge between two points that creates a potential difference (electromotive force or voltage) which can move electrons from one point to another (a current flow).

Also, you have learned that scientists have defined units of measure for the quantities of charge, potential difference, and current. Recall that the unit of charge is the coulomb (6.25×10^{18} electrons); the unit of electromotive force, or potential difference, is the volt; and the unit of current flow, or current, is the ampere. Of the electrical parameters described, you will frequently use the units for current and electromotive force (i.e., the ampere and the volt).

A third electrical quantity you will often use is **resistance.** If you have tried to sandpaper a piece of wood, you have experienced *physical resistance*.

Current flow through a conductor encounters molecular resistance to its flow. The amount of resistance to current flow depends on a number of factors. One factor is the type of material through which the current is flowing. Other factors are the dimensions of the conductor wire, including its cross-sectional area and its length, and temperature. An electrical pump (emf or voltage) causes current to flow through the resistance in its path.

As you have probably reasoned, electrical resistance is the opposition shown to current flow. Again, an electrical unit has been named after a scientist, Georg Simon Ohm; thus, the unit of resistance is the **ohm.**

The ohm of resistance has been defined in several ways. One definition states an ohm is the resistance of a column of pure mercury with a cross-sectional dimension of 1 millimeter squared and a length of 106.3 centimeters at a temperature of zero degrees centigrade. Another definition is an ohm is the amount of resistance that develops 0.24 calories of heat when one ampere of current flows through it. *The definition for ohm that is most useful is:*

> *An ohm of resistance is the amount of electrical resistance that limits the current to one ampere when one volt of electromotive force is applied.*

BASIC ELECTRICAL CIRCUIT
Generalized Description

Earlier in the chapter, a simple description was given of an electrical circuit. Recall that an electrical circuit has three basic ingredients: the source, the means to carry electricity from one point to another point, and a load. Before thoroughly examining this basic circuit, we will discuss the difference between a closed and an open circuit.

The Closed Circuit

You have learned that if a path is provided and when a potential difference exists between two points, electrons move from the point of excess electrons (negative polarity point) to the point of electron deficiency (positive polarity point). One method of providing this path is to connect a conductor wire between the points, thus providing a route through which electrons move and establish current flow. A closed circuit is a complete, unbroken path through which electrical current flows whenever voltage is applied to that circuit, Figure 1–32.

The Open Circuit

You probably already understand that if a closed circuit is an unbroken path for electron flow, then an open circuit has a break in the path for current

**FIGURE 1–32   A closed circuit provides an
unbroken path for current flow.**

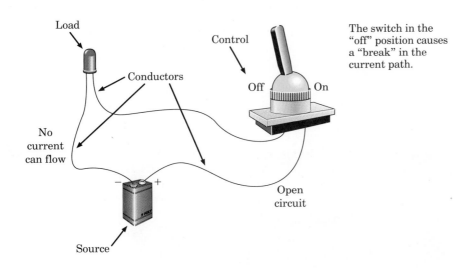

**FIGURE 1–33   An open circuit has a break somewhere in the
current path.**

flow, Figure 1–33. This break may be either a desired (designed) break or
an undesired (unplanned) break. The most common method for purposely
opening a circuit is an electrical switch, such as the one you use to turn the
lights on or off. The switch is a fourth element of the basic electrical circuit,
and many circuits contain such a control component, or circuit. In Figure
1–33 the switch is the control component.

(a)

(b)

(c)

**FIGURE 1–34   Direction of current flow through a circuit: (a) Schematic;
(b) Pictorial; and (c) Photo (Photo by Michael A. Gallitelli)**

PARTS OF A CIRCUIT	EXAMPLE OF EACH	IMPORTANT ELECTRICAL QUANTITIES	
1) SOURCE	POWER COMPANY GENERATORS	EMF	(VOLTS)
		RESISTANCE	(OHMS)
2) TRANSPORTING MEANS	CONDUCTOR WIRES		
		CURRENT	(AMPERES)
3) LOAD	LIGHT BULB/S		
4) (CONTROL)	(SWITCH)		

FIGURE 1–35   Important facts about a basic electrical circuit. As emf increases, current increases. As resistance increases, current decreases.

The Basic Electrical Circuit Summarized

The basic electrical circuit has 1) a source, 2) a means of conducting electron movement (current), 3) a load, and may have 4) a control element, such as a switch.

In order for current to flow (electron movement) through the circuit, there must be a voltage source and a closed circuit or a complete current-conducting path. If the conductor wire(s) were accidentally connected across the source with no other load component, the wire (with its small resistance) would become the load. Depending on the resistance of the wire and the amount of source voltage, this undesired low resistance condition (sometimes called a short circuit) would probably result in the wire overheating and melting, thus breaking the circuit.

Applying a voltage (electromotive force) to a closed circuit causes current to flow through the circuit conductors and load from the source's negative side through the circuit and back to the source's positive side, Figure 1–34. The amount of current (amperes) that flows depends on the value of voltage (volts) applied to the circuit and how much resistance (ohms) the circuit offers to current flow, Figure 1–35. Remember these points because they will become more important as your studies proceed.

SUMMARY

▶ Matter occupies space, has weight, and sometimes can be tangible to one or more of our five senses.

▶ Matter's physical states include solids, liquids, and gases.

▶ Matter's chemical states include elements, compounds, and mixtures.

▶ Elements are comprised of only one type of atom. But compounds are unique combinations of different elements' atoms that result in a new matter or substance.

▶ A mixture does not cause chemical or physical changes in the elements being mixed.

▶ The smallest particle into which a compound can be divided and retain its basic characteristics is a molecule. Elements can be divided to the atom level and retain their inherent characteristics.

▶ The atom is the fundamental building block of matter. Special combinations of atoms, called molecules, are the building blocks of compounds.

▶ Bohr's model of an atom pictures the central portion of the atom, the nucleus, being comprised of protons and neutrons. Around this nucleus is a particle called the electron, which orbits the nucleus like planets orbit the sun.

▶ The electron theory helps illustrate electricity and behavior of electrons in matter and relates to the atom model with its protons, neutrons, and electrons.

▶ The atom's proton has a positive electrical charge, while the orbiting electrons have a negative electrical charge. The neutrons in the center of the atom add mass to the atom but are electrically neutral.

▶ Protons and neutrons are approximately 1800 times heavier than electrons.

▶ The net charge of an atom is typically neutral, since there are equal numbers of electrons and protons and their electrical charges offset each other.

▶ The atomic number of an element is the number of protons in the nucleus of each of its atoms.

▶ The atomic weight of an element is a comparison of its weight to that of carbon-12.

▶ Electrons travel around the nucleus of an atom in shells that are at different distances from the nucleus. The first ring (shell) carries a maximum of 2 electrons; the second shell outward a maximum of 8; the next shell, 18, and the fourth shell, 32. The outermost electron ring, valence electrons, determines if the atom is electrically and chemically active or stable and is full when it has 8 electrons.

▶ If the outer shell of an atom contains 8 electrons, it is stable. If the outer ring has less than 4 electrons, it is usually a conductor. If the valence electrons number 4, it is a semiconductor material.

▶ Materials, such as copper, that have only 1 outermost ring electron per atom are good conductors. These electrons, sometimes called free electrons, are easily moved within the material from atom to atom.

▶ An atom that has lost or gained electrons is called an ion. If it has lost electron(s), it is a positive ion. If it has gained electron(s), it is a negative ion.

▶ Ions and/or movement of electrons within a material are caused by various external and internal energies or forces. Several examples of energy that can electrically unbalance atoms or move electrons are friction, chemical, heat, and light.

▶ The law of electrical charges is that unlike charges attract and like charges repel each other.

▶ Static electricity is usually associated with nonconductor materials and is often caused by friction between them.

▶ Polarity designates differences between two points or objects relative to each other, or opposites. For example, electrical polarities of "−" or "+" show which point has excess electrons and electron deficiency relative to each other. Magnetic polarity indicates differences between magnetic poles (North and South). Also, electrical polarity shows direction of current flow through an electrical component or circuit. For example, electron movement (current) moves from the source's negative side to the source's positive side through the components and circuitry connected to that source.

▶ Coulomb's law of charges states, the force of attraction or repulsion between two charges is directly related to the product of the two charges and is inversely related to the square of the distance between them.

▶ The unit of charge is the coulomb, which represents 6.25×10^{18} electrons of charge.

▶ Electrical potential is the potential to perform the electrical work of moving electrons. It is sometimes termed electromotive force.

▶ The unit of measure for potential difference (electromotive force) is the volt.

▶ Several types of energy that establish and maintain potential differences between two points (i.e., create a voltage source) are chemical, heat, light, magnetic, and mechanical energies.

► Organized electron movement, or electron flow, is known as current flow or current.

► Current flow caused by electron movement is from the negative side of the voltage source to the positive side of the voltage source through the circuit connected to the source, if the circuit is a closed circuit.

► The unit of current is the ampere, which represents a rate of electron flow of one coulomb per second.

► A formula that relates the amount of current flow (in amperes) to the amount of charge (in coulombs) moved in a given amount of time (in seconds) is $I = Q/T$. The formula may also be transposed to read $Q = I \times T$.

► Three important electrical quantities of measure are:

1. volt, the unit of potential difference, or amount of potential difference causing one ampere of current to flow through a resistance of one ohm.

2. ampere, the unit of current flow, or amount of current flow represented by a rate of charge flow of one coulomb per second.

3. ohm, the unit of resistance, or amount of electrical resistance limiting the current to one ampere with one volt applied.

► A basic electrical circuit has an electrical source (emf), a way to transport electrons from one point to another point, and a load, which uses the electrical energy. Another part of many basic circuits is a control device, such as a switch.

► A closed circuit has an unbroken path for current flow: from one side of the source, through the circuitry and back to the other side of the source.

► An open circuit does not provide a continuous path for current flow. The path is broken, either by design (using a switch) or by accident. The open circuit presents an infinite resistance path for current.

► A short circuit is an undesired, low resistance path for current flow. If the short is across both terminals of a voltage source, it frequently causes undesired effects, such as melted wires, fire, and other damage to the voltage source.

► In the basic electrical circuit, the voltage source provides the electromotive force that moves electrons and causes current flow. The circuit conductors and components provide resistance to current, thereby limiting the current. The various techniques of controlling electrons and electron flow to perform desired effects form the basis of electronic science.

FORMULAS AND CALCULATOR SEQUENCES SUMMARY

FORMULA 1–1 $F = k \dfrac{Q_1 \times Q_2}{d^2}$

k value, $\boxed{\times}$, Q_1 value, $\boxed{\times}$, Q_2 value, $\boxed{=}$, $\boxed{\div}$, distance value, $\boxed{x^2}$, $\boxed{=}$

FORMULA 1–2 $I = \dfrac{Q}{T}$

charge in coulombs, $\boxed{\div}$ time in seconds, $\boxed{=}$

REVIEW QUESTIONS

1. Define the term matter.
2. List the three physical states of matter and give examples of each.
3. List the three chemical states of matter and give examples of each.
4. Define the term element and give two examples of elements.
5. Define compound and give two examples of compounds.
6. Sketch a hydrogen atom. Identify each particle.
7. On your hydrogen atom sketch, designate the electrical charge of each particle.
8. a. Which has more weight or mass, the electron or the proton?
   b. Approximately how many times heavier is the one compared to the other?
9. Define the term free electron.
10. a. Define the term ion.
    b. What causes an ion to be classified as a positive ion?
11. Name four energy types that can be used to purposely create electrical imbalances in atoms of materials.
12. How many valence electrons do each of the following materials have?
    a. A conductor.
    b. A semiconductor.
    c. An insulator.

13. Describe the basic law of electrical charges.
14. Describe Coulomb's law of electrical charges.
15. If the distance between two charges is tripled and the charge strength of each charge remains the same, what happens to the force of attraction or repulsion between them?
16. In an electrical circuit, what is potential difference?
17. What unit of measure expresses the amount of electromotive force present between two identified points in a circuit?
18. In an electrical circuit, what is current flow?
19. a. What basic unit of measure expresses the value of current in a circuit?
    b. What is the value of current flow if 20 coulombs of charge move past a given point in 5 seconds?
20. Describe the term resistance in terms of voltage and current, and give the unit of measure.
21. Describe the difference between a closed circuit and an open circuit.
22. List three or more elements of a basic electrical circuit.

TEST YOURSELF

1. Draw a diagram illustrating an electrically balanced atom that has 30 protons in its nucleus and whose first three shells are full, with the fourth (last) shell having two electrons.
2. Label all electrical charges in your diagram.
3. What kind of material is pictured in your diagram?

4. Is this material a conductor, semiconductor, or insulator?

5. What is its atomic number?

6. Define the term polarity as used in electrical circuits.

7. Draw an electrical circuit showing voltage source, conductor wires, and load. Show the polarity of that source, and, by using an arrow, show the direction of current flow through the circuit. (Assume a closed circuit.)

Electrical Quantities and Components

2

KEY TERMS

Ampere (A; mA; μA)

Block diagrams

Conductance *(G)*

Current *(I)*

Ohm (Ω; kΩ; MΩ)

Resistance *(R)*

Resistor (R)

Resistor color code

Schematic diagrams

Siemens (S; mS)

Switch (S)

Volt (V; mV; μV)

Voltage *(V)*

OUTLINE

Basic electrical units and abbreviations

Using the metric system to help

Conductors and their characteristics

Resistors

Resistor color code

Measuring voltage, current, and resistance with meters

Diagrams used for electronic shorthand

CHAPTER PREVIEW

This chapter will further your study of the basic electrical quantities which are vital to your knowledge as a technician. For example, commonly used metric prefixes and powers of ten related to electrical quantities are presented. Information regarding basic characteristics and types of conductors and resistors is presented. You will also study and learn the resistor color code. (NOTE: Interpreting the color code is one of the most practical and most frequently used skills ever acquired by technicians).

How to use meters to measure voltage, current, and resistance is also presented. Of course, application of this knowledge, and practice at doing this, is acquired by performing laboratory exercises that require making actual meter measurements.

Finally, the chapter introduces you to the technicians' and engineers' "shorthand" method of presenting electrical circuits. Both block diagram and schematic diagram techniques are presented.

OBJECTIVES

After studying this chapter, you should be able to:

• List the units of measure for charge, potential (emf), current, resistance, and conductance and give the appropriate abbreviations and symbols for each

• Use metric system terms and abbreviations to express subunits or multiple units of the primary electrical units

• List the factors that affect the resistance of a conductor

• Recognize common types of conductors

• Use a wire table to find conductor resistance for given lengths

• Recognize and/or draw the diagrammatic representations for conductors that cross and electrically connect, and that cross and do not connect

• Give the characteristics of several types of resistors

• Use the resistor color code

• Explain how to connect meters to measure voltage, current, and resistance

• Recognize and/or draw the diagrammatic symbols for elemental electronic components or devices

• Interpret basic facts from block and schematic diagrams

• List key safety habits to be used in laboratory work

BASIC ELECTRICAL UNITS AND ABBREVIATIONS

Charge

Recall that the basic unit of charge is the coulomb and that the letter "Q" (or q) represents charge. However, you need to know that the abbreviation for the coulomb is **C.** (Remember that one C equals 6.25×10^{18} electrons.) Notice that this is a capital "C" and not a lower case "c." If the unit abbreviation or symbol for electronic measurement is named after an individual, such as coulomb, volt, ampere, or ohm, the abbreviation is a capital letter and not a lower case letter. Keep this in mind and get in the habit of correctly writing unit abbreviations or symbols.

Potential

The unit for potential difference, or electromotive force, is the **volt,** named after Alessandro Volta. The abbreviation, or symbol, for this unit is **V. Voltage** is expressed in volts. Recall that one volt equals the amount of electromotive force (emf) that moves a current of one ampere through a resistance of one ohm.

Current

The unit of measure for current flow is the **ampere.** The abbreviation, or symbol, for this basic unit of measure is **A.** Don't confuse the letter I (thought of as intensity of electron movement) that represents the general term **current** with the A used for the unit of current. Remember that one ampere equals an electron flow of one coulomb per second past a given point.

Resistance

Resistance is another electrical parameter that uses two letters: "R" represents the general term resistance and the Greek letter omega (Ω) represents the unit of resistance, the **ohm**. Remember that one ohm equals the resistance that limits the current to one ampere with one volt applied. [NOTE: Figure 2–1 shows some "multimeters" used to measure V, A, and Ω, the units of measure we have just discussed.]

Conductance

Another electrical parameter is **conductance.** It sometimes is defined as the comparative ease with which current flows through a component or circuit. Conductance is the opposite of resistance. The unit of conductance is the **siemens (S)** named after the scientist Ernst von Siemens. The abbreviation for the general term conductance is G, and the unit of measure is the siemens (S). Conductance, in siemens, is defined as the reciprocal of resistance in ohms.

$$G(\text{in siemens}) = \frac{1}{R} \text{ or } G = \frac{1}{R}(\text{S})$$

(a)   (b)

FIGURE 2–1   Multimeters used to measure voltage, current, and resistance: (a) An analog volt-ohm-milliammeter (VOM) (Courtesy of Simpson Electric Company); (b) A digital multimeter (DMM) (Courtesy of John Fluke Mfg. Co., Inc.)

These units of measure and abbreviations, or symbols, are important, so a chart has been provided, Figure 2–2. Study this chart until you have memorized all the units and their abbreviations.

USING THE METRIC SYSTEM TO HELP

Some Familiar Metrics

Many people have been exposed to the metric system. The meter (approximately 39 inches), centimeter (one one-hundredth of a meter), millimeter (one one-thousandth of a meter), and kilometer (1,000 meters) are not uncommon terms. As you can see, the prefix to "meter" identifies a multiple or subdivision of the basic unit. In other words, the meter is the basic unit and the prefix centi equals 1/100, milli equals 1/1,000, and kilo equals 1,000 times the basic unit.

This technique of using metric prefixes is also useful in identifying multiples or subdivisions of basic electrical units. For example, in electronic circuits it is common to use thousandths or millionths of an ampere, or thousands of ohms of resistance. Although the subunits and multiple-units in Figure 2–3 do not represent all prefixes encountered in electronics, they do represent those you will be immediately applying. Study Figure 2–3 until you learn each of the metric terms and their abbreviations or symbols.

There is one other bit of related knowledge that can be of great help to you in practical work. That is, mathematical ways to represent each of these metric prefixes. These mathematical representations are helpful in study-

BRIEF DEFINITION OF QUANTITY	ELECTRICAL QUANTITY OR PARAMETER	BASIC UNIT OF MEASURE	ABBREVIATION OR SYMBOL FOR UNIT
Excess or deficiency of electrons	Charge (Q)	Coulomb	C $(6.25 \times 10^{18}$ electrons)
Force able to move electrons	Potential Difference (emf)	Volt (Force that moves one coulomb of charge per second through one ohm of resistance)	V
Progressive flow of electrons	Current (I)	Ampere (An electron flow rate of one coulomb per second)	A
Opposition to current flow	Resistance (R)	Ohm (A resistance that limits current to a value of one ampere with one volt applied)	Ω
Ease with which current can flow through a component or circuit	Conductance (G)	Siemens (The reciprocal of resistance, or, $\frac{1}{R}$)	S

FIGURE 2–2   Important electrical units' abbreviations and symbols

ing electronic circuits, because simple calculations can be used to analyze, verify, or predict circuit behavior.

Since the subunit and multiple-unit prefixes in Figure 2–3 are based on a decimal system (multiples or submultiples of 10), it is convenient to express these prefixes in powers of ten. Recall that:

10^0 = (ten to the zero power) = 1
10^1 = (ten to the 1st power) = 10
10^2 = (ten to the 2nd power) = $10 \times 10 = 100$
10^3 = (ten to the 3rd power) = $10 \times 10 \times 10 = 1,000$
10^{-1} = (ten to the negative 1st power) = 1/10 = 0.1
10^{-2} = (ten to the negative 2nd power) = 1/100 = 0.01
10^{-3} = (ten to the negative 3rd power) = 1/1,000 = 0.001

METRIC TERM	SYMBOL	MEANING	TYPICAL USAGE WITH ELECTRONIC UNITS
pico	p	one millionth of one millionth of the unit	picoampere (pA)
nano	n	one thousandth of one millionth of the unit	nanoampere (nA) nanosecond (ns)
micro	μ	one millionth of the unit	microampere (μA) microvolt (μV)
milli	m	one thousandth of the unit	milliampere (mA) millivolt (mV)
kilo	k	one thousand times the unit	kilohms (kΩ) kilovolts (kV)
mega	M	one million times the unit	megohm(s) (MΩ)

FIGURE 2–3 **Some commonly used metric units**

Figure 2–4 shows how powers of ten relate to the electrical units and prefixes discussed earlier. **Learn these because you will use them frequently.**

Study Figure 2–4. You can surmise that powers of ten can simplify writing very large or very small numbers. For example, instead of writing 3.5 microamperes (μA) as 0.0000035 amperes, you can state it as 3.5×10^{-6} amperes. The power of ten tells us how many decimal places and which direction the decimal moves to state the number as a unit. Thus, 3.5×10^{-6} indicates the decimal moves six places to the left when stated as the basic unit (i.e., 0.0000035 amperes). This is three and one-half millionths of an ampere.

Another example is 5.5 megohms (MΩ) of resistance simplified as 5.5×10^6 ohms. How would this be stated as a number representing the basic unit? If you said the power of ten indicates moving the decimal six places to the right, you are correct. This yields 5,500,000 ohms.

Obviously, it is easier to manipulate numbers using powers of ten rather than using the bulky multiple decimal places involved in the small and large numbers typically associated with electrical/electronic units. You will get more practice using powers of 10 in the next chapter.

NUMBER	PWR OF 10	TERM	SAMPLE ELECTRONIC TERM
0.000000000001	10^{-12}	pico	pA(1×10^{-12} ampere)
0.000000001	10^{-9}	nano	nA(1×10^{-9} ampere)
0.000001	10^{-6}	micro	μA(1×10^{-6} ampere)
0.001	10^{-3}	milli	mA(1×10^{-3} ampere)
1,000	10^3	kilo	kΩ(1×10^3 ohms)
1,000,000	10^6	mega	MΩ(1×10^6 ohms)

FIGURE 2–4 **Powers of ten related to metric and electronic terms**

In addition to practice using powers of 10, you will learn that related skills, such as using scientific and engineering notation and using a scientific calculator can greatly simplify manipulation of the numbers associated with electronic circuitry.

✔ IN-PROCESS LEARNING CHECK I

Fill in the blanks as appropriate.

1. Charge is represented by the letter _____. The unit of measure is the _____, and the abbreviation is _____.

2. The unit of potential difference is the _____. The symbol is _____.

3. The abbreviation for current is _____. The unit of measure for current is the _____, and the symbol for this unit is _____.

4. The abbreviation for resistance is _____. The unit of measure for resistance is the _____, and the symbol for this unit is _____.

5. Conductance is the _____ with which current can flow through a component or circuit. The abbreviation is _____. The unit of measure for conductance is the _____, and the symbol for this unit is _____.

6. How many microamperes does 0.0000022 amperes represent? _____ How is this expressed as a whole number times a power of 10? _____

7. What metric prefix represents one-thousandth of a unit? _____ How is this expressed as a power of 10? _____

8. Using a metric prefix how would you express 10,000 ohms? _____ How is this expressed as a whole number times a power of 10? _____

CONDUCTORS AND THEIR CHARACTERISTICS

Functions of Conductors

You have already learned that the function of conductors is to carry electrical energy from one point to another. More explicitly, conductors provide a path for the flow of electrical current between components or circuits. As you can observe at home, at work, or at school, there are numerous types of conductors. One type is used to transport power into your home. Another type brings the TV signal from the antenna to your TV set. Still another style of conductors delivers telephone signals. The heavy battery cables in your autombile represent yet another type of conductor.

Types of Conductors

The type of conductor required for a given job is determined by the nature of the application and the amount of electrical energy it must carry. Some ways of classifying conductors include the:

• Type of metal used (silver, copper, gold, aluminum)

• Physical dimensions of the wire itself (its "size")

• Form of the metal conductor/s (solid, stranded, or braided)

• Number of conductors used or packaged together

• Insulation characteristics (uninsulated, insulated, and type of insulation used)

Several examples of conductor types are shown in Figure 2–5.

Resistance of Solid Conductors

The ideal conductor would have zero resistance, and thus would not dissipate any of the electrical energy being transported. In reality, however, all conductors do have some resistance to the flow of electrons.

PHYSICAL FACTORS INFLUENCING CONDUCTOR RESISTANCE.
The resistance of a given conductor is related to the following factors:

1. The type material making up the conductor. (The more conductive the material, the lower its resistance per given physical dimensions.)

2. The length of the conductor. (The longer it is, the higher its total resistance from one end to the other.)

3. The cross-sectional area of the conductor. (The greater the cross-sectional area, the lower its resistance per given length.)

4. The temperature of the conductor. (Typically, for metal conductors, as temperature increases, the resistance increases.)

RESISTIVITY. The "resistivity" (or "specific resistance") of a given material refers to its characteristic resistance (in ohms) per standard length and cross-sectional area, at a specified temperature. In SI units, resistivity is

FIGURE 2–5 **Examples of several types of conductors: (a) Auto battery cable, (b) Coaxial cable, (c) TV twin lead, (d) Stranded, (e) Solid (Photos by Michael A. Gallitelli)**

expressed in terms of meters for length, and square meters for cross-sectional area. Different material each have their own resistivity characteristic. For example, in SI units, silver has a resistivity of 16 nano-ohms per meter; copper, a resistivity of 17 nano-ohms per meter; gold, a resistivity of 24 nano-ohms per meter; and aluminum, a resistivity of 27 nano-ohms per meter. These are expressed in nano-ohms per meter for simplicity of reading, rather than having to state that silver has a resistivity of 0.000000016 ohms/meter, and so on. The symbol used for resistivity is ρ (the Greek letter rho).

In everyday technician work, you are more likely to be dealing with conductors in terms of diameter in mils (thousandths of an inch), area in circular mils (CM), and length in feet. A circular mil describes the cross-sectional area of a round conductor having a diameter of one-thousandth of an inch (1 mil diameter). The cross-sectional area in circular mils is thus found as:

$$\boxed{\text{Formula 2–1} \quad A = d^2}$$

where A = cross-sectional area in circular mils
d = diameter of wire in mils
See Figure 2–6.

 diameter value (in mils), $\boxed{\times^2}$

EXAMPLE:

If a wire has a diameter of 4 mils (0.004″), what is its cross-sectional area in circular mils?

Answer: $A = d^2$. $A = 4^2$. $A = 16$ CM

PRACTICE PROBLEMS I

What is the circular-mil cross-sectional area of a conductor whose diameter is 2.5 mils?

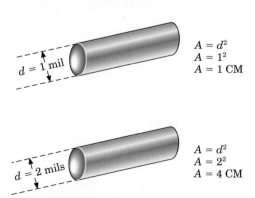

$d = 1$ mil
$A = d^2$
$A = 1^2$
$A = 1$ CM

$d = 2$ mils
$A = d^2$
$A = 2^2$
$A = 4$ CM

FIGURE 2–6   Cross-sectional area of conductors in circular mils (CM)

Because resistivity (ρ) of a material is a measure of a material's resistance per standard length and cross-sectional area (at a given temperature), ρ can be expressed in terms of circular-mil ohms per foot. That is, CM-Ω/ft.

CALCULATING THE RESISTANCE OF A CONDUCTOR.

A general formula for finding the resistance a wire (assuming a given temperature) is:

| Formula 2–2 | $R = \dfrac{\rho l}{A}$ |

where R = resistance in ohms
ρ = resistivity in CM-Ω/ft (for the type material in conductor)
l = length in feet
A = cross-sectional area of conductor in CM

 Resistivity (CM-Ω/ft), $\boxed{\times}$, length (ft), $\boxed{\div}$, area (CM), $\boxed{=}$

EXAMPLE:

What is the resistance of a copper wire that is 250 feet in length and has a cross-sectional area of 39.75 CM? (NOTE: ρ for copper is 10.4 CM-Ω/ft.)

Answer: $R = \dfrac{\rho l}{A} = \dfrac{(10.4 \text{ CM-}\Omega/\text{ft})(250 \text{ ft})}{39.75} = 65.4 \ \Omega$

PRACTICE PROBLEMS II

What is the resistance of a copper wire that is 300 feet in length and has a cross-sectional area of 2,048 CM? (Remember: ρ for copper is 10.4 CM-Ω/ft.)

SUMMARY.

The resistance a given conductor exhibits to current flow is directly related to its resistivity characteristic and its length, and inversely

- Same type wire
- Same diameter for 1 and 2
- Conductor 2 is twice the length of conductor 1
- Conductor 2's resistance is twice that of Conductor 1, from one end to the other

- Same type wire
- Conductors 3 and 4 are the same length
- Conductor 4 has twice the diameter of 3 or ≈ 4 times the cm area
- Conductor 4's resistance is one-fourth that of conductor 3 from end to end

FIGURE 2–7 Relationship of wire resistance to conductor diameter and length

related to its diameter squared (its cross-sectional area). That is, the larger the wire (of a given type material), the lower its resistance per given length. The longer the wire, the higher its resistance per given cross-sectional area. See Figure 2–7. These are important facts to remember as you deal with conductors and conductive paths as a technician.

Wire Sizes

There is a standardized sizing of wires by "wire gage numbers." This standard is referred to as the American Standard Wire Gage, or AWG. Figure 2–8 shows a tool some technicians use to measure the AWG size of wires. Wire sizes quickly communicate relative size information about conductors. For example, if a wire is categorized as an AWG #14 wire, its cross-sectional area in CM will be 4,107 CM. If the wire is sized as a #18 wire, its cross-sectional area in CM will be close to 1,624 CM. [NOTE: The higher the AWG number, the smaller the wire is in diameter (and hence, the less the cross-sectional area it has).] It is interesting to know that for each change of three wire-gage sizes, the CM area doubles or halves, depending on which direction you move. For example, a #11 wire has approximately twice the cross-sectional area of a #14 wire. This means its resistance per unit length is approximately half that of #14 wire.

Wire Table

A very useful source of information regarding conductor sizes and their resistances is a wire table, Figure 2–9. The wire table displays data regarding AWG wire numbers, their cross-sectional areas and their resistance in ohms per given length into one concise list or matrix. Most wire tables show the standard resistance per given length in terms of ohms per thousand feet.

FIGURE 2–8   A wire-gage tool
(Photo by Michael A. Gallitelli)

AWG WIRE SIZE	DIAMETER IN MILS	CIRCULAR-MIL AREA	OHMS PER 1,000 FT AT 25°C	CURRENT CAPACITY @ 700 CM/A
---	--------	--------	--------	--------
10	101.9	10,380	1.018	14.8
11	90.7	8,234	1.284	11.8
12	80.8	6,530	1.619	9.33
13	72.0	5,178	2.042	7.4
14	64.1	4,107	2.575	5.87
15	57.1	3,257	3.247	4.65
16	50.8	2,583	4.094	3.69
17	45.3	2,048	5.163	2.93
18	40.3	1,624	6.510	2.32
19	35.9	1,288	8.210	1.84
20	32.0	1,022	10.35	1.46
21	28.5	810	13.05	1.16
22	25.3	642	16.46	0.918
23	22.6	510	20.76	0.728
24	20.1	404	26.17	0.577
25	17.9	320	33.00	0.458
26	15.9	254	41.62	0.363
27	14.2	202	52.48	0.288
28	12.6	160	66.17	0.228
29	11.3	127	83.44	0.181
30	10.0	101	105.2	0.144
31	8.9	80	132.7	0.114
32	8.0	63	167.3	0.090
etc.	etc.	etc.	etc.	etc.

NOTE: Current-carrying capacity varies greatly, depending on environmental conditions. For example, a single #14 wire in open air can carry up to 32 amperes. In conduit, it can only carry a maximum of 17 amperes. At 700 CM/ampere, it can only carry about 5.87 amperes.

FIGURE 2–9 Sample partial wire table

PRACTICE PROBLEMS III

1. Refer to Figure 2–9 and determine the resistance of 500 feet of #12 wire.

2. What gage wire has a cross-sectional area of about 10,380 CM?

3. What is the resistance in ohms per thousand feet of the wire in question 2?

As you might reason, the size of the conductor also dictates the maximum current it can safely carry without overheating and being destroyed. Some wire tables will also show this maximum current-carrying capacity for stated conditions, as well as providing the resistance-per-given length data. For a given type conductor, the larger the diameter of the wire, the greater current-carrying capacity it has. For example, as stated earlier, the cross-sectional area of the #11 wire is about double that of #14 wire. This means that the #11 wire can safely handle about double the current of the

#14 wire. The general rule of thumb is that for every three wire sizes changed, the current-carrying capacity of the wire will double, or halve, depending on which direction you move in the wire sizes.

✔ IN-PROCESS LEARNING CHECK II

1. The greater the resistivity of a given conductor, the (higher, lower) _____ its resistance will be per given length.

2. The circular-mil area of a conductor is equal to the square of its _____ in _____.

3. The smaller the diameter of a conductor, the (higher, lower) _____ its resistance will be per given length.

4. The higher the temperature in which a typical metal conductor must operate, the (higher, lower) _____ will be its resistance.

5. Wire tables show the AWG wire _____, the cross-sectional area of wires in _____, the resistance per given length of copper wire, and the operating _____ at which these parameters hold true.

RESISTORS

Uses

Resistors limit current and help determine current values for a given circuit. This current-controlling function is one of the important roles that resistors perform in electronics. Another function, which you'll study later, is dividing voltage. Both of these purposes are important.

Types

Resistors are designed to perform these functions under a variety of operating conditions, including various voltage, current, and power levels. Each situation demands certain characteristics of this component. Due to these varying demands, two basic categories of resistors have been designed:

1. Carbon, or composition resistors, Figure 2–10

2. Wire-wound resistors, Figure 2–11

The power-handling capability of any resistor is related to its physical size and ability to dissipate heat. You will be studying the unit of power,

FIGURE 2–10 Carbon resistor
physical appearance

FIGURE 2-11 Wire-wound resistors

called the watt, in the next chapter. Typically, power ratings for carbon resistors range from one-fourth watt to two watts. Wire-wound resistors have a power-handling capability in a range from about 5 watts to 100 watts or more.

Familiarity with these two basic categories of resistors will be gained in laboratory applications. As a technician, the important characteristics to consider when selecting or replacing resistors include value in ohms, wattage (power) rating, voltage rating, accuracy of value (or tolerance rating), and physical dimensions.

Resistor Types and Symbols

Resistors may come in "fixed," "semifixed," or "variable" resistance forms. A fixed resistor is one in which the resistance between its terminals or component leads cannot be changed externally. See Figure 2–12a. A semifixed resistor has one contact, or terminal, that can be adjusted to any given spot along the resistance element, and affixed to that position, as shown in Figure 2–12b. Thus the resistance value can be set between this movable terminal and the end terminals at a desired value. Once adjusted, the movable element is not normally moved again. The variable resistor, however, has a movable contact that can be moved anywhere along the resistance element. In the variable resistor, the movable contact has a shaft attached to purposely enable easy changing of the contact's setting along the resistance element, Figure 2–12c.

Figure 2–13 shows the "schematic symbols" used to represent two of these types. As you will soon learn, schematic symbols are the technician's shorthand for representing electrical components and circuits on paper. You

(**a**) Fixed resistor

(**b**) Semifixed resistor

(**c**) Variable resistor

FIGURE 2–12   Fixed, semifixed, and variable resistor constructions

Fixed resistor symbols

Generic variable resistor
symbol and/or rheostat
symbol

Potentiometer-type variable
resistor symbol

Potentiometer connected
as a two-terminal
rheostat device

FIGURE 2-13   Schematic symbols for fixed and variable resistors

will become very familiar with these, and other electronic diagram symbols
as you continue in your studies.

Resistor Construction

Carbon, or composition resistors, are composed of carbon or graphite mixed
with powdered insulating material. The mixture controls the amount of
resistance exhibited for a given dimension of the material. Carbon fixed
resistors come in a tubular shape, Figure 2–14. Notice the colored stripes on
this tubular-shaped resistor. You will be studying this color-coding system
in the next section of this chapter. Carbon resistive elements are also found
in variable resistors, Figure 2–15.

Wire-wound resistors generally are made of special resistance wire,
wrapped around a ceramic or insulator core. The amount of resistance of the
element is controlled by the size of the wire, the type of material in the wire,
and the length of the wire wrapped around the form. Refer again to Figure
2–11. Wire-wound resistors are manufactured as fixed, semi-fixed, or variable resistances.

Some variable resistors, called "potentiometers," have resistive elements which can be either carbon or wire-wound elements, Figure 2–12c.
The position of the "wiper-arm" contact touching the circular carbon or
wire-resistive element determines the amount of resistance from the center
(wiper-arm) terminal to the terminals at each end of the potentiometer,
Figure 2–15. You will see in your further studies that the potentiometer is

**FIGURE 2–14   Carbon resistor
with color coding**

FIGURE 2–15 Inside view of a potentiometer carbon element and wiper-arm contact

FIGURE 2–16 High current-carrying capacity wire-wound variable resistor. This resistor can be wired as a two-terminal rheostat or as a three-terminal potentiometer.

a valuable component for voltage-dividing functions. Volume controls for radios and TVs are a common application for potentiometers.

Wire-wound resistors are often used where very low values of resistance are needed or where high current-carrying capacity is required. In Figure 2–16, you see a large wire-wound variable resistor, typically used as a rheostat.

RESISTOR COLOR CODE

A skill used continuously by electronics technicians is that of decoding the **resistor color code.** Because many resistors are so physically small, it is difficult to print their value in a readable size. A system of color coding has been adopted and standardized by the Electronics Industries Association (EIA); thus, technicians everywhere understand the code.

You should be aware that there are also resistors which can, and do, have their values labeled on them, and do not therefore need to have color coding. For those that do use the color code, the band system of color coding consists of three, four, or five bands (or stripes) of color around the tubular-shaped resistors. The first three stripes, when decoded, indicate the resistor's rated resistance value, in ohms. The fourth stripe, when present, indicates the tolerance percentage, or amount of deviation from the rated value that the resistor's actual value may vary in manufacture. If there is not a fourth band, the tolerance is automatically assumed to be ±20%. When a fifth band is present, it is used to indicate a reliability factor. This rating is based on the percentage failure rate for 1,000 hours of service.

Significance of Color Band Placement

Look at Figure 2–17 and observe the placement of these color bands. The first color is considered to be the one closest to the end of the resistor. The other color bands progress toward the center of the resistor body. The key to decoding the color code is to learn the significance of each band position, and of each color.

Refer again to Figure 2–17 and observe that the band positions are interpreted as follows:

1. The first band (closest to the end of the resistor) is the first significant digit in the resistor's ohmic value rating.

A = First significant figure of resistance value

B = Second significant figure of resistance value

C = Decimal multiplier for resistance value

D = Tolerance rating (in percentage)

E = Reliability factor (in percentage, based on
                    1,000 hours of operation)

FIGURE 2-17   Color band significance

2. The next, or second band, represents the second significant digit in the resistor's ohmic value rating.

3. The third band indicates the multiplier, or number of zeros that should follow the first two numbers in order to know the resistor's rated ohmic value.

4. The fourth band provides the percentage tolerance information. That is how much, in percentage, the resistor can acceptably vary from its rated, color-coded value and still be within the manufacturer's specifications.

5. The fifth band, when present, indicates the failure rate (in percentage) per 1,000 hours of service. This is sometimes called the reliability factor.

Significance of Colors

For resistors that are greater than 10 ohms in value, the ten colors used to represent the values zero through nine (0–9) are used and are fairly easy to remember by means of a memory-aid sentence. As you can see in Figure 2–18, the color code starts with the dark colors at zero, and progresses toward lighter colors as the numbers increase. The figure shows how the sentence "*Big Boys Race Our Young Girls, But Violet Generally Wins*" can be used to remember the color code. These colors represent the sequential numbers zero through nine. The first letter of each word in the sentence is the same as the first letter of each color. Learn these!

Tolerance Band Colors

Gold and silver are frequently used tolerance band colors. Gold indicates ±5% tolerance from rated value. For example, if a resistor is color coded as

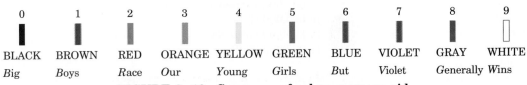

FIGURE 2–18   Sequence of colors memory aid

a 100-ohm resistor with ±5% tolerance rating, its actual value could be anywhere between 95 and 105 ohms, and still be within the manufacturer's rating limits.

Silver represents a tolerance of ±10%. In this case, a resistor color coded 100 ohms could actually have a value between 90 ohms and 110 ohms and still be considered good.

As stated earlier, if there is no fourth band, the tolerance rating is ±20% of rated value.

Use of Gold and Silver Colors as Multipliers

Low value resistors (below 10 ohms) require a special multiplier value for the third band. This indicates that the numbers decoded from the first two color bands are to be multiplied by 0.1 or 0.01 to find the value for the resistor. A gold third band indicates the multiplier is one-tenth (0.1). For example, if the first two color bands are red and violet (27), then a gold third band would indicate a resistor value of 0.1×27, or 2.7 ohms.

A silver third band indicates a multiplier of one-hundredth (0.01). This means if the resistor is color coded brown-black-silver (in that order), the resistor value is 0.1 ohm ($0.01 \times 10 = 0.1$).

A summary chart showing the significance of various colors used in the color code is shown in Figure 2–19.

Examples

Let's see how to apply the combination of color information and band position to decode some resistor values and tolerances. Also, we'll look at one case where a reliability factor is given. Notice in Figure 2–20a that the first band is yellow, representing the number 4. The second band is violet, representing 7. The third band is orange, indicating a multiplier of 1,000, which adds three zeros to the 47. Therefore, the resistor color-coded value is

COLOR		NUMBER IT REPRESENTS (IN 1ST AND 2ND BANDS)	DECIMAL MULTIPLIER IT REPRESENTS (IN 3RD BAND)	TOLERANCE IT REPRESENTS (IN 4TH BAND)	RELIABILITY FACTOR IT REPRESENTS (IN 5TH BAND)
Black	■	0	1 (no zeros)	—	—
Brown	■	1	10 (1 zero)	—	1%
Red	■	2	100 (2 zeros)	—	0.1%
Orange	■	3	1,000 (3 zeros)	—	0.01%
Yellow	▢	4	10,000 (4 zeros)	—	0.001%
Green	■	5	100,000 (5 zeros)	—	—
Blue	■	6	1,000,000 (6 zeros)	—	—
Violet	■	7	10,000,000 (7 zeros)	—	—
Gray	▢	8	100,000,000 (8 zeros)	—	—
White	▢	9	1,000,000,000 (9 zeros)	—	—
Gold	▨	—	0.1	5%	—
Silver	▨	—	0.01	10%	—

FIGURE 2–19   Color significance summary chart

**FIGURE 2–20   (a) 47 kΩ, 10%
tolerance resistor; (b) 10 kΩ, 5%
tolerance resistor; (c) 2.2 kΩ, 10%
tolerance, 1% reliability factor**

47,000 ohms (or 47 kΩ). The fourth band is silver, indicating a 10% tolerance rating. This means that the resistor's actual value can be 47,000 ohms,
plus or minus 4,700 ohms. That is, its value should be between 47,000 +
4,700 ohms (51,700 ohms) and 47,000 ohms − 4,700 ohms (42,300 ohms).

In Figure 2–20b, the color code is brown, black, orange, gold, in that
order. Decoded, this value is 10,000 ohms, ±5%. In Figure 2–20c, the five-
band color code is red, red, red, silver, brown. The value of the resistor is
therefore 2, 2, and two zeros, or 2,200 ohms. The tolerance is ±10%. The
reliability factor is 1%. (See Figure 2–19.)

Now you try a couple!

PRACTICE PROBLEMS IV

1. What is the resistor value and tolerance of the resistor shown?

Orange White Orange Silver

2. What is the resistor value, tolerance, and reliability rating of the
   resistor shown?

Brown Gray Orange Gold Red

Special Precision Resistors

Although the majority of resistors you will work with use the color coding as
presented thus far, there are some precision resistors with tolerances of one
to two percent that use a special five-band color-code technique. In these
cases, the first three bands are value digits and the fourth band is the
multiplier. The fifth band, then, provides the tolerance rating. For these

special resistors, a gold fifth band signifies a 1% tolerance rating and a silver fifth band indicates a 2% tolerance.

Other Exceptions

We mentioned earlier that some resistors are labeled rather than color coded. Some labeled resistors use a cryptic numbering and lettering system. The letter R indicates the decimal point position for resistors under 100 ohms. For example, if a resistor is labeled 2R7, its value is 2.7 ohms. Letters are also used to designate tolerance ratings, where the letter F means ±1%, $G = ±2\%$, $J = ±5\%$, $K = ±10\%$, and $M = ±20\%$. So if the label shows 2R7K, the resistor is a 2.7-ohm resistor with 10% tolerance.

For resistor values requiring more than three numbers, plus a multiplier, the first three labeled numbers are followed by a fourth number, which is the multiplier, indicating the number of zeros to follow the preceding numbers. An example of this might be a resistor labeled 4702K. In this case, the value is 47,000 ohms and the tolerance is ±10%.

MEASURING VOLTAGE, CURRENT, AND RESISTANCE WITH METERS
The Importance of Knowing How to Measure *V, I,* and *R*

As a technician, you will frequently measure voltage, current, and resistance in electrical or electronic circuits. These measurements are sometimes used to set up or adjust circuit conditions for proper operation. Measure-

(a)   (b)

**FIGURE 2-21   (a) Digital meter (Courtesy of John Fluke Mfg. Co., Inc.)
(b) Analog meter (Courtesy of Simpson Electric Company)**

ments are often used to check that a circuit is operating properly or to troubleshoot circuits that are not operating properly. The skill to make measurements correctly will become a valuable tool to you. In this section, we will introduce you to the most elemental concepts of measuring V, I, and R. Mastery of this measurement skill will only come with actual practice.

Types of Meters

Two general categories of meters you will use in making measurements of V, I, and R are analog meters and digital meters, Figure 2–21. You will learn more about the significance of the terms analog and digital as you progress in your studies. For now, you can see that the scale on the digital meter is directly readable, just by reading the numbers in the readout portion of the meter. The analog meter, however, takes some practice in interpreting values in terms of where the pointer rests along the continuous scale. You should strive to become adept at using either type of meter, since both kinds are found in industry.

Symbols for Meters

In Figure 2–22 you can see both the general schematic symbols and the pictorial symbols we will use in this book to represent analog and digital voltmeters, current meters (ammeters and milliammeters), and the ohmmeter (which measures resistance).

Making a Voltage Measurement with a Voltmeter

To measure the voltage between two points, you simply connect the voltmeter across (or in parallel with) the two points at which you wish to measure the potential difference, Figure 2–23. Notice, when measuring dc voltage, it is necessary to connect the negative test lead from the meter to the

ANALOG METER PICTORIALS AND SCHEMATIC SYMBOLS

DIGITAL METER PICTORIALS

FIGURE 2-22   Pictorial and schematic meter symbols

(a)

(b)

(c)

FIGURE 2–23   Measuring voltage "across" a component
(Photo by Michael A. Gallitelli)

more negative point of the circuit and the positive test lead to the more
positive point in the circuit. You can see in the figure that the meter is
reading the value of voltage across one component in the circuit (in this
case, component number 2).

Making a Current Measurement with an Ammeter

An illustration of how to connect an ammeter or a milliammeter to measure
circuit current is shown in Figure 2–24. Observe that the circuit has to be
broken in order for the meter to be put in the circuit's current path for
measurement purposes. You can also see that in this dc circuit, the meter

(d)

FIGURE 2–24   Measuring current "through" a circuit
(Photo by Michael A. Gallitelli)

leads must be connected so the current enters the meter's negative terminal, travels through the meter, and then exits the meter from the meter's positive terminal. You will learn more details about the concept of polarity a little later.

Making a Resistance Measurement with an Ohmmeter

Very important *cautions* to be observed when using an ohmmeter are that *power should be off the circuit and the component to be measured should be disconnected from the remainder of the circuit!* For example, to measure the resistance of a component, as shown in Figure 2–25:

1. Turn the power off the circuit by opening the switch.

2. Disconnect one resistor lead from the remainder of the circuit.

3. Make the resistance measurement by connecting the ohmmeter across both leads of the resistor.

KEY POINTS REGARDING USING METERS.   As indicated earlier, you can only gain practical experience in using meters to make measurements by actually making measurements. When learning and practicing this skill, be sure to observe all the safety precautions to protect yourself and the test equipment.

NOTICE: Power is off due to open switch. Component
being measured is isolated from the rest of the
circuit by disconnecting one of its leads (at "✕").

(a)

(b)
FIGURE 2–25   Ohmmeter measurement
(Photo by Michael A. Gallitelli)

SAFETY HINTS

1. Turn power off circuit to be tested when connecting meter leads.

2. Appropriately connect meter test leads to circuit. Observe proper polarity if measuring dc voltage or current.

3. Make sure meter is in correct mode (ac, dc, V, A, Ω).

4. Make sure range selected on meter is high enough to handle the value that is about to be tested.

5. a. If measuring V or I, turn on power and take reading.
   b. If measuring R, *do not turn on power.*

6. Turn off power and remove meter test leads from circuit.

7. If measuring I, reconnect circuit, as appropriate.

1. When measuring voltage, connect the voltmeter test leads to the two points at which you wish to know the potential difference, in volts. (That is, connect across, or in parallel, with the component or points to be checked).

2. When measuring current, break the circuit and insert the current meter leads in line (in series) with the current path through which you are trying to measure the current.

3. If measuring dc voltage or current, be sure to observe the direction of current through the circuit, and connect the positive and negative test leads accordingly. (Observe polarity.)

4. When measuring resistance, be sure to *turn off the power* to the circuit, isolate the component to be measured, then make the resistance measurement, as appropriate.

✓ IN-PROCESS LEARNING CHECK III

1. In the diagram shown, if you wanted to measure the dc voltage across component #1, at which point in the circuit would you attach the meter's negative test lead? At which point would you connect the meter's positive test lead?

Meter negative lead connected to Point _____.
Meter positive lead connected to Point _____.

2. In the diagram shown, if you wanted to measure the dc circuit current and you were going to break the circuit at point "X," to which point would you connect the ammeter's negative lead? At which point would you connect the ammeter's positive lead?

Meter negative lead connected to Point _____.
Meter positive lead connected to Point _____.

3. In the diagram shown, if you wanted to measure resistance, what is the first thing you would do, before connecting the meter across the component to be measured?

DIAGRAMS USED FOR ELECTRONIC SHORTHAND

Types of Diagrams

During your experiences in electronics you will encounter various types of diagrams which can be helpful. Several of these types include **block diagrams, schematic diagrams,** chassis layout diagrams, and others. For now, we'll focus on the block diagram and the schematic diagram since these are the kind you will be using most.

Some Basic Symbols

Electronic circuits are frequently represented by either block diagrams, schematic diagrams, or both. In order to interpret and use schematic diagrams, it is necessary to learn some basic electronic symbols used in schematics. Because diagrams can be made to represent an infinite variety of circuits and configurations in a concise manner, schematic diagrams are the electronic technician's or engineer's shorthand.

    At this point, we'll introduce you to some of the common schematic symbols you will be using almost immediately. Additional schematic symbols will be introduced throughout the text, as you need to know them.

Become familiar with the symbols; the details of the represented compo-nents' operation and applications will unfold as your learning continues.

Examine Figure 2–26 and observe a few of the important electronic symbols. Don't be worried about committing these to rote memory at this time. You will automatically learn these by repeated usage as you continue in your studies.

Block Diagrams

Block diagrams are useful because they are simple and conveniently por-tray the "functions" relating to each block. Recall that we used a simple block diagram in an earlier chapter when a simple electrical system was discussed.

Refer to Figure 2–27 (see page 64) and see if you can understand the concept of a block diagram. In essence, a block diagram is created by ana-lyzing a system: breaking it into meaningful subsystems, which are drawn as individual blocks; then, interconnecting the subsystem blocks in a way that portrays flow and interaction between the blocks.

Note that the diagram quickly conveys what the system does. That is, this system distributes electrical power from a source to several loads. Also, it graphically shows the direction of flow of the electrical energy between blocks. Additionally, it very simply displays what each part of the system does. The power source supplies power, the distribution components control and allocate portions of the power to the various loads, and the loads use the power delivered to them.

Now, see if you can interpret the block diagram of a system most ev-eryone is familiar with.

PRACTICE PROBLEMS V

Look at Figure 2–28 (see page 64) and answer the following ques-tions:

1. What is the purpose of this system?

2. In which direction does the TV signal flow on the diagram, from right-to-left or from left-to-right?

3. What subsystem receives signals from the TV antenna?

4. What are the two types of outputs from the TV receiver circuitry?

Schematic Diagrams

Now that you have briefly seen the usefulness of block diagrams, let's move on to the schematic diagrams. In schematic diagrams you apply the knowl-edge of electrical symbols. Because the symbols are standardized, it is easy for others working in electronics to "decode" the information that the orig-inal designer of the schematic is conveying. The information is concise and useful in communicating the detailed electrical connections in the circuitry, the types and values of components, and other important information.

COMPONENT TYPES	PICTORIAL REPRESENTATIONS	SCHEMATIC SYMBOLS	CHAPTER REFERENCES
Conductors (connected)			Chapter 1 and following
Conductors (not connected)			Chapter 1 and following
Cell			Chapters 1, 2, & 9
Battery			Chapters 1, 2, & 9
Switch (SPST) Single-pole Single-throw			Chapters 1, 2, & following
Switch (SPDT) Single-pole Double-throw			Chapter 2
Switch (DPST) Double-pole Single-throw			Chapter 2
Switch (DPDT) Double-pole Double-throw			Chapter 2
Switch Rotary type			Chapter 2
Switch (NOPB) Push-button type (normally open)		NOPB	Chapter 2
Switch (NCPB) Push-button type (normally closed)		NCPB	Chapter 2

FIGURE 2–26 Various components and their schematic symbols (Photos c and d courtesy of Eveready Battery Co.; photos g, h, i, j, and k courtesy of Grayhill, Inc.; photo m courtesy of Johanson General Electric)

COMPONENT TYPES	PICTORIAL REPRESENTATIONS	SCHEMATIC SYMBOLS	CHAPTER REFERENCES
Fixed Resistors			Chapter 2 and following
Variable Resistor			Chapter 2 and following
Voltmeter			Chapter 2 and following
Ammeters Milliammeter and Microammeter			Chapter 2 and following
Ohmmeter			Chapter 2 and following
Fuses		or	Chapter 2
Circuit Breaker			Chapter 2
Fixed Inductors		AIR CORE IRON CORE	Chapter 14 and following
Variable Inductor			Chapter 14 and following
Fixed Capacitors			Chapter 18 and following
Variable Capacitor			Chapter 18 and following

FIGURE 2–26 (cont.)   Various components and their schematic symbols (Photos n and p courtesy of Simpson Electric Company; photo q courtesy of Bussman Division, Cooper Industries; photo r courtesy of Square D Co.; photo u courtesy of Mallory Capacitor Co.; photo v courtesy of Johanson General Electric)

Look at the sample pictorial and schematic diagrams in Figure 2–29 and notice the following facts:

1. Each component has an individualized callout. For example, one resistor is called R_1, another is called R_2, and so on. The switch is labeled S_1. If there were more switches, the series would carry on with S_2, and so on.

FIGURE 2–27   Sample block diagram

FIGURE 2–28   Sample block diagram problem

(**a**) Pictorial diagram

(**b**) Schematic diagram

(**c**)

FIGURE 2–29   Sample pictorial and schematic diagrams (Photo by Michael A. Gallitelli)

2. In the pictorial, you can see that the resistor color-code bands are shown. In this schematic, the resistor values are given. If there were other components with important value data, their values would also be shown on the diagram.

3. In both the pictorial and the schematic diagrams, the battery terminals are marked with + and −, as appropriate. This marking of polarity can be given at each component; however, since you will study more about polarity in the next chapter, we will wait until then to use these polarity markings throughout the diagram.

4. Finally, an extremely important feature of all schematic, pictorial, or other wiring diagrams is that every circuit interconnection is shown, allowing the reader to understand every component's electrical relationship to every other component in the circuit.

Also, on many schematic diagrams, the values of electrical parameters such as voltage, current, resistance, and so on are called out. These data help immensely when technicians are trying to troubleshoot a circuit.

PRACTICE PROBLEMS VI

Look at Figure 2–30 and answer the following questions:

1. How many different kinds of components are there in the circuit? Name them.

2. What is the highest value resistor?

(a) Pictorial

(b) Schematic

FIGURE 2–30   Diagrams for Practice Problems VI

3. What kind of switch is used, a SPST or a SPDT?

4. Is the voltage source a cell or a battery?

PRACTICE PROBLEMS VII

Look at Figure 2–31 and answer the following questions:

1. How many resistors are in the circuit?

2. Which resistor has the lowest resistance value? What is its value?

3. How many switches are in the circuit?

4. What is the value of the circuit applied voltage?

5. Which switch is opened and which is closed?

Did you do well? As you can see, schematics can provide a large amount of information in a little space. Your appreciation for the value of schematics will undoubtedly continue to grow with your experiences in electronics.

(**a**) Pictorial diagram

(**b**) Schematic diagram

FIGURE 2–31   Sample pictorial and schematic diagrams for Practice Problems VII

! **SAFETY HINTS**

You may have already been exposed to some safety hints about electricity and electronics. Nevertheless, here are a few facts and tips that can serve either as information or as reinforcement of earlier learning.

Background Facts

Electrical shock occurs when your body provides a path for current between two points having a difference of potential. Generally, the amount of injury a person receives from a shock depends on the amount of current forced through the path of current. A range of effects from minor discomfort, severe muscular contractions, severe burns, ventricular fibrillation (erratic heart action), and death can occur. Many times the injury caused by involuntary jerking or muscular contractions is worse than the shock. Even very low values of current, if they flow through your body under the right conditions, can be fatal. Therefore, use extreme caution, no matter what value of voltage or current you are working with.

A higher voltage causes more current to flow through a circuit. The higher the resistance path, the lower the current. Therefore, anything that lowers the voltage on those circuits or any means of increasing resistance of possible current paths through the body is helpful.

Some General Shop or Laboratory Safety Tips

1. *Remove power* from circuits before working on them, as appropriate.

2. If you must work on a circuit with power applied, *work with only one hand at a time.* This will prevent touching two points in the circuit simultaneously and providing a current path from one hand, through your body, to the other hand. Also, avoid touching two points with any part of your body. For example, your hand touching a point in the circuit, and your arm leaning against the circuit chassis, ground, or another point in the circuit can provide "shocking results."

3. *Insulate your body* from ground so a path is not provided from one point (through your body) to ground.

4. Use only *tools with insulated handles.*

5. *Know where the closest switch or main breaker switch is located* that will turn off power to the system. Make sure everyone working near you also knows this.

6. *Remove metal jewelry.* Dangling jewelry can not only make electrical contact, but, if you are working on "rotating machinery," can entangle you with the machinery.

7. When working on circuits with capacitors that store charge even after the power has been removed, *discharge the capacitors* prior to working on the circuit.

8. *Ground all equipment that should be grounded,* including equipment chassis, cabinets, and so forth.

Safety When Working With Hot Soldering Irons

1. *Position hot soldering irons properly* to prevent accidentally laying your arm, hands, or other objects on them.

2. *Wear safety glasses* when soldering.

3. *Be careful when shaking or wiping hot solder off a soldering iron.* Hot or molten solder splatters easily and causes serious burns.

NOTE: Other safety hints will be provided at strategic times during your training. Don't just read these safety hints; *follow them!*

SUMMARY

▶ Charge (Q or q) is measured in coulombs. The abbreviation for coulomb is C. A typical sub-unit of the coulomb is the μC, or microcoulomb, meaning one-millionth of a coulomb.

▶ Potential difference, or voltage, is represented by the letter V. The unit of measure is the volt, also abbreviated "V." The submultiples and multiple of voltage commonly found include mV (millivolt), μV (microvolt), and kV (kilovolt).

▶ Current (I) is measured in amperes. The abbreviation for the unit of measure is A. Typical submultiples of this unit are mA (milliampere), μA (microampere), and occasionally pA (picoampere).

▶ Conductance uses the letter G and the unit of measure is the siemens (S). The commonly found submultiples are mS (millisiemens) and μS (microsiemens).

▶ Resistance uses the letter R and the unit of measure is the ohm (Ω). Typical multiples of this unit are the kΩ (kilohm) and the MΩ (megohm).

▶ Several components found in many circuits are resistors, switches, batteries (or voltage sources), and the conductor wires. Know the standard symbols used for each.

▶ Commonly used metric subunit prefixes are pico, or micro micro, meaning one millionth of one millionth (one trillionth) of a unit; nano, meaning one thousandth of a millionth (a billionth) of a unit; micro, meaning one mil-lionth of a unit; and milli, meaning one thousandth of a unit.

▶ Powers of ten for the above subunit prefixes are 10^{-12} for pico (or micro micro), 10^{-9} for nano, 10^{-6} for micro, and 10^{-3} for milli units.

▶ Common metric multiple-unit prefixes are kilo (1,000 times the unit) and mega (1,000,000 times the unit).

▶ Symbols for all of the above metric prefixes are p (pico), n (nano), μ (micro), m (milli), k (kilo), and M (mega).

▶ Conductors can vary in the materials they are made from, in physical size, in length, and in overall makeup. Solid and stranded wires are two basic types of conductors. Other differences between conductor types may relate to how they are packaged and the type of insulation used.

▶ There is a standard wire-gaging system called the American Standard Wire Gage, abbreviated AWG. The lower the gage number, the larger the wire.

▶ The cross-sectional area of conductors is measured in circular mils (CM). One mil equals one-thousandth of an inch. A one-CM wire has a diameter of 1 mil, or 0.001 inch.

▶ The larger a conductor's size; i.e., diameter and cross-sectional area, the lower its resistance, and the higher its current-carrying capacity. The greater the length of any given conductor, the greater its resistance from end to end.

▶ Resistivity (specific resistance) of any given type material is a measure of that material's resistance per standard length and cross-sectional area. The formula for resistivity is $R = \rho l/A$.

▶ A wire table is a convenient source of information regarding a conductor's resistance per given length and size, as well as its current-carrying capacity for specified conditions.

▶ Resistors perform several useful functions in electronic circuits: They limit current and divide voltage.

▶ Several types of resistors are carbon or composition resistors, film resistors, and wire-wound resistors. Another way to categorize resistors is fixed, semifixed, and variable resistors.

▶ Carbon or composition resistors normally are manufactured with power dissipation ratings from 1/4 to 2 watts. Wire-wound resistors generally are rated to handle power levels from 5 watts to over 100 watts.

▶ Because carbon resistors are typically very small, a color-coding system has been devised to indicate resistor values.

▶ The sentence *"Big Boys Race Our Young Girls, But Violet Generally Wins"* can help you remember the ten colors in the code: Black, Brown, Red, Orange, Yellow, Green, Blue, Violet, Gray, and White.

▶ Block diagrams illustrate the major functions within a circuit or system and convey the flow of signals, electrical power, and so forth. General flow is typically shown from left to right.

▶ Schematic diagrams provide more information than block diagrams about types of components in the circuit/s, their values, and specific connections and related electrical parameter values.

▶ Block diagrams provide an overview of the system, whereas schematic diagrams provide the information needed to enable circuit analysis, circuit repairs or maintenance, and/or desired circuit modifications.

▶ Safety rules relating to electrical circuits, hot soldering irons, and rotating machinery should always be followed.

FORMULAS AND CALCULATOR SEQUENCES SUMMARY

FORMULA 2–1 $A = d^2$
diameter value (in mils), $\boxed{\times^2}$

FORMULA 2–2 $R = \dfrac{\rho l}{A}$
Resistivity (CM-Ω/ft), $\boxed{\times}$, length (ft), $\boxed{\div}$, area (CM), $\boxed{=}$

REVIEW QUESTIONS

1. Convert 3 mV to an equivalent value in V.

2. If a value of current is 12 µA, this can be stated as a value of 12×10 to what power amperes? What is this same value of current when stated as a decimal value of amperes?

3. What submultiple of a milli unit is a micro unit?

4. How many kΩ are there in a MΩ?

5. Draw a schematic diagram illustrating a voltage source (battery), two resistors, and a

SPST switch all connected end-to-end. Illustrate the first resistor connected to the source's negative side, the other resistor connected to the source's positive side, and the switch between the resistors.

6. ___ is the symbol for a fixed _____.

7. ___ is the symbol for a variable _____.

8. ___ is the symbol for a _____.

9. ___ is the symbol for a _____.

10. Convert the following units:
    a. 15 milliamperes to amperes.
    b. 15 milliamperes to 15 times a power of ten.

    c. 5,000 volts to kilovolts.
    d. 5,000 volts to 5.0 times a power of ten.
    e. 0.5 megohms to ohms.
    f. 100,000 ohms to megohms.
    g. 0.5 ampere to milliamperes.

11. Name two or more types of resistors and discuss their physical construction.

12. Give the resistance value and tolerance rating for the following color-coded resistors: (NOTE: The first color shown is the first band; the second color represents the second band, and so forth.)
    a. Yellow, violet, yellow, and silver.
    b. Red, red, green, and silver.
    c. Orange, orange, black, and gold.
    d. White, brown, brown, and gold.
    e. Brown, red, gold, and gold.

13. Name at least three important parameters, or factors, to consider when selecting a resistor for a given circuit.

14. If a resistor has a fifth color band, what does this band designate?

15. List five or more safety rules to follow when working on electrical circuits and/or around hot soldering irons or rotating machinery.

CHALLENGE QUESTIONS

1. In your own words, describe why a resistor color code is necessary.

2. If the environmental temperature conditions for a certain conductor's operation increase, describe what happens to the following parameters:
    a. Its resistance
    b. Its current-carrying capacity

3. Explain in your own words what the reliability rating on certain resistors means.

4. Describe in your own words the procedures and precautions for:
    a. Measuring dc voltage across a specified component in a multi-component circuit.
    b. Measuring circuit current.
    c. Measuring the value of a specified resistor in a multiresistor circuit.

TEST YOURSELF

1. Draw a block diagram illustrating a simple power distribution system. Use at least three blocks, and label all blocks and diagram features.

2. Draw a schematic diagram showing two resistors, one meter, two switches, and a battery. Label all components.

3. List the color coding for the following resistors:
   a. 1,000 ohms, 10% tolerance.
   b. 27 kilohms, 20% tolerance.
   c. 1 megohm, 10% tolerance.
   d. 10 ohms, 5% tolerance.

PERFORMANCE PROJECTS CORRELATION LISTING

Suggested performance projects that correlate with topics in this chapter are:

CHAPTER TOPIC	PERFORMANCE PROJECT		PROJECT NUMBER
Making a Voltage Measurement with a Voltmeter	Voltmeters	(Use & Care of Meters series)	1
Making a Current Measurement with an Ammeter	Current Meters	(Use & Care of Meters series)	2
Making a Resistance Measurement with an Ohmmeter	Ohmmeters	(Use & Care of Meters series)	3
Resistor Color Code	Resistor Color Code	(Ohm's law series)	4

Ohm's Law

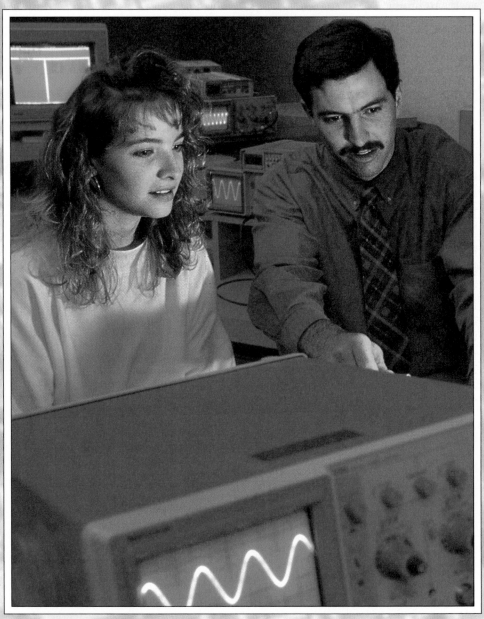

Courtesy of Wellman Thermal System Corp., © 1992 Loren Santow

3

KEY TERMS

Conventional current

Energy

Ohm's law

Polarity

Power

Power dissipation

Watt (W)

Watthour (Wh)

Wattsecond (Ws)

Work

OUTLINE

Ohm's law and relationships between electrical quantities

Rearranging to find the unknown quantity

Powers of ten and scientific notation

Sample application of metric prefixes and powers of ten

Some basic rules relating to powers of ten

Direction of current flow

Polarity and voltage

Comparison of circuit current directions from DC and AC sources

Work, energy, and power

Measuring electrical energy consumption

Commonly used versions of Ohm's law and power formulas

CHAPTER PREVIEW

This important chapter will provide a thorough knowledge of Ohm's law. Ohm's law is fundamental to your career as a technician. It is the basis for calculations of electrical quantities in all types of electrical and electronic circuits. You will have opportunity to combine the knowledge you have already acquired with some new and very practical circuit analysis information.

Additionally, you will begin to use the scientific calculator to solve circuit problems. In this chapter, you will only use the rudimentary functions, such as multiplying, dividing, or squaring numbers. Also, you will see how to use the calculator to find the scientific and engineering notation related to powers of ten. As you progress in the text, you will learn to use many other time-saving functions of the scientific calculator. We will explain these functions as the need for them unfolds.

OBJECTIVES

After studying this chapter, you should be able to:

• Explain the relationships of current, voltage, and resistance

• Use Ohm's law to solve for unknown circuit values

• Illustrate the direction of current flow and polarity of voltage drops on a schematic diagram

• Use metric prefixes and powers of ten to solve Ohm's law problems

• Use a calculator to solve circuit problems and express numbers with scientific or engineering notation

• Explain power dissipation

• Use appropriate formulas to calculate values of power

OHM'S LAW AND RELATIONSHIPS BETWEEN ELECTRICAL QUANTITIES

Georg Simon Ohm, after whom the unit of resistance is named, observed and documented important relationships between three fundamental electrical quantities. He verified a relationship that indicated that electrical current in a circuit was *directly related* to the voltage applied and *inversely related* to the circuit's resistance. Recall that if two items are *directly related,* as one item increases, the other item also increases, and it increases proportionately to the first item's increase. If one item goes down, the other items goes down, proportionately. *Inversely related* means that as one item increases, the other item decreases, proportionately, and vice versa. Also recall that in defining some of the electrical quantities, the relationships in **Ohm's law** were implied in the definitions. For example, an ohm is that amount of resistance that limits current to one ampere when one volt is applied.

The Relationship of Current to Voltage with Resistance Constant

These relationships are stated in a formula known as Ohm's law, where:

Current (amperes) = Potential difference (volts) / Resistance (ohms)

Formula 3–1	$I = \dfrac{V}{R}$; $\left(\text{Amperes} = \dfrac{\text{Volts}}{\text{Ohms}}\right)$

 voltage value, $\boxed{\div}$, resistance value, $\boxed{=}$

EXAMPLE

If a circuit has 10 volts applied voltage and a resistance of 5 ohms, what is the value of current in amperes?

$$I = \frac{V}{R}, \text{ where } I = \frac{10\text{ V}}{5\ \Omega} = 2\text{ A (See Figure 3–1)}$$

As you can see from the formula, if V increases and R remains unchanged, then a large value V divided by the same value R will produce a larger number for I. This means that with a given resistance value, if voltage is increased, then current must increase also, and it will *increase* in direct proportion to the voltage change. Compare the current in Figure 3–1a with the current in Figure 3–1b. As you can see, when voltage increased from 10 V to 15 V and the resistance remained at 5 Ω, the current increased from 2 A to 3 A. That is directly proportional to the voltage change, isn't it? As voltage increased to one and one-half times the original value, current also increased to one and one-half times its original value.

On the other hand, if voltage is decreased while the resistance remains the same, the current will *decrease* proportionately with the voltage change. Compare the current in Figure 3–1c with the current shown for the circuit in Figure 3–1a. Again, you can see a direct proportion. The circuit current in Figure 3–1c is half the value of 2 A, shown in Figure 3–1a. This is due to the fact the applied voltage in the circuit of Figure 3–1c (5 V) is half of the 10-V applied voltage, shown in Figure 3–1a. This type relationship may be referred to as a "linear proportionality."

We conclude, then, that *current is directly related to voltage* for a given resistance value. If the circuit applied *voltage increases* for a given resistance value, *current increases* in direct proportion to the increase in voltage. If *voltage decreases* for a given resistance value, *current* through the circuit will *decrease* in direction proportion to the voltage change.

The Relationship of Current to Resistance with Voltage Constant

In Figure 3–2a, the voltage is 10 V and the resistance value is 10 Ω. Thus, the current value is 1 A.

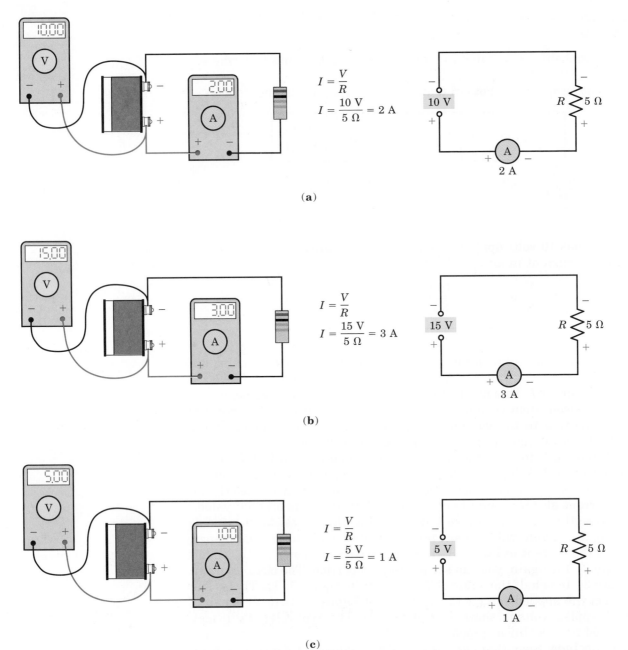

$$I = \frac{V}{R}$$

$$I = \frac{10 \text{ V}}{5 \text{ }\Omega} = 2 \text{ A}$$

(a)

$$I = \frac{V}{R}$$

$$I = \frac{15 \text{ V}}{5 \text{ }\Omega} = 3 \text{ A}$$

(b)

$$I = \frac{V}{R}$$

$$I = \frac{5 \text{ V}}{5 \text{ }\Omega} = 1 \text{ A}$$

(c)

FIGURE 3–1   Ohm's law and the relationship of current to voltage with resistance constant

Recall, in Figure 3–1a, when the resistance was 5 Ω, and the applied voltage was 10 V, current was 2 A. By *increasing* the resistance to 10 Ω, with the same value of voltage (that is, 10 V), the circuit current *decreases* to 1 A. In other words, with V remaining the same and R increasing, the I decreases proportionately, showing an inverse relationship between I and R. Current is *inversely related to resistance*.

In Figure 3–2b, you can see that if voltage is kept constant at 10 V and the resistance is decreased to 5 Ω, the current will increase proportionately.

$$I = \frac{V}{R}$$

$$I = \frac{10\ V}{10\ \Omega} = 1\ A$$

(a)

$$I = \frac{V}{R}$$

$$I = \frac{10\ V}{5\ \Omega} = 2\ A$$

(b)

$$I = \frac{V}{R}$$

$$I = \frac{10\ V}{15\ \Omega} = 0.66\ A$$

(c)

FIGURE 3–2   Ohm's law and the relationship of current to resistance with voltage constant

Likewise, in Figure 3–2c, it is obvious that if the resistance is increased in value, for a given applied voltage, the circuit current will decrease in proportion to the resistance change.

The direct relationship between voltage *(V)* and *(I),* and the inverse relationship between resistance *(R)* and current *(I)* are illustrated with the Ohm's law formula in Figure 3–3.

Now that we've shown you some examples, you can use the knowledge you have gained regarding Ohm's law, and direct and inverse relationships on the following problems.

Since $I = \dfrac{V}{R}$, then:

$$
\begin{array}{llll}
\text{If} & \dfrac{V\uparrow}{R\rightarrow} & \text{then } I\uparrow & (I \text{ directly related to } V. \\
 & & & \text{If } V \text{ doubles, } I \text{ doubles, if} \\
 & & & R \text{ is not changed)} \\[2em]
\text{If} & \dfrac{V\downarrow}{R\rightarrow} & \text{then } I\downarrow & (I \text{ directly related to } V. \\
 & & & \text{If } V \text{ halves, } I \text{ halves, if} \\
 & & & R \text{ is not changed)}
\end{array}
$$

(a)

$$
\begin{array}{llll}
\text{If} & \dfrac{V\rightarrow}{R\uparrow} & \text{then } I\downarrow & (I \text{ inversely related to } R. \\
 & & & \text{If } R \text{ doubles, } I \text{ halves, if} \\
 & & & V \text{ is not changed)} \\[2em]
\text{If} & \dfrac{V\rightarrow}{R\downarrow} & \text{then } I\uparrow & (I \text{ inversely related to } R. \\
 & & & \text{If } R \text{ halves, } I \text{ doubles, if} \\
 & & & V \text{ is not changed)}
\end{array}
$$

(b)

**FIGURE 3–3   Direct and inverse relationships of
voltage, current, and resistance:
(a) Direct; (b) Inverse**

PRACTICE PROBLEMS I

1. Determine the circuit current for a circuit that has 150 V of
   applied voltage and a resistance of 10 Ω. (To solve, write down the
   Ohm's law formula, then substitute values in the formula as
   appropriate.)

2. If the voltage in problem 1 decreased to 75 V, but the resistance
   remained at 10 Ω, what would be the new circuit current?

3. Suppose that both voltage and resistance doubled for the circuit
   in problem 1:
   a. What would happen to the circuit current? Would it increase,
      decrease, or remain the same?
   b. Use Ohm's law and calculate the actual value resulting from
      these changes.

PRACTICAL NOTES

To solve problems, get in the habit of writing down the appropriate
formula, substituting the knowns into the formula, then solving for the
unknown. Also, if the problem is stated in words (without a diagram), draw
a simple diagram illustrating the circuit in question. It will help you think
clearly!

So far, you have looked at only one arrangement of Ohm's law: $I = V/R$. Now, a quick review of transposing an equation will be given because this process isolates the unknown quantity of interest on one side of the equation in order to simplify finding its value.

Recall that an equation is a statement of equality between two quantities, or sets of quantities. The equality is expressed by the "=" sign. For example, the Ohm's law equation of $I = V/R$ says that the current is equal to the voltage divided by the resistance. Notice that the I is shown on one side of the equation, isolated from the other factors. To find its value, it is only necessary to carry out the operations on the other side of the equation. If both V and R are known, then their values are substituted into the equation to solve for I.

REARRANGING TO FIND THE UNKNOWN QUANTITY

How can the $I = V/R$ equation be transposed to solve for V or R? Before doing that, let's look at two basic premises relating to equations.

1. *Multiplying* or *dividing* both sides of an equation by the *same* factor does not change the equality.

2. By using the correct value factor, cancellation on one side of the equation can be made to provide the desired change in the equation by isolating the desired unknown.

EXAMPLE

Use the formula for Ohm's law and solve for V and for R.

$$I = \frac{V}{R}$$

To isolate V, multiply both sides of the equation by R, and cancel factors, as appropriate.

$$\text{Therefore, } I \times R = \frac{V}{R} \times R. \text{ Then } I \times R = \frac{VR}{R}.$$

Formula 3–2	$V = I \times R$; (Volts = Amperes × Ohms)

 current value, $\boxed{\times}$, resistance value, $\boxed{=}$

Now, I and R are on one side of the formula and R is no longer a denominator in a fraction.

How can we isolate R? If both sides are divided by I, then the I on the right side of the equation will cancel because I divided by $I = 1$.

$$V = IR. \text{ Then, } \frac{V}{I} = \frac{\cancel{I} \times R}{\cancel{I}}$$

Formula 3–3	$R = \dfrac{V}{I}; \left(\text{Ohms} = \dfrac{\text{Volts}}{\text{Amperes}} \right)$

 voltage value, $\boxed{\div}$, current value, $\boxed{=}$

As you can see, either multiply or divide *by the factor you are trying to cancel,* depending on whether that factor is in the numerator or the denominator of the equation. Multiply both sides of the equation by that factor, if you are trying to cancel the factor in a denominator. Divide both sides of the equation by the factor you are trying to cancel, if the factor is in the numerator.

You have now been exposed to all three variations of Ohm's law formula.

$$I = \frac{V}{R}$$

$$V = I \times R$$

$$R = \frac{V}{I}$$

This review of transposing formulas to isolate unknowns will help as other similar occasions arise. A useful memory aid often helps students remember these three variations of Ohm's law, Figure 3–4.

Simply look at (or cover up) the desired quantity (I, V, or R) and observe the relationships of the other two quantities. To remember the correct version of the formula:

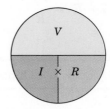

Look at I, the aid shows $\dfrac{V}{R}$

$$I = \frac{V}{R}$$

Look at V, the aid shows $I \times R$
$$V = I \times R$$

Look at R, the aid shows $\dfrac{V}{I}$

$$R = \frac{V}{I}$$

FIGURE 3–4   Ohm's law memory aid

Restating these three Ohm's law variations:

$$I = \frac{V}{R} \text{ or Amperes} = \frac{\text{Volts}}{\text{Ohms}}$$

$$V = I \times R \text{ or Volts} = \text{Amperes} \times \text{Ohms}$$

$$R = \frac{V}{I} \text{ or Ohms} = \frac{\text{Volts}}{\text{Amperes}}$$

Knowing any two of the three electrical quantities, you can solve for the unknown value by using the appropriate version of the Ohm's law formula. This capability is important!

In the practical world of electronics, you usually will not be dealing with just the basic units of voltage, amperes, and ohms. You will frequently work with millivolts, milliamperes, microamperes, kilohms, and megohms.

POWERS OF TEN AND SCIENTIFIC NOTATION

Powers of Ten

By way of review, observe Figure 3–5, which again shows the relationships between decimal numbers, powers of ten, metric terms, and electronic terms that are expressed using metric prefixes. Recall that in the preceding chapter, it was indicated that very small and very large electrical quantities can more easily be written and worked with by using powers of ten. In this chapter, we want you to begin to use the powers-of-ten mathematical tool because you will have some actual circuit problems to solve.

Scientific Notation

Using just powers of 10 we can express a given decimal number in numerous ways.

EXAMPLE

Using powers of ten, the decimal number 250 can be expressed as:

250×10^0, or
25×10^1, or
2.5×10^2, or
0.25×10^3, and so forth.

NUMBER	POWER OF 10	TERM	SAMPLE ELECTRONIC TERM
0.000000000001	10^{-12}	pico	pA (1×10^{-12} ampere)
0.000000001	10^{-9}	nano	nA (1×10^{-9} ampere)
0.000001	10^{-6}	micro	μA (1×10^{-6} ampere)
0.001	10^{-3}	milli	mA (1×10^{-3} ampere)
1,000	10^3	kilo	kΩ (1×10^3 ohms)
1,000,000	10^6	mega	MΩ (1×10^6 ohms)

FIGURE 3–5   Powers of ten related to metric and electronic terms

A related system of notation, called "scientific notation," uses the powers of ten to advantage, and adds the idea of converting any number to a number between 1 and 10 times the appropriate power of ten.

EXAMPLES

The decimal number 250 expressed in scientific notation becomes:

$$2.5 \times 10^2$$

The number 1,500 expressed in scientific notation becomes:

$$1.5 \times 10^3$$

The idea of scientific notation is to always convert the number into a small, easy to handle number between 1 and 10, times a power of ten.

Now, let's have you practice a little with powers of 10 and scientific notation.

PRACTICE PROBLEMS II

Convert the following decimal numbers to scientific notation:

1. 25,000
2. 12.335

3. 0.100
4. 0.0015

Convert the following numbers to their decimal equivalents:

5. 235×10^2
6. 0.001×10^6

7. 10×10^3
8. 15×10^{-3}

Scientific Notation Mode

As was stated earlier, powers of 10 and scientific notation can be easily used, when using a calculator, as well as when performing calculations without the calculator. Most "scientific" calculators have a scientific notation mode (sometimes abbreviated on the keyboard as "Sci"). Also, scientific calculators have an exponent $\boxed{\text{EE}}$ or $\boxed{\text{EXP}}$ key, which allows easy utilization of powers of 10. See Figure 3–6.

EXAMPLE

To find the scientific notation expression for any given decimal number:

1. Set your calculator to the *scientific notation mode*. (See your particular calculator's instruction manual to perform this appropriately.)

2. Once the scientific notation mode has been selected, simply *"input" the decimal number* on the calculator number keypad. For example, input the number 25647.

**FIGURE 3–6   Photo closeup of a typical
scientific calculator (Courtesy of
Texas Instruments)**

3. Then *push* $=$. The calculator readout should show you the scientific
notation for the decimal number you inputted. Does it? In this case,
the result should be 2.5647^{04} or 2.5647 04, indicating 2.5647×10^4.

Now you try a couple on your own.

PRACTICE PROBLEMS III

Use a calculator and convert the following numbers to scientific
notation form:

1. 0.00234   2. 12.55   3. 1,245.50

After completing the scientific notation problems, take the calculator out of
the scientific notation mode. (See your calculator instruction book, if necessary, to see how to do this).

Using the [EE] or [EXP] Key

Now let's try to use the exponent [EE] or [EXP] key in conjunction with powers of ten.

1. Input a number with the calculator keypad.

2. Push the [EE] or [EXP] button.

3. Input the power of 10 you want to raise that number to.

4. Press the = sign.

You will see that number expressed in terms of a power of ten number.

EXAMPLES

Determine the value of (a) 235 × 10 to the 3rd power and (b) 235 × 10 to the minus 2nd power.

(a) Input 235, [EE] or [EXP], input 3, [=]

The readout indicates that 235×10^3 equals 235,000.

If you want to use a negative power of ten, you simply input the negative sign prior to expressing the number for the power of ten.

(b) Input 235, [EE] or [EXP], plus/minus [+/−] or [±], input 2, [=]

The result indicates that 235×10^{-2} equals 2.35. Now you try the following problems.

PRACTICE PROBLEMS IV

Use the calculator and the [EE] or [EXP] key to find the decimal values for the following:

1. Determine the value of the number 1.456 times ten to the 6th power.

2. Determine the value of the number 333 times ten to the minus 4th power.

3. Determine the value of the number 1,500,000 times ten to the minus 6th power.

4. Determine the value of the number 10 times ten to the 5th power.

SAMPLE APPLICATION OF METRIC PREFIXES AND POWERS OF TEN

Suppose that a circuit has 10,000 ohms of resistance with 150 volts of applied voltage. What is the current?

$$I = \frac{V}{R}$$

$$I = \frac{150 \text{ V}}{10,000 \text{ }\Omega} \left(\text{or } \frac{150 \text{ V}}{10 \text{ k}\Omega} \right)$$

$$I = 0.015 \text{ A (or 15 mA)}$$

It is cumbersome to divide units by thousands of units (kilounits), isn't it? It turns out that when you divide units (volts) by kilounits (kΩ), the answer is in milliunits (mA). That is a convenient thing to remember, and you will use it often.

However, how would this work if we used powers of ten?

$$I = \frac{V}{R}$$

$$I = \frac{150 \text{ V}}{10 \times 10^3 \text{ }\Omega}$$

$$I = 15 \times 10^{-3} \text{ A (or 15 mA)}$$

Did you notice that when 10^3 is brought from the denominator to the numerator, it becomes 10^{-3}? Suppose we had divided one by one thousand. How many times would one thousand go into one? The answer is one one-thousandth time. Recall that one thousand equals 10^3 and one one-thousandth equals 10^{-3}. Therefore, when 150 volts are divided by 10,000 ohms, the answer is in thousandths of an ampere. Stated mathematically:

$$\frac{1}{1,000} = 0.001 \text{ or one thousandth (or } 1 \times 10^{-3}) \text{ and}$$

$$\frac{150}{10,000} = 0.015 \text{ or 15 thousandths (or } 15 \times 10^{-3})$$

SOME BASIC RULES RELATING TO POWERS OF TEN

Since you will be using powers of ten frequently, we will summarize some rules.

Rules for Moving Decimal Places Using Powers of Ten

To express a decimal number as a whole number times a power of ten, move the decimal point to the right and count the number of places moved from the original point. The number of places counted (or moved) is the appropriate negative power of ten.

EXAMPLE

The number 0.00253 can be expressed using powers of ten in any of the following ways:

(1) 253×10^{-5} (2) 2530×10^{-6} (3) 25.3×10^{-4} and so on.

To express a large number as a smaller one times a power of ten, move the decimal point to the left and count the number of places moved from the original point. The number of places counted is the appropriate positive power of ten.

EXAMPLE

The number 42,853 can be expressed using powers of ten in any of the following ways:

(1) 4.2853×10^{4} (2) 42.853×10^{3} (3) 4285.3×10^{1} and so on.

Rule for Multiplication Using Powers of Ten

When multiplying powers of ten, add the exponents (algebraically). (NOTE: Laws of exponents are applicable when using powers of ten.)

EXAMPLE

$$10^{2} \times 10^{2} = 10^{4}$$
$$2 \times 10^{4} \times 2 \times 10^{-3} = 4 \times 10^{1}$$

Rule for Division Using Powers of Ten

When dividing powers of ten, subtract the exponents (algebraically).

EXAMPLE

$$\frac{4 \times 10^{-2} \ \text{(numerator)}}{2 \times 10^{-4} \ \text{(denominator)}} = 2 \times 10^{-2} \times 10^{4} = 2 \times 10^{2}$$

$$\frac{5 \times 10^{3}}{2.5 \times 10^{6}} = 2 \times 10^{3} \times 10^{-6} = 2 \times 10^{-3}$$

NOTE: Change the sign of the denominator power of ten and add algebraically to the numerator power of ten exponent. If the power of ten exponent in the denominator is the same sign and same value as the numerator, they cancel.

Rules for Multiplication and Division Combined

The rules are: a. Multiply all factors in the numerator.
                   b. Multiply all factors in the denominator.
                   c. Divide.

EXAMPLE

$$\frac{15 \times 10^2 \times 10 \times 10^4}{2 \times 10^{-2} \times 12.5 \times 10^{-2}} = \frac{150 \times 10^6}{25 \times 10^{-4}} = 6 \times 10^{10}$$

Rule for Addition of Powers of Ten

Change all factors to the same power of ten and add the coefficients. The answer retains the same power of ten as the added factors.

EXAMPLE

$$(2 \times 10^2) + (5 \times 10^3) = (0.2 \times 10^3) + (5 \times 10^3) = 5.2 \times 10^3$$

Examples of coefficients

Rule for Subtraction of Powers of Ten

Change all powers of ten to the same power of ten and subtract the coefficients.

EXAMPLE

$$(5 \times 10^3) - (2 \times 10^2) = (5 \times 10^3) - (0.2 \times 10^3) = 4.8 \times 10^3$$

See if you can apply Ohm's law, metric prefixes, and powers of ten.

✓ IN-PROCESS LEARNING CHECK I

1. Refer to the diagram below and determine the circuit current. $I = $ _____ amperes, or _____ milliamperes.

2. Refer to the circuit below and determine the circuit applied voltage.
   $V =$ _____ volts.

3. Refer to the circuit below and determine the value of circuit resistance.
   $R =$ _____ ohms, or _____ kilohms.

4. a. If R increases and V remains the same in Problem 1, does I increase, decrease, or remain the same?
   b. If I increases and R remains the same in Problem 2, would the voltage applied be higher, lower, or the same as stated in the problem?
   c. In Problem 3 if I doubles but the V remains the same, what must be true regarding R?

DIRECTION OF CURRENT FLOW
Electron Flow Approach

In Chapter 1, there was a brief discussion about electrons (negative in charge) moving to a positively charged point in a circuit. When circuit diagrams are shown, it is quite common to show the direction of current flow through the circuit by means of arrows. In the cases shown in Figure 3–7, the arrows are used to show the direction of the electron current flow. *For*

FIGURE 3–7   Method of showing direction of electron current flow with arrows

this text, this is the standard that will be used for indicating direction of currents through circuits. This is coherent with our discussion of the fact that current flow is comprised of the movement of free electrons. Because electrons are negative in charge, they are repelled from negative points in circuits and are attracted toward more positive points; thus, electron current flow moves from negative-to-positive locations in components or circuits external to the voltage source.

Conventional Current Flow Approach

Prior to the development of the electron theory, from which we derive our electron-flow approach to analyzing electron movement, there was a theory that assumed that current flow was in the direction that a positive charge would drift or move through a circuit. Obviously, this asusmption would show current flowing from positive-to-negative points in a circuit. This difference in representing direction of current flow is referred to as **conventional current.** See Figure 3–8 for examples of how diagrams would represent this type current.

Since many engineering formulas and thought processes were created back when this theory was prevalent, numerous engineering texts represent current flow in this manner. Also, this is the manner that has been used in creating certain electronic component symbols. For these reasons, we want you to be aware of the term conventional current. Be assured that regardless of whether electron current flow or conventional current flow is used in circuit analysis, the answers come out the same! As indicated earlier, this text will adhere to the electron current flow approach, as indicated in Figure 3–7.

POLARITY AND VOLTAGE

The concept of polarity, or one point in a component or circuit being considered more negative or positive than another point, was discussed earlier. Recall that the more negative point was labeled with a negative sign, and the less negative (or positive point) was labeled with a positive sign. This technique indicates there is a potential difference, or voltage, between the two points.

To label the polarity of voltages across components in a circuit diagram, note the direction of current flow, then label the point where the current enters the component as the negative end. The other end of the component

FIGURE 3–8   Arrows showing direction of "conventional current"

FIGURE 3–9   Illustrating polarity of voltages

(where the current exits) is labeled as the positive point. This process is true for all components in the circuit, except the voltage source. In that case, the negative side of the source is the source of the electron flow, and the positive end of the source is the end where the electrons return after traveling through the external circuit, Figure 3–9.

> **PRACTICAL NOTES**
>
> You should have observed throughout our discussions that current flows through a component and/or circuit. Voltage does **not** "go" or "flow." It is a difference of potential between points, so think of voltage as being **across,** or **between** two points.

COMPARISON OF CIRCUIT CURRENT DIRECTIONS FROM DC AND AC SOURCES

So far, we have only mentioned dc (direct current) sources, and the resulting direction of current flow through circuits connected to them. At this point, we want to introduce you to the fact that there is also another type of voltage and current known as "ac." The ac stands for alternating current. An example of this type of voltage and current is that supplied to your home from the power company. An ac source provides voltage that periodically alternates in direction. The output waveform from such a source is known as a "sine wave." See Figure 3–10a. A complete "cycle" of a sine wave has occurred at the moment the waveform starts repeating itself. For example, in the waveform shown in Figure 3–10a, the voltage starts from zero level, moves to a maximum positive level, and returns back to zero. *Then,* the source's output voltage *reverses* direction, or polarity. The output voltage now moves to a maximum negative point, and then, back to zero. At this point, the voltage waveform starts the same sequence of events all over again. This begins the second cycle. See Figure 3–10b.

One-half of the complete cycle is called an "alternation"; therefore, there are two alternations in a cycle of ac, Figure 3–10a. As you can see, there is one positive alternation and one negative alternation in a cycle. You will be studying this type waveform in more depth in a later chapter. For now, we just want you to see the general differences between dc and ac.

Notice the contrast between dc and ac voltage as shown in Figure 3–11. The dc voltage remains at one polarity. This means the current is always traveling through the circuit in one direction, Figure 3–11a. In the ac

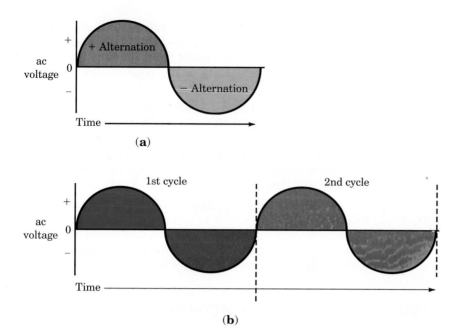

FIGURE 3–10   Representation of ac sine-wave voltages

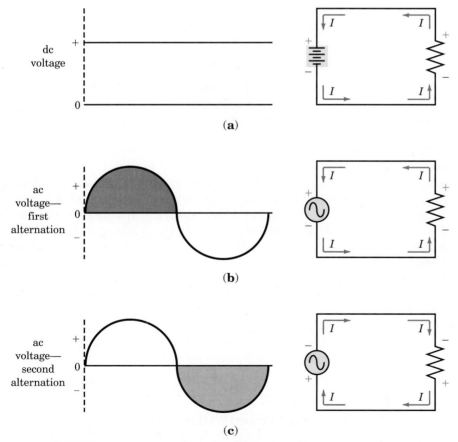

FIGURE 3–11   Contrasts between dc and ac voltages and circuit currents

circuit, you see that during one alternation of the ac cycle, current travels through the circuit in one direction, Figure 3–11b. (Incidentally, you probably noticed that the electrons are traveling from the negative side of the source, through the external circuit, back to the positive side of the source, just like they did in the dc circuit.) Obviously, when the source reverses polarity (on the next alternation), the electrons will reverse their direction of travel through the circuit. However, once again, the electrons leave the negative side of the source, travel through the external circuit, and return to the positive side of the source, Figure 3–11c. This indicates that the polarity of the ac source voltage applied to the circuit determines the direction of the electron current flow through the circuit. As the polarity of source voltage reverses, so does the direction of current flow through the circuit, Figures 3–11b and c.

Later in your studies you will learn some more specific facts and terms related to ac voltages and signals. These facts will be presented at the particular time you will be best able to apply them.

WORK, ENERGY, AND POWER

Work is the expenditure of energy. **Energy** is the ability to do work, or that which is expended in doing work. **Power** is the rate of using energy. A formula for mechanical work is:

Formula 3–4	Work (Foot pounds) = Force (Pounds) × Distance (Feet)

force value, $\boxed{\times}$, distance value, $\boxed{=}$

The basic unit of mechanical work is the foot poundal. At sea level, 32.16 foot poundals equals 1 foot pound. Mechanical power is the rate of doing work. For example, the "horsepower" equals a rate of 550 foot pounds per second.

Electrical energy is the ability to do electrical work. The unit of energy used for electrical energy is the joule (J). For electrical applications, the joule is defined as the energy required to move one coulomb of electrical charge between two points having a difference of potential of one volt.

Electrical power is the rate at which electrical energy is used. The unit of electrical power is the **watt,** named after the scientist James Watt. The abbreviation for watt or watts is **W.** *A watt of electrical power is defined as electrical energy expended at the rate of one joule per second.* Since joules can be related to energy, and energy can be related to force multiplied by distance, it can be determined that 746 watts is equivalent to one horsepower of mechanical power. As we already stated, electrical power in watts represents the time rate of expending electrical energy in performing electrical work; for example, the work of causing current flow (moving electrons) through an electrical resistance. Power can be related to electrical energy in the following manner:

Formula 3–5	Power (Watts) $= \dfrac{\text{Energy (Joules)}}{\text{Time (Seconds)}}$

energy value, $\boxed{\div}$, time value, $\boxed{=}$

EXAMPLE

If 100 joules of energy are used in 10 seconds, determine the power used.

$$P = \frac{100}{10} = 10 \text{ W}$$

PRACTICE PROBLEMS V

1. What is the power, in watts, for a circuit in which 25 joules of energy are being used in 2 seconds?

2. What is the power, in milliwatts, in a circuit where 1 joule is used in 100 seconds?

MEASURING ELECTRICAL ENERGY CONSUMPTION

By finding how much power has been used over a period of time, we can determine the total electrical energy that has been consumed. You should be aware that the symbol for both work and energy is *W;* however, the SI unit for both is the joule (J).

(NOTE: Don't get the symbol *W* for work or energy confused with the abbreviation for the unit of power, the watt (W).

The electric power company charges you for the amount of electrical energy you use for a given period of time. For example, they typically charge you for the total number of "kilowatthours" of electrical energy you have used for that month. When you use one kilowatthour of energy, you are using 1,000 watts for one hour. For example, if you burn ten 100-watt bulbs for one hour, you are using one kilowatthour's worth of energy. If you use a 500-watt iron for two hours, you are using one kilowatthour of energy, and so on.

| Formula 3–6 | Energy (Watthours) = Power (Watts) × Time (Hours) |

 power value, $\boxed{\times}$, time value, $\boxed{=}$

The time can be measured in seconds or hours. When the time is expressed in seconds, the unit of energy consumption is called the **wattsecond (Ws).** When the time is expressed in hours, the unit of energy consumption is called the **watthour (Wh),** or, when convenient due to the amount of power involved (as in the case of the power company), the kilowatthour (kWh).

EXAMPLES

1. A power consumption of 200 watts for 3 hours = 600 Wh of electrical energy consumption.

$$W = P \times T; \ W = 200 \times 3 = 600 \text{ Wh (or, 0.6 kWh)}$$

2. If 450 watts are being used over a period of 9 hours, find the energy consumption.

$$W = P \times T; W = 450 \times 9 = 4,050 \text{ Wh (or 4.05 kWh)}$$

PRACTICE PROBLEMS VI

How many kilowatthours of energy are used by a 250-watt bulb if it is left on for 8 hours?

The Basic Power Formula

Relating electrical power to practical circuits:

• A joule is the energy used to move one coulomb of charge between two points with a one volt potential difference.

• An ampere is the rate of charge movement of one coulomb per second.

• A watt is the rate of doing work at one joule per second.

Thus, one watt of power equals one ampere times one volt. The formula is:

Power (in watts) = Voltage (in volts) × Current (in amperes)

| Formula 3–7 | $P = V \times I$ (Watts = Volts × Amperes) |

 voltage value, $\boxed{\times}$, current value, $\boxed{=}$

EXAMPLE

If a circuit has 50 milliamperes of current flow and an applied voltage of 10 volts, how much power does the circuit dissipate? Using the basic formula for power:

$$P = V \times I$$
$$P = 10 \text{ V} \times 50 \text{ mA}$$
$$P = 10 \times 50 \times 10^{-3} = 500 \times 10^{-3} \text{ W or 500 mW}$$

PRACTICE PROBLEMS VII

1. If a circuit has an applied voltage of 150 V, and the circuit resistance limits the current to 10 mA; how much power does the circuit dissipate?

2. If the circuit voltage for Problem 1 remained at 150 V, but the circuit resistance were doubled:
   a. What would happen to circuit current?
   b. What would the circuit power dissipation be?

✓ IN-PROCESS LEARNING CHECK

1. Electron current flow is considered to flow from the (negative, positive) _____ side of the source, through the external circuit, back to the (negative, positive) _____ side of the source.

2. "Conventional" current is considered to flow from the (negative, positive) _____ side of the source, through the external circuit, back to the (negative, positive) _____ side of the source.

3. This text uses the (electron, conventional) _____ current flow approach.

4. One of the basic differences between dc and ac is that current in a dc circuit always flows in _____ direction; whereas, current flow in an ac circuit _____ direction for each alternation of the ac voltage.

5. Work is the _____ of energy; energy is the ability to do _____, or that which is expended in doing _____. Power is the rate of using _____.

6. Power in watts is equal to energy in _____ divided by time in _____.

7. Energy = power (in watts) × time. A common measure of energy consumption used by power companies for billing customers is the _____.

Variations of the Basic Power Formula

Just as the Ohm's law formula can be rearranged, so can the power formula. The formula we started with was $P = V \times I$. Two other arrangements are $V = P/I$ and $I = P/V$. Incidentally, the same style of visual aid used for the Ohm's law formula will work here, Figure 3–12.

Other practical variations of the power formula are developed by combining knowledge of Ohm's law and the power formulas. For example, since $V = I \times R$ (Ohm's law) and $P = V \times I$, the formula can be combined to show that $P = I^2 \times R$ or $I^2 R$. You have substituted for V in the Ohm's law formula $P = V \times I$; therefore $P = (I \times R) \times I$.

Formula 3–8	$P = I^2R$ (Watts = Amperes squared × Ohms

 current value, $\boxed{\times^2}$, $\boxed{\times}$, resistance value, $\boxed{=}$

Using this same idea, you can derive the formulas that state:

$$P = \frac{V^2}{R}, \qquad P = VI, \qquad P = V \times \left(\frac{V}{R}\right)$$

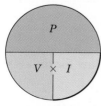

Again, by looking at the desired parameter (or covering it up) and seeing the arrangement of the other two parameters, the correct formula is evident. For example, if you know V and I, to find power take $V \times I$. If you know the power and the current, simply divide P by I to get V, and so forth.

FIGURE 3–12   Power formula memory aid

| Formula 3–9 | $P = \dfrac{V^2}{R} \left(\text{Watts} = \dfrac{\text{Volts squared}}{\text{Ohms}} \right)$ |

 voltage value, $\boxed{\times^2}$, $\boxed{\div}$, resistance value, $\boxed{=}$

You have simply substituted the Ohm's law formula for I into the $P = V \times I$ power formula. In order to solve for any desired electrical quantity, each of the power formulas can be rearranged using the transposition and cancelling techniques learned earlier in this chapter.

$$P = V \times I, \qquad I = \frac{P}{V}, \qquad V = \frac{P}{I}$$

$$P = I^2 \times R, \qquad I = \sqrt{\frac{P}{R}}, \qquad R = \frac{P}{I^2}$$

$$P = \frac{V^2}{R}, \qquad V = \sqrt{PR}, \qquad R = \frac{V^2}{P}$$

COMMONLY USED VERSIONS OF OHM'S LAW AND POWER FORMULAS

The most important equations to remember are

$$I = \frac{V}{R}, \qquad V = IR, \qquad \text{and } R = \frac{V}{I} \text{ (Ohm's law formulas)}$$

$$P = VI, \qquad P = I^2R, \qquad \text{and } P = \frac{V^2}{R} \text{ (power formulas)}$$

By understanding and knowing these Ohm's law and power formulas and by becoming adept at applying them to practical electronic problems, you will be on the way to becoming a good technician.

For a change of pace from simply reading about these formulas, we now want you to see how these formulas can be applied to practical problems. Follow the steps carefully as several sample problems and solutions are shown. You will be doing this sort of thing on your own throughout the remainder of your training.

EXAMPLE

Refer to Figure 3–13 for the following.

1. In Diagram 1, find the voltage *(V)*. Ohm's law formula for voltage is $V = I \times R$.
   Substitute the known values for I and R.

FIGURE 3–13   Ohm's law example problems

Therefore, $V = 5 \times 10^{-3}$ ampere $\times 10 \times 10^3$ ohms. $V = 50$ volts
Remember: When multiplying powers of ten, add their exponents algebraically:
10^{-3} and 10^3 add to 10^0, which equals 1. Also, milliunits multiplied by kilounits result in units, and, in this case, mA $\times$ kΩ result in volts.

2. In Diagram 2, find the circuit current (I). Ohm's law formula for current is $I = V/R$.
    Substitute the known values for V and R.

$$I = \frac{200 \text{ V}}{100 \text{ k}\Omega} = \frac{200 \times 10^0}{100 \times 10^3} = 2 \times 10^{-3} = 2 \text{ mA}$$

Remember: When dividing powers of ten, subtract the exponents algebraically. Since the voltage did not show a power of ten, show it as 200×10^0 [$10^0 = 1$]. To subtract the denominator 10^3 from the numerator 10^0, change the sign of the denominator power of ten, then add 10^{-3} to 10^0, which equals 10^{-3}. Also, units divided by kilounits result in milliunits. In this case, V/kΩ results in milliamperes (mA).

3. In Diagram 3, find the resistance (R). Ohm's law formula for resistance is $R = V/I$.
    Substitute the known values for V and I.

$$R = \frac{100 \text{ V}}{5 \text{ mA}} = \frac{100}{5 \times 10^{-3}} = 20 \text{ k}\Omega$$

Remember: When moving a power of ten in the denominator by dividing it into the numerator, change its sign and add the exponents as appropriate. The 10^{-3} for the milliamperes becomes the 10^3 for kilohms in the answer. Also, in this problem units divided by milliunits result in kilounits; thus, volts divided by milliamperes result in the answer as kilohms.

PRACTICE PROBLEMS VIII

FIGURE 3–14   Ohm's law practice problems

1. Find the applied voltage in the circuit of Figure 3–14a.
2. Find the circuit current for the circuit of Figure 3–14b.
3. Find the circuit resistance for the circuit of Figure 3–14c.

In summary, when applying Ohm's law and power formulas:

• look at the circuit to determine the knowns and unknown(s)

• select and write the appropriate formula to use

• substitute the knowns into the formula

• use metrics and powers of ten to manage large and small numbers

• remember that units divided by milliunits equals kilounits and that units divided by kilounits equals milliunits

• remember that milliunits multiplied by kilounits equals units

• recall various expressions for Ohm's law

Solving Problems Using Both the Ohm's Law and Power Formulas

Technicians are often required to combine their knowledge of Ohm's law and the power formulas in order to find the electrical quantities of interest. The following examples provide an idea about how you need to logically address problems of this nature.

EXAMPLE

Refer to Figure 3–15. Find the unknown values shown.

1. In Figure 3–15a, which formula would be easiest to use first? Observe that values for V and R are known.

FIGURE 3–15   Combination Ohm's law power formulae problems

$$I = \frac{V}{R} = \frac{100\ V}{10\ \Omega} = 10\ A$$

Knowing the current, you can solve the remaining unknown.

$P = V \times I = 100\ V \times 10\ A = 1{,}000$ watts (or 1 kilowatt)

Also, since V, I, and R are known, you can use the other power formulas, as well.

2. In Figure 3–15b, which formula might be used first? In this case, voltage and current are known values, but resistance is unknown. If the current meter does not drop any appreciable voltage, you could assume that V_R equalled the applied voltage and first solve for power. In many problems, you can start at more than one place, depending on the known value(s). Let's start with the power formula.

$P = V \times I = 300\ V \times 2 \times 10^{-3}\ A = 600 \times 10^{-3}\ W$ (or 600 mW)

Now to solve for the other unknowns, assume 300 V across R, where $V_R = 300\ V$.

$$R = \frac{V}{I} = \frac{300\ V}{2 \times 10^{-3}A} = 150 \times 10^3\ \Omega \text{ (or 150 k}\Omega)$$

3. In Figure 3–15c, a good starting point is to first solve for I since you know the values of V and R.

$$I = \frac{V}{R} = \frac{250\ V}{50\ k\Omega} = \frac{250}{50 \times 10^3} = 5 \times 10^{-3}\ A \text{ (or 5 mA)}$$

Now that voltage and current are known, it is easy to solve for the other unknown.

$P = V \times I = 250\ V \times 5 \times 10^{-3}\ A = 1{,}250 \times 10^{-3}\ W$ or 1.25 W

FIGURE 3–16

In summary, now you should know:

- how to find the known value(s) from the given diagram or facts

- where there are several starting places it is best to look for the formula that might be the easiest to use and start there

- how to combine Ohm's law knowledge with power formulas

PRACTICE PROBLEMS IX

Try the problems in Figure 3–16 to see if you can apply logic and appropriate formulas. Don't cheat yourself by looking ahead.

Now check yourself to see how well you did! You will be getting more practice, both in lab and within this text, as you proceed in your training.

If you have difficulty in solving these, take a few minutes to review appropriate parts of this chapter to find the formulas and the needed guidance or hints.

SUMMARY

▶ Current is directly related to voltage and inversely related to the circuit resistance. Ohm's law states this as $I = V/R$.

▶ If the other factors in the equation are known, the techniques of transposition allow rearrangement of the equation to solve for an unknown factor. For example, by rearranging the Ohm's law formula, two other useful equations can be derived, namely $V = I \times R$ and $R = V/I$.

▶ Powers of ten and scientific notation are useful when working with either large or small numbers in formulas. When using powers of ten, the rules relating to exponents are applied. Scientific notation involves simply converting any decimal number to a number between 1 and 10, times a power of ten.

▶ Scientific calculators are very useful to simplify working electrical problems. The Scientific Notation mode and the [EE] or [EXP] keys are helpful aids when using calculators to solve problems.

▶ Two good habits to form when solving electrical circuit problems are: 1) Draw a diagram of the circuit, if one is not given; 2) Write down any pertinent formulas that will be used; then, substitute knowns into them appropriately to solve for the unknowns.

▶ The direction of electron current flow through a circuit, external to the voltage source, is from the source's negative side, through the circuit, and back to the source's positive side. Arrows are frequently drawn on schematic diagrams to illustrate this. The direction electrons would move through a circuit is the standard being used in this textbook.

▶ Conventional current, as used in some engineering textbooks, is considered to flow in the direction that a positive charge would move through the circuit. That is, from positive to negative.

▶ When labeling the polarity of voltage across a given component in a circuit, the component end where the current enters is labeled the negative end, and the end where the current exits is labeled the positive end.

▶ Electrical power is the rate of using electrical energy to do electrical work. Power is generally dissipated in a circuit or in a component in the form of heat. One watt of power is the performance of electrical work at the rate of 1 joule per second.

▶ A joule is the energy used in moving one coulomb of charge between two points that have a difference of potential of one volt between them.

▶ The formulas commonly used to calculate electrical power, or power dissipation are: $P = V \times I$; $P = I^2R$, and $P = V^2/R$

▶ Electrical energy usage is computed by multiplying the power used times the amount of time it was used ($W = P \times T$). Units of measure for usage of electrical energy are: watt-second (Ws); watthour (Wh); and for large amounts of energy usage, kilowatthour (kWh).

FORMULAS AND CALCULATOR SEQUENCES SUMMARY

FORMULA 3–1 $I = \dfrac{V}{R}\left(\text{Amperes} = \dfrac{\text{Volts}}{\text{Ohms}}\right)$
voltage value, $\boxed{\div}$, resistance value, $\boxed{=}$

FORMULA 3–2 $V = I \times R$ (Volts = Amperes × Ohms)
current value, $\boxed{\times}$, resistance value, $\boxed{=}$

FORMULA 3–3 $R = \dfrac{V}{I}\left(\text{Ohms} = \dfrac{\text{Volts}}{\text{Amperes}}\right)$
voltage value, $\boxed{\div}$, current value, $\boxed{=}$

FORMULA 3–4 Work (Foot pounds) = Force (pounds) × Distance (Feet)
force value, $\boxed{\times}$, distance value, $\boxed{=}$

FORMULA 3–5 Power (Watts) = $\dfrac{\text{Energy (Joules)}}{\text{Time (Seconds)}}$
energy value, $\boxed{\div}$, time value, $\boxed{=}$

FORMULA 3–6 Energy (Watthours) = Power (Watts) × Time (Hours)
power value, $\boxed{\times}$, time value, $\boxed{=}$

FORMULA 3–7 $P = V \times I$ (Watts = Volts × Amperes)
voltage value, $\boxed{\times}$, current value, $\boxed{=}$

FORMULA 3–8 $P = I^2 R$ (Watts = Amperes squared × Ohms)
current value, $\boxed{x^2}$, $\boxed{\times}$, resistance value, $\boxed{=}$

FORMULA 3–9 $P = \dfrac{V^2}{R}\left(\text{Watts} = \dfrac{\text{Volts squared}}{\text{Ohms}}\right)$
voltage value, $\boxed{x^2}$, $\boxed{\div}$, resistance value, $\boxed{=}$

REVIEW QUESTIONS

1. State the three forms of:
   a. The Ohm's law formula
   b. The formula for electrical power

2. State the following numbers in scientific notation form:
   a. 33,456
   b. 25
   c. 1,055,000
   d. 10

3. State the following numbers in the appropriate number times 10^3:
   a. 5,100
   b. 47,000
   c. 0.001
   d. 39

4. If voltage is doubled and resistance is halved in a circuit, what is the relationship of the new current to the original current?

5. If current triples in a circuit, but resistance remains unchanged, what happens to the power dissipated by the circuit?

6. a. Draw a schematic diagram showing a voltage source, a current meter, and a resistor.
   b. Assume the voltage applied is 30 V and the current meter indicates 2 mA of current. Calculate the circuit resistance and circuit power dissipation.

c. Label all electrical parameters, the polarity of voltage across the resistor and across the voltage source, and show the direction of current flow.

7. a. Define the terms work, energy, and power.
   b. Give the units used to measure work, energy, and power.

Refer to Figure 3–17 when answering questions 8 through 15:

FIGURE 3–17

8. In Figure 3–17, if the value of V is assumed to be 50 V and the value of R is assumed to be 33 kΩ, what value of current will the meter M read?

9. If the current meter (M) in Figure 3–17 is reading 3 mA, and R has a value of 12 kΩ, what is the value of circuit applied voltage?

10. Indicate whether the following items would increase, decrease, or remain the same if the value of voltage in question 9 were doubled:
    a. V has _____
    b. R will _____
    c. I will _____

11. If current through the circuit of Figure 3–17 is 50 mA and circuit voltage is 41 V, what is the value of R?

12. In Figure 3–17, what is the voltage applied if the circuit power dissipation is 100 mW and the circuit current is 12.5 mA?

13. For the circuit of Figure 3–17, what is the circuit current if the power dissipated by the 10-Ω resistor is 100 W?

14. If current in the circuit of Figure 3–17 decreases to one-third its original value, and the original power dissipation was 180 W, what is the new power dissipation?

15. Assume an applied voltage of 100 V for the circuit of Figure 3–17. What value of R is needed to limit circuit current to 8.5 mA?

CHALLENGE QUESTIONS

1. Determine the electrical power equivalent to 3.5 horsepower.

2. In your own words, describe the relationships and/or proportionalities between the following electrical parameters:
   a. Power related to current (for a given resistance)
   b. Voltage related to current (for a given resistance)
   c. Energy (W) related to power (for a given time)

3. In a given circuit, if power dissipation has increased to eight times original power, while resistance has decreased to half its original value, what must have happened to the circuit applied voltage?

TEST YOURSELF

1. a. Draw a schematic diagram showing a source, a resistor, and a current meter.
   b. Assume the power dissipation is 1,000 watts and resistance is 10 ohms. Transpose the $P = I^2R$ formula and solve for the value of I.
   c. Calculate the circuit applied voltage.
   d. Label all parameters on the circuit diagram, including the direction of current flow and the polarity of voltages across each component.

2. Using powers of ten, change the following as directed:
   a. $\dfrac{10^7 \times 10^{-4}}{10^5 \times 10^2 \times 10^{-2}}$ to a whole number.
   b. $10^3 \times 10^5 \times 10^6 \times 10^{-2}$ equals 10 to what power?
   c. Add: $(7.53 \times 10^4) + (8.15 \times 10^3) + (225 \times 10^1)$
   d. Subtract: $(6.25 \times 10^3) - (0.836 \times 10^2)$

PERFORMANCE PROJECTS CORRELATION LISTING

Suggested performance projects that correlate with topics in this chapter are:

CHAPTER TOPIC	PERFORMANCE PROJECT	PROJECT NUMBER
Relationship of Current to Voltage with Resistance Constant	Relationship of I & V with R Constant (Ohm's law series)	5
Relationship of Current to Resistance with Voltage Constant	Relationship of I & R with V Constant (Ohm's law series)	6
The Basic Power Formula	Relationship of P to V with R Constant (Ohm's law series)	7
Variations of the Basic Power Formula	Relationship of P to I with R Constant (Ohm's law series)	8

PART II

BASIC

CIRCUIT

ANALYSIS

Series Circuits

Courtesy of International Business Machines Corporation

4

KEY TERMS

Battery

Cell

Kirchhoff's voltage law

Open circuit

Series circuit

Short circuit

Series-dropping resistor

Voltage-divider action

OUTLINE

Definition and characteristics of a series circuit

Resistance in series circuits

Voltage in series circuits

Kirchhoff's voltage law

Using a scientific calculator

Power in series circuits

Effects of opens and troubleshooting hints

Effects of shorts and troubleshooting hints

Designing a series circuit to specifications

Special applications

Critical thinking and the SIMPLER troubleshooting sequence

Introduction to troubleshooting skills

The SIMPLER sequence for troubleshooting

Troubleshooting levels

CHAPTER PREVIEW

Series circuits are found in many systems and subsystems. Examples are light switch circuits and automobile ignition switch circuits. It is important that you learn the characteristics of the series circuit. That knowledge will aid you in analyzing series circuits whether they are stand-alone circuits or portions of more complex circuits.

In studying series circuits, you will apply Ohm's law and power formulas. Also, you will learn how Kirchhoff's voltage law helps you analyze series circuits. Insight will be gained about the effects of opens and shorts in series circuits. You will see how voltage sources are used in series circuits. Also, voltage dividers and the concept of reference points will be examined. The final portion of this chapter will present a useful troubleshooting method that will be valuable to you throughout your training, and more importantly, in your career.

You will be introduced to a special troubleshooting sequence called the SIMPLER sequence. By solving a troubleshooting problem called Chapter Challenge, you will gain experience using the SIMPLER sequence. The system for performing the procedure and solving the Chapter Challenge will be discussed. One sample solution to the challenge troubleshooting problem is presented at the end of the chapter so you can compare your thinking process against the sample solution.

OBJECTIVES

After studying this chapter, you should be able to:

- Define the term **series circuit**
- List the primary characteristics of a series circuit
- Calculate the total resistance of series circuits using two different methods
- Calculate and explain the voltage distribution characteristics of series circuits
- State and use **Kirchhoff's voltage law**
- Calculate power values in series circuits
- Explain the effects of **opens** in series circuits
- Explain the effects of **shorts** in series circuits
- List troubleshooting techniques for series circuits
- Design series circuits to specifications
- Series-connect voltage sources for desired voltages
- Analyze a **voltage divider** with reference points
- Calculate the required value of a **series-dropping resistor**
- Use the SIMPLER troubleshooting sequence to solve the Chapter Challenge problem

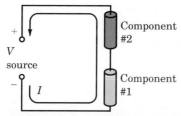

**FIGURE 4–1   Components
connected end-to-end in a
two-component series circuit**

DEFINITION AND CHARACTERISTICS OF A SERIES CIRCUIT

A **series circuit** is any circuit having only one path for current flow. In other words, two or more electrical components or elements are connected so the same current passes through all the connected components.

This situation exists when components are connected end-to-end in the circuit external to the source, Figure 4–1. This is termed a two-component series circuit. Also, notice there is only *one possible path* for current flow from the source's negative side, through the external circuit, and back to the source's positive side.

Key Characteristics

Excluding the source, how many components are in series in Figure 4–2? If you said four, you are correct! The circuit in Figure 4–3 is a six-component series circuit. In all these cases, the important characteristics to remember are there is only one path for current flow and the current is the same through all parts of a series circuit. As a technician, you will have many opportunities to apply this knowledge.

RESISTANCE IN SERIES CIRCUITS

Knowing the characteristics that identify a series circuit helps you analyze the important parameters of these circuits.

(a)   (b)
FIGURE 4–2   Four-resistor series circuit

**FIGURE 4–3
Six-resistor series circuit**

First, let's analyze the series circuit in terms of its resistance to current flow. In previous chapters we have limited circuits to a single resistor, and the circuit's resistance to current flow was obvious. Now, we are connecting two or more resistors in tandem or in series. How does this affect the circuit resistance?

Series Circuit Total Resistance Formula

If the circuit current must sequentially flow through all the resistors since there is only one path for current, then the total resistance to current flow will equal the *sum* of all the resistances in series. Once again, in series circuits, the total resistance (R_T or R total) equals the sum of all resistances in series.

| Formula 4–1 | $R_T = R_1 + R_2 \ldots + R_n$ |

R_1 value, $\boxed{+}$, R_2 value, $\boxed{+}$, . . . , $\boxed{=}$

R_n represents the last of the remaining resistor values.

EXAMPLE

Refer to Figure 4–4 for an application of this formula. This illustrates that to find a series circuit's total resistance, simply add all the individual resistance values.

PRACTICE PROBLEMS I

1. Referring back to Figure 4–2, what is the circuit's total resistance if R_1 and R_2 are 10 kΩ resistors and R_3 and R_4 are 27 kΩ resistors?

2. The values of the resistors in Figure 4–4 are changed so that R_1 = 2.7 kΩ, R_2 = 5.1 kΩ, and R_3 = 8.2 kΩ. What is the new total resistance value for the circuit?

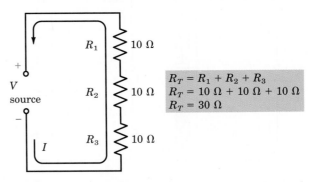

FIGURE 4–4   **Application of series-circuit total resistance formula**

The Ohm's Law Approach

Another important method to determine total resistance *in any circuit,* including series circuits, is Ohm's law. Recall that the Ohm's law formula to find resistance is $R = V/I$. To find total resistance, this formula becomes:

Formula 4–2	$R_T = \dfrac{V_T}{I_T}$.

 Total voltage value, $\boxed{\div}$, total current value, $\boxed{=}$

If there is any means of determining the value of total voltage applied to the series circuit *and* the total current (which is the same as the current through any part of the series circuit), Ohm's law can be used to find total resistance by using $R_T = V_T/I_T$.

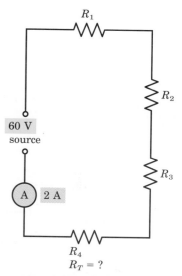

EXAMPLE

Look at Figure 4–5 and use Ohm's law to find total resistance. Your answer should be 30 ohms because 60 volts divided by 2 amperes equals 30 ohms. Does this show the value of each of the resistors in the circuit? No, it only reveals the circuit's total resistance.

FIGURE 4–5   Finding R_T by Ohm's law

EXAMPLE

Now look at Figure 4–6. Again, it is possible to use Ohm's law to determine the circuit's total resistance, where total resistance equals total voltage divided by total circuit current.

$$R_T = \frac{V_T}{I_T} = \frac{70 \text{ V}}{2 \text{ A}} = 35 \ \Omega$$

Because we know the voltage drop across R_1 and know that the current through all parts of a series circuit is the same value , we can also determine the value of R_1. In this case, R_1 equals the voltage across R_1 divided by the 2 amperes of current through R_1. Thus, R_1 equals 30 volts divided by 2 amperes. $R_1 = 15$ ohms.

PRACTICE PROBLEMS II

1. Again refer to Figure 4–6, if $V_T = 100$ V, $I_T = 1$ A, and $V_I = 47$ V, what is the value of R_1? What is the value of R_T?

2. Assume that the electrical parameters in the circuit of Figure 4–6 are changed so that $V_T = 42$ V and $I_T = 4$ mA. What is the new value of R_T?

3. In the circuit described in problem 2, R_2 and R_3 are equal value resistors and $R_1 = 2.7$ kΩ. What is the value of R_2?

FIGURE 4–6   Finding one R value by Ohm's law

We have discussed and illustrated two important facts about series circuits. First, *current* is the same throughout all parts of a series circuit. Second, the total circuit *resistance* equals the sum of all the resistances in series, which indicates total resistance must be greater than any one of the resistances. We will now study voltage, a third important electrical parameter in series circuits.

VOLTAGE IN SERIES CIRCUITS

To help you understand how voltage is distributed throughout a series circuit, refer to Figure 4–7.

FIGURE 4–7   Voltage distribution in series circuits

Individual Component Voltages

Since there is only one path for current, the current *(I)* through each resistor $(R_1, R_2,$ and $R_3)$ must have the same value since it is the same current, Figure 4–7. From the study of Ohm's law, you know the value of voltage dropped by R_1 must equal its *I* times its *R,* or

Formula 4–3	$V_1 = I_{R_1} \times R_1$

 current value, $\boxed{\times}$, resistor #1 value, $\boxed{=}$

EXAMPLE

In this case, the current is 2 mA and the resistance of R_1 is 100 kΩ; therefore, V_1 = 200 volts, Figure 4–7a. (Remember that milliunits times kilounits equals units). Voltage dropped by R_2 must equal 2 mA times 27 kΩ, or 54 volts, Figure 4–7b. The voltmeter, which is measuring the voltage across R_3, would indicate 20 volts because 2 mA $\times$ 10 kΩ = 20 volts, Figure 4–7c.

PRACTICE PROBLEMS III

1. In the circuit shown in Figure 4–7c, the voltmeter reads 30 V. What are the values of V_A, I_T, V_{R_2}, and V_{R_1}?

2. In the circuit shown in Figure 4–7c, the voltmeter reads 25 V. What are the values of V_A, I_T, V_{R_2}, and V_{R_1}?

PRACTICAL NOTES

In review, several observations are made.

1. The largest value resistor in a series circuit drops the largest voltage and the smallest value resistor drops the least voltage. These drops occur because the current is the same through each of the resistors and each resistor's voltage drop equals *I* times its *R.*

2. Since the same current passes through all components in series, a given component's voltage drop equals *the same percentage or fraction* of the total circuit applied voltage as its resistance value is of the total circuit resistance.

3. Relating the above concepts, when comparing two specific components' voltage drops, the ratio of voltage drops is the same as the ratio of their individual resistances.

Refer back to the voltages calculated for the circuit in Figure 4–7. Did the largest resistor value drop the highest voltage? Yes, R_1 dropped 200 volts compared to 54 volts for R_2 and 20 volts for R_3. Is the comparative ratio of voltage drops between resistors the same as the ratio of their resistances? Yes, R_1 resistance value is 10 times that of R_3 and V_1 is 10 times V_3. Also note that R_2 voltage drop is 2.7 times that of R_3, which is the same as their resistance ratios.

The concept of ratios of voltage drops equaling the ratios of resistances can be used to compare the voltage drops of any two components in the series circuit. For example, if the R values and applied voltage are known, it is possible to find all of the voltages for each component throughout the circuit using the proportionality technique. Naturally, you can also determine individual voltages around the circuit by solving for total resistance, then total current (V_T/R_T), and then determining each $I \times R$ drop with $I \times R_1$ for V_1, $I \times R_2$ for V_2, and so forth.

The Voltage Divider Rule

You have just seen that the amount of voltage dropped by a given resistor in a series circuit is related to its resistance value compared to the others in the circuit. The voltage divider rule shows how in a series circuit, any given resistor's voltage drop relates to its resistance compared to the circuit's total resistance. Using this rule, you can determine a series resistor's voltage drop, without having to know the circuit current, if you know the applied voltage and circuit total resistance. Here's the simple formula:

$$\text{Formula 4-4} \quad \bigg| \quad V_X = \frac{R_X}{R_T} \times V_T$$

where: V_X = the voltage drop across the selected resistor
R_x = the resistance value of the selected resistor
R_T = the circuit total resistance
V_T = the circuit applied voltage

 R_X value, $\boxed{\div}$, R_T value, $\boxed{\times}$, V_T value, $\boxed{=}$

PRACTICAL NOTES

This type calculation is very simple using a calculator. You should get into the habit of using your calculator for MOST of your calculations (if your program promotes it at this stage of your learning). In some cases, however, the resistor values and the voltage applied value are such that you can almost do this type calculation in your head. In fact, here's a hint! It is a good idea to learn to approximate answers in your head for any problem that you can, just as a double check on your computations. Approximating is a powerful skill to learn and practice! It can often save you from embarrassing results. If you inadvertently make a mathematical error along the way, your approximation will show you that the erroneous results are not logical. Then, you can check back through your procedure, and quickly find the mistake.

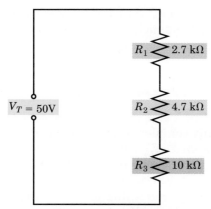

**FIGURE 4–8   Voltage-divider
rule sample circuit**

EXAMPLE

Refer to Figure 4–8 and note how we can find the voltage dropped across R_2 by simply substituting the knowns into the formula. (In this case, R_2 is the value to be used for R_X in the formula.)

$$V_{R_2} = \frac{R_2}{R_T} = V_T; \; V_{R_2} = \frac{4.7 \text{ k}\Omega}{17.4 \text{ k}\Omega} \times 50; \; V_{R_2} = 0.27 \times 50 = 13.5 \text{ V}$$

PRACTICE PROBLEMS IV

Now you use the voltage divider formula to find the following answers.

1. For the circuit of Figure 4–8, what is the voltage dropped by R_1?

2. For the circuit of Figure 4–8, what is the voltage dropped by R_3?

3. Sum the rounded voltages of R_1, R_2, and R_3. Do they add up to close to the applied voltage of 50 volts?

As you can see, the voltage divider rule can be very handy! You will get more opportunities to use this rule as you proceed in your training.

Finding the Value of Applied Voltage

Again referring to Figure 4–7, let's examine some ways to find the circuit applied voltage. One obvious way is the Ohm's law expression where $V_T = I_T \times R_T$. In this circuit, V total (or V applied) = 2 mA times R_T. As you can easily determine, R_T in this circuit equals 100 kΩ + 27 kΩ + 10 kΩ = 137 kΩ total resistance. Therefore, circuit applied voltage equals 2 mA × 137 kΩ = 274 volts.

Another way to find applied voltage is to add all the individual voltage drops to find total voltage, just as we add all the individual resistances to find total resistance. This yields the same answer where 200 volts + 54 volts + 20 volts = 274 volts. This method suggests an important concept called Kirchhoff's voltage law.

KIRCHHOFF'S VOLTAGE LAW

Kirchhoff's voltage law states the *arithmetic sum* of the voltages around a single circuit loop (any complete closed path from one side of the source to the other) *equals V applied (V_A)*. It also says the *algebraic sum* of all the loop voltages, including the source or applied voltage, must *equal zero*. That is, if you observe the polarity and value of voltage drops by the circuit elements and the polarity and value of the voltage source, and add the complete loop's values algebraically, the result is zero. For our purposes, the arithmetic sum approach will be most frequently used. It illustrates how Kirchhoff's voltage law helps determine unknown circuit parameters. Refer to Figure 4–9 as you study the following example.

EXAMPLE

If the V_A is 50 volts and V_2 is 20 volts, how can Kirchhoff's voltage law help determine V_1? If the sum of voltages (*not* counting the source) must equal V applied, then:

$$V_A = V_1 + V_2$$

Since we know V_A and V_2, transpose to solve for V_1.

Therefore, $V_1 = V_A - V_2 = 50\ \text{V} - 20\ \text{V} = 30\ \text{V}$

$$V_A = V_1 + V_2$$

FIGURE 4–9   Using Kirchhoff's voltage law where $V_A = V_1 + V_2$

If individual voltage drops were known but not the applied voltage, find V applied by adding the individual voltage drops. In this case 30 V + 20 V = 50 V. This agrees with Kirchhoff's voltage law that the arithmetic sum equals the applied voltage. Kirchhoff's voltage law can help you find unknown voltages in series circuits through either addition or subtraction, as appropriate. And if there are more than two series components:

Formula 4–5	$V_T = V_1 + V_2 \ldots + V_n$

V_1 value, $\boxed{+}$, V_2 value, $\boxed{+}$, ... , $\boxed{=}$

where V_n represents the last of the remaining voltage values.

In Figure 4–10, observe the polarity of voltages around the closed loop. To indicate polarities used in Kirchhoff's voltage law, trace the loop from the source's positive side (Point A), through the resistors and back to the source's negative side (Point B). Consider any voltage a positive voltage whose + point is reached first, and vice versa. In this case, the first voltage reached is +20 V; the next voltage is +30 V, and the source negative terminal is −50 V. Adding these voltages yields:

$$(+20) + (+30) + (-50) = 0; \text{ or } 20 + 30 - 50 = 0$$

Tracing the other direction through the circuit gives:

$$(-30) + (-20) + (+50) = 0; \text{ or } -30 + -20 + 50 = 0$$

In either case, the algebraic sum of the voltage drops *and* the voltage source around the entire closed loop equals zero.

Likewise, the arithmetic sum of the *voltage drops* around a given loop must equal the value of the *applied voltage,* regardless of which direction is used to trace the loop.

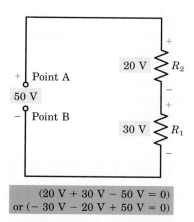

(20 V + 30 V − 50 V = 0)
or (− 30 V − 20 V + 50 V = 0)

FIGURE 4–10   Kirchhoff's algebraic sum example

PRACTICE PROBLEMS V

Refer again to Figure 4–10. Using the proportionality of voltage drops concept and Kirchhoff's voltage law, if V_1 is 45 V, what are the values of V_2 and V_A?

IN-PROCESS LEARNING CHECK I

Fill in the blanks as appropriate.

1. The primary identifying characteristic of a series circuit is that the _____ is the same throughout the circuit.

2. The total resistance in a series circuit must be greater than any one _____ in the circuit.

3. In a series circuit, the highest value voltage is dropped by the _____ value resistance, and the lowest value voltage is dropped by the _____ value resistance.

4. In a two-resistor series circuit, if the applied voltage is 210 volts and one resistor's voltage drop is 110 volts, what must the voltage drop be across the other resistor? _____ V.

5. Answer each part of this question with increase, decrease, or remain the same. In a three-resistor series circuit, if the resistance value of one of the resistors increases, what happens to the circuit total resistance? _____ To the total current? _____ To the adjacent resistor's voltage drop? _____

6. What is the applied voltage in a four-resistor series circuit where the resistor voltage drops are 40 V, 60 V, 20 V, and 10 V, respectively?

A Quick Review

Current is the same throughout a series circuit; total resistance equals the sum of all the individual resistances in series, and voltage distribution around a series circuit is directly related to the resistance distribution, since current is the same through all components and each component's voltage drop = current × its resistance. Kirchhoff's voltage law states that the arithmetic sum of voltage drops equals the voltage applied, or the algebraic sum of all the voltage drops and the voltage source equal zero.

USING A SCIENTIFIC CALCULATOR

Try solving this problem with a calculator, following these instructions:

Problem: Find R_T, I_T, and V_T for a series circuit containing three 20-kΩ resistors where $V_1 = 10$ V, Figure 4–11.

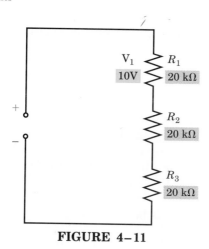

FIGURE 4–11

Step 1: Turn calculator on and clear.

Step 2: Find R_T where $R_T = R_1 + R_2 + R_3$, by entering 20,000 + 20,000 + 20,000, then press the ☐= button. The answer should be 60,000. (Of course, this is 60,000 Ω, or 60 kΩ.)

Step 3: Find I by finding the I through R_1, where $I_1 = V_1/R_1$. Clear calculator by pushing the ☐CE/C button. Input as follows: Punch in 10 (for V_1), then the ☐÷ symbol button, then 20,000 (for R_1), then the ☐= button. The answer should be 0.0005. (Of course, this is 0.0005 A or 0.5 mA.)

Step 4: Find V_T by using the formula $V_T = I_T \times R_T$. You may clear the calculator (or leave the 0.0005 you had just calculated, since you need it in this calculation). At any rate, be sure that 0.0005 is showing, then push the ☐× symbol button, then input 60,000 (for the R_T value). Push the ☐= button. The answer should be 30, representing 30 volts.

NOTE: This approach is appropriate for most calculators. Some brands, depending on their notation approach, will not use the exact same sequence of key strokes.

PRACTICE PROBLEMS VI

Try this one on your own, using a calculator. For the circuit shown in Figure 4–12, find the parameters called for.

FIND
$V_T =$ _____
$P_3 =$ _____
$R_1 =$ _____
$R_2 =$ _____
$R_3 =$ _____

FIGURE 4–12

Now that you've used the elemental method of entering numeric data; let's look at three other ways you can enter data and use a scientific calculator to advantage. These ways include:

1. Using the ☐EE or ☐EXP key on a scientific calculator (along with powers of ten).

2. Using the �託EE or ⌷EXP key on a scientific calculator that is in the "engineering mode" (along with powers of ten).

3. Using the ⌷EE or ⌷EXP key on a scientific calculator that is in the "scientific notation mode" (along with powers of ten).

All three of these methods will save you from having to enter numerous zeros for very large or very small numbers. For our particular example problem, these approaches save you from having to put in all the zeros for large resistance values, or all the zeros after the decimal for the small current value. Any one of these approaches can be very helpful for most of your electronic calculations. Once you have learned these techniques, use the one that suits your needs best.

SPECIAL NOTATIONS REGARDING SCIENTIFIC CALCULATORS
There are numerous brands and types of scientific calculators on the market. Functions such as squaring a number, finding square roots, finding the arc tangent, finding antilogarithms, and so on may differ slightly from calculator to calculator. For the types of calculations we perform throughout this text, we want to mention some common variations you may see or use. If you are using a different type scientific calculator than the one that was used in creating calculator sequences in our text (CALCULATOR #1), the following equivalent keystrokes chart may be helpful to you.

TASK DESCRIPTION	CALCULATOR #1 KEYSTROKE(S) TI 36X SOLAR	CALCULATOR #2 KEYSTROKE(S) HP 20S	CALCULATOR #3 KEYSTROKE(S) (OTHERS)
Finding the reciprocal	1/x	1/x	2nd, 1/x
Squaring a number	x^2	↰ , x^2	2nd, x^2
Taking a square root	$\sqrt{x}$	$\sqrt{x}$	$\sqrt{x}$
Finding the arctan	2nd, tan	↰ , ATAN	INV, tan
Finding the antilog of a natural logarithm (ϵ)	2nd, e^x	e^x	e^x
Converting from polar to rectangular form: Finding the x-axis and y-axis (j) values:	phasor value, X ⇅ Y, angle value, 2nd, (P ▷ R) (Displays x-axis value) Follow above sequence with X ⇅ Y key again (Displays y-axis (j) value)	phasor value, INPUT, angle value, ↗,→R (Displays y-axis (j) value) Follow above sequence with ↰ , SWAP (Displays x-axis value)	phasor value, P-R key, angle value, = key (Displays x-axis value) Follow above sequence with X-Y key (Displays y-axis (j) value)
Converting from rectangular to polar form: Finding the phasor value and the angle	x-axis value, X ⇅ Y, y-axis (j) value, 3rd, v(R ▷ P) Displays the phasor value Press the X ⇅ Y key again (Displays the angle value)	x-axis value, INPUT, y-axis (j) value, ↰ , →P, (Displays angle value) Follow above sequence with ↰ , SWAP (Displays the phasor value)	x-axis value, R-P key, y-axis (j) value, =, (Displays phasor value) Follow the above sequence with X-Y key, (Displays the angle value)

The significant key you will use on the keyboard in conjunction with the three modes mentioned may be called ⌷EE on some calculators or ⌷EXP on others. Let's see how this system works.

Using the [EE] (or [EXP]) Key with Powers of Ten for Metric Units

Using the same sample problem as before; that is, finding R_T, I_T, and V_T, where there are three series 20-kΩ resistors and $V_1 = 10$ V:

Step 1: Turn calculator on and clear by pressing the [AC/ON] key.
Step 2: Find R_T, (where $R_T = R_1 + R_2 + R_3$) by entering the following:
   input 20, [EE] (or [EXP]), 3,
      (the display reads 20^{03} or 20. 03, meaning 20×10^3)
   then input [+] ,
      (the display now reads 20000)
   then input 20, [EE] (or [EXP]), 3, a second time
      (the display once again reads 20^{03} or 20. 03, showing the number you just put in)
   then input [+] ,
      (the display now reads 40000)
   then input 20, [EE] (or [EXP]), 3, a third time
      (the display will read 20^{03} or 20. 03, again representing the number you just entered)
   then enter [=] ,
      (the display will read 60000. This indicates that the value of R_T is 60,000 Ω, or 60 kΩ)
Step 3: Find I by finding the I through R_1, (where $I_1 = V_1/R_1$).
   Clear the calculator by pressing the [CE/C] key.
   Enter 10 (for the V_1 value),
   then enter [÷] ,
   then enter 20, [EE] (or [EXP]), 3 (for the 20-kΩ R_1 value)
   then press the [=] key.
   The display reads 0.0005. (This indicates that current is 0.5 mA.)
Step 4: Find V_T by using the formula $V_T = I_T \times R_T$.
   Clear the calculator with the [CE/C] button.
   Enter the value of I_T: 0.5, [EE] (or [EXP]), [+/−] , 3
      (The display should read 0.5^{-03} or 0.5 −03.)
   then press the [×] key,
   then input 60, [EE] (or [EXP]), 3
   then press the [=] key.
   The display reads 30. (This indicates that $V_T = 30$ volts.)

PRACTICE PROBLEMS VII

Now you try this technique to find the answers called for in the circuit shown in Figure 4−13.

Using the Engineering Mode with Powers of Ten for Metric Units

The engineering mode provides a notation that again uses powers of tens. Since most science and engineering calculations use metric measurements

FIND
$R_T = ?$
$I_T = ?$
$V_{R_2} = ?$

FIGURE 4-13

which typically have *exponents* that are in multiples or submultiples of three, the data input and answers appear as a number, times a power of ten, with a multiple of three exponent. For example, in electronics, kilohms and kilowatts are expressed in terms of 10^3. Millivolts and milliamperes are expressed in terms of 10^{-3}. Microamperes are stated as 10^{-6}. Nanoamperes are indicated as 10^{-9}, and so on.

Using the example problem in Figure 4-11; that is, finding R_T, I_T, and V_T, with three series 20-kΩ resistors and $V_1 = 10$ V:

Step 1: Turn calculator on and clear by pressing the $\boxed{\text{AC/ON}}$ key.

Step 2: Press any buttons required to get your calculator into the *engineering mode*. See your calculator's instruction manual for this. (NOTE: For the TI 36X solar calculator, the key sequence to get into the engineering mode is $\boxed{\text{3rd}}$, ENG. When you are in this mode, the calculator readout will read $0.^{00}$ or 0.00, prior to your entering new data.)

Step 3: Find R_T in the *engineering mode* as follows:
Be sure your calculator is in the *engineering mode* and the display reads $0.^{00}$ or 0. 00.
Enter 20, $\boxed{\text{EE}}$ (or $\boxed{\text{EXP}}$), 3 (the readout reads $20.^{03}$ or 20. 03),
then press $\boxed{+}$ (the readout will still read $20.^{03}$ or 20. 03),
then enter 20, $\boxed{\text{EE}}$ (or $\boxed{\text{EXP}}$), 3 a second time (the readout will still read $20.^{03}$ or 20. 03),
then enter $\boxed{+}$ (the readout will read $40.^{03}$ or 40. 03),
then enter 20, $\boxed{\text{EE}}$ (or $\boxed{\text{EXP}}$), 3 a third time (the readout will read $20.^{03}$ or 20.03),
then press the $\boxed{=}$ sign (the readout now reads $60.^{03}$ or 60. 03).
This indicates 60×10^3, or 60,000 Ω, or 60 kΩ.

Step 4: Find I_T in the *engineering mode* by entering the following (where $I_T = I_1$ and $I_1 = V_1/R_1$):

Be sure your calculator is in the *engineering mode* and the display reads 0.^{00} or 0. 00.

Enter 10 (for the V_1 value),

then press the $\boxed{\div}$ key,

then enter 20, $\boxed{\text{EE}}$ (or $\boxed{\text{EXP}}$), 3,

then press the $\boxed{=}$ key.

The display should show 500^{-06} or 500. -06.

(This indicates 500×10^{-6}, or 0.0005 A, which is the same as 0.5 mA. $I_T = 0.5$ mA.)

Step 5: Find V_T in the *engineering mode* by using the formula $V_T = I_T \times R_T$.

Be sure the calculator is in the *engineering mode* and the display reads 0.^{00} or 0. 00.

Enter the I_T value by entering 0.5, $\boxed{\text{EE}}$ (or $\boxed{\text{EXP}}$), $\boxed{+/-}$, 3,

then press the $\boxed{\times}$ key,

then enter 60, $\boxed{\text{EE}}$ (or $\boxed{\text{EXP}}$), 3,

then press the $\boxed{=}$ key.

The display reads 30.^{00} or 30. 00. (This means 30×10^0, which means that $V_T = 30$ V.)

PRACTICE PROBLEMS VIII

Get some practice in using the *engineering mode* by finding the required data in the circuit shown in Figure 4–14.

FIND
$R_T = ?$
$I_T = ?$
$V_T = ?$

FIGURE 4–14

Using the Scientific Notation Mode with Powers of Ten for Metric Units

As you know, scientific notation expresses numerical values as a number between 1 and 10, times the appropriate power of ten. Performing the same calculations as we have been using, the sequence of keys are as follows (NOTE: Make sure the calculator is *cleared* and you are in the *scientific notation mode* for each of the following calculations. For the TI 36X solar calculator, the key sequence to get into this notation mode is: 3rd , SCI.):

To find R_T for our example circuit:
Enter 20, EE (or EXP), 3, + , 20, EE (or EXP), 3, + , 20, EE (or EXP), 3, then press = . The display reads $6.^{04}$ or 6. 04, meaning 6×10^4, or 60,000. Again, this is interpreted as 60,000 Ω, or 60 kΩ for our circuit.

To find I_T (which is equal to I_1):
Enter 10, ÷ , 20, EE (or EXP), 3, then press = . The display $5.^{-04}$ or 5. −04, which is interpreted as 5×10^{-4}, or 0.5 mA.

To find V_T:
Enter 5, EE (or EXP), +/− , 4, × , 60, EE (or EXP), 3, then press = . The display reads $3.^{01}$ or 3. 01, meaning 3×10^1, or 30 V.

PRACTICE PROBLEMS IX

Get some practice in using the *scientific notation mode* by finding the required data in the circuit shown in Figure 4–15.

5V

10 kΩ
R_1

R_2 1 kΩ

V_A

R_3 10 kΩ

R_4 1 kΩ

R_5

3.9 kΩ

FIND
$R_T = ?$
$I_T = ?$
$V_T = ?$

FIGURE 4–15

PRACTICAL NOTES

As you continue to work more electronic circuit problems, it is good to remember that:

• Units divided by Kilo units = Milli units. This is often used when finding current: $I = V/R$; where V/kΩ = mA.

• Units divided by Milli units = Kilo units. This is often used when finding resistance: $R = V/I$; where V/mA = kΩ.

• Milli units times Kilo units = Units. This is often used when finding voltage: $V = I \times R$; where mA × kΩ = V.

POWER IN SERIES CIRCUITS

The last major electrical quantity to be discussed in relation to series circuits is power. Recall that power dissipated by a component or circuit (or supplied by a source) is calculated with the formulas $P = V \times I$, $P = I^2R$, or $P = V^2/R$. You have learned that in series circuits the current (I) is the same through all components. Therefore, the I^2 factor in the $P = I^2R$ formula is the same for each component in series and the largest R value dissipates the most power. Conversely, the component with the least R value dissipates the least power (I^2R), Figure 4–16.

Individual Component Power Calculations

Individual component power dissipations can be found if any two of the three electrical parameters for the given component are known. For example, if the component's resistance and the current through it are known, use the $P = I^2R$ formula. If the component's voltage drop and the current are known, use the $P = V \times I$ formula. Knowing the voltage and resistance allows you to use the $P = V^2/R$ version of the power formula.

As implied above, the individual power dissipations around a series circuit are *directly related* to the resistance of each element, just as the

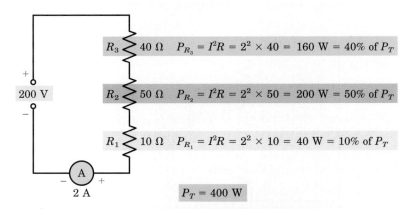

FIGURE 4–16   **Largest R dissipates most power, and smallest R dissipates least power.**

voltage distribution is directly related to each element's resistance. A specific resistor's power dissipation is the same percentage of the circuit's total power dissipation as its resistance value is of the total circuit resistance (R_T). For example, if its resistance is one-tenth (1/10) the total circuit resistance, it dissipates 10% of the total power. If it is half the R_T, it dissipates 50% of the total power, and so forth, Figure 4–16.

Total Circuit Power Calculations

Total power dissipation (or power supplied to the circuit by the power source) is determined by adding all the individual power dissipations, if they are known. See Formula 4–6.

| Formula 4–6 | $P_T = P_1 + P_2 \ldots + P_n$ |

P_1 value, $+$, P_2 value, $+$, . . . , $=$

EXAMPLE

Refer to Figure 4–16. Using Formula 4–6:

$$P_T = P_{R_1} + P_{R_2} + P_{R_3} = 40 \text{ W} + 200 \text{ W} + 160 \text{ W} = 400 \text{ W}$$

PRACTICE PROBLEMS X

Refer to Figure 4–17.

Find the values for:

$P_1 =$ _____ watts		$V_1 =$ _____ volts
$P_2 =$ _____ watts		$V_2 =$ _____ volts
$P_3 =$ _____ watts		$V_3 =$ _____ volts
$P_T =$ _____ watts		$R_3 =$ _____ % of R_T
$I_T =$ _____ amperes		$P_3 =$ _____ % of P_T

R_1 10 Ω

90 V R_2 10 Ω $R_T = R_1 + R_2 + R_3$
$R_T = 10 \text{ Ω} + 10 \text{ Ω} + 10 \text{ Ω}$
$R_T = 30 \text{ Ω}$

R_3 10 Ω

FIGURE 4–17

EFFECTS OF OPENS AND TROUBLESHOOTING HINTS

Opens

Thus far, you have studied how to analyze current, resistance, voltage, and power in series circuits. At this point we will introduce some practical situations that might occur in *abnormal* series circuits. The two extreme conditions are opens and shorts.

First, what are the effects of an open in a series circuit? An **open circuit** occurs anytime a break in the current path occurs somewhere within the series circuit. Since there is only one path for current flow, if a break (or open) occurs *anywhere* throughout the current path, there is no current flow. In effect, the circuit resistance becomes "infinite" and I_T (circuit current) must decrease to zero.

If an open causes circuit resistance to increase to infinity and circuit current to decrease to zero, how does this affect the circuit voltages and power dissipations?

EXAMPLE

In Figure 4–18, assume that R_3 is physically broken in half, causing an open or break in the current path. Since there is no complete path for current flow, what would the current meter read? The answer is zero mA.

If there is no current, what is the $I \times R$ drop of R_1? The answer is zero volts since zero times any value of R equals zero. And the answer for V_2 is also zero for the same reason. This means that the difference of potential across each of the good resistors (R_1 and R_2) is zero. That is, R_1 has zero voltage drop and R_2 has zero voltage drop. In effect, the positive side of the source potential is present anywhere from the source itself around the circuit to point A. The negative side of the source potential is present at every point along the circuit on the opposite side of the break all the way to point B.

What is the potential difference across the break, or open? If your answer is 100 volts (or V_A), you are correct! This indicates there will be no $I \times R$ drops across the *good* components in the circuit, and the circuit applied voltage appears across the open portion of the circuit, regardless of where the open appears in the circuit.

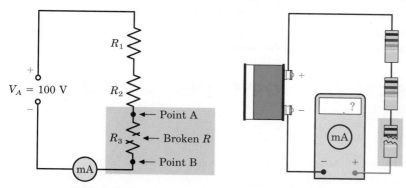

FIGURE 4–18   An open in a series circuit

An example of a purposely opened circuit is the light switch circuit in your home. When you turn the light on, you *close* an open set of contacts on the switch, which is in series with the light. When you turn the light off, you *open* those switch contacts in series with the light; therefore, you open the path for current flow, causing the applied voltage to appear across the opened switch contacts and zero volts to appear across the light element(s). If you measure the voltage across the switch with the lights *off,* you will measure the source voltage, or *V* applied. CAUTION: Don't try this! The voltage in a 120-volt lighting circuit is dangerous.

PRACTICE PROBLEMS XI

1. In a series circuit having a 100-V source and comprised of four 27-kΩ resistors, indicate whether the following parameters would increase, decrease, or remain the same if one of the resistors opened.
   a. Total resistance would _____.
   b. Total current would _____.
   c. Voltage across the unopened resistors would _____.
   d. Voltage across the opened resistor would _____.
   e. Total circuit power dissipation would _____.

2. Would the conditions you have described for question 1 be true no matter which resistor opened? _____

TROUBLESHOOTING HINTS

When you suspect an open in a series circuit, you can measure the voltage across each component. The component, or portion of the circuit across which you measure *V* applied, is the opened component, or circuit portion. *Caution:* To measure these voltages, power must be applied to the circuit. Use all safety precautions possible.

An alternative to this "power-on" approach is to use *"power-off"* resistance measurements. The two points in the circuit at which you measure "infinite ohms" (with the circuit completely disconnected from the voltage/power source) is the opened portion of the circuit. Resistance of the good components should measure their rated values.

EFFECTS OF SHORTS AND TROUBLESHOOTING HINTS

Shorts

What are the effects of a short in a series circuit? First, what is a short? A **short circuit** may be defined as an undesired very low resistance path in or around a given circuit. An example is when someone drops a metal object across two wires connected to the output terminals of a power source. In this case you short all the other circuits connected to the power source terminals and provide a *very low* resistance path for current flow through the new metal current path.

A short may occur across one component, several components, or the complete circuit (as in our dropped metal object example). Regardless, consider a short to be a virtually zero resistance path that disrupts the normal operation of the circuit.

EXAMPLE

Observe Figure 4–19. Notice that the undesired low resistance path is a piece of bare wire with a virtually zero resistance that has fallen across the leads of R_1. *Without* the short, the circuit total resistance would equal 20 kΩ ($R_1 + R_2$). Therefore, with 20 V applied voltage, the current is 1 mA (20 V/20 kΩ). *With* the short, the resistance from point A to point B is close to zero ohms because R_1 is shorted). As far as the voltage source is concerned, "it sees" only 10 kΩ resistance of R_2 as total circuit resistance. Thus, current is 20 V/10 kΩ, or 2 mA (double what it would be without the short). The short causes R_T to decrease to half its normal value and current to double since the V applied is unchanged.

This circuit analysis points to some important generalizations. 1) If any portion of a series (or any other type circuit) is shorted (or simply decreases in resistance), the total circuit resistance decreases, and 2) The circuit's total current increases, assuming the applied voltage remains unchanged.

What happens to the other circuit parameters if a short (or decreased resistance) occurs in a series circuit? Using our sample circuit in Figure 4–19: Let's think through it.

1. R_1 becomes zero ohms.

2. R_T decreases, causing circuit current (I_T) to increase.

FIGURE 4–19 Effects of a short in a series circuit

3. Since I has increased and R_2 is still the same value, the voltage across R_2 increases. In this case, it will increase to V applied (20 V) since there is no other resistor in the circuit to drop voltage.

4. V_1 (the voltage across R_1) decreases to zero volts since any value of I times zero ohms equals zero volts.

5. The total circuit power dissipation increases, since I has doubled and V applied has remained unchanged ($P = V \times I$).

6. Power dissipated by the good resistor (R_2) increases, since its R is the same, but the current through it has doubled ($P = I^2 R$).

7. Of course, the power dissipated by R_1 decreases to zero.

What happens if the *total* circuit is shorted instead of part of it? This means both R_1 and R_2 are shorted and the short is across the power supply terminals. The voltage source is looking at zero ohms, and it would try to supply "infinite" current. In reality this cannot happen. Probably the power supply fuse would blow, and/or the power supply and circuit wires/conductors might be damaged. However, the generalizations for the partially shorted circuit would also apply. That is, circuit resistance goes down, total current increases (until a fuse blows or the power supply "dies"), and the voltage across the shorted portion of the circuit decreases to zero.

PRACTICE PROBLEMS XII

1. If a series circuit having a 200-V source is composed of a 10-kΩ, a 27-kΩ, a 47-kΩ and a 100-kΩ resistor, and the 47-kΩ resistor shorts, what will the voltages be across each of the resistors listed? (Don't forget to use the "draw-the-circuit" hint to help you in your analysis.)

2. With the short present across the 47-kΩ resistor, do the voltage drops across the other resistors increase, decrease, or remain the same?

TROUBLESHOOTING HINTS

If circuit fuses blow, components are too hot, or smoke appears, there is a good chance that a component, circuit, or portion of a circuit may have acquired a short. Again, measuring resistances helps locate the problem that requires removing power from the circuit to be tested. It is best to remove power from the circuit until the fault can be cleared; therefore, voltage measurement techniques must be used judiciously, if at all. Applying power to circuits of this type often causes other components or circuits to be damaged because excessive current may pass through the nonshorted components or circuits.

With power removed from the circuit, it is possible to use an ohmmeter to measure the resistance values of individual components, or of selected circuit portions. If the normal values are known, it is obvious which component(s), or portion(s) of the circuit have the very low resistance value. Often, there will be visual signs of components or wires that have overheated, leaving a trail that indicates where the excessive current flowed as a result of the short.

In special cases, if voltage measurements are used, voltage drops across the good components are higher than normal (due to the higher current), and voltage drops across the shorted component(s) or circuit portion(s) are close to zero.

If there are current meters or light bulbs in the circuit, the meter readings are higher than normal and/or the unshorted light bulbs glow brighter than normal.

A Special Troubleshooting Hint

A simple but useful technique used in troubleshooting circuits that have sequential "in-line" components or subcircuits is called the divide-and-conquer approach, or split technique. This technique can save many steps in series situations where one of many components or subcircuits in the sequential line is the malfunctioning element.

The technique is to make the first test *in the middle* of the circuit. The results tell the technician which half of the circuit has the problem. For example, Figure 4–20 represents a twelve-light series Christmas tree circuit with an open bulb. Recall, if one light becomes open, all the lights go out since the only path for current has been interrupted. Note in Figure 4–20, the first test is made from one end of the "string" to the "mid-point," in this case between bulbs 6 and 7. If making a voltage test with power on and the meter indicates V applied, then the open bulb must be between bulb 1 through bulb 6. If V measures 0 volts, then the problem is in the other half of the circuit between bulb 7 through bulb 12. With *one* check, half of the circuit has been eliminated as the possible area of trouble.

Series Christmas tree lights (10-volt bulbs)

$V_A = 120$ V (if using voltage checks)

FIGURE 4–20   Divide-and-conquer troubleshooting technique—1st step (NOTE: Meter "M" may be either a voltmeter, if voltage is applied to the circuit under test, or an ohmmeter, if the circuit is disconnected from the power source.)

FIGURE 4–21   Divide-and-conquer troubleshooting technique—2nd step (NOTE: Meter "M" may be either a voltmeter, if voltage is applied to the circuit under test, or an ohmmeter, if the circuit is disconnected from the power source.)

Incidentally, this technique can be used with the power-off resistance measurement approach. With the circuit *disconnected* from the power source, if the first R measurement shows infinite resistance between bulbs 1 and 6, the problem is in that half of the circuit. If the reading is a low resistance, then the problem is in the other half of the circuit between bulb 7 through bulb 12.

After dividing the abnormally operating circuit's suspect area in half by making the first measurement made at the circuit mid-point, the next step is to divide the remaining suspect area of the circuit in half again by making the second check from one end of the suspect section to its mid-point; thus, narrowing down to a quarter of the circuit the portion that remains in question, Figure 4–21. This splitting technique can continue to be used until only two components or subcircuits remain. Then each is checked, as appropriate, to find *the* malfunctioning component.

The divide-and-conquer approach can be used in any in-line (linear) system of components or circuits where current flow, power, fluids, or signals must flow sequentially from one component or subcircuit to the next and is useful in troubleshooting electrical, electronic, hydraulic, and many other systems having the in-line condition.

PRACTICAL NOTES

In review, solve circuit problems with the following steps:

1. Collect all known values of electrical parameters.

2. If no circuit diagram is given, draw a diagram and write known quantities on the diagram.

3. Solve the first part of the problem at a point where you have, or can easily find, sufficient knowns to solve for an unknown. Then proceed by solving the remaining portions of the problem, as appropriate.

DESIGNING A SERIES CIRCUIT TO SPECIFICATIONS

Let's apply this in a sample design problem, then you can try one to see whether you have learned the process.

EXAMPLE

Design a three-resistor series circuit where two of the resistors are 10 kΩ, the total circuit current equals 2 mA, and the applied voltage equals 94 V.

The first step is to collect the knowns, which were given to you. The next step is to draw the circuit and label the knowns, Figure 4–22. The third step is to start at a point where sufficient knowns are available to solve for a desired unknown.

To apply the third step, observe that in the circuit we drew, we know both the I and the R values for both R_2 and R_3. This makes it easy to find their voltage drops.

$$V_2 = I \times R_2 = 2 \text{ mA} \times 10 \text{ k}\Omega = 20 \text{ V}$$
$$V_3 = I \times R_3 = 2 \text{ mA} \times 10 \text{ k}\Omega = 20 \text{ V}$$

The unknown is the value of R_1, which completes our design problem.

If the voltage drop of R_1 can be found, it is easy to find its resistance value. This is because we know the current must be 2 mA and its resistance must equal its voltage drop divided by 2 mA. It is convenient to use Kirchhoff's voltage law to find V_1. According to Kirchhoff, V_A must equal $V_1 + V_2 + V_3$. Thus, transposing to solve for the unknown V_1:

$$V_1 = V_A - (V_2 + V_3) = 94 \text{ V} - (20 \text{ V} + 20 \text{ V}) = 54 \text{ V}$$

Therefore, $R_1 = \dfrac{54 \text{ V}}{2 \text{ mA}} = 27 \text{ k}\Omega$.

Another way to solve R_1's value is to find R_T, where $R_T = V_T/I_T$. Thus, $R_T = 94 \text{ V}/2 \text{ mA} = 47 \text{ k}\Omega$. Knowing total resistance equals the sum of all

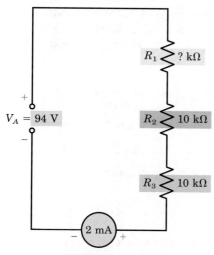

FIGURE 4–22

individual resistances in a series circuit and R_2 plus R_3 equals 20 kΩ, R_1 must make up the difference of 27 kΩ to provide a total resistance of 47 kΩ. The more formal way to illustrate this is to write the R_T formula, then transpose it to solve for R_1. Thus,

$$R_T = R_1 + R_2 + R_3$$

$$R_1 = R_T - (R_2 + R_3) = 47 \text{ kΩ} - (10 \text{ kΩ} + 10 \text{ kΩ}) = 27 \text{ kΩ}$$

Does our design meet the specifications? Yes, it is a three-resistor series circuit where two of the resistors are 10 kΩ, the total circuit current is 2 mA, and the applied voltage equals 94 V.

PRACTICE PROBLEMS XIII

Try this next design problem yourself. Complete it on a separate sheet of paper.

Problem: Design a three-resistor series circuit where R_1 is 20 kΩ and drops two-fifths of V applied, R_2 drops 1.5 fifths of V applied, and R_3 has a voltage drop equal to V_2. Assume V applied is 50 V. Draw the circuit and label all V, R, P, and I parameter values.

SPECIAL APPLICATIONS

Voltage Sources in Series

Voltage sources can be connected in series to provide a higher or lower total (resultant) voltage than one of the sources provides alone. The *resultant*

Series-aiding sources

6.0 V battery

1.5 V cell

R 1 Ω   7.5 V

A

7.5 A

Kirchhoff's algebraic sum around closed loop = 0 V
Going clockwise from point A =
+ 7.5 V − 1.5 V − 6.0 V = 0 V

(a)

Series-opposing sources

6.0 V battery

1.5 V cell

R 1 Ω   4.5 V

A

4.5 A

Kirchhoff's again
from point A (clockwise)
+ 4.5 V + 1.5 V − 6.0 V = 0 V

(b)

FIGURE 4–23   Series-connected voltage sources

voltage of more than one voltage in series depends on the values of each voltage *and* whether they "series-aid" or "series-oppose" each other. Figure 4–23a is a series-aiding type connection. Figure 4–23b is an example of a series-opposing arrangement.

Note that in Figure 4–23, you see the terms **cell** and **battery.** A cell is a single voltaic device which converts chemical energy to electrical energy. The symbols associated with a single cell are shown in each circuit in the figure as the 1.5 V sources. A familiar example of cells are the cells you put into your flashlight. Typically, these are 1.5-V cells.

A battery is simply two or more cells interconnected and put into one package. In the figure, the battery symbol is shown in each circuit related to the 6.0-volt sources. A familiar example of a battery is the 12-volt automobile battery. It is comprised of six 2-volt cells, interconnected and packaged in one battery case.

Note, in Figure 4–23a, that the "resultant" voltage applied to the circuit from the series-aiding sources equals the sum of the two sources. In this case 6.0 V + 1.5 V = 7.5 volts. The resulting current through the resistor is 7.5 amperes,

$$I = \frac{V}{R} = \frac{7.5 \text{ V}}{1 \text{ }\Omega} = 7.5 \text{ A}$$

Ways to know these sources are series-aiding are 1) the negative terminal of one source is connected to the positive terminal of the next, and 2) both sources try to produce current in the same direction through the circuit.

In Figure 4–24, what is the total voltage applied to the circuit? What is the current value? If you answered 27 V for the resultant voltage and 0.5 mA for the current, you are correct.

As you can see in Figure 4–25, series-opposing sources are connected so that: 1) the negative terminal of one source is connected to the negative

FIGURE 4–24 **Series-aiding sources have the following features: (a) Sources are in series. (b) Negative terminal of one source connects to positive terminal of next, and so on. (c) All three sources try to produce current in the same direction through the circuit. (d) Verifiable using Kirchhoff's law.**

Pt. **A** (clockwise) = + 2.25 V + 2.25 V + 4.5 V − 9.0 V = 0 V

FIGURE 4–25 **Series-opposing sources have the following features: (a) Sources are in series. (b) Negative terminal of one source connects to negative terminal of next (or positive connects to positive), and so on. (c) Both sources try to produce current in opposite directions through the circuit. (d) Verifiable using Kirchhoff's law.**

terminal of the next, or positive-to-positive; and 2) the sources try to produce current through the circuit in opposite directions.

To determine the resultant or equivalent voltage of series-opposing sources, *subtract* the smaller voltage from the larger voltage.

EXAMPLE

In Figure 4–25, 9.0 V minus 4.5 V equals a resultant of 4.5 V applied to the circuit connected to the series-opposing sources.

A final point about the concept of series-aiding and series-opposing voltages is that voltage drops can be series-aiding or series-opposing just like voltage sources are series-aiding or series-opposing. If the voltage drops are such that series-connected polarities are − to +, or + to −, *and* the current through the series components is in the same direction, they are series-aiding voltage drops. For example, V_{R_1} and V_{R_2} in Figure 4–25 are series-aiding. Conversely, if the polarities of series component voltage drops are − to −, or + to +, they are series-opposing voltage drops.

PRACTICE PROBLEMS XIV

1. Refer to Figure 4–24. If Source #1 were reversed in polarity (turned around in this circuit), what would be the value of V_{R_1}?

2. Refer to Figure 4–25. If the 4.5-V battery were replaced by a 6-V battery (using the same polarity connections into the circuit), what would be the value of V_{R_2}?

Simple Voltage Dividing Action and Reference Points

VOLTAGE DIVIDING.   As you have learned, voltage drops around a series circuit are proportional to the resistance distribution since the current is the same through all components. You can *select* the values of resistors in series to "distribute" or "divide" the applied voltage in any desired fashion. Therefore, combining this concept with the series-aiding voltage drop idea, we can create a simple **voltage-divider action** using a series circuit, Figure 4–26. Notice the polarity of voltage drops, and the difference of potential between the various points identified in the circuit.

Equal resistance values are used for R_1, R_2, and R_3, and the voltage source is equally divided by the three resistors. Also, since the voltage drops are series-aiding, the voltages at the various points are cumulative.

PRACTICE PROBLEMS XV

Referring again to Figure 4–26, assume the parameters are changed so source voltage = 188 V, R_1 = 20 kΩ, R_2 = 27 kΩ, and R_3 = 47 kΩ. Indicate the voltages between the following points: D to C; C to B; B to A; and D to B.

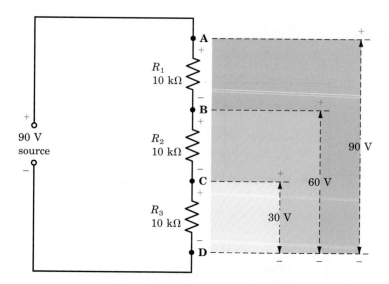

FIGURE 4-26 Example of voltage-divider action

Sample Applications for Voltage Dividers

VOLUME CONTROLS. You will recall that in an earlier chapter we mentioned that potentiometers were often used as voltage-dividing devices. In Figure 4–27, you can see that if the wiper-arm on the potentiometer is set at point A, the total voltage coming from the preceeding stage in an amplifier or radio circuit is fed to the next stage. If the wiper arm is midway on the resistive element, half the voltage is applied to the next stage, and so on. This is the technique used in radio and TV receivers, and in audio amplifiers to control the volume you hear from the system speaker(s). Obviously, the higher the "divided voltage" that is fed to the next (audio amplifier) stage after the potentiometer, the louder the sound you hear.

TRANSISTOR BIAS CIRCUIT. Later in your studies you will learn appropriate details about transistor amplifier stages. For now, you simply need to see that one very practical and commonly used application for a voltage divider is to supply appropriate voltage levels to various parts of a transistor, Figure 4–28.

REFERENCE POINT(S). Recall in an earlier chapter we discussed the idea of something "with respect to" or "in reference to" something else. The example used was "John is taller than Bill," or "Bill is shorter than John."

In electronic circuits it is common to relate electrical parameters, like voltage, to some common reference point in the circuit. The point frequently used as an electrical reference point in circuits is called "chassis ground," and the electrical schematic symbol is /̅7̅ , Figure 4–29.

In Figure 4–29, point C is where the chassis ground is connected. How is this used as a reference point to describe the voltages along the "divider"?

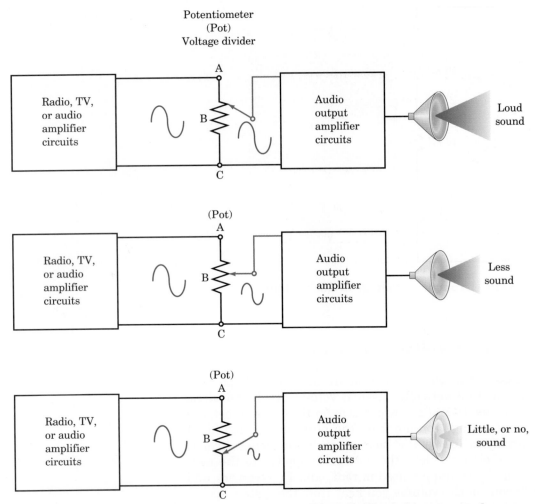

FIGURE 4–27   Potentiometer acting as voltage divider/volume control

The voltages with respect to this ground reference point can be described as follows:

Point D with respect to chassis ground = −30 V
Point C with respect to chassis ground =   0 V
Point B with respect to chassis ground = +30 V
Point A with respect to chassis ground = +60 V

Suppose that the chassis ground reference point is moved from point C to point B, Figure 4–30. How are the voltages described with respect to the reference point?

Point D with respect to chassis ground = −60 V
Point C with respect to chassis ground = −30 V
Point B with respect to chassis ground =   0 V
Point A with respect to chassis ground = +30 V

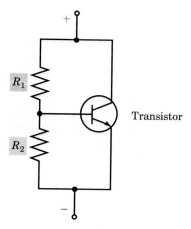

The voltages applied to the various elements of the transistor depend upon the value of the applied voltage and the ratio of the voltage divider resistances R_1 and R_2.

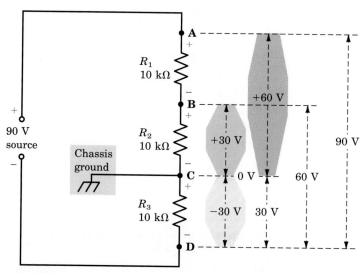

FIGURE 4–29   Chassis ground as a reference point at point C. The 30 V, 60 V, and 90 V indicators show the cumulative voltage drops. The −30 V, +30 V, and +60 V indicators show voltages with respect to point C, where ground reference is connected.

Polarity

Again, the polarity and value of voltage are described in terms of its polarity and value with respect to the designated reference point. For example, in Figure 4–29, the voltage at point C with respect to the voltage at point D is +30 V. Conversely, the voltage at point D with respect to the voltage at

FIGURE 4–30   Chassis ground as a reference point
at point B

point C is −30 V. (Note the direction and amount of current is the same in both of the examples.)

A more in-depth study of various types of voltage dividers will be given in the chapter on series-parallel circuits. The idea of reference points, as just discussed, will be used throughout your studies and your career as a technician.

PRACTICE PROBLEMS XVI

Assume the circuit components and parameters are changed in Figure 4–30 so that V_A now equals 50 V; R_1 now equals 12 kΩ; R_2 now equals 33 kΩ; and R_3 now equals 15 kΩ. Assume that the chassis ground reference point is still point B and that the polarity of the source is the same as in Figure 4–30. Using the voltage-divider rule, and your recently gained knowledge of reference points, determine the following: (*Hint:* Redraw the circuit so you will have something to look at. Use a calculator, if possible.)

1. What is the voltage across R_2?
2. What is the voltage at the top of R_2 with respect to point C?
3. What is the voltage across R_1?
4. What is the voltage at point C with respect to point A?

The Voltage-Dropping Resistor

One final topic relating to some special applications of series circuits is the **series-dropping resistor.**

Often, there is a need to supply a specific voltage at a specific current to an electrical load. If the "fixed" power supply voltage available is higher than the desired voltage value, a series-dropping resistor is used that drops the voltage to the desired level, Figure 4–31.

EXAMPLE

The current through the electrical load and the series-dropping resistor must be the same since they are in series. To calculate the value of the dropping resistor, the value of voltage it must drop, and the value of current the load requires, provide the knowns to solve the problem. In this case, the dropping resistor must drop 60 V, and the load requires 30 mA of current. Therefore, the dropping resistor must have 30 mA passing through it *when* it is dropping 60 V. Ohm's law indicates the resistor value must be $R = V/I$, or the dropping resistor value = 60 V/30 mA = 2 kΩ. Naturally, this design calculation technique can be used to select resistor values for an infinite variety of dropping-resistor applications and is another example of the many special applications for series circuits.

Dropping resistor must drop 60 V with 30 mA of current through it.

R dropping

100 V source

Electrical load requires 40 V at 30 mA

FIGURE 4–31   Example of a series-dropping resistor

PRACTICAL NOTES

The power rating of the dropping resistor must be greater than the power it is required to dissipate. Usually, a rating *of at least two times* the actual dissipation is chosen.

It should be pointed out that a voltage drop is the difference in voltage between two points caused by a loss of pressure (or emf) as current flows through a component offering opposition to current flow. An example is the potential difference caused by a voltage drop across a resistor.

PRACTICE PROBLEMS XVII

1. What is the minimum power rating for the dropping resistor, Figure 4–31?

2. What should the value of the dropping resistor be if the applied voltage were 150 V and the load requirements remained 40 V at 30 mA? What is the minimum power rating for the dropping resistor?

THE SIMPLER SEQUENCE FOR TROUBLESHOOTING

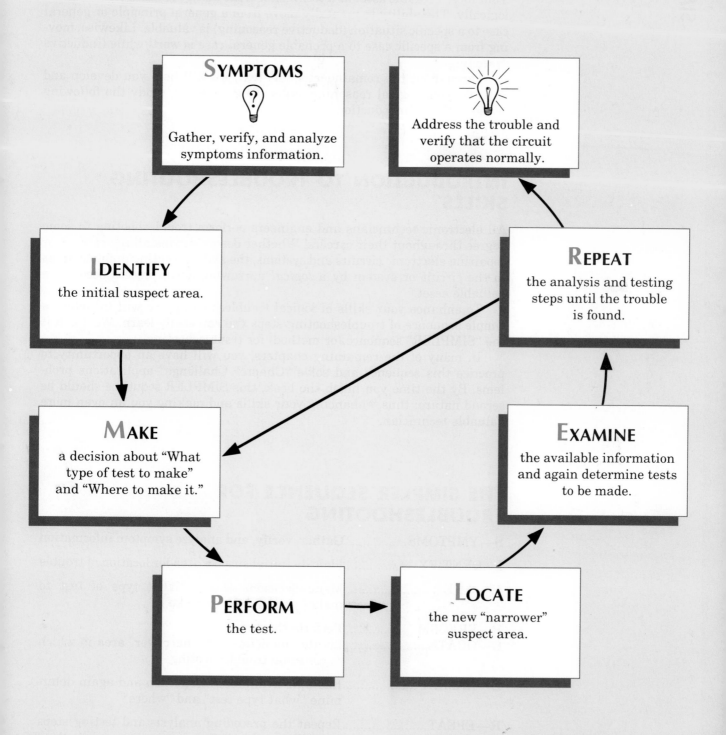

SYMPTOMS

Gather, verify, and analyze symptoms information.

Address the trouble and verify that the circuit operates normally.

IDENTIFY

the initial suspect area.

REPEAT

the analysis and testing steps until the trouble is found.

MAKE

a decision about "What type of test to make" and "Where to make it."

EXAMINE

the available information and again determine tests to be made.

PERFORM

the test.

LOCATE

the new "narrower" suspect area.

CRITICAL THINKING AND THE SIMPLER TROUBLESHOOTING SEQUENCE

Your most valuable asset as a technician is the ability to think critically and logically. The ability to mentally move from a general principle or general case to a specific situation (deductive reasoning) is valuable. Likewise, moving from a specific case to a probable general case is worthwhile (inductive reasoning).

Throughout the remainder of this text, we will help you develop and enhance your logical reasoning skills. To get started, study the following section called "Introduction to Troubleshooting Skills."

INTRODUCTION TO TROUBLESHOOTING SKILLS

All electronic technicians and engineers perform troubleshooting to some degree throughout their careers! Whether designing, installing, testing, or repairing electronic circuits and systems, the skill of locating problem areas in the circuit or system by a *logical* narrowing down process becomes a valuable asset.

To enhance your skills of logical troubleshooting, we will introduce a simple sequence of troubleshooting steps you can easily learn. We'll call it the "SIMPLER" sequence (or method) for troubleshooting.

In many of the remaining chapters, you will have an opportunity to practice this sequence and solve "Chapter Challenge" applications problems. By the time you finish the book, this SIMPLER sequence should be second nature; thus, enhancing your skills and making you an even more valuable technician.

THE SIMPLER SEQUENCE FOR TROUBLESHOOTING

S—YMPTOMS.................Gather, verify, and analyze symptom information

I—DENTIFY....................Identify initial suspect area for location of trouble

M—AKE...........................Make decisions about: "What type of test to make" and "Where to make it"

P—ERFORM ....Perform the test

L—OCATE.......................Locate and define new "narrower" area in which to continue troubleshooting

E—XAMINE....................Examine available information and again determine "what type test" and "where"

R—EPEAT.......................Repeat the preceding analysis and testing steps until the trouble is found

Further Information About Each Step in the Sequence

1. **SYMPTOMS** may be provided or collected from a number of sources, such as:
   a. What *the user* of the circuit or system tells you.
   b. *Your five senses* frequently can give you strong clues to the trouble. Sight and hearing allow evaluation of the circuit or system output indicators; e.g., normal or abnormal picture, normal or abnormal sound, system meter readings. Sight also allows you to see obviously overheated/burned components. Smell helps find overheated or burned components. (Once you have smelled certain types of overheated or burned components, you will never forget the smell!) Touch can also be used but should be used *very cautiously!* Bad burns result from touching a component that is running exceedingly hot! (Some components run that way normally, others do not.)
   c. By using your evaluation of the *easy-to-check indicators*. For example, if a TV has sound but no picture, or a radio receiver has hum, but you can't tune in radio stations. Most systems have easy-to-get-at switches and controls that aid in narrowing down possible trouble areas before you begin in-depth troubleshooting. For example, TVs have channel selectors, radios have tuning and volume controls, and electronic circuits frequently have switches and variable resistances.
   d. By *comparing actual operation to normal operation and "norms."* "Norms" are normal characteristics of signals, voltages, currents, etc., that are known by experience or determined by referring to diagrams and documentation about the circuit or system being checked. This means you, as a technician, should know how to read and thoroughly understand block and schematic diagrams. (In specific cases, you may have to trace the actual circuit and draw the diagrams yourself.)

2. **IDENTIFY** the initial suspect area by:
   a. Analyzing all the symptom data.
   b. From the analysis of the symptom data, determining all the possible sections of a system, or components within a circuit, that might cause or contribute to the problem.
   c. Circling, or in some way marking, the initial suspect area.

3. **MAKE** a decision about "what kind of test" and "where" by:
   a. Looking at the initial suspect area, and determining where a test should be made to narrow down the suspect area most efficiently. And, determining what type of test would be most appropriate to make. (NOTE: Many times the "where" will dictate what type of test is to be made. At other times, the easiest type of test to make, which yields useful information, is the determinant of both the "where" and "what kind" of tests.)
   b. Typically, you begin with general tests, such as looking at obvious indicators, manipulating switches and controls, and similar measures. As you narrow down the suspect area, your tests become more precise, such as voltage, current, or resistance measurements. Finally, substituting a known good component for the suspected bad

one confirms your analysis. (In some cases, eradicating the problem is done by soldering, or removing a short, or moving a wire rather than changing a component.)

4. **PERFORM** what your "where-and-what-kind-of-test" analysis has indicated (i.e., what kind of test and where), and you are provided with new information and insight to use in your quest to narrow down and find the problem.

5. **LOCATE** the questionable area of the circuit or system using information you have. Again, you will identify this new smaller area by circling or bracketing it so you won't make needless checks outside the logical area to be analyzed.

6. **EXAMINE** the collected symptom information, test results, and other data you have. Now make a decision about the next test that will provide further meaningful information.

7. **REPEAT** the analysis, testing, and narrowing down process as many times as required, and you will eventually find the fault in the circuit or system. At this point, you have successfully used the SIMPLER sequence!

Using the SIMPLER Sequence: An Example

The information given to you is:

A basic circuit is comprised of a voltage source, conducting wires, a switch, and a light bulb (with its socket). It is known that the voltage source and conductor wires have been previously tested and they are good.

The complaint (which reveals symptom information) is the light bulb does not light, even when the switch is flipped to the "on" position.

1. The **symptom** information gathered is that the light bulb doesn't light! You verify this by connecting the circuit. Analysis of the symptom information can be done by mentally visualizing the circuit, or drawing the circuit diagram. It is then apparent that if the source and the conductor wires are good, the initial suspect areas are the switch, light bulb, and socket.

2. Now you mentally (or on paper) **identify** the portion of the circuit containing the switch, the light bulb, and the socket (by circling or bracketing), Figure 4–32. This becomes the "initial suspect area" for locating the trouble.

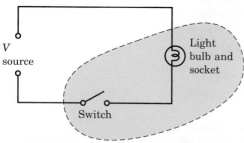

FIGURE 4–32 Initial suspect area

FIGURE 4–33 Checking the light bulb first

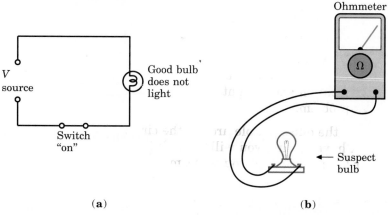

**FIGURE 4–34   Results of test: (a) Good bulb does not light.
(b) Suspect bulb has continuity.**

3. You now **make** a decision about "what type of test to make" and "where to make it." Some of the possibilities are to: a) make a voltage test with a voltmeter across the light bulb socket; b) turn the switch to the "off" position and replace the bulb with a known good bulb, then turn the switch to the "on" position to see if the replacement bulb lights; c) disconnect the power source, unscrew the bulb, and make an ohmmeter test of the suspect bulb to see if it has continuity; d) measure the voltage across the switch terminals with the switch open and with it closed to see if the voltage changes; and e) disconnect the power source and use an ohmmeter to check the switch in both the "off" and "on" positions to see if it is operating properly. Since changing the bulb would be reasonably easy to do and bulbs do fail frequently, you decide to make the light bulb test first, Figure 4–33.

4. When you **perform** this test, you find that the "good" replacement bulb does not light, even with the power source connected to the circuit and the switch in the "on" position. Or if you tested the "suspect" bulb using the ohmmeter check approach, you would find the bulb checks good, Figure 4–34.

5. You can now **locate** the new smaller suspect area that includes only the socket and switch, Figure 4–35.

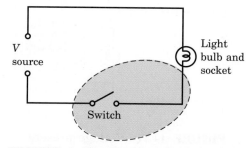

FIGURE 4–35   Narrowed suspect area

FIGURE 4–36 **Voltage check at**
light socket

6. When you **examine** all the accumulated information, you decide the most likely suspect between the socket and switch might be the switch since it is somewhat of a mechanical device which fails more frequently. Three ways you can check into this possibility are to: a) disconnect the power source from the circuit and use an ohmmeter to check the switch continuity in both the "off" and "on" positions; b) measure the voltage at the light socket to see if voltage is present with the switch in the "on" position; and c) measure voltage across the switch in both the "off" and "on" positions to see if there is voltage across the switch when in the "off" position and zero volts present when the switch is in the "on" position. After examining the information, your decision about "what type of test" and "where" is to check voltage at the light socket since this is an easy check to make, Figure 4–36.

7. **Repeating** the analysis *and* testing procedures, you make the test at the bulb socket with the switch "on." Voltage is present, Figure 4–37. This indicates the switch is good. With this analysis, your remaining suspect is the light socket. In testing the socket (either by ohmmeter continuity testing with the power off or by replacing with a known good socket), you find the socket is indeed bad. You have found the problem! The solution is obvious: Replace the socket. After replacing the socket, you verify the accuracy of your troubleshooting by testing the circuit. Your reward is that the circuit works properly, Figure 4–38.

FIGURE 4–37 **Voltage present**
at socket

FIGURE 4–38   The circuit works

Although this hypothetical case is simple, it illustrates the concepts of the SIMPLER troubleshooting sequence. Practice using the sequence because it will help you become a logical troubleshooter!

TROUBLESHOOTING LEVELS

There are two basic levels of troubleshooting technicians may have to perform. One is troubleshooting to the "block" or "module" level. The other is troubleshooting to the single component level.

The block level requires knowledge of the normal "inputs" and "outputs" used for each block or module in the total system. These inputs and/or outputs may be audio or video signals, certain voltage levels, certain current values, and so forth.

To troubleshoot to the block level, we'll refer to the "inputs/outputs" theory, where the "inputs" refer to block inputs and the "outputs" refer to the block output(s). If the input quantities (signals, voltages, etc.) check normal for a given block but the output quantities are abnormal, the problem is probably in that block.

On the other hand, if the input quantities are not normal, you trace backward to where the input is supposed to come from until the place of abnormality is found. In some cases, the abnormality can be caused by the input circuitry of the block being tested, rather than from blocks feeding this block. Your job is to isolate the block causing the problem and to replace it. See A Block-level Troubleshooting Example that follows.

Single-component level troubleshooting requires knowledge of normal parameters throughout the circuitry within each block of the system. By isolating and narrowing down (using the SIMPLER sequence approach), you will eventually narrow the problem to the bad component(s) within a module or block. Replacing the bad component(s) solves the problem. See A Component-level Troubleshooting Example that follows.

The SIMPLER Sequence and Troubleshooting Levels

The SIMPLER sequence can be applied to both the block and component levels of troubleshooting. You will have a chance to try both troubleshooting levels in the Chapter Challenge problems you will find throughout this book.

Now, try the first Chapter Challenge, found at the end of this chapter, to begin practicing these concepts.

A BLOCK-LEVEL TROUBLESHOOTING EXAMPLE

If you were told there was no sound from the speaker, Figure 4–39, even though someone was talking into the microphone, block-level troubleshooting scenario might be as follows:

Symptoms   No sound from speaker.

Identify   initial suspect area anywhere in the system for this situation. The trouble could be in the speaker block, amplifier block, or microphone block.

Make   decisions about test (what kind of test and where). Probably make test about in the middle of the system (at the amplifier input). That way you have cut the area of possible trouble in half with your first test. (Remember the "divide-and-conquer" technique explained with Figure 4–21.)

Perform   the test: As you can see from the diagram, the signal into the amplifier is OK (the "input" is all right for this block).

Locate   the new smaller suspect area. Your first test has eliminated the microphone as a possible trouble spot. The remaining possible trouble area is the amplifier block and the speaker block.

Examine   available information and make new test decisions. Now see if the amplifier output is normal. The "type" of check is a signal check; the "where" is at the amplifier output.

Repeat   testing and analysis procedure. A check at the amplifier output shows no signal. (The "output" signal is bad.) If the input to the amplifier is good and the output of the amplifier is bad, the chances are the trouble is in the amplifier block. This is troubleshooting to the block level. If we were to go inside the block to find the component in the block causing the trouble, that would be troubleshooting to the component level.

A COMPONENT-LEVEL TROUBLESHOOTING EXAMPLE

FIGURE 4-40

If you were given the facts in Figure 4-40, and told the voltage source had been checked as being O.K., component-level troubleshooting might be done as follows:

Symptoms   No voltage across R_2, even though circuit input voltage is normal.

Identify   initial suspect area. The trouble could be with any one of the resistors or with the interconnecting wires in the circuit, since the symptom could be caused by R_2 being shorted or by an open elsewhere in the circuit.

Make   decisions about test (what type of test and where). Since voltage tests are easy and yield much information for small effort, the "what kind of test" is answered with a voltage test. Since you were told that the voltage source is O.K., you decide to check the voltage across either one of the two remaining resistors to determine if current is flowing anywhere in the circuit.

Perform   the test. You decide to make the voltage test across R_1. The result of the test is that V_{R_1} is found to be 6 volts. This is abnormal for this circuit!

Locate   the new smaller suspect area. The voltage test across R_1 leads you to believe it may be the suspect! (Recall in a series circuit applied voltage appears across the open portion of the circuit, and zero volts appear across the good components).

Examine   available information and make new test decisions. The logical new test might be to disconnect the power source and use an ohmmeter to check the value of resistance exhibited by R_1. The "what kind of test" is a resistance check. The "where to make the test" decision is across R_1.

Repeat   testing and analysis procedure. When R_1 is tested with the ohmmeter, it measures infinite resistance, or "open." R_1 is indeed the culprit. This is verified by replacing R_1 and energizing the circuit. When this is done, the voltage drops measured across each of the resistors is 2 volts. This is troubleshooting down to the component level.

SUMMARY

▶ The definition and characteristics of a series circuit state that the circuit components are connected so that there is only one current path; therefore, the current is the same through all components in series.

▶ Current in a series circuit is calculated by Ohm's law, using either total voltage divided by total resistance

$$I = \frac{V_T}{R_T}$$

or the voltage drop across one of the components divided by its resistance; for example,

$$I = \frac{V_1}{R_1}$$

▶ Total resistance in a series circuit equals the sum of the individual resistances ($R_T = R_1 + R_2 \ldots + R_n$). R_T can also be calculated if V_T and I are known using

$$R_T = \frac{V_T}{I_T}$$

▶ Because I is the same throughout all components in series, each component's voltage drop ($I \times R$) is proportional to its resistance compared to the total resistance. The largest value resistance drops the highest percentage or fraction of V applied; the smallest value resistance drops the smallest portion of the total voltage.

▶ The voltage-divider rule is a convenient way of finding individual voltage drops in series circuits where the applied voltage and total resistance are known. The formula states:

$$V_X = \frac{R_X}{R_T} \times V_T$$

▶ Calculators are very useful for most calculations in electronics. Calculators may be used in several different modes of operation. One method is simply entering the literal numbers. Other methods use the advantages of a "scientific calculator" by application of powers of ten. The engineering mode deals with numbers times powers of 10 with exponents having multiples of three. The scientific notation mode translates every number into a number between 1 and 10 times a power of ten. The engineering mode is the mode you will probably use most.

▶ V applied must equal the sum of the voltage drops around the closed loop. That is, if there are three components around the series string, V applied must equal the sum of the three components' voltage drops. If there are four components, V applied equals the sum of the four individual voltage drops.

▶ Kirchhoff's voltage law states the arithmetic sum of the voltages around a single circuit loop equals V applied. Also, Kirchhoff's voltage law can be stated as the algebraic sum of all the loop voltages, including the source, must equal zero. It is important when utilizing Kirchhoff's law(s) to notice the voltage polarities.

▶ The power distribution throughout a series circuit is directly related to the resistance distribution. For example, if a component's R is 25% of the total R, it dissipates 25% of the total power supplied by the source.

▶ Total power dissipated by a series circuit (or any other type circuit) equals the sum of all the individual power dissipations. Total power can be calculated by $P_T = V_T \times I_T$, or $P_T = P_1 + P_2 \ldots + P_n$.

▶ A break or open anywhere along the path of current in a series circuit causes R_T to increase to infinity; I_T to decrease to zero; voltage drops across unopened components or circuit portions to decrease to zero; and circuit applied voltage to be felt across the open portion of the circuit. P_T decreases to zero since I equals zero.

▶ A short (very low resistance path) across any component, or portion, of a series circuit causes R_T to decrease; I_T to increase; voltage drops across the normal portions of the circuit to increase; and voltage dropped across the shorted portion of the circuit to decrease to virtually zero. A short also causes P_T to increase and the power dissipated by the remaining unshorted components to increase.

▶ When designing a circuit or solving for unknown quantities in an electrical circuit, it is good to: 1) collect all known values; 2) draw a diagram and label known value(s); 3) solve at any point where enough knowns are available to solve for an unknown. (NOTE: Usually, Ohm's law and/or Kirchhoff's law(s) will be used in the solution).

▶ When troubleshooting a series circuit with an open condition, voltage measurements will reveal zero voltage drop across good components (or circuit portions) and V applied across the open points. Take resistance measurements only with the circuit disconnected

from the source. Then normal components will reveal normal readings, and the opened component/portion will reveal an infinite ohms reading.

▶ When troubleshooting a series circuit with a short condition, use *resistance* measurements with the power source disconnected from the circuit. Normal components (or circuit portions) will measure normal values of R, while the shorted component (or circuit portion) will measure very low, or zero R.

▶ Series circuits are useful for several special applications, such as series-aiding and series-opposing voltage sources, simple voltage-dividing action, and series-dropping an available voltage down to a desired level.

FORMULAS AND CALCULATOR SEQUENCES SUMMARY

FORMULA 4–1 $R_T = R_1 + R_2. . .+ R_n$
R_1 value, $\boxed{+}$, R_2 value, $\boxed{+}$, . . . , $\boxed{=}$

FORMULA 4–2 $R_T = \dfrac{V_T}{I_T}$
total voltage value, $\boxed{\div}$, total current value, $\boxed{=}$

FORMULA 4–3 $V_1 = I_{R_1} \times R_1$
current value, $\boxed{\times}$, resistor #1 value, $\boxed{=}$

FORMULA 4–4 $V_X = \dfrac{R_X}{R_T} \times V_T$
R_X value, $\boxed{\div}$, R_T value, $\boxed{\times}$, V_T value, $\boxed{=}$

FORMULA 4–5 $V_T = V_1 + V_2 . . . + V_n$
V_1 value, $\boxed{+}$, V_2 value, $\boxed{+}$, . . . , $\boxed{=}$

FORMULA 4–6 $P_T = P_1 + P_2. . . + P_n$
P_1 value, $\boxed{+}$, P_2 value, $\boxed{+}$, . . . , $\boxed{=}$

REVIEW QUESTIONS

NOTE: While working on these questions, don't forget to draw and label diagrams, as appropriate. It is a good habit to form.

1. If a series string of two equal resistors draws 5 amperes from a 10-volt source, what is the value of each resistor?

2. What is the current through a series circuit comprised of one 10-kΩ, one 20-kΩ, and one 30-kΩ resistor? Assume the circuit applied voltage is 40 V.

3. Draw a diagram of a series circuit containing resistors having values of 50 Ω, 40 Ω, 30 Ω, and 20 Ω, respectively. Label the 50-Ω resistor as R_1, the 40-Ω resistor as R_2, and so forth. Calculate the following and appropriately label your diagram:
   a. What is the value of R_T?
   b. What is V applied if I equals 2 A?
   c. What is the voltage drop across each resistor?
   d. What is the value of P_T?
   e. What value of power is dissipated by R_2? By R_4?
   f. What fractional portion of V_T is dropped by R_4?
   g. If R_3 increases in value while the other resistors remain the same, would the following parameters increase, decrease, or remain the same?
      (1) Total resistance
      (2) Total current
      (3) V_1, V_2, and V_4
      (4) P_T
   h. If R_2 shorted, would the following parameters increase, decrease, or remain the same?
      (1) Total resistance
      (2) Total current
      (3) V_1, V_3, and V_4

4. According to Kirchhoff's voltage law, if a series circuit contains components dropping 10 V, 20 V, 30 V, and 50 V, respectively, what is V applied? What is the algebraic sum of voltages around the complete closed loop, including the source?

5. Draw a diagram showing how you would connect three voltage sources to acquire a circuit applied voltage of 60 V, if the three sources equaled 100 V, 40 V, and 120 V, respectively.

6. a. Calculate the P_T and P_1 values for the circuit shown in Figure 4–41.
   b. What is the V_T?
   c. What is the R_T?
   d. How many times greater is P_4 than P_1?
   e. If R_4 shorted what would the value of P_T become? Of P_1?

FIGURE 4–41

7. What precaution should be taken when troubleshooting a series circuit suspected of having a short? Briefly explain why this precaution should be taken.

8. If a series circuit's current suddenly drops to zero, what has happened in the circuit?

9. If certain components are overheating, circuit current has increased, and the voltage drops across R_2 and R_3 have drastically decreased, which of the following has occurred?
   a. The total circuit has been shorted.
   b. The total circuit has been opened.
   c. R_1 has shorted.
   d. R_2 and R_3 have shorted out.
   e. None of the above.

10. Find the value of V_A for the series circuit shown in Figure 4–42.
    a. Assume that R_3 and R_4 are equal. What is the value of R_3?
    b. What is the value of R_2?

FIGURE 4–42

11. Draw a simple three-resistor series voltage divider that provides equal voltage division of a 180-V source and draws 2 mA of current.

Referring to Figure 4–43, calculate and answer questions 12–19:

12. What value of voltage will be read by the voltmeter?

13. What value of current will be read by the milliammeter?

14. What is the voltage drop across R_2?

15. What is the voltage drop across R_3?

16. What is the voltage drop across R_4?

17. If R_3 were to short out, what would be the voltage reading on the voltmeter?

18. Which resistor in the circuit of Figure 4–43 dissipates the most power? The least power?

19. If V_A doubled for the original circuit of Figure 4–43, would the percentage of applied voltage dropped by R_3 increase, decrease, or remain the same?

20. In a three-resistor series circuit, if V applied is 200 V, V_1 is 40 V, and V_2 is 90 V, what is the value of V_3?

21. In a three-resistor series circuit, if V_1 is 25 V, V_2 is 50 V, and V_3 is twice V_1, what is the value of V applied?

22. What basic concept is often used in troubleshooting a defective series circuit that has a number of components in series?

23. For the conditions shown in the circuit in Figure 4–44, to what value is R_1 adjusted?
    a. What is the value of I_T?
    b. What percentage of total power is dissipated by R_1?
    c. If R_1 were set to the middle of its R range, what would be the value of I_T?
    d. If R_1 were set at the least value possible, what value of voltage would be indicated by the voltmeter in the circuit?

FIGURE 4–44

24. a. Find R_T and V_T for the circuit shown in Figure 4–45.
    b. What value will meter M_1 indicate?
    c. What value will meter M_3 indicate?
    d. What is the value of I_T for this circuit?
    e. If all R values remain the same, but P_T doubles, what is the new value of V_T? What is the new I_T?

FIGURE 4–43   **Review questions problem circuit**

FIGURE 4–45

25. Refer to Figure 4–45. If all R values are as shown, but P_{R_5} equals 45 mW:
    a. What is the value of P_{R_2}?
    b. What is the value of P_T?
    c. What is the value of I_T?

CHALLENGE QUESTIONS

1. In your own words, explain the difference between scientific notation and engineering notation.

2. List at least two other applications for simple voltage dividers, other than those mentioned in the chapter.

3. In your own words, express the key facts you have learned in this chapter regarding:
   a. Current in series circuits
   b. Voltage distribution around a series circuit
   c. Power distribution around a series circuit
   d. The effects of an open component on circuit current and voltage distribution
   e. The effects of a shorted series component on circuit current and voltage distribution
   f. At least two means of gathering symptom information regarding a malfunctioning circuit
   g. The "divide-and-conquer" rule

TEST YOURSELF

1. Draw a circuit diagram of a series circuit where V_3 is three times the value of V_1 and V_2 is twice V_1. Assume V applied is 60 V, R_T is 120 kΩ, and calculate:
   a. Value of R_1.
   b. Values of V_1, V_2, and V_3.
   c. Values of I and P_T.

2. Determine the value of a series-dropping resistor needed to drop a source voltage of 100 V to the appropriate value for a load rated as 50 V at 5 mA. Draw the circuit and label all components and electrical parameters.

PERFORMANCE PROJECTS CORRELATION CHART

Suggested performance projects in the Laboratory Manual that correlate with topics in this chapter are:

CHAPTER TOPIC	PERFORMANCE PROJECT	PROJECT NUMBER
Resistance in Series Circuits	Total Resistance in Series Circuits	9
Definition and Characteristics of a Series Circuit	Current in Series Circuits	10
Voltage in Series Circuits	Voltage Distribution in Series Circuits	11
Power in Series Circuits	Power Distribution in Series Circuits	12
Effects of Opens in Series Circuits and Troubleshooting Hints	Effects of an Open in Series Circuits	13
Effects of Shorts in Series Circuits and Troubleshooting Hints	Effects of a Short in Series Circuits	14

NOTE: It is suggested that after completing the above projects, the student should be required to answer the questions in the "Summary" at the end of this section of projects in the Laboratory Manual.

THE SIMPLER SEQUENCE FOR TROUBLESHOOTING

Following the **SIMPLER** troubleshooting sequence, find the problem with this circuit. As you follow the sequence, record your responses to each step on a separate sheet of paper.

CHALLENGE CIRCUIT 1

SYMPTOMS Gather, verify, and analyze symptom information. (Look at the "Starting Point Information.")

IDENTIFY Identify initial suspect area for the location of the trouble.

MAKE Make a decision about "What type of test to make" and "Where to make it." (To simulate making a decision about each test you want to make, select the desired test from the "TEST" column listing below.)

PERFORM Perform the test. (Look up the result of the test in Appendix C. Use the number in the "RESULTS" column to find the test result in the Appendix.)

LOCATE Locate and define a new "narrower" area in which to continue troubleshooting.

EXAMINE Examine available information and again determine "what type of test' and "where."

REPEAT Repeat the preceding analysis and testing steps until the trouble is found. What would you do to restore this circuit to normal operation? When you have solved the problem, compare your results with those shown on the following pages.

STARTING POINT INFORMATION

1. Circuit diagram
2. $V_T = 6$ V
3. Measured $I_T = 1.5$ mA

TEST   Results in Appendix C

V_{A-B}......................................(5)
V_{B-C}......................................(86)
V_{C-D}......................................(42)
Res. of R_1 ....(16)
Res. of R_2 ....(62)
Res. of R_3...............................(3)
R_T...(36)

CIRCUIT

1

Meter lead

Source lead

Symptoms

AUTO-RANGE DIGITAL MULTIMETER

From source +

From source −

Symptoms The total current is too low for voltage applied. This implies that the total resistance has increased, meaning that one or more resistor(s) has changed value.

Identify initial suspect area: R_1, R_2, and R_3 (i.e., total circuit).

Make test decision: Check voltage across R_2 (middle of circuit).

Perform 1st Test: Look up the test result. V_{B-C} is 1.5 V.

Locate new suspect area: R_1 and R_3. NOTE: V_{B-C} would be greater than 2 V if R_2 had increased and the other resistors had not.

Examine available data.

Repeat analysis and testing:

CHALLENGE CIRCUIT 1

Check voltage across R_1. V_{A-B} is 1.5 V.

Check voltage across R_3. V_{C-D} is 3 V.

Disconnect the V source and check resistance of R_3. The result is that R_3 is 2 kΩ, which is abnormal.

Replace R_3 with a good 1-kΩ resistor and note the current. When this is done, the circuit checks out normal. Each resistor drops 2 V and the circuit current measures 2 mA.

1st Test

2nd Test

3rd Test

4th Test

5th Test

Parallel Circuits

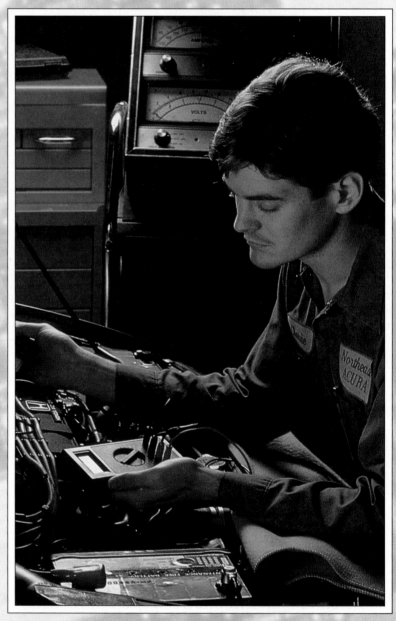

Courtesy of Cleveland Institute of Electronics, Inc.

5

KEY TERMS

Assumed voltage method

Current divider circuit

Equivalent circuit resistance

Kirchhoff's current law

Parallel branch

Parallel circuit

Product-over-the-sum method

OUTLINE

Definition and characteristics of a parallel circuit

Voltage in parallel circuits

Current in parallel circuits

Kirchhoff's current law

Resistance in parallel circuits

Methods to calculate total resistance (R_T)

Power in parallel circuits

Effects of opens in parallel circuits and troubleshooting hints

Effects of shorts in parallel circuits and troubleshooting hints

Similarities and differences between series and parallel circuits

Designing a parallel circuit to specifications

Sources in parallel

Current dividers

CHAPTER PREVIEW

Most electrical and electronic circuits, no matter how simple or complex, contain portions that can be examined using parallel circuit analysis. Of course, some circuits contain only parallel circuit arrangements. For example, the lights and wall outlets in your home generally use parallel circuitry to connect several lights or outlets on one circuit, Figure 5–1. Many automobile accessories (heaters and radios) are connected to the battery using parallel circuit methods, Figure 5–2.

In this chapter, you will again apply Ohm's law and the power formulas. Also, Kirchhoff's current law and some *vital contrasts* between parallel and series circuits will be examined. Several approaches to solve for total resistance in parallel circuits will be learned and applied. Finally, information on circuit troubleshooting techniques will be presented.

OBJECTIVES

After studying this chapter, you should be able to:

- Define the term **parallel circuit**
- List the characteristics of a parallel circuit
- Determine voltage in parallel circuits
- Calculate the total current and branch currents in parallel circuits
- Compute total resistance and branch resistance values in parallel circuits using at least three different methods
- Determine conductance values in parallel circuits
- Calculate power values in parallel circuits
- List the effects of opens in parallel circuits
- List the effects of shorts in parallel circuits
- Describe troubleshooting techniques for parallel circuits
- Use current divider formulas
- Use the SIMPLER troubleshooting sequence to solve the Chapter Challenge problem

DEFINITION AND CHARACTERISTICS OF A PARALLEL CIRCUIT

Recall from the last chapter the important characteristic defining a series circuit is only one path for current exists. Therefore, the current through all components is the same. Because of this fact, the voltage distribution throughout the circuit (voltage across each of the components) depends on each component's resistance value (i.e., the highest R value dropping the highest voltage or the lowest R value dropping the lowest voltage).

Lights in a home are connected in parallel.

Wall outlets in a home are connected in parallel.

FIGURE 5–1   Parallel circuits are common in a home.

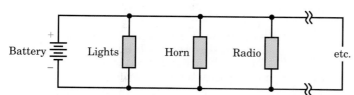

FIGURE 5–2   Automotive accessories are often connected in parallel across the battery. NOTE: Series switches in circuits are not shown.

The significant features of **parallel circuits** are: 1) *the voltage across all parallel components must be the same,* Figure 5–3; and 2) *there are two or more paths (branches) for current flow,* Figure 5–4. Because the voltage is the same across each **parallel branch,** the current distribution throughout the circuit depends on each branch's resistance value. Each branch current is *inverse* to its resistance value, which means the higher the resistance value, the lower the branch current, Figure 5–5.

Thus, a parallel circuit is defined as one where the voltage across all (parallel branch) components is the same and where there are two or more branches for current flow.

FIGURE 5–3   Voltage in parallel circuits is the same across all components, where $V_{R_1} = V_A$ and $V_{R_2} = V_{R_1} = V_A$.

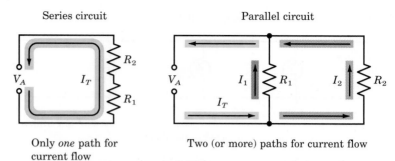

Series circuit

Only *one* path for
current flow

Parallel circuit

Two (or more) paths for current flow

**FIGURE 5–4   A basic difference between series and
parallel circuits**

$V = 10$ V $10\ \Omega$ R_1 $5\ \Omega$ R_2

$I_1 = \dfrac{V}{R_1} = 1$ A $I_2 = \dfrac{V}{R_2} = 2$ A

**FIGURE 5–5   In parallel circuits,
current through each branch is
inversely proportional to each
branch's resistance.**

VOLTAGE IN PARALLEL CIRCUITS

If you know the voltage applied to a parallel circuit or the voltage across any
one of the branches, you automatically know the voltage across any one or
all of the parallel branches. This voltage is the same across all branches of
a parallel circuit, Figure 5–3.

PRACTICE PROBLEMS I

Again refer to Figure 5–3. If R_2 is 27 kΩ and current through R_2 is
2 mA, what is the value of V_{R_1}? Of V_A?

CURRENT IN PARALLEL CIRCUITS
Branch Currents and Total Current

Refer to Figure 5–6 and note the following points as current and Kirchhoff's current law are discussed:

• Total circuit current leaves from the source's negative side and travels
  along the conductor until it reaches the circuit "junction" at point A.

FIGURE 5–6 Kirchhoff's current law states that current coming to a point must equal the current leaving that same point.

- At point A, a portion of the total current travels through branch resistor R_1 (current path I_1) on its way back to the source's positive side. The remaining portion travels on to the circuit junction at point B.

- At point B, current "splits" and a portion travels through R_2 (current path I_2) and the remainder through R_3 (current path I_3) on its way back to the source's positive side.

- At point C, currents I_2 and I_3 rejoin and travel to junction D. Thus, current value from point C to D must equal $I_2 + I_3$.

- At point D, current $I_2 + I_3$ joins current I_1 and this total current travels back to the source's positive side. NOTE: *The total current equals the sum of the branch currents.* This is an important fact; remember it! Mathematically:

Formula 5–1	$I_T = I_1 + I_2 + I_3 \cdots + I_n$

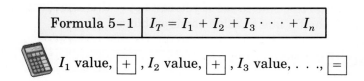
I_1 value, $\boxed{+}$, I_2 value, $\boxed{+}$, I_3 value, . . ., $\boxed{=}$

PRACTICE PROBLEMS II

Assume a five-branch parallel circuit. What is the formula for I_T? Which resistor passes the most current? Which resistor passes the least current?

KIRCHHOFF'S CURRENT LAW

The previous detailed statements express **Kirchhoff's current law.** Simply stated, this law is *the value of current entering a point must equal the value of current leaving that same point.*

In parallel circuit branch currents, since V across each branch is the same, the current divides through the branches in an inverse relationship to the individual branch resistances. That is, the highest R value branch allows the least current and the lowest R value branch allows the highest branch current.

FIGURE 5–7   Current is inverse to branch R value.

EXAMPLE

In using Ohm's law to solve for branch currents, this fact becomes apparent, Figure 5–7. It suggests an interesting proportionality approach in thinking about branch currents in parallel circuits, Figure 5–8. You can see that if branch 1 has twice the resistance of branch 2, it will have half the current value through it. Conversely, branch 2 will have twice the current value of branch 1. If branch 1 were to have one-fourth the R value of branch 2, then it would have four times the current of branch 2, and so forth.

Now you know voltage is the same across each parallel branch, current to and from any given circuit point must be the same, branch currents are inverse to their branch resistances, and total current equals the sum of the branch currents. It's time to see if you can apply this knowledge and Ohm's law and Kirchhoff's law to solve some problems.

PRACTICE PROBLEMS III

Look at Figure 5–9 and, without looking ahead, determine the values for V_2, V_1, V_A, I_1, and I_T. Read the following explanation.

To find V_2 use Ohm's law, where $V_2 = I_2 \times R_2 = 1 \text{ A} \times 50 \text{ }\Omega = 50 \text{ V}$. Since the voltage must be the same across all parallel branches, then V_1 and V_A must also equal 50 V in this circuit. Since you know V_1 and R_1, then

FIGURE 5–8   Looking at branch 1 via the proportionality approach: If branch 1 has twice the R of branch 2, then I_1 = half of I_2 (or $I_2 = 2 \times I_1$). If branch 1 has one-fourth the R of branch 2, then $I_1 = 4 \times I_2$ (or I_2 is one-fourth of I_1).

FIND
$V_2 = ?$
$V_1 = ?$
$V_A = ?$
$I_1 = ?$
$I_T = ?$

FIGURE 5–9

Ohm's law will again enable you to find the answer for I_1. That is, I_1 must equal V_1/R_1, or 5 A. And I_T equals the sum of the branch currents; thus, 1 A + 5 A = 6 A.

Using Ohm's law, what must the value of R_T be? The answer is

$$R_T = \frac{V_T}{I_T} = \frac{50\text{ V}}{6\text{ A}} = 8.33\ \Omega$$

It is interesting to note the total circuit resistance is *less* than either branch's resistance. With respect to the power source, the two branch resistances of 10 Ω and 50 Ω could be replaced by one resistor with a value of 8.33 Ω and it would still supply the same total current and circuit power. Later in the chapter you will study several methods to determine total circuit resistance for parallel circuits. However, at this point, remember that total resistance of parallel circuits is always less than the least branch resistance.

PRACTICE PROBLEMS IV

1. Refer to Figure 5–10 and with your knowledge of parallel circuits, solve for the unknowns. Don't cheat yourself by looking up the answers prior to working the problems.

2. Again referring to Figure 5–10, assume the parameters are changed to: $V_A = 125$ V, $I_T = 10.64$ mA, $I_1 = 4.63$ mA, and $R_3 = 56$ kΩ. Find R_1, I_2, I_3, R_2, and the value of R_T.

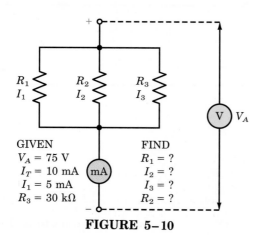

GIVEN
$V_A = 75$ V
$I_T = 10$ mA
$I_1 = 5$ mA
$R_3 = 30$ kΩ

FIND
$R_1 = ?$
$I_2 = ?$
$I_3 = ?$
$R_2 = ?$

FIGURE 5–10

RESISTANCE IN PARALLEL CIRCUITS

Again, total resistance of parallel circuits is always *less than the least branch resistance.* Let's examine the reason for this and learn several useful ways to solve for total resistance in parallel circuits.

Since each parallel branch provides another current path where the voltage/power source must supply current, then the source must provide more current for each additional branch in a parallel circuit. Since the V is the same from the source, but the I supplied by the source is higher for every branch added in parallel, then Ohm's law ($R = V/I$) tells us that the total or **equivalent circuit resistance** must decrease every time another branch (or current path) is added to the parallel circuitry. For example, if a circuit comprises only the lowest R value branch, the total resistance equals that value. If *any* value R branch is added in parallel with that branch, another current path has been provided. Therefore, total current supplied by the source increases, source voltage remains constant, and total or equivalent circuit resistance decreases. Moving forward from that thought, let's look at several methods available to determine parallel circuit total resistance.

METHODS TO CALCULATE TOTAL RESISTANCE (R_T)

The Ohm's Law Method

Recall the formula $R_T = V_T/I_T$ to find total resistance.

EXAMPLE

Look at Figure 5–11. The total resistance according to Ohm's law is

$$R_T = \frac{100 \text{ V}}{20 \text{ mA}} = 5 \text{ k}\Omega$$

A logical extension of this law for parallel circuits is that each branch resistance can be solved by knowing branch voltage and branch current.

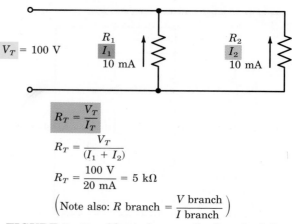

$$R_T = \frac{V_T}{I_T}$$

$$R_T = \frac{V_T}{(I_1 + I_2)}$$

$$R_T = \frac{100 \text{ V}}{20 \text{ mA}} = 5 \text{ k}\Omega$$

$$\left(\text{Note also: } R \text{ branch} = \frac{V \text{ branch}}{I \text{ branch}} \right)$$

FIGURE 5–11   Ohm's law method to find R_T

This is stated as

$$R \text{ branch} = \frac{V \text{ branch}}{I \text{ branch}}$$

Again, refer to Figure 5–11. Since we know the voltage is the same across all branches and must equal V applied, then

$$R_1 = \frac{100 \text{ V}}{10 \text{ mA}} = 10 \text{ k}\Omega$$

$$R_2 = \frac{100 \text{ V}}{10 \text{ mA}} = 10 \text{ k}\Omega$$

$$R_T = \frac{100 \text{ V}}{20 \text{ mA}} = 5 \text{ k}\Omega$$

Note this result of $R_T = 5$ kΩ agrees with our earlier statement that total circuit R must be less than any one of the branch resistances in parallel.

Assume the following changes in parameters for the circuit of Figure 5–11. $V_T = 50$ V, $I_1 = 1.06$ mA, $I_2 = 1.85$ mA. Again, total resistance can be found using Ohm's law. (NOTE: Rounding values to two decimal places, the answers will be close, but not absolutely precise in value.)

$$R_T = \frac{V_T}{I_T} = \frac{50 \text{ V}}{(1.06 \text{ mA} + 1.85 \text{ mA})} = \frac{50 \text{ V}}{2.91 \text{ mA}} = 17.18 \text{ k}\Omega$$

$$R_1 = \frac{V_1}{I_1} = \frac{50 \text{ V}}{1.06 \text{ mA}} = 47 \text{ k}\Omega$$

$$R_2 = \frac{V_2}{I_2} = \frac{50 \text{ V}}{1.85 \text{ mA}} = 27 \text{ k}\Omega$$

PRACTICE PROBLEMS V

1. Again, refer to Figure 5–11 and apply Ohm's law. If $I_1 = 15$ mA and $R_1 = 2$ kΩ, what are the values of V_{R_1} and V_T?
2. If $R_2 = 10$ kΩ, what is the value of R_T?

The Reciprocal Method

Since total current in a parallel circuit equals the sum of the branch currents, we can derive a reciprocal formula for resistance using this fact plus Ohm's law. Since $I_T = I_1 + I_2 \ldots + I_n$, and Ohm's law states $I_T = V/R_T$ and each branch current equals its V over its R, then:

$$\frac{V}{R_T} = \frac{V}{R_1} + \frac{V}{R_2} + \frac{V}{R_3} \ldots + \frac{V}{R_n}$$

Recall R_n represents the last resistor value in the sequence. Since V is common in parallel circuits and is the same across each of the resistances, we divide by V in each case and derive the formula for $\dfrac{1}{R_T}$.

Formula 5–2	$\dfrac{1}{R_T} = \dfrac{1}{R_1} + \dfrac{1}{R_2} + \dfrac{1}{R_3} \cdots + \dfrac{1}{R_n}$

R_1 value, $\boxed{1/x}$, $\boxed{+}$, R_2 value, $\boxed{1/x}$, $\boxed{+}$, R_3 value, $\boxed{1/x}$, . . . , $\boxed{=}$

This formula plus a couple simple math steps is used to find total or equivalent resistance of the circuit (R_T or R_e) for any number of branch resistances.

EXAMPLE

If a three-resistor circuit has a 10-Ω branch, a 15-Ω branch, and a 20-Ω branch, the total, or equivalent resistance, is found as follows:

$$\frac{1}{R_T} = \frac{1}{10} + \frac{1}{15} + \frac{1}{20} = \frac{6}{60} + \frac{4}{60} + \frac{3}{60} = \frac{13}{60}$$

To find R_T:

$$\frac{1}{R_T} = \frac{13}{60}$$

Cross-multiply.

$$60 = 13\, R_T.$$

Then divide.

$$R_T = \frac{60}{13} = 4.62\ \Omega$$

In the real world of electronics, rarely will the resistor values be as convenient to deal with as in our example. We chose those values to show you the concepts of using the reciprocal method. We now want to give you an example with more typical resistor values, and also promote the idea of your using a calculator to find R_T (or $R_{equivalent}$) for a parallel circuit. Since your calculator has a reciprocal function by using the "1/x" button (or x^{-1} button on some calculators), it makes calculations of this sort very easy.

EXAMPLE

If a four-resistor parallel circuit is comprised of resistors having values of 22 Ω, 39 Ω, 56 Ω, and 82 Ω, the total resistance of the circuit may be found, using a calculator, as follows:

Step 1: Clear the calculator by pressing the $\boxed{\text{AC/ON}}$ button. The readout should show 0.

Step 2: Enter 22, then press the $\boxed{1/x}$ button, then the $\boxed{+}$ button. The readout should show 0.04545.

Step 3: Enter 39, then the $\boxed{1/x}$ button, then the $\boxed{+}$ button. The readout should show 0.07109.

Step 4: Enter 56, then the $\boxed{1/x}$ button, then the $\boxed{+}$ button. The readout should show 0.0889.

Step 5: Enter 82, then the $\boxed{1/x}$ button, then the $\boxed{=}$ button. The readout should show 0.1011.

Step 6: Press the $\boxed{1/x}$ button to find the circuit total resistance. $R_T = 9.886$ Ω (rounded to two places, $R_T = 9.89$ Ω, or approximately 10 Ω.

(Is this less than the least branch R? Yes, and it does seem reasonable, using the mental approximating technique.)

Again, R equivalent is less than the least branch. Another variation of the reciprocal method is:

Formula 5–3	$R_T = \dfrac{1}{\dfrac{1}{R_1} + \dfrac{1}{R_2} + \dfrac{1}{R_3} \cdots + \dfrac{1}{R_n}}$

 R_1 value, $\boxed{1/x}$, $\boxed{+}$, R_2 value, $\boxed{1/x}$, $\boxed{+}$, R_3 value, $\boxed{1/x}$, . . . , $\boxed{=}$, $\boxed{1/x}$

PRACTICAL NOTES

As you have seen in the example, the reciprocal method is very easy with calculators. Practice this method until it is second nature to you. In most cases, it is the easiest method to use.

PRACTICE PROBLEMS VI

1. Assume a parallel circuit has branch resistances of 10 kΩ, 27 kΩ, and 20 kΩ, respectively. Use the reciprocal method and find R_T.

2. Assume a parallel circuit has branch resistances of 1.5 kΩ, 3.9 kΩ, 4.7 kΩ, and 5.6 kΩ, respectively. Again, use the reciprocal method and find the circuit R_T.

The Conductance Method

Recall resistance is the opposition shown to current flow, and conductance, conversely, is the ease with which current passes through a component or circuit. To calculate conductance use the reciprocal of resistance, $G = 1/R$, where G is conductance in siemens (S) and R is resistance in ohms (Ω).

EXAMPLE

If a circuit has a resistance of 10 Ω, it exhibits one-tenth S of conductance. Conversely, if a circuit has one-tenth S of conductance, it must have 10 Ω of resistance. That is, $R = 1/G$, where $R = 1/0.1 = 10$ Ω. Total conductance of

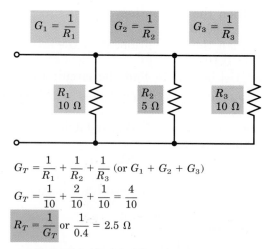

FIGURE 5–12   The conductance
method to find R_T

parallel branches is found by adding all conductances of parallel branches,
Figure 5–12.

| Formula 5–4 | $G_T = G_1 + G_2 + G_3$ |

G_1 value, $\boxed{+}$, G_2 value, $\boxed{+}$, G_3 value, $\boxed{=}$

In other words, $\dfrac{1}{R_T}$ (or G_T) $= \dfrac{1}{R_1} + \dfrac{1}{R_2} + \dfrac{1}{R_3}$.

PRACTICE PROBLEMS VII

Refer to Figure 5–13, and use the conductance method to find the
circuit total resistance.

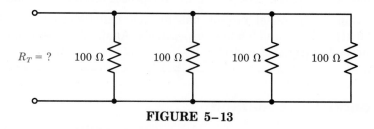

FIGURE 5–13

The Product-Over-The-Sum Method

One popular method to solve for total resistance in parallel circuits is the
product-over-the-sum method. This method is limited to solving total
resistance values for two-resistor circuits, or when circuits have more than
two resistors, solving them, two resistors at a time. The mathematical formula is derived from the basic reciprocal resistance (conductance) formulas.

That is

$$G_T = \frac{1}{R_T} = \frac{1}{R_1} + \frac{1}{R_2}$$

Since resistance is the reciprocal of conductance, the formula can be re-stated as:

$$R_T = \frac{1}{G_T}, \text{ or } R_T = \frac{1}{\dfrac{1}{R_1} + \dfrac{1}{R_2}}$$

Formula 5–5	$R_T = \dfrac{R_1 \times R_2}{R_1 + R_2}$

 R_1 value, $\boxed{\times}$, R_2 value, $\boxed{\div}$, $\boxed{(}$, R_1 value, $\boxed{+}$, R_2 value, $\boxed{)}$, $\boxed{=}$

EXAMPLE

An illustration of this formula is in the two-resistor parallel circuit of Figure 5–14.

What would be the total circuit resistance if $R_1 = 100$ ohms and $R_2 = 50$ ohms? In this case, the product is $100 \times 50 = 5000$, and the sum is $100 + 50 = 150$. Therefore, $R_T = 5000/150$, and total resistance = 33.33 ohms. This product-over-the-sum method works for any two branch values.

EXAMPLE

Let's assume the resistor values in Figure 5–14 are changed so that one resistor's value is 3.9 kΩ and the other resistor's value is 6.8 kΩ.

$$R_T = \frac{R_1 \times R_2}{R_1 + R_2} = \frac{(3.9 \times 10^3) \times (6.8 \times 10^3)}{(3.9 \times 10^3) + (6.8 \times 10^3)} = \frac{26.52 \times 10^6}{10.7 \times 10^3}$$

$$R_T = 2.48 \times 10^3 \text{ or } 2.48 \text{ k}\Omega$$

Our mental approximation, quick-check technique here might be $4 \times 7 = 28$, and $28 \div 4 + 7$ (or 11) would be approximately 2.5 or so. Our answer makes sense! Now you try a couple of practice problems!

$R_T = ?$ R_1 20 Ω R_2 20 Ω

$$R_T = \frac{R_1 \times R_2}{R_1 + R_2}$$

$$R_T = \frac{20 \times 20}{20 + 20}$$

$$R_T = \frac{400}{40}$$

$$R_T = 10 \ \Omega$$

FIGURE 5–14 The product-over-the-sum method to find R_T

PRACTICE PROBLEMS VIII

1. If the circuit in Figure 5–14 has resistances of 30 Ω and 20 Ω, respectively, solve for R_T using the product-over-the-sum method.

2. Assume the values of resistors in Figure 5–14 are changed to 10 kΩ and 33 kΩ, respectively. Use the product-over-the-sum approach and determine the new value of R_T or (or R_e) for the new circuit.

This method works well for two-branch circuits. How might a similar technique be used for a three-branch circuit? The answer is simple: Use the formula on two branches, *then use these results* from that calculation as being one equivalent resistance in parallel with the third resistor, and repeat the process. Of course, you can also use the reciprocal or conductance approach, which was recommended earlier, where

$$\frac{1}{R_T} = \frac{1}{R_1} + \frac{1}{R_2} + \frac{1}{R_3} \text{ or, better yet, } R_T = \frac{1}{\dfrac{1}{R_1} + \dfrac{1}{R_2} + \dfrac{1}{R_3}}$$

The following example shows how the product-over-the-sum approach might be used for a three-branch circuit.

EXAMPLE

In Figure 5–15 three branch resistances are in parallel. Since each branch has a resistance of 3 Ω, what is the circuit total resistance?

Using the product-over-the-sum formula for the first two branches yields:

$$\frac{R_1 \times R_2}{R_1 + R_2} = \frac{3 \times 3}{3 + 3} = \frac{9}{6} = 1.5 \ \Omega$$

indicating that the equivalent resistance of branches 1 and 2 = 1.5 Ω. However, there is another branch involved! The equivalent resistance of 1.5 Ω is

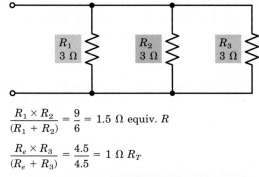

$$\frac{R_1 \times R_2}{(R_1 + R_2)} = \frac{9}{6} = 1.5 \ \Omega \text{ equiv. } R$$

$$\frac{R_e \times R_3}{(R_e + R_3)} = \frac{4.5}{4.5} = 1 \ \Omega \ R_T$$

FIGURE 5–15 Using product-over-the-sum method with more than two branches

in parallel with branch R_3 (3 Ω). Repeating the product-over-the-sum formula for this situation gives:

$$\frac{R_e \times R_3}{R_e + R_3} = \frac{1.5 \times 3}{1.5 + 3} = \frac{4.5}{4.5} = 1\ \Omega$$

Note that using the reciprocal approach would also yield:

$$\frac{1}{R_T} = \frac{1}{3} + \frac{1}{3} + \frac{1}{3} = \frac{3}{3} = \frac{1}{1} = 1\ \Omega$$

PRACTICE PROBLEMS IX

If the resistances in Figure 5−15 are 47 Ω, 100 Ω, and 180 Ω, respectively, what is the value of R_T? Use the product-over-the-sum approach to calculate.

What would happen if there were four branches? To find the equivalent resistance of any two branches, use the product-over-the-sum formula. Then find the equivalent resistance of the other two branches in the same way, and use the product-over-the-sum formula for the two equivalent resistances to find total circuit resistance.

A Useful Simplifying Technique

You have undoubtedly observed in Figure 5−14 described earlier, which had two 20-Ω resistors in parallel, that the total $R = 10\ \Omega$. This number is half the value of one of the equal branch resistances. What do you suppose the total resistance of a three-branch parallel circuit is if the circuit has three equal branches of 30 Ω each? If you said 10 Ω again, you are right. That is, two equal parallel branches are equivalent to half the R value of one of the equal branches, three equal branches are equivalent to one-third the value of one of the equal branches, and so on. The idea is to mentally *combine* equal resistance branches to find their equivalent resistance. Doing this mental process reduces the complexity of the problem.

Look at Figure 5−16. By realizing that R_2 and R_4 can be combined as an equivalent resistance of 10 Ω, then combined with the two remaining 10-Ω

Step 1 = Combine R_2 and R_4 to yield 10 Ω.
Step 2 = Knowing that three 10 Ω resistors
        in parallel yields 1/3 of 10 Ω
        as their R_T. Then $R_T = 3.33\ \Omega$.

FIGURE 5−16   A useful simplifying technique

branches, you can see the total circuit resistance will be one-third of 10 Ω, or 3.33 Ω.

This same concept can be used whenever branches of a multi-branched parallel circuit have equal values. Combine equal branches in a way to simplify to the fewest remaining branches. Then use the most appropriate method to solve for total circuit resistance, such as product-over-the-sum method, reciprocal method, Ohm's law, etc.

NOTE: The unequal values of resistances used in many circuits do not readily lend themselves to using this simplifying technique. However, whenever you can use it to help, do it! It will simplify the computation efforts.

The Assumed Voltage Method

One technique to solve for total resistance of parallel circuits containing miscellaneous "mismatched" resistance values is the **assumed voltage method.** You'll frequently find problems of this sort on FCC radio licensing tests.

The technique uses a two-step approach:

1. Assume an uncomplicated or easy circuit applied voltage value to determine each branch current, should the "assumed voltage" actually be applied to the circuit. For example, choose a value that is an even multiple of all the branch resistor values (the "least common denominator" value).

2. Determine the branch currents by dividing the assumed voltage by each branch R value. Then add branch currents to find total current. Now *divide "assumed voltage" by calculated total current to find total circuit resistance.*

EXAMPLE

Refer to Figure 5–17. Following the preceding steps, the solution for this problem is as follows:

1. Assume 300 V applied voltage since it is easily divided by each of the branch R values.

2. Solve each branch current using Ohm's law. Branch currents are 12 A, 6 A, 3 A, and 2 A, respectively.
   Total current = sum of branch currents = 23 A
   Thus, $R_T = \dfrac{300 \text{ V}}{23 \text{ A}} = 13.04 \ \Omega$

In essence, this technique often allows you to mentally perform the first step without paper and pencil to find the total current. All that remains is one division problem, V_T/I_T, to find total circuit resistance.

A good assumed voltage is 300 V due to R values.

I_1 = 12 A I_2 = 6 A I_3 = 3 A I_4 = 2 A

I_T = sum of branch currents = 23 A

$R_T = \dfrac{300\ V}{23\ A} = 13.04\ \Omega$

FIGURE 5–17   The assumed voltage method to find R_T

In Figure 5–17, if the R values were changed to 10, 13, 20, and 26 Ω respectively, what would be the total circuit resistance? Try using the assumed voltage method to find the answer.

Your answer should be 3.77 Ω. One value of voltage that could have been assumed is 260 V, since it is evenly divisible by all the R values in the circuit. NOTE: It does not matter if you assumed another value of voltage applied, your answer will be the same!

A Useful Parallel Circuit Resistance Design Formula

As a technician, there may be times when you will want to purposely decrease the resistance of an existing circuit by placing a resistor in parallel with it. Or you may need a resistor of some unavailable specific value. In either case, you may have a number of resistors, but none are the right value. This happens frequently! How can we determine the unknown resistor value to be paralleled with a given (known) value to achieve the desired resultant resistance? (NOTE: We interchangeably use the terms R_T (R total) and R_e (R equivalent) to indicate the resultant total resistance of paralleled resistances. Be familiar with both terms as you will see them used frequently.)

The answer is a formula derived from the product-over-the-sum method:

Formula 5–6 $R_u = \dfrac{R_k \times R_e}{R_k - R_e}$

Where R_u = R unknown

R_e = equivalent R of paralleled resistors

R_k = known R value that will be paralleled by R unknown achieving the desired R equivalent.

R known value, $\boxed{\times}$, R equivalent value, $\boxed{\div}$, $\boxed{(}$, R known value, $\boxed{-}$, R equivalent value, $\boxed{)}$, $\boxed{=}$

FIGURE 5–18   A practical circuit design formula

FIGURE 5–19

EXAMPLE

What value of R is needed in parallel with a 10-ohm resistor to achieve a total resistance of 6 ohms? See the solution in Figure 5–18.

PRACTICE PROBLEMS X

Use this technique and solve the problem in Figure 5–19.

POWER IN PARALLEL CIRCUITS
Total Power

Total power for *any* resistive circuit (series, parallel, or combination) is computed by adding all individual power dissipations in the circuit, or by using one of the power formulas, $P_T = V_T \times I_T$; $P_T = I_T^2 \times R_T$; or

$$P_T = V_T^2/R_T$$

Likewise, power dissipated by any one resistive component is calculated using that particular component's parameters. That is, its voltage drop, current through it, and its resistance value.

   Look at the circuit in Figure 5–20 as you read the next section about power in parallel circuits.

Power Dissipated by Each Branch

To find a branch power dissipation, use the parameters applicable to that branch. In the case of branch 1 (R_1), $P = 100 \text{ V} \times I_1$. To find I_1, divide branch

$$P_1 = V_1 \times I_1 \qquad P_2 = V_2 \times I_2 \qquad P_3 = V_3 \times I_3$$
$$P_1 = 100 \text{ V} \times 1 \text{ mA} \quad P_2 = 100 \text{ V} \times 4 \text{ mA} \quad P_3 = 100 \text{ V} \times 5 \text{ mA}$$
$$P_1 = 100 \text{ mW} \qquad P_2 = 400 \text{ mW} \qquad P_3 = 500 \text{ mW}$$

$$P_T = 100 \text{ mW} + 400 \text{ mW} + 500 \text{ mW} = 1000 \text{ mW (or 1 W)}$$
or
$$P_T = 100 \text{ V} \times 10 \text{ mA} = 1000 \text{ mW (or 1 W)}$$

FIGURE 5–20 Power in parallel circuits

V (100 V) by branch R (100 kΩ). The answer is 1 mA. Thus, $P_1 = 100$ V × 1 mA = 100 mW.

$$\text{Branch 2: Branch } I = \frac{100 \text{ V}}{25 \text{ k}\Omega} = 4 \text{ mA}$$

$$\text{Branch } P = 100 \text{ V} \times 4 \text{ mA} = 400 \text{ mW}$$

$$\text{Branch 3: Branch } I = \frac{100 \text{ V}}{20 \text{ k}\Omega} = 5 \text{ mA}$$

$$\text{Branch } P = 100 \text{ V} \times 5 \text{ mA} = 500 \text{ mW}$$

EXAMPLE

If we added all the individual power dissipations in the circuit in Figure 5–20, the total power = 100 mW + 400 mW + 500 mW = 1 W. Using the power formula, since V_T equals 100 volts and I_T equals 10 mA, the total power is 100 V × 10 mA = 1000 mW (or 1 W). In either case, the total power supplied to and dissipated by the circuit is calculated as 1 W.

PRACTICE PROBLEMS XI

What is the total power in the circuit of Figure 5–20 if $V_T = 150$ V?

Relationship of Power Dissipation to Branch Resistance Value

Refer again to Figure 5–20 and note that the *lowest* branch R dissipates the *most* power; the *highest* value R branch dissipates the *least* power. This is because V is the same for all branches, and current through each branch is *inverse* to its R value. **This is the opposite of what you learned about**

series circuits. In *series circuits*, the largest value R dissipates the most power because its I is the same as all others in the circuit; however, its V is directly proportional to its R value.

Soon we will look at other interesting similarities and differences between series and parallel circuits that you should learn. However, before we do that, let's look at some variations from the norm when parallel circuits develop opens or shorts.

EFFECTS OF OPENS IN PARALLEL CIRCUITS AND TROUBLESHOOTING HINTS

Opens

Analysis of one or more opened branches in a parallel circuit is simple. Refer to Figure 5–21.

1. Normal operation for this circuit is: $V_T = 100$ V; $I_T = 40$ mA; $P_T = 4000$ mW (4 W); each branch $I = 10$ mA; and each branch $P = 1000$ mW (1 W).

2. If branch 2 opens (R_2), then total current decreases to 30 mA, since there is no current through branch 2. Under these conditions, the following is true:

   • Total resistance increases.

   • Total current decreases (30 mA rather than 40 mA).

   • Total voltage and branch voltages remain unchanged.

   • Total power decreases (100 V $\times$ 30 mA = 3000 mW, or 3 W).

FIGURE 5–21   Effects of an open in a parallel circuit

• Current through the good (unopened) branches remains unchanged since both their V and R values stay the same.

• Also, since voltage and current through unopened branches are unaffected, individual branch power dissipations stay the same, *except* for the opened branch or branches.

• Since current decreases to zero through the opened branch, its power dissipation also drops to zero. The voltage across the opened branch, however, stays the same.

PRACTICAL NOTES

Some generalizations worth noting about an open in *any* circuit (series, parallel, or combination) are:

1. An open causes total circuit resistance to increase.

2. An open causes total circuit current to decrease.

3. An open causes total circuit power to decrease.

Incidentally, opens can be caused either by *components* opening, or by *conductor paths* opening. On printed circuit boards, a cracked conductor path or poor solder joint can cause an open. On "hard-wired" circuitry, a broken wire (or poor solder joint) can cause an open conductor path, Figure 5–22.

FIGURE 5–22 Open circuits can be caused by broken conductor paths or poor solder joints.

TROUBLESHOOTING HINTS

Since an open in a parallel branch causes total resistance to increase, total current to decrease, and current through the opened branch to decrease to zero, troubleshooting hints help isolate and identify these facts.

With power removed from the circuit, you can measure its total resistance with an ohmmeter. If the measured value indicates a higher than normal value for the circuit, then there is a good chance an open has occurred or at least some part of the circuit has opened or increased in R value. If it is convenient to electrically "lift" (electrically isolate) one end of each branch one at a time from the remainder of the circuit as you watch the meter, you can spot the troubled branch. For example, if lifting one end of a branch from the circuit causes a change in R (increase), the branch just lifted is probably good. If lifting of the branch causes no change in the R reading, you have probably found the branch with the problem. Since it was already opened, opening it again didn't change anything!

If you prefer using "power-on" current readings, the first indication of trouble you will see is a lower-than-normal total circuit current. If you can conveniently connect the meter to measure each branch's current individually, the branch with zero current is the open branch. Other branches should measure normal values.

EFFECTS OF SHORTS IN PARALLEL CIRCUITS AND TROUBLESHOOTING HINTS

Shorts

A shorted branch in a parallel circuit is a serious condition. Since each branch is connected directly across the voltage source, a shorted branch puts a resistance of zero ohms, or close to zero ohms, directly across the source. Refer to Figure 5–23.

1. Normal operation for this circuit is the same as described earlier for Figure 5–21, where $V_T = 100$ V; $I_T = 40$ mA; $P_T = 4000$ mW (4 W); each branch $I = 10$ mA; and each branch $P = 1000$ mW (1 W).

2. If branch 2 (R_2) shorted, then the following things would probably happen:

   • I_T would try to drastically increase until a fuse blew, or the power supply "died"!

   • V_T would drastically decrease! Due to greatly increased $I \times R$ drop across the source's internal resistance, the output terminal voltage would decrease. It would probably drop to nearly zero volts with most power sources.

   • P_T would drastically increase until the power source quit, or a fuse blew!

   • Current through the other branches would decrease to zero, or close to it, because the V has dropped to almost zero, or zero.

Normal operation

(Reading 100 V)

V_T
100 V
from dc
source

(Reading 40 mA)

(a)

First instant after R_2 is shorted

(Reading 0 V)

Short

V_T
drops
to near
zero

(Meter "pinned" at full-scale deflection)

(b)

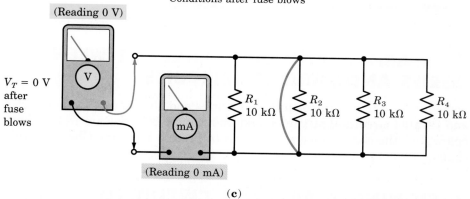

Conditions after fuse blows

(Reading 0 V)

$V_T = 0$ V
after
fuse
blows

(Reading 0 mA)

(c)

**FIGURE 5–23   Effects of a short in a parallel circuit. R_T decreases and I_T
increases until the fuse blows or the power source quits.**

- Total circuit resistance would drastically decrease! Remember the
  total resistance will be less than the least branch R.

- If there were any power supply fuses or house circuit fuses associated
  with the faulty circuit, they would probably blow. Also, there might be
  some smoke rising from some sections of the circuit or power supply.

PRACTICAL NOTES

Some generalizations about a short in *any* type of circuit (series, parallel, or combination) are:

1. A short causes total circuit resistance to decrease.

2. A short causes total circuit current to increase. (At least momentarily!)

3. A short causes total circuit power to increase. (At least momentarily!)

4. A short likely causes circuit voltage to decrease. (Probably to zero volts!)

TROUBLESHOOTING HINTS

First *disconnect* the circuit in question from the power source! Resistance checks can then be made on the circuitry without further damage either to the circuitry, power source, connecting wires, and so forth. Power-on troubleshooting should *not* be done for shorted conditions.

Again, the best method to find the troubled area is to isolate each branch one at a time. By electrically lifting one end of each branch from the remaining circuitry, an ohmmeter check of the branch resistance can be made. Normal *R* values will be found for the good branches. Close to zero ohms will be measured across the shorted branch or component.

Once the branch is located, a visual inspection of components, wires, printed circuit paths, and so on generally reveals the problem. If there is an internal short within the component, an *R* measurement of the component reveals this problem.

SIMILARITIES AND DIFFERENCES BETWEEN SERIES AND PARALLEL CIRCUITS

Some interesting and important similarities and differences between series and parallel circuits that you should know and remember are in the chart below. Use the chart on page 187 as an In-Process Learning Check for this chapter.

DESIGNING A PARALLEL CIRCUIT TO SPECIFICATIONS

EXAMPLE

Assume you want to create a four-branch parallel circuit so that branches 1 and 3 have equal current values, and branches 2 and 4 also have equal current values but their current values are twice that of branches 1 and 3. Further assume that branch 1 has an R value of 10 kΩ and its current value is 10 mA.

COMPARING SERIES AND PARALLEL CIRCUITS

SERIES CIRCUIT CHARACTERISTICS	PARALLEL CIRCUIT CHARACTERISTICS
R_T = Sum of Rs	R_T = Less than lowest R in parallel
$I_T = \dfrac{V_T}{R_T}$	$I_T = \dfrac{V_T}{R_T}$
I_T = Same current throughout the circuit	I_T = Sum of branch currents, where branch current $$(I_b) = \dfrac{V_b}{R_b}$$ Branch I is inverse to branch R.
$V_T = I_T \times R_T$	$V_T = I_T \times R_T$
V_T = Sum of all IR drops	V_T = Same as across each branch
$V_{R_X} = I_{R_X} \times R_X$; Any given R's voltage drop is related to the ratio of its R to R total. $$V_{R_X} = \left(\dfrac{R_X}{R_T}\right) \times V_T$$	Voltage across all resistors is same value and equals the circuit applied voltage.
P dissipated by a given R is *directly related* to its R value. Largest R in circuit dissipates most power; smallest R dissipates least power.	P dissipated by a given R is *inverse* to its R value. Largest R dissipates least power, smallest R value dissipates most power.
$P_T = V_T \times I_T$; or sum of all Ps	$P_T = V_T \times I_T$; or sum of all Ps
Opens cause 0 current; 0 V across good components; and V_T (applied circuit voltage) to appear across open.	*Opens* cause no change in good branches; opened branch has zero current, but V across all branches stays the same.
Shorts cause I to increase; V to decrease across short; V to increase across remaining components; and total circuit power to increase.	*Shorts* cause I_T and P_T to increase (at least momentarily); V to decrease across all branches; and current through good branches to decrease.

The design problem involves determining all the circuit parameters and drawing the schematic for this circuit. Label all component values and all V, R, and I electrical parameters for each branch and for the total circuit.

Recall the way to begin is to draw a diagram and label all the knowns. Then proceed in the most appropriate manner to solve the unknowns based on the knowns.

See Figure 5–24a for this first step, which is drawing a diagram and labeling all knowns.

The second step is to use the knowns to solve for the unknowns, then label the diagram with the appropriate parameters.

• Since branch 1's R and I values are known, it's easy to find its voltage value. Thus, $V = I \times R = 10$ mA $\times$ 10 kΩ = 100 V.

• Since this is a parallel circuit, 100 V is the voltage across all the branches as well as the V applied value.

• Because branches 3 and 1 have the same current value, and since we know the V applied to both branches is the same, it is obvious branch 3 resistance value must be the same as branch 1. So we now know branch 3 R_3 = 10 kΩ, I_3 = 10 mA, and V_3 = 100 V.

Step 1: Draw diagram. Label and/or list "knowns." Step 2: Use knowns to solve unknowns. Label diagram.

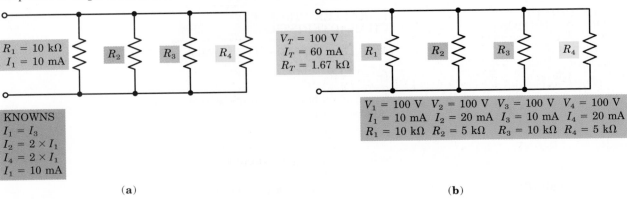

(a) (b)

FIGURE 5–24

- Because branches 2 and 4 have twice the current value as branches 1 and 3, it is obvious their current values must be 20 mA each. Since the V across each branch is 100 V, each branch's R value is

$$\frac{100 \text{ V}}{20 \text{ mA}} = 5 \text{ k}\Omega$$

- Solving the total circuit current and resistance are simple, and you already know the V applied value. Total current (I_T) equals the sum of the branch currents = 60 mA.

$$R_T = \frac{V_T}{I_T} = \frac{100 \text{ V}}{60 \text{ mA}} = 1.67 \text{ k}\Omega$$

Incidentally, you can use the R values and solve for the total resistance. The two 10-kΩ branches are the equivalent of 5 kΩ. This 5 kΩ in parallel with two other 5-kΩ branches is equivalent to 5 kΩ divided by 3, or 1.67 kΩ. Remember the technique with equal value branches?

- The last task is to label all appropriate parameters on the diagram, Figure 5–24b.

PRACTICE PROBLEMS XII

Try this problem yourself. Design and draw a three-resistor parallel circuit so that:

1. V applied to the circuit is 120 V.
2. The total circuit power dissipation is 1.44 W.
3. Branch 1 has half the current value of branch 2 and one-third the current value of branch 3.

Label your diagram with all $V, I, R,$ and P values for the circuit and for each branch.

12 – V auto batteries

FIGURE 5–25   Equal voltage sources in parallel can deliver more current and power than one source alone.

SOURCES IN PARALLEL

Voltage and power sources may be connected in parallel to provide higher current delivering capacity (and power) to circuits or loads (at a given voltage) than can be delivered by only one source. The sources should be equal in output terminal voltage value when using this technique. A common application of this method is the "paralleling" of batteries or cells, Figure 5–25.

CURRENT DIVIDERS

You are aware that parallel circuits cause the total current to divide through that circuit's branches. You also know the current divides in inverse proportion to each branch resistance value. These characteristics of parallel circuits offer the possibility of current dividing "by design," where one purposefully controls the ratios of branch currents by choosing desired values of resistances. Also, this current-dividing trait gives another method to find current through branches if only the total resistance, branch resistances, and total current are known.

General Formula for Parallel Circuit with Any Number of Branches

A general formula to find any given branch's current is:

Formula 5–7	$I_X = \dfrac{R_T}{R_X} \times I_T$

  R_T value, $\boxed{\div}$, R_x value, $\boxed{\times}$, I_T value, $\boxed{=}$

where I_X is a specified branch's current value; R_T is the parallel circuit total resistance; and R_X is the value of resistance of the specified branch. Refer to Figure 5–26 for example applications of this formula.

GIVEN
$R_T \approx 0.55$ kΩ
$I_T \approx 5.5$ mA

Find approximate values for I_1, I_2, and I_3

$$I_1 \approx \left(\frac{R_T}{R_1}\right) \times I_T = \left(\frac{0.55 \text{ k}\Omega}{1 \text{ k}\Omega}\right) \times 5.5 \text{ mA} \approx 3.0 \text{ mA}$$

$$I_2 \approx \left(\frac{R_T}{R_2}\right) \times I_T = \left(\frac{0.55 \text{ k}\Omega}{2 \text{ k}\Omega}\right) \times 5.5 \text{ mA} \approx 1.5 \text{ mA}$$

$$I_3 \approx \left(\frac{R_T}{R_3}\right) \times I_T = \left(\frac{0.55 \text{ k}\Omega}{3 \text{ k}\Omega}\right) \times 5.5 \text{ mA} \approx 1.0 \text{ mA}$$

FIGURE 5–26   A general current-divider example

PRACTICE PROBLEMS XIII

1. Refer again to Figure 5–26. Assume that $R_1 = 27$ kΩ, $R_2 = 47$ kΩ, $R_3 = 100$ kΩ, $R_T = 14.64$ kΩ, and $I_T = 3.42$ mA. Find each of the branch currents using the divider formula, as appropriate.

2. Assume the circuit of Figure 5–26 is changed so that $I_T = 4$ A, $R_1 = 10$ Ω, $R_2 = 5.6$ Ω, and $R_3 = 2.7$ Ω.
   a. Find R_T.
   b. Use the general current divider formula to find the current through each branch.

Simple Two-Branch Parallel Circuit Current Divider Formula

If there are only two branches, Figure 5–27, a formula is available to solve for each branch current, knowing I_T and branch R values:

$$\boxed{\text{Formula 5–8} \quad I_1 = \frac{R_2}{R_1 + R_2} \times I_T}$$

 R_2 value, $\boxed{\div}$, $\boxed{(}$, R_1 value, $\boxed{+}$, R_2 value, $\boxed{)}$, $\boxed{\times}$, I_T value, $\boxed{=}$

$$I_1 = \frac{10 \text{ k}\Omega}{11 \text{ k}\Omega} \times 50 \text{ mA} = 0.909 \times 50 \text{ mA} = 45.45 \text{ mA}$$

$$\boxed{\text{Formula 5–9} \quad I_2 = \frac{R_1}{R_1 + R_2} \times I_T}$$

 R_1 value, $\boxed{\div}$, $\boxed{(}$, R_1 value, $\boxed{+}$, R_2 value, $\boxed{)}$, $\boxed{\times}$, I_T value, $\boxed{=}$

$$I_2 = \frac{1 \text{ k}\Omega}{11 \text{ k}\Omega} \times 50 \text{ mA} = 0.0909 \times 50 \text{ mA} = 4.55 \text{ mA}$$

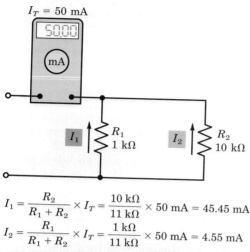

$$I_1 = \frac{R_2}{R_1 + R_2} \times I_T = \frac{10 \text{ k}\Omega}{11 \text{ k}\Omega} \times 50 \text{ mA} = 45.45 \text{ mA}$$

$$I_2 = \frac{R_1}{R_1 + R_2} \times I_T = \frac{1 \text{ k}\Omega}{11 \text{ k}\Omega} \times 50 \text{ mA} = 4.55 \text{ mA}$$

FIGURE 5–27 Two-resistor current
divider example

NOTE: There is an inverse relationship of current to branch Rs. Thus, when solving for I_1, R_2 is in the numerator; when solving for I_2, R_1 is in the numerator.

PRACTICE PROBLEMS XIV

1. Refer again to Figure 5–27 and assume the values of R_1 and R_2 change to 100 kΩ and 470 kΩ, respectively. Also, assume the total current is 2 mA. Find the values of current through each branch using the appropriate two-branch divider formula.

2. Assume that the circuit in Figure 5–27 is changed so that total current measures 20 mA, R_1 equals 1.8 kΩ, and R_2 equals 8.2 kΩ. Using the two-resistor current divider formula, determine the currents through each branch.

PRACTICAL NOTES

You have learned that whenever *any* resistance value is placed in parallel with an existing circuit, it alters the circuit resistance and current because another branch has been added; thus providing another current path.

In a later chapter, we will discuss how this feature impacts the types and uses of measurement devices, such as voltmeters. For now, we only want to alert you to an important observation about parallel circuits.

As already stated, no matter what value of R is paralleled with (or "bridged across") an existing circuit, the circuit R will decrease. But the question is *how much* change will there be?

R_T without 1 MΩ connected = 5 kΩ
R_T with 1 MΩ connected = 4.975 kΩ

FIGURE 5–28   Change in R_T of a given circuit when adding 1 MΩ branch

R_T without 1 kΩ connected = 5 kΩ
R_T with 1 kΩ connected = 0.833 kΩ

FIGURE 5–29   Change in R_T of a given circuit when adding 1 kΩ branch

Look at Figure 5–28. Note the impact (change) that adding a 1 megohm resistor branch has on the circuit resistance. As you can see, adding a *high* R in parallel causes a small change in total circuit R.

Look at Figure 5–29. Note the change in R_T caused by introducing a 1-kΩ resistor in parallel with the same original circuit. Adding a *low R* in parallel with an existing circuit causes a much more significant change in the circuit total resistance (and total current) than paralleling an existing circuit with a high R value.

The amount of change caused by adding a new branch R is related to *both* the R value of the "new" R branch *and* to the total circuit resistance of the original circuit. If the original circuit's equivalent resistance is low initially, adding a low value R in parallel will not have nearly the effect that it does when the original circuit has a high R_T.

SUMMARY

▶ A parallel circuit is one in which there are two or more paths (or branches) for current flow, and all the branches have the same voltage applied across them.

▶ Total circuit current divides through the parallel branches in an inverse ratio to the branch resistances.

▶ Total circuit current in a parallel circuit equals the sum of the branch currents.

▶ Kirchhoff's current law states the value of current entering a point must equal the value of current leaving that same point in the circuit.

▶ The total resistance of a parallel circuit must be less than the least resistance branch R value.

▶ Several methods to find the total resistance of parallel circuits are:

1. *Ohm's law*

$$R_T = \frac{V_T}{I_T}$$

2. *reciprocal resistance method*

$$\frac{1}{R_T} = \frac{1}{R_1} + \frac{1}{R_2} \cdots + \frac{1}{R_n} \text{ or } \frac{1}{\frac{1}{R_1} + \frac{1}{R_2} \cdots + \frac{1}{R_n}}$$

3. *conductance method*

$$R_T = \frac{1}{G_T}$$

4. *product-over-the-sum method*

$$R_T = \frac{R_1 \times R_2}{R_1 + R_2}$$

5. *assumed voltage method.*

▶ One technique to solve for parallel circuit total resistance is to combine equal branch R values into their equivalent resistance value, whenever possible. For example, two 10-kΩ resistors in parallel equal 5-kΩ equivalent resistance (or half of one equal branch R value). Three 10-kΩ resistors in parallel equal 3.33-kΩ equivalent resistance (or one-third of one equal branch R value).

▶ A useful formula to find the R value needed in parallel with a given R value to obtain a desired equivalent (total resistance) value is

$$R_u = \frac{R_k \times R_e}{R_k - R_e}$$

where R_u is the unknown R value being calculated; R_e is the resultant total resistance desired from the "paralleled" resistors; and R_k is the known value resistor across which the unknown value resistor is paralleled to achieve the desired equivalent, or total parallel circuit resistance. Also, you can use

$$\frac{1}{R_e} - \frac{1}{R_k} = \frac{1}{R_u}$$

▶ Total power supplied to, or dissipated by, a parallel circuit is found with the power formulas $P_T = V_T \times I_T$, $I_T^2 \times R_T$, or V_T^2/R_T, or by summing the power dissipations of all the branches.

▶ Power dissipated by any branch of a parallel circuit is computed with the power formulas by substituting the specified branch V, I, and/or R parameters into the formulas, as appropriate.

▶ Branch power dissipations are such that the highest R value branch dissipates the least power, and the lowest R value branch dissipates the most power.

▶ An open branch in a parallel circuit causes total circuit current (and power) to decrease; total circuit R to increase; current through the unopened branches to remain the same, and current through the opened branch to decrease to zero. The circuit voltage remains the same.

▶ A shorted branch in a parallel circuit may cause damage to the circuit wiring and/or to the circuit power supply and/or may cause fuse(s) to blow. Until a fuse blows or the circuit power source fails, the short causes total circuit current to drastically increase; circuit R to decrease to zero ohms; and power demanded from the source to greatly increase. When a fuse blows or the source fails, circuit V, I, and P decrease to zero.

▶ When troubleshooting parallel circuits with an open branch problem, isolate the bad branch by power-off resistance checks or by power-on branch-by-branch current measurements. Another technique is to lift each branch electrically from the remainder of the circuit while monitoring circuit resistance or circuit current. The branch that causes no change is the bad branch.

▶ When troubleshooting parallel circuits with a shorted branch problem, *turn the power off and remove the circuit from the power source terminals before performing power-off checks to isolate the problem*. Branch-by-branch resistance measurements usually locate the problem. Also, you can electrically lift and reconnect each branch sequentially while monitoring the total circuit resistance.

▶ There are several significant *differences* between parallel and series circuits. Current is *common* in series circuits; voltage is *common* in parallel circuits. *Voltage* divides around a series circuit in *direct* relationship to the R values. *Current* divides throughout a parallel circuit in *inverse* relationship to the branch R values. In series circuits, the largest R dissipates the most power; in parallel circuits, the smallest R dissipates the most power. In series circuits, the total resistance equals the sum of all resistances. In parallel circuits, the total circuit resistance is *less* than the smallest branch resistance value.

▶ Voltage/power sources can be connected in parallel to provide higher current/power at a given voltage level.

FORMULAS AND CALCULATOR SEQUENCES SUMMARY

FORMULA 5–1 $I_T = I_1 + I_2 + I_3 \ldots + I_N$

I_1 value, $\boxed{+}$, I_2 value, $\boxed{+}$, I_3 value, . . . , $\boxed{=}$

FORMULA 5–2 $\dfrac{1}{R_T} = \dfrac{1}{R_1} + \dfrac{1}{R_2} + \dfrac{1}{R_3} \ldots + \dfrac{1}{R_n}$

R_1 value, $\boxed{1/x}$, $\boxed{+}$, R_2 value, $\boxed{1/x}$, $\boxed{+}$, R_3 value, $\boxed{1/x}$, . . . , $\boxed{=}$

FORMULA 5–3 $R_T = \dfrac{1}{\dfrac{1}{R_1} + \dfrac{1}{R_2} + \dfrac{1}{R_3} \ldots + \dfrac{1}{R_n}}$

R_1 value, $\boxed{1/x}$, $\boxed{+}$, R_2 value, $\boxed{1/x}$, $\boxed{+}$, R_3 value, $\boxed{1/x}$, . . . , $\boxed{=}$, $\boxed{1/x}$

FORMULA 5–4 $G_T = G_1 + G_2 + G_3$

G_1 value, $\boxed{+}$, G_2 value, $\boxed{+}$, G_3 value, $\boxed{=}$

FORMULA 5–5 $R_T = \dfrac{R_1 \times R_2}{R_1 + R_2}$

R_1 value, $\boxed{\times}$, R_2 value, $\boxed{\div}$, $\boxed{(}$, R_1 value, $\boxed{+}$, R_2 value, $\boxed{)}$, $\boxed{=}$

FORMULA 5–6 $R_u = \dfrac{R_k \times R_e}{R_k - R_e}$

R known value, $\boxed{\times}$, R equivalent value, $\boxed{\div}$, $\boxed{(}$, R known value, $\boxed{-}$, R equivalent value, $\boxed{)}$, $\boxed{=}$

FORMULA 5–7 $I_x = \dfrac{R_T}{R_x} \times I_T$

R_T value, $\boxed{\div}$, R_x value, $\boxed{\times}$, I_T value, $\boxed{=}$

FORMULA 5–8 $I_1 = \dfrac{R_2}{R_1 + R_2} \times I_T$

R_2 value, $\boxed{\div}$, $\boxed{(}$, R_1 value, $\boxed{+}$, R_2 value, $\boxed{)}$, $\boxed{\times}$, I_T value, $\boxed{=}$

FORMULA 5–9 $I_2 = \dfrac{R_1}{R_1 + R_2} \times I_T$

R_1 value, $\boxed{\div}$, $\boxed{(}$, R_1 value, $\boxed{+}$, R_2 value, $\boxed{)}$, $\boxed{\times}$, I_T value, $\boxed{=}$

REVIEW QUESTIONS

1. Refer to Figure 5–30 and find R_T.
   Circle the method you used to find the answer:
   Ohm's law
   Reciprocal method
   Product-over-the-sum method
   Assumed voltage method

FIGURE 5–30

2. Refer to Figure 5–31 and find R_T.
   Describe the method(s) you used to solve this problem.

Recall that crossed wires are not connected since there is not a dot!

FIGURE 5–31

3. Refer to Figure 5–32 and find I_2.
   Describe the method(s) you used to solve this problem.

FIGURE 5–32

4. Assume in Figure 5–33 that each of the six resistors is 150 kΩ and determine the following circuit parameters:
   R_T, I_T, and P_T;
   Current reading of meter 1
   Current reading of meter 2

FIGURE 5–33

5. Refer to Figure 5–34 and using the information provided, answer the following with:
   "I" for increase
   "D" for decrease
   "RTS" for remain the same
   a. If R_2 opens,
      I_1 will _____ P_T will _____
      I_2 will _____ V_1 will _____
      I_3 will _____ V_T will _____
      I_T will _____
   b. Assume that R_3 shorts internally. Prior to fuse or power supply failure,
      I_1 will _____ I_3 will _____
      I_2 will _____   total circuit current will _____.

FIGURE 5–34

6. Refer to Figure 5–35 and find the value of R_1.

FIGURE 5–35

7. Refer to Figure 5–36. If I_T is 18 mA, what are the values of I_1 and I_2?

FIGURE 5–36

8. Refer to Figure 5–37. What is the value of R_1?

FIGURE 5–37

9. Refer to Figure 5–38. What is the value of P_{R_1}?

FIGURE 5–38

10. Refer to Figure 5–39 and find the following parameters:

$R_T = $ _____
$I_T = $ _____
$P_T = $ _____
$I_1 = $ _____
$I_2 = $ _____
$I_3 = $ _____
$I_4 = $ _____
$I_5 = $ _____
$P_1 = $ _____
$P_2 = $ _____
$P_3 = $ _____
$P_4 = $ _____
$P_5 = $ _____

FIGURE 5–39

11. Refer to Figure 5–40 and answer the following:
    a. What is current reading of M_1?
    b. What is current reading of M_2?
    c. What is current reading of M_3?
    d. What is current reading of M_4?
    e. What is current reading of M_5?
    f. What is the power dissipation of R_1?
    g. What is the power dissipation of R_2?
    h. What is the power dissipation of R_3?
    i. What is total power dissipation of the circuit?
    j. If R_2 opens, what is new value of P_T?

FIGURE 5–40

FIGURE 5-41

Refer to this circuit, Figure 5-41, when answering questions 12 through 16.

12. With nothing connected between points A and B, show your "step-by-step" work and solve for R_T using:
    a. The reciprocal method
    b. The conductance method
    c. The assumed voltage method
    d. The Ohm's law method

13. What resistance (R_X) needs to be connected between points A and B for the circuit total resistance to equal 10 kΩ?

14. Assume the R_X of the preceding problem is left connected to the circuit. What happens to the circuit R_T if R_4 is doubled in value and R_2 is halved in value? Would the answer be the same if R_X were removed?

15. With the R_X of problem 13 still connected:
    a. Which resistor in the circuit dissipates the most power?
    b. What is the value of power dissipated by this resistor?

16. With the R_X of problem 13 *disconnected*:

a. Which resistor in the circuit dissipates the least power?
b. What is the value of power dissipated by this resistor?

Refer to Figure 5-42 as you answer questions 17 through 25.

17. What is the value of V_T?

18. What is the value of R_1?

19. What is the value of current through R_2?

20. What is the value of R_2?

21. Which resistor dissipates the most power?

22. Which resistor dissipates the least power?

23. What would the total current be if R_2 were changed to a 1.5-kΩ resistor (assuming V_T remained the same)?

24. If R_1 were to increase in value (due to aging and use), what would happen to the total power supplied by the source?

25. For the conditions described in question 24, what would happen to the power dissipated by R_3?

FIGURE 5-42

CHALLENGE QUESTIONS

1. What is the purpose of performing mental approximations, whenever possible, when performing circuit calculations?

2. What is one possible application of a current divider circuit?

3. Assume a circuit has three parallel resistors. If R_1 equals 12 kΩ, R_2 equals 15 kΩ, and R_3 equals 18 kΩ, determine the total resistance using a calculator and the techniques defined below:
   a. Use a calculator in the engineering mode and determine R_T by the reciprocal method. Show the answer as it is expressed in the calculator readout.
   b. Use a calculator in the scientific notation mode and determine R_T by the reciprocal method. Show the answer as it is expressed in the calculator readout.

4. Because of the resistor values given, what approach might have been easily used for mental approximation of the R_T value?

TEST YOURSELF

1. Draw a circuit diagram of a three-branch parallel circuit with the largest R having a value of 100 kΩ. This current divider circuit should have values that cause the total circuit current to divide as follows:
   Branch 1's current is half of branch 2 and branch 2's current is one-fifth of branch 3.
   Label all R values on your diagram.

2. What value of R is needed in parallel with a 50-kΩ resistor to have a resultant equivalent resistance of 10 kΩ?

PERFORMANCE PROJECTS CORRELATION CHART

Suggested performance projects in the Laboratory Manual that correlate with topics in this chapter are:

CHAPTER TOPIC	PERFORMANCE PROJECT	PROJECT NUMBER
Voltage in Parallel Circuits	Voltage in Parallel Circuits	17
Current in Parallel Circuits & Kirchhoff's Current Law	Current in Parallel Circuits	16
Resistance in Parallel Circuits	Equivalent Resistance in Parallel Circuits	15
Power in Parallel Circuits	Power Distribution in Parallel Circuits	18
Effects of Opens in Parallel Circuits & Troubleshooting Hints	Effects of an Open in Parallel Circuits	19
Effects of Shorts in Parallel Circuits & Troubleshooting Hints	Effects of a Short in Parallel Circuits	20

NOTE: It is suggested that after completing the above projects, the student should be required to answer the questions in the "Summary" at the end of this section of projects in the Laboratory Manual.

Following the SIMPLER troubleshooting sequence, find the problem with this circuit. As you follow the sequence, record your responses to each step on a separate sheet of paper.

CHALLENGE CIRCUIT 2

STARTING POINT INFORMATION

1. Circuit diagram
2. $V_T = 10$ V
3. Measured $I_2 = 2.1$ mA
4. Measured $I_T = 6.7$ mA
5. Rs measured with remainder of circuit isolated from points being tested

SYMPTOMS Gather, verify, and analyze symptom information. (Look at the "Starting Point Information.")

IDENTIFY Identify initial suspect area for the location of the trouble.

MAKE Make a decision about "What type of test to make" and "Where to make it." (To simulate making a decision about each test you want to make, select the desired test from the "TEST" column listing below.)

PERFORM Perform the test. (Look up the result of the test in Appendix C. Use the number in the "RESULTS" column to find the test result in the Appendix.)

LOCATE Locate and define a new "narrower" area in which to continue troubleshooting.

EXAMINE Examine available information and again determine "What type of test" and "where."

REPEAT Repeat the preceding analysis and testing steps until the trouble is found. What would you do to restore this circuit to normal operation? When you have solved the problem, compare your results with those shown on the following pages.

TEST   Results in Appendix C

TEST	Results
$V_{\text{A-D}}$	(83)
$V_{\text{B-D}}$	(11)
$V_{\text{E-H}}$	(22)
$V_{\text{F-G}}$	(95)
$V_{\text{I-L}}$	(52)
$V_{\text{J-K}}$	(75)
$V_{\text{A-B}}$	(8)

TEST	Results
$V_{\text{C-D}}$	(56)
$V_{\text{I-J}}$	(103)
$R_{\text{A-B}}$	(46)
$R_{\text{C-D}}$	(90)
$R_{\text{I-J}}$	(14)
$R_{\text{K-L}}$	(29)

CIRCUIT

2

Meter lead

Source lead

Symptoms The current through R_2 is normal, but the total current is low for this circuit. This suggests that the total R is higher than normal.

Identify initial suspect area: Branch 1 (points A to D) and branch 3 (points I to L) are suspect areas since branch 2 seems to be operating normally.

Make test decision: Check the voltage between points A and D (branch 1).

Perform Look up the test result. V_{A-D} is 10 V, which is normal.

Locate new suspect area: Points A to B and C to D in branch 1 and all of branch 3 are the new suspect areas.

Examine available data.

Repeat analysis and testing:

The V_{B-D}. V_{B-D} is 0 V, which is abnormal. This could result from a bad connection between points A and B or between points C and D.

Check V_{A-B}. V_{A-B} is 0 V, which is normal.

Check V_{C-D}. V_{C-D} is 10 V, which is abnormal.

Isolate branch 1 and check the resistance between points C and D. R_{C-D} is infinite ohms, indicating an open. (NOTE: This meter shows a blinking 30.00 to indicate infinite ohms. Other digital multimeters may use different readouts to indicate infinite ohms.)

1st Test

From source +

From source −

CHALLENGE CIRCUIT 2

Make a solid connection between points C and D and note the total current reading. It should be 10.4 mA if all the branches are operating properly. That is, the result of a good connection is that total current is approximately 10.4 mA. When this is done, the circuit operates normally.

2nd Test

3rd Test

4th Test

5th Test

6th Test

SPECIAL NOTE: You could have found the problem more quickly by calculating what the total current should be, then determining the amount of current missing was equal to the value branch 1 should have. Then you would have immediately known the problem was with branch 1. You still would have to perform isolation tests within that branch to locate the open.

Series-Parallel Circuits

Courtesy of John Fluke Mfg. Co., Inc.

6

KEY TERMS

Bleeder current

Bleeder resistor

Bridge circuit

Bridging or shunting

Ground reference

In-line

Load

Load current

Loaded voltage divider

Loading effect

Series-parallel circuit

Subcircuit

Wheatstone bridge circuit

OUTLINE

What is a series-parallel circuit?

Approaches to recognize and analyze series and parallel circuit portions

Total resistance in series-parallel circuits

Current in series-parallel circuits

Voltage in series-parallel circuits

Power in series-parallel circuits

Effects of opens in series-parallel circuits and troubleshooting hints

Effects of shorts in series-parallel circuits and troubleshooting hints

Designing a series-parallel circuit to specifications

Loaded voltage dividers

The Wheatstone bridge circuit

CHAPTER PREVIEW

Most electronic devices, equipment, and systems are comprised of series-parallel circuitry. Consumer products, such as high fidelity amplifiers, televisions, computers, and virtually all electronic products are filled with series-parallel circuits. A combination of both series and parallel connections form circuits that are coupled, or combined, to perform a desired task or function.

In this chapter, you will examine various configurations of these "combinational" circuits and will become familiar with analyzing them, using the principles you have already learned about series and parallel circuits. Also, you will see how a change in one part of a circuit has a greater or lesser effect on another portion of the circuit, depending on total circuit configuration and component values. Troubleshooting concepts will be examined. Your understanding of these circuits will be used by designing a simple series-parallel circuit to specifications. Finally, special applications of series-parallel circuitry will be studied.

OBJECTIVES

After studying this chapter, you should be able to:

- Define the term **series-parallel circuit**
- List the primary characteristic(s) of a series-parallel circuit
- Determine the total resistance in a series-parallel circuit
- Compute total circuit current and current through any given portion of a series-parallel circuit
- Calculate voltages throughout a series-parallel circuit
- Determine power values throughout a series-parallel circuit
- Analyze the effects of an open in a series-parallel circuit
- Analyze the effects of a short in a series-parallel circuit
- Design a simple series-parallel circuit to specifications
- Explain the **loading effects** on a series-parallel circuit
- Calculate values relating to a **loaded voltage divider**
- Make calculations relating to **bridge circuits**
- Use the SIMPLER troubleshooting sequence to solve the Chapter Challenge problem

WHAT IS A SERIES-PARALLEL CIRCUIT?

Definition

A **series-parallel circuit** contains a combination of both series-connected and parallel-connected components. There are both **in-line** series current paths and "branching-type" parallel current paths, Figure 6–1.

Characteristics

Series components in a series-parallel circuit may be in series with other *individual* components, or with other *combinations* of components, Figure 6–2.

Parallel components in a series-parallel circuit may be in parallel with other *individual* components, or with other *combinations* of components, Figure 6–3.

(**a**) Pictorial

(**b**) Schematic

I convergence point
(Current comes back together into one path)

Parallel connected components

R_1

"In-line" current path

+
Source
−

R_2

R_3

I

I branch point
(Current branches into two different paths)

FIGURE 6–1   A basic series-parallel circuit

R_1 and R_2 are in series with each other, the source, and subcircuit B.

R_5 and R_6 are in series with each other.

(**a**) Schematic diagram (actual circuit)

(**b**) Subcircuits block diagram

FIGURE 6–2   Series components in a series-parallel circuit

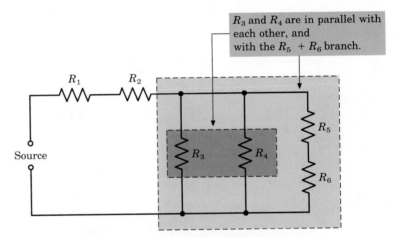

R_3 and R_4 are in parallel with each other, and with the $R_5 + R_6$ branch.

**FIGURE 6–3   Parallel components in a
series-parallel circuit**

APPROACHES TO RECOGNIZE AND ANALYZE SERIES AND PARALLEL CIRCUIT PORTIONS

1. Start analysis at the portion of the circuit farthest from the source and work back toward the source to identify those components or circuit portions that are in series and in parallel.

2. Trace common current paths to identify components in series, Figure 6–4. Components, or combinations of components with common current are in series with each other.

3. Observe voltages shared (in common) to identify components in parallel, Figure 6–5. Components, or combinations of components with the same voltage connection points (at both ends) are in parallel with each other.

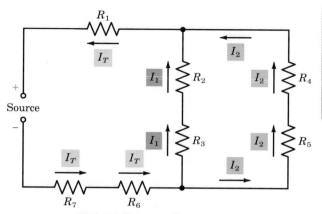

Starting "away from the source" and working back:
1. R_4 and R_5 have common current (I_2) and are in series with each other.
2. R_2 and R_3 have common current (I_1) and are in series with each other.
3. R_1, R_6, and R_7 have common current (I_T) and they are in series with each other, and with the total parallel combination of the $R_2 + R_3$ branch in parallel with the $R_4 + R_5$ branch.

FIGURE 6–4   Tracing common current paths to identify series circuit portions

NOTE: Points **A** and **B** are electrically the same point and points **C** and **D** are electrically the same point. Therefore, the voltage from point **B** to point **C** is the same voltage as the voltage from point **A** to point **D**.

Branch 2 ($R_4 + R_5$) is across the same voltage points as branch 1 ($R_2 + R_3$). Therefore, branch 1 and branch 2 are in parallel with each other.

FIGURE 6–5   Observing common voltage points to identify parallel circuit portions

4. Observe current branching and converging points to identify components, or combinations of components that are in parallel with each other, Figure 6–6.
   a. A point where current splits, or branches, is one end of a parallel combination of components, Figure 6–6, points A and B.
   b. The point where those same currents converge (or rejoin) is the other end of that same parallel combination, Figure 6–6, points C and D.

5. In summary, in the series portions of the circuit, current is common and the voltages (and resistances) are additive. In the parallel portions of the circuit, voltage is common and the branch currents are additive.

FIGURE 6–6   Observing current branching and converging points to identify parallel circuit portions

---

PRACTICE PROBLEMS I

Refer to Figure 6–7 and identify *individual* or single components that:

1. Carry total current.

2. Are in parallel with each other. That is, have the same voltage across them.

3. Are in series with each other. That is, have the same current through them.

Let's move to some analysis of series-parallel circuit electrical parameters.

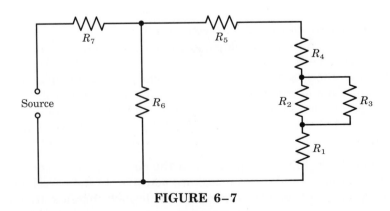

FIGURE 6–7

TOTAL RESISTANCE IN SERIES-PARALLEL CIRCUITS

If total circuit current and circuit applied voltage are known, then solving for total circuit resistance is easy.

Using Ohm's Law

Simply use Ohm's law, where $R_T = V_T/I_T$. If source voltage (V_S) and total current are unknown, then apply the concepts discussed in the preceding section.

Using the "Outside Toward the Source" Approach

1. Start at the end of the circuit farthest from the source and work toward the source, identifying and solving the series, the parallel, and the series-parallel combinations, Figure 6–8. You will apply series circuit rules for series portions, and parallel circuit rules for parallel portions in the circuitry.

2. First, identify and solve parallel component segments. The common voltage and the current branching and converging concepts are true for R_2 and R_3; therefore, these two components are in parallel, Figure 6–8a. As you recall, the equivalent resistance of two equal value resistors in parallel is equal to half the value of one equal branch R. Thus, R_e of R_2 and $R_3 = 5\ \Omega$. Naturally, you can also solve it with the product-over-the-sum formula. That is:

$$R_e = \frac{R_2 \times R_3}{R_2 + R_3} = \frac{(10\ \Omega) \times (10\ \Omega)}{(10\ \Omega) + (10\ \Omega)} = \frac{100\ \Omega}{20\ \Omega} = 5\ \Omega$$

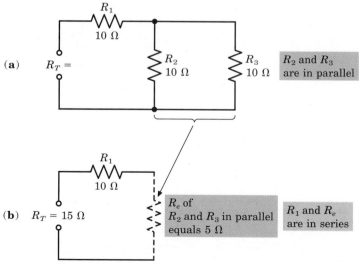

FIGURE 6–8 Analyzing a circuit to find R_T

3. Next, work toward the source to solve the series and/or series-parallel combinations. Total current splits and rejoins as it flows through the parallel combination of R_2 and R_3. Total current also flows through R_1. Thus, R_1 and the parallel combination of R_2 and R_3 are in series with each other, Figure 6–8b. Therefore, total resistance of R_1 in series with the parallel combination of R_2 and R_3 equals 10 Ω + 5 Ω = 15 Ω. And the circuit total resistance is 15 Ω.

EXAMPLE

Referring to Figure 6–9, let's look at another example of using the "outside-in" approach. This time we'll use more typical resistor values that aren't quite as easy to analyze as our previous example.

1. Starting with the outside branch, you can see that R_7 and R_8 are in parallel with each other, and their R_e is in series with both R_6 and R_9. Using the calculator's reciprocal function and several simple key-strokes, (1.5, $\boxed{\text{EE}}$ (or $\boxed{\text{EXP}}$), 3, $\boxed{1/x}$, $\boxed{+}$, 2.2, $\boxed{\text{EE}}$ (or $\boxed{\text{EXP}}$), 3, $\boxed{1/x}$, $\boxed{=}$, $\boxed{1/x}$), we quickly find that the 1.5-kΩ and 2.2-kΩ parallel resistors have an R_e = 891.9 Ω or 0.892 kΩ. Let's round that to its approximate value of 0.9 kΩ. Now, adding the values of R_6 + R_e (of R_7 and R_8 in parallel) + R_9, gives us 3.9 kΩ + 0.9 kΩ + 4.7 kΩ = 9.5 kΩ, which is the resistance of the outer branch.

2. Working into the next branch, we can see that the resistance between points A and B will be the equivalent resistance of the outer branch in parallel with the series resistors R_4 and R_5. R_4 + R_5 = 10 kΩ + 5.6 kΩ = 15.6 kΩ. The equivalent resistance of this 15.6 kΩ in parallel with the outer branch value of 9.5 kΩ can quickly be calculated with

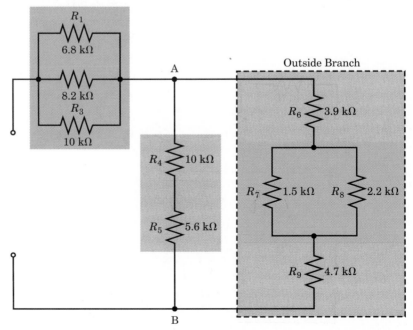

FIGURE 6–9

the calculator: 15.6, $\boxed{\text{EE}}$ (or $\boxed{\text{EXP}}$), 3, $\boxed{1/x}$, $\boxed{+}$, 9.5, $\boxed{\text{EE}}$ (or $\boxed{\text{EXP}}$), 3, $\boxed{1/x}$, $\boxed{=}$, $\boxed{1/x}$. The answer is 5.9 kΩ for the value of resistance between points A and B.

3. Now, all that remains is to find the equivalent resistance of R_1, R_2, and R_3, which are in parallel with each other, and add that R_e to the resistance between points A and B, and we will have solved the total resistance for this circuit. Once again, using our calculator and the reciprocal method: 6.8, $\boxed{\text{EE}}$ (or $\boxed{\text{EXP}}$), 3, $\boxed{1/x}$, $\boxed{+}$, 8.2, $\boxed{\text{EE}}$ (or $\boxed{\text{EXP}}$), 3, $\boxed{1/x}$, $\boxed{+}$, 10, $\boxed{\text{EE}}$ (or $\boxed{\text{EXP}}$), 3, $\boxed{1/x}$, $\boxed{=}$, $\boxed{1/x}$. The R_e of the three resistors = 2.7 kΩ. This 2.7 kΩ is in series with $R_{\text{A}-\text{B}}$, which is 5.9 kΩ. Thus, R_T for this circuit equals 2.7 kΩ + 5.9 kΩ = 8.6 kΩ.

The Reduce-and-Redraw Approach

A useful method to analyze series-parallel circuits is the reduce-and-redraw approach. This technique simplifies circuit analysis.

EXAMPLE

Look at Figure 6–10 as R_T is discussed for this series-parallel combination circuit.

FIGURE 6–10   Simplifying by the reduce-and-redraw technique

1. R_3 and R_4 are in parallel with each other and have an equivalent resistance that we'll call R_{e_1}. Thus, $R_{e_1} = 5$ kΩ. R_{e_1}, the resistance of R_3 and R_4 in parallel equals

$$\frac{(10 \text{ k}\Omega)(10 \text{ k}\Omega)}{(10 \text{ k}\Omega) + (10 \text{ k}\Omega)} = \frac{100 \times 10^6}{20 \times 10^3} = 5 \times 10^3 = 5 \text{ k}\Omega \text{ (Figure 6–10b)}$$

Of course, a simpler way is to remember that two equal branch resistances in parallel have an equivalent resistance equal to half of one of the branch resistances.

2. R_5 and R_6 are in series with each other and with R_{e_1} of parallel resistors R_3 and R_4. See Figure 6–10c and the calculations below.

The resistance of $R_{e_1} = 5$ kΩ
The resistance of $R_5 + R_6$ in series $= 10$ k$\Omega + 10$ k$\Omega = 20$ kΩ
The resistance of $R_{e_1} + R_5 + R_6 = 25$ kΩ

3. R_2 (10 kΩ) is in parallel with the total 25-kΩ combination of components R_3 through R_6 ($R_{e_1} + R_5 + R_6$). See Figures 6–10c and Figure 6–10d and the following calculations.

$$R_{e_2} = \frac{(10 \text{ k}\Omega)(25 \text{ k}\Omega)}{(10 \text{ k}\Omega) + (25 \text{ k}\Omega)} = \frac{250 \times 10^6}{35 \times 10^3} = 7.14 \times 10^3 = 7.14 \text{ k}\Omega$$

4. The overall combination of components R_2 through R_6 (R_{e_2}) is in series with R_1. See Figure 6–10d and the following calculations.

Total circuit resistance $(R_T) = R_1 + R_{e_2} = 10$ k$\Omega + 7.14$ k$\Omega = 17.14$ kΩ

PRACTICE PROBLEMS II

Try practicing the reduce-and redraw method as you perform the following tasks.

1. Find the total resistance of the circuit shown in Figure 6–11.

FIGURE 6–11

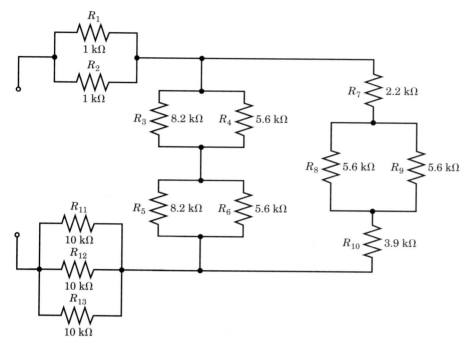

FIGURE 6–12

2. Find the total resistance of the circuit shown in Figure 6–12.

IN-PROCESS LEARNING CHECK I

Try solving the total resistance problems in Figures 6–13a and 6–13b. Use the outside toward the source approach to solve Figure 6–13a and use the reduce and redraw technique to solve the circuit in Figure 6–13b.

CURRENT IN SERIES-PARALLEL CIRCUITS

Examine Figure 6–14 as we now discuss current in a series-parallel circuit. One can start by finding the total circuit current, then work from that point to determine current distribution throughout the circuit using Ohm's law, Kirchhoff's laws, current divider formulas, and/or other techniques.

It is easier if Ohm's law is used to find total current, where $I_T = V_S/R_T$. In this circuit, the source voltage (V_S) is 175 volts and the total resistance is 175 Ω. For practice, verify the R_T value using the techniques you learned in the preceding section.

Now that R_T and V_S are known, it's easy to solve for I_T.

$$I_T = \frac{V_S}{R_T} = \frac{175 \text{ V}}{175 \text{ }\Omega} = 1 \text{ A}$$

(a)

(b)

FIGURE 6–13 Solve for R_T values in each circuit.

What about the distribution of current throughout the circuit? For this circuit, distribution of current is easy to find because of the values used. One example of a thought process that can be used is as follows:

1. R_1 and R_7 must have 1 ampere through them, since they are in series with the source and the remaining circuitry. In other words, I_T passes through both these resistors, Figure 6–14.

2. Since R_2 (250 Ω) and the outer branch (250 Ω) are equal resistance value branches, the total current equally divides between them. Thus, current through R_2 = 0.5 ampere, and current through the outer branch must equal 0.5 ampere, Figure 6–14.

3. It is apparent in tracing current through the outer branch that all of the 0.5 ampere must pass through both R_5 and R_6. However, this 0.5 ampere evenly splits between R_3 and R_4 because they are equal value parallel resistances. This means current through R_3 must be 0.25 ampere, and current through R_4 is also 0.25 ampere. In other words, the 0.5 ampere evenly splits at point C into two 0.25-ampere branch currents and rejoins at point D, Figure 6–14.

To find circuit resistance, you frequently start at a point farthest from the source and work back toward the source. *But when analyzing circuit current, you often start at the source with total current, then work outward through*

FIGURE 6–14   Solve for current values throughout a series-parallel circuit.

the circuit, analyzing the current distribution. In this case, the circuit currents are determined to be:

> I_T; I through R_1; and I through R_7 = 1 ampere.
> I through R_2; I through R_5; and R_6 = 0.5 ampere.
> I through R_3 and I through R_4 each = 0.25 ampere.

Of course, several other methods, or combinations of methods can be used to provide solutions, such as Ohm's law, Kirchhoff's voltage and current laws, and current divider formulas. In those cases where the values are not as convenient, you will probably have to use these other methods.

EXAMPLE

Having solved for total current (1 ampere), you can refer to Figure 6–14 and apply known values to solve for the unknown parameters.

1. I_T passes through both R_1 and R_7. We can use Ohm's law to find the voltage drops. Thus,

$$V_1 = I_T \times R_1 = 1 \text{ A} \times 25 \text{ }\Omega = 25 \text{ V}$$
$$V_7 = I_T \times R_7 = 1 \text{ A} \times 25 \text{ }\Omega = 25 \text{ V}$$

2. According to Kirchhoff's voltage law, the sum of the voltages around any closed loop must equal V applied. This means the sum of $V_1 + V_{A-B} + V_7$ must equal 175 V.
Thus, 25 V + V_{A-B} + 25 V = 175 V

$$50 \text{ V} + V_{A-B} = 175 \text{ V}$$
$$V_{A-B} = 175 \text{ V} - 50 \text{ V} = 125 \text{ V}$$

3. Kirchhoff's current law indicates the current leaving point A must equal the current entering point A. Stated another way, the total of

the currents through R_2 and the outer branch must equal I_T in this case.

a. Use Ohm's law to find current through R_2.

$$I \text{ through } R_2 = \frac{V_2}{R_2} = \frac{125 \text{ V}}{250 \text{ } \Omega} = 0.5 \text{ A}$$

b. Using Kirchhoff's current law:

$$I \text{ through outer branch} = I_T - I \text{ (through } R_2\text{)};$$
$$= 1 \text{ A} - 0.5 \text{ A} = 0.5 \text{ A}$$

4. Ohm's law and Kirchhoff's voltage law are now used to solve the voltage drops across the resistors in the outer branch and the currents through R_3 and R_4.

   Knowing the outer branch current is 0.5 A:
   a. $V_6 = 0.5 \text{ A} \times R_6 = 0.5 \text{ A} \times 100 \text{ } \Omega = 50 \text{ V}$
   b. $V_5 = 0.5 \text{ A} \times R_5 = 0.5 \text{ A} \times 100 \text{ } \Omega = 50 \text{ V}$

   Kirchhoff's voltage law indicates that:
   c. $V_{\text{C-D}} = V_{\text{A-B}} - (V_5 + V_6) = 125 \text{ V} - (50 \text{ V} + 50 \text{ V}) = 25 \text{ V}$

   Ohm's law can now be used to solve for the currents through R_3 and R_4.
   d. $I \text{ through } R_3 = \dfrac{V_3}{R_3} = \dfrac{25 \text{ V}}{100 \text{ } \Omega} = 0.25 \text{ A}$
   e. $I \text{ through } R_4 = \dfrac{V_4}{R_4} = \dfrac{25 \text{ V}}{100 \text{ } \Omega} = 0.25 \text{ A}$

PRACTICE PROBLEMS III

1. For practice in solving for current distribution throughout a series-parallel circuit, refer to Figure 6–15. Solve for I_T, I_1, I_2, and I_3.

FIGURE 6–15

FIGURE 6–16

2. Refer to Figure 6–16 and find the values of I_T, I_1, I_2, and I_3.

VOLTAGE IN SERIES-PARALLEL CIRCUITS

Voltage distribution throughout a series-parallel circuit is such that: 1) voltage across components, or circuit portions that are in series with each other are additive, 2) voltage across components, or circuit portions that are in parallel must have equal values.

As you would expect, a given component's voltage drop in a series-parallel circuit is contingent on its circuit location and its resistance value. The location dictates what portion of the circuit's total current passes through it. Consequently, location of any given component within the series-parallel circuit has direct influence on its $I \times R$ drop.

Refer to Figure 6–17 as the voltage distribution characteristics are analyzed with two methods: the Ohm's law method and resistance and voltage divider techniques.

FIGURE 6–17   Analyzing voltages in a series-parallel circuit

EXAMPLE

Ohm's Law Method

Solve circuit R and I values, then use Ohm's law to compute voltages.

1. $R_{\text{C–D}} = \dfrac{R_4 \times R_5}{R_4 + R_5} = \dfrac{(10 \times 10^3) \times (10 \times 10^3)}{(10 \times 10^3) + (10 \times 10^3)}$

   $= \dfrac{100 \times 10^6}{20 \times 10^3} = 5 \times 10^3 = 5 \text{ k}\Omega$

2. $R_3 + R_{\text{C–D}} + R_6 = 5 \text{ k}\Omega + 5 \text{ k}\Omega + 5 \text{ k}\Omega = 15 \text{ k}\Omega$

3. $R_{\text{A–B}} = \dfrac{(R_2)(R_3 + R_{\text{C–D}} + R_6)}{(R_2) + (R_3 + R_{\text{C–D}} + R_6)}$

   $= \dfrac{(15 \times 10^3) \times (15 \times 10^3)}{(15 \times 10^3) + (15 \times 10^3)} = \dfrac{225 \times 10^6}{30 \times 10^3} = 7.5 \times 10^3 = 7.5 \text{ k}\Omega$

4. $R_T = R_1 + R_{\text{A–B}} = 7.5 \text{ k}\Omega + 7.5 \text{ k}\Omega = 15 \text{ k}\Omega$

5. $I_T = \dfrac{V_T}{R_T} = \dfrac{15 \text{ V}}{15 \text{ k}\Omega} = 1 \text{ mA}$

6. $V_1 = I_T \times R_1 = 1 \text{ mA} \times 7.5 \text{ k}\Omega = 7.5 \text{ V}$

7. Current splits evenly between branches R_2 and the R_3 through R_6 branch since each branch R equals 15 kΩ. This means that:

   $$V_2 = 0.5 \, I_T \times R_2 = 0.5 \text{ mA} \times 15 \text{ k}\Omega = 7.5 \text{ V}$$
   $$V_3 + V_{\text{C–D}} + V_6 \text{ equals } 7.5 \text{ V}$$

8. $V_3 = 0.5 \, I_T \times R_3 = 0.5 \text{ mA} \times 5 \text{ k}\Omega = 2.5 \text{ V}$

9. Current splits evenly between R_4 and R_5; thus, each carries 0.25 mA of current.

   $$V_4 = 0.25 \, I_T \times R_4 = 0.25 \text{ mA} \times 10 \text{ k}\Omega = 2.5 \text{ V}$$
   $$V_5 = 0.25 \, I_T \times R_5 = 0.25 \text{ mA} \times 10 \text{ k}\Omega = 2.5 \text{ V}$$

10. $V_6 = 0.5 \, I_T \times R_6 = 0.5 \text{ mA} \times 5 \text{ k}\Omega = 2.5 \text{ V}$

EXAMPLE

Resistance and Voltage Divider Techniques

1. Observing the circuit, you can see that:
   a. R_4 and R_5 are in parallel; thus, $V_4 = V_5$.
   b. R_e of $R_{4-5} = 5$ kΩ.
   c. R_3, R_{4-5}, and R_6 are in series and together form a 15-kΩ branch of the circuit. Furthermore, each of these drops one-third the voltage present from point A to point B, since each of the three entities represents 5 kΩ.
   d. This 15-kΩ branch is in parallel with R_2, another 15-kΩ branch. Thus, R_e from point A to point B equals 7.5 kΩ.

e. R_1 is also 7.5 kΩ and is in series with the equivalent 7.5 kΩ found from point A to point B. Thus V_1 and V_{A-B} are equal, and each equals half the voltage applied (Kirchhoff's voltage law).

2. Applying the principles of voltage dividers:

a. $V_{A-B} = \dfrac{R_{A-B}}{R_T} \times V_T = \dfrac{7.5\text{ k}\Omega}{15\text{ k}\Omega} \times 15\text{ V} = 7.5\text{ V}$

b. $V_{C-D} = \dfrac{R_{4-5}}{R_3 + R_{4-5} + R_6} \times V_{A-B} = \dfrac{5\text{ k}\Omega}{15\text{ k}\Omega} \times 7.5\text{ V} = 2.5\text{ V}$

c. $V_3 = V_{4-5} = V_6$

d. $V_2 = V_{A-B}$

e. $V_1 = \dfrac{R_1}{R_T} \times V_T = \dfrac{7.5\text{ k}\Omega}{15\text{ k}\Omega} \times 15\text{ V} = 7.5\text{ V}$

PRACTICAL NOTES

An approach often used to simplify notation for indicating that components are in parallel with each other is to use two vertical bars, or short vertical lines between component designators. For example, in the circuit of Figure 6–17 to show that R_4 and R_5 are in parallel, we could state $R_4 \| R_5$. If we were going to use this type of notation, we could indicate the resistors and their connections, which make up the resistance at points A–B, as follows:

$$R_{A-B} = R_2 \| (R_3 + R_4 \| R_5 + R_6)$$

EXAMPLE

Let's look at one more example of analyzing a series-parallel circuit, using the Ohm's law method. Solve for circuit R and I values, then compute voltages, as appropriate, for the circuit of Figure 6–18.

FIGURE 6–18

1. $R_{C-D} = R_4 \| (R_5 + R_6)$

$$R_{C-D} = \frac{R_4 \times (R_5 + R_6)}{R_4 + (R_5 + R_6)} = \frac{(3.9 \times 10^3) \times (4.9 \times 10^3)}{(3.9 \times 10^3) + (4.9 \times 10^3)}$$

$$R_{C-D} = \frac{19.11 \times 10^6}{8.8 \times 10^3} = 2.17 \text{ k}\Omega$$

2. $R_3 + R_{C-D} = 4.7 \text{ k}\Omega + 2.17 \text{ k}\Omega = 6.87 \text{ k}\Omega$

3. $R_{A-B} = R_2 \| (R_3 + R_{C-D}) = \dfrac{(5.6 \times 10^3) \times (6.87 \times 10^3)}{(5.6 \times 10^3) + (6.87 \times 10^3)} = \dfrac{38.47 \times 10^6}{12.47 \times 10^3}$

$\qquad = 3.09 \text{ k}\Omega$

4. $R_T = R_1 + R_{A-B} = 12 \text{ k}\Omega + 3.09 \text{ k}\Omega = 15.09 \text{ k}\Omega$

5. $I_T = \dfrac{V_S}{R_T} = \dfrac{15 \text{ V}}{15 \text{ k}\Omega} \approx 1 \text{ mA}$

6. $I_{R_1} = 1 \text{ mA}; V_{R_1} = I_{R_1} \times R_1 = 1 \text{ mA} \times 12 \text{ k}\Omega = 12 \text{ V}$

7. $I_{R_2} = \dfrac{V_{R_2}}{R_2} = \dfrac{3 \text{ V}}{5.6 \text{ k}\Omega} \approx 0.54 \text{ mA}$

8. $I_{R_3} = I_T - I_{R_2} = 1 \text{ mA} - 0.54 \text{ mA} \approx 0.46 \text{ mA}$
$\quad V_{R_3} = I_{R_3} \times R_3 = 0.46 \text{ mA} \times 4.7 \text{ k}\Omega \approx 2.16 \text{ V}$

9. $I_{R_4} = \dfrac{V_{R_4}}{R_4} = \dfrac{V_S - (V_{R_1} + V_{R_3})}{R_4} = \dfrac{15 \text{ V} - 14.16 \text{ V}}{3.9 \text{ k}\Omega}$

$\quad I_{R_4} = \dfrac{0.84 \text{ V}}{3.9 \text{ k}\Omega} \approx 0.22 \text{ mA}$

10. $I_{R_5} = I_{R_6} = \dfrac{V_{C-D}}{R_5 + R_6} = \dfrac{0.84 \text{ V}}{4.9 \text{ k}\Omega} \approx 0.17 \text{ mA}$
$\quad V_{R_5} = I_{R_5} \times R_5 = 0.17 \text{ mA} \times 2.7 \text{ k}\Omega \approx 0.46 \text{ V}$
$\quad V_{R_6} = I_{R_6} \times R_6 = 0.17 \text{ mA} \times 2.2 \text{ k}\Omega \approx 0.37 \text{ V}$

PRACTICE PROBLEMS IV

1. Analyze the voltage and current parameters throughout the circuit of Figure 6–19. List the voltage dropped by each resistor and
the current through each resistor.

FIGURE 6–19

FIGURE 6–20

2. Analyze the voltage and current parameters throughout the circuit of Figure 6–20. Again, list the voltage dropped by each resistor and the value of current through each resistor.

POWER IN SERIES-PARALLEL CIRCUITS

Analysis of power dissipations throughout a series-parallel circuit employs the same techniques used in series or parallel circuits. Important points to remember are:

1. Total power $= V_T \times I_T$ (or $I_T^2 \times R_T$; or, V_T^2/R_T)

2. Total power also equals the sum of all individual power dissipations.

3. Individual component power dissipations are calculated using the individual component's parameters. That is, $V \times I$; I^2R; or V^2/R values.

EXAMPLE

Notice the power dissipations throughout the circuit shown in Figure 6–21. If you take time to analyze this circuit completely, you find that the circuit's total resistance is 30 Ω. This means that the circuit's total current is 30 V/30 Ω, or 1 A. Current divides evenly in the two outer branches. Therefore, current through R_4 and $R_6 = 0.5$ A, and current through R_6 and R_7 is also 0.5 A. As you can see from the circuit diagram, the total 1 A current also evenly splits through R_2 and R_3, with each carrying 0.5 A.

With the current through each of the resistors known, you can then find the voltage drops across each component, and the power dissipated by each component, Figure 6–21. Since we know current, resistance, and voltage, any one of the power formulas ($P = V \times I$; $P = I^2 R$; or $P = V^2/R$) might be used to compute each of the component's power dissipations. For example: $P_{R_7} = V \times I = 10$ V $\times$ 0.5 A $= 5$ W; $P_{R_1} = V \times I = 10$ V $\times$ 1 A $= 10$ W; and so on.

FIGURE 6–21   Effect of location on power, current, and voltage

PRACTICAL NOTES

Note that:

1. In series circuits, the largest value resistor dissipates the most power, and the lowest value resistor dissipates the least power.

2. In parallel circuits, the smallest resistance branch dissipates the most power, and the largest resistance branch dissipates the least power.

3. In series-parallel circuits, the resistor dissipating the most power *may or may not* be the highest value resistor in the circuit. The resistor dissipating the lowest power *may or may not* be the lowest value resistor in the circuit.

The power dissipated is determined by the position of the given resistor in the total circuit configuration. An illustration of this is in Figure 6–21. Note that:

- R_7 is twice the value of R_1 but dissipates only half the power.

- R_1 and R_2 are equal value resistors but each dissipates a different value of power.

To make sure you are learning to analyze the key parameters for a series-parallel circuit, perform the following In-Process Learning Check.

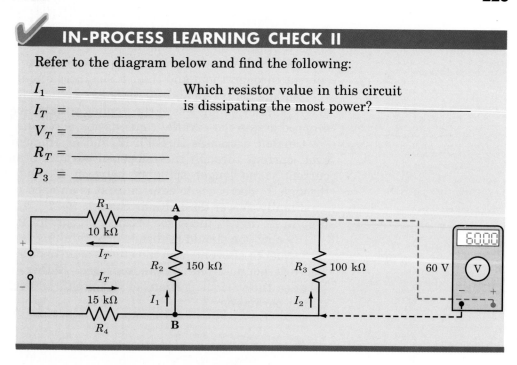

IN-PROCESS LEARNING CHECK II

Refer to the diagram below and find the following:

I_1 = _____   Which resistor value in this circuit
I_T = _____   is dissipating the most power? _____
V_T = _____
R_T = _____
P_3 = _____

EFFECTS OF OPENS IN SERIES-PARALLEL CIRCUITS AND TROUBLESHOOTING HINTS

Effects

As you have seen, the electrical location of a given component in a series-parallel circuit has great influence on all its electrical parameters. The location of an open in the circuit configuration also has a strong influence on all the parameters throughout the circuit.

Looking at the circuit used in In-Process Learning Check II, you can see that if either R_1 or R_4 opens, then the path for total current is broken. Thus, all other components' voltage drops, currents, and power dissipations decrease to zero!

On the other hand, if R_3 opens, total current decreases, since the total resistance increases from 85 kΩ to 175 kΩ. But current does not decrease to zero. Thus, individual component voltages, currents, and power dissipations change; however V applied remains the same.

Here is a **special thinking exercise** for you. Again refer to the In-Process Learning Check circuit and assume R_2 opens. Will the following parameters increase, decrease, or remain the same?

V_T will _____ I_T will _____
V_1 will _____ I through R_1 will _____
V_2 will _____ I through R_2 will _____
V_3 will _____ I through R_3 will _____
V_4 will _____ I through R_4 will _____
P_T will _____ P_3 will _____ P_4 will _____

Total power and total current should decrease because total resistance increases while V applied remains the same.

Voltage drops across R_1 and R_4 also decrease, since they now have less current through them and their resistances remain unchanged.

Voltage across R_2 and R_3 increases because the series components (R_1 and R_4) are dropping less of the applied voltage, leaving more voltage to be dropped across the parallel combination of R_2 and R_3, points A–B.

Current decreases through R_1 and R_4 (remember I_T decreased); however, current through the unopened parallel branch, R_3, increases since current is no longer splitting between R_2 and R_3. Obviously, current through R_2 decreases to zero, since it is an open path.

Finally, power dissipations change. P_T, P_1, and P_4 decrease due to the drop in I_T, and P_3 increases because its current increases.

By now you should realize logical thinking is required when analyzing the sequence of events and the effects of an open in a series-parallel circuit. There is not just one simple approach. Rather, think of the interaction between the various portions of the circuit on each other and on the total circuit parameters.

TROUBLESHOOTING HINTS

Because the component's position in the circuit is important due to the interaction it has on the remaining circuitry, the hints given here are somewhat general.

If an open occurs in *any* component in the circuit, total resistance increases; consequently, total current decreases. If you measure (or monitor) total current and it is lower than normal with normal applied voltage, an open, or at least an increased resistance value, has occurred somewhere in the circuit.

NOTE: Two typical ways to determine total current in a circuit are shown in Figure 6–22.
a. Insert current meter in series with source and the circuitry.
b. Measure voltage across a known resistor value that carries total current, and use Ohm's law (V/R) to determine current.

PRACTICAL NOTES

A convenient resistance value for this approach is a 1-kΩ resistor, if available, because its voltage drop equals the number of mA through it. In other words,

$$\frac{10V}{1k\Omega} = 10mA$$

(See Figure 6–22.)

1. If an open occurs in any component in series with the source, total R increases to ∞, and total I decreases to zero. As you learned in series circuits, V applied appears across the opened series component and zero volts are dropped across the good components.

2. If an open occurs in any component other than those in series with the source, current measurements and *common sense* will isolate the problem.

3. If current measurements cannot be easily performed, you can make a number of voltage checks throughout the circuit. Observe where voltages are higher and lower than normal, then logically determine where the open is located.

(**a**) Pictorial   (**b**) Schematic

FIGURE 6–22   Two ways of checking total current

EFFECTS OF SHORTS IN SERIES-PARALLEL CIRCUITS AND TROUBLESHOOTING HINTS

Effects

Again, the location of the faulty component or circuit portion in the total circuit configuration controls what happens to the parameters of the remaining good components.

Refer to Figure 6–23. If the shorted component is in a line that carries total current, that is, in series with the source, then: 1) total current increases, 2) voltage across the shorted component drops to zero, and 3) all other voltages throughout the circuit increase (except V_T).

Refer to Figure 6–24. If the shorted component is elsewhere in the circuit, then: 1) total current still increases since total R decreases; however,

FIGURE 6–23   **Effects of a shorted series (in-line) element**

If R_1 shorts: R_T ↓ (from 10 kΩ to 7.5 kΩ)
V_T → (stays at 100 V)
I_T ↑ (from 10 mA to 13.33 mA)
V_{R_1} ↓ (to zero)
$V_{R_2 - R_5}$ ↑ (due to I ↑ and V_{R_1} ↓)

FIGURE 6–24   **Effect of shorted element in the combinational portion of a series-parallel circuit**

If R_3 shorts: R_T ↓ (from 10 kΩ to 8.5 kΩ)
V_T → (stays at 100 V)
I_T ↑ (from 10 mA to 11.76 mA)
V_{R_1} ↑ (from 25 V to 29.4 V)
$V_{R_3-R_4}$ ↓ (to zero V, due to short)
V_{R_2} and V_{R_5} ↓ (due to increased drop V_{R_1} with V_T staying the same)

2) voltage across the shorted component *and* components *directly* in parallel with the shorted component decreases to zero, since the shorted component is shunting the current around components in parallel with it; and 3) other components' voltage drops may increase or decrease, depending on their location.

TROUBLESHOOTING HINTS

Some general thoughts about troubleshooting shorts in series-parallel circuits are:

1. A short of *any* component in the circuit causes total resistance to decrease and total current to increase.

2. A higher than normal total current (or lower than normal R_T) implies that either a short or a lowered R has occurred somewhere in the circuit.

3. Voltage measurements across the circuit components reveal close to zero volts across the shorted component and those directly in parallel with it. The good components may have either higher or lower than normal voltage levels, depending on their individual positions within the circuit. Voltage measurement is one simple way to locate the component or combination of components that are shorted.

4. Once the shorted portion of the circuit is located, resistance measurements will isolate the exact component or circuit portion containing the short. Resistance measurement involves turning the power off during measurement and lifting one end of each branch from the circuit.

Let's see if your analysis skills are developing. Try the troubleshooting problems in the following In-Process Learning Check.

IN-PROCESS LEARNING CHECK III

1. Refer to Figure 6–25. Determine the defective component and the nature of its defect.

2. Refer to Figure 6–26. Indicate which component or components might be suspect and what the trouble might be.

3. Refer to Figure 6–27. Indicate which voltages will change and in which direction, if R_2 increases in value.

NOTE: Voltages indicated are from specified test points to chassis ground reference.

FIGURE 6–25

NOTE: Voltages are indicated from specified test points to chassis ground reference.

FIGURE 6–26

FIGURE 6–27

DESIGNING A SERIES-PARALLEL CIRCUIT TO SPECIFICATIONS

For a design problem, let's assume we are designing and drawing the circuit diagram for a simple lighting circuit that contains four light bulbs, each of which normally operates at 10 volts. Using four lights and three SPST switches, configure the circuit so switch 1 acts as an on-off switch for all the lights; switch 2 controls two of the lights' on-off modes (assuming switch 1 is on), and switch 3 controls the remaining lights' on-off conditions (assuming switch 1 is on). The source voltage is 20 volts.

EXAMPLE

The thought processes to solve this problem are as follows:

1. Since the source equals 20 V and the bulbs are 10-V bulbs, it is logical to assume two bulbs are in series across the 20-V source.

2. Since there are four lights, it is also logical to assume there are two, two-bulb branches connected to the 20-V source voltage.

3. Since switch 1 controls the on-off condition of the total circuit connected to the source, it is logical that this switch is in series with the source and connected between the source and the remaining circuitry.

4. Since the circuit specification is to control each of the two bulb branches independently, switches 2 and 3 must be placed in series with the bulbs in each branch, respectively.

Using the preceding knowns, it is possible to draw the circuit diagram, Figure 6–28.

PRACTICE PROBLEMS V

See if you can design a series-parallel circuit that meets the following conditions:

Given six 10-kΩ resistors and a 45-V source, design and draw the circuit diagram where:

$V_1 = 20$ V; $V_2 = 5$ V; $V_3 = 5$ V; $V_4 = 5$ V; $V_5 = 5$ V; $V_6 = 20$ V

LOADED VOLTAGE DIVIDERS

Two applications of series-parallel circuits commonly encountered by electronic technicians are **loaded voltage dividers** and the **Wheatstone bridge circuit.**

Frequently in electronic circuits, a requirement is to provide different levels of voltage for various portions of the circuit or system. Also, these different portions may demand different load currents. Voltage dividers are

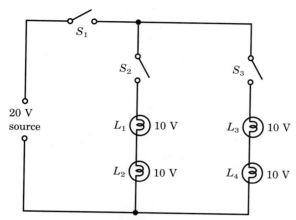

FIGURE 6–28   Sample design problem circuit

often used with power supplies to provide and distribute these different levels of voltages at the different current values demanded.

Since loads connected to voltage dividers are frequently functional circuits rather than fixed-value resistors, load requirements (i.e., current demands by the loads) often change during operation. For this reason, voltage dividers need to be designed so that reasonable variations in load current demands do not unreasonably upset the voltage levels at the various divider output points. When this is true, we say that the voltage divider has been designed so that it is "stiff." In order to be a stiff voltage divider, the design is such that the bleeder current through the divider itself (without load currents) is at least ten times the load current that will be demanded.

Refer to Figure 6–29. Recall in the Series Circuits chapter we learned voltage values at different points along the series circuit are controlled by the ratios of the resistances throughout the series string. These voltage dividers are called unloaded voltage dividers because no external circuit loads (current-demanding components or devices) are connected to the voltage divider circuit.

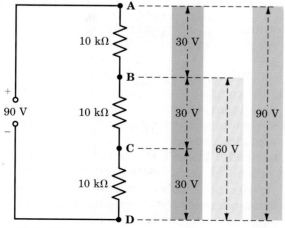

FIGURE 6–29   Unloaded voltage divider

FIGURE 6–30 Terms related to loaded voltage dividers

To help in our discussion of loaded voltage dividers, it is appropriate to define some important terminology used in our discussion. Refer to Figure 6–30 as you study these definitions.

1. **Load** is a component/device or circuit drawing current from a power source.
2. **Load current** is the current required by the components or circuits that are connected to the power source and/or its output voltage divider.
3. R_L is the resistance value of the load resistor or component.
4. **Ground reference** is an electrical *reference point* used in defining or measuring voltage values in a circuit.

Frequently the electrical ground reference point in a circuit is chassis ground, which is the common metal conducting path through which many components or circuits are electrically connected to one side of the power supply. In printed circuits, it is a common metal conductor *path* where many components are electrically connected, or "commoned." In circuits built on a metal chassis, the metal chassis is often the "return" path, or common conductor path; hence, the term "chassis ground." The symbol for chassis ground is ⌁ . In electrical wiring circuits, such as house wiring, "earth ground" is a common reference and connecting point. The symbol for earth ground is ⏚ .

PRACTICAL NOTES

Safety must be observed when dealing with equipment where the equipment chassis is used as the "common" circuit return path. In many (or most) cases, it is advisable to make sure that the chassis is also connected to a good earth ground via some means to prevent inadvertent shock or life-threatening danger due to differences of potential between equipment chassis and earth ground. Earth ground connections are frequently made via electrical connection made to a water pipe, or a metal rod driven into the ground. The ac electrical power system normally has one of the conductor paths "neutralized" by its being connected to an electrical earth ground.

There are some pieces of equipment that are designed to be operated with a "floating ground," which is not connected to earth ground. That is the chassis is NOT to be connected to earth ground. In those cases, it is a desirable safety habit to use an "isolation transformer" to isolate the ac power source earth ground from the chassis common circuit return path. You will learn more about transformers in a later chapter. We simply wanted to CAUTION you to be careful when working with circuits and equipment that might shock (or kill) you if you touch both the chassis and an earth ground at the same time. Operate the equipment properly, and be aware of whether the chassis is properly grounded or isolated.

You must be aware of the ground conditions that exist for the equipment you are using, working on, or testing. It is always a good idea to check for differences of potential between equipment and earth grounds using a voltmeter. If the chassis is "hot" with respect to earth or power wiring grounds, you must determine whether you should even work on it at all, if it needs to be repaired prior to working on it, or whether the power plug needs to be reversed.

5. **Bleeder current** is considered the fixed current through a resistive network, or a bleeder resistor connected across the output of the power supply. Many power supplies use a bleeder resistor that draws a fixed minimum current from the supply to help regulate the power supply output voltage; that is, keeping it more constant under varying load conditions.

6. **Bleeder resistor** is the resistor, or resistor network, connected in parallel with the power supply circuitry, that draws the bleeder current from the source. Another common function of a bleeder resistor is to discharge power supply capacitors after the circuit is turned off, for safety purposes. You will learn more about this in later studies regarding power supply circuits.

7. **Potentiometer** (Figure 6–31) is a three-terminal resistive device used as a voltage-dividing component. One terminal is at one end of the resistive element, a second contact is at the other end of the resistive element, and the third contact is a "wiper arm" that can move to any position along the resistive element. The position of the wiper arm

FIGURE 6–31   The potentiometer (Photo courtesy of Bourns, Inc.)

along the resistive element determines the resistance, and thus the voltage from it to either end contact.

With these terms in mind, let's look at a brief analysis of a typical loaded voltage divider circuit.

Two-Element Divider with Load

Observe Figures 6–32a and 6–32b as you study the following discussion. Let's think through the circuit.

What causes voltage between point A and ground reference to decrease from 25 volts to about 16.7 volts when load resistor R_L is connected? Before

FIGURE 6–32   Two-element divider circuit

the load was connected, the two-series-connected equal value resistors divided the input voltage equally (25 volts each) since they were passing the same current.

When the load resistor was connected, a new branch path for current was created through R_L. The current through R_L must also pass through R_1 to return to the source's positive side. Now R_1 has the original unloaded voltage divider current through it, *plus* the load current of R_L.

If the current through R_1 increases and its R value is unchanged, then its $I \times R$ drop must increase. In this case it increased from 25 volts to about 33.33 volts. Thus, only 16.66 volts remain to be dropped from point A to ground compared to 25 volts previously dropped before the load was connected.

Three-Element Divider with Multiple Loads

Observe in Figure 6–33a and 6–33b the changes caused by connecting a load to a three-element voltage divider are similar to those changes that occur in a two-element divider.

Let's look at the three-element loaded voltage divider in Figure 6–33b from the perspective of Ohm's law and Kirchhoff's voltage law.

R_T is solved by using your knowledge of series and parallel circuits. R_L is in parallel with the series combination R_{2-3}. This indicates the R_L value of 50 kΩ is in parallel with 50 kΩ so the equivalent resistance from point A to ground equals 25 kΩ. This 25 kΩ is in series with the R_1 value of 25 kΩ yielding a total circuit resistance of 50 kΩ.

(a)   (b)

FIGURE 6–33 Comparison of loaded and unloaded voltage dividers
(Effect of connecting a load to a voltage divider circuit)

Solve I_T using Ohm's law, where

$$I_T = \frac{V_T}{R_T} = \frac{75 \text{ V}}{50 \text{ k}\Omega} = 1.5 \text{ mA}$$

Since I_T passes through R_1, $V_1 = I_T \times R_1 = 1.5$ mA $\times$ 25 kΩ = 37.5 volts.

Since point A to ground is composed of two equal 50-kΩ branches (R_L and R_{2-3}), total current evenly divides between the branches. This means 0.75 mA passes through R_L and 0.75 mA passes through R_{2-3}. The R_L voltage drop = 0.75 mA $\times$ 50 kΩ = 37.5 V; R_2 voltage drop = 0.75 mA $\times$ 25 kΩ = 18.75 V; R_3 voltage drop also = 0.75 mA $\times$ 25 kΩ = 18.75 V; therefore, V_{2-3} = 37.5 V.

Of course, by Kirchhoff's voltage law (the sum of the voltage drops around a closed loop must equal V applied), you know the voltage from point A to ground has to be 37.5 V. Since the calculated R_1 drop is 37.5 V, the remainder of V applied has to drop from A to ground reference.

Let's move one step further in studying loaded voltage dividers by performing a design and analysis problem to achieve specified voltage levels and load currents.

EXAMPLE

Refer to Figure 6–34 for the desired parameters of the circuit to be designed. Note the voltage-dividing system provides voltage and current levels as follows:

Load 1 (R_{L_1}) = 25 V at 12.5 mA
Load 2 (R_{L_2}) = 75 V at 25 mA
Load 3 (R_{L_3}) = 225 V at 50 mA

Bleeder current (I_B) = 400 mA (to provide a "stiff" divider)

For additional information we'll make one task to determine the resistance values of loads 1, 2, and 3. Using the parameters provided, this is not critical to the design tasks but will be of interest.

Since you know the desired voltage across each load and the currents passing through them, simply apply Ohm's law to find load resistance values.

$$R_{L_1} = \frac{25 \text{ V}}{12.5 \text{ mA}} = 2 \text{ k}\Omega$$

$$R_{L_2} = \frac{75 \text{ V}}{25 \text{ mA}} = 3 \text{ k}\Omega$$

$$R_{L_3} = \frac{225 \text{ V}}{50 \text{ mA}} = 3.5 \text{ k}\Omega$$

The second task, a design task, is to determine the values of voltage divider resistors R_1, R_2, and R_3 that will provide the appropriate voltages to the loads.

FIGURE 6–34   Analyzing and designing a loaded voltage divider circuit

Again refer to Figure 6–34 as we examine the design task as follows:

1. Looking at the circuit, you can see that none of the load currents pass through R_3. This means only the given bleeder current (I_B) of 400 mA is passing through R_3. Because its voltage must equal the load 1 voltage of 25 V since they are in parallel, then

$$R_3 = \frac{25 \text{ V}}{400 \text{ mA}} = 62.5 \ \Omega$$

2. The R_2 current must equal the 400-mA bleeder current, *plus* the 12.5-mA load 1, or 412.5 mA. The voltage dropped by R_2 must equal 75 V minus the voltage dropped by R_3. Thus, V_{R_2} equals 50 V. Therefore,

$$R_2 = \frac{50 \text{ V}}{412.5 \text{ mA}} = 121 \ \Omega$$

3. The R_1 current must equal the bleeder current plus load 1 current plus load 2 current. Therefore, the current through R_1 equals 400 mA (bleeder current) plus 12.5 mA (I_{L_1}) plus 25 mA (I_{L_2}) equals 437.5 mA. Thus, R_1 voltage drop must equal 225 V minus the drop from point B to ground, or 225 V − 75 V = 150 V. This means

$$R_1 = \frac{150 \text{ V}}{437.5 \text{ mA}} = 343 \ \Omega$$

Look at Figure 6–35 for the completed design.

FIGURE 6–35   Voltage divider problem solution

PRACTICE PROBLEMS VI

1. Draw the circuit of a three-resistor voltage divider with three loads connected as they were in our example. Let the top divider resistor be R_1, the middle one R_2, and the bottom one R_3.

2. Determine the value of each of the load resistances and find the values for the divider resistors R_1, R_2, and R_3 assuming the following:
   a. Source voltage = 100 V
   b. Bleeder current = 200 mA
   c. Load #1 needs 10 V and demands 10 mA of current
   d. Load #2 needs 30 V and demands 20 mA of current
   e. Load #3 needs 100 V and demands 40 mA of current

3. Completely label your drawing showing all resistances, voltages, and currents.

A Practical Voltage Divider Variation

Look at the circuit in Figure 6–36. With the potentiometer's wiper arm at the top (position 1), the 20-kΩ R_L is in parallel with total potentiometer resistance of 30 kΩ, yielding an equivalent resistance of 12 kΩ from point A to ground. The circuit voltage-dividing action is dividing the 200 V across the 40-kΩ total circuit resistance consisting of the 28-kΩ resistor and the 12-kΩ parallel circuit equivalent resistance in series. Therefore, R_L drops twelve-fortieths of 200 V, or 60 V. The 28-kΩ resistor drops the remaining 140 V.

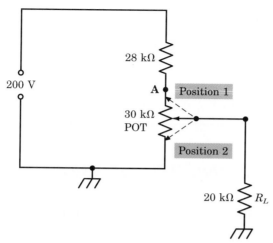

The potentiometer enables varying V across R_L from 0 to 60 V

FIGURE 6–36   Example of potentiometer varying voltage output

On the other hand, if the wiper arm is at position 2 (or ground), both ends of R_L will connect to ground. Therefore, the voltage across R_L is zero V. This means that with the potentiometer, we can vary the voltage across R_L from zero to sixty volts. As you will see throughout your electronics experience, this "variable-voltage-divider" capability of the potentiometer is convenient.

PRACTICE PROBLEMS VII

Refer to Figure 6–37 and answer the following:

1. What is the value of voltage across R_L? How much power is dissipated by R_L?

2. What is the value of R_L?

3. What is the value of R_2?

4. If V_{R_L} suddenly decreases due to a change in the R_2 value, does this indicate R_2 has increased or decreased in value?

FIGURE 6–37

5. If the decrease in V_{R_L} in question 4 was caused by a change in R_1 value, would it indicate R_1 has increased or decreased in value?

6. If R_L is disconnected from the voltage divider in the circuit, what is the new value of V_{R_2}?

7. What happens to V_{R_L} if R_1 opens? If R_2 shorts?

THE WHEATSTONE BRIDGE CIRCUIT

Various types of special series-parallel circuits, called **bridge circuits,** are used to make measurements in electronic circuits, Figure 6–38.

Note that zero volts appear between points A and B when the bridge is balanced. You will see this balance *only* when the ratio of the resistances in the left arm and in the right arm of the bridge are equal.

For example, observe Figure 6–38. In the left arm of the bridge, $R_1 = 10$ kΩ and $R_2 = 20$ kΩ. In the right arm of the bridge, $R_3 = 50$ kΩ and $R_4 = 100$ kΩ. If V applied is 150 V, then R_1 drops one-third the voltage across the left arm of the bridge, or 50 V (since it is one-third of the left arm's total resistance). R_2 drops the other two-thirds of the 150 volts, or 100 V.

What about the other arm of the bridge? The same 150 V is applied across that arm's 150 kΩ, with R_3 (50 kΩ) dropping one-third the applied voltage, or 50 V, and R_4 dropping two-thirds the applied voltage (R = two-thirds the right arm's total R), or 100 V.

Note the voltage differential between points A and B is zero. This is because point A is at a potential of +100 volts with respect to the power supply's negative terminal. Point B is also at a +100 volts with respect to the power supply's negative terminal. Thus, the difference of potential between points A and B is zero volts, since they are at the same potential.

This balanced condition (zero volts between points A and B) exists whenever the resistance ratios of the top resistor to the bottom resistor in the left and right arm of the bridge are the same. That is,

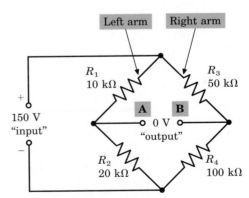

FIGURE 6–38   A balanced bridge circuit exists when $\dfrac{R_1}{R_2} = \dfrac{R_3}{R_4}$.

Formula 6–1	Bridge balanced when: $\dfrac{R_1}{R_2} = \dfrac{R_3}{R_4}$

 R_1 value, $\boxed{\div}$, R_2 value, $\boxed{=}$, (check for equaling) R_3 value, $\boxed{\div}$,
R_4 value, $\boxed{=}$

NOTE: The values of the resistors do not matter. It's the *ratio* of the resistances in each arm that determines the voltage distribution throughout that arm. Recall the series circuit concepts. If both arms have equal voltage distribution, then the midpoint of each arm is at the same potential and the bridge is balanced.

The **Wheatstone bridge circuit,** a special application of the bridge circuit, measures unknown resistance values. When the bridge circuit is balanced, the output terminals (points A and B) have zero potential difference between them. On the other hand, if the bridge circuit is unbalanced, there is a difference of potential between the output terminals. Refer to Figure 6–39 as you read the following discussion.

One method of using the Wheatstone bridge circuit to determine an unknown resistance is to:

1. Carefully select R_1 and R_2 to be matched (equal) values.

2. Place a sensitive "zero-center-scale" current meter between points A and B, which are the output terminals of the bridge circuit.

3. Use a calibrated variable resistance, such as a "resistance decade box," as R_3. NOTE: This type variable resistance has calibrated dials to enable you to read its resistance setting, Figure 6–40.

4. Place the unknown resistor (R_X) in the same arm of the circuit as the calibrated variable resistance.

5. Since R_1 and R_2 are equal value in the left arm, then the bridge is balanced when the calibrated variable resistance and the unknown resistance equal each other in the right arm. In this case, the sensitive

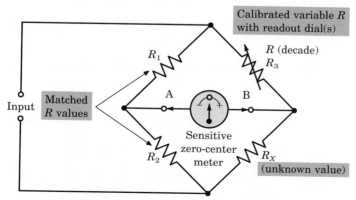

**FIGURE 6–39   Concepts of the Wheatstone
bridge circuit**

FIGURE 6-40 Decade resistor box used as a calibrated R value (Courtesy of Clarostat Mfg. Company, Inc.)

current meter's pointer is pointing at "0" in the center of the scale. If the bridge is unbalanced, the pointer points to the left of center, indicating current through the meter in one direction, or to the right of center, indicating current in the other direction. The pointer's direction is dependent on the polarity of the voltage difference between points A and B, as determined by the comparative voltages at points A and B in the bridge network.

6. Once the calibrated variable resistance is adjusted for zero current reading (balanced), the value of the unknown resistor is read directly from the calibrated variable resistance's dial(s).

Another method to determine the unknown resistance value is to use the known resistance values and the concepts of the balanced bridge circuit.

1. Recall balance occurs when $\dfrac{R_1}{R_2} = \dfrac{R_3}{R_4}$.

2. With these relationships, it is mathematically shown that:

$$R_1 \times R_4 = R_2 \times R_3 \text{ (cross-multiplying factors in formula)}$$

3. If R_4 is the unknown resistance, (R_X), the formula can be rearranged to solve R_X as follows:

Formula 6-2	$R_X = \dfrac{R_2 \times R_3}{R_1}$

 R_2 value, $\boxed{\times}$, R_3 value, $\boxed{\div}$, R_1 value, $\boxed{=}$

4. Look at Figure 6-41 with the values shown to see if the steps described above will work.

1. Balance: $\dfrac{R_1}{R_2} = \dfrac{R_3}{R_4}$; therefore $\dfrac{10}{20} = \dfrac{30}{R_X}$
2. Rearranging: $R_1 \times R_4 = R_2 \times R_3 \therefore 10\,R_X = 600\ \Omega$
3. Solving for R_X: $R_X = \dfrac{R_2 \times R_3}{R_1} = \dfrac{20 \times 30}{10} = \dfrac{600}{10} = 60\ \Omega$

FIGURE 6–41   The calculation approach may be used to solve for the R_X value in a bridge circuit.

5. Verifying results by substituting our answer into the original balanced bridge formula: $\dfrac{R_1}{R_2} = \dfrac{R_3}{R_X} = \dfrac{10}{20} = \dfrac{30}{60} = \dfrac{1}{2}$. (It checks out!)

Some Applications for Bridge Circuits

In industry, there are numerous types of sensors and control circuits used to control the operation of machinery and manufacturing systems. For example, sensors for detecting changes in temperature, flow, pressure, or changes in electrical conditions are all quite common. The output of one of these sensors can be used as one leg of a bridge circuit, with the bridge circuit "zeroed" at the desired reference level for temperature, pressure, flow, and so on, Figure 6–42. Any change in the condition being sensed will then cause the bridge to become unbalanced. This differential at the output terminals of the bridge is then typically used as an input to circuits or devices that eventually translate the change into a readable output indication, such as a digital readout, for monitoring purposes, or a feedback control signal, Figure 6–42. This system condition change can then be managed, either manually, or in many cases automatically, to bring the system back to the desired operational characteristics.

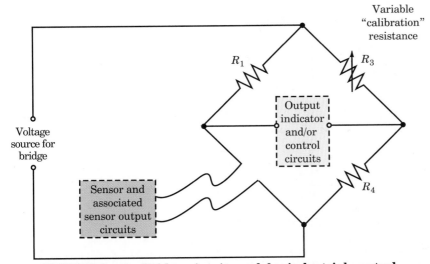

FIGURE 6–42   Bridge circuit used for industrial control

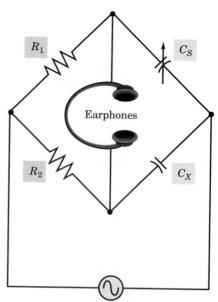

Audio signal source

**FIGURE 6–43   Bridge circuit
used to measure unknown
capacitor value**

We have already named one common application of the Wheatstone
bridge; that is, measuring the value of an unknown resistance. Variations of
the bridge circuit can also be used to measure the values of other types of
components, such as capacitors, about which you will learn details in a later
chapter. Again, the component to be measured is put into one leg of the
bridge, Figure 6–43. When the bridge is balanced, by adjusting the "standard capacitor" (C_S) to the same value as the unknown capacitor's value
(C_X), the audio tone in the earphones is "nulled" or minimized.

Still another application of a modified bridge circuit, which has been
used for years by the power and telephone companies, is the Murray loop.
The Murray loop is used to locate where an undesired conductor ground has

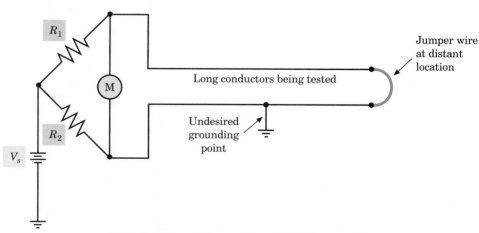

FIGURE 6–44   Example of the Murray loop

occurred along the miles of lines used to transport power or telephone messages, Figure 6–44. By knowing the resistance per foot of the conductors, and using the outgoing and return wires in a loop that forms part of the bridge circuit, the precise location of the ground fault can be determined. A service person can then be sent to the location to perform the necessary repair.

PRACTICAL NOTES

Because series-parallel circuits are combinational circuits that are connected in an infinite variety of configurations, no single analysis technique is applicable. Rather, it is necessary to break the circuit down into its series and parallel portions for analysis.

In studying this chapter, you have learned the series portions of series-parallel circuits are analyzed using series-circuit concepts, where 1) current is the same through series components and/or circuit portions, and 2) voltage drops on series components are proportional to their resistance ratios.

Also you have learned that the parallel portions of the series-parallel circuits abide by the rules of parallel circuits, where 1) voltage is the same across parallel components or combinations, and 2) current divides in inverse proportion to the parallel branch resistances.

Furthermore, you have observed that the position (location) of a given component or circuit portion in the circuit influences all the parameters throughout the circuit. Therefore, to troubleshoot series-parallel circuits, be sure to isolate individual circuit portions using concepts appropriate to that circuit portion. Then move to the next portion, and so forth, until the faulty circuit portion is found. Isolation techniques are then used in that portion to isolate the specific faulty component or device within that portion.

SUMMARY

▶ A series-parallel circuit is a combination of series and parallel connected components and/or circuit portions.

▶ Series components or circuit portions are analyzed using the rules of series circuits.

▶ Parallel components or circuit portions are analyzed using the rules of parallel circuits.

▶ Points throughout the circuit where current divides and rejoins help identify parallel portions.

▶ Points throughout the circuit that share common voltage also help identify components or circuit portions that are in parallel.

▶ Paths in the circuitry that carry the same current help identify the series circuit elements.

▶ Analysis of total circuit resistance is approached by starting away from the source, then solving and combining results as one works back toward the source. (NOTE: If V_T and I_T are known, this is unnecessary. Use $R_T = V_T/I_T$ formula).

▶ Analysis of current throughout the circuit is approached by finding the total current value. Then working outward from the source, analyze currents through various circuit portions.

▶ Power dissipations throughout the circuit are analyzed by finding the appropriate I, V, and R parameters. Then use the appropriate power formulas, such as $V \times I$, I^2R, or V^2/R.

▶ A short, or decreased value of resistance, *anywhere* in the circuit causes R_T to decrease and I_T to increase.

▶ An open, or increased value of resistance, *anywhere* in the circuit causes R_T to increase and I_T to decrease.

▶ Two special applications of series-parallel circuits are voltage dividers and bridge circuits used for various purposes.

FORMULAS AND CALCULATOR SEQUENCES SUMMARY

FORMULA 6–1   Bridge balanced when: $\dfrac{R_1}{R_2} = \dfrac{R_3}{R_4}$

R_1 value, $\boxed{\div}$, R_2 value, $\boxed{=}$, (check for equaling) R_3 value, $\boxed{\div}$, R_4 value, $\boxed{=}$

FORMULA 6–2 $R_x = \dfrac{R_2 \times R_3}{R_1}$

R_2 value, $\boxed{\times}$, R_3 value, $\boxed{\div}$, R_1 value, $\boxed{=}$

REVIEW QUESTIONS

1. Refer to Figure 6–45 and determine R_T.

FIGURE 6–45

2. Refer to Figure 6–46 and find:

I_T _____ V_3 _____ I_2 _____
V_2 _____ V_5 _____ I_3 _____

FIGURE 6–46

3. Refer to Figure 6–47 and answer the following with I, for increase; D, for decrease; or RTS, for remains the same. Assume that R_5 shorted:

R_T will _____
R_4 will _____
V_1 will _____
V_3 will _____
P_T will _____

FIGURE 6–47

4. In Figure 6–48, what is the maximum total current possible?

FIGURE 6–48

5. In Figure 6–49, what are the minimum and maximum voltages possible from point A to ground, if the setting of the potentiometer is varied? From _____ V to _____ V

FIGURE 6–49

6. In Figure 6–50, I_T is 10 amperes. Find R_1 and P_{R_2}.

FIGURE 6–50

7. In Figure 6–51, find the value of R_4 and I_2.

FIGURE 6–51

8. In Figure 6–52, find the resistance between points A and B.

FIGURE 6–52

9. In Figure 6–53, will the light get brighter, dimmer, or remain the same after the switch is closed?

FIGURE 6–53

10. Find the applied voltage in Figure 6–54.

FIGURE 6–54

11. Accurately copy the diagram, Figure 6–55, on a separate sheet of paper. When you are done, draw arrows on the diagram showing the direction of current through each resistor. Using this circuit, assume all resistors have a value of 100 kΩ. Answer the follow-

ing questions. (Round answers to whole numbers.)

a. Find R_T.
b. Find I_T.
c. Find V_{R_1}.
d. Find V_{R_2}.
e. Find V_{R_3}.
f. Find V_{R_4}.
g. Find V_{R_5}.
h. Find V_{R_6}.
i. Find V_{R_7}.
j. Find V_{R_8}.

FIGURE 6-56

FIGURE 6-55

12. Refer to the circuit in Figure 6-56 and answer the following questions:

a. If R_1 increases, what happens to the current through R_5?

b. If R_7 decreases, what happens to the voltage across R_9?

c. If R_4 shorts, what happens to the voltage across R_2? Across R_8?

d. If all resistors are 10 kΩ, what is the value of R_T? (All conditions are normal!)

e. If the circuit applied voltage is 278.5 volts, what is the voltage drop across R_1?

f. If the circuit applied voltage is reduced to half the value shown in question e, what is the power dissipation of R_9?

g. If the circuit applied voltage is 557 volts, which resistor(s) dissipate the most power?

h. Regardless of the applied voltage, which resistor(s) dissipate the least power?

i. For the conditions defined in question g, what power is dissipated by the resistor dissipating the most power? What power is dissipated by the resistor dissipating the least power?

FIGURE 6-57   Circuit for Review Questions 13-21

Refer to Figure 6-57 and answer questions 13-21.

13. Prior to connecting the voltmeter between point A and ground, what is the voltage across R_1?

14. Prior to connecting the voltmeter between point A and ground, what is the voltage across R_2?

15. Assume the voltmeter is being used on its 10-volt range and the meter's resistance on that range is 200,000 ohms. When the meter is connected to the circuit to measure the voltage between point A and ground, what voltage will the meter indicate across R_2?

16. For the conditions described in question 15, and based on Kirchhoff's voltage law, what must be the voltage drop across R_1 at that time?

17. Does connecting the meter to the circuit change the circuit's electrical parameters? Explain.

18. Explain the circuit conditions if the voltmeter were connected across R_1 rather than R_2.

19. With the meter connected across just one of the resistors, what is the total circuit current?

20. What is the total circuit current without the meter connected?

21. At what points in the circuit could a meter be connected and not affect the voltage drops across R_1 and R_2?

Refer to Figure 6–58 and answer questions 22–25.

22. Determine the values of R_{L_1}, R_{L_2}, and R_{L_3}.

23. Determine the values for the divider resistors R_1, R_2, and R_3.

24. Would the voltage divider used in this circuit be considered a "stiff" divider?

25. If the required conditions for R_{L_2} were changed so that the desired parameters were 150 V at 10 mA, would the value of R_1 in the divider need to be increased, decreased, or kept the same?

FIGURE 6–58 Circuit for Review Questions 22–25

CHALLENGE QUESTIONS

1. In your own words, describe the key characteristics of series-parallel circuits.

2. Name two methods of identifying portions of combination (series-parallel) circuits in which components or branches are in parallel.

3. Name two methods of identifying portions of combination circuits in which components are in series.

4. Describe a generally useful approach for analyzing a combination circuit in terms of the circuit's total resistance.

5. Briefly define the sequence of keystrokes or steps needed to perform calculation of $R_{equivalent}$ of three parallel resistors, using the reciprocal function of a calculator. Assume that you want to use the "engineering mode" of the calculator for this task.

6. In your own words, describe the difference between an "unloaded" and a "loaded" voltage divider circuit.

7. Explain how using a voltmeter to measure voltage across part of a series portion of a circuit causes circuit conditions to change somewhat.

FIGURE 6–59

Perform appropriate analysis of the circuit shown in Figure 6–59 and answer the following statements with I for increase, D for decrease, or RTS for remain the same:

8. If R_2 decreased in value by a great amount:
   a. The voltage across R_3 would _____
   b. The voltage across R_1 would _____
   c. The current through the load connected in parallel with R_4 would _____

9. If R_2 "opened":
   a. The voltage across the load resistor connected to the bottom of R_1 (the "outer loop" load) would _____
   b. The voltage across R_1 would _____

10. Briefly explain how an industrial sensor's output may be used in conjunction with a bridge circuit to help control industrial processes.

TEST YOURSELF

1. Determine the values of R_1, R_2, R_3, and R_4 for the circuit, Figure 6–59.

2. Determine the value of R_X in the circuit, Figure 6–60: Assume the bridge is balanced by the variable R set at 100 kΩ.

FIGURE 6–60

PERFORMANCE PROJECTS CORRELATION CHART

Suggested performance projects in the Laboratory Manual that correlate with topics in this chapter are:

CHAPTER TOPIC	PERFORMANCE PROJECT	PROJECT NUMBER
Resistance in S-P Circuits	Total Resistance in S-P Circuits	21
Current in S-P Circuits	Current in S-P Circuits	22
Voltage in S-P Circuits	Voltage Distribution in S-P Circuits	23
Power in S-P Circuits	Power Distribution in S-P Circuits	24
Effects of Opens in S-P Circuits	Effects of an Open in S-P Circuits	25
Effects of Shorts in S-P Circuits	Effects of a Short in S-P Circuits	26

NOTE: After completing above projects, perform Summary at end of Series-Parallel Circuits performance projects section in the Laboratory Manual.

THE SIMPLER SEQUENCE FOR TROUBLESHOOTING

Following the SIMPLER troubleshooting sequence, find the problem with this circuit. As you follow the sequence, record your responses to each step on a separate sheet of paper.

CHALLENGE CIRCUIT 3

SYMPTOMS Gather, verify, and analyze symptom information. (Look at the "Starting Point Information.")

IDENTIFY Identify initial suspect area for the location of the trouble.

MAKE Make a decision about "What type of test to make" and "Where to make it." (To simulate making a decision about each test you want to make, select the desired test from the "TEST" column listing below.)

PERFORM Perform the test. (Look up the result of the test in Appendix C. Use the number in the "RESULTS" column to find the test result in the Appendix.)

LOCATE Locate and define a new "narrower" area in which to continue troubleshooting.

EXAMINE Examine available information and again determine "what type of test" and "where."

REPEAT Repeat the preceding analysis and testing steps until the trouble is found. What would you do to restore this circuit to normal operation? When you have solved the problem, compare your results with those shown on the following pages.

STARTING POINT INFORMATION

1. Circuit diagram

2. I_T measures 1.4 mA

3. V_{R_1} measures 14 V

4. Resistances are measured with remainder of circuit being isolated from points being tested.

TEST   Results in Appendix C

V_{R_1}	(49)
V_{R_2}	(80)
V_{R_3}	(12)
V_{R_4}	(39)
V_{R_5}	(61)
V_{R_6}	(88)

Res. of R_1	(25)
Res. of R_2	(54)
Res. of R_3	(93)
Res. of R_4	(37)
Res. of R_5	(68)
Res. of R_6	(33)

CIRCUIT

3

Source lead

Meter lead

Symptoms

From source −

From source +

Symptoms The total current is a little low. Since current should be close to 1.7 mA, R_T is higher than it should be.

Identify initial suspect area: All resistors are initial suspect areas since if one or more resistor increased in value, the symptom would be present. However, the most likely suspect resistors and associated circuitry are related to R_2 and R_3, plus R_5 and R_6. This is because an open condition for any resistor branch would change the circuit resistance just enough to cause the symptom.

Make test decision: Check the resistance of R_2 (with it isolated).

Perform Look up the test result. R_2 is 10 kΩ, which is normal.

Locate new suspect area: R_3, R_5, and R_6.

Examine available data.

Repeat analysis and testing:

R_6
10 kΩ

R_5
10 kΩ

R_4
10 kΩ

+

V_T
50 V

−

R_2
10 kΩ

R_3
10 kΩ

mA

R_1
10 kΩ

CHALLENGE CIRCUIT 3

Measure the resistance of R_3. R_3 is 10 kΩ, which is normal.

Measure the resistance of R_5. R_5 is 10 kΩ, which is normal.

Measure the resistance of R_6. R_6 is infinite ohms. NOTE: This meter shows 30.00 MΩ when measuring an open (or an R value from 30 MΩ to ∞).

Replace R_6 and note the circuit parameters. When this is done, the circuit operates properly.

1st Test

2nd Test

3rd Test

4th Test

5th Test

From source −

From source +

SPECIAL NOTE: Another way to check is to lift one end of each resistor and observe whether total current changed. The resistor you lift without having the total current change is the open one, since opening an open does not change the circuit conditions.

Basic Network Theorems

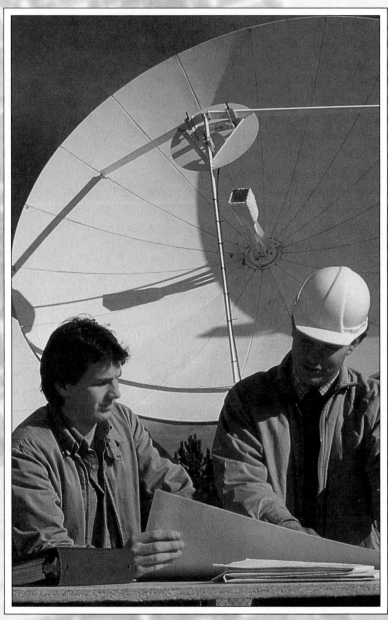

7

KEY TERMS

Bilateral resistance

Constant current source

Constant voltage source

Efficiency

Impedance

Linear network

Maximum power transfer theorem

Multiple-source circuit

Network

Norton's theorem

Steady-state condition

Superposition theorem

Thevenin's theorem

OUTLINE

Some important terms

Maximum power transfer theorem

Superposition theorem

Thevenin's theorem

Norton's theorem

Converting Norton and Thevenin equivalent parameters

CHAPTER PREVIEW

In this chapter, several commonly used electrical network theorems will be discussed. These include **maximum power transfer, superposition, Thevenin's,** and **Norton's** theorems. For our purposes, a network is defined as a combination of components that are electrically connected. Theorems are ideas or statements that are not obvious at first, but that we can prove using some more accepted premises. When the theorems are proved this way, they also become laws. The theorems you will learn about in this chapter can be proven in theory and in practice.

The theorems in this chapter deal with **bilateral resistances** that are in **linear networks** under **steady-state conditions,** not during transient switch on or off moments.

Using the maximum power transfer theorem, you can know what value of load resistance (or impedance) allows maximum transfer of power from a source to its load, (e.g., what speaker impedance in your stereo allows transfer of maximum power from amplifier to speaker). Using the superposition theorem, you can readily analyze circuits having more than one source.

One of the advantages in using Thevenin's theorem is once the Thevenin equivalent circuit has been determined, you can easily predict electrical parameters for a variety of load resistances (R_L) without having to recalculate the whole network each time.

Norton's theorem provides another approach to simplify complex networks and make circuit parameter analysis easier.

OBJECTIVES

After studying this chapter, you should be able to:

- State the **maximum power transfer theorem**
- Determine the R_L value needed for maximum power transfer in a given circuit
- State the **superposition theorem**
- Solve circuit parameters for a circuit having more than one source
- State **Thevenin's theorem**
- Determine V_L and I_L for various values of R_L connected across specified points in a given circuit or network
- State **Norton's theorem**
- Apply Norton's theorem in solving specified problems
- Convert between Norton and Thevenin equivalent parameters

SOME IMPORTANT TERMS

In the Chapter Preview we mentioned several terms that should be defined at this point. We indicated that the theorems you will study in this chapter relate to circuits that are linear networks having bilateral resistances that are being operated under steady-state conditions. These are rather high-sounding words, but their meanings are quite simple. Let's briefly look at definitions for each term:

- **Bilateral resistance**—a resistance having equal resistance in either direction. That is, the resistance is the same for current passing either way through the component. This type component produces the same voltage drop across it, no matter which direction the current flows through it.

- **Linear network**—a circuit whose electrical behavior does not change with different voltage or current levels or values.

- **Steady-state condition**—the condition where circuit values and conditions are stable or constant. The opposite of transient conditions, such as switch-on or switch-off conditions, when values are changing from one state or level to another.

MAXIMUM POWER TRANSFER THEOREM

In many electronic circuits and systems, it is important to have maximum transfer of power from the source to the load. For example, in radio or TV transmitting systems, it is desired to transfer the maximum power possible from the last stage of the transmitter to the antenna system. In your high fidelity amplifier, you want maximum power transferred from the last amplifier stage to the speaker system. This is usually accomplished by proper "**impedance** matching," which means the load resistance or impedance matches the source resistance or impedance. NOTE: In dc circuits, such as you have been studying, the term for opposition to current flow is resistance. In ac circuits, which you will soon study, the term for opposition to current flow is impedance.

If it were possible to have a voltage source with zero internal resistance but able to maintain constant output voltage (even if a load resistance of zero ohms were placed across its output terminals), the source theoretically could supply infinite power to the load, where $P = V \times I$. In this case, V would be some value and I would be infinite. Thus, $P = V$ value $\times$ infinite current = infinite power in watts.

Since these conditions are not possible, the practical condition is that there is an optimum value of R_L where maximum power is delivered to the load. Any higher or lower value of R_L causes less power to be delivered to the load.

The Basic Theorem

The **maximum power transfer** theorem states maximum power is transferred from the source to the load (R_L) when the resistance of the load *equals* the resistance of the source. In other words, it is a series circuit. In practical

**FIGURE 7–1   Maximum power transfer theorem
(NOTE: R_T seen by source is the series total
of $r_{int} + R_L$.)**

circuits, the internal resistance (r_{int}) of the source causes some of the source's power to be dissipated within the source, while the remaining power is delivered to the external circuitry, Figure 7–1.

PRACTICAL NOTES

It is important to realize that all power sources have some internal resistance, for which we are using the designation r_{int} in our discussions. We are showing you the r_{int} as an internal source series resistance, which alters the source output voltage delivered to the load, and limits the maximum amount of current that can be delivered by the source, Figure 7–1.

Refer to the circuits in Figure 7–2a, b, and c. Assuming the same source for all three circuits, notice the power delivered to the load in each case.

Observe in the first circuit, Figure 7–2a, the load resistance (R_L) equals the internal resistance of the source (r_{int}). The power delivered to the load is 10 watts.

In the circuit of Figure 7–2b, the load resistance is less than the source's internal resistance. The power delivered to the load is 8.84 watts.

In the third circuit, Figure 7–2c, R_L is greater than r_{int}, and the power delivered to the load is 7.5 watts. You can make calculations for a large number of R_L values to confirm that maximum power is delivered when $R_L = r_{int}$.

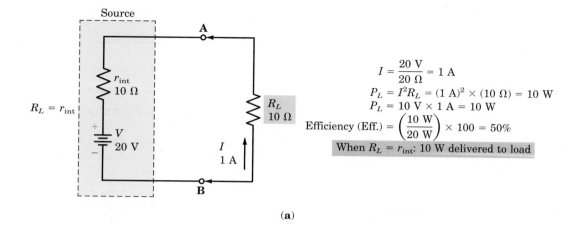

$$I = \frac{20 \text{ V}}{20 \text{ }\Omega} = 1 \text{ A}$$
$$P_L = I^2 R_L = (1 \text{ A})^2 \times (10 \text{ }\Omega) = 10 \text{ W}$$
$$P_L = 10 \text{ V} \times 1 \text{ A} = 10 \text{ W}$$
$$\text{Efficiency (Eff.)} = \left(\frac{10 \text{ W}}{20 \text{ W}}\right) \times 100 = 50\%$$

When $R_L = r_{\text{int}}$: 10 W delivered to load

(**a**)

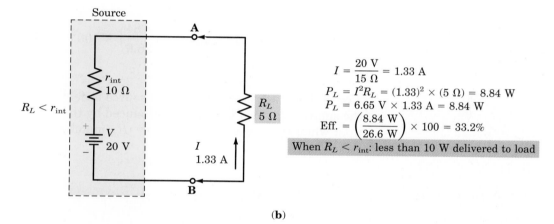

$$I = \frac{20 \text{ V}}{15 \text{ }\Omega} = 1.33 \text{ A}$$
$$P_L = I^2 R_L = (1.33)^2 \times (5 \text{ }\Omega) = 8.84 \text{ W}$$
$$P_L = 6.65 \text{ V} \times 1.33 \text{ A} = 8.84 \text{ W}$$
$$\text{Eff.} = \left(\frac{8.84 \text{ W}}{26.6 \text{ W}}\right) \times 100 = 33.2\%$$

When $R_L < r_{\text{int}}$: less than 10 W delivered to load

(**b**)

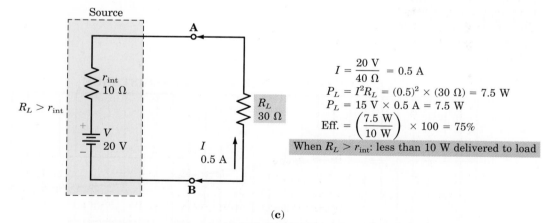

$$I = \frac{20 \text{ V}}{40 \text{ }\Omega} = 0.5 \text{ A}$$
$$P_L = I^2 R_L = (0.5)^2 \times (30 \text{ }\Omega) = 7.5 \text{ W}$$
$$P_L = 15 \text{ V} \times 0.5 \text{ A} = 7.5 \text{ W}$$
$$\text{Eff.} = \left(\frac{7.5 \text{ W}}{10 \text{ W}}\right) \times 100 = 75\%$$

When $R_L > r_{\text{int}}$: less than 10 W delivered to load

(**c**)

FIGURE 7–2   Verification of maximum power transfer theorem

Efficiency Factor

Formula 7–1	Efficiency (%) $= \dfrac{P_{out}}{P_{in}} \times 100$

P_{out} value, $\boxed{\div}$, P_{in} value, $\boxed{\times}$, 100, $\boxed{=}$

If you calculate the **efficiency** factor, based on what percentage of total power generated by the source is delivered to the load, you will find efficiency equals 50% when R_L equals r_{int}.

In Figure 7–2a, $V = 20$ V and $I = 1$ A. Thus, P_T (or P_{in}) = 20 V × 1 A = 20 W. The load is dissipating $(1 \text{ A})^2 (10 \ \Omega) = 10$ W, or half of the 20 watts.

In Figure 7–2b, $V = 20$ V and $I = 1.33$ A. Thus, $P_T = 20$ V × 1.33 A = 26.6 W, but only 8.84 W is delivered to the load. Therefore,

$$\text{Efficiency} = \left(\frac{8.84}{26.6}\right) \times 100 = 33.2\%$$

On the other hand, in Figure 7–2c, $P_T = 10$ W and $P_L = 7.5$ W. Thus, 75% of the power generated by the source is delivered to the load. However, the smaller value of total power produced by the source causes P_L to be less than when there was 50% efficiency (R_L matched r_{int}).

PLOT OF LOAD POWER VS. LOAD RESISTANCE

Maximum power transferred when $R_L = R_{int}$

Load power in watts

Load resistance in ohms

FIGURE 7–3

Where is the remainder of the power consumed in each case? If you said by the internal resistance of the source, you are right! In other words, when there is maximum power transfer, half the total power of the source is dissipated by its own internal resistance; the other half is dissipated by the external load. When the efficiency is about 33%, Figure 7−2b, 67% of the total power is dissipated within the source. When efficiency is 75%, Figure 7−2c, 25% of the total power is dissipated within the source, while 75% is delivered to the load.

Summary of the Maximum Power Transfer Theorem

- Maximum power transfer occurs when the load resistance equals the source resistance.

- Efficiency at maximum power transfer is 50%.

- When R_L is greater in value than r_{int}, efficiency is greater than 50%.

- When R_L is less than r_{int}, efficiency is less than 50%.

To visually see the effects on load power and efficiency when R_L is varied, using a given source, observe the graphs in Figures 7−3 and 7−4.

FIGURE 7−4

PRACTICAL NOTES

You may or may not have noticed the amount of voltage at the output terminals of the source varies with load current. This is due to the internal *IR* drop across r_{int}. That is, higher currents result in greater internal *IR* drop and less voltage available at the output terminals of the source. In fact, one way to obtain an idea of a source's internal resistance is to measure its open circuit output terminal voltage, then put a load that demands current from the source across its terminals. The change in output terminal voltage from no-load (zero current drain) to loaded conditions allows calculation of r_{int}.

Formula 7–2	$r_{int} = \dfrac{V \text{ (no load)} - V \text{ (loaded)}}{I \text{(load)}}$

V (no load) value, $\boxed{-}$, V (loaded) value, $\boxed{=}$, $\boxed{\div}$,

I (load) value, $\boxed{=}$

If V (no-load) is 13 V and V (loaded) is 12 V with 100 A (load) current, then:

$$\frac{13 \text{ V} - 12 \text{ V}}{100 \text{ A}} = 0.01 \ \Omega$$

A practical example of this approach to determine if a source's internal resistance is higher than it should be is the automotive technician's technique of checking a battery's output voltage at no-load, then under heavy load. If the terminal voltage of the battery drops too much under load, its internal resistance is too high, and the battery is either defective or needs charging.

FIGURE 7–5

PRACTICE PROBLEMS I

Try solving the maximum power transfer problems in Figure 7–5.

SUPERPOSITION THEOREM

Frequently, you will analyze circuitry having more than one source. The superposition theorem is a useful tool in these cases. The **superposition theorem** states that in linear circuits having more than one source, the voltage across or current through any given element equals the algebraic sum of voltages or currents produced by each source acting alone with the other sources disabled, Figure 7–6. NOTE: To disable voltage sources for purposes of calculations, consider them shorted. To disable current sources, consider them opened.

EXAMPLE

A circuit containing two voltage sources—**a multiple-source circuit**—is illustrated in Figure 7–6a. In Figures 7–6b through 7–6d, the step-by-step application of the superposition theorem is shown. *Important:* **Noting the direction of current flow and polarity of voltage drops is important for each step.**

In Figure 7–6b, the 60-V source is considered disabled (shorted out). Notice this effectively places R_3 in parallel with R_2. This means the R_1 value of 2 kΩ is in series with the parallel combination of $R_2 - R_3$ across the 30-V source. Thus, total R is 3 kΩ. Total I is 10 mA. R_1 passes the total 10 mA, whereas, the 10 mA is equally divided through R_2 and R_3 with 5 mA through each one. Using Ohm's law, the voltages are computed as: $V_1 = +20$ V with respect to point A; $V_2 = -10$ V with respect to point A; and $V_3 = +10$ V with respect to point B.

In Figure 7–6c, the 30-V source is considered disabled. In this case, R_1 and R_2 are in parallel with each other and in series with R_3 across the 60-V source. Notice the voltage drops are such that $V_1 = +20$ V with respect to point A; $V_2 = +20$ V with respect to point A; and $V_3 = +40$ V with respect to point B.

In Figure 7–6d, you see the algebraically computed results. Recall V_1 is +20 V with respect to A with the 60-V source disabled and +20 V with the 30-V source disabled. This means the voltages are additive and the resultant voltage across R_1 equals +40 V with respect to point A. This same type addition also determines the resultant value of V_2 equals +10 V with respect to point A, and the resultant value of V_3 equals +50 V with respect to point B.

In a similar manner, directions and values of the currents through each component are determined, using the algebraic summation technique. For example, note in Figure 7–6b, current flows from point B toward point A through resistor R_2 and has a value of 5 mA. In Figure 7–6c, current flows

Circuit with two sources

(a)

Assuming 60-V source shorted out

NOTE: If you were going to physically do this, you would *remove* the 60-V source and replace it with a conductor.

Step 1: Assume 60-V source is replaced by a short (0Ω).

Step 2: Analyze R_T of circuit:

a. $R_2 \| R_3 = \dfrac{2\text{ k}\Omega \times 2\text{ k}\Omega}{2\text{ k}\Omega + 2\text{ k}\Omega} = \dfrac{4 \times 10^6}{4 \times 10^3} = 1\text{ k}\Omega$

b. R_e of $R_2 \| R_3$ is in series with R_1

c. $R_T = R_1 + R_e = 2\text{ k}\Omega + 1\text{ k}\Omega = 3\text{ k}\Omega$

Step 3: Solve for total current and branch currents:

a. $I_T = \dfrac{30\text{ V}}{3\text{ k}\Omega} = 10\text{ mA}$

b. Since each branch = 2 kΩ, current divides equally:

$I_{R_2} = 5\text{ mA}; I_{R_3} = 5\text{ mA}$

Step 4: Use Ohm's law and solve the voltage drops:

a. $V_1 = I_T \times R_1 = 10\text{ mA} \times 2\text{ k}\Omega = +20\text{ V}$ (with respect to point A)

b. $V_2 = I_{R_2} \times R_2 = 5\text{ mA} \times 2\text{ k}\Omega = -10\text{ V}$ (with respect to point A)

c. $V_3 = I_{R_3} \times R_3 = 5\text{ mA} \times 2\text{ k}\Omega = +10\text{ V}$ (with respect to point B)

FIGURE 7–6 **Verification of superposition theorem**
(b)

from point A toward point B through R_2 and has a value of 10 mA. This means the resultant current through R_2 with both sources active is 5 mA traveling from point A toward point B. You get the idea! Refer to Figure 7–6d to see the other resultant values, as appropriate.

EXAMPLE

If the polarity of each of the sources in Figure 7–6 is reversed (while the circuit configuration remains the same), the results are like those in Figure 7–7.

The same analysis is used. That is, disable each voltage source, one at a time by viewing them as shorted. Then determine the currents and voltages throughout the circuit, using Ohm's law and so on.

Assuming 30-V source shorted out

Resultant circuit Vs and Is with two sources

NOTE: To physically perform this step, you would *remove* the 30-V source and replace it with a "short."

Step 5: Assume 30-V source is replaced by a short (0Ω).

Step 6: Analyze R_T of circuit:

a. $R_1 \| R_2 = \dfrac{2 \text{ k}\Omega \times 2 \text{ k}\Omega}{2 \text{ k}\Omega + 2 \text{ k}\Omega} = \dfrac{4 \times 10^6}{4 \times 10^3} = 1 \text{ k}\Omega$

b. R_e of $R_1 \| R_2$ is in series with R_3

c. $R_T = R_3 + R_e = 2 \text{ k}\Omega + 1 \text{ k}\Omega = 3 \text{ k}\Omega$

Step 7: Solve for total current and branch currents:

a. $I_T = \dfrac{60 \text{ V}}{3 \text{ k}\Omega} = 20 \text{ mA}$

b. Since each branch = 2 kΩ, current divides equally:
$I_{R_2} = 10 \text{ mA}; I_{R_3} = 10 \text{ mA}$

Step 8: Use Ohm's law and solve the voltage drops:

a. $V_1 = I_1 \times R_1 = 10 \text{ mA} \times 2 \text{ k}\Omega = +20 \text{ V}$ (with respect to point A)

b. $V_2 = I_{R_2} \times R_2 = 10 \text{ mA} \times 2 \text{ k}\Omega = +20 \text{ V}$ (with respect to point A)

c. $V_3 = I_T \times R_3 = 20 \text{ mA} \times 2 \text{ k}\Omega = +40 \text{ V}$ (with respect to point B)

(c)

Step 9: Algebraically add voltages and currents from the two sets of computations to find resultant component electrical parameters:

a. $V_1 = +20 \text{ V} + 20 \text{ V} = +40 \text{ V}$ (with respect to point A)

b. $V_2 = -10 \text{ V} + 20 \text{ V} = +10 \text{ V}$ (with respect to point A)

c. $V_3 = +10 \text{ V} + 40 \text{ V} = +50 \text{ V}$ (with respect to point B)

d. I_{R_1} resultant current = 20 mA (away from point A)

e. I_{R_2} resultant current = 5 mA (away from point A)

f. I_{R_3} resultant current = 25 mA (away from point B)

(d)

FIGURE 7–6   Continued

With the 60-V source assumed as shorted, V_1 equals -20 V with respect to point A; V_2 equals $+10$ V with respect to point A; and V_3 equals -10 V with respect to point B, Figure 7–7b.

With the 30-V source assumed as shorted, V_1 equals -20 V with respect to point A; V_2 equals -20 V with respect to point A; and V_3 equals -40 V with respect to point B, Figure 7–7c.

Combining the results of the two assumed conditions, V_1 equals -40 V with respect to point A; V_2 equals -10 V with respect to point A; and V_3 equals -50 V with respect to point B, Figure 7–7d.

As you can see, when only the polarity of the sources changes, the resultant direction of currents and polarities of the voltage drops reverses but their values remain the same. Compare Figure 7–6d and 7–7d.

FIGURE 7–7   Superposition theorem example: reversed polarities

Summary of the Superposition Theorem

- Ohm's law is used to analyze the circuit, using one source at a time.
- Then the final results are determined by algebraically superimposing the results of all the sources involved.

PRACTICE PROBLEMS II

For practice in using the superposition theorem, try the following problems:

1. Solve for the resultant current(s) and voltage(s) for the circuit in Figure 7–8.

2. Assume only the 150-volt source is reversed in polarity and solve for the resultant current(s) and voltage(s). **Caution: Only one source is being reversed!**

FIND
I through R_1
V across R_1, with respect to Point **A**
I through R_2
V across R_2, with respect to Point **A**
I through R_3
V across R_3, with respect to Point **A**

FIGURE 7–8

IN-PROCESS LEARNING CHECK I

Fill in the blanks as appropriate.

1. For maximum power transfer to occur between source and load, the load resistance should _____ the source resistance.

2. The higher the efficiency of power transfer from source to load, the _____ the percentage of total power is dissipated by the load.

3. Maximum power transfer occurs at _____ % efficiency.

4. If the load resistance is less than the source resistance, efficiency is _____ than the efficiency at maximum power transfer.

5. To analyze a circuit having two sources, the superposition theorem indicates that Ohm's law _____ be used.

6. What is the key method in using the superposition theorem to analyze a circuit with more than one source?

_____

7. Using the superposition theorem, if the sources are considered "voltage" sources, are these sources considered shorted or opened during the analysis process? _____ .

8. When using the superposition theorem when determining the final result of your analysis, the calculated parameters are combined, or superimposed, _____ .

THEVENIN'S THEOREM

Thevenin's theorem is used to simplify complex **networks** so the calculation of voltage(s) across any two given points in the circuit (and consequently, current(s) flowing between the two selected points) is easily determined.

Once the Thevenin equivalent circuit parameters are known, solving circuit parameters for *any value* of R_L merely involves solving a two-resistor series circuit problem. A **constant voltage source** is assumed. This is a source whose output voltage remains constant under varying load current (or current demand) conditions.

As you recall, voltage distribution was considered when you studied series circuits. Thevenin's theorem deals with a simple equivalent series circuit to simplify voltage distribution analysis. Let's see how this is applied.

The Theorem Defined

One version of **Thevenin's theorem** states that any **linear** two-terminal **network** (of resistances and source[s]) can be replaced by a simplified equivalent circuit consisting of a single voltage source (called V_{TH}) and a single series resistance (called R_{TH}), Figure 7–9.

V_{TH} is the open-terminal voltage felt across the two load terminals (A and B) with the load resistance removed. R_{TH} is the resistance seen when looking back into the circuit from the two load terminals (without R_L connected) with voltage sources replaced by their internal resistance. For simplicity, we assume them shorted.

Refer to Figures 7–10a, b, c, and d. Figure 7–10a shows the original circuit from which we will determine the Thevenin equivalent circuit, a process called "Thevenizing" the circuit.

Figure 7–10b illustrates how the value for V_{TH} is determined. Notice R_L is considered opened (between points A and B). V_{TH} is considered the value of voltage appearing at network terminals A and B (with R_L removed). For example, if Ohm's law is used, R_T is 100 Ω; V_T is 100 V; thus I_T = 1 A. Each of the 25-Ω series resistors drops 25 V (1 A × 25 Ω). The 1 A evenly splits between the two 100-Ω parallel resistors. Thus, voltage between points A and B = 0.5 A × 100 Ω, or 50 V. Therefore, V_{TH} = 50 V.

Figure 7–10c shows the process to find R_{TH}. Mentally short the source and determine the circuit resistance by looking toward the source from the two network points A and B. With the source shorted, the two 100-Ω parallel resistors are effectively in parallel with 50 Ω (25 Ω + 25 Ω). The equivalent resistance of two 100-Ω resistors is 50 Ω, and that 50 Ω in parallel with the remaining 50-Ω branch yields an R_T of 25 Ω. This is the value of R_{TH}.

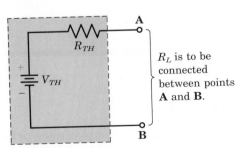

FIGURE 7–9   Thevenin's theorem
equivalent circuit

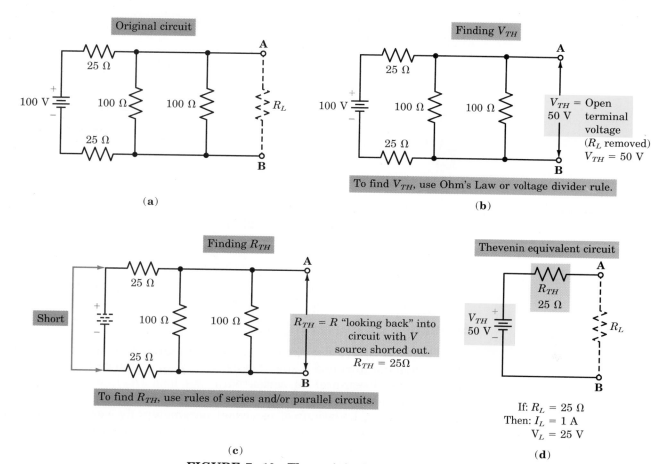

FIGURE 7–10 Thevenin's theorem illustration

Figure 7–10d shows the Thevenin equivalent circuit of a 50-V source (V_{TH}) and the series R_{TH} of 25 Ω. It is thus possible to determine the current through any value of R_L and the voltage across R_L by analyzing a simple two-resistor series circuit.

EXAMPLE

If R_L has a value of 25 Ω, what are the values of I_L and of V_L? If you answered 1 A and 25 V, you are correct!
Using formula 7–3, solve this way:

$$I_L = \frac{50 \text{ V}}{25 \text{ Ω} + 25 \text{ Ω}} = \frac{50 \text{ V}}{50 \text{ Ω}} = 1\text{A}$$

Formula 7–3	$I_L = \dfrac{V_{TH}}{R_{TH} + R_L}$

 V_{TH} value, $\boxed{\div}$, $\boxed{(}$, R_{TH} value, $\boxed{+}$, R_L value, $\boxed{)}$, $\boxed{=}$

V_L is calculated by using Ohm's law $(I_L \times R_L)$, or the voltage divider rule where:

Formula 7–4	$V_L = \dfrac{R_L}{R_L + R_{TH}} \times V_{TH}$

 R_L value, $\boxed{\div}$, $\boxed{(}$, R_L value, $\boxed{+}$, R_{TH} value, $\boxed{)}$, $\boxed{=}$, $\boxed{\times}$, V_{TH} value, $\boxed{=}$

Using formula 7–4, solve this way:

$$V_L = \frac{25\ \Omega}{25\ \Omega + 25\ \Omega} \times 50\ \text{V} = \frac{25\ \Omega}{50\ \Omega} \times 50\ \text{V} = 0.5 \times 50\ \text{V} = 25\ \text{V}$$

Summary of Thevenin's Theorem

The Thevenin equivalent circuit consists of a source, V_{TH}, (often called the Thevenin generator) and a single resistance, R_{TH}, (often called the Thevenin internal resistance) in series with the load (R_L), which is across the specified network terminals. Therefore, this equivalent circuit is useful to simplify calculations of electrical parameters for various values of R_L since it reduces the problem to a simple two-resistor series circuit analysis.

EXAMPLE

Let's assume that we are to put a load resistance (R_L) across the output terminals (points A and B) of a Wheatstone bridge circuit, Figure 7–11. Further, we would like to determine the V and I parameters associated with the load resistor. By using Thevenin's theory, we can define the simple Thevenin equivalent circuit for the bridge. Once that is found, simple two-resistor series circuit analysis techniques make it easy to find the V and I parameters for any given R_L value.

Refer to Figure 7–11 as you study the following steps.

1. Find V_{TH} (the difference of potential between points A and B, with the load resistor removed):

$$V_{TH} = V_A - V_B = \left(\frac{R_2}{R_1 + R_2}\right) \times V_S - \left(\frac{R_4}{R_3 + R_4}\right) \times V_S$$

$$V_{TH} = \left(\frac{220\ \Omega}{320\ \Omega}\right) \times 12\ \text{V} - \left(\frac{470\ \Omega}{800\ \Omega}\right) \times 12\ \text{V}$$

$$V_{TH} = 8.25\ \text{V} - 7.05\ \text{V} = 1.2\ \text{V}$$

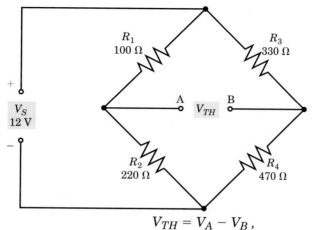

$$V_{TH} = V_A - V_B,$$

where (using the voltage divider method): $V_A = \left(\dfrac{R_2}{R_1 + R_2}\right) \times V_S$

$$V_B = \left(\dfrac{R_4}{R_3 + R_4}\right) \times V_S$$

$$V_{TH} = 1.2 \text{ V}$$

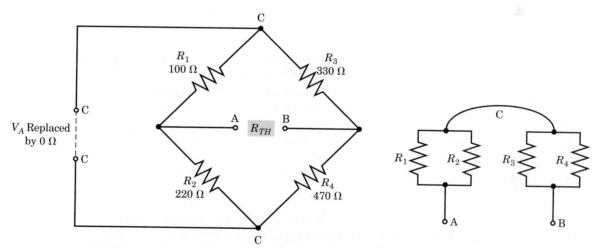

FIGURE 7–11   Analyzing a bridge circuit with Thevenin's theorem

2. Find R_{TH}: By assuming source to be zero ohms:

$R_{TH} = R_e$ of $R_1 \parallel R_2 + R_e$ of $R_3 \parallel R_4 \cong 263 \ \Omega$, as shown below.

$$R_{TH} = \frac{R_1 \times R_2}{R_1 + R_2} + \frac{R_3 \times R_4}{R_3 + R_4}$$

$$R_{TH} = \frac{100 \ \Omega \times 220 \ \Omega}{100 \ \Omega + 220 \ \Omega} + \frac{330 \ \Omega \times 470 \ \Omega}{330 \ \Omega + 470 \ \Omega}$$

$$R_{TH} = 68.75 \ \Omega + 193.875 \ \Omega = 262.625 \ \Omega$$

FIGURE 7–12   Thevenin equivalent circuit

3. Draw the equivalent circuit. See Figure 7–12.

Let's assume that an R_L value of 270 Ω is to be connected between points A and B of the bridge circuit. What is the value of V_{R_L} and of I_L?

$$V_{R_L} = \left(\frac{R_L}{R_{TH} + R_L}\right) \times V_{TH} = \left(\frac{270\ \Omega}{263\ \Omega + 270\ \Omega}\right) \times 1.2\ \text{V} \approx 0.61\ \text{V}$$

$$I_L = \frac{V_{TH}}{R_T} = \frac{1.2\ \text{V}}{533\ \Omega} = 2.25\ \text{mA}$$

PRACTICAL NOTES

In a practical circuit, you can remove the load resistor (R_L) and measure the open-terminal voltage at those two points with a voltmeter to find V_{TH}. Then you can remove the power supply from the circuit and short the points where its output terminals are connected. Next, you can use an ohmmeter to measure the circuit resistance (at the open-terminal points) by looking at where the source points are shorted. Thus, the measured value of R is R_{TH}.

As you have seen, one example of a practical application of Thevenin's theorem is in analyzing an unbalanced bridge circuit. Try applying your Thevenin's theorem knowledge to the following practice problems.

FIGURE 7–13

PRACTICE PROBLEMS III

1. Refer again to Figure 7–10. What are the values of I_L and V_L, if R_L is 175 Ω?

2. In Figure 7–10, what value of R_L must be present if I_L is 125 mA?

3. Assume that the parameters in our sample bridge circuit, Figure 7–11, were changed so that $V_A = 18$ V, $R_1 = 220$ Ω, $R_2 = 330$ Ω, $R_3 = 470$ Ω, and $R_4 = 560$ Ω. If R_L is 470 Ω, what are the values of I_L and V_{R_L}?

4. In Figure 7–13, what are the values of V_{TH} and R_{TH}? What are the values of I_L and V_L, if R_L equals 16 kΩ?

NORTON'S THEOREM

Norton's theorem is another method to simplify complex networks into manageable equivalent circuits for analysis. Again, the idea is to devise the equivalent circuit so parameter analysis at two specified network terminals (where R_L is connected) is easy.

Thevenin's theorem uses a voltage-related approach with a voltage source/series resistance equivalent circuit where voltage divider analysis is applied. On the other hand, Norton's theorem uses a current-related approach with a current source/shunt resistance equivalent circuit where a current-divider analysis approach is applied.

The Theorem Defined

Norton's theorem says any linear two-terminal network can be replaced by an equivalent circuit consisting of a single (**constant**) **current source** (called I_N) and a single shunt (parallel) resistance (called R_N), Figure 7–14.

FIGURE 7–14   Norton's theorem
equivalent circuit

FIGURE 7–15   Norton's I_N = Current from A to B, if R_L shorted (0 Ω). (In this case, I_N would equal 100 V/50 Ω = 2 A.)

Refer to Figure 7–15, points A and B. I_N is the current available from the Norton equivalent current source. I_N is determined as the current flowing between the load resistance terminals, if R_L were shorted (i.e., if R from points A to B is zero ohms). R_N is the resistance of the Norton equivalent shunt resistance. This value is computed as the R value seen when looking in the network with the load terminals open (R_L not present) and sources of *voltage* replaced by their R_{int}, Figure 7–16.

This method is the same one used to calculate R_{TH} in Thevenin's theorem. In fact, the R_{TH} and R_N values are the same. The only difference is that in Norton's theorem, this resistance value is considered in parallel (or shunt) with the I_N equivalent current source. In Thevenin's theorem, this resistance value is considered in series with the V_{TH} equivalent voltage source.

Observe the similarities and differences between the two types of equivalent circuits in Figure 7–17b and 7–17c. Note these equivalent circuits are derived from the same original circuit used to explain Thevenin's theorem earlier in the chapter.

Correlating the prior thinking of the Thevenin analysis to the Norton analysis, let's assume the same 25-Ω R_L (load resistance) used in Figure 7–10d, and apply the Norton equivalent circuit approach.

First, let's see how the Norton equivalent circuit parameters, Figure 7–17c, are derived. Analysis starts with the original circuit, Figure 7–18a.

I_N equals current between points A and B with R_L shorted. This short also shorts the two 100-Ω parallel resistors, so the source sees a total circuit R of 50 Ω (the two 25-Ω resistors in series, through the short). Thus, current between points A and B is 100 V/50 Ω, or 2 A (I_N = 2 A), Figure 7–18b.

FIGURE 7–16   Norton's R_N = R_{A-B} with R_L removed and source considered as a short. (In this case, R_N = 25 Ω.)

(a)

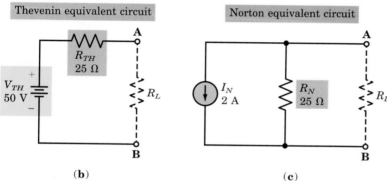

(b)   (c)

**FIGURE 7–17   Comparisons of Thevenin and Norton
equivalent circuits**

R_N equals R looking back into the network with points A and B open
and with *voltage* source(s) shorted. R_N thus equals the equivalent resistance
of the two 100-ohm parallel resistors (or 50 Ω) in parallel with the two 25-Ω
series resistors' branch (or 50 Ω). This yields a resultant circuit equivalent
resistance of 25 Ω (R_N = 25 Ω), again see Figure 7–18b.

Now, refer to Figure 7–18c to see how this equivalent circuit is used to
determine electrical parameters associated with various values of R_L.

If an R_L value of 25 Ω is placed from terminal A to B, it is apparent the
Norton equivalent circuit current of 2 A equally divides between R_N and
R_L, since they are equal value parallel branches. This means I_L = 1 A. V_L
is easily calculated using Ohm's law where, $I_L \times R_L$ = 1 A × 25 Ω = 25 V.
It is interesting to note the I_L and V_L computed using Norton's theorem are
exactly the same values as those calculated for a load resistance of 25 Ω
when using Thevenin's theorem for the same original circuit, Figure 7–10.
It makes sense that the same R_L connected to the same circuit has the same
operating parameters, regardless of the method used to analyze these parameters!

___

EXAMPLE

Suppose the R_L value changes to 75 Ω. The Norton equivalent circuit is the
same. That is, I_N = 2 A and R_N = 25 Ω. With the higher value of R_L, I_L
obviously is a smaller percentage of the I_N value.

Using the current divider rule, which is derived from the fact current divides inverse to the resistance of the branches, we can determine I_L from the following formula:

$$\text{Formula 7-5} \quad\bigg|\quad I_L = \frac{R_N}{R_N + R_L} \times I_N$$

R_N value, $\boxed{\div}$, $\boxed{(}$, R_N value, $\boxed{+}$, R_L value, $\boxed{)}$, $\boxed{=}$, $\boxed{\times}$, I_N value, $\boxed{=}$

With $R_L = 75\ \Omega$ and $R_N = 25\ \Omega$,

$$I_L = \frac{25\ \Omega}{25\ \Omega + 75\ \Omega} \times 2\ A = 0.25 \times 2\ A = 0.5\ A$$

To find V_L, again use Ohm's law where $V_L = 0.5\ A \times 75\ \Omega = 37.5\ V$

Original circuit

(a)

Norton equivalent circuit

(b)   (c)

I_L by current divider rule:

$$I_L = \frac{R_N}{R_N + R_L} \times I_N$$

$$I_L = \frac{25}{(25 + 25)} \times 2 = 0.5 \times 2 = 1\ A$$

FIGURE 7-18   Application sample of Norton's theorem

PRACTICE PROBLEMS IV

In Figure 7–18a, assume R_L is 60 Ω. What are the values of I_L and V_L?

Summary of Norton's Theorem

Norton's equivalent circuit is comprised of a constant current source (I_N) and a shunt resistance called R_N. When a load (R_L) is connected to this equivalent circuit, the current (I_N) divides between R_N and R_L according to the rules of parallel circuit current division (i.e., inverse to the resistances). The simple two-resistor current divider rule is applied to determine the current through the load resistance. Try the following Norton's theorem problems.

PRACTICE PROBLEMS V
---

1. Use Norton's theorem and determine the values of I_L and V_L for the circuit in Figure 7–18a, if R_L has a value of 100 Ω.

2. Refer again to Figure 7–18a. Will I_L increase or decrease if R_L changes from 25 Ω to 50 Ω? Will V_L increase or decrease?

CONVERTING NORTON AND THEVENIN EQUIVALENT PARAMETERS

Since we have been discussing the similarities and differences between the two theorems, it is appropriate to show you some simple conversions that can be done between the two.

You already know R_{TH} and R_N are the same value for a given network or circuit. You may have noticed if you multiply $I_N \times R_N$, the answer is the same as the value of V_{TH}.

Let's summarize these thoughts to show you the conversions possible. Finding Thevenin parameters from Norton parameters:

$$R_{TH} = R_N$$
$$V_{TH} = I_N \times R_N$$

Finding Norton parameters from Thevenin parameters:

$$R_N = R_{TH}$$
$$I_N = \frac{V_{TH}}{R_{TH}}$$

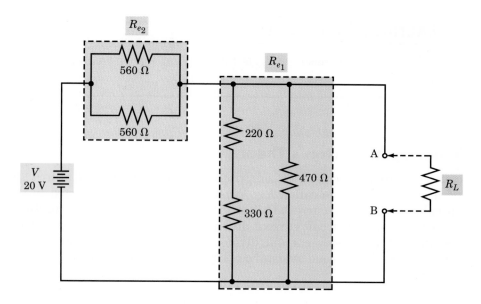

Thevenin's theorem:

$$V_{TH} = \left(\frac{R_{e_1}}{R_{e_1} + R_{e_2}}\right) \times 20\text{ V} = \left(\frac{253\ \Omega}{253\ \Omega + 280\ \Omega}\right) \times 20\text{ V} \cong 9.5\text{ V}$$

$$R_{TH} = \frac{R_{e_1} \times R_{e_2}}{R_{e_1} + R_{e_2}} = \frac{253\ \Omega \times 280\ \Omega}{253\ \Omega + 280\ \Omega} = 133\ \Omega$$

Norton's theorem:

$$I_N = \frac{V}{R_{e_2}} = \frac{20\text{ V}}{280\ \Omega} = 71\text{ mA}$$

$$R_N = \frac{R_{e_1} \times R_{e_2}}{R_{e_1} + R_{e_2}} = 133\ \Omega$$

CONVERSIONS
Thevenin from Norton:

$$R_{TH} = R_N$$
$$133\ \Omega = 133\ \Omega$$
$$V_{TH} = I_N \times R_N$$
$$V_{TH} = 71 \times 10^{-3} \times 133\ \Omega \cong 9.5\text{ V}$$

Norton from Thevenin:

$$R_N = R_{TH}$$
$$133\ \Omega = 133\ \Omega$$
$$I_N = \frac{V_{TH}}{R_{TH}}$$
$$I_N = \frac{9.5\text{ V}}{133\ \Omega} \cong 71\text{ mA}$$

FIGURE 7–19

EXAMPLE

Let's look briefly at a sample circuit, analyze it with Thevenin's theorem and with Norton's theorem, then plug values into our simple conversion formulas. This will serve as a quick review of the two theorems, plus illustrate the simplicity of conversion from one set of parameters to the other. See Figure 7–19.

SUMMARY

▶ The theorems studied apply to two-terminal linear networks that have bilateral resistances or components.

▶ The maximum power transfer theorem indicates maximum power is transferred from source to load when the load resistance equals the source resistance. Under these conditions the efficiency is 50%.

▶ The superposition theorem says electrical parameters are computed in linear circuits with more than one source by analyzing the parameters produced by each source acting alone, then superimposing each result by algebraic addition to find the circuit parameters with multiple sources.

▶ Thevenin's theorem simplifies any two-terminal linear network into an equivalent circuit representation of a voltage source (V_{TH}) and series resistance (R_{TH}).

▶ Steps for Thevenizing a circuit are:

1. Open (or remove R_L) from the network circuitry.

2. Determine the open-circuit V at points where R_L is to be connected. This is the value of V_{TH}.

3. Mentally short the circuit voltage source and determine resistance by looking into the network (from R_L connection points, with R_L disconnected). This is the value of R_{TH}.

4. Draw the Thevenin equivalent circuit with source = V_{TH} and series resistor of the R_{TH} value.

5. Make calculations of I_L and V_L for the desired values of R_L (where R_L is as-

sumed to be across the equivalent circuit output terminals) by using two-resistor series circuit voltage analysis.

▶ Voltage distribution in a Thevenin equivalent circuit is computed using Ohm's law and/or the voltage divider rule.

▶ Norton's theorem simplifies any two-terminal linear network into an equivalent circuit representation of a current source (I_N) and a parallel or shunt resistance (R_N).

▶ Steps for Nortonizing a circuit are:

1. Determine the value for I_N by finding the value of current flowing through R_L if it were zero ohms.

2. Find the value for R_N by assuming the R_L is removed from the circuit (R_L = infinite ohms) and any voltage sources are zero ohms. Calculate the circuit resistance by looking from the R_L connection points toward the source(s).

3. Draw the Norton equivalent circuit as a current source of a value equal to I_N and a resistance in parallel with the source equal to R_N.

4. Compute R_L parameters, as appropriate, by assuming the load resistance at the output terminals of the equivalent circuit, using concepts of two-resistor parallel circuit current division analysis.

▶ Current division in a Norton equivalent circuit is computed using Ohm's law and/or the current divider rule.

FORMULAS AND CALCULATOR SEQUENCES SUMMARY

FORMULA 7–1 Efficiency (%) = $\dfrac{P_{out}}{P_{in}} \times 100$

P_{out} value, $\boxed{\div}$, P_{in} value, $\boxed{\times}$, 100, $\boxed{=}$

FORMULA 7–2 $r_{int} = \dfrac{V \text{ (no load)} - V \text{ (loaded)}}{I \text{ (load)}}$

V (no load) value, $\boxed{-}$, V (loaded) value, $\boxed{=}$, $\boxed{\div}$, I (load) value, $\boxed{=}$

FORMULA 7–3 $I_L = \dfrac{V_{TH}}{R_{TH} + R_L}$

V_{TH} value, $\boxed{\div}$, $\boxed{(}$, R_{TH} value, $\boxed{+}$, R_L value, $\boxed{)}$, $\boxed{=}$

FORMULA 7–4 $V_L = \dfrac{R_L}{R_L + R_{TH}} \times V_{TH}$

R_L value $\boxed{\div}$, $\boxed{(}$, R_L value, $\boxed{+}$, R_{TH} value, $\boxed{)}$, $\boxed{=}$, $\boxed{\times}$, V_{TH} value, $\boxed{=}$

FORMULA 7–5 $I_L = \dfrac{R_N}{R_N + R_L} \times I_N$

R_N value, $\boxed{\div}$, $\boxed{(}$, R_N value, $\boxed{+}$, R_L value, $\boxed{)}$, $\boxed{=}$, $\boxed{\times}$, I_N value, $\boxed{=}$

REVIEW QUESTIONS

1. Name at least two applications for the maximum power transfer theorem.

2. In Figure 7–20, use the superposition theorem, and determine the voltage across R_1 and R_2.

3. Refer again to Figure 7–20. Assume the polarity of V_{S_2} is reversed. What is the value of V_{R_1} and V_{R_2}?

4. In Figure 7–21, use Thevenin's theorem and determine V_{TH}, R_{TH}, I_L, and V_L. Assume R_L is 10 kΩ.

FIGURE 7–20

FIGURE 7–21

5. Refer again to Figure 7–21. Assume R_L changes to 27 kΩ. What are the values of V_{TH}, R_{TH}, I_L, and V_L?

6. Draw and label the values for the Norton equivalent circuit for the circuit in question 4.

7. Draw and label the values for the Norton equivalent circuit for the circuit in question 5.

8. Define the term "a bilateral linear network."

9. State the sequence of steps used in "Thevenizing" a circuit.

10. State the sequence of steps used in "Nortonizing" a circuit.

11. What is the maximum power that can be delivered to the load in Figure 7–22?

FIGURE 7–22

12. If the efficiency of power transfer from source to load is 60%, should the load resistance be increased or decreased to enable maximum power transfer?

13. If the voltage at the source terminals of a circuit is 100 volts without load and 90 volts when 10 amperes are drawn from the source, what value is the load resistance to afford maximum power transfer?

14. For the circuit in Figure 7–23, what are the values of R_{TH} and V_{TH}?

FIGURE 7–23   Find R_{TH} and V_{TH}.

15. For the circuit in Figure 7–24, what are the values of R_N and I_N?

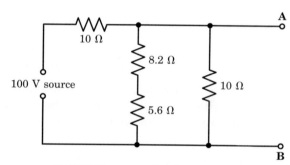

FIGURE 7–24   Find R_N and I_N.

16. Convert Norton parameters to Thevenin parameters, given I_N = 3 A and R_N = 10 Ω.

17. Convert Thevenin parameters to Norton parameters, given R_{TH} = 15 Ω and V_{TH} = 25 V.

18. Describe the most important application of the superposition theorem.

19. Which theorem relates most closely to voltage divider analysis techniques?

20. Which theorem relates most closely to current divider analysis techniques?

CHALLENGE QUESTIONS

1. Explain why it is important for a voltage source to have as low an internal resistance (r_{int}) as possible.

2. For a given power source, why is it important for the load's opposition to current flow to be equal to that of the source?

3. a. Give the main reason it is important to observe the direction of current flow through components when using the superposition theorem to analyze a two-source circuit.
   b. Describe which component(s) in the circuit are especially impacted by the fact described in question 3a.

TEST YOURSELF

1. In your own words, verbally explain the significance of R_{TH} in using Thevenin's Theorem, and explain how it is determined.

2. In your own words, verbally explain the significance of V_{TH} in using Thevenin's Theorem, and explain how it is found.

3. In your own words, verbally explain the significance of R_N in using Norton's Theorem, and explain how it is found.

4. In your own words, verbally explain the significance of I_N in using Norton's Theorem, and explain how it is found.

5. a. Perform appropriate research to enable listing at least one type circuit or application in which there may be two or more sources.
   b. Which network theorem would you use to help you in analyzing such a circuit?

PERFORMANCE PROJECTS CORRELATION CHART

Suggested performance projects in the Laboratory Manual that correlate with topics in this chapter are:

CHAPTER TOPIC	PERFORMANCE PROJECT	PROJECT NUMBER
Maximum Power Transfer	Maximum Power Transfer Theorem	29
Thevenin's Theorem	Thevenin's Theorem	27
Norton's Theorem	Norton's Theorem	28

NOTE: After completing above projects, perform the Summary at end of performance projects section in the Laboratory Manual.

THE SIMPLER SEQUENCE FOR TROUBLESHOOTING

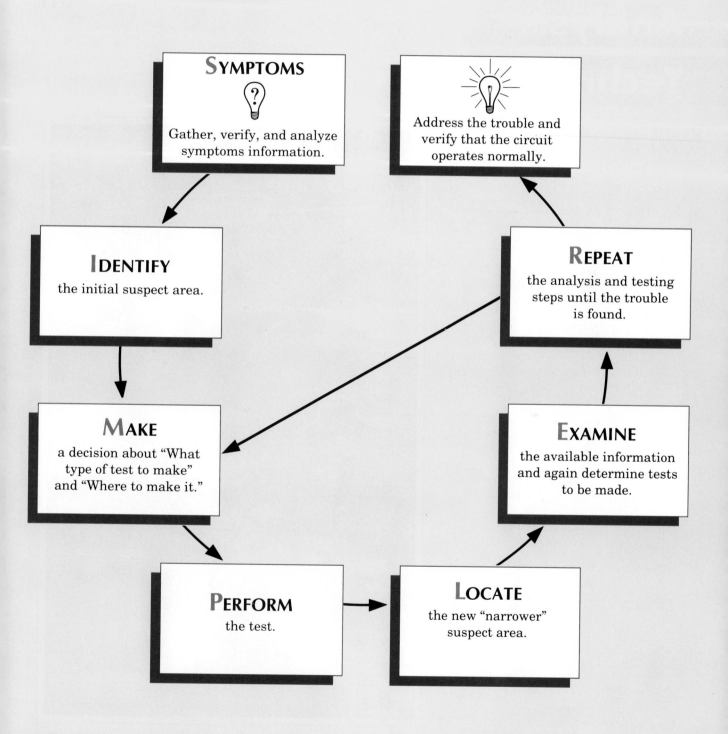

SYMPTOMS

Gather, verify, and analyze symptoms information.

Address the trouble and verify that the circuit operates normally.

IDENTIFY

the initial suspect area.

REPEAT

the analysis and testing steps until the trouble is found.

MAKE

a decision about "What type of test to make" and "Where to make it."

EXAMINE

the available information and again determine tests to be made.

PERFORM

the test.

LOCATE

the new "narrower" suspect area.

Network Analysis Techniques

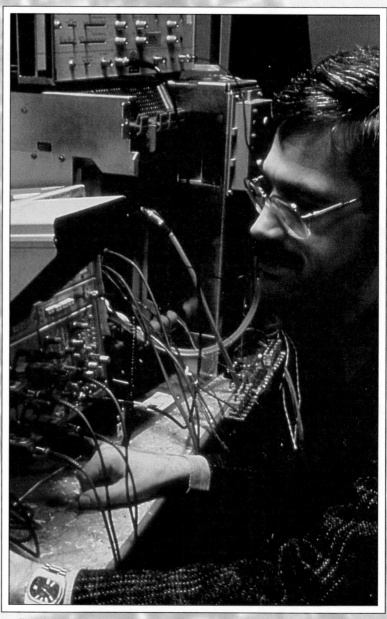

Courtesy of International Business Machines Corporation

8

KEY TERMS

Assumed mesh current(s)

Delta-Wye conversion

Loop

Loop voltages

Mesh

Node

Wye-Delta and Delta-Wye conversions

OUTLINE

Assumed loop or mesh current analysis approach

Nodal analysis approach

Conversions of Delta and Wye networks

285

CHAPTER PREVIEW

As you have already learned, circuit and network analysis enable you to find the various currents and voltages throughout the circuit, or network, being analyzed. In this chapter, two other approaches will be introduced for solving network parameters. One method uses Kirchhoff's voltage law as a basis for writing equations for electrical quantities around a closed loop (also sometimes called a mesh). These equations then enable finding current and voltage parameters throughout the network. Another method uses Kirchhoff's current law as a basis for writing "nodal" equations, which also enable solving network parameters. Each method has advantages and disadvantages. The techniques presented in this chapter enable you to analyze complex networks in which connections between components are such that you cannot simply use basic series-parallel circuit analysis methods.

This chapter will also introduce some techniques for converting two specific types of networks from one form to the other when this would be advantageous. One type network is called the wye (Y) or (T) network configuration. The other is known as the delta (Δ) or pi (π) configuration. These networks, or circuit configurations, are used in a number of places in electrical power systems and in electronic circuitry. Knowledge regarding these circuit configurations is therefore very useful, and has value to you as a technician.

OBJECTIVES

After studying this chapter, you should be able to:

• Define the terms mesh, loop, and node

• Analyze a single-source circuit using a loop procedure

• Use the assumed mesh currents approach to find voltage and current parameters for each component in a network having two sources

• Use the nodal analysis approach to find voltage and current parameters for each component in a network having two sources

• Convert from delta (Δ) circuit configuration parameters to wye (Y) circuit configuration parameters, and vice versa

ASSUMED LOOP OR MESH CURRENT ANALYSIS APPROACH

Background Information

A **loop** is considered to be a closed, or complete, path within a circuit. A **mesh** is also considered to be an unbroken, or closed, loop. Observe in Figure 8–1a that two closed loops are shown in the circuit with arbitrary mesh current assignments and identifiers. Figure 8–1b illustrates the commonly accepted "single-window frame" definition for meshes and mesh currents.

(**a**) Mesh currents—generalized loop definition

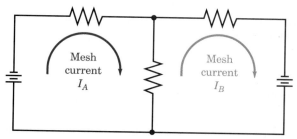

(**b**) Mesh currents—single "window-frame" definition

**FIGURE 8–1 Examples of assumed
mesh currents**

For each mesh, or loop, we will use an **assumed mesh current** direction in which the mesh current does NOT branch, as an actual circuit current would. These arbitrarily assigned currents, for analysis purposes, are considered to flow only in their own loop, or mesh. (This nonbranching aspect simplifies matters, and is one advantage of the mesh analysis method). Kirchhoff's voltage law equations will then be written, using the assumed mesh current designators.

The version of Kirchhoff's voltage law that will be used stipulates that the sum of the voltage drops around a closed loop must equal the voltage source(s). That is: $V_s = V_1 + V_2 \ldots + V_n$. From the results of these equations, we will then be able to compute actual circuit voltages and currents throughout the network.

Polarity Assignments

It is important to note directions of currents and polarities of voltages in the analysis steps. The general rules to follow during the analysis include:

1. Use the arbitrary convention that assumed loop currents will be shown circulating in a clockwise direction through each loop.

2. Assume any voltage drop created across a resistor by its own assumed loop or mesh current to be positive in that mesh's equation. Any voltage drop across a resistor due to another loop or mesh current is considered negative for that mesh's equation.

3. For source voltage polarity assignment (the algebraic sign to be used in equations), use the arbitrary convention that if the assumed mesh

current is returning to the source into its positive terminal, the source voltage will be considered positive. If the assumed mesh current is returning (going into) the negative terminal, it will be considered negative.

Steps to Use the Loop/Mesh Analysis Approach

1. Label the assumed mesh currents on the diagram. (Use the arbitrary clockwise direction convention.)

2. When assigning polarities to sources in the Kirchhoff equations, consider the source voltage positive if the assumed mesh current is entering its positive terminal, and negative if the assumed mesh current enters its negative terminal.

3. Write the Kirchhoff equation for each loop's voltage drops. Assume voltages across resistors in a given loop to be positive when caused by that mesh's assumed current, and negative when caused by an adjacent mesh's assumed current.

4. Solve the resulting equations using simultaneous equation methods.

5. Verify the results using Ohm's law, Kirchhoff's law, and so on, as appropriate.

EXAMPLE

Let's try this system out with a single-source bridge circuit, and see how it works. For the given circuit, let's find the value of voltage between points A and B.

Step 1. Draw assumed loop/mesh currents on the diagram. See Figure 8–2.

Step 2. Determine polarity of source polarity assignment(s). In this case, both mesh current I_A and mesh current I_B are shown entering the positive terminal of V_S. Thus, the source voltage will have a positive sign in our equations.

Step 3. Write the Kirchhoff equation for each loop's voltage drops:

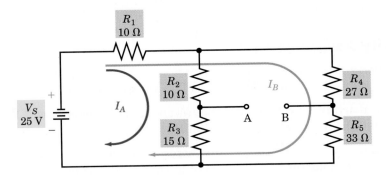

Kirchhoff's voltage law equation 1: $35\,I_A + 10\,I_B = 25$ V
Kirchhoff's voltage law equation 2: $10\,I_A + 70\,I_B = 25$ V
$$I_A = 0.638 \text{ A}$$
$$I_B = 0.266 \text{ A}$$

FIGURE 8–2

a. Loop A:

$$(I_A + I_B)R_1 + I_A R_2 + I_A R_3 = V_S$$

Switching positions of terms for ease of notation and manipulation:

$$10(I_A + I_B) + 10I_A + 15I_A = 25$$
$$35I_A + 10I_B = 25 \text{ (equation 1)}$$

b. Loop B:

$$(I_A + I_B)R_1 + I_B R_4 + I_B R_5 = V_S$$

Switching positions of terms for ease of notation and manipulation:

$$10(I_A + I_B) + 27(I_B) + 33(I_B) = 25$$
$$10I_A + 70\ I_B = 25 \text{ (equation 2)}$$

Step 4.
$$\text{Equation 1:} \quad 35I_A + 10I_B = 25$$
$$\text{Equation 2:} \quad 10I_A + 70I_B = 25$$

Multiplying equation 1 by -7 in order to cancel the I_B term and solve for I_A:

$$-245I_A - 70I_B = -175$$
$$10I_A + 70I_B = 25$$

Adding to cancel the I_B term:

$$-235I_A = -150$$
$$I_A = 0.638 \text{ A}$$

Substituting 0.638 for I_A into equation 2 in order to solve for I_B:

$$10(0.638) + 70I_B = 25$$
$$6.38 + 70I_B = 25$$
$$70I_B = 18.62$$
$$I_B = 0.266 \text{ A}$$

Thus:

$$I_A = 0.638 \text{ A}$$
$$I_B = 0.266 \text{ A}$$

Using mesh currents to solve voltage drops:

$$V_1 = (I_A + I_B) \times 10 = 0.904 \times 10 = 9.04 \text{ V}$$
$$V_2 = I_A \times 10 = 0.638 \times 10 = 6.38 \text{ V}$$
$$V_3 = I_A \times 15 = 0.638 \times 15 = 9.57 \text{ V}$$
$$V_4 = I_B \times 27 = 0.266 \times 27 = 7.18 \text{ V}$$
$$V_5 = I_B \times 33 = 0.266 \times 33 = 8.78 \text{ V}$$

Point A is $+9.57$ V with respect to point C in the circuit. Point B is $+7.18$ V with respect to point C in the circuit. Thus, point A is $+2.39$ V with respect to point B.

Step 5. Checking the voltage drops around the loops to see if it verifies our calculations:

Loop A: $V_1 + V_2 + V_3 = 9.04 + 6.38 + 9.57 = 24.99$ V, which checks out against our 25-V source.

Loop B: $V_1 + V_4 + V_5 = 9.04 + 7.18 + 8.78 = 25$ V, which also checks out against our 25-V source.

FIGURE 8–3

PRACTICE PROBLEMS I

Refer to Figure 8–3, and using the mesh analysis approach, solve
the following:

1. I_A
2. I_B
3. V_1
4. V_2
5. V_3
6. V_4
7. V_5
8. V_{A-B}

EXAMPLE

Now, let's try the mesh approach with a circuit containing two sources.
Refer to Figure 8–4 as you follow the steps for this example.

Step 1. Label the assumed mesh currents on the diagram, using the arbitrary clockwise convention. See Figure 8–4.

Step 2. Determine polarity of sources. By observation you can see that the
mesh currents are such that current goes into the positive terminal
of the 15-volt source, thus it will be designated as a positive number. The mesh current goes into the negative terminal of the 12-volt
source, thus it will be signed as a negative number in equations.

Step 3. Write the Kirchhoff voltage equation for each loop's voltage drops:

$$(3.9I_A + 5.6I_A) - 5.6I_B = 15 \text{ V}$$
$$9.5I_A - 5.6I_B = 15 \text{ V} \qquad \text{(equation 1)}$$
$$-5.6I_A + (5.6I_B + 6.8I_B) = -12 \text{ V}$$
$$-5.6I_A + 12.4I_B = -12 \text{ V} \qquad \text{(equation 2)}$$

Kirchhoff's voltage law equation 1: $9.5\,I_A - 5.6\,I_B = 15$ V
Kirchhoff's voltage law equation 2: $-5.6\,I_A + 12.4\,I_B = -12$ V
$$I_A = 1.377 \text{ A}$$
$$I_B = 0.346 \text{ A}$$

FIGURE 8–4

Step 4: In order to cancel the I_B terms and solve for I_A, we multiply equation 1 by the factor 2.214 (which is derived from 12.4 being divided by 5.6). This yields:

$$21I_A - 12.4I_B = 33.21$$
$$-5.6I_A + 12.4I_B = -12$$

Adding to cancel the I_B term produces:

$$15.4I_A = 21.21$$
$$I_A = 1.377 \text{ A}$$

To find I_B we substitute 1.377 A for I_A in the second equation, which yields:

$$-5.6(1.377) + 12.4I_B = -12$$
$$-7.71 + 12.4I_B = -12$$
$$12.4I_B = -4.29$$
$$I_B = -0.346 \text{ A}$$

The negative sign of the outcome tells us that the actual current is in the opposite direction from that which we assumed in creating the mesh currents.

Now using the values of 1.377 A for I_A and 0.346 A for I_B, we can calculate circuit parameters quite easily.

$$V_1 = I_A \times 3.9 = 5.37 \text{ V}$$
$$V_2 = (I_A + I_B) \times 5.6 = (1.377 + 0.346) \times 5.6 = 9.65 \text{ V}$$
$$V_3 = I_B \times 6.8 = 0.346 \times 6.8 = 2.35 \text{ V}$$

Step 5. Checking the loop voltages as a verification of our calculations, we find:
Loop A: $V_1 + V_2$ should equal V_{S_1}: 5.37 + 9.65 = 15.02 V (which verifies loop A)
Loop B: $V_2 + V_3$ should equal V_{S_2}: 9.65 + 2.35 = 12 V (which verifies loop B)

FIGURE 8–5

PRACTICE PROBLEMS II

For further practice in using the mesh approach, refer to Figure 8–5 and perform the required work to find the following circuit parameters.

1. I_A
2. I_B
3. V_1 and the current through R_1
4. V_2 and the current through R_2
5. V_3 and the current through R_3

✓ IN-PROCESS LEARNING CHECK I

1. A loop is considered to be a closed, or complete _____ within a circuit.

2. In the loop or mesh approach, it is important to label the assumed _____ currents on the diagram you use for analysis.

3. When assigning polarities to sources in the Kirchhoff equations, a source voltage is considered (negative, positive) _____ if the assumed mesh current enters the positive terminal of the source.

4. When writing Kirchhoff equations for each loop's voltage drops, the component voltage is considered (negative, positive) _____ when caused by its own mesh current, and (negative, positive) _____ when caused by an adjacent mesh's current flow.

NODAL ANALYSIS APPROACH

Background Information

A **node** is a current junction point of two or more components in a circuit; that is, a point where currents enter and leave. Kirchhoff's current law can be used to write equations related to these nodes in circuits. Nodes where three or more components are joined are sometimes identified as "major nodes."

For nodal analysis, one major node is specified as the "reference node." Typically, a node that is common to as many parts of the network as possible is chosen as the reference node. Voltages within the circuit are then designated for each of the other nodes in the circuit, relative to the reference node. A Kirchhoff current equation is written for each of the nodes, other than the reference node. For each current term, we use the appropriate V_R/R expression. This is in contrast to our having used the appropriate $I \times R$ expression for voltage terms in the mesh/loop analysis approach.

Steps to Use the Node Analysis Approach

1. Find the number of major nodes that are in the circuit, then select the one major node you will use as the reference node.

2. Designate currents into and out of all major nodes, with the exception of the reference point node.

3. Write Kirchhoff Current Law (KCL) equations for currents into and out of each major node, using the V_R/R approach to represent each current. The number of equations required is generally one less than the number of major nodes.

4. Incorporate known values of voltage and resistance in the equations, and solve for the unknown node voltages.

5. Based on the results of step 4, solve for current and voltage values for each component, as appropriate.

6. Verify results by using Kirchhoff's voltage law for each loop in the circuit.

EXAMPLE

Refer to the circuit of Figure 8–6 as you observe the following steps in a sample nodal analysis.

Step 1. Find the major nodes and select the reference node. Observe in the circuit that only two major nodes exist. One is at the top of R_2, and the other is at other end of R_2, which happens to be ground reference in this case. Let's let the ground reference be the reference node. The voltage across R_2 will then be our voltage of interest, against which other parameters will be determined.

Step 2. Designate currents into and out of major nodes. See Figure 8–6 again.

Step 3. Write the Kirchhoff current equations for currents into and out of these major nodes. For the loop with the 15-V source:

$$I_1 + I_2 = I_3$$

(NOTE that the way our circuit components are labeled, current I_2 is the current through R_3, and current I_3 is the current through R_2 and the reference node.)

NOTE: I_1 is expressed as V_{R_1}/R_1
I_2 is expressed as V_{R_3}/R_3
I_3 is expressed as V_{R_2}/R_2

FIGURE 8–6

$$\frac{V_{R_1}}{R_1} + \frac{V_{R_3}}{R_3} = \frac{V_{R_2}}{R_2}$$

(NOTE: V_{R_2} is the voltage of special interest, since it is associated with the reference node.)

Step 4. Since we know the V_{S_1} value, we can state that:

$$V_{R_1} + V_{R_2} = 15 \text{ V or } V_{R_1} = 15 - V_{R_2}$$

For the loop with the 12-V source, we can state that:

$$V_{R_3} + V_{R_2} = 12 \text{ V or } V_{R_3} = 12 - V_{R_2}$$

Now, incorporating these terms into our V/R current statements:

$$\frac{15 - V_{R_2}}{10} + \frac{12 - V_{R_2}}{1} = \frac{V_{R_2}}{5}$$

Multiplying each term by 10 to remove the denominators:

$$\left(\frac{\cancel{10}}{1}\right) \times \frac{15 - V_{R_2}}{\cancel{10}} + \left(\frac{10}{1}\right) \times \frac{12 - V_{R_2}}{1} = \left(\frac{\cancel{10}}{1}\right) \times \frac{V_{R_2}}{5}$$

$$15 - V_{R_2} + 10(12 - V_{R_2}) = 2\, V_{R_2}$$

$$15 - V_{R_2} + 120 - 10\, V_{R_2} = 2\, V_{R_2}$$

$$135 = 13\, V_{R_2}$$

$$V_{R_2} = \frac{135}{13}$$

$$V_{R_2} = 10.38 \text{ V}$$

Step 5. Therefore:

$$I_{R_2} = \frac{10.38}{5} = 2.076 \text{ A}$$

From this, we can continue with:

$$V_{R_3} = 12 - 10.38 = 1.62 \text{ V}$$

$$I_{R_3} = \frac{1.62}{1} = 1.62 \text{ A}$$

$$V_{R_1} = 15 - 10.38 = 4.62 \text{ V}$$

$$I_{R_1} = \frac{4.62}{10} = 0.462 \text{ A}$$

Step 6. Verifying results:

$$V_{R_1} + V_{R_2} = V_{S_1}$$
$$4.62 + 10.38 = 15 \text{ V (It checks!)}$$
$$V_{R_2} + V_{R_3} = V_{S_2}$$
$$10.38 + 1.62 = 12 \text{ V (It checks!)}$$

EXAMPLE

The same procedures we have been showing you will work even if the component values are not as convenient to work with as the ones in our first example. Follow along in this example, which does not use a least common denominator to eliminate the denominators, but rather simply uses one of the denominators as the factor to move things from the denominator to the numerator. The steps used in the solution process are essentially the same as those used in the preceding example, so we will not detail the step numbers.

Refer to Figure 8–7 as you follow the procedure shown.

$$I_1 + I_2 = I_3$$

FIGURE 8–7

(Again, note that I_2 is R_3's current, and I_3 is R_2's current, where the R_2 parameters are the ones of special interest, since this is the major reference node part of the circuit.)

$$\text{Currents Restated: } \frac{V_{R_1}}{R_1} + \frac{V_{R_3}}{R_3} = \frac{V_{R_2}}{R_2}$$

15-V loop: $V_{R_1} + V_{R_2} = 15$ V or $V_{R_1} = 15 - V_{R_2}$

12-V loop: $V_{R_3} + V_{R_2} = 12$ V or $V_{R_3} = 12 - V_{R_2}$

Substituting these terms into the $I_1 + I_2 = I_3$ equation:

$$\frac{15 - V_{R_2}}{3.9} + \frac{12 - V_{R_2}}{6.8} = \frac{V_{R_2}}{5.6}$$

Multiplying each term by 6.8 to move the denominators:

$$1.74(15 - V_{R_2}) + (12 - V_{R_2}) = 1.214 \, V_{R_2}$$

$$(26.1 - 1.74 \, V_{R_2}) + (12 - V_{R_2}) = 1.214 \, V_{R_2}$$

$$38.1 = 3.95 \, V_{R_2}$$

$$V_{R_2} = 9.64 \text{ V}$$

Using the reference node voltage, we find the other parameters as follows:

$$V_{R_1} = 15 - 9.64 = 5.36 \text{ V}$$

$$I_{R_1} = \frac{5.36 \text{ V}}{3.9 \text{ } \Omega} = 1.37 \text{ A}$$

$$V_{R_3} = 12 - 9.64 = 2.36 \text{ V}$$

$$I_{R_3} = \frac{2.36 \text{ V}}{6.8 \text{ } \Omega} = 0.347 \text{ A}$$

$$\text{Current through } R_2 = \frac{9.64 \text{ V}}{5.6 \text{ } \Omega} = 1.72 \text{ A or}$$

$$I_1 + I_2 = I_3 = 1.37 \text{ A} + 0.347 \text{ A} = 1.717 \text{ A}$$

PRACTICE PROBLEMS III

Refer to Figure 8–8 and find the following circuit parameters, using the nodal analysis method.

1. V_{R_1}
2. V_{R_2}
3. V_{R_3}
4. Current through R_1
5. Current through R_2
6. Current through R_3

FIGURE 8–8

IN-PROCESS LEARNING CHECK II

1. A node is a _____ junction point for two or more components in a circuit.
2. A node where three or more components are joined is sometimes called a _____ _____.
3. For nodal analysis, one node is chosen as the _____ _____.
4. In nodal analysis, a Kirchhoff _____ law equation is written for each major node, with the exception of the reference point node.

CONVERSIONS OF DELTA AND WYE NETWORKS

As a technician, you will occasionally deal with two circuit configurations found in numerous applications: the Wye or Tee configuration, shown in Figure 8–9, and the Delta, or Pi configuration, shown in Figure 8–10. For example, both types of connections are found in ac power applications. The pi configuration is found in many filter circuits used in electronic systems. As you can see, the names of these configurations are derived from the shapes they form. The delta configuration has the triangular shape

(**a**) Delta [Δ] configuration

(**a**) Wye [Y] configuration

(**b**) Tee [T] configuration

FIGURE 8–9 Commonly used Y and T circuit configurations

(**b**) Pi [π] configuration
FIGURE 8–10
Commonly used Δ and π configurations

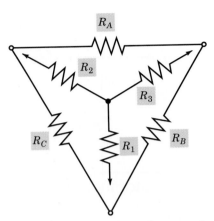

**FIGURE 8–11   Superimposed
Δ and Y configurations with
components labeled**

of the Greek letter delta (Δ). The wye configuration is often represented in the shape of the letter Y.

There are times when it is convenient and helpful to be able to convert from one circuit configuration to the other. One example of simplifying calculations using this technique is the analysis of a bridge circuit, which we will show you a little later. To do this conversion, or transformation, we need to be able to derive equivalent circuits from one configuration to the other. So, let's look a little closer at each of these common network connection configurations, and how they might be converted, or transformed.

For ease of reference, we will label the components and their connection points as shown in Figure 8–11. Notice that we have superimposed one configuration on the other, so you can easily see the cross-referencing between components when transformation formulas are used. The wye components are labeled as R_1, R_2, and R_3; the delta components are labeled as R_A, R_B, and R_C, respectively.

Converting a Wye Network to an Equivalent Delta Network

The object of our conversion is to find out what values the delta components must be in order to provide the same electrical load to the source as the original wye network does. In other words, we are looking for the delta equivalent to the wye circuit. The equivalent circuit will present the same resistance, and draw the same current from the source as the original wye network does. It should be noted, however, that voltages between connection points will differ for each configuration.

The equations used to convert from wye to delta networks (**Wye-Delta Conversion**) are:

1. To find R_A's value:

Formula 8–1	$R_A = \dfrac{(R_1 \times R_2) + (R_2 \times R_3) + (R_3 + R_1)}{R_1}$

$\boxed{(}$, R_1 value, $\boxed{\times}$, R_2 value, $\boxed{)}$, $\boxed{+}$, $\boxed{(}$, R_2 value, $\boxed{\times}$, R_3 value, $\boxed{)}$, $\boxed{+}$, $\boxed{(}$, R_3 value, $\boxed{\times}$, R_1 value, $\boxed{)}$, $\boxed{=}$, $\boxed{\div}$, R_1 value, $\boxed{=}$

2. To find R_B's value:

Formula 8–2	$R_B = \dfrac{(R_1 \times R_2) + (R_2 \times R_3) + (R_3 \times R_1)}{R_2}$

 $\boxed{(}$, R_1 value, $\boxed{\times}$, R_2 value, $\boxed{)}$, $\boxed{+}$, $\boxed{(}$, R_2 value, $\boxed{\times}$, R_3 value, $\boxed{)}$, $\boxed{+}$, $\boxed{(}$, R_3 value, $\boxed{\times}$, R_1 value, $\boxed{)}$, $\boxed{=}$, $\boxed{\div}$, R_2 value, $\boxed{=}$

3. To find R_C's value:

Formula 8–3	$R_C = \dfrac{(R_1 \times R_2) + (R_2 \times R_3) + (R_3 \times R_1)}{R_3}$

 $\boxed{(}$, R_1 value, $\boxed{\times}$, R_2 value, $\boxed{)}$, $\boxed{+}$, $\boxed{(}$, R_2 value, $\boxed{\times}$, R_3 value, $\boxed{)}$, $\boxed{+}$, $\boxed{(}$, R_3 value, $\boxed{\times}$, R_1 value, $\boxed{)}$, $\boxed{=}$, $\boxed{\div}$, R_3 value, $\boxed{=}$

Observe that the numerator is the same in all three equations, and it is comprised of the sum of the products of all the wye resistors, two at a time. The denominator in each case is the resistor in the wye configuration that is across from, or opposite to, the resistor value being sought for the delta equivalent circuit.

EXAMPLE

Find the delta equivalent circuit to the Y configuration shown in Figure 8–12. To make it a little easier to follow, we have again used our superim-

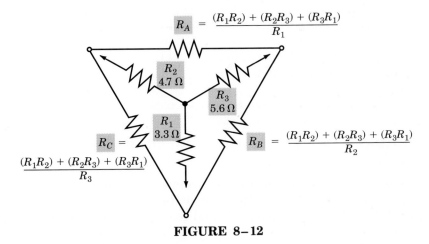

FIGURE 8–12

posed diagrams technique. NOTE: Get out your calculator and verify our calculations as you track our presentation.

Assume that $R_1 = 3.3\ \Omega$, $R_2 = 4.7\ \Omega$, and $R_3 = 5.6\ \Omega$.

$$R_A = \frac{(R_1 R_2) + (R_2 R_3) + (R_3 R_1)}{R_1}$$

$$R_A = \frac{(3.3 \times 4.7) + (4.7 \times 5.6) + (5.6 \times 3.3)}{3.3} = \frac{60.31}{3.3} = 18.28\ \Omega$$

The numerator is the same for all three equations. All we need to do is divide the 60.31 by the appropriate resistor value from the Y network to find the resistor of interest in our equivalent delta network. Thus:

$$R_B = \frac{60.31}{4.7} = 12.83\ \Omega$$

$$R_C = \frac{60.31}{5.6} = 10.77\ \Omega$$

PRACTICAL NOTES

On occasion, you will find that the same value resistors are used for R_1, R_2, and R_3 in the wye configuration. This simplifies conversion calculations. For example, if each of the wye resistors is a 10-Ω resistor, then the resulting numerator in our conversion formula (the sum of the products part) is 300. The denominator is 10. The resulting answer is that the delta resistor must equal 30 Ω, or three times the Y resistor. This is true for all three legs. It is also true that the equivalent delta resistor will always be three times its related Y resistor value, as long as all the wye network resistors are equal in value.

This fact can also be used in reverse order. When converting from a symmetrical value delta configuration to a wye configuration, the wye resistors will each be one-third the value of their related delta resistors. In other words, if, in our previous example, the delta resistors had been 30 Ω each, the wye equivalent circuit would have had resistors equal to 10 Ω each.

In summary, when converting from wye to delta (where all resistors are equal in value), use three times each wye resistor value for each delta resistor. When converting from delta to wye (where all resistors are equal in value), make each wye resistor one-third the value of each of the delta resistors.

Converting a Delta Network to an Equivalent Wye Network

As an interesting way to find if the conversion techniques will work, let's look at the formulas for converting from delta to wye, insert the values we just calculated for the delta network, and see if, after conversion, we arrive back at the same values given in our original wye network. The equations used to convert from delta to wye networks (**Delta-Wye Conversion**) are:

Formula 8–4	$R_1 = \dfrac{R_B \times R_C}{R_A + R_B + R_C}$

 R_B value, $\boxed{\times}$, R_C value, $\boxed{\div}$, $\boxed{(}$, R_A value, $\boxed{+}$, R_B value, $\boxed{+}$, R_C value, $\boxed{)}$, $\boxed{=}$

Formula 8–5	$R_2 = \dfrac{R_C \times R_A}{R_A + R_B + R_C}$

 R_C value, $\boxed{\times}$, R_A value, $\boxed{\div}$, $\boxed{(}$, R_A value, $\boxed{+}$, R_B value, $\boxed{+}$, R_C value, $\boxed{)}$, $\boxed{=}$

Formula 8–6	$R_3 = \dfrac{R_A \times R_B}{R_A + R_B + R_C}$

R_A value, $\boxed{\times}$, R_B value, $\boxed{\div}$, $\boxed{(}$, R_A value, $\boxed{+}$, R_B value, $\boxed{+}$, R_C value, $\boxed{)}$, $\boxed{=}$

Notice in these equations, the denominator is the same in every case. That is, the denominator equals the sum of the resistances in the delta network. Notice also that the numerator for each equation shows the product of the two delta resistors that connect to one end of the wye resistor of interest in our superimposed networks drawing, or the product of the two delta resistors that are adjacent to the wye network resistor of interest.

EXAMPLE

Let's insert our computed delta resistor values into these formulas and see if we come close to the original wye circuit values.

$$R_1 = \frac{R_B \times R_C}{R_A + R_B + R_C} = \frac{12.83 \times 10.77}{18.28 + 12.83 + 10.77} = \frac{138.18}{41.88} = 3.3 \ \Omega$$

For these equations, the denominator will be constant, so we only need to determine the numerator in each case, then divide by 41.88. Therefore,

$$R_2 = \frac{R_C \times R_A}{41.88} = \frac{10.77 \times 18.28}{41.88} = \frac{196.88}{41.88} = 4.7 \ \Omega$$

$$R_3 = \frac{R_A \times R_B}{41.88} = \frac{18.28 \times 12.83}{41.88} = \frac{234.53}{41.88} = 5.6 \ \Omega$$

Observe that our values do check out and verify the techniques!

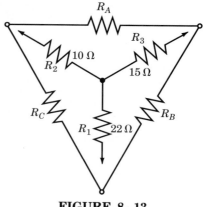

FIGURE 8–13

PRACTICE PROBLEMS IV

1. Perform the necessary conversion from wye to delta configurations for Figure 8–13.

2. Use the values you computed for the delta resistor values in question 1 and determine the values of the wye network resistors, using the delta-wye conversion equations.

3. Did the answers you got in problem 2 appropriately correspond with the original circuit parameters shown in Figure 8–13?

✔ IN-PROCESS LEARNING CHECK III

1. Due to their visual shapes and electrical connections:
   a. The wye circuit configuration is sometimes also called the _____ configuration.
   b. The delta circuit configuration is sometimes also called the _____ configuration.

2. Delta and wye circuit configurations are often found in _____ power applications.

3. The value in knowing how to determine equivalent circuits, or convert from delta-to-wye and wye-to-delta circuits is that the equivalent circuits will present the same resistance and draw the same _____ from the source feeding the circuit.

4. The voltage between connection points in wye and delta configurations (is, is not) _____ the same.

FIGURE 8–14   Sample bridge circuit

Bridge Circuit Analysis Using Conversion Techniques

Observe the bridge circuit of Figure 8–14. When shown as a bridge circuit, it appears that the network is essentially two delta networks, back-to-back. Simple series-parallel circuit analysis techniques cannot be used here. If we convert the bottom delta portion of this circuit to a wye configuration, it will simplify the analysis of the circuit considerably.

In Figure 8–15, you see that we have superimposed the proposed wye network in the bottom delta of the bridge diagram. By converting this delta circuit portion to an equivalent wye circuit, we can redraw the circuit as shown in Figure 8–16. As stated earlier, this lends itself to much easier analysis.

For simplification in transferring the concepts of the examples given earlier to this situation, we have assigned the resistors designations similar to those used in our preceding examples.

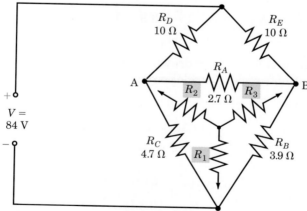

FIGURE 8–15   Sample bridge circuit analysis

FIGURE 8–16   Redrawn bridge network equivalent with bottom delta transformed to wye

EXAMPLE

If it were required that we find the R_T and I_T of the circuit, we might use the transformation techniques as follows.

Converting the bridge circuit delta portion to an equivalent wye:

$$R_1 = \frac{R_B \times R_C}{R_A + R_B + R_C} = \frac{3.9 \times 4.7}{2.7 + 3.9 + 4.7} = \frac{18.33}{11.3} = 11.3 \ \Omega$$

$$R_2 = \frac{R_C \times R_A}{11.3} = \frac{4.7 \times 2.7}{11.3} = \frac{12.69}{11.3} = 1.12 \ \Omega$$

$$R_3 = \frac{R_A \times R_B}{11.3} = \frac{2.7 \times 3.9}{11.3} = \frac{10.53}{11.3} = =0.93 \ \Omega$$

Now the bridge circuit can be redrawn as shown in Figure 8–17. The analysis is now that of a series-parallel circuit.

$$R_T = (R_D + R_2) \parallel (R_E + R_3) + R_1$$

$$R_T = (11.12 \parallel 10.93) + 11.3$$

$$R_T = 5.5 \ \Omega + 11.3 \ \Omega = 16.8 \ \Omega$$

$$I_T = \frac{V}{R_T} = \frac{84 \text{ V}}{16.8 \ \Omega} = 5 \text{ A}$$

FIGURE 8–17

PRACTICE PROBLEMS V

For the circuit of Figure 8–15, assume that the values are changed so that $R_A = 7\ \Omega$, $R_B = 14\ \Omega$, and $R_C = 21\ \Omega$. Use the delta-to-wye transformation techniques to find the following parameters:

1. R_T
2. I_T

SUMMARY

▶ A loop is a complete path in a circuit, from one side of the source, through a series of circuit components, back to the other side of the source.

▶ A mesh is an unbroken, closed loop within a network.

▶ Mesh currents are assumed currents (and current directions) used in conjunction with Kirchhoff's voltage law to analyze networks. Mesh currents are different than actual circuit currents, in that they are assumed to not branch.

▶ A common convention regarding labeling mesh currents on a diagram is that they are shown in a clockwise direction.

▶ Voltages across resistors in a network that are caused by their own mesh current are considered positive. Voltage drops due to another mesh current are considered negative.

▶ An arbitrary convention regarding assignment of source polarities in a network is that if a mesh current returns to the positive side of the source, the source is denoted as a positive term in the equations. If the mesh current returns to the negative terminal of the source, the source is signed as negative in the equations.

▶ For mesh analysis, voltages are expressed as $I \times R$ terms.

▶ Simultaneous equation techniques are used to eliminate one unknown from the resulting Kirchhoff equations. This enables solving for the other unknown factor.

▶ A node is a point where two or more circuit components join and where when circuit current is flowing, circuit currents may branch. A major node is where three or more components join.

▶ Nodal analysis uses Kirchhoff's current law to define parameters in the network. Currents to be used in the analysis, however, are expressed in terms of V/R.

▶ Delta and wye circuit configurations are quite common to many electrical and electronic circuits. The ability to analyze these circuits is enhanced by knowing how to convert, or transform, values of one type network to the other.

▶ In the conversion formulas for converting from wye to equivalent delta circuit values, the numerator is the common factor used in determining each of the three resistor values.

▶ In the conversion formulas for converting from delta to equivalent wye circuit values, the denominator is the common factor used in determining each of the three resistor values.

▶ In symmetrical wye circuits (those where all three legs of the wye have equal value resistors), the delta equivalent resistances will be three times the value of the wye resistors.

▶ In symmetrical delta circuits (those where all three sides of the delta have equal value resistors), the wye equivalent resistances will be one-third the value of the delta resistors.

FORMULAS AND CALCULATOR SEQUENCES SUMMARY

FORMULA 8–1 $R_A = \dfrac{(R_1 \times R_2) + (R_2 \times R_3) + (R_3 \times R_1)}{R_1}$

$\boxed{(}$, R_1 value, $\boxed{\times}$, R_2 value, $\boxed{)}$, $\boxed{+}$, $\boxed{(}$, R_2 value, $\boxed{\times}$, R_3 value, $\boxed{)}$, $\boxed{+}$, $\boxed{(}$, R_3 value, $\boxed{\times}$, R_1 value, $\boxed{)}$, $\boxed{=}$, $\boxed{\div}$, R_1 value, $\boxed{=}$

FORMULA 8–2 $R_B = \dfrac{(R_1 \times R_2) + (R_2 \times R_3) + (R_3 \times R_1)}{R_2}$

$\boxed{(}$, R_1 value, $\boxed{\times}$, R_2 value, $\boxed{)}$, $\boxed{+}$, $\boxed{(}$, R_2 value, $\boxed{\times}$, R_3 value, $\boxed{)}$, $\boxed{+}$, $\boxed{(}$, R_3 value, $\boxed{\times}$, R_1 value, $\boxed{)}$, $\boxed{=}$, $\boxed{\div}$, R_2 value, $\boxed{=}$

FORMULA 8–3 $R_C = \dfrac{(R_1 \times R_2) + (R_2 \times R_3) + (R_3 \times R_1)}{R_3}$

$\boxed{(}$, R_1 value, $\boxed{\times}$, R_2 value, $\boxed{)}$, $\boxed{+}$, $\boxed{(}$, R_2 value, $\boxed{\times}$, R_3 value, $\boxed{)}$, $\boxed{+}$, $\boxed{(}$, R_3 value, $\boxed{\times}$, R_1 value, $\boxed{)}$, $\boxed{=}$, $\boxed{\div}$, R_3 value, $\boxed{=}$

FORMULA 8–4 $R_1 = \dfrac{R_B \times R_C}{R_A + R_B + R_C}$

R_B value, $\boxed{\times}$, R_C value, $\boxed{\div}$, $\boxed{(}$, R_A value, $\boxed{+}$, R_B value, $\boxed{+}$, R_C value, $\boxed{)}$, $\boxed{=}$

FORMULA 8–5 $R_2 = \dfrac{R_C \times R_A}{R_A + R_B + R_C}$

R_C value, $\boxed{\times}$, R_A value, $\boxed{\div}$, $\boxed{(}$, R_A value, $\boxed{+}$, R_B value, $\boxed{+}$, R_C value, $\boxed{)}$, $\boxed{=}$

FORMULA 8–6 $R_3 = \dfrac{R_A \times R_B}{R_A + R_B + R_C}$

R_A value, $\boxed{\times}$, R_B value, $\boxed{\div}$, $\boxed{(}$, R_A value, $\boxed{+}$, R_B value, $\boxed{+}$, R_C value, $\boxed{)}$, $\boxed{=}$

REVIEW QUESTIONS

Refer to Figure 8–4. Assume that all three resistors now equal 10 Ω. Assume V_{S_1} is changed to 20 V and V_{S_2} is changed to 16 V. Use the mesh analysis approach to determine the parameters called for in the following questions.

1. I_A
2. I_B
3. V_{R_1}
4. V_{R_2}
5. V_{R_3}
6. I_{R_1}
7. I_{R_2}
8. I_{R_3}

Refer to Figure 8–6. Assume that all three resistors equal 4.7 Ω. Assume V_{S_1} is changed to 21 V and V_{S_2} is changed to 14 V. Use the nodal analysis approach to determine the parameters called for in the following questions.

9. I_1
10. I_2
11. I_3
12. V_{R_1}
13. V_{R_2}
14. V_{R_3}

Refer to Figure 8–11. Assume that $R_1 = 10$ Ω, $R_2 = 12$ Ω, and $R_3 = 15$ Ω. Use the wye-to-delta conversion formulas to find the values of:

15. R_A
16. R_B
17. R_C

CHALLENGE QUESTIONS

1. Explain the basic difference between a mesh current and an actual circuit current.

2. Define the terms major node and reference node.

3. Perform research and find where in industrial plants you might commonly find delta and wye configurations of electrical circuits.

TEST YOURSELF

1. Define the terms mesh, loop, and node.

2. When would mesh or nodal analysis be used rather than standard series-parallel analysis of a network?

3. How many equations must be written for a three-mesh circuit in order to find the circuit parameters?

PERFORMANCE PROJECTS CORRELATION CHART

CHAPTER TOPIC	PERFORMANCE PROJECT	PROJECT NUMBER
Assumed Loop or Mesh Current Analysis Approach	Loop or Mesh Analysis	30
Nodal Analysis Approach	Nodal Analysis	31
Conversions of Delta and Wye Networks	Wye-Delta and Delta-Wye Conversions	32

PART

PRODUCING &

MEASURING

ELECTRICAL

QUANTITIES

Cells and Batteries

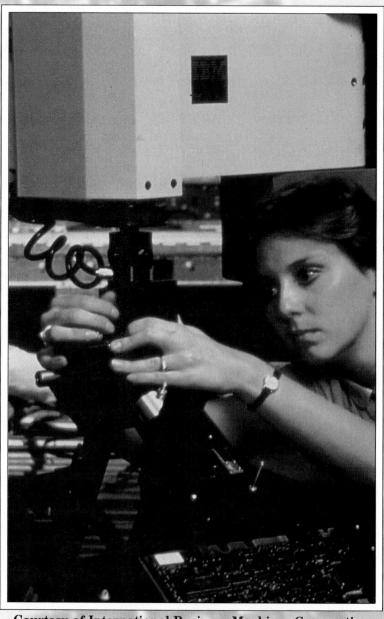

Courtesy of International Business Machines Corporation

9

KEY TERMS

Amalgamation

Ampere-hour (Ah) rating

Battery

Cell

Depolarizer

Electrode

Electrolyte

Local action

Polarization

OUTLINE

Chemical action in a cell

The common dry cell

Operating characteristics of dry cells

Rechargeable (secondary) cells and batteries

Battery capacity ratings

Data on recharging rechargeable batteries

Smart chargers

Construction and operational characteristics of the lead-acid battery

Battery charging methods

Examples of other types of primary and secondary cells

CHAPTER PREVIEW

In this chapter you will study common types of cells and batteries; learn about their operation; and examine some of their basic features. Also, some important facts about their care and usage are presented.

Cells and **batteries** are important sources of dc voltage and current that operate a variety of electrical and electronic circuits and devices. Everything, from automated cameras, burglar alarms, and emergency lighting systems, to portable radio and TV receivers, is made possible by various types of cells and batteries, Figure 9–1.

A battery is two or more electrically connected cells usually contained in a single unit. Cells and batteries depend on the process of changing chemical energy into electrical energy by separating and/or producing charges.

Two general categories of these dc sources are nonrechargeable primary cells and rechargeable secondary cells. The common flashlight battery is an example of a nonrechargeable primary cell and the cells in a rechargeable automobile storage battery are examples of secondary cells.

OBJECTIVES

After studying this chapter, you should be able to:

- Describe the difference between primary and secondary cells
- Briefly explain chemical action in a cell
- List physical and operating features of dry cells
- List physical and operating features of secondary cells
- Describe precautions to take with dry cells and wet cells
- List a variety of primary and secondary cells
- List typical electrical parameters of at least three primary and secondary cells
- Use the SIMPLER troubleshooting sequence to solve the Chapter Challenge problem

CHEMICAL ACTION IN A CELL

An early version of a cell that converts chemical energy to electrical energy is called a voltaic cell, named in honor of Alessandro Volta. A voltaic cell consists of two different metal plates (copper and zinc) immersed in a diluted acid solution (like a weak sulfuric acid solution), Figure 9–2. These metal plates are the **electrodes** and the acid solution is the **electrolyte.**

The chemical action occurring under these conditions is briefly described as follows:

1. When sulfuric acid dissolves in water, it splits into positive hydrogen ions and negative sulfate ions. (Recall from Chapter 1 that positive ions are atoms that have lost electrons and are positively charged enti-

FIGURE 9–1   Some common items that use batteries as a power source: (a) 35-mm camera (Courtesy of Pentax Corp); (b) portable radio (Courtesy of SONY Corp. of America); (c) portable television (Photo by Michael A. Gallitelli)

ties. Negative ions are atoms that have gained electrons and are negatively charged entities).

2. When the zinc plate is immersed in the acid solution, the acid attacks the zinc. This action causes zinc ions to separate from the plate, leaving the zinc plate with an excess of electrons. The zinc plate has a net negative charge. NOTE: When the zinc plate wears away, the cell can no longer operate, and the cell is dead.

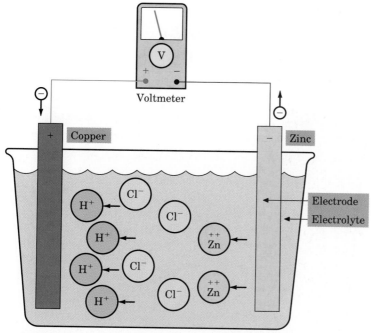

FIGURE 9–2   A simple voltaic cell consists of two different metal plates (electrodes) immersed in a diluted acid solution (electrolyte).

3. The positive hydrogen ions draw electrons from the copper plate, leaving the copper plate positively charged.

4. Because of the chemical actions described, the zinc plate is negative with respect to the copper plate (and/or the copper plate is positive with respect to the zinc plate). Therefore, a difference of potential exists between the electrodes. This electromotive force (emf) causes current to flow between the electrode terminals, if a conducting path is connected between them, Figure 9–3.

5. The open circuit difference of potential in primary cells depends on the metals used as the electrodes.

6. The amount of deliverable current depends upon the physical size of the electrodes.

In summary, the concepts discussed show that chemical reactions produce electromotive force. This principle of changing chemical energy to electrical energy is fundamental to the operation of all cells and batteries.

THE COMMON DRY CELL

Using the materials previously described, the useful life of the voltaic cell is very short. Therefore, it is not a practical type of cell.

A more useful primary cell is the carbon-zinc cell (also known as the LeClanche cell). This cell is commonly termed a dry cell.

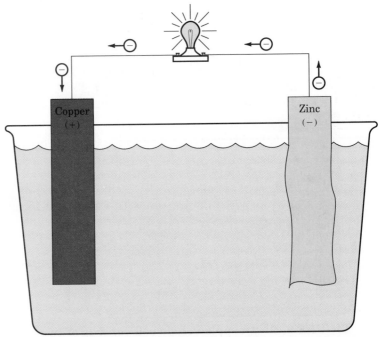

FIGURE 9–3   A cell produces current through the conducting path between electrodes.

Construction

The basic construction of a dry cell is shown in Figure 9–4. The key components are:

1. the positive electrode, which is carbon;

2. the negative electrode, which is zinc; and

3. the electrolyte, which is a solution of ammonium chloride called sal ammoniac that forms a paste-like substance.

Polarization

As the zinc electrode dissolves during the chemical process, hydrogen bubbles form on the positive carbon electrode that is detrimental to the cell's operation. This process is called **polarization.**

To overcome polarization, a **depolarizer** agent is introduced that combines its oxygen with the hydrogen bubbles to form water. This prevents the buildup of the undesired bubbles. A common depolarizing agent is manganese dioxide.

Local Action

Another operational problem confronted in the manufacture of dry cells is **local action.** Due to the manufacturing process, most zinc has impurities (such as carbon). Such impurities are suspended in the electrolyte close to the zinc and create independent cell actions. This undesired "voltaic" action dissolves the zinc.

To reduce local action, a thin coating of mercury is put on the zinc by a process called **amalgamation.** Amalgamation seals the impurities, separating them from the electrolyte to prevent local action.

Shelf Life

An unused dry cell has an indeterminant shelf life, since some local action occurs and the electrolyte paste will eventually dry up, depending on storage conditions.

Sealing compound
Cardboard container
Zinc casing
Blotting paper
Granulated carbon, manganese dioxide, and ammonium chloride
Carbon

FIGURE 9–4   The construction of a carbon-zinc (LeClanche) cell, commonly termed a dry cell

The cell becomes worn-out when the zinc, electrolyte, or depolarizer is consumed. This is observed when output voltage, under normal load current conditions, is appreciably lower than it should be.

OPERATING CHARACTERISTICS OF DRY CELLS

During a dry cell's useful life, voltage output ranges from about 1.4 volts to 1.6 volts with 1.5 volts being the nominal voltage.

Current capability typically ranges from a few milliamperes for small cells with small electrodes to 1/8 ampere or more for large cells, such as a #6 dry cell.

As a cell or battery ages, its internal resistance increases. Consequently, at a given current level, its terminal voltage decreases accordingly. Recall in the previous chapter we described how you can determine the internal resistance of a cell or battery by dividing the difference between no-load and full-load terminal voltage by the load current at full load.

PRACTICE PROBLEMS I

1. What is the internal resistance of a dry cell having a no-load terminal voltage of 1.5 V and a full-load terminal voltage of 1.25 V with a load current of 100 mA?

2. According to our nominal normal range values discussed earlier, is this cell considered good, fair, or bad in performance?

PRACTICAL NOTES

As indicated earlier, drying of the sal ammoniac electrolyte paste and local action shorten the shelf life of dry cells. Also, trying to draw too much current and/or drawing high current for extended periods will shorten operating life.

To lengthen shelf life, dry cells and batteries should not be stored in hot environments and should not be operated at currents that are not within normal specified operating ranges.

If terminal voltage appreciably drops from the nominal rated value, the internal resistance of the cell has increased due to its deteriorated operating condition. To insure proper operation of the circuit(s) using the cell as a source, replace the cell when this condition is apparent.

IN-PROCESS LEARNING CHECK I

Fill in the blanks as appropriate.

1. The carbon-zinc type dry cell is composed of a negative electrode made of _____ and a positive electrode made of _____. The electrolyte solution in this type cell is _____ _____, called sal _____.

2. The process which causes hydrogen bubbles to form on the positive electrode, and which is detrimental to the cell's operation is called _____.

3. Manganese dioxide is often added to the cell to prevent the process mentioned in question 2. This agent is sometimes known as a _____ agent.

4. Another problem that can occur in dry cells is that of little batteries being formed due to impurities suspended in the cell's electrolyte near the zinc electrode. This type activity is known as _____ _____. To reduce this undesired activity, mercury is put on the zinc by a process called _____.

5. The shelf life of dry cells is (enhanced, hurt) _____ by storing the cells in hot conditions.

6. Nominal cell terminal voltage of dry cells is about _____ V.

RECHARGEABLE (SECONDARY) CELLS AND BATTERIES

Lead-Acid Batteries

The most commonly used rechargeable battery is, of course, the lead-acid type, used in automotive and other services, Figure 9–5. In its automotive application, it partially discharges at a high rate (during starting of the automobile), then is recharged by the automobile's alternator system while the car is running. While the car is not running, certain accessories may operate off the battery. For example, the clock runs continuously. Radios, tape decks, and so on may also be operated when the car's engine is off. The nominal cell voltage for the common 12-volt lead-acid type battery is 2 volts per cell.

FIGURE 9–5   Automotive lead-acid battery

Deep-Cycle Batteries

Another popular, high-power, lead-acid type battery is called the "deep-cycle" battery. These are typically used for powering electrical fishing motors and for other recreational devices. They are typically the size of a medium automotive battery and may furnish from 1,000 to 1,200 watt-hours per charge in normal operating temperatures. If cared for properly, this type battery may operate satisfactorily for over 200 charge/discharge cycles.

Some Other Applications

Lead-acid batteries are often used in other applications. They provide power for mobile and portable transmitters and transceivers, and for electric vehicles and wheel chairs. Special, glass-encased, stationary lead-acid batteries, or banks of batteries, provide backup power for telephone company installations. You will find this type battery being used almost anywhere considerable dc power is required, particularly where a portable or mobile source is needed.

Nickel-Cadmium (Ni-Cad) Cells

FIGURE 9-6 The nickel-cadmium battery (Courtesy of Eveready Battery Co., Inc.)

The nickel-cadmium (ni-cad) cell is probably the most frequently used, all-purpose, rechargeable dry cell available today, Figure 9-6. The nominal voltage, per cell, for this type is 1.2 volts/per cell. Properly used, these cells/batteries are capable of more than 500 charge/discharge cycles. This type cell or battery should never be discharged to less than 1.0 volt per cell, as a cell can reverse polarity. This can cause a short circuit and/or ruptured seals.

Catalog listings for this battery indicate capacities in the range of 110 to over 1,000 milliampere-hour ratings. This type battery has a long shelf life, and is typically used in applications where there is cyclic rather than continuous use. A slow, "trickle" charging method is recommended for the recharging action.

Nickel-Metal-Hydride Cell

This newer type rechargeable cell has some features that differ from the more commonly known ni-cad cells. Its main advantage is that it can deliver about 50-60% greater capacity per given size. However, it has several disadvantages at our present state of technology. Some of its disadvantages are that it is more costly than ni-cads, its cycle life is shorter, and it cannot be fast-charged. For example, you cannot charge it at a 1C rate, and there are some low-temperature charging concerns. For this reason, you cannot use the same charger to charge this type cell as you would use to charge a ni-cad.

BATTERY CAPACITY RATINGS

The capacity ratings of cells and batteries are related to the continuous current that can be supplied by the cell or battery over a specified period of time. This common rating is often referred to as the **ampere-hour (Ah)**

FIGURE 9-7

rating. This rating is often expressed as the C rating. For example C/12 would indicate the current available continuously for 12 hours. The value for this C rating is dependent upon the discharge rate. Figure 9–7 shows a graph that illustrates the relationship of discharge rate to the ampere-hour (C) rating for two sizes of lead-acid storage batteries. As you can see, the higher the discharge rate, the lower the ampere-hour, or C rating for a given battery.

Typical capacity ratings may be in the range of a few milliampere-hours for very small batteries (such as used in hearing aids and other low-power devices) to over 100 ampere-hours for deep-cycle storage batteries.

EXAMPLE

A battery is rated as C/15. In this case, the C indicates the current continuously available for 15 hours. As shown in our previous graph, if the discharge rate is changed, then the ampere-hour rating (C) changes.

From the graph shown in Figure 9–7, we can see that at a discharge rate of 10 amperes, the C rating for the type 27 lead-acid battery is about 83. The product of amperes and hours (or C rating), then, would indicate that this battery could deliver 10 amperes for about 83/10, or for 8.3 hours. That is 10 A × 8.3 h = 83 Ah.

PRACTICE PROBLEMS II

1. What continuous level of current can be delivered over a 20-hour period by a battery rated at 150 ampere-hours?

2. Using the graph, what would be the C rating for a type 24 battery being dischaged at a rate of 12.5 amperes?

DATA ON RECHARGING RECHARGEABLE BATTERIES

The full-charge rating *(C)* of a battery is expressed in ampere-hours. Since there is not 100% efficiency in recharging a battery, it will take more hours to fully charge the battery than that simply indicated by the *C* rating. For example, if you are charging a battery at one-tenth its *C* rating current (0.1 C, or the 10-hour rate), it may take over 12 hours to fully charge the battery, rather than just the 10 hours indicated by the 0.1 C charging rate.

You should also be aware that charging methods used for lead-acid and ni-cad type cells and batteries should be different. In order to fully charge a ni-cad battery, it should be charged using a "constant-current" method at a rate of about 0.1 C (one-tenth full current rating) level. Typically, this would be about 50 mA for an AA-size cell. This charge rate is high enough to recharge the cells, yet not so high as to create overcharging damage. NOTE, many ni-cads are manufactured so as to accept a certain level of overcharging, but overcharging still should be minimized.

For gel-filled (nonspill, sealed, any-mounting-position) lead-acid batteries, voltage-limited charging is best, since this type cannot accept overcharging. Charging should typically be done at C/10 or C/20 until each cell's voltage reaches about 2.3 volts. At this point, the charging current should taper to simply a maintenance of charge level. Deep-cycle, lead-acid batteries are also best charged at a slow rate.

Some automotive lead-acid batteries, and some ni-cad cells/batteries may be safely "quick-charged" at higher charging rates, but the manufacturer's recommendations should be followed in doing this.

SMART CHARGERS

Many quick-charge systems simply use a "timing" system, without regard to the condition of the cells or batteries being charged. There are occasions where this technique can actually harm the cell or battery.

New *Smart Chargers,* which *sample* and *test* the condition of the cell or battery being charged, are now on the market in the communications field. The charger then cycles on and off, appropriately, to optimally charge the cell or battery without overcharging or damaging it.

CONSTRUCTION AND OPERATIONAL CHARACTERISTICS OF THE LEAD-ACID BATTERY

The key components include:

1. the positive plate(s), which is lead peroxide (PbO_2) (brown);
2. the negative plate(s), which is spongy lead (Pb) (gray);
3. the electrolyte, which is dilute sulphuric acid (H_2SO_4);
4. nonconducting separators; and
5. hard rubber or plastic enclosure.

Charged Condition(s)

When the automotive lead-acid battery is fully charged, the conditions include:

1. Positive plate(s) are lead peroxide and dark brown.

2. Negative plate(s) are spongy lead and gray.

3. The electrolyte specific gravity is between 1.275 and 1.30.

4. Closed-circuit (normal load) voltage per cell is about 2 volts. (Open-circuit voltage per cell is about 2.2 volts without load when the battery is fully charged).

Discharging Condition(s)

As the lead-acid battery is discharging, the battery's chemical actions cause:

1. The positive plates to deteriorate, resulting in formulation of lead sulfate and water, which further dilutes the electrolyte;

2. The negative plates to deteriorate, resulting in formulation of lead sulfate; and

3. The electrolyte to weaken and decrease in specific gravity.

Discharged Condition(s)

When the lead-acid battery has fully discharged, typical conditions are:

1. Both positive and negative plates are changed to lead sulfate.

2. The creation of water weakens the sulfuric acid electrolyte so the specific gravity (measured with a hydrometer) drops below 1.150.

3. Closed-circuit voltage per cell drops to about 1.75 volts.

4. If the battery remains in an uncharged condition for too long, the sulfate hardens and the chemical reversal (via charging action) becomes impossible. This is called sulfation.

BATTERY CHARGING METHODS

Overview

Charging any partially or fully discharged secondary battery involves forcing current through it in the reverse direction (from the normal current flow) when it is acting as a source.

Look at Figure 9–8. Notice the difference in current flow directions when the battery is a source (providing current to a load) and when the battery is the load for the charging device or circuit. NOTE: To accomplish the charging action, the charger voltage must be greater than the battery voltage; the negative output terminal of the charger is connected to the negative battery terminal; and the positive charger output terminal is connected to the battery positive terminal. That is, the connections are *negative-to-negative* and *positive-to-positive*.

FIGURE 9–8   Directions of
current through battery during
charge and discharge

Two Charging Methods

The constant voltage method, Figure 9–9, applies a constant charging voltage to the battery. When the battery voltage, which is series-opposing to the
charger voltage, is low, the charging current is high. As the battery increases its charge, thus its terminal voltage, the battery voltage almost
equals the charging source. Thus, the charging current's value tapers downward as the battery charges and its series-opposing voltage nears that of the
charger voltage.

The constant current method, Figure 9–10, essentially charges the battery at a constant current rate (usually the battery's normal discharge rate)
for about 90% of the charging time. During the remaining time, the rate

Higher charging current initially when battery's
series-opposing voltage is lower.

Charging current tapers downward in value as battery
charges, and its series-opposing voltage nears that of
the charger voltage.

NOTE: A constant voltage source is one that has very
low internal R; thus, varying values of load current
won't cause source output voltage to change much.

FIGURE 9–9   Constant voltage method of
charging a battery

Charged at near normal discharge rate for about 90% of charging cycle, then reduced to about 40% of normal discharge rate for the remainder of the charging time.

NOTE: A constant current source has a high internal R compared to R of load; thus, load has little effect on circuit current.

FIGURE 9–10   Constant current method of charging a battery

reduces to about 40% of normal discharge rate. For example, a 100-ampere-hour battery initially charges at a rate of about 12 amperes, then reduces to about 5 amperes for the remaining period of charge.

EXAMPLES OF OTHER TYPES OF PRIMARY AND SECONDARY CELLS

Electronic parts catalogs list a variety of primary and secondary cells and batteries. We will briefly mention several common types whose application depend on inherent features, such as size, weight, current capacity, voltage, longevity, and durability.

Examples of Rechargeable Secondary Cells and Batteries

THE EDISON CELL.   The Edison (nickel-iron) cell is frequently used in heavy-duty industrial applications. This cell is lighter and more durable than the lead-acid cell. Although it endures many rechargings, it is more expensive than the lead-acid cell.

OTHER TYPES.   Other rechargeable secondary cells and batteries are lead-calcium, silver-zinc, and silver-cadmium.

Examples of Nonrechargeable Primary Cells and Batteries

Primary cells and batteries are used in a variety of applications, such as watches, cameras, computer memory protection circuits, and radios.

ALKALINE CELL.   The alkaline cell has a longer shelf life than the standard carbon-zinc dry cell, Figure 9–11.

FIGURE 9–11   The alkaline cell (Photo by Michael A. Gallitelli)

FIGURE 9–12   The mercury cell

FIGURE 9–13   The lithium cell (Courtesy of Eveready Battery Co., Inc.)

MERCURY CELL.   The mercury is often used in voltage reference applications because of the stable output voltage (typically 1.35 volts per cell) during its lifetime, Figure 9–12.

ZINC-CHLORIDE CELL.   The main feature of the zinc-chloride cell is high current capacity.

LITHIUM CELL.   Figure 9–13 shows an example of a lithium cell. Because of their long life, lithium cells are commonly used in watches, cameras, calculators, computers, and similar devices. Although they are expensive, their long life and energy density make them practical.

SOLAR CELL.   The solar cell is a special category of primary cells used in space vehicles, calculators, and so forth. It has a voltage of about 0.26 volts per cell and a very low current rating.

   Refer to Figure 9–14 for a listing of primary and secondary cells and batteries and their nominal output voltage per cell.

PRIMARY CELLS/BATTERIES (NONRECHARGEABLE)		SECONDARY CELLS/BATTERIES (RECHARGEABLE)	
TYPE	**NOMINAL _V_**	**TYPE**	**NOMINAL _V_**
Carbon-Zinc	1.5	Lead-Acid	2.2
Alkaline	1.5	Nickel-Iron	1.4
Mercury	1.35	Nickel-Cadmium	1.2
Silver-Oxide	1.5		
Lithium	3.0		

FIGURE 9–14   Chart of primary and secondary cells/batteries with nominal voltage values

Final Comments About Cells and Batteries

Selection of the right cell or battery for the job depends on the requirements of the application relative to *size, current drains, voltage, frequency of replacement,* and *cost.*

Recall from an earlier chapter the possibilities of connecting cells or batteries, Figure 9–15a, in series to increase voltage, Figure 9–15b, and of connecting like-value voltage sources in parallel to increase current delivery, Figure 9–15c. When performing either process, it is recommended that the cells and batteries have the same manufacturer's electrical ratings.

Also, recall that as a cell or battery deteriorates, its internal resistance increases. This means that for a given current drain, its output terminal voltage decreases due to the increased internal IR drop at the given current value. NOTE: Some *newer types* of batteries typically *maintain* their *voltage* output then *suddenly die.* Examples are mercury (primary) cells and nickel-cadmium (secondary) cells.

Shelf life of cells is greatly affected by temperature. Most dry cells operate near 70 degrees F, so long exposure to higher temperatures drastically shortens their life. Conversely, batteries stored for long periods at low temperatures (e.g., 0 degrees F) have a good shelf life.

Single cell = 1.5 V (nominal)

(a)

Cells in series provide higher voltage out
(series-aiding connections)

(b)

Cells in parallel can provide higher current
to the load at a voltage = to that of one cell

(c)

FIGURE 9–15   Advantages of multiple-cell connections

PRACTICAL NOTES

Caution: *Do not leave dry cells or batteries in equipment for long periods without checking them!* With age, most cells and batteries leak electrolyte or acid that damages battery holder contacts, circuitry, or equipment. Therefore, remove a dead or degenerated cell or battery, even if a replacement is not available. Do not leave a dead cell or battery in the equipment.

Finally, do not recharge batteries at too high a charging rate; be cautious with sparks or flames near a charging battery; and, of course, be careful to prevent spillage of the electrolyte acid!

SUMMARY

▶ Cells and batteries (combinations of cells) are important sources of dc voltage and current, especially for portable equipment (e.g., emergency back-up equipment).

▶ Primary cells are *not rechargeable,* have a variety of voltage ratings, current capabilities, expected shelf life ratings, and costs, and are produced in many sizes and shapes.

▶ Secondary cells and batteries *are rechargeable.* Secondary cells or batteries (such as lead-acid batteries and ni-cad batteries) are available in a variety of ratings, sizes, shapes, and costs.

▶ The common dry cell is composed of a positive carbon electrode, a negative zinc electrode, and an electrolyte of ammonium chloride (and water), called sal ammoniac.

▶ During a dry cell's normal operating chemical process, hydrogen bubbles form around the positive electrode. This polarization degenerates the battery's output. To reduce polarization, a depolarizing agent is placed in the dry cell during manufacture. A common depolarizing agent is manganese dioxide.

▶ Local action also causes degeneration in dry cells. Impurities in the negative zinc electrode suspend in the electrolyte near the zinc, creating their own voltaic action and eating away the zinc. To reduce local action, the inside of the zinc container (negative electrode) is covered with a thin coating of mercury;

thus preventing most potential for local action.

▶ Shelf life of cells and batteries depends on storage conditions and chemical properties.

▶ Voltage ratings of cells and batteries depend on the electrode material. Current ratings largely depend on the size (active area) of the electrodes.

▶ When using dry cells, care should be taken with their storage. Also, frequently check their condition, and make replacements, as appropriate.

▶ One common secondary (rechargeable) cell is the lead-acid cell. In this cell, the positive plates are lead peroxide, the negative plate(s) are spongy lead, and the electrolyte is diluted sulfuric acid.

▶ Batteries are rated according to the amount of current they deliver for a specified time. For example, a battery rated at 120 ampere-hours can nominally deliver 15 amperes for 8 hours, or 6 amperes for 20 hours. Small batteries and cells are rated in milliampere-hours.

▶ Rechargeable batteries are charged at high rates for short periods, or at lower rates for longer periods to restore them to full charge. The preferred method is to charge at lower rates for longer periods, decreasing the risk of cell or battery damage as well as preventing excessive gassing.

▶In addition to the common carbon-zinc primary cell, many other types exist, such as alkaline, mercury, zinc-chloride, and lithium cells.

▶Common types of rechargeable batteries are the lead-acid, nickel-iron (Edison), and nickel-cadmium.

REVIEW QUESTIONS

1. Briefly define positive ion.
2. In a carbon-zinc cell, which element is the positive electrode and which is the negative electrode?
3. Briefly describe the process of amalgamation.
4. What term describes the undesired collection of hydrogen bubbles around a dry cell's positive electrode?
5. Name a common depolarizing agent.
6. Name a common rechargeable cell used for many small portable electronic devices.
7. What type of cell is frequently a reference voltage source due to its stable output voltage?
8. Name two advantages and one disadvantage of the lead-acid cell.
9. What is the ampere-hour rating of a battery that delivers 12.5 amperes for 8 hours without its output voltage excessively dropping?
10. Using the constant voltage method, describe the process and features of charging action when a battery is recharged.

TEST YOURSELF

1. Explain the polarity relationships between charger and battery when recharging a battery.
2. Using the constant current method, explain the process when a battery is recharged.
3. Refer to an electronic parts catalog and find the typical ampere-hour or milliampere-hour ratings for:
   a. Ni-cad cells
   b. Alkaline dry cells
4. What is the internal resistance of a battery whose no-load voltage is 12 volts, but whose terminal voltage drops to 11 volts when a load current of 20 amperes is demanded?
5. Refer to an electronic parts catalog. Find and record the voltage and current for the following cells:
   a. AAA
   b. AA
   c. A
   d. C
   e. D

6. What primarily determines the voltage-delivering capability of a single cell?

7. What primarily determines the current-delivering capability of a single cell?

8. Why does a cell's output voltage decrease under load, particularly when the cell is going bad?

9. What are some common applications of the modern lithium cell?

10. What makes the lithium cell so useful for the applications you named in question 9?

Following the SIMPLER troubleshooting sequence, find the problem with this circuit. As you follow the sequence, record your responses to each step on a separate sheet of paper.

CHALLENGE CIRCUIT 4

SYMPTOMS Gather, verify, and analyze symptom information. (Look at the "Starting Point Information.")

IDENTIFY Identify initial suspect area for the location of the trouble.

MAKE Make a decision about "What type of test to make" and "Where to make it." (To simulate making a decision about each test you want to make, select the desired test from the "TEST" column listing below.)

PERFORM Perform the test. (Look up the result of the test in Appendix C. Use the number in the "RESULTS" column to find the test result in the Appendix.)

LOCATE Locate and define a new "narrower" area in which to continue troubleshooting.

EXAMINE Examine available information and again determine "what type of test" and "where."

REPEAT Repeat the preceding analysis and testing steps until the trouble is found. What would you do to restore this circuit to normal operation? When you have solved the problem, compare your results with those shown on the following pages.

STARTING POINT INFORMATION

1. Circuit Diagram
2. Cells are 1.5 V cells
3. With S_1 closed, $V_R = 5$ V

TEST   Results in Appendix C

(With S_1 Open)
V_{A-B} . . . .(18)
V_{B-C} . . . .(1)
V_{C-D} . . . .(51)
V_{D-E} . . . .(99)
V_{A-F} . . . .(77)

(With S_1 Closed)
V_{A-B} . . . .(7)
V_{B-C} . . . .(50)
V_{A-C} . . . .(81)
V_{C-D} . . . .(112)
V_{D-E} . . . .(73)
V_{C-E} . . . .(10)
V_{A-F} . . . .(105)

CIRCUIT

Meter lead

Meter lead

4

Symptoms With the switch closed, V_R is 5 volts. However, if all the 1.5-volt cells were operating properly, it should be 6 volts. Since the resistor drops 5 volts with the switch closed, there is circuit current and circuit continuity. This implies a worn cell in the group.

Identify initial suspect area: All the cells.

Make test decision: If we use the divide-and-conquer (splitting) technique, we can check the voltage from A to C or from C to E. This narrows the suspect area to two cells, either C_1 and C_2, or C_3 and C_4. Let's start with the first pair by checking voltage from point A to C with S_1 closed to cause circuit current. That is, the batteries are being tested under load.

Perform The result of measuring V_{A-C} is 2 V, which is an abnormal reading for two 1.5-V cells in series.

Locate new suspect area: Cells C_1 and C_2.

Examine available data.

Repeat analysis and testing: Check the voltage across either C_1 or C_2 with S_1 closed.

CHALLENGE CIRCUIT 4

Check voltage across C_1. V is 1.5 V, which is normal.

Check voltage across C_2. V is 0.5 V, which is abnormal, suggesting that cell C_2 is probably the bad cell.

Replace cell C_2 with a good cell, and observe voltage across R and S_1 closed. V_R is 6 V, which is normal.

1st Test

2nd Test

3rd Test

4th Test

Magnetism and Electromagnetism

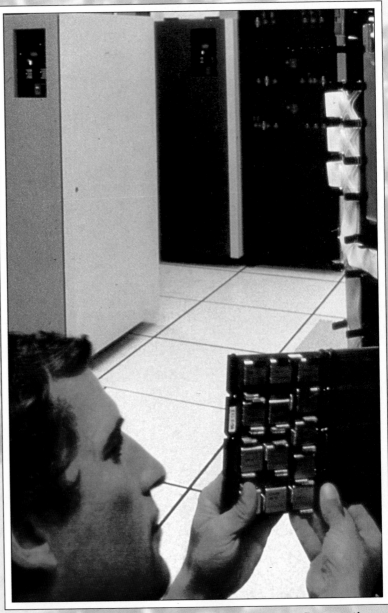

Courtesy of International Business Machines Corporation

10

KEY TERMS

Ampere-turns (AT)

Electromagnetism

Faraday's law

Flux

Flux density

Induction

Left-hand rules

Lenz's law

Lines of force

Magnetic field

Magnetic field intensity (H)

Magnetic polarity

Magnetism

Magnetomotive force

Permeability (μ)

Relative permeability (μ_r)

Reluctance

Residual magnetism

Saturation

Tesla

Weber

OUTLINE

Background information

Fundamental laws, rules, and terms to describe magnetism

Elemental electromagnetism

Some practical applications of magnetism

Important magnetic units, terms, symbols, and formulas

Practical considerations about core materials

The B-H curve

The hysteresis loop

Induction and related effects

Faraday's law

Lenz's law and reciprocal effects of motors and generators

Other topics related to magnetism

CHAPTER PREVIEW

Of all the phenomena in the universe, one of the most fascinating is magnetism and its unique relationship to electricity.

Magnetic storage media have become extremely important in today's world. Computers use magnetic "floppy disks" and "hard disks" to store tremendous amounts of data. Audio and video recorders use magnetic tape to store magnetic patterns that are transformed to sound and visual intelligence, which we can enjoy at our choosing.

Knowledge of magnetism is important for understanding a variety of industrial measurement and control devices and systems. Some examples include electrical meters, relays, solenoids, magnetic switches, inductors, transformers, and motors and generators, just to name a few.

In achieving your goal to become a technician, you will have the opportunity to study a number of these devices and components. In this chapter, you will focus on basic principles related to magnetism, some units of measure used with magnetic circuits, and important relationships between magnetic and electrical phenomena.

OBJECTIVES

After studying this chapter, you should be able to:

- Define **magnetism, magnetic field, magnetic polarity,** and **flux**
- Draw representations of magnetic fields related to permanent magnets
- State the magnetic attraction and repulsion law
- State at least five generalizations about magnetic lines of force
- Draw representations of fields related to current-carrying conductors
- Determine the polarity of electromagnets using the **left-hand rule**
- List and define at least five magnetic units of measure, terms, and symbols
- Draw and explain a B-H curve and its parameters
- Draw and explain a hysteresis loop and its parameters
- Explain motor action and generator action related to magnetic fields
- List the key factors related to induced emf
- Briefly explain the relationships of quantities in **Faraday's law**
- Briefly explain **Lenz's law**

BACKGROUND INFORMATION

Early History

Early discovery of magnetism happened by people noticing the unique characteristics of certain stones called lodestones. These stones (magnetite material), which were found in the district of Magnesia in Asia Minor, attracted small bits of iron. It was also noted that certain materials stroked by lodestones became magnetized.

Materials or substances that have this unique feature are called magnets, and the phenomenon associated with magnets is called **magnetism.** Materials that are attracted by magnets are called magnetic materials.

As early as the eleventh century, the Chinese used magnets as navigational aids, since magnets that are suspended, or free to move, align approximately in a north-south geographic direction.

Today we have materials, called permanent magnets, that retain their magnetism for long periods. Examples of permanent magnets include hard iron or special iron-nickel alloys.

Magnetic materials that lose their magnetism after the magnetizing force is removed are termed temporary magnets.

Later Discoveries

Centuries later the relationship between electricity and magnetism was discovered. In the 1800s several scientists made important observations. Hans Oersted noted that a free-swinging magnet would react to a current-carrying wire, if they were in close proximity. As long as current passed through the wire the reaction occurred. But when there was no current, there was no interaction between the magnet and wire. This indicated a steady current caused a steady magnetic effect near the current-carrying conductor. This "field of influence" would eventually be called a **magnetic field.**

Later in the same century, Joseph Henry and Michael Faraday discovered interesting phenomena that occurred when changing current levels or moving magnetic fields were present. Henry determined when current changed levels in a conductor, current was induced in a nearby conductor (having a closed current path). Faraday determined a permanent magnet moving in the vicinity of a conductor induced current in the wire (if there existed a complete path for current through the conductor). It should also be stated that even without a closed current path, voltage is induced.

All of these discoveries are very important! You will see these scientists' names again since several units of magnetic measure are named in their honor.

FUNDAMENTAL LAWS, RULES, AND TERMS TO DESCRIBE MAGNETISM

When a magnet is created, there is an area of magnetic influence near the magnet called a magnetic field. A magnetic field is established either by alignment of internal magnetic forces within a magnetic material, Figure 10–1, or by organized movement of charges (current flow) through conductor materials. This magnetic field is composed of magnetic **lines of force,**

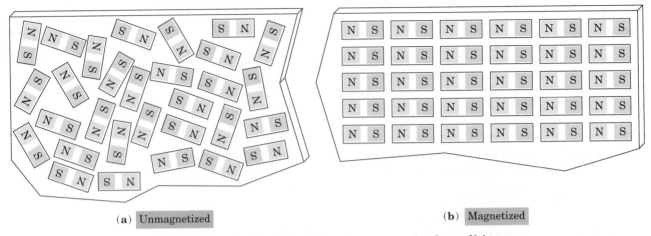

(a) Unmagnetized   (b) Magnetized

FIGURE 10–1   Magnetized and unmagnetized conditions

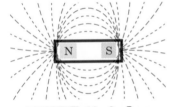

FIGURE 10–2   Iron filings displaying magnetic field pattern of a bar magnet

sometimes called **flux lines.** Thus, magnetic flux refers to all the magnetic lines of force associated with that magnet. The symbol for flux is the Greek letter Phi (ϕ). Each magnetic field line is designated by a unit called the maxwell (Mx). That is, one magnetic field line equals one maxwell, or 50 magnetic field lines equal 50 maxwells.

Another basic unit associated with flux is the **weber.** Whereas the maxwell equals one line of flux, the weber (Wb) equals 10^8 lines of flux, or 100 million flux lines. Thus, a microweber equals 100 lines, or 100 Mx.

Another fundamental term related to flux is **flux density,** which refers to the number of lines per given unit area. We'll discuss flux density later in the chapter.

If a nonmagnetic material, like cardboard, is laid on top of a bar magnet (or other magnets), sprinkled with small iron filings, then tapped or vibrated, the iron filings position themselves in a pattern illustrating the effect of the magnetic field around the magnet, Figure 10–2.

A diagram frequently used to show this effect is in Figure 10–3. Notice that at each end of the bar magnet, the lines of force are concentrated, but

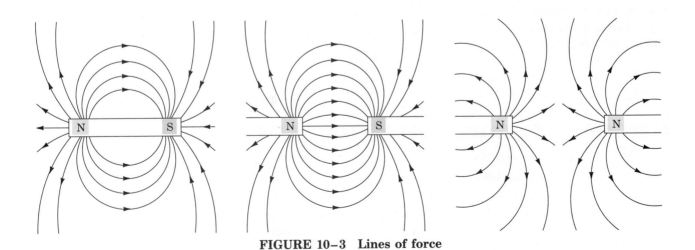

FIGURE 10–3   Lines of force

**FIGURE 10–4   Like poles repel each
other, and unlike poles attract
each other.**

as they get farther from the ends, they spread. As you can see, the ends of
the bar magnets (where the lines of force are concentrated) are labeled "N"
and "S" and are the "poles" of the magnet. That is, the N and S indicate
North-seeking and South-seeking poles.

To represent directions of the lines of force, it is customary to say the
flux lines exit the North pole and reenter the South pole of the magnet.
Inside the magnet, of course, the flux lines travel from the South pole to the
North pole. Each flux line is an unbroken loop, or ring.

In Figure 10–4 you can see two bar magnets used to illustrate that *like
poles repel each other, and unlike poles attract each other.* This is an important law, so remember it!

Rules Concerning Lines of Force

Along with the law defined above, you need to know the following important
generalizations related to *lines of force* (flux lines), Figure 10–5.

1. Magnetic lines are continuous, Figure 10–5a.

2. Magnetic lines flow from North to South Pole outside the magnet and
   from South to North inside the magnet, Figure 10–5a.

3. Magnetic lines take the shortest or easiest path, Figure 10–5b.

4. Lines in the same direction repel each other but are additive, strengthening the overall field. Lines in opposite directions attract and cancel
   each other, weakening the overall field, Figure 10–5c.

5. Magnetic lines penetrate nonmagnetic materials, Figure 10– 5d.

6. Lines of force do not cross each other, Figure 10–5e.

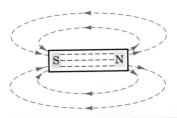

Magnetic lines are continuous and are assumed to flow
from North to South poles external to the magnet
and from South to North poles inside the magnet.

(a)

Magnetic lines tend to take the shortest or easiest path.

(b)

Lines in same
direction repel each other,
but are additive,
strengthening overall field.

Lines in opposite directions
attract and cancel each other,
weakening overall field.

(c)

Glass

Magnetic lines penetrate nonmagnetic materials.

Lines of force don't cross each other.

(d)   **(e)**

FIGURE 10–5   Rules about lines of force

Some Practical Points About Permanent Magnets

Before discussing electromagnetism, here are some practical points about
permanent magnets.

1. They are not really permanent, because they lose their magnetic
   strength over time, or if heated to a high temperature, or if physically
   pounded upon, and so on.

2. They can be configured into different shapes in addition to bar magnets. One common shape is termed a horseshoe magnet due to its appearance, Figure 10–6.

3. The field lines' shape depends on the easiest path for flux, Figure
   10–7.

4. Appropriate methods to store magnets help preserve their field
   strengths for longer periods. For example, store bar magnets so oppo-

**FIGURE 10–6
Horseshoe
magnet**

**FIGURE 10–7
Shape of field
lines depends
on easiest
path for flux.**

**FIGURE 10–8   Preserving the
strength of magnets**

site poles are together, or use a "keeper bar" of magnetic material
across the ends of a horseshoe magnet, Figure 10–8.

5. A magnet induces magnetic properties in a nearby object of magnetic
   material, if the object is in the path of the original magnet's field lines,
   Figure 10–9.

NOTE: Polarity of "induced" magnetism places "unlike" poles next to each other.
This explains how magnets attract metal objects.

FIGURE 10–9   Induced magnetism

✓ IN-PROCESS LEARNING CHECK I

Fill in the blanks as appropriate.

1. Small magnets that are suspended and free to move align in a _____ and _____ direction.

2. Materials that lose their magnetism after the magnetizing force is removed are called _____ _____. Materials that retain their magnetism after the magnetizing force is removed are called _____ _____.

3. Can a wire carrying dc current establish a magnetic field? If it does, is this field stationary or moving?

4. A law of magnetism is like poles _____ each other and unlike poles _____ each other.

5. The maxwell is a magnetic unit that represents how many lines of force?

6. The weber is a magnetic unit that represents how many lines of force?

7. Are magnetic lines of force continuous or noncontinuous?

8. Lines of force related to magnets exit the _____ pole of the magnet and enter the _____ pole.

9. Do nonmagnetic materials stop the flow of magnetic flux lines through themselves?

10. Can magnetism be induced from one object to another object?

ELEMENTAL ELECTROMAGNETISM

Direction of Field Around a Current-Carrying Conductor

Recall that Hans Oersted discovered a current-carrying wire has an associated magnetic field. A relationship between the current direction through the wire and the magnetic field direction around the current-carrying conductor is established while moving the compass around the conductor and noting the effect on the compass needle.

Note in Figure 10–10a you are looking at the end of wire (black dot), and the current is coming toward you. In Figure 10– 10b you get an overview of the relationship of current direction with the magnetic field around the current-carrying conductor. Both of these figures illustrate one of the **left-hand rules.**

The Left-Hand Rule for Current-Carrying Conductors

Grasping the wire with your left hand so your extended thumb points in the direction of current flow, the magnetic field around the conductor is in the direction of the fingers that are wrapped around the conductor. (NOTE: For those who use "conventional current," where the direction of current flow is considered in the opposite direction, the "right-hand" rule is used instead of this left-hand rule).

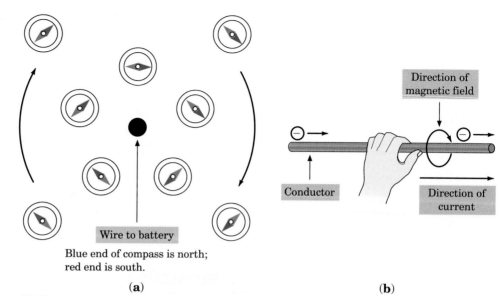

Blue end of compass is north;
red end is south.

(a)   **(b)**

**FIGURE 10–10   (a) Field around current-carrying conductor; (b) Left-hand
rule for current-carrying conductor**

EXAMPLE

Another way to illustrate this rule is shown in Figure 10–11. In this illustration the current (and your thumb) is coming toward you from the conductor with the dot, Figure 10–11a. The dot is like looking at the tip of an arrow coming toward you. In this case, the magnetic field is in a clockwise direction.

In the other picture, Figure 10–11b, the conductor has a plus sign, indicating current is going away from you into the page. This is similar to looking at the tail feathers of an arrow. In this case, the direction of magnetic field is counterclockwise.

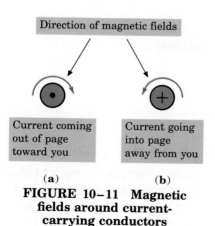

**FIGURE 10–11   Magnetic
fields around current-
carrying conductors**

FIGURE 10–12   Force between
adjacent current-
carrying conductors

FIGURE 10–13

PRACTICE PROBLEMS I

1. Draw a picture of two parallel conductors with the direction of current going away from you in both cases.

2. Using arrows, show the directions of the magnetic fields around each conductor.

Force Between Parallel Current-Carrying Conductors

Look at Figure 10–12a. Notice that when two parallel current-carrying wires have current in the same direction, the flux lines between the wires are in opposite directions. This situation causes the flux lines to attract and cancel each other, and the magnetic field pattern is modified since lines of force cannot cross. This indicates the two adjacent wires move toward the weakened portion of the field or toward each other.

On the other hand, if the two wires have current in opposite directions, Figure 10–12b, the lines of force between the two conductors move in the same direction. Hence, the flux lines repel each other and the wires move apart.

PRACTICE PROBLEMS II

Indicate whether the conductors move together (toward each other) or apart in Figure 10–13a, b, and c.

FIGURE 10–14
Stronger
"combined" field

Field Around a Multiple-Turn Coil

Individual fields of conductors carrying current in the same direction modify to a larger and stronger field, as shown in Figure 10–14.

When a single current-carrying wire forms multiple loops (such as in a coil), there are a number of fields that combine from adjacent turns of the coil, producing a large field through and around the coil, Figure 10–15.

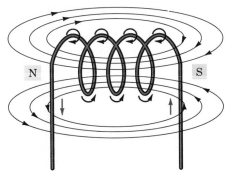

FIGURE 10–15   Strong field of coil and establishment of magnetic polarity

(a)   (b)

FIGURE 10–16   Left-hand rule to determine magnetic polarity of an electromagnet

Also notice in Figure 10–15 the magnet established by this coil has North and South magnetic poles, like a permanent bar magnet. The **magnetic polarity** of this electromagnet is determined by another left-hand rule.

Left-Hand Rule To Determine Polarity of Electromagnets

Grasping the coil so your fingers are pointing in the same direction as the current passing through the coil, the extended thumb points toward the end of the coil that is the North pole, Figure 10–16.

PRACTICE PROBLEMS III

See if you can determine the magnetic polarity of the coils in Figure 10–17a, b, and c.

(a)   (b)   (c)

FIGURE 10–17

Factors Influencing the Field Strength of an Electromagnet

From the previous discussions, you should surmise the number of turns influences the strength of the electromagnet. That is, the higher the number of turns per unit length of coil, the greater the strength of the magnetic field, Figure 10–18a.

Also, you might have reasoned the amount of current through the conductor directly affects the magnetic strength, since a higher current indicates more moving charge per unit time, Figure 10-18b. These are correct assumptions. Furthermore, the larger the cross-sectional area of the coil, the greater the strength, Figure 10–18c.

Another factor influencing strength is the characteristics of the material in the flux-line path. If the material in the flux path is magnetic, the strength is greater compared with a path using nonmagnetic material, Figure 10–18d.

The length of the magnetic path also influences strength. For example, suppose we have a 100-turn coil with a given length of "X." Now if we spread that 100 turns over a length twice as great (2X), the strength reduces because the turns are farther apart, thus the effective magnetizing force is less, Figure 10– 18e.

To summarize, field strength is influenced by:

1. number of turns on the coil;

2. value of current through coil;

3. cross-sectional area of coil;

4. type of core (magnetic characteristics of material in flux path); and

5. length of the coil (number of turns per unit length).

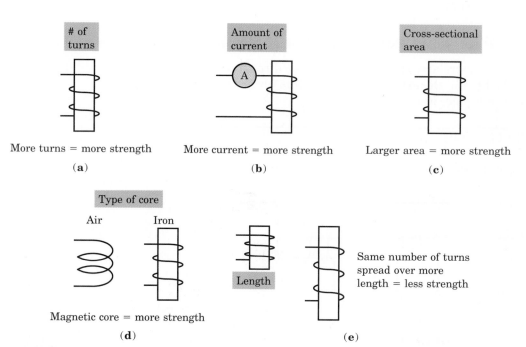

FIGURE 10–18   Some factors that determine strength of an electromagnet

IN-PROCESS LEARNING CHECK II

1. For a given coil dimension and core material, what two factors primarily affect the strength of an electromagnet? _____ and _____.

2. The left-hand rule for determining the polarity of electromagnetics states that when the fingers of your left hand point (in the same direction, in the opposite direction) _____ as the current passing through the coil, the thumb points toward the (North, South) _____ pole of the electromagnet.

3. Adjacent current-carrying conductors that are carrying current in the same direction tend to (attract each other, repel each other) _____.

4. If you grasp a current-carrying conductor so that the thumb of your left hand is in the direction of current through the conductor, the fingers "curled around the conductor" (will indicate, will not indicate) _____ the direction of the magnetic field around the conductor.

5. When representing an end view of a current conductor pictorially, it is common to show current coming out of the paper via a (dot, a cross, or plus sign) _____.

Highlighting Magnetism and Electromagnetism Facts Discussed Thus Far:

1. Like poles repel, and unlike poles attract, again see Figures 10–4 and 10–19a.

2. Amount of repulsion or attraction depends on the strength and proximity of the poles involved. Repulsion or attraction is directly related to the poles' strength and inversely related to the *square* of their separation distance, Figure 10–19a.

Like poles repel

Electrical circuit

$$I = \frac{V}{R}$$

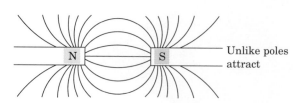

Unlike poles attract

Amount of repulsion or attraction depends on strength of each magnet and is inverse to distance squared.

$$F = \frac{M_1 M_2}{d^2}$$

Magnetic circuit

$$\phi = \frac{mmf}{\Re}$$

(a)   (b)

FIGURE 10–19 Highlights

Formula 10–1	Force $= \dfrac{\text{Pole 1 Strength} \times \text{Pole 2 Strength}}{d^2}$

 pole 1 value, $\boxed{\times}$, pole 2 value, $\boxed{\div}$, $\boxed{(}$, distance value, $\boxed{\times^2}$, $\boxed{)}$, $\boxed{=}$

3. Strength of magnet is related to number of flux lines in its magnetic field per unit cross-sectional area.

4. Magnetic lines outside of the magnet travel from North pole to South pole, again see Figure 10–5a and 10–19a.

5. Characteristics of the flux path determine how much flux is established by a given magnetizing force. Therefore, if the path is air, vacuum, or any nonmagnetic material, less flux is established. If the path is a magnetic material, flux lines are more easily established. These concepts indicate that when air gaps are introduced into a magnetic path, the magnetic strength of the field is diminished.

6. The shape or configuration of the magnetic field flux lines is controlled by features of the materials in their path. That is, lines of force follow the path of least opposition, called **reluctance** in magnetic circuits, again see Figure 10–5b.

7. An analogy can be drawn between Ohm's law in electrical circuits and the relationships of a magnetic circuit. In Ohm's law, the amount of current through a circuit is directly related to the amount of electromotive force (V) applied to the circuit and inversely related to the circuit's opposition to current flow (R). In a magnetic circuit, the number of flux lines through the magnetic circuit is directly related to the value of the magnetizing force establishing flux and inversely related to the circuit's opposition to those flux lines. This opposition is the reluctance of the path, and the symbol is $\mathcal{R}$, Figure 10–19b.

FIGURE 10–20　Several magnetic media (Photo by Michael A. Gallitelli)

SOME PRACTICAL APPLICATIONS OF MAGNETISM

Magnetic Storage Media

In Figure 10–20 you see several commonly used magnetic storage media elements.

Information on the computer diskettes is stored in structured track and sector formats. Figure 10–21. This allows the computer to selectively access and recover data at any desired location in the "magnetically memorized" data, by virtue of that data's location on the diskette.

On audio cassette tapes, of course, sound waves have been translated into electrical signals at the recording tape head, which in turn causes

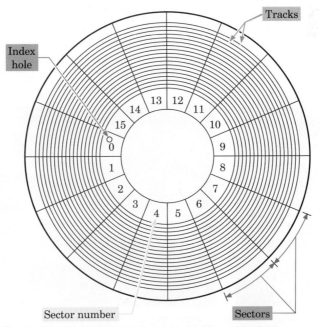

FIGURE 10–21   Computer "floppy" disks typical structure and format

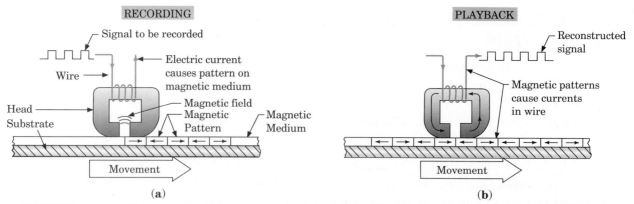

FIGURE 10–22   **Using magnetism to record and play back signals (audio, video, or digital data)**

magentic patterns to be induced onto the magnetic tape, Figure 10–22a. When the recorded tape is run over the playback head of the recorder, the reverse process happens. That is, the magnetic patterns on the tape induce electrical signals into the pickup head, Figure 10–22b. These signals are then processed, amplified, and eventually become sound waves, which activate your ear drums.

Video cassette tape recording systems use a technique similar to, but somewhat more complicated than, the audio recording system. Both audio and video signals are magnetically stored on a videotape's magnetic coating. Once recorded, you can enjoy both seeing and hearing the results by playing them back on your VCR.

Industrial Components and Devices

In industry (and in many home appliances, as well), there are a myriad of applications for relays, solenoids, and magnetic switches, Figure 10–23. Although solid-state switching devices have begun replacing electromechanical relays and solenoids in low-power circuitry, relays and solenoids

FIGURE 10–23   **Common industrial electromagnetic devices (Photos a and b courtesy of Guardian Electric Mfg. Co.; photo c by Michael A. Gallitelli)**

are still in wide use. The contacts on relays are used to make and break (open and close) circuits as the relay is energized by current flowing through its electromagnetic coil, or is deenergized by current ceasing to flow through its coil, Figure 10–24.

The movable element (plunger, or movable core) in a solenoid is typically used to physically move something. It changes electrical energy into linear, or straight-line, motion, Figure 10–25. The travel of this movable core, then, can be used to do everything from shifting cycles in your washing machine, to operating valves, or positioning items on a production line in industry.

As a technician in industry, you will frequently use metering devices or instruments, which also depend on magnetism and electromagnetism for

(a)

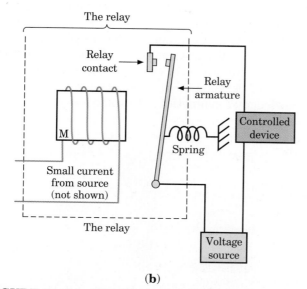

(b)

FIGURE 10–24   Simplified diagrams of relays in use
(a) Automotive starter system contactor relay (b) Single-pole,
single-throw relay operated by low power but controlling
higher-power device

FIGURE 10–25   Several frame-style solenoids (Courtesy of Guardian Electric Mfg. Co.)

their operation. At home or at work, you will hear speakers. Most speakers depend upon the interaction of a fixed (permanent magnet) and a changing current through a movable electromagnet (the voice coil) to cause the speaker cone to move the air in the room, Figure 10–26. This movement of air, of course, sets up the sound waves you hear.

Motors and generators represent another large application area for the use of magnetism and electromagnetism. A motor changes electrical energy into rotary motion. In contrast, recall that the solenoid changes electrical energy into straight-line motion.

Later in this book, you will be studying the basic operation of transformers. Transformers are widely found in consumer products, delivery of electrical power, and industrial applications. They represent another important application of magnetism and electromagnetism.

In depth details of all these magnetic devices will be taught at the appropriate times in your studies. For now, we wanted you to see just a few of the applications of permanent and electromagnets and realize how important the knowledge of magnetics is to you as a technician or engineer.

FIGURE 10–26   Permanent magnet dynamic speaker

IMPORTANT MAGNETIC UNITS, TERMS, SYMBOLS, AND FORMULAS

Now, let's study some units of measure for the magnetic characteristics discussed earlier. Recall a single flux line (or magnetic line of force) is a maxwell, and 10^8 flux lines equal a Weber (Wb). Also, although magnetizing force was presented, we did not specifically define it.

Before thoroughly discussing magnetic units, some background about systems of units would be helpful. Most of us daily use units of measure pertaining to dimensions, volume, weight, speed, and so on. In the field of electronics, it is critical that some meaningful system of units exists.

Before 1960 there were several systems of units related to electricity, electronics, and magnetism. One system was the common unit of the volt, ampere, and watt; another system dealt with electrostatics; and a third system explained magnetics. Both electrostatics and magnetics used a system called the *cgs* system, where *c* was distance in centimeters, *g* was weight in grams, and *s* was time in seconds.

However, since the 1960s, a system has been accepted that uses the *Meter* for distance, the *Kilogram* for weight, the *Second* for time, and the *Ampere* (necessary for electrical systems) for current measure. This system is abbreviated as the *MKSA*. Since this system has been adopted internationally, it is called the International System, abbreviated "SI," *Système internationale d'unités*.

Examples of the units, symbols, and formulas are in Figure 10–27. You will need to refer to this chart several times as you study the next section.

The Basic Magnetic Circuit

Since technicians do not typically have to deal with all the systems of units presented in Figure 10–27, we will not dwell on this subject. However, you should be aware of these units and have an understanding of the magnetic circuit. Repeating some of the ideas stated earlier, the magnetic equivalent of Ohm's law, sometimes called Rowland's law, is shown in Figure 10–28.

Relationships Among *B, H,* and Permeability Factors

From our previous discussion, it is possible to note H (magnetic field intensity or mmf per unit length) is the factor producing flux per unit area (or flux density, B) within a given magnetic medium.

It is further observed the amount of B a given H produces is directly related to the permeability of the flux path. For a vacuum only, it is stated that:

Formula 10–2	$\mu_0 \; = \; \dfrac{B}{H}$

 B value, $\boxed{\div}$, H value, $\boxed{=}$

TERM OR QUANTITY	SYMBOL OR ABBREVIATION	SI UNIT AND FORMULA	CGS UNIT AND FORMULA
Flux (lines)	ϕ	weber (Wb) $= \dfrac{\text{number lines}}{10^8}$	Maxwell (Mx) number lines in field
Flux Density (Magnetic flux per unit cross-sectional area at right angles to the flux lines)	B	$\dfrac{\text{webers}}{\text{sq meter}} = \textbf{tesla}$ (T) $B = \dfrac{\phi\,(\text{mks})}{A\,(\text{mks})}$ or teslas $= \dfrac{\text{Wb}}{\text{sq mtr}}$	$\dfrac{\text{lines}}{\text{sq cm}}$ $B = \dfrac{\phi\,(\text{cgs})}{A\,(\text{cgs})}$ or gauss $= \dfrac{\text{Mx}}{\text{sq cm}}$
Magnetomotive Force (That which forces magnetic lines of force through a magnetic circuit)	MMF	Ampere-turns or $\textbf{AT} = NI$	F (cgs) $= 0.4\pi NI$ or F (gilberts) $= 1.26 \times N \times I$
Magnetic Field Intensity (Magnetomotive force per unit length)	H	Ampere-turns per meter $\dfrac{NI}{\text{length}}$ or $\dfrac{AT}{\text{meter}}$	H (oersteds) $= \dfrac{0.4\pi NI}{\text{length}}$ or $\dfrac{1.26\,NI}{1\text{ cm}}$
Permeability (Ability of a material to pass, conduct, or concentrate magnetic flux; analogous to conductance in electrical circuits)	μ	Vacuum or air; $\mu_o = B/H$. B is teslas; H is AT/meter where free space is considered to have an "absolute" permeability of: $4\pi \times 10^{-7}$ or 12.57×10^{-7}.	Free space: $\mu_o = B/H$. B is in gauss; H is in oersteds; In cgs system, $\mu_o = 1$.
Relative Permeability (Not constant because it varies with the degree of magnetization)	μ_r	Relative permeability of a material is a ratio. Thus, $\mu_r =$ $\dfrac{\text{Flux density with core material}}{\text{Flux density with vacuum core}}$ where, flux density in the core material is: $B = \mu_o\mu_r H$ teslas, and absolute permeability of core materials is: $\mu = B/H = \mu_o\mu_r$ (SI units)	Same concept in both systems

FIGURE 10–27 Magnetic units, symbols, and formulas chart

OHM'S LAW	ROWLAND'S LAW	EXPLANATION
$R = \dfrac{V}{I}$, where: Resistance in ohms equals electromotive force in volts divided by current in amperes.	$\mathscr{R} = \dfrac{F(mmf)}{\phi}$, where: Reluctance (the magnetic circuit's opposition to the establishment of flux) equals the magnetomotive force (F) divided by the flux (ϕ). Also, it can be shown that: $\mathscr{R} = \dfrac{l}{\mu A}$ where l = length of the path in meters; μ = permeability of the path A = cross-sectional area of the path perpendicular to flux, in meters.	Reluctance equals magnetomotive force divided by flux. Flux equals magnetomotive force divided by reluctance. $F\ (mmf) = \phi \times \mathscr{R}$ or flux times reluctance.

FIGURE 10–28 Rowland's Law

Formula 10–3	$B = \mu_o \times H$

 μ_o value, $\boxed{\times}$, H value, $\boxed{=}$

Formula 10–4	$H = \dfrac{B}{\mu_o}$

 B value, $\boxed{\div}$, μ_o value, $\boxed{=}$

Referring again to Figure 10–27 and other related discussions, we will now apply this information to solve some problems dealing with magnetic parameters. This will give you a better understanding of the relationships.

EXAMPLES

Problem 1: What is the **magnetomotive force,** if a current of 5 amperes passes through a coil of 100 turns?

Answer:

Formula 10–5	$mmf = NI$

 number of turns value, $\boxed{\times}$, current value, $\boxed{=}$

Thus, $mmf = 100 \times 5 = 500$ ampere-turns

Problem 2: If the length of the coil in problem 1 is 0.1 meters, what is the **magnetic field intensity?**

Answer:

Formula 10–6	$H = \dfrac{AT}{\text{meter}}$

 value of current (amperes), $\boxed{\times}$, number of turns, $\boxed{\div}$, number of meters, $\boxed{=}$

Thus, $H = \dfrac{500}{0.1} = 5000$ ampere-turns per meter

Problem 3: What is the **flux density** (SI units) if there are 10,000 lines of force coming from a North magnetic pole face (surface) having a cross-sectional area of 0.002 square meters?

Answer:

Formula 10–7	$B = \dfrac{\phi\ (\text{Wb})}{A\ (\text{m}^2)}$

 $\boxed{(}$, number of lines, $\boxed{\div}$, 10^8, $\boxed{)}$, $\boxed{\div}$,
area in meters squared, $\boxed{=}$

$$\text{Thus, } B = \left[\frac{\dfrac{10^4}{10^8}}{2} \times (10^{-3}\text{m}^2)\right]$$

$$= \frac{10^{-4}}{0.002} = \frac{0.0001}{0.002} = 0.05 \text{ T}$$

NOTE: Stated in micro units, $\dfrac{100\ \mu\text{Wb}}{2{,}000\ \mu\text{m}} = 0.05 \text{ T}$

Problem 4: What is the **"absolute"** permeability (SI) of a core material if its relative permeability (μ_r) is equal to 2,000?

Answer:

Formula 10–8	$\mu = \mu_o \times \mu_r$

 μ_o value, $\boxed{\times}$, μ_r value, $\boxed{=}$

Thus, $\mu = (12.57 \times 10^{-7}) \times 2{,}000 = 2{,}514 \times 10^{-6}$

Problem 5: What is the **flux density** (B) in tesla (T) if there are 100 webers per square meter?

Answer:

$$B = \frac{100\ \text{Wb}}{1\text{m}^2} = 100 \text{ T}$$

Problem 6: If a magnetic circuit has an air gap with a length of 2 mm and a cross-sectional area of 5 cm^2, what is the gap's **reluctance?**

Answer:
Since $\mu_r = 1$ for air, then:

$$R = \frac{l}{\mu A} = \left[\frac{2 \times 10^{-3}}{12.57 \times 10^{-7} \times 5 \times 10^{-4}}\right]$$

$$= \frac{2 \times 10^{-3}}{62.85 \times 10^{-11}}$$

$$= (2000 \times 10^{-6})(62.85 \times 10^{-11})$$

$$= 31.82 \times 10^5 \text{ SI units}$$

Problem 7: If a given coil has a flux density of 3,000 webers per square meter using unknown core material and it has a flux density of 1,000 webers per square meter using air as its core, what is the **relative permeability** of the "brand x" material?

Answer:

$$\mu_r = \frac{\text{flux density with unknown core}}{\text{flux density with air core}}$$

$$= \frac{3,000}{1,000}$$

$$= 3$$

PRACTICAL CONSIDERATIONS ABOUT CORE MATERIALS

Since many of the components and devices you will be studying use principles related to electromagnetism, we will now consider the practical aspects of the core materials used in these components. You know that for a given magnetizing force, a coil's magnetic field strength is weaker if its core is air than if its core has a ferromagnetic substance, Figure 10–29. Also, we have indicated that certain magnetic properties of a material are not constant but vary with degrees of magnetization. Therefore, there is not a constant relationship between the magnetizing force field intensity (H) and the flux density (B) produced.

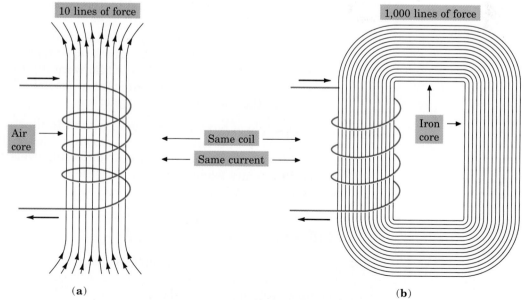

FIGURE 10–29   Effects of core material(s)

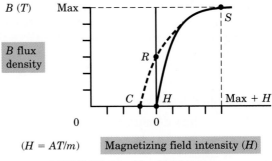

FIGURE 10–30   *B-H* curve

THE *B-H* CURVE

Manufacturers of magnetic materials often provide data about their materials to help select materials for a given purpose. One presentation frequently used is a *B-H* curve, or a magnetization curve, Figure 10–30. Notice in Figure 10–30 that as the magnetizing intensity (H) increases, the flux density (B) increases but not in a linear fashion. This graph provides several important points.

1. During magnetization (the solid line), a point is reached when increasing H (by increasing current through the coil) does not cause a further significant flux density increase. This is due to the core reaching "saturation," represented at point S on the graph. At this point, it can't contain more flux lines.

2. During demagnetization (with current decreasing to zero through the coil), the dotted line indicates some other interesting characteristics:
    a. When the H reaches zero (no current through coil), some flux density (B) remains in the core due to **residual** or "remanent" magnetism. The amount is indicated by the vertical distance from the graph baseline to point R.
    b. To remove the residual magnetism and reduce the magnetism in the core to zero, it takes a magnetizing intensity in the opposite direction equal to that represented horizontally from point 0 to point C. This value indicates the amount of "coercive" force required to remove the core's residual magnetism.

An example of how the *B-H* curve is used to illustrate important magnetic features of different materials is shown in Figure 10–31. For example, notice different materials saturate at different levels of flux density (B). **Saturation** occurs when a further increase in H does not result in an appreciable increase in B.

THE HYSTERESIS LOOP

Another useful graph is the hysteresis loop. It portrays what happens when a core is saturated with current in one direction through its coil, then the current direction is reversed and increased until the core is saturated in the opposite polarity. If this is continually repeated, a hysteresis loop graph,

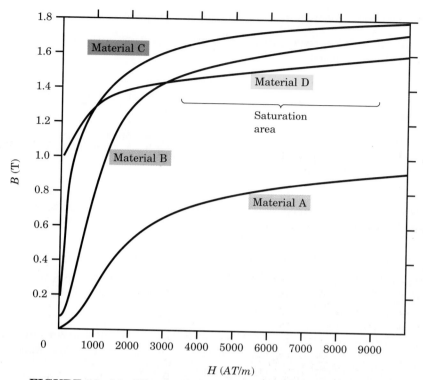

FIGURE 10–31   **Magnetization curves for different metals**

similar to that in Figure 10–32, can be created that enables meaningful analysis of the material in question. This is typically done by application of an alternating current source to the coil. NOTE: "Hysteresis" means "lagging behind." As this graph demonstrates, the flux buildup and decay lags behind the changes in the magnetizing force.

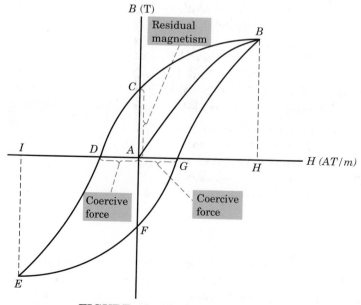

FIGURE 10–32   **Hysteresis loop**

Important information provided by this graph includes:

1. Points A to B represent the initial magnetization curve from zero H until the core reaches saturation.

2. Points B to C represent the flux density decay as H changes from its maximum level back to zero. The vertical distance from A to C is the amount of residual magnetism in the core.

3. Points C to D represent the flux density decay back to zero as the opposite polarity H is applied. The horizontal distance from A to D represents the coercive force.

4. Points D to E represent the buildup of flux density to core saturation in the opposite polarity from the initial magnetization. Of course, the total horizontal distance from point A to I shows the H required for this saturation. The vertical axis of the graph shows the amount of flux density (B) involved.

5. Points E to F show decay of flux density as H changes from its maximum level of this polarity back to zero. The residual magnetism is represented from points A to F.

6. Points F to G illustrate the decay of flux density back to zero due to the coercive force of the reversed H overcoming the residual magnetism. The coercive force is from points A to G.

7. Points G back to B complete the loop, displaying the increase of B to saturation in the original direction. For this segment of the loop, the vertical distance from point B to H is the flux density value at saturation.

— To ac power supply

FIGURE 10–33 Example of demagnetizing coil

8. The area inside the hysteresis loop represents the core losses from this cycle of magnetizing, demagnetizing, magnetizing in the opposite polarity, and demagnetizing the core. These losses are evidenced by the heating of the core as a result of the "magnetic domains" continuously having to be realigned with alternating current (ac) applied to the coil. We'll discuss core losses in greater detail at a later point in the book.

Incidentally, you may wonder how material that has been magnetized can be demagnetized. One method is to put the material through this hysteresis loop cycle while gradually decreasing the H swing. The result would be shown as a shrinking area under the hysteresis loop. One practical way to accomplish a decrease in H swing is to use a large ac demagnetizing coil so the material is placed within the coil. Gradually pull the material away from the ac energized coil, Figure 10–33. You may have seen a TV technician use a "degaussing" coil on color picture tubes to demagnetize them. A color picture tube that has become magnetized will have a degraded picture. Demagnetizing it usually helps.

IN-PROCESS LEARNING CHECK III

1. A B-H curve is also known as a _____ curve.

2. B stands for flux _____.

3. H stands for magnetizing _____.

4. The point where increasing current through a coil causes no further significant increase in flux density is called _____.

5. The larger the area inside a "hysteresis loop" the (smaller, larger) _____ the magnetic losses represented.

INDUCTION AND RELATED EFFECTS

Motor Action

When talking about motor action, we refer here to converting electrical energy to mechanical energy. You have studied the law that states lines of force in the same direction repel each other, and lines of force in opposite directions attract and cancel each other. One practical application of this phenomenon is motor action. The simplest example of this is a current-carrying conductor placed in a magnetic-field environment.

Look at Figure 10–34 and note the following:

1. The direction of the magnet's flux lines—North to South.

2. The direction of the current-carrying conductor's flux lines (Use the left-hand rule for current-carrying conductors.)

3. The interaction between the conductor's flux lines and the magnet's lines of force is such that the lines at the top of the conductor are repelled (lines in the same direction repel); the lines at the bottom of the

FIGURE 10–34   Motor action
on a single conductor

conductor are attracted (lines in opposite directions attract); and the
field is weakened at the bottom. Therefore, the conductor will be
pushed down. In other words, the lines at the top aid each other and
the lines at the bottom cancel each other.

4. The amount of force that pushes the conductor down depends on the
strength of both fields. The higher the current through the conductor,
the stronger its field. And the higher the flux density of the magnet,
the stronger its field.

5. Another method to cause stronger motor action is to make the single
conductor into a coil (in the form of an armature) with many turns,
thus increasing the ampere-turns and the strength of its field,
Figure 10–35.

6. Refer to Figure 10–36b. When the current through the conductor is
reversed or the magnetic poles of the magnet are switched, the force
(motor action) is in the opposite direction of that in Figure 10–36a.
When both current and poles are simultaneously reversed, Figure
10–36d, the force remains in the original direction, prior to being reversed, as that shown in Figure 10–36a.

FIGURE 10–35   Concept of stronger motor
action with an armature coil

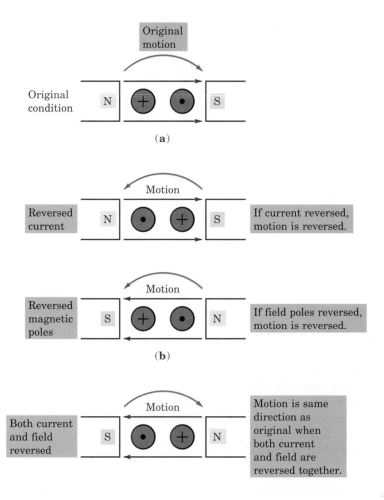

FIGURE 10–36   Effects of changing direction of current or field

Generator Action

When talking about generator action, we refer to converting mechanical energy to electrical energy. Throughout the chapter, we have alluded to a relationship between electricity and magnetism. Electromagnetic **induction** clearly illustrates this linkage. As you will recall, Faraday observed there is an induced emf (causing current if there is a path) when a conductor cuts across magnetic flux, or conversely, when the conductor is cut by lines of force. In essence, *when there is relative motion* between the two (i.e., the conductor and the magnetic lines of flux), an emf is induced.

Observe Figure 10–37 and note the following:

1. The direction of the magnet's field: North to South.

2. The physical direction of the conductor movement with respect to the magnetic field of the magnet: Upward.

**FIGURE 10–37   Converting
mechanical energy into
electrical energy**

3. The fact that when the conductor cuts this field in the direction as indicated, the direction of induced current is toward you, Figure 10–37a. In fact, the left-hand rule for current-carrying conductors (studied earlier) would indicate the direction of current, as shown. For example, if you aim your fingers in the direction of the flux (N to S) and bend your fingers in the same direction the magnet's field is bent, then the induced current is in the same direction as your extended thumb.

4. If the conductor moves down into the field, rather than up, the induced current is in the opposite direction, Figure 10–37b. (Try the left-hand rule again!)

5. The faster we move the conductor relative to the field, the higher the induced current (and/or emf), since more lines are being cut per unit time.

You may be wondering how cutting flux lines with a conductor causes induced emf, thus producing current flow (assuming a closed current path). Conceptually, it is briefly explained by indicating the magnetic fields involved with each of the electrons in the wire are caused to align, causing electron movement, which is current flow. When the conductor is open (with no complete path for current flow), one end of the conductor will be caused to have an excess of electrons (negatively charged); the other end has a deficiency of electrons (positively charged). Thus, we have an induced emf. If a closed path for current is provided, this induced emf causes current to flow.

What factors affect the amount of induced voltage? The key factors are:

1. amount of flux

2. number of turns linked by the flux

3. angle of cutting the flux

4. rate of relative motion

All of these factors relate to the number of flux lines cut per unit time.

FARADAY'S LAW

Faraday's law states *the amount of induced emf depends on the rate of cutting the flux (ϕ)*. The formula is:

Formula 10–9	$V_{ind} = \dfrac{\Delta\phi}{\Delta T}$

 change in flux value, $\boxed{\div}$, amount of time value, $\boxed{=}$

where, V_{ind} is the induced emf, and $\Delta\phi/\Delta T$ indicates the *rate* of cutting the flux. Using the SI system, if a single conductor cuts one weber of flux (10^8 lines) in one second, the induced voltage is 1 volt, a two-turn coil would have 2 volts induced, and so forth. (NOTE: The symbol Δ is the fourth letter in the Greek alphabet called delta, which, for our purposes, represents the amount of change in, or a small change in). Sometimes, the lower case letter "d" is used to indicate delta rather than the triangular symbol. Thus:

Formula 10–10	$V_{ind} = N\,(\text{\# of turns}) \times \dfrac{d\phi\,(\text{Wb})}{dt\,(\text{s})}$

 number of turns, $\boxed{\times}$, $\boxed{(}$, number of webers cut, $\boxed{\div}$, time to cut, $\boxed{)}$, $\boxed{=}$

EXAMPLE

If a 100-turn coil is cutting 3 webers per second ($d\phi/dt$ in formula), then the induced voltage is $100 \times 3 = 300$ volts. In essence, 3×10^8 lines of flux are cutting the 100-turn coil each second, resulting in 300 volts of induced voltage.

Although we will not thoroughly discuss generators, it is important you know that if a loop of wire is rotated within a magnetic field, Figure 10–38, the maximum voltage is induced when the conductor(s) cut at right angles to the flux lines. Voltage will incrementally decrease as the loop rotates at less than right angles, reaching zero when the loop conductor moves parallel to the lines of flux (in the vertical position), Figure 10–39. The key point is when the conductor moves parallel to the lines of flux (an angle of 0 degrees), no voltage is induced; when the conductor moves at right angles to the flux lines, maximum voltage is induced; and when the conductor moves at angles between zero and 90 degrees, some value between zero and the maximum value is induced.

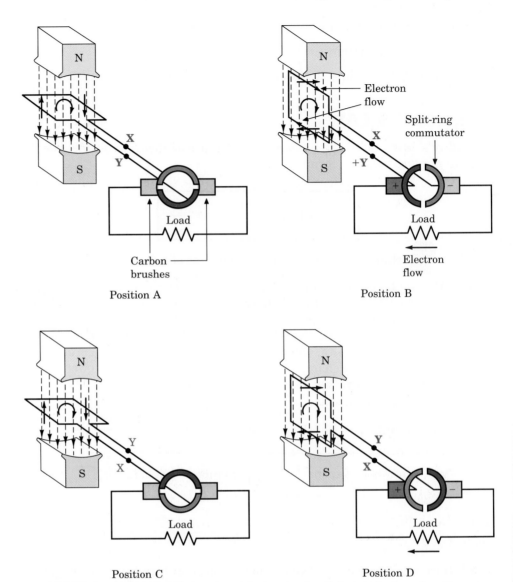

FIGURE 10–38   **Regions of maximum and minimum induced voltage**

LENZ'S LAW AND RECIPROCAL EFFECTS OF MOTORS AND GENERATORS

Lenz's Law

As you have already noted, the law of energy conservation is you don't get something for nothing! In the case of motor action in the generator, to generate electricity with the generator we have to supply mechanical energy that turns the armature shaft and overcomes the force of the opposing motor effect that results from the induced current and its resultant field(s).

Lenz's law states *the direction of an induced voltage, or current, is such that it tends to oppose the change that caused it.* Another way of stating this

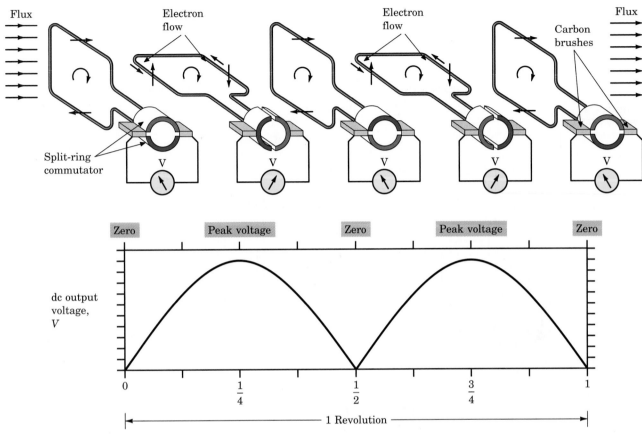

FIGURE 10–39   Variation in output voltage through one rotation of loop

law is when there is a change in the flux linking a circuit, an emf/current is induced that establishes a field that opposes the change.

Look at Figure 10–40 and use the left-hand rules for current-carrying conductors *and* for electromagnets. Notice when the magnet is moved *down* in the coil, the induced current establishes a North pole at the top end of the coil, facing (and repelling) the North pole of the magnet, whose flux linking the coil caused the original induced current. In other words, as we force the magnet down into the coil, the induced current produces a field to oppose the motion, as stated by Lenz. Conversely, when we pull the magnet out of the coil, the induced current is in the reverse direction, establishing a South pole at the top of the coil and tending to oppose retraction of the magnet. Again, the magnetic polarity can be defined in accordance with Lenz's law.

Motor Effect in a Generator

The generator converts mechanical energy (armature shaft rotation) into electrical energy (generator's output voltage). As you know, when the generator's conductors (armature windings) move through the magnetic field of the generator's field magnets, voltage is induced according to Faraday's law. This induced voltage causes current through the conductors, assuming a

(a)   **(b)**

FIGURE 10-40   Lenz's law

closed path for current. The *direction* of the resulting induced current is such that the magnetic field established around the conductor(s), resulting from the induced current, *opposes* the motion causing it, Figure 10-41.

Generator Effect in a Motor

In Figure 10-42, you can see the direction of current we are passing through the conductor causes the loop to rotate clockwise due to appropriate

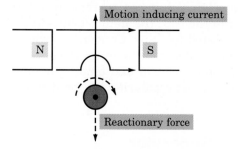

Induced current (coming toward you) results from upward mechanical movement.

Induced current sets up magnetic field around conductor which opposes upward motion that induced it in the first place.

FIGURE 10-41   Motor effect in a generator

FIGURE 10–42   Generator effect in a motor is such that the induced current due to cutting flux will be in opposite direction of current causing motor to rotate.

motor action. You can also observe when the conductor rotates clockwise, the wire loop's left side moves up into the flux and the wire loop's right side moves down into the flux. Since we have movement between a conductor and flux lines, there is an induced voltage/current.

This induced current is in the opposite direction from the current that is causing the motor's armature to move clockwise. This phenomenon is often termed generator effect in a motor and is related to Lenz's law.

OTHER TOPICS RELATED TO MAGNETISM

Several topics not previously discussed are now presented since you now have some general understanding of magnetism. These topics are:

1. the toroidal coil form (closed magnetic path);

2. magnetic shields;

3. shaping fields across a gap via pole pieces;

4. the Hall effect; and

5. special classifications of materials.

Toroidal Coil Form

Notice in Figure 10–43 the toroidal coil form resembles a doughnut and is made of soft iron, ferromagnetic, or other low-reluctance, high-permeability material. The key feature is this coil effectively confines the lines of flux

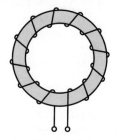

Flux is virtually contained within core with very little "leakage" flux

FIGURE 10–43   Toroid coil

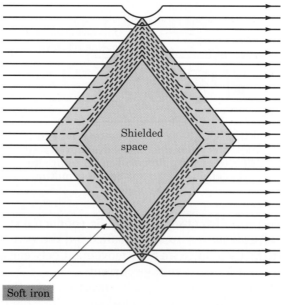

**FIGURE 10–44   Concept of
the magnetic shield**

within itself (because there are no air gaps); therefore, it is a very efficient carrier of flux lines. It allows little "leakage flux" (flux outside the desired path, generally in the surrounding air), thus does not create magnetic effects on nearby objects. At the same time, its internal field is affected very little by other magnetic fields near it.

It is interesting to note there are no poles, as such, in the toroid configuration. However, if one takes a slice out of the doughnut, creating an air gap, North and South poles are established at opposite sides of the air gap.

Magnetic Shields

Remember magnetic lines of force penetrate nonmagnetic materials. To protect components, devices, and so forth from stray or nearby magnetic fields, a magnetic shield is used, Figure 10–44. A high-permeability, soft iron enclosure is useful for the shielding function. Actually, it is simply diverting the lines of flux through its low-reluctance path, thus protecting anything contained inside of it. One application for this type shielding is in electrical measuring instruments or meter movements.

Shaping Fields Via Pole Pieces

Notice in Figure 10–45 the use of pole pieces. These pieces control the shape of the field across the air gap to afford a more linear field with almost equal flux density across the area of the gap. This control is beneficial for many applications.

FIGURE 10–45   Concept of pole pieces linearizing a field

The Hall Effect

Figure 10–46 illustrates the Hall effect. Named after its discoverer, E. H. Hall, the phenomenon states when a current-carrying material is in the presence of an external magnetic field, a small voltage develops on two opposite surfaces of the conductor.

As the illustration indicates, this effect is observed when the positional relationships of the current, the field's flux lines, and the surfaces where the difference of potential develops are such that:

1. Flux is perpendicular to direction of current flow through the conductor. (NOTE: The amount of voltage developed is directly related to the flux density, B).

2. If current is traveling the length of the conductor, the Hall voltage (v_H) develops between the sides or across the width of the conductor.

Indium arsenide is a semiconductor material used in devices, such as gaussmeters, to measure flux density via the Hall effect. This material develops a relatively high value of v_H per given flux density. Since the amount of Hall-effect voltage developed directly relates to the flux density where the current-carrying conductor is immersed, a meter connected to this probe can be calibrated to indicate the flux density of the probed field, Figure 10–47.

FIGURE 10–46   The Hall effect

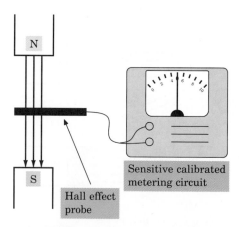

FIGURE 10–47   Using the Hall effect to measure flux density

Special Classifications of Materials

In this chapter, we have primarily been discussing ferromagnetic materials that typically have high permeabilities and can be strongly magnetized. Examples include iron, steel, nickel, and various alloys. Other classifications of materials worth noting are:

1. *Paramagnetic* materials that have weak magnetic properties. Examples are aluminum and chromium.

2. *Diamagnetic* materials that have permeabilities of less than one and align perpendicular to the direction of magnetic fields. Examples are mercury, bismuth, antimony, copper, and zinc.

3. *Ferrites* that are powdered and compressed materials with high permeabilities but high electrical resistance that minimizes I^2R/eddy current losses when used in high-frequency applications. Examples are nickel ferrite, nickel-cobalt ferrite, and yttrium-iron garnet. Note that eddy currents are those induced into the core material that cause undesired I^2R losses within the core.

SUMMARY

▶ A magnet is an object that attracts iron, steel, or certain other materials. It can be natural (magnetite or lodestone) or produced in iron, steel, and other materials in a form called a permanent magnet. Also, a magnet can exist in soft iron and other materials as a temporary magnet that requires a current-carrying coil around it.

▶ Magnetic materials, such as iron and nickel, are attracted by magnets and can be made to have magnetic properties.

▶ Magnets that are suspended and free to move align in an approximate North/South direction. The end of the magnet seeking the earth's North pole is the North-seeking pole. The other end of the magnet is the South-seeking pole.

▶ A magnetic field composed of magnetic lines of force called flux surrounds a magnet. This region is where the magnetic forces act.

▶ Lines of force are continuous; travel outside the magnet from the magnet's North pole to the magnet's South pole; and flow inside the magnet from the South pole to the North pole. They travel the shortest path or the path with least opposition to the flow of flux, and when traveling in the same direction, repel each other. When traveling in opposite directions, they attract each other. Furthermore, lines of force do not cross each other and will penetrate nonmagnetic materials.

▶ A current-carrying conductor or wire is surrounded by a magnetic field. The strength of the magnetic field is directly related to the value of current passing through it.

▶ A coil of wire carrying current develops a magnetic field that is stronger than a straight conductor with the same current. The magnetic field becomes stronger if an iron core is inserted in the coil, making it an electromagnet.

▶ A basic law of magnetism is like poles repel each other and unlike poles attract each other.

▶ Permanent magnets come in a variety of shapes; for example, bar-shaped (rectangular), horseshoe-shaped, and disc-shaped. Permanent magnets are not really permanent since they lose their magnetism over time.

▶ Induction of magnetism occurs when a magnetized material is brought in proximity to or rubbed across some magnetic material.

▶ The left-hand rule to determine direction of the magnetic field surrounding a current-carrying conductor states when the wire is grasped in the left hand with the extended thumb in the direction of current flow through the wire, then the fingers are wrapped around the conductor in the same direction as the magnetic field.

▶ The left-hand rule to determine the polarity of electromagnets (i.e., which end is the North pole and which end is the South pole) states when the coil is grasped with the fingers pointing in the same direction as the current, the extended thumb points toward the end of the coil that is the North pole.

▶ Factors influencing the strength of an electromagnet are: the number of turns on the coil; the coil's length; the coil's value of current;

the type of core; and the coil's cross-sectional area.

▶ Several important terms and units relating to magnetism include: flux; weber; tesla; magnetomotive force; magnetic field intensity; permeability; and relative permeability.

▶ Comparing magnetic circuit parameters with electrical circuits and Ohm's law can be performed by stating that the amount of flux flowing through a magnetic circuit is directly related to the magnetomotive force and inversely related to the path's reluctance (opposition) to flux.

▶ Core materials differ in their magnetic and electrical features. For example, some materials saturate at lower levels of magnetization.

▶ An ideal core material has very high permeability; loses all its magnetism when there is no current flow in the coil; does not easily saturate; and has low I^2R loss due to eddy currents.

▶ An important relationship exists between flux density produced by a given magnetic field intensity and the permeability of the material involved. This relationship stated for the permeability of a vacuum is:

$$\mu_o = \frac{B}{H}, \text{ and } B = \mu_o \times H, \text{ and } H = \frac{B}{\mu_o}$$

where: μ_o is the absolute permeability of a vacuum
B is flux density and
H is the magnetic field intensity

▶ When an ac passes through an electromagnet's coil, there is a lag between the magnetizing force and the flux density produced. This is illustrated by the *B-H* curve and the hysteresis loop. These graphs also illustrate residual magnetism and the coercive force needed to overcome it.

▶ Faraday's law states the amount of voltage induced when there is relative motion between conductor(s) and magnetic flux lines is such that when 10^8 lines of flux are cut by one conductor (or vice versa) in one second, one volt is induced. In other words, induced voltage relates to the number of lines of force cut per unit time.

▶ Lenz's law states when there is an induced voltage (or current), the direction of the induced voltage or current opposes the change causing it.

▶ A motor converts electrical energy into mechanical energy.

▶ A generator converts mechanical energy into electrical energy.

FORMULAS AND CALCULATOR SEQUENCES SUMMARY

FORMULA 10–1 Force $= \dfrac{\text{Pole 1 Strength} \times \text{Pole 2 Strength}}{d^2}$

pole 1 value, $\boxed{\times}$, pole 2 value, $\boxed{\div}$, $\boxed{(}$, distance value, $\boxed{x^2}$, $\boxed{)}$, $\boxed{=}$

FORMULA 10–2 $\mu_o = \dfrac{B}{H}$

B value, $\boxed{\div}$, H value, $\boxed{=}$

FORMULA 10–3 $B = \mu_o \times H$

μ_o value, $\boxed{\times}$, H value, $\boxed{=}$

FORMULA 10–4 $H = \dfrac{B}{\mu_o}$

B value, $\boxed{\div}$, μ_o value, $\boxed{=}$

FORMULA 10–5 $mmf = NI$

number of turns value, $\boxed{\times}$, current value, $\boxed{=}$

FORMULA 10–6 $H = \dfrac{AT}{\text{meter}}$

value of current (amperes), $\boxed{\times}$, number of turns, $\boxed{\div}$, number of meters, $\boxed{=}$

FORMULA 10–7 $B\,(\text{T}) = \dfrac{\phi(\text{Wb})}{A(\text{m}^2)}$

$\boxed{(}$, number of lines, $\boxed{\div}$, 10^8 , $\boxed{)}$, $\boxed{\div}$, area in meters squared, $\boxed{=}$

FORMULA 10–8 $\mu = \mu_o \times \mu_r$

μ_o value, $\boxed{\times}$, μ_r value, $\boxed{=}$

FORMULA 10-9 $V_{ind} = \dfrac{\Delta\phi}{\Delta T}$

change in flux value, $\boxed{\div}$, amount of time value, $\boxed{=}$

FORMULA 10-10 $V_{ind} = N(\text{\# of turns}) \times \dfrac{d\phi(Wb)}{dt}$

number of turns, $\boxed{\times}$, $\boxed{(}$, number of webers cut, $\boxed{\div}$, time to cut, $\boxed{)}$, $\boxed{=}$

REVIEW QUESTIONS

NOTE: Redraw illustrations on separate paper to answer questions involving diagrams.

1. Briefly define magnetism, flux, field, pole, and magnetic polarity.

2. Draw a typical field pattern for the magnetic situations in Figure 10–48. Indicate direction(s) of flux lines with arrows.

FIGURE 10-48

3. State the law regarding magnetic attraction and repulsion.

4. Draw magnetic field lines around the conductors in Figure 10–49 and indicate direction with arrows.

FIGURE 10-49

5. Indicate the North and South poles of the current-carrying coil in Figure 10–50.

FIGURE 10-50

6. List five statements about the behavior of magnetic lines of force.

7. What term indicates the ease with which a material passes, conducts, or concentrates flux?

8. What term indicates the opposition to magnetic lines of force?

9. Using SI units, define the terms, expressions, or units for flux and flux density. Show symbols, if appropriate.

10. Using SI units, define the terms, expressions, or units for magnetomotive force, magnetic field, and field intensity. Show symbols, if appropriate.

11. Using SI units, define the terms, expressions, or units for permeability and relative permeability. Show symbols, if appropriate.

12. How many lines of force are represented by the weber? (NOTE: Express as a whole number, not a power of ten).

13. What is the difference between magnetomotive force and magnetic field intensity?

14. If the permeability of a given flux path is unchanged and H doubles, what happens to B?

15. What happens to the reluctance of a given flux path if its length triples while the magnetic circuit's μ and A remain unchanged?

16. If current doubles through a given electromagnet's coil, does reluctance increase, decrease, or remain the same? (Assume the magnet is in its linear operating range.)

17. For the situation in question 16, will flux density increase, decrease, or remain the same?

18. What is meant by saturation in a magnetic path?

19. How does a magnetic shield give the effect of shielding something from magnetic lines of force?

20. What does a gaussmeter measure?

CHALLENGE QUESTION

1. In your own words, define what is meant by the term magnetic polarity.

TEST YOURSELF

1. Draw a typical hysteresis loop and explain how saturation, residual magnetism, and coercive force are shown on the graph.

2. If an air gap is 2 mm in length and has a cross-sectional area of 10 cm^2, what is the air gap's reluctance?

3. If a pole piece has a cross-sectional area of 0.005 m^2 and is emitting 1 μWb of flux:
   a. How many lines of flux are involved?
   b. What is the flux density, expressed in teslas?

4. How much induced emf results if a coil of 500 turns is cut at a rate of 10 μWb per second?

THE SIMPLER SEQUENCE FOR TROUBLESHOOTING

DC Measuring Instruments

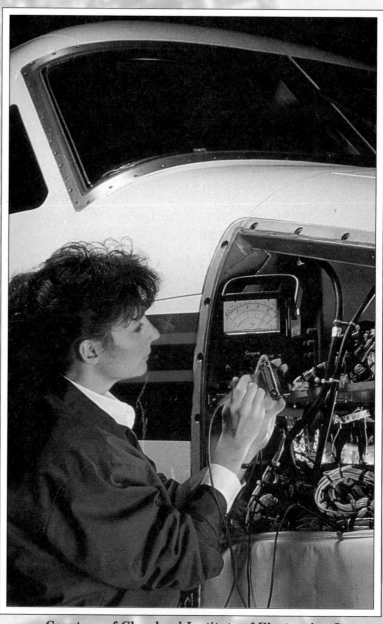

Courtesy of Cleveland Institute of Electronics, Inc.

11

KEY TERMS

Backoff ohmmeter scale

Core

Damping (electrical)

Frame/movable coil

Full-scale current

$I_M(I_{fs})$

I_S

Loading effect

Meter shunt (current meter)

Moving-coil (d'Arsonval) movement

Multimeter

Multiplier resistor

Ohms-per-volt rating

Pointer

Pole pieces

R_M

R_{Mult}

Range selector (switch)

Scale

Sensitivity rating

Series-type ohmmeter circuit

Springs

Stops

Zero-adjust control

OUTLINE

General requirements of basic meter movements

The d'Arsonval (moving coil) movement

Current meters

Current meter shunts

Voltmeters

Calculating the multiplier values

Ohmmeters

VOMs, DMMs, and other related devices

The Digital Multimeter (DMM)

Troubleshooting hints

CHAPTER PREVIEW

Lord Kelvin can be paraphrased as saying, "If you can measure what you are speaking about, you know something about it. If you can't, your knowledge is very limited." Someone else has indicated that measuring instruments are like special eyes to technicians, engineers, and scientists. The dc measuring instruments you will study in this chapter are invaluable aids to people working in electronics. These instruments are indispensable for designing, troubleshooting, servicing, and monitoring equipment operation. Some samples of these instruments are in Figure 11–1.

There is a technical distinction between the terms instrument and meter. Instruments measure the value of present quantities. Meters, such as the watt-hour meter, measure and register quantities with respect to time. For convenience, we will refer to the measurement devices discussed using either term.

The digital multimeter has become the most used device by technicians and engineers for measuring voltage, current, resistance, and other electrical quantities. There are still a number of areas in which the analog-type meter has advantages over the digital meters. Advantages and disadvantages of each type instrument will be discussed further in the chapter.

At a later time in your studies, you will study the theory and principles of operation of the digital meter. However, a portion of this chapter is dedicated to having you apply the electrical principles and laws you have studied thus far, for analyzing the operation of typical analog-type instruments. You will use and practice your basic knowledge of magnetism, series circuits, parallel circuits, and series-parallel circuits. In doing this, you will see practical applications of this knowledge, and also gain a basic understanding of how dc analog meters operate.

In this chapter, you will examine the most frequently used moving-coil movement, called the d'Arsonval movement. You will see how this movement is incorporated in measuring instrument circuits that measure current, voltage, and resistance. Also, you will study the proper use of these instruments; the instruments' effects on the circuit(s) being tested; and some hints about their practical applications.

OBJECTIVES

After studying this chapter, you should be able to:

- Describe the **d'Arsonval movement**

- List key parts and functions of a movement

- Define linear deflection, full-scale sensitivity, shunt, multiplier, **sensitivity rating, loading effect,** VOM, and DMM

- Explain the proper methods to connect and use current meters, voltmeters, and ohmmeters

- Calculate the values required for **current meter shunts**

- Calculate the values required for **voltmeter multipliers**
- Briefly explain calibration of series-type ohmmeters
- Describe at least two special-purpose measuring devices
- List troubleshooting applications for current meters, voltmeters, and ohmmeters
- Use the SIMPLER troubleshooting sequence to solve the Chapter Challenge problem

GENERAL REQUIREMENTS OF BASIC METER MOVEMENTS

Three essentials most analog meter movements contain are:

1. A moving element that reacts to the magnitude of the electrical parameter being measured.

2. An indicating function that shows how much reaction has occurred.

3. A damping function that aids the moving element to reach its appropriate resting place without undue overshoot or back and forth mechanical oscillation.

THE D'ARSONVAL (MOVING COIL) MOVEMENT

The **d'Arsonval movement** contains parts meeting the above requirements and is a device that reacts to the value of current through it.

FIGURE 11–1   Examples of measuring instruments: a. Volt-ohm-milliammeter; b. Benchtop digital multimeter (Photos courtesy of Simpson Electric Company)

(**a**) Moving-coil movement

(**b**) "Phantom" view of commercial permanent-magnet moving-coil movement

(**c**) Moving element with pointer and springs

FIGURE 11–2   Key parts of a moving-coil movement

Refer to Figure 11–2 and note the following key parts and their functions in this movement.

1. *Permanent magnet*—establishes magnetic field.

2. ***Pole pieces***—linearize and strengthen magnet's field

3. ***Core***—soft iron core concentrates, strengthens magnetic field and, along with pole pieces, produces a uniform air gap between the permanent magnet's pole pieces and itself.

4. *Movable coil and support frame*—light aluminum frame provides support and damping action for coil; movable coil creates electromagnetic field when current passes through it.

5. *Spindle and jewel bearings*—spindle supports movable coil and coil frame; jewel bearings provide a low-friction bearing surface to allow turning action of movable coil mechanism. (NOTE: Sapphire bearings are not shown.)

6. *Springs*—provide electrical contact or current path to coil and mechanical torque, restoring coil to starting position when current is not present; also provide a force against which the movable coil acts.

7. **Pointer** *and counterbalance*—provide means of indicating amount of deflection.

8. **Scale**—calibrated in appropriate units gives meaning to pointer deflection.

9. **Stops**—provide left and right limits to pointer deflection. (NOTE: Stops are not shown.)

Principle of Operation

The operation of the **moving-coil movement** is based on the motor action occurring between the permanent magnet's field and the field established by the movable coil when current passes through it, Figures 11–2 and 11–3, as appropriate.

The moving element function, described earlier, involves many parts—the moving coil, the light aluminum coil frame, the pole pieces, the magnetic core piece, the spindle, the springs, and the sapphire bearings on which the spindle rotates. You will recall the strength of an electromagnet depends on the number of turns involved (per given length), the current through those turns, and the permeability features of the flux path. Since the number of turns and the flux path are known values, the only variable, in this case, is the amount of current through the movable coil. The higher current produces a stronger motor action; thus, the moving coil moves further.

Observe in Figures 11–2a and 11–3 the polarity of the electromagnetic field set up by current passing through the moving coil is such that the electromagnet's North pole is adjacent to the permanent magnet's North

**FIGURE 11–3   Polarity required for proper
direction of motor action**

pole, and the electromagnet's South pole is adjacent to the permanent magnet's South pole. This means there is a force of repulsion. Thus, the movable coil moves in a clockwise direction when current passes in the correct direction.

It also is evident that current in the opposite direction through the coil reverses the polarity of the coil's field, causing attraction rather than repulsion. Thus, the movable coil moves in a counterclockwise direction, possibly damaging the fragile meter movement. Note that proper operation of this movement depends on dc current (current in only one direction) and proper polarity, or direction of that current. The indicating function is performed by the combination of the pointer and calibrated scale.

The **damping** function is really an electrical damping provided by the light aluminum coil frame. Recall that when a conductor moves within or links a magnetic field and has relative motion, a current-producing emf is induced. The direction of the induced current establishes a magnetic field that opposes the motion causing the induced current in the first place (Lenz's law). This action causes a mechanical damping effect on the movement of the movable-coil-pointer assembly. The result is the pointer does not significantly overshoot its appropriate resting place, nor does it oscillate back and forth or above and below the appropriate deflection location for long periods.

Advantages and Disadvantages of the Moving-Coil Movement

Primary advantages of the moving-coil movement are:

1. Linear scale, including scale divisions, is easy to read.

2. Construction provides a self-contained magnetic shielding effect.

3. It can be very accurate.

4. It can be quite sensitive since it has a large number of turns on coil and a strong permanent magnet.

Some disadvantages of the moving-coil movement are:

1. It is quite fragile so will not stand rough treatment.

2. It measures only dc parameters.

3. It is very expensive.

✔ IN-PROCESS LEARNING CHECK I

Fill in the blanks as appropriate.

1. Three essential elements in an analog meter movement include a _____ element, an _____ means, and some form of _____.

2. The purpose of the permanent magnet in a d'Arsonval meter movement is to establish a fixed _____ _____.

3. The movable element in a moving coil meter moves due to the interaction of the fixed _____ _____ and the magnetic field of the movable coil when current passes through the coil.

4. Electrical conduction of current in a movable coil meter movement is accomplished through the _____, which also provides a desired (mechanical magnetic) _____ force against which the movable coil must act.

5. The element in a moving-coil movement producing electrical damping is the _____ _____ _____.

CURRENT METERS

How Current Meters are Connected

The moving-coil movement is primarily a current indicator. Its basic action depends on the magnitude of the current through its moving coil. Thus, to measure current, the circuit to be tested must be broken, and the current meter inserted in series with the circuit path where it is desired to measure current value (being sure to observe proper polarity), Figure 11–4. As you already know, current is the same throughout a series circuit path. Hence, the meter indicates the current through that portion of the circuit where it is inserted.

Sensitivity

Current meters come in a variety of measurement ranges, Figure 11– 5. Two common methods to define the **sensitivity** of current meters are the amount of current required to cause full-scale deflection and the number of millivolts dropped by the coil at a current equal to full-scale deflection. Sensitive basic movements (for example, the laboratory galvanometer) measure current values in the range of microamperes; other movements measure current in milliamperes.

1. Turn power off.
2. Break part of circuit to be tested.
3. Insert current meter in series.
4. Observe polarity.

FIGURE 11–4 Proper connection of current meters

FIGURE 11-5   Meters come in different ranges. (Photos courtesy of Simpson Electric Co.)

However, there is a limit to the range of measurement for which a basic movement can be designed. Physical factors, such as the size of wire, the weight of the coil frame, and the size of springs, limit the range when designing a basic movement. As you will see in a later discussion, using meter shunts helps solve this problem.

What factors cause a meter to be more or less sensitive? The primary factors are the number of turns on the coil and the permanent magnet's field strength. For example, when a meter deflects full-scale with 10 microamperes of current through the coil, the meter has a full-scale sensitivity (*fs*) of 10 microamperes. When it takes 100 microamperes to cause full-scale deflection, the full-scale (*fs*) sensitivity rating is 100 microamperes and so on.

Another physical factor influencing meter sensitivity is the force of the hairsprings. In order to have a very sensitive movement (one taking a small current value causing full-scale deflection of the pointer), it takes a strong permanent magnet and a coil with many turns of fine wire, and hairsprings without much force.

On the other hand, to have a less sensitive meter, a movable coil using less turns of larger diameter wire is used. Also, using a weaker permanent magnet and/or hairsprings with more force produces a less sensitive meter.

Typically, basic movements are produced in ranges from a few microamperes to a few milliamperes. When measuring currents higher than this, the use of a parallel (current bypass) path, called a **meter shunt,** is necessary.

The number of millivolts dropped by the meter movement at full-scale current adds yet another useful rating because it involves *both* the **full-scale current (I_M)** and the resistance of the wire in the movable coil **(R_M)** of the meter movement. For example, if the meter is a 1-mA meter (*fs* = 1 mA) and the meter resistance (R_M) is 10 Ω, then when the meter deflects a full-scale value of 1 mA, the meter movement is dropping a voltage equal to $I_M \times R_M = 1 \text{ mA} \times 10 \text{ Ω} = 10 \text{ mV}$. Therefore, this movement is rated as a 10 mV, 0-1-mA movement.

Trade-Offs

What are the trade-offs of basic meter movements? Two important trade-offs are the desired sensitivity and *the desire not to disturb the tested circuit more than necessary.*

A more sensitive meter has a smaller wire and greater number of turns on the movable coil. Small diameter wire has a higher resistance per unit length than larger wire. This means that the more sensitive the meter, the higher is its own internal resistance called R_M. Consequently, a meter with higher resistance has more effect on the tested circuit, since its R_M is a larger percentage of the test circuit's total resistance. Ideally, a current meter should have zero internal resistance so as not to disturb the circuit tested. But zero resistance is impossible.

PRACTICAL NOTES

A common rule of thumb about current meter resistance versus the resistance of the circuit tested is the current meter should not be more than about 1% of the resistance of the tested circuit.

EXAMPLE

For example, if the circuit through which you are trying to measure current has a resistance of 10,000 ohms, the current meter should not have a resistance greater than 100 ohms to ensure that the tested circuit is not altered too much when the current meter is inserted.

CURRENT METER SHUNTS

The Basic Shunt

Current meter shunts extend the range of current that a given meter movement measures. The shunt is simply a low-resistance current path connected in parallel (or shunt) with the meter movement (the moving coil). The shunt effectively causes some percentage of the current to bypass the meter through the parallel path with the remaining percentage passing through the meter to cause deflection, Figure 11–6.

IMPORTANT PARAMETERS. To calculate the required shunt resistance value that extends the meter's range, several electrical parameters are of importance. These parameters are:

1. the full-scale current for the meter movement (I_{fs}, or I_M);

2. the resistance of the meter movement (R_M);

(Meter current)

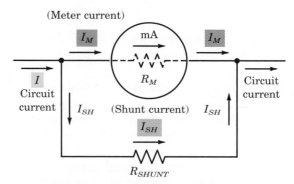

Circuit current "splits." A percentage of current passes through meter while remaining percentage passes through shunt. Thus,
$I_M + I_{SH}$ = circuit current.

FIGURE 11–6   Concept of a "shunt" resistor path to extend meter range

3. the desired maximum current, or the extended meter range value to be indicated by full-scale deflection (I);

4. the current that passes through the shunt (I_S or $I - I_M$); and

5. the resistance of the shunt (R_S).

METHODS USED TO CALCULATE R_{SHUNT} VALUE.   More than one approach can be used to calculate the R value needed to shunt a given meter movement to increase its measurement range. However, all the techniques evolve from variations of Ohm's law, Figure 11–7.

One method is to use *Ohm's law* in its basic form, where:

$$\text{Formula 11–1} \quad R_S = \frac{V_S}{I_S}$$

 voltage across shunt, $\boxed{\div}$, current value to pass through shunt, $\boxed{=}$

NOTE: V_S is the same as the parallel voltage V_M (or $I_M R_M$).

I_S is the difference between the highest current to be measured (the extended range full-scale value) and the movement's full-scale current value ($I_S \times I - I_M$).

Substituting in Formula 11–1:

$$\text{Formula 11–2} \quad R_S = \frac{I_M R_M}{(I - I_M)}$$

 I_M value, $\boxed{\times}$, R_M value, $\boxed{\div}$, $\boxed{(}$, I (extended range) value, $\boxed{-}$,
I_M , $\boxed{)}$, $\boxed{=}$

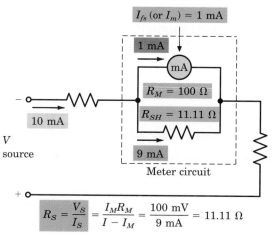

$$R_S = \frac{V_S}{I_S} = \frac{I_M R_M}{I - I_M} = \frac{100 \text{ mV}}{9 \text{ mA}} = 11.11 \; \Omega$$

FIGURE 11–7   Basic shunt resistance calculation

Formula 11–3	$R_S = \dfrac{I_M R_M}{I_S}$

 I_M value, $\boxed{\times}$, R_M value, $\boxed{\div}$, I_S value, $\boxed{=}$

Incidentally, the $I_M R_M$ product is the same as the meter's millivolt rating spoken of earlier, assuming full-scale current value in milliamperes. The $(I - I_M)$ portion in Formula 11–2 is really the current value that must pass through the shunt (I_S) in Formula 11–3. Therefore, if you know the millivolt rating of the meter and the current value passing through the shunt,

$$R_S = \frac{\text{millivolt rating}}{I_S}.$$

EXAMPLE

Given a 0–1 mA movement with a meter resistance of 100 Ω, what value of shunt R is needed to make the meter act as a 0–10 mA instrument?

Answer: $R_S = \dfrac{I_M R_M}{(I - I_M)}$

$$R_S = \frac{(1 \times 10^{-3}) \times 100}{(10 \times 10^{-3}) - (1 \times 10^{-3})}$$

$$R_S = \frac{(100 \times 10^{-3})}{(9 \times 10^{-3})}$$

$$R_S = 11.11 \; \Omega$$

Another approach is to use the *inverse proportionality concept.* That is, branch currents are inverse to branch resistances.

For example, if it is desired to have nine times the current through branch 2 as there is current through branch 1, then make the R of branch 2 equal to one-ninth that of branch 1. The inverse relationship for our case is stated as:

$$
\boxed{\;\text{Formula 11–4} \quad \left| \quad \frac{R_S}{R_M} = \frac{I_M}{I_S} \right.\;}
$$

 R_S value, $\boxed{\div}$, R_M value, $\boxed{=}$, (ratio should compare to), I_M value, $\boxed{\div}$, I_S value, $\boxed{=}$

NOTE: The ratio of the *shunt* resistance *to* the *meter* resistance is the same as the ratio of the *meter* current *to* the *shunt* current. Notice the inverse relationship on the two sides of the equation.

Another formula expressing the ratio concept is:

$$
\boxed{\;\text{Formula 11–5} \quad \left| \quad R_S = \frac{R_M}{N - 1} \right.\;}
$$

 R_M value, $\boxed{\div}$, $\boxed{(}$, no. of times range is to be multiplied, $\boxed{-}$, 1 , $\boxed{)}$, $\boxed{=}$

where R_S = R of shunt, R_M = R of meter, and N = number of times range is being multiplied. In other words, 1 mA meter whose increased range is 10 mA, so N equals 10.

EXAMPLE

Using the same meter and desired current range as in the previous problem, substitute values in the inverse relationship equation as follows:

$$\frac{R_S}{R_M} = \frac{I_M}{I_S}$$

$$\frac{R_S}{100\ \Omega} = \frac{1\ \text{mA}}{9\ \text{mA}} \quad \text{or} \quad \begin{array}{l} R_S = \text{one-ninth of } 100\ \Omega \\ R_S = 11.11\ \Omega \end{array}$$

Using this same concept, if you want to double the basic range of a given meter, what is the shunt resistance value? If you said the shunt resistance (R_S) equals the meter resistance (R_M), you are correct! In this case, we want the total current to equally split between the shunt path and the meter movement path; thus, the R values must be equal. Since you now understand the inverse proportionality method, try the following problem.

Special "shorting-type" switch makes next
contact before "breaking" with previous contact.

**FIGURE 11–8   Switchable shunts provide
multiple ranges with one meter.**

PRACTICE PROBLEMS 1

What shunt resistance value is needed to extend the range of a
0–1-mA meter to 0–100 mA. Assume the meter movement's R is
100 Ω.

A Basic Multiple-Range Meter Circuit

Notice in Figure 11–8 we have several shunts that can be selectively
switched and connected with the meter. Obviously, since each shunt is a
different value, the meter circuit can measure different values of current,
depending on which shunt is connected to the circuit.

PRACTICAL NOTES

1. *A special shorting-type switch is used to switch ranges.* This type of
   switch, with its very long wiper contact, assures the meter is never
   without a shunt! In other words, when changing ranges, this
   switch makes contact with the next switch contact before it breaks
   contact with the contact it is leaving. Of course, the reason we
   don't want to leave the meter without a shunt when switching
   between ranges is that when measuring current in a circuit having current greater than the basic movement's full-scale current
   rating, the meter is damaged or ruined by excessive current
   through the movement during that moment when it is unshunted.

2. *The shunt value for each range is determined in the same manner
   as you determined the single shunt,* Figure 11–9 for sample calculations.

$$R_{S_1} = \frac{200 \text{ mV}}{9 \text{ mA}} = 22.22 \ \Omega$$

$$R_{S_2} = \frac{200 \text{ mV}}{49 \text{ mA}} = 4.08 \ \Omega$$

$$R_{S_3} = \frac{200 \text{ mV}}{99 \text{ mA}} = 2.02 \ \Omega$$

**FIGURE 11–9   Sample calculation of
shunts for three ranges**

Current Meter Key Points Capsule

1. Moving-coil movements are essentially dc current indicators.

2. The minimum current measured is based on the sensitivity of the basic meter movement.

3. The sensitivity of a basic meter movement is determined by the number of turns in the moving coil, the strength of the magnetic field, and other related mechanical factors, such as the springs, the weight of the coil frame, and so forth.

4. Currents higher than the basic movement's full-scale current rating can be measured if appropriate values of resistance are connected in shunt (or parallel) with the basic movement.

5. Current meters are generally linear devices. That is, when 50% of full-scale current value flows through the meter movement, deflection is mid-scale on the meter. When 25% of full-scale current is measured, the pointer moves one-fourth full-scale distance on the scale.

6. To measure current, meters must be connected in series, observing proper polarity, and must have adequate range.

7. Because current meters are connected in series, the lower the meter resistance (R_M) used in making the measurement, the less the actual circuit operating conditions will be changed and the more useful the reading will be.

VOLTMETERS

The Purpose of the Multiplier Resistor

Another extremely useful application for which movements can be adapted is that of measuring voltage, Figure 11–10. Remember that the basic meter

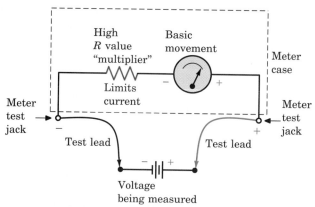

FIGURE 11–10   **The multiplier resistor**

movement is a current indicator. By adding select value resistors, called **multiplier resistors, R_{Mult},** in series with the meter movement, it is possible to measure voltage. The multiplier resistor limits the current through the movement. The amount of meter deflection is directly related to the voltage applied to the meter circuit. For example, a voltage equaling one-third the voltage range causes one-third scale deflection. As you will soon see, different multiplier values provide different voltage ranges.

Connecting the Voltmeter to the Circuit Being Tested

Refer to Figure 11–11. Notice that although the voltmeter circuit is a series circuit ($R_M + R_{Mult}$), the voltmeter is always connected *in parallel* with the two points across which it is desired to measure the potential difference, or voltage. Because the voltmeter is connected in parallel with the portion of the circuit being tested, it is **not** *necessary to break (open) the circuit,* as is done when inserting a standard current-measuring instrument for measuring current.

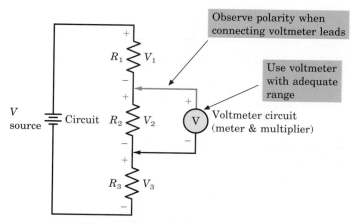

Voltmeter is measuring V_2 value and is connected in parallel with voltage to be measured.

FIGURE 11–11   **Connecting a voltmeter to make voltage measurements**

It is necessary, however, to *observe polarity* when measuring dc voltages, or the meter deflection will be in the wrong direction, as it is for the basic current meter.

As indicated earlier in the chapter, *a sufficiently high range* is needed to assure no more than full-scale deflection of the meter; otherwise, damage will result. Put another way, you are asking for bent pointers and maximum smoke if an adequate range is not used! Again, look at Figure 11–11 to see these points illustrated.

CALCULATING THE MULTIPLIER VALUES

Using our familiar 0–1-mA meter with a resistance (R_M) of 100 Ω, let's see how we can design a voltmeter that will measure 10 V, 50 V, and 100 V.

Since the multiplier resistor's purpose is to limit the current through the meter, it makes sense that the multiplier resistor (R_{Mult}) value should limit the current through the meter to its full-scale current rating at the voltage range we want to design. That is, to have a 10-volt range, the series value of $R_M + R_{Mult}$ should limit current to 1 mA when it is measuring 10 V. This is a calculation for a two-resistance series circuit using Ohm's law. As you read the following examples, refer to Figure 11–12.

EXAMPLES

For the 10-volt range, then:

(Total R of meter circuit) | Formula 11–6 | $R_{Mult} + R_M = \dfrac{V \text{ (range)}}{I \text{ (full-scale)}}$

 V range value, $\boxed{\div}$, I full-scale value, $\boxed{=}$

(Multiplier R) | Formula 11–7 | $R_{Mult} = \dfrac{V \text{ (range)}}{I \text{ (fs)}} - R_M$

 V range value, $\boxed{\div}$, I full-scale value, $\boxed{-}$, R_M value, $\boxed{=}$

Therefore, $R_{Mult} = \dfrac{10 \text{ V}}{1 \text{ mA}} - 100 \text{ Ω} = 10{,}000 - 100 = 9{,}900 \text{ Ω}$

For the 50-volt range:

(Total R of meter circuit) $R_{Mult} + R_M = \dfrac{V}{I}$

(Multiplier R) $R_{Mult} = \dfrac{V}{I} - R_M$

$R_{Mult} = \dfrac{50 \text{ V}}{1 \text{ mA}} - 100 \text{ Ω} = 50{,}000 - 100 = 49{,}900 \text{ Ω}$

$$R_1 = \frac{10 \text{ V}}{1 \text{ mA}} - 100 \text{ }\Omega = 10{,}000 \text{ }\Omega - 100 \text{ }\Omega = 9{,}900 \text{ }\Omega$$

$$R_2 = \frac{50 \text{ V}}{1 \text{ mA}} - 100 \text{ }\Omega = 50{,}000 \text{ }\Omega - 100 \text{ }\Omega = 49{,}900 \text{ }\Omega$$

$$R_3 = \frac{100 \text{ V}}{1 \text{ mA}} - 100 \text{ }\Omega = 100{,}000 \text{ }\Omega - 100 \text{ }\Omega = 99{,}900 \text{ }\Omega$$

FIGURE 11–12 Examples of multiplier calculations

For the 100-volt range:

$$(\text{Total } R \text{ of meter circuit}) \; R_{Mult} + R_M = \frac{V}{I}$$

$$(\text{Multiplier } R) \; R_{Mult} = \frac{V}{I} - R_M$$

$$R_{Mult} = \frac{100 \text{ V}}{1 \text{ mA}} - 100 \text{ }\Omega = 100{,}000 - 100 = 99{,}900 \text{ }\Omega$$

PRACTICE PROBLEMS II

Given a 50-µA movement with R_M value of 2,000 Ω, what value of multiplier resistor is needed to create a voltmeter range of 250 V?

Approximating Multiplier Values

An interesting point you have probably observed is that the meter's resistance is a very small percentage of the total meter circuit R ($R_M + R_{Mult}$). Therefore, not much error is introduced if we indicate that R_{Mult} is essentially equal to

$$\frac{V \text{ (desired range)}}{I(fs)}$$

Range Selector Switch

Observe in Figures 11–13a and 11–13b that a **range selector switch** selects the desired voltage measurement range by switching in the appropriate multiplier resistor(s). No matter what range is selected, the multiplier resistance limits current to the meter's full-scale current value when the voltage applied to the total metering circuit (meter plus multiplier) equals the voltage range value.

FIGURE 11–13   **The range selector switch**

Also note that either the individual R multiplier or the cumulative R multipliers approach can be used. Most commercial multi-range meters use the cumulative R type circuitry. Refer again to Figure 11–13b.

PRACTICAL NOTES

Caution About Accuracy

There are many times when measurement accuracy is *critical!* In certain sensitive circuits, a percentage point of inaccuracy can make a meaningful difference—the difference between operating or not operating, in some cases. For that reason, we want to remind you: ***Always be aware of the test instruments' accuracy as related to the acceptable level of accuracy needed to measure the circuit being tested!***

Meter scales have calibrations that either directly match or are a convenient multiple or submultiple of the various selectable ranges. Refer to Figure 11–14.

Voltmeter Ohms-Per-Volt $\left(\dfrac{\Omega}{V}\right)$ Ratings

Previously we discussed the full-scale current- and millivolt-type ratings used with the basic current meter movements.

Another useful rating, related particularly to voltmeters, is the **ohms-per-volt rating** (Ω/V), Figure 11–15. This rating indicates how many ohms

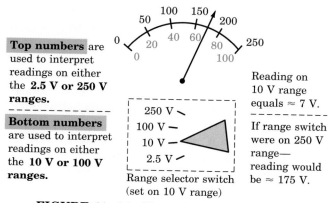

Top numbers are used to interpret readings on either the **2.5 V or 250 V ranges.**

Bottom numbers are used to interpret readings on either the **10 V or 100 V ranges.**

Reading on 10 V range equals ≈ 7 V.

If range switch were on 250 V range—reading would be ≈ 175 V.

Range selector switch (set on 10 V range)

FIGURE 11–14 **Meter scales correlate to selectable ranges.**

of meter circuit resistance ($R_M + R_{Mult}$) must be present to limit current to the meter's full-scale current rating value *for each volt* applied. In other words,

$$R = \frac{1\text{ V}}{I(fs)}$$

or the reciprocal of the full-scale current rating provides the ohms-per-volt rating.

EXAMPLE

A 1-mA movement has an Ω/V rating of $1/1 \times 10^{-3}$, or 1,000 ohms per volt. A sensitive 50-μA movement has an ohms-per-volt rating of $1/50 \times 10^{-6}$, or 20,000 ohms per volt.
A 10-mA movement has a rating equal to the reciprocal of 0.010, or $1/10 \times 10^{-3} = 100$ ohms per volt.

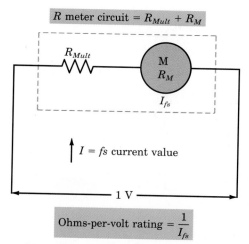

R meter circuit $= R_{Mult} + R_M$

R_{Mult}

M
R_M

I_{fs}

$I = fs$ current value

1 V

Ohms-per-volt rating $= \dfrac{1}{I_{fs}}$

FIGURE 11–15 **The ohms-per-volt rating**

Greater sensitivity of the basic movement means a higher ohms-per-volt rating. Two observations are worth noting related to this rating.

One, if you know the ohms-per-volt rating, it's easy to calculate the multiplier R value for any desired voltmeter range, neglecting the small value of R_M. However, *be careful!* If a 20,000-Ω/V movement has a 2,000-Ω meter resistance, which is typical, a 10% error is introduced on the 1-V range by this technique. One example of using this ballpark or approximating technique is as follows. If the meter rating is 10,000-Ω/V, and you want to create a 100-V range, multiply the rating by 100. The answer is 100 × 10,000, or 1 MΩ. You can see the trend!

Another more important fact is this rating gives you a good idea of how much meter loading effect will occur when you make voltage measurements. The next section discusses this concept in more detail. It is important, so study it!

PRACTICE PROBLEMS III

1. What is the ohms-per-volt rating of a meter that uses a 20-microampere movement?

2. What is the full-scale current of a meter rated at 2,000 ohms per volt?

Voltmeter Loading Effect

Since a voltmeter is connected in parallel with the component or the circuitry it is measuring, it would be ideal if the voltmeter circuitry had *infinite* resistance. This would mean none of the normal circuit operating conditions is altered with the voltmeter connected. But this is not the case! The voltmeter circuit has some finite value of resistance, and it alters circuit conditions! This situation is called **loading effect,** Figure 11–16.

It is apparent that the higher the voltmeter circuit's resistance, the less loading effect it has on the tested circuit. That is, the higher the ohms-per-volt rating of the voltmeter circuit *or* the higher the voltage range, the less

FIGURE 11–16   Voltmeter loading effect

FIGURE 11–17   Sample conditions of loading effect

the loading effect. Again, greater sensitivity of the basic movement means a higher ohms-per-volt rating. This situation correlates with the voltmeter causing less loading effect on the tested circuit.

Also, a higher voltage range results in a higher multiplier resistance, which causes less voltmeter loading effect, Figure 11– 17.

The greater the voltmeter circuit resistance ratio to the component or circuit resistance being tested, the more accurately the measurement reflects circuit conditions; that is, the less the circuit conditions change with the voltmeter attached. Stated another way, not only is the meter circuit resistance important, *but* the tested circuit's resistance characteristics are equally important! A higher circuit resistance prior to voltmeter connection results in greater disturbance of normal operating parameters when the meter is connected! In other words, low-resistance circuits are disturbed less by the meter connection than high-resistance circuits, Figure 11–18.

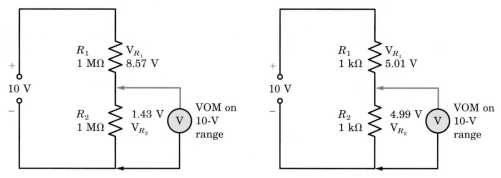

FIGURE 11–18   Variation in loading effect due to differences in circuits being tested

One practical rule of thumb is to be sure the resistance of the voltmeter circuit (on the range being used) is at least ten times the resistance of the circuit portion being tested by the voltmeter. For practical purposes, commercial voltmeters range from 20,000-Ω/V sensitivity ratings for common multimeters, or VOMs (volt-ohm-milliammeters), to 10 megohms, or higher input circuit resistances for the voltmeter circuitry common in digital measuring devices, Figure 11–19. Obviously, meters with multi-megohm input resistances cause negligible loading effect on lower-resistance circuits to be tested.

FIGURE 11–19   A DMM's higher meter circuitry resistance means less loading effect. (Courtesy of Simpson Electric Co.)

Voltmeter Key Points Capsule

1. A voltmeter is a combination of a current meter and a series current-limiting resistor, called a multiplier. The multiplier resistance limits current through the voltmeter when voltage is applied to the circuit.

2. A common sensitivity rating system for voltmeters is the ohms-per-volt rating that expresses how much resistance is needed to limit current through the meter to full-scale value when one volt is applied to the voltmeter circuitry.

3. The multiplier resistance value needed to achieve a given range voltmeter is determined by using Ohm's law and simple series-circuit techniques.

$$(R_M + R_{Mult}) = \frac{V \text{ desired range}}{I_{fs}}$$

4. Linear scales are usually present on dc voltmeters.

5. To measure voltage, the voltmeter is connected in parallel with the component or circuit portion where the difference of potential is to be measured. When measuring dc voltage, it is necessary to observe polarity. Also, care must be taken to assure that the voltage range of the metering circuit is sufficient to handle the voltage to be measured.

6. Because the voltmeter is connected in parallel, the higher the metering circuit resistance, the less it changes the circuit conditions, and the more accurately the voltmeter reading shows the actual circuit operating conditions.

OHMMETERS

NOTE: Although **series-type ohmmeter circuits** are becoming less common, due to the availability of digital multimeter devices, we will study the series-type ohmmeter briefly so you can gain a basic understanding of ohmmeter concepts.

FIGURE 11–20 Typical ohmmeter elements

The ohmmeter is yet another application for dc meter movements. An ohmmeter circuit has three main elements—a battery (generally housed in the meter case), the meter movement, and some current-limiting resistance, Figure 11–20.

Because resistance measurements are made with the power off on the tested component or circuit, the internal battery becomes the current source during testing. Obviously, the meter reacts to the amount of current flowing, and the current-limiting resistance controls the amount of current flowing through the meter.

Adjusting, Using, and Reading the Ohmmeter

The most common ohmmeter (used in analog instruments) uses the series-type ohmmeter circuit. Refer to Figure 11–21 and notice that when the test leads are not touching each other or are not connected to an external current path, there is an "open" circuit. Therefore, there is no current through the meter, and the **pointer** is at its left-hand resting place.

On the ohmmeter scale, zero current indicates infinite resistance between the test probes. The scale is marked with the infinity symbol (∞) on the left end of the scale. When the test leads are touched together, there are zero ohms between the test leads, and there is a complete current path. That is, current flows through the meter. The amount of current is controlled by adjusting the variable resistor (called the **zero-adjust control**) so there is exactly full-scale deflection on the current meter. The right end of the ohmmeter scale is marked with a "0" calibration mark, indicating that when full-scale current is flowing, there are zero ohms between the test probes.

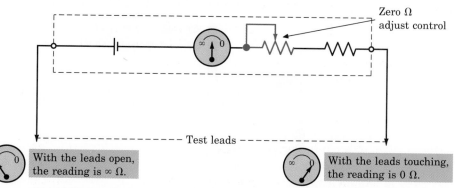

FIGURE 11–21 Series-type ohmmeter circuit characteristics

R_X value	% fs	Reading
R_X = 0 Ω	100%	0 Ω
R_X = 1000 Ω	75%	1000 Ω
R_X = 3000 Ω	50%	3000 Ω
R_X = 6000 Ω	33.3%	6000 Ω
R_X = 277 kΩ	1%	Hard to read. Looks close to ∞ Ω

FIGURE 11–22   Ohmmeter deflection for various R_X values

It is apparent when a resistance value between 0 and ∞ ohms is placed between the test leads, some amount of current flows that is less than full-scale current value and more than zero current. Therefore, the scale is calibrated to show this resistance value. Series circuit analysis is used to show what percentage of full-scale deflection occurs for various resistance values, Figure 11–22.

Note the ohmmeter scale is a nonlinear scale! This nonlinear scale contrasts with most ammeter and voltmeter scales that have evenly spaced or linear scales. A nonlinear ohmmeter scale is sometimes termed a **backoff ohmmeter scale,** since it is a "right-to-left" scale. In other words, zero is on the right end of the scale, and higher resistance readings are to the left. Again, see Figure 11–21.

Typically, commercial multimeters (volt-ohm-milliammeters or VOMs) have more than one range for measuring resistance that enables reading both low and high R values. Notice in Figure 11–23 the ohmmeter range switch shows ranges expressed in terms of R times some factor. For example, in the $R \times 1$ range, the ohmmeter scale is read directly. In the $R \times 100$ range position, the technician should multiply the scale reading by 100 to interpret the value of resistance being measured. The method for multiple ranges involves switching to different metering circuit shunt resistances, which is necessary because when measuring high resistances, little current flows; thus high meter sensitivity is needed. When measuring low resis-

FIGURE 11–23   Switchable ohmmeter ranges

tances, much current flows; thus, the meter can be shunted, as appropriate. Also, it is common to find a switching of internal voltage source(s), so when measuring high resistances the voltage supplied to the circuit is higher.

Ohmmeter Key Points Capsule

1. The series-type ohmmeter is a dc movement, an internal dc source, and a series current-limiting resistance connected in series. The test probes, when connected to the external component(s) being tested, place the unknown value of R being measured in series with the metering circuit.

2. The commonly used series-type ohmmeter uses a backoff (right-to-left) scale that is nonlinear. The lower value resistance scale calibrations (on the right end of the scale) are spread apart more than the higher resistance value calibrations, which appear on the left end of the scale.

3. *Ohmmeters should NEVER be used to measure components or circuit portions that have power applied.* (Most of the time, if that is done, the meter will be damaged). The ohmmeter's internal battery supplies the necessary voltage and current to make the measurement(s).

4. When using the series-type ohmmeter, the test probes are touched together and the zero-adjust control is adjusted so there is full-scale current through the meter and the ohmmeter scale reads "zero" resistance. After "zeroing" the meter and with the power off, the test probes are then connected across the component or components being tested to read their resistance value(s).

5. Most ohmmeters have more than one resistance range to measure a wider range of resistance values. The range selector switch indicates R times some factor at each switch position. For example, if the meter indicates 100 on the scale and the range selector switch is in the $R \times 10$ position, the actual resistance being measured is 10×100, or 1,000 Ω. When changing ranges, be sure to re-zero the meter on the new range setting.

VOMs, DMMs, AND OTHER RELATED DEVICES
Background Information

As indicated throughout the chapter, most test instruments are designed so they have multiple ranges for the electrical quantity they are measuring. Figure 11–24 shows an example of a panel meter designed to measure one type of electrical quantity. This type of meter is found in electronic equipment, such as radio transmitters and power supplies, and frequently has only one range.

As an electronics technician, you will probably use devices called **multimeters** in your troubleshooting and designing tasks. These meters not only have multiple ranges for each type of electrical quantity to be measured, but can measure more than one of the basic electrical quantities (I, V, or R). Let's take a brief look at several types of popular multimeters.

**FIGURE 11–24   Single-range
panel meter (Courtesy of
Simpson Electric Co.)**

VOMs

The VOM (volt-ohm-milliammeter) is a popular multimeter used for many
years by technicians, Figure 11–25. The advantages and disadvantages of
the VOM are:

ADVANTAGES	DISADVANTAGES
• Compact and portable • Does not require an external power source • Measures dc current, dc voltage resistance, and ac voltage • Typically has multiple ranges for each electrical quantity it measures • Is analog in nature, so is *good* for adjusting, tuning, or peaking tuned circuits	• Can cause significant meter loading effects on circuit being tested • Less accurate than a DMM • Analog readout scale can introduce "parallax" reading errors if pointer is viewed from an angle

THE DIGITAL MULTIMETER (DMM)

The digital multimeter (DMM), Figure 11–26, is probably the piece of test
equipment most used by technicians today. There are a number of applications where the analog multimeter (VOM) is still the best device to use.
However, the DMM's accuracy, expanded measurement capabilities, ease of
reading, and lighter loading of the circuit under test make it a prime choice
in most cases.

Some DMMs provide polarity-indicated readouts and autoranging features. If the polarity being measured is negative, the readout indicates this.
In contrast, a VOM would have its pointer hitting and/or pushing against
the scale's left-side limit peg, if the polarity were opposite the way you
connected the test leads. Certainly, meter damage could result. When used
in the autoranging mode, a DMM will automatically select the range with
the best resolution, without the technician having to manually select the
correct range.

Many DMMs also offer measurement capabilities beyond simply measuring voltage, current, or resistance. Some allow measurement of ac fre-

FIGURE 11–25   The popular VOM
(volt-ohm-milliammeter) (Courtesy
of Simpson Electric Co.)

(a)

(b)

FIGURE 11–26   (a) The DMM (digital multimeter): (b) a portable DMM
(a. Courtesy of John Fluke Mfg. Co., Inc. b. Courtesy of Simpson Electric Co.)

quency, capacitor values, semiconductor diode condition, and so on. You will better understand the significance of these capabilities after you have studied ac signals and components used in ac.

Some popular DMMs also are capable of remembering important measurement data that you want to read at a later time. For example, there may be a "MIN MAX" mode on the meter. This means that the meter will remember the lowest and highest readings over a period of time, take the average, and provide that data in its readout. Also, some meters can store a reading in memory, then compare it to subsequent readings, and display the difference between the memorized reading and the subsequent reading. For safety sake, some meters have a "hold" the reading feature. This allows you to keep your eyes on the test probes when working in dangerous or difficult places for making measurements. You can then read the display when you have removed the test probes and are better able to fix your eyes on the meter readout.

Some of the advantages and disadvantages of the DMM are:

ADVANTAGES	DISADVANTAGES
• Digital readout usually prevents a reading error • Can have "autoranging" that prevents destroying a meter by having it on the wrong range • Can have "autopolarity" that prevents problems from connecting meter to test circuit with wrong polarity • Can have excellent accuracy (plus or minus 1%) • Portable and uses internal batteries	• Can be physically larger than equivalent VOM (not in all cases) • Can be more costly to buy and operate • Due to "step-type" digital feature rather than "smooth" analog action, are *not as good* for adjusting "tuned" circuits or "peaking" tunable responses • Frequency limitations

Related Devices

There are special probes used with multimeters and special self-contained metering devices that you should know about.

FIGURE 11-27   The high-voltage probe (Courtesy of Amprobe Instruments)

HIGH-VOLTAGE PROBE. Basic multimeters cannot generally measure very high voltages in the multikilovolt range; however, a high-voltage probe will enable this to be done, Figure 11–27. In essence, the probe consists of a high-value resistance that acts as a multiplier resistance, external to the meter. The probe's resistor(s) are designed to withstand high voltages. One typical application of this probe is to measure high voltages on TV picture tubes.

CLAMP-ON CURRENT PROBE. Another accessory found on some multimeters is the clamp-on current probe that measures ac current, Figure 11–28a. When ac current levels are in the ampere ranges, this device reacts to the magnetic fields surrounding the current-carrying conductors and translates this activity into current readings. The key advantage to this device is that it measures current *without having to break the circuit* when inserting the metering circuit. The clamp-on probe simply surrounds the current-carrying conductor without breaking the circuit. NOTE: This device *does not measure dc* current!

OTHER DEVICES. There are a variety of special clamp-on measurement devices described in many electronic distributor catalogs. You will find the clamp-on meters we've been describing, as well as clamp-on volt-ohm-milliammeters (Figure 11–28b), clamp-on digital wattmeters, and "power factor" meters. (You will learn about "power factor" meters in later chapters).

(a)

(b)

FIGURE 11–28 Clamp-on probes (Photos courtesy of Amprobe Instruments)

PRACTICAL NOTES

Measuring instruments are a vital tool to technicians when they are troubleshooting circuits that are operating abnormally. Electrical parameters indicate whether a component, circuit portion, or a total circuit is operating normally, abnormally, or is dead.

In a *qualitative* sense, the measurements indicate whether the values measured are normal, higher than normal, or lower than normal.

In a *quantitative* sense, the measurements indicate specific values for the parameters being measured. These values are then interpreted as normal or abnormal.

To troubleshoot any circuit or system, the technician must know "what should be." That is, he or she *must know the norms*. This is gained through experience, knowledge, or reference to appropriate technical documentation, such as schematics.

Although the following discussion is not exhaustive or comprehensive, the hints and techniques can be useful to you during your training and throughout your career as a technician. Study and apply these items, as appropriate.

TROUBLESHOOTING HINTS

General Information

Use test instruments properly:

1. **Current meters**—*Turn power off circuit.* Insert meter in series, making sure to observe proper range and polarity. Turn power on circuit to be tested. Be sure the current metering circuit (meter or meter and shunt) does not have too high a resistance, causing undue or unaccounted-for changes in circuit conditions.

2. **Voltmeters**—Using appropriate safety precautions to prevent shock, connect the meter in parallel with desired test points, making sure to observe proper range and polarity. For safety's sake, only use one hand at a time in making voltage measurements. Be sure the meter and range do not cause unaccounted for meter loading because of too low a meter resistance.

3. **Ohmmeters**—*Turn power off circuit* where component or circuit portion to be measured resides. Make appropriate zero-adjust calibration with test probes shorted. Disconnect at least one component lead. Connect test probes across component or circuit portion to be measured. Be sure ohmmeter batteries are good.

Techniques for Voltmeter Measurements

1. **Connecting one lead to common or ground**—Many electronic circuits have a common or ground reference point, such as the chassis or a common conductor "bus" where all voltages are referenced. By con-

necting one lead of the voltmeter (frequently the black or negative lead) to this ground reference, measurements are made throughout the circuit by moving the other lead to the various points where voltage measurements are desired, Figure 11–29. Frequently, the black lead on a meter has an alligator clip, making it easy to attach the lead to an electrically common point in the circuit.

2. **Using voltage checks to determine current**—Because it is often awkward and time consuming to break a circuit when installing a current meter, technicians frequently use voltage measurements to determine current. For example, if the measured voltage drop across a 1-kΩ resistor is 15 V, it is simple (Ohm's law) to deduce the current through that resistor must be 15 mA. That is,

$$I = \frac{V}{R} = \frac{15 \text{ V}}{1 \text{ k}\Omega} = 15 \text{ mA}$$

Obviously, Ohm's law enables you to determine current regardless of the R value where you are measuring the voltage. This technique prevents you from having to a) turn off the circuit; b) break the circuit; c) insert a current meter; d) make the measurement, then turn circuit off again; and e) reconnect the circuit when the current meter is removed.

3. **Using voltage measurements to find opens**—Recall from a previous chapter that when an open is in a series circuit, the applied voltage appears across the open. This means the voltmeter is a very effective

When + voltmeter lead is at:
Top of R_3, reading = 10 V
Top of R_2, reading = 20 V
Top of R_1, reading = 30 V

FIGURE 11–29   Making voltage measurements with respect to ground reference

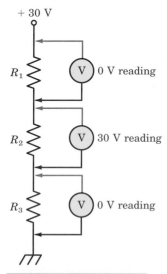

+ 30 V

R_1 (V) 0 V reading

R_2 (V) 30 V reading

R_3 (V) 0 V reading

Readings when R_2 is open!

FIGURE 11–30 Voltage measurements can find "opens."

(a)

Schematic symbol for this type phone jack
(b)

FIGURE 11–31 Shorting-type jack (Photo reprinted with permission from Perozzo, *Practical Electronics Troubleshooting* by Delmar Publishers Inc., Albany, NY)

means to find the circuit's open portion. The series circuit applied voltage appears across the open component(s), and zero volts are measured across the remaining component(s), Figure 11–30.

4. **Using voltage measurements to find shorts**—Again recall that a zero-volt *IR* drop appears across a shorted component or circuit portion. Voltage measurements quickly show this.

Techniques for Current Meter Measurements

1. **Total current measurement as a circuit condition indicator**—Total current value gives a general indication of circuit condition. If the total current is higher than normal, a lower-than-normal total circuit resistance is indicated. If the total current is lower than normal, a higher-than-normal total circuit resistance is indicated. Additional isolation techniques are then used to find the problem.

2. **Special jacks for current measurement**—Some circuits or systems have built-in closed-circuit jacks where the current meters are connected to measure current in the circuit, Figure 11–31.

   When the current meter is not present, the jack is shorted, providing a closed-current path. But when the current meter is present, the current must pass through the meter. (NOTE: This current-jack situation is uncommon.) More typically, voltage measurements determine current(s). In transmitters and other similar circuits where monitoring current is important, permanently mounted and installed current meters are used.

3. **Clamp-on metering for ease of measurement**— Previously we discussed the advantage of not breaking the circuit when using this type of current measurement. The limitations of this type instrument are the limited available ranges and the limited physical situations where it is possible to connect such a meter.

Techniques for Ohmmeter Measurements

1. **Continuity checks to find opens**—One frequently used check is the continuity check. The ohmmeter indicates ∞ ohms resistance if there is no "continuity," or an open between the points where the probes are connected. Of course, an open is one extreme condition that can be measured.

   A continuity check is very useful for checking fuses, light bulbs, switches, circuit wires, and circuit board paths. For example, all of these items will show very low or almost zero ohms resistance across their terminals, if they are normal. If the measurement shows infinite ohms, the item is obviously open or blown.

   In fact, sometimes visual checks are used to confirm where the open is occurring. You can see the broken filament in an automobile light bulb and a broken copper path on the circuit board!

2. **Ohmmeter checks to find shorts**—Ohmmeter checks are valuable for finding shorted components or circuit portions. It is apparent when measuring a component or circuit resistance that when the resistance

is zero ohms, there is a short. In the case of a closed switch, this is the normal situation. However, in cases where there should be a measurable resistance value, you have located a problem. By physically lifting one end of each component or circuit portion from the remaining circuitry in order to electrically isolate it, and making appropriate measurements, you can determine which component is shorted.

3. **Checking continuity of long cables or power cords**—Frequently, it is handy to verify whether a cable has a break in one of its conductors, or a short between conductors. Examples are TV twin leads, coaxial cables for antenna installations, and cables in computer network systems. Note Figure 11–32 and see one popular technique used to verify continuity in such cables. Rather than having an ohmmeter with 200-foot-long test leads, the technician creates temporary shorts on the conductors at one end of the cable. Then, using the ohmmeter at the other end, determines if there is continuity throughout the cable length for both conductors. After making the measurement, the technician removes the temporary short to establish normal operation. Incidentally, when the temporary short is removed, there should *not* be continuity. If there is, this indicates an undesired low resistance, or shorted path between conductors.

4. **Measuring resistances with the ohmmeter**— Reiterating, *never* connect an ohmmeter to a circuit with power on it! Remove the power plug!

WATCH OUT FOR "SNEAK PATHS"!   One frequent error made by technicians is not sufficiently isolating the component or circuit portion to be tested from the remaining circuitry. Any unaccounted current path that remains electrically connected to the specific path you want to measure can change the measured R values from the "expected" to the "unexpected."

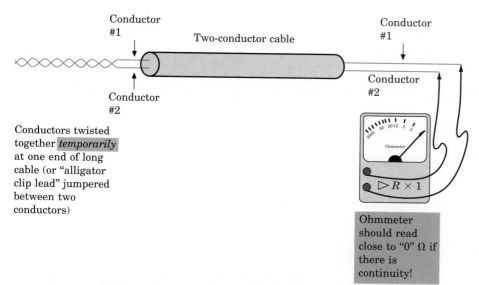

FIGURE 11–32   Checking continuity of long cables

Where possible, it is best to completely remove, or at least lift one end of the component being tested from the remaining circuit.

Another common "sneak path" occurs when the technician inadvertently allows his or her fingers to touch the test probes. This puts their body resistance in parallel with whatever is being measured. Where it is impossible to isolate the component or circuit portion to be tested, but "in-circuit" tests are required, have a schematic diagram, or good knowledge about the circuitry so you can accurately interpret the measurements.

Naturally, when measuring resistances, use all of the normal operating procedures. That is, perform the zero-adjust calibration for the range to be used, remove power from the circuit or component being tested, and use the proper range. If resistance measurements indicate improper values, proceed with additional isolating procedures, as appropriate.

Closing Comments About Measuring Instruments

In this chapter, we have focused on dc measurement instruments. We did mention, however, that most multimeters have additional built-in circuitry enabling the measurement of ac voltages.

Also you should know there are several other types of meter movement designs that react to ac quantities, which were not discussed in this chapter. These movements include the attraction-type moving-iron meter movement, the repulsion-type moving-iron movement, and the electrodynamometer movement (frequently used in wattmeters). It is beyond the scope of this chapter to discuss how each of these types operate, but we did want you to be aware of them.

As indicated throughout the chapter, meter movements are very fragile instruments. It is important that you give them the proper care to prevent physical abuse, and that you use proper operating procedures to prevent electrical damage.

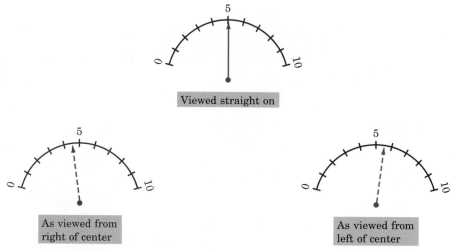

FIGURE 11–33   "Parallax" error

Many instruments are manufactured with electrical protection circuitry built in. A variety of semiconductor circuits, from simple diode protection to more elaborate circuits, are examples of these circuits.

We have mentioned that technicians should be aware of factors affecting accuracy when making measurements. Meter loading effects were mentioned as one of these factors. Use simple series-parallel circuit analyses to determine the *voltmeter's* effect on the tested circuit. Use series circuit analyses to assess the *current meter's* effect on a given circuit.

Another error is that of parallax when reading scales on analog meters, Figure 11–33. Some instruments have a mirror on the scale so the technician will not read the pointer from an angle that might cause a misinterpreted reading, Figure 11–34. By making sure the real pointer is superimposed over the reflection, the technician will be reading the scale at the angle that yields the best accuracy.

Another point is the basic accuracy rating of the instrument itself. Typically, meters are rated from 1–3% accuracy at full-scale deflection for analog instruments and from 0.1–1% (or plus/minus one digit) on digital meter readouts. It is essential technicians understand the limitations of the instruments used and have knowledge of the circuits or systems they are testing to properly interpret what is being measured. As a technician, you will find that *properly used and properly interpreted* circuit parameter measurements (enabled by the test instruments we have been discussing) are essential to the performance of your job. This is true whether the task be in designing, troubleshooting, servicing, or monitoring the operation of circuits and systems.

Mirror → strip

The technician makes sure the "real pointer" is superimposed over the reflection of the pointer when reading the meter to minimize the parallax reading error.

FIGURE 11–34   Mirrored scale for minimizing parallax error

SUMMARY

▶ A moving-coil meter movement (often called the d'Arsonval movement) contains several key elements—permanent magnet, pole pieces, soft iron core, movable coil and support frame, spindle and bearings, springs, pointer and counterbalance, calibrated scale, and stops.

▶ The moving-coil movement depends on a force of repulsion between the permanent magnet's fixed magnetic field and the movable coil's magnetic field when current passes through the movable coil.

▶ Electrical damping prevents the instrument's pointer from oscillating above and below the actual reading; develops from the effects of induced current in the movable coil frame; and is analyzed by Lenz's law.

▶ Moving-coil meters are current indicators and can be manufactured with sensitivities ranging from microamperes to milliamperes.

▶ Movement sensitivity relates to the strength of the permanent magnet, the number of turns on the moving coil, and other factors.

▶ Current meter sensitivities are usually expressed in full-scale current, but are also expressed in the amount of voltage dropped by the coil with full-scale current value passing through it, (e.g., 1-mA, 50-mV movement).

▶ When using the moving-coil movement as a current meter, the instrument is connected in series with the circuit portion where it is desired to measure current flow. Thus, breaking or opening the circuit, inserting the meter, and observing proper polarity and range are required. (*Caution:* Turn power off circuit before breaking circuit).

▶ The minimum current value an instrument measures is determined by its basic movement full-scale current sensitivity. An instrument can measure higher currents by using current meter shunts. A shunt is a parallel-resistance path that divides the circuit current. Thus, a percentage of the current passes through the meter and the remainder passes through the shunt.

▶ Parallel circuit analysis (current-divider) techniques determine the shunt resistance value for a desired current range. Inverse proportionality of current dividing in parallel circuits is a useful technique for this analysis, where

$$\frac{R_S}{R_M} = \frac{I_M}{I_S}$$

▶ When multiple ranges are available on the instrument for current measurements (selectable shunts), the switch is normally a shorting-type switch to assure the basic meter movement is never without a shunt between switch positions.

▶ A moving-coil meter movement can measure voltage by adding a series multiplier resistance. The multiplier R value and the meter's resistance must limit the current to the movement's full-scale current value for the voltage range desired. Ohm's law and series-circuit techniques determine the multiplier value. (For all practical purposes, the value of

$$R_{Mult} = \frac{V \text{ (desired range)}}{I \text{ (full-scale)}}$$

To be accurate, subtract the meter's resistance value from this answer.)

▶ Multiple ranges for voltage measurement are obtained by using a switch and contacting different multiplier resistance values at each switch position.

▶ Voltmeters are often rated by ohms-per-volt sensitivity. This rating indicates how many ohms must be in the metering circuit to limit current to the meter's full-scale current value when one volt (1 V) is applied. The formula for this is

$$\frac{\Omega}{V} \text{ (sensitivity)} = \frac{1 \text{ V}}{I_{fs}}$$

▶ Because voltmeters are connected in *parallel* with the portion of circuitry under test, the meter circuit's resistance alters the test circuit's operating conditions. This is called meter loading. The higher the meter circuit's input resistance, or Ω/V rating, the less loading effect it will cause.

▶ Most analog ohmmeters use series-type circuitry, whereby the meter's resistance, the internal voltage source, and the resistance under test are all in series. Therefore, the ohmmeter scales read from right to left, with low R values appearing on the right side and higher R values appearing on the left side. This scale is sometimes called a backoff scale. Also, the scale calibrations for ohms are nonlinear, but the scale calibrations for current and voltage are linear.

▶ Factors to be observed when using dc instruments include: use the proper range; connect the meter to the circuit in a proper fashion; observe proper polarity; and, with the ohm-

meter, make sure power is removed from the component or circuit being tested.

▶ Multimeters measure more than one type of electrical parameter, such as voltage, current, and resistance.

▶ A VOM typically has lower R input than a DMM. Therefore, a VOM causes more meter loading effects when measuring voltage.

▶ A number of special probes and devices are available for electrical measurements. Some of these include clamp-on current probes, thermocouples, clamp-on multimeters, and high-voltage probes.

▶ Using dc measuring instruments to troubleshoot actually means using common sense, properly applying Ohm's law, and having some knowledge of how the tested circuit or system operates.

FORMULAS AND CALCULATOR SEQUENCES SUMMARY

FORMULA 11–1 $R_S = \dfrac{V_S}{I_S}$

voltage across shunt, $\boxed{\div}$, current value to pass through shunt, $\boxed{=}$

FORMULA 11–2 $R_S = \dfrac{I_M R_M}{(I - I_M)}$

I_M value, $\boxed{\times}$, R_M value, $\boxed{\div}$, $\boxed{(}$, I (extended range) value, $\boxed{-}$, I_M , $\boxed{)}$, $\boxed{=}$

FORMULA 11–3 $R_S = \dfrac{I_M R_M}{I_S}$

I_M value, $\boxed{\times}$, R_M value, $\boxed{\div}$, I_S value, $\boxed{=}$

FORMULA 11–4 $\dfrac{R_S}{R_M} = \dfrac{I_M}{I_S}$

R_S value, $\boxed{\div}$, R_M value, $\boxed{=}$, (ratio should compare to), I_M value, $\boxed{\div}$, I_S value, $\boxed{=}$

FORMULA 11–5 $R_S = \dfrac{R_M}{N-1}$

R_M value, $\boxed{\div}$, $\boxed{(}$, no. of times range is to be multiplied, $\boxed{-}$, 1 , $\boxed{)}$, $\boxed{=}$

FORMULA 11–6 $R_{Mult} + R_M = \dfrac{V\ (\text{range})}{I\ (\text{full-scale})}$

V range value, $\boxed{\div}$, I full-scale value, $\boxed{=}$

FORMULA 11–7 $R_{Mult} = \dfrac{V\ (\text{range})}{I(fs)} - R_M$

V range value, $\boxed{\div}$, I full-scale value, $\boxed{-}$, R_M value, $\boxed{=}$

REVIEW QUESTIONS

REMINDER: Don't forget to use your calculator for these problems. The more you use it, the more proficient and efficient you will become in performing calculations.

 1. A current meter is rated as a 2-mA, 200-mV meter movement. What is its full-scale current rating? What is the resistance of the movement?

2. It is desired to increase the range of a 5-mA movement to 500 mA. What value of shunt resistance should be used? (Assume R_M = 20 Ω)

3. Briefly explain the action of a shorting-type switch. Under what circumstances should it be used in dc measuring instruments?

4. Should a current meter have a high or low resistance? Why?

5. Draw a three-resistor parallel circuit with a dc source. Show where and how current meters are connected to measure I_T and the current through the middle branch. Indicate polarities, as appropriate.

6. A current meter is rated at 3 mA, 150 mV. What multiplier resistance value is needed to use this meter as a 0—30-V meter? Draw and label the circuit.

7. If the full-scale current rating of a meter is 40 μA, what is its ohms-per-volt sensitivity

rating as a voltmeter? What is the value of $R_M + R_{Mult}$ if the voltage range selected is the 500-volt range?

8. Refer to Figure 11–35 and indicate what the voltmeter reading will be if the meter has Ω/V sensitivity rating of 20,000 Ω/V. (Assume that you are using the 0–10-V range). What would the R_2 voltage drop be without the voltmeter present?

FIGURE 11–35

9. What is the value of the metering circuit resistance for a current meter having a 1-mA, 100-mV rating that is shunted to measure 0–2 mA?

10. If a voltmeter is reading 22 V on the 0–50-V range, what percentage of full-scale current is flowing through the meter movement? What would the reading be if 75% of full-scale current were flowing?

11. Should a voltmeter circuit have a high or low resistance? Why?

12. Draw the basic circuit of a series-type ohmmeter. Show the test probes and an unknown value resistor to be measured. (R_x)

13. For the circuit in question 12, if the meter is a 1-mA, 100-Ω movement and the ohmmeter voltage source is a 1.5-V cell, what would be the calibration value at mid-scale on the ohmmeter scale? At quarter scale?

14. Current meters are connected in _____ with the component or circuit to be tested. Voltmeters are connected in _____ with the component or circuit to be tested. For ohmmeters using the backoff scale, the resistance under test is in _____ with the meter, its current limiting resistor, and the internal voltage source.

15. Why is it easier to use voltmeter checks, rather than current meter checks when troubleshooting?

16. Why are pole pieces used in dc moving-coil movements?

17. What is meant by damping relative to dc moving-coil meter movements?

18. How is damping achieved in a moving-coil instrument?

19. What are two distinct advantages of a DMM compared with an analog VOM?

20. For what situation is an analog-type instrument easier to use than a digital-type instrument?

TEST YOURSELF

1. If a voltmeter with a sensitivity rating of 1,000 Ω/V that is set on the 0–50-V range is connected across one resistor of a series circuit comprised of two 100-kΩ resistors, what will be the voltmeter reading? (Assume a 100-V dc source.)

2. In question 1, what would be the reading if the meter had a sensitivity of 100,000 Ω/V and the 0–100-V range were used?

3. In either question 1 or 2, if two identical meters, each set on the same voltage range were simultaneously connected across R_1 and R_2 respectively (i.e., one meter measuring V_1, the other measuring V_2), what voltage reading would each show? (Still assume 100 V applied voltage).

PERFORMANCE PROJECTS CORRELATION CHART

Suggested performance projects that correlate with topics in this chapter are:

CHAPTER TOPIC	PERFORMANCE PROJECT	PROJECT NUMBER
Current Meter Shunts	The Basic Current Meter Shunt	33
Voltmeters	The Basic Voltmeter Circuit	34
Ohmmeters	The Series Ohmmeter Circuit	35

After completing the above projects, perform the "Summary" checkout at the end of the performance projects section.

THE SIMPLER SEQUENCE FOR TROUBLESHOOTING

Following the SIMPLER troubleshooting sequence, find the problem with this circuit. As you follow the sequence, record your responses to each step on a separate sheet of paper.

CHALLENGE CIRCUIT 5

B_1
M_1
r_m
5 kΩ pot R_1
Test probes

SYMPTOMS Gather, verify, and analyze symptom information. (Look at the "Start Point Information.")

IDENTIFY Identify initial suspect area for the location of the trouble.

MAKE Make a decision about "What type of test to make" and "Where to make it." (To simulate making a decision about each test you want to make, select the desired test from the "TEST" column listing below.)

PERFORM Perform the test. (Look up the result of the test in Appendix C. Use the number in the "RESULTS" column to find the test result in the Appendix.)

LOCATE Locate and define a new "narrower" area in which to continue troubleshooting.

EXAMINE Examine available information and again determine "what type of test' and "where."

REPEAT Repeat the preceding analysis and testing steps until the trouble is found. What would you do to restore this circuit to normal operation? When you have solved the problem, compare your results with those shown on the following pages.

STARTING POINT INFORMATION

1. Circuit diagram
2. B_1 is a 1.5-V cell
3. Meter is a 1-mA movement
4. There is no meter response when the test probes are touched together to "zero" on the ohmmeter

TEST   Results in Appendix C

(With test probes not touching)
V_{A-B} . . . . (44)
V_{B-C} . . . . (115)
V_{C-D} . . . . (9)
V_{D-E} . . . . (91)
V_{E-F} . . . . (40)
V_{F-A} . . . . (101)
R_{C-E} . . . . (13)
R_{C-D} . . . . (59)
R_{D-E} . . . . (108)
R_{F-A} . . . . (84)

(With test probes touching)
V_{A-B} . . . . (64)
V_{B-C} . . . . (27)
V_{C-D} . . . . (97)
V_{D-E} . . . . (20)
V_{E-F} . . . . (110)
V_{F-A} . . . . (47)

CIRCUIT

5

Symptoms

Symptoms Information. When the ohm-meter test probes are touched to zero the ohmmeter, there is not meter response. This indicates a bad meter, a lack of circuit continuity somewhere, or a dead battery.

Identify initial suspect area: The total circuit, since a discontinuity can be caused anywhere in the circuit.

Make test decision: Since the battery is the only item in the circuit having a definite shelf life, let's first check the battery. Measure V between points A and B with the probes open.

Perform 1st Test: The result of measuring V_{A-B} is 1.5 V, which is normal. NOTE: There is no picture shown measuring V_{A-B}. With the test probes open, there is no load current drawn from the battery, thus it might measure normal without load and not completely good when under load. However, this tells us the battery is probably not totally dead, which it would have to be to match our symptom information.

Locate new suspect area: The remainder of the circuit, since discontinuity can be anywhere along the current path.

Examine available data.

Repeat analysis and testing: By checking continuity between points C and E (with test leads not touching), we can check the continuity of the two circuit elements, R_1 and the test-lead/test-probe.

CHALLENGE CIRCUIT 5

Check R_{C-E}. R is infinite ohms, indicating a problem. (NOTE: This meter shows "1" when measuring infinite ohms.)

Check R_{C-D} to verify that R_1's resistive element or wiper arm is not causing the discontinuity. R is about 1,500 ohms. However, moving the wiper arm causes R to change, which implies everything is normal.

Check R_{D-E}. R is infinite ohms, indicating the discontinuity is between these points. A further check shows the test-lead wire is broken inside the test-probe plastic handle.

Repair the wire connection to the test probe. Then, try to zero the ohmmeter. Now it works.

2nd Test

3rd Test

4th Test

5th Test

Basic AC Quantities

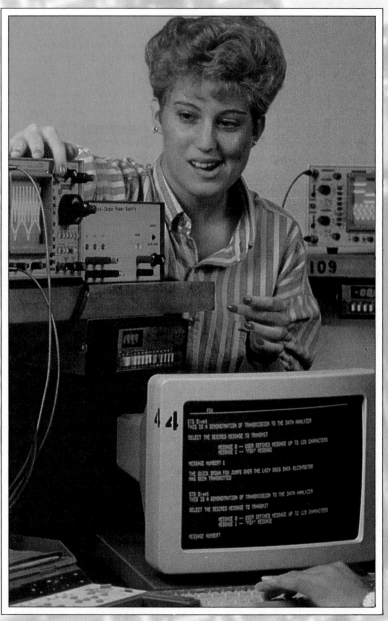

(Courtesy of DeVry Institutes)

KEY TERMS

Ac

Alternation

Average value

Coordinate system

Cosine

Cycle

Duty cycle

Effective value (rms)

Frequency (f)

Hertz (Hz)

Instantaneous value

Magnitude

Peak value (V_p or maximum value, V_{max})

Peak-to-peak value (p-p)

Period (T)

Phase angle (θ)

Phasor

Quadrants

Rectangular wave

Sawtooth wave

Sine

Sine wave

Sinusoidal quantity

Square wave

OUTLINE

Background information

Generating an AC voltage

Rate of change

Introductory information summary regarding AC sine-wave signals

Some basic sine-wave waveform descriptors

Period and frequency

Phase relationships

Important AC sine-wave current and voltage values

The purely resistive AC circuit

Other periodic waveforms

CHAPTER PREVIEW

Previous chapters have discussed dc voltages, currents, and related quantities. As you have learned, direct current (dc) flows in one direction through the circuit. Your knowledge of dc and your practice in analyzing dc circuits will help you as you now investigate alternating current (ac)—current that periodically reverses direction.

In this chapter, you will become familiar with some of the basic terms used to define important ac voltage, current, and waveform quantities and characteristics. Knowledge of these terms and values, plus their relationships to each other, form the critical foundation on which you will build your knowledge and skills in working with ac, both as a trainee and as a technician.

Studying this chapter, you will learn how ac is generated; how it is graphically represented; how its various features are described; and how specific important values, or quantities are associated with these features.

In the next chapter, you will study a powerful test instrument, called the oscilloscope, which will enable you to visually observe the ac characteristics you will study in this chapter.

OBJECTIVES

After studying this chapter, you should be able to:

- Draw a graphic illustrating an **ac** waveform
- Define **cycle, alternation, period, peak, peak-to-peak,** and **effective value (rms)**
- Compute effective, peak, and peak-to-peak values of ac voltage and current
- Explain **average** with reference to one-half cycle of sine-wave ac
- Explain **average** with reference to nonsinusoidal waves
- Define and calculate **frequency** and period
- Draw a graphic illustrating phase relationships of two **sine waves**
- Describe the phase relationships of V and I in a purely resistive ac circuit
- Label key parameters of nonsinusoidal waveforms

BACKGROUND INFORMATION
The Difference Between DC and AC

Whereas direct current is unidirectional and has a constant value, or magnitude, alternating current, as the name suggests, alternates in direction and is bidirectional. Also, ac differs from dc in that it is continuously changing in amplitude or value, as well as periodically changing in polarity.

In Figure 12–1, you see how both a simple dc circuit and a simple ac circuit may be represented diagrammatically. You are already familiar with the symbols in the dc circuit shown in Figure 12–1a. Notice the symbol used to represent the ac source in the ac circuit shown in Figure 12–1b. You will be seeing this symbol used throughout the remainder of the book. The wiggly line in the center of the symbol represents a sine-wave waveform, which you will study shortly.

Figure 12–1 also highlights the comparisons and contrasts between dc and ac voltage characteristics, using their graphical representations. As you proceed in the chapter, you will be focusing your attention on learning important specifics about the sine wave and other ac waveforms.

FIGURE 12–1   Graphic comparing dc and ac waveforms

CHARACTERISTIC DESCRIPTOR	IS CHARACTERISTIC OF:	
	AC	DC
Current in one direction only	No	Yes
Current periodically alternates in direction one-half the time in each direction	Yes	No
Voltage can be "stepped up" or "stepped down" by transformer action	Yes	No
Is an efficient means to transport power over long distances	Yes	No
Useful for all purposes	Yes	No

FIGURE 12–2 Similarities and differences between dc and ac power

Alternating Current Used for Power Transmission

Alternating current (**ac**) has certain advantages over dc when transporting electrical power from one point to another point. A key advantage is its ability to be "stepped up" or "stepped down" in value. Again, you will study more about this later in the text. These advantages are why power is brought to your home in the form of ac rather than dc.

Some interesting contrasts between dc and ac are shown in Figure 12–2. This ac electrical power is produced by alternating-current generators. A simple discussion about generating ac voltage will be covered shortly. However, some groundwork will now be presented.

Defining Angular Motion and the Coordinate System

As we deal with ac quantities, it is helpful to understand how angular relationships are commonly depicted. Refer to Figures 12–3a (angular motion) and 12–3b (the coordinate system) as you study the following discussions.

Mathematical and Graphic Concepts About AC

1. Angular motion is defined in terms of a 360° circle or a four-quadrant coordinate system.

2. The zero degree (starting or reference plane) is a horizontal line or arrow extending to the right, Figure 12–3a. NOTE: Other terms you might see for this line or arrow are a rotating radius and a vector called a **phasor**. A **vector** identifies a quantity that has (or illustrates) both magnitude *and* direction.

3. Angles increase in a positive, counterclockwise (CCW) direction from the zero reference point. In other words, the horizontal phasor to the right is rotated in a CCW direction until it is pointing straight upward, indicating an angle of 90° (with respect to the 0° reference point).

   • When the phasor continues in a CCW rotation until it is pointing straight to the left, that point represents 180° of rotation.

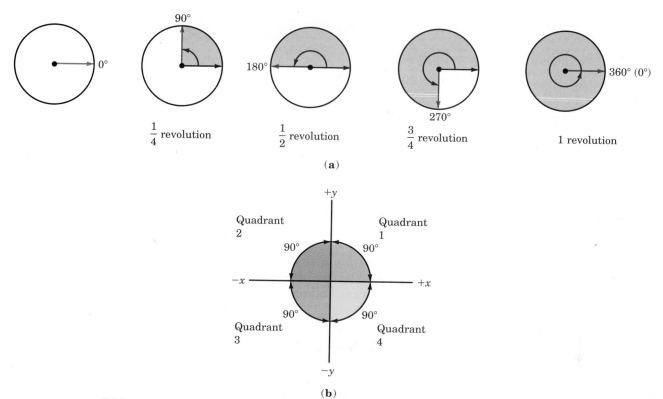

$\frac{1}{4}$ revolution $\frac{1}{2}$ revolution $\frac{3}{4}$ revolution   1 revolution

(a)

(b)

FIGURE 12–3   (a) Describing angular motion; (b) The coordinate system

- When rotation continues until the phasor is pointing straight downward, the angle at that point is 270°.

- When the phasor completes a CCW rotation, returning to its starting point of zero degrees, it has rotated through 360°.

- When the phasor is halfway between 0° and 90°, the angle is 45°. When the phasor is halfway between 90° and 180°, the angle is 135°.

4. The **coordinate system,** Figure 12–3b, consists of two perpendicular lines, each representing one of the two axes. The horizontal line is the x-axis. The vertical line is the y-axis. It is worth your effort to learn the names and directions of these two axes— they will be referenced often during your technical training. Remember the x-axis is horizontal and the y-axis is vertical.

5. There are four **quadrants** associated with this coordinate system. Again, look at Figure 12–3b and note that the angles between 0° and 90° are in quadrant 1. Angles between 90° and 180° are in quadrant 2. Angles between 180° and 270° are in quadrant 3; and angles between 270° and 360° (or 0°) are in quadrant 4.

6. When the phasor moves in a clockwise (CW) direction, the various angle positions are described as negative values. For example, the location halfway between 360° (0°) and 270° (−90°) in quadrant 4 (the same location as the +315° point) can be described as either −45° or +315°.

1. Define the difference between dc and ac.
2. Which direction is the reference, or zero-degree position when describing angular motion?
3. Is the y-axis the horizontal or vertical axis?
4. Is the second quadrant between 0 and 90 degrees, 90 and 180 degrees, 180 and 270 degrees, or 270 and 360 degrees?
5. Define the term vector.
6. Define the term phasor.

Keeping in mind these thoughts about defining angles in degrees and a reference point on the coordinate system, let's move to a brief explanation of how an ac sine-wave voltage is generated.

GENERATING AN AC VOLTAGE

Refer to Figure 12–4a and note the elementary concepts of an ac generator. The simplified generator consists of a magnetic field through which a conductor is rotated. As you would expect when the maximum number of magnetic field flux lines are being cut (or linked) by the rotating conductor, maximum voltage is induced. Recall that if 10^8 flux lines are cut by a single

(a)

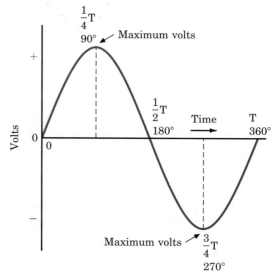

$$T = \text{Period} = \frac{1}{f} = \text{Time for 1 cycle}$$

(b)

FIGURE 12–4 Basic concepts of an ac generator

conductor in one second, the induced voltage is one volt. Conversely, when minimum (or zero) flux lines are cut, minimum (or zero) voltage is induced.

Slip rings, (sometimes called collector rings), are electrical contacting devices designed to make contact between stationary and rotating electrical junctures. As you can see in Figure 12–4, the brushes are in a stationary position, and make contact between the external load resistance and the slip rings. The slip rings are connected to, and rotate along with, the rotating conductors. The outside surfaces of the slip rings, however, are in continuous sliding electrical contact with the brushes. The conductors, which are rotating through the magnetic field, are connected so that the same end of each conductor is always connected to the same slip ring.

Also, observe that the slip rings and brushes that connect the external circuit load to the rotating loop so each end of the load is always connected to the same end of the rotating conductor. Thus, if the polarity of the induced voltage in the conductor changes, the voltage felt by the load through the brushes and the slip rings also changes.

Figure 12–4b shows the graphical representation of the changing magnitude and polarity of generator output felt by the load throughout one complete 360-degree rotation of the conductor loop(s). As we stated earlier, this graphic representation of the changing amplitude and polarity of the output voltage over time is called a waveform. In this case, the waveform is that of one full **cycle** of an ac sine-wave signal. A cycle is often defined as one complete sequence of a series of recurring events. There are many examples of cycles. The 24-hour day, the 12-hour clock face, one revolution of a wheel, and so on. In Figure 12–4b, you can see how at any given instant or point throughout the cycle, the signal's amplitude can be related to the angular degrees of rotation of the generator. Figure 12–5 again relates sine-wave amplitude and polarity to degrees throughout 360 degrees. For a sine wave, you can see that maximum positive amplitude occurs at 90 degrees. Maximum negative amplitude occurs at 270 degrees. And, the sine-wave signal is at zero level at 0, 180, and 360 degrees. (Assuming we start the waveform analysis at the zero point, with the signal progressing in a positive direction.)

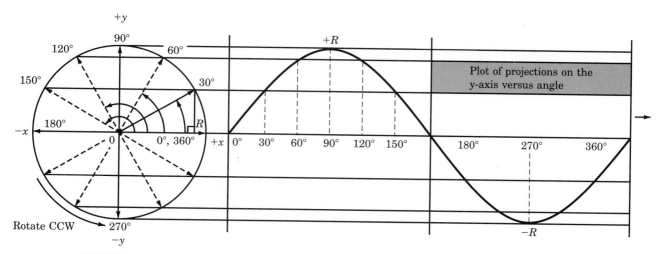

FIGURE 12–5 **Projections of sine-wave amplitudes in relation to various angles**

PRACTICAL NOTES

As you will see in upcoming studies, we do not always have to analyze an ac sine-wave cycle as starting at zero amplitude level, and a cycle being completed at zero amplitude. We can define one complete cycle as starting at any point along the sine wave, and being completed at the next point along the waveform that is at the same amplitude and changing in the same direction. (That is, the point at which the sequence of recurring events starts again).

The Sine Wave

This unique, single-frequency waveform reflects a quantity that constantly changes in amplitude and periodically reverses in direction.

Arc = Radius =
1 Radian

57.3°

Radius
of circle

**FIGURE 12–6
Illustration of the radian
angular measurement**

Radians of Angular Measure

You have been introduced to the fact that sine-wave signals are often analyzed in terms of angular measure. The radian is another angular measure of importance, as many of the mathematics associated with ac analysis often use this measure.

As you might surmise from the word *radian* itself, there is a relationship between the term radius (of a circle) and the term radian (used in circular angular measure). One definition for a radian (sometimes abbreviated "rad") is: "In a circle, the angle included within an arc equal to the radius of the circle." In Figure 12–6, you can see that the angle included by an arc of this length is about 57.3 degrees.

The thing that makes the radian of particular interest, mathematically, is that there are exactly 2π radians in 360 degrees. If you divide 360 degrees by 57.3 degrees, you will see that it would take 6.28 (or 2π) radians to make up a complete 360-degree circle. Observe Figure 12–7, and note the rela-

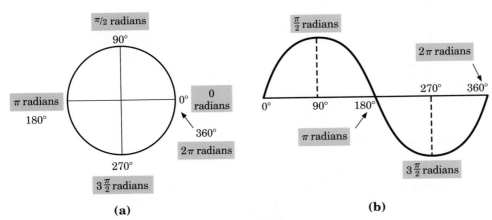

(a)   (b)

FIGURE 12–7   Degrees and radians of angular measure

For angle *A:*
1. The side opposite the right angle is called the *hypotenuse* (c).
2. The side opposite angle *A* is called the *opposite side* (a).
3. The side of angle *A* which is not the hypotenuse is called the *adjacent side* (b).

Sine (sin) of angle $= \dfrac{\text{opposite side}}{\text{hypotenuse}}$; $\sin A = \dfrac{a}{c}$

Cosine (cos) of angle $= \dfrac{\text{adjacent side}}{\text{hypotenuse}}$; $\cos A = \dfrac{b}{c}$

Tangent (tan) of angle $= \dfrac{\text{opposite side}}{\text{adjacent side}}$; $\tan A = \dfrac{a}{b} = \dfrac{\sin A}{\cos A}$

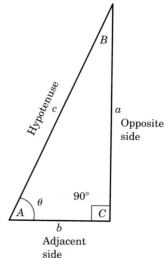

FIGURE 12–8   The right triangle

tionships between the degrees and radians around the circle. It is important to remember that one complete cycle of ac (360 degrees) equals 2π radians, that half a cycle (180 degrees) equals 1π radian, and so on. See Figure 12–7b. The waveform in Figure 12–7b is called a **sine wave** because the amplitude (amount) of voltage at any given moment is directly related to the sine function, used in trigonometry. That is, the trigonometric sine (sin) value for any given angle (θ or theta), represented along the sine wave, relates directly to the voltage amplitude at that same angle. NOTE: The sine of a given angle, using a right triangle as the basis of analysis, is equal to the ratio of the opposite side to the hypotenuse, Figure 12–8.

Also observe in Figure 12–8 that we have shown the side relationships that are important in computing the cosine (cos) and the tangent (tan) trig function values for the selected acute angle in the right triangle. As you can see, the cosine (cos) equals the ratio of the side that is adjacent (to the angle of interest) to the hypotenuse. The tangent (tan) equals the ratio of the side opposite to the side adjacent for the angle of interest. You will be studying and using these trig functions in later studies in this text, along with the sine (sin) function, which has been introduced in this chapter.

PRACTICAL NOTES

To find the sine of an angle using your calculator, first be sure your calculator is in the degrees ("DEG") mode of operation, rather than the "RAD" or "ERAD" mode. Then simply input the value of the angle (for example, 30, for 30 degrees), press the $\boxed{\text{SIN}}$ function button, and readout the value of the sine. In our example of 30 degrees, the value of the sine is

0.5. The key sequence again is: (the angle value), then, the trig function key. The answer shows the value of the trig function.

To go the other way; that is, if you know that the ratio of the opposite side to the hypotenuse equals 0.5, and you want to find the angle whose sine value is 0.5: input 0.5, then press the $\boxed{2^{nd}}$ and $\boxed{SIN^{-1}}$ keys. On some calculators, you press $\boxed{INV}$, then $\boxed{SIN}$. By pressing the $\boxed{2^{nd}}$ and $\boxed{SIN^{-1}}$ (or $\boxed{INV}$, $\boxed{SIN}$ keys, you are telling the calculator you want to know the angle whose sin is 0.5. When you do this, you should see 30 in the readout of the calculator. This means the angle whose sine is 0.5 is 30 degrees. The key sequence again is: (the trig function value), the $\boxed{2^{nd}}$ or ($\boxed{INV}$) key, then the trig function key (i.e., the $\boxed{SIN^{-1}}$ or $\boxed{SIN}$ key). The answer shows the value of the angle.

You can use the same techniques to find the value for cosines and tangents and/or the angles involved. That is, to find the trig function value when the angle is known: input the value of the angle, then press the correct trig function key to get your answer. In the case of the cosine, if you input 30 (for 30 degrees), then press the $\boxed{COS}$ button, the readout will indicate that the value of the cosine for an angle of 30 degrees is 0.866. If you input 30, then press the $\boxed{TAN}$ button, you find the tangent of an angle of 30 degrees is 0.577. Going the other way, input the trig function value, then the $\boxed{2^{nd}}$ or $\boxed{INV}$ key, then the appropriate trig function key (i.e., $\boxed{SIN^{-1}}$ or $\boxed{SIN}$ key, as appropriate). The answer is the angle involved.

Refer to Figure 12−9 as you read the following discussion. For simplicity, we will equate the angle of rotation from the starting point of the ac generator's mechanical rotation (where no flux lines are being cut by the rotating conductor) to the angle of the sine-wave output voltage waveform. For example, at the 0-degree point of rotation, no flux lines are being cut and the output voltage is zero (position a). If you find the sine value of 0° on a "trig" table, or use the "sin" function on a calculator, you will find it to be zero, and the output voltage at this time is zero volts.

At 45° into its rotation, the conductor cuts sufficient flux lines to cause output voltage to rise from 0 to 70.7% of its maximum value. The sine value for 45° is 0.707.

As the conductor rotates to the 90° position (position b), it cuts maximum lines of flux per unit time; hence, has maximum induced voltage. The sine function value for 90° is 1.00. For a sine wave, 100% of maximum value occurs at 90°.

As the conductor rotation moves from 90° toward 180°, fewer and fewer magnetic lines of force per unit time are being cut; hence, induced voltage decreases for maximum (positive) value back to zero (position c).

Continuing the rotation from 180° to 270°, the conductor cuts flux in the opposite direction. Therefore, the induced voltage is of the opposite polarity. Once again, voltage starts from zero value and increases in its opposite polarity **magnitude** until it reaches maximum negative value at 270° of rotation (position d).

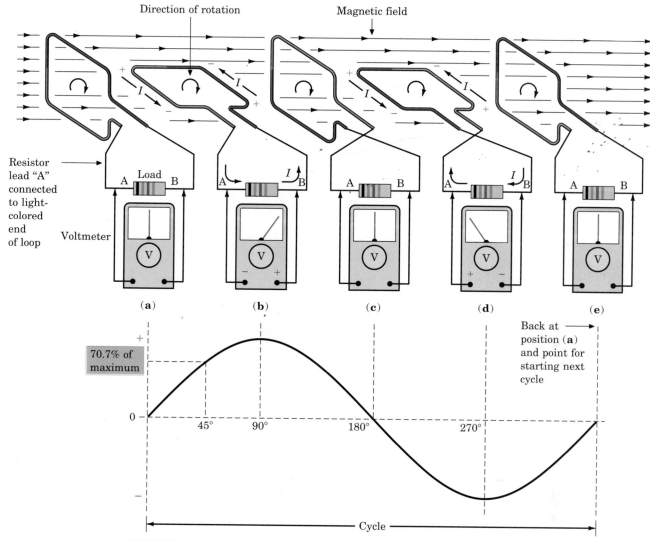

FIGURE 12-9   Relationship of flux lines cut in relation to output

From 270° degrees to 360°, the output voltage decreases from its maximum negative value back to zero (position e). This complete chain of events of the sine wave is a cycle. If rotation is continued, the cycle is repeated.

(NOTE: A cycle can start from *any* point on the waveform. That particular cycle continues until that same polarity and magnitude point is reached again. At that point, repetition of the recurring events begins again, and a new cycle is started.)

RATE OF CHANGE

Starting at one of the zero points, voltage increases from zero to maximum positive, then back to zero, on to maximum negative value, and then back to zero. As you can see, the sine wave of voltage continuously changes in amplitude and periodically reverses in polarity.

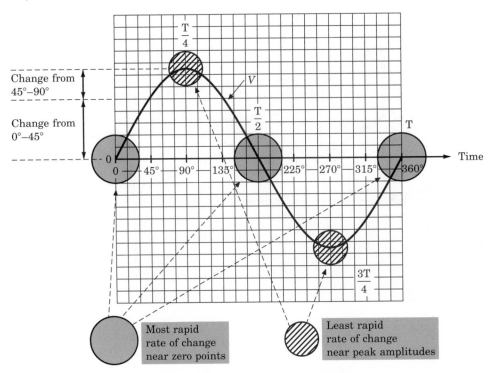

T = time for one cycle. This is known as the "period" of the waveform.

FIGURE 12–10   Different rates of change of voltage at various parts of sine wave

Let's look at another interesting aspect of the sine wave of voltage. The sine-wave waveform is nonlinear. That is, there is not an equal change in amplitude in one part of the waveform for a given amount of time, compared with the change in amplitude for another equal time segment at a different portion of the waveform. Therefore, the *rate of change* of voltage is different over different portions of the sine wave.

EXAMPLE

Notice in Figure 12–10 the voltage is either increasing or decreasing in amplitude at its most rapid rate near the zero points of the sine wave, (i.e., the largest amplitude *change* for a given amount of time).

Also, notice that near the maximum positive and negative areas of the sine wave, specifically, near the 90° and 270° points, the rate of change of voltage is minimum. This rate of change has much less *change* in amplitude per unit time. Knowledge of this rate of change (with respect to time) will be helpful to you in some later discussions, so try to retain this knowledge.

INTRODUCTORY INFORMATION SUMMARY REGARDING AC SINE-WAVE SIGNALS

1. A sine-wave signal is an ac signal whose amplitude characteristics can be related to the trigonometric sine function.

2. An ac sine-wave signal continuously varies in amplitude and periodically changes polarity.

3. One complete cycle of an ac signal can be measured in terms of 360 degrees and/or 2π radians (a radian = 57.3 degrees).

4. When a signal's zero-degree starting point is considered to be at zero amplitude, it is common to think of that signal's maximum positive point occurring at 90 degrees, the next zero point being at 180 degrees, the maximum negative point being at 270 degrees, and the cycle being completed at 360 degrees, at zero amplitude, ready to start the next cycle.

5. One cycle of a sine-wave signal can be measured from any starting point along the waveform to the next point along the waveform that matches it, and starts another similar waveform sequence in terms of amplitude and direction variations.

6. The "rate of change" of the sine-wave waveform is greatest at the zero amplitude points, and least at the maximum amplitude points of the waveform.

PRACTICE PROBLEMS I

1. Use a calculator or trig table and determine the sine of 35 degrees.

2. Indicate on the following drawing at what point cycle 1 ends and cycle 2 begins, if the starting point for the first cycle is point A.

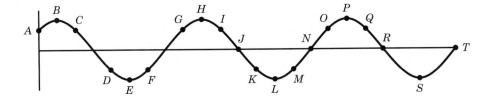

3. Which points on the drawing above represent areas of maximum rate of change of the sine wave(s)?

SOME BASIC SINE-WAVE WAVEFORM DESCRIPTORS

Refer to Figure 12–11. The horizontal line (x-axis) divides the sine wave into two parts—one above the zero reference line, representing positive values, and one below the zero reference line, representing negative values. Each half of the sine wave is an **alternation.** Naturally, the upper half is the "positive alternation," or positive half cycle. The lower half of the wave is the "negative alternation," or negative half cycle. With zero as the starting point, an alternation is the variation of an ac waveform from zero to a maximum value, then back to zero (in either polarity).

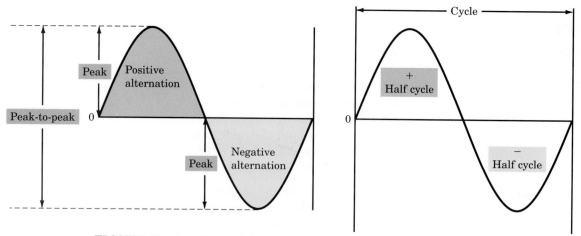

FIGURE 12–11   Several important descriptors of sine-wave ac

Examine Figure 12–11 again, and learn what is meant by positive peak, negative peak, and peak-to-peak values of the sine wave. Also, note in this illustration how a cycle is depicted.

Generalizing, a cycle, whether in electronics or any other physical phenomena, is one complete set of recurring events. Examples of this are one complete revolution of a wheel, the 24-hour day (representing one complete revolution of the earth), and the 11-year sunspot cycle.

PERIOD AND FREQUENCY
Periodic Waves

Periodic waves are those that repeat themselves regularly in time and form. Most ac signals and waves that you will study as a technician fall into this class. The period, or time of interest, for such waves is the time it takes for one complete cycle of that waveform. Let's see how the sine wave we have been discussing is defined in this regard.

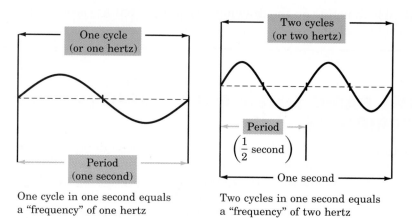

One cycle in one second equals   Two cycles in one second equals
a "frequency" of one hertz   a "frequency" of two hertz

(a)   (b)

FIGURE 12–12   Comparing two frequencies

Period

The ac sine-wave waveform has been described in terms of amplitude and polarity and has been referenced to various angles. It should be emphasized that the sine-wave occurs over a period of time—it does not instantaneously vary through all its amplitude values. The household ac, with which you are familiar, takes 1/60 of a second to complete a cycle. Look at Figure 12–12 and observe how the sine-wave illustrations have been labeled. Notice that time is along the horizontal axis. In Figure 12–12a, a one-cycle-per-second signal is represented. If this waveform represented one cycle of the 60-cycle ac house current, the time, or **period** (*T*) of a cycle would be 1/60 of a second, rather than the one second shown for the one-cycle signal. Learn this term period! Again, period is the time required for one cycle.

Frequency

The number of cycles occurring in one second is the **frequency** (*f*) of the ac. Refer again to Figure 12–12. The period of one cycle is inversely related to frequency. That is, the period equals the reciprocal of frequency, or 1/*f*. The letter *T* represents period; that is, *T* is time of one cycle. Therefore:

Formula 12–1	$T(\text{s}) = \dfrac{1}{f\,(\text{Hz})}$

 frequency in hertz, $\boxed{1/x}$

PRACTICAL NOTES

Use the $\boxed{1/x}$ (or x^{-1}) reciprocal function of your calculator for this type of calculation.

Restated, the period of one cycle (in seconds) equals 1 divided by the frequency in cycles per second. Rather than saying "cycles per second" when talking about frequency, we use a unit called the **Hertz (Hz),** named for the famous scientist Heinrich Hertz. For example, 1 Hz equals one cycle per second; 10-Hz equal 10 cycles per second; 1 kHz equals 1,000 cycles per second; and 1 MHz equals 1,000,000 cycles per second. Notice the abbreviation for Hertz (Hz) *starts* with a capital H and is abbreviated Hz to prevent confusion with another electrical quantity you will study later—the Henry, which is abbreviated with a capital H.

EXAMPLE

1. The period of a 100-Hz sine wave is $\dfrac{1}{100}$ of a second, $\left(T = \dfrac{1}{f}\right)$.

2. The period of a 1,000-Hz signal is $\dfrac{1}{1,000}$ of a second, $\left(T = \dfrac{1}{f}\right)$.

Relationship of Frequency and Period

Converse to calculating the period of one cycle by taking the reciprocal of the frequency, (i.e., $T = 1/f$), the formula is transposed to find the frequency, if you know the time it takes for one cycle. The formula is:

$$\text{Formula } 12\text{--}2 \quad f\,(\text{Hz}) \; = \; \frac{1}{T\,(\text{s})}$$

 time for one cycle in seconds , $\boxed{1/x}$

This indicates the period is inverse to the frequency, and vice versa. That is, the higher (or greater) the frequency, the less time required for one cycle (the shorter the period), or the longer (or greater) the period, the lower the frequency must be. In other words, if $T\uparrow$, then $f\downarrow$; and if $f\uparrow$, then $T\downarrow$. NOTE: again the $\boxed{1/x}$ function on your calculator is useful.

EXAMPLE

1. A signal with a period of 0.0001 seconds indicates a signal with a frequency of 10,000 Hz, $\left(f = \dfrac{1}{T}\right)$.

2. A signal with a period of 0.005 seconds indicates a signal with a frequency of 200 Hz, $\left(f = \dfrac{1}{T}\right)$.

IN-PROCESS LEARNING CHECK II

1. Determine the frequency of an ac signal whose period is 0.0001 second.
2. What is the period for a frequency of 400-Hz?
3. What time does one alternation for a frequency of 10-kHz take?
4. Does the y-axis represent time or amplitude in waveforms?
5. For waveforms, at what *angular points* does the amplitude equal 70.7% of the positive peak value? At what angular points is the sine wave at a 70.7% level during the negative alternation?
6. In the coordinate system, in which quadrants are all the angles represented by the positive alternation of a sine wave?
7. What is the period of a frequency of 15-MHz?
8. What frequency has a period of 25 ms?
9. As frequency increases, does T increase, decrease, or remain the same?
10. The longer a given signal's period, the ____ the time for each alternation.

PHASE RELATIONSHIPS

Alternating current (ac) voltages and currents can be *in-phase* or *out-of-phase* with each other by a difference in angle. This is called **phase angle** and is represented by the Greek letter theta (θ). When two sine waves are the same frequency and their waveforms pass through zero at different times, and when they do not reach maximum positive amplitude at the same time, they are out-of-phase with each other.

On the other hand, when two sine waves are the same frequency, and when their waveforms pass through zero and reach maximum positive amplitude at the same time, they are in-phase with each other.

Look at Figure 12–13 and note the following:

1. Sine waves a and b are in-phase with each other because they cross the zero points and reach their maximum positive levels at the same time (shown on the x-axis). Zero points are at times t_0, t_2, and t_4, and the positive peak at time t_1.

2. Sine waves b and c are out-of-phase with each other because they do not pass through the zero points and reach peak values at the same time. In fact, sine wave b reaches the peak positive point (at t_1) of its waveform 90 degrees before sine wave c does (at T_2). It can be said that sine wave c "lags" sine wave b (and also sine wave a) by 90°. Conversely, it can be said that sine wave b "leads" sine wave c by 90°. In other words, when viewing waveforms of two *equal-frequency* ac sig-

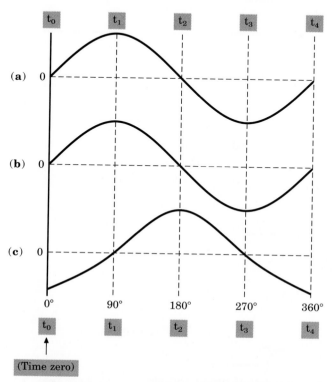

FIGURE 12–13   Phase relationships

nals, the signal reaching maximum positive first is leading the other signal, in phase. "First" refers to the peak that is closest to the y-axis; thus, showing an occurrence at an earlier time.

Obviously, ac signals can be out-of-phase by different amounts. This is illustrated by sine-wave waveforms and **phasor** diagrams, Figure 12–14. Figure 12–14a shows two signals that are 45° out-of-phase. Which signal is leading the other one by 45°? The answer, of course, is A is leading B by 45°.

In Figure 12–14b, waveforms A and B are 180° out of phase. If these signals were of equal amplitude, they would cancel each other out.

In Figure 12–14c, you see two signals that are 90° out-of-phase.

Notice how the phasors in Figure 12–14 illustrate both magnitude and relative phase for several conditions. Observe that the *length* of the phasors can represent *amplitude* values of the two waveforms. Both phasors must represent the same relative point or quantity for each of the two waveforms being compared.

PRACTICE PROBLEMS II

1. Illustrate two equal-frequency sine waves that are 60° out of phase with each other. Label the leading wave as A, and the lagging wave as B. Also, show the A waveform starting at the zero-degree point!

2. Assume the signals in problem 1 are equal in amplitude. Draw and label a diagram using phasors to show the conditions described in problem 1. Also, show the B phasor as the horizontal vector!

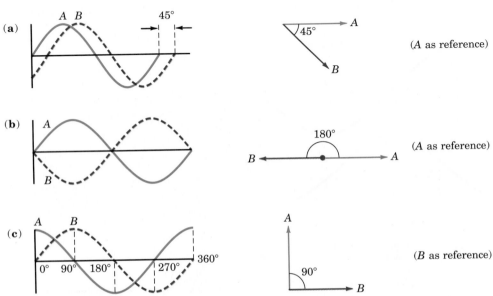

FIGURE 12–14 Diagrams depicting out-of-phase signals

IMPORTANT AC SINE-WAVE CURRENT AND VOLTAGE VALUES

The importance of learning the relationships between the various values used to measure and analyze ac currents and voltages cannot be overemphasized! These values are defined in this section. Learn these relationships well, since you will use them extensively in your training and work from this point on.

Brief Review of Terms

Look at Figure 12–15 and recall the maximum positive or negative amplitude (height) of one alternation of a sine wave is called **peak value.** For voltage, this is called V_p, or V_M; for current, this is called I_p, or I_M.

Also, remember the total amplitude from the peak positive point to the peak negative point is called **peak-to-peak** value, sometimes abbreviated as $V_{p\text{-}p}$ when speaking of voltage values.

In other words, for a symmetrical waveform:

$$V_p = \frac{1}{2} \times V_{p\text{-}p}, \text{ and } V_{p\text{-}p} = 2 \times V_p$$

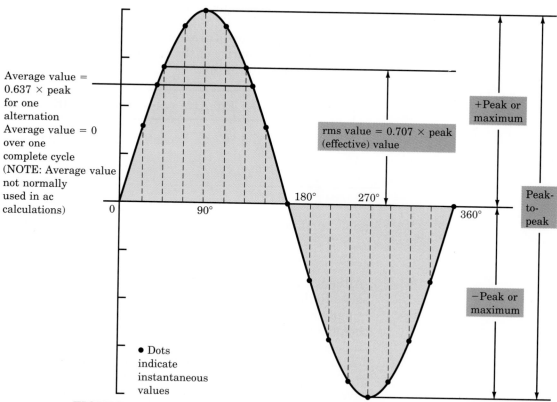

FIGURE 12–15   **Important values related to sine-wave waveforms**

Some New Terms and Values

Refer again to Figure 12–15 as you study the following discussions:

1. Since a given amount of current through a component in either direction causes the same amount of voltage drop, power dissipation, heating effect, and so on, the effects of an ac voltage or current are analyzed on a single alternation. For example, the positive alternation causes the same amount of current through a component (like a resistor) as the negative alternation, and vice versa.

2. The most frequently used value related to ac is called the **effective value (rms).** This is *that value of ac voltage or current producing the same heating effect as an equal value of dc voltage or current.* That is, it produces the same power:

$$I^2 R \text{ or } \frac{V^2}{R}$$

For example, an effective value of ac current of two amperes through a given resistor produces the same heating effect as two amperes of dc current through that same resistor.

Effective value is also commonly termed the root-mean-square value. The root-mean-square term comes from the mathematical method used to find its value. It is derived from taking the square root of the mean (average) of all the squares of the sine values. Thus, effective value (rms) is 0.707 times maximum or peak value. Again, the reason effective value is greater than half of the maximum value relates to the nonlinearity of the rate of change of voltage or current throughout the alternation.

Formula 12–3	Effective value (rms) = 0.707 × Peak value

0.707, $\boxed{\times}$, peak value of voltage, $\boxed{=}$

EXAMPLE

Again, assuming 1-V peak value for the waveform in Figure 12–15, $V_{rms} = 0.707 \times 1 = 0.707$ V

---

3. Although **average value** only has meaningful significance as it relates to the subject of rectification of ac (changing of ac to dc), we will simply introduce you to the concepts here. This awareness will then become useful in your later studies. If you compute the average amplitude or level under the waveform outline of *one-half cycle* or one alternation of an ac waveform, the result is the **average value** equals 0.637 (63.7%) of the maximum (or peak) value. Again, the reason the average value is more than 50% of the peak value is because the rate of change is slower near peak value than it is near the zero level. That

means the voltage level is above 50% of maximum level for a longer time than it is below 50% of maximum level.

In other words, the sine wave is a nonlinear waveform. Therefore, the average is not half of the peak value. *Incidentally, when you "average" over the entire cycle, rather than for one alternation, the result is zero,* since the positive and negative alternations are symmetrical, equal in amplitude and time, and opposite in polarity. However, computing only one alternation or half-cycle, the formula is:

| Formula 12–4 | Average value = 0.637 × Peak value |

 0.637, $\boxed{\times}$, peak value of voltage, $\boxed{=}$

EXAMPLE

If the peak value in Figure 12–15 were 1 V, V_{avg} (one-half cycle) = 0.637 × 1.0 = 0.637 V.

PRACTICAL NOTES

Most voltmeters and current meters used by technicians are calibrated to measure effective, or rms, ac values. This includes VOMs, DMMs, and clip-on current and volt meters.

In ac, power (heating effect) is computed in purely resistive components or circuits by using the effective voltage or current values. That is, $P = V_{\text{rms}} \times I_{\text{rms}}$, or $P = I^2R$, or $P = V^2/R$.

Summary of Relationships for Common AC Values (For Sinusoidal Waveforms)

1. Effective value (rms) = 0.707 times peak value, or 0.3535 times peak-to-peak value.

2. Average value = 0.637 times peak value. (NOTE: it also = 0.9 times effective value, since 0.637 is approximately 9/10 of 0.707.)

3. Peak value = maximum amplitude within one alternation. Since effective value (rms) is 0.707 of peak value, it follows that peak value equals 1/0.707, or is 1.414 times effective value. That is, $V_p = 1.414 \times$ rms value. (NOTE: it also equals 0.5 times peak-to-peak value.)

4. Peak-to-peak value = two times peak value and is the total sine wave amplitude from positive peak to negative peak. It follows then, that effective value (rms) is 0.3535 of the peak-to-peak value (or half of 0.707). That is, V_{p-p} equals 1/0.3535, or peak-to-peak value equals 2.828 times effective value.

With the knowledge of these various relationships, try the practice problems below.

PRACTICE PROBLEMS III

1. The peak value of a sine-wave voltage is 14.14 volts. What is the rms voltage value?

2. For the voltage described in question 1, what is the V_{avg} value over one alternation?

3. Approximately what percentage of effective value computed in question 1 is the average value computed in question 2?

4. Does frequency affect the rms value of a sine-wave voltage?

5. Are the results from calculating ac values from the negative alternation the same as ac values calculated from the positive alternation?

6. The peak-to-peak voltage of a sine-wave signal is 342 volts. Calculate the following:
   a. ac rms (effective) voltage value ____
   b. peak voltage value ____
   c. instantaneous voltage value at 45° ____
   d. instantaneous voltage value at 90° ____
   e. ac average voltage value (over one-half cycle) ____

THE PURELY RESISTIVE AC CIRCUIT

Earlier, we discussed in-phase and out-of-phase ac quantities. In purely resistive ac circuits, the ac voltage across a resistance is in-phase with the current through the resistance, Figure 12–16. It is logical that when maximum voltage is applied to the resistor, maximum current flows through the resistor. Also, when the instantaneous ac voltage value is at the zero point, zero current flows and so forth. Ohm's law is just as useful in ac circuits as it has been in dc circuits. (NOTE: **instantaneous values** are expressed as "e" or "v" for instantaneous voltage and "i" for instantaneous current.)

EXAMPLE

Applying Ohm's law to the circuit in Figure 12–16, if the rms ac voltage is 100 V, what is the effective value of ac current through the 10-Ω resistor? If you answered 10 amperes, you are correct! That is,

$$I = \frac{V}{R} = \frac{100 \text{ V}}{10 \ \Omega} = 10 \text{ A}$$

In ac circuits where voltage and current are in-phase, the calculations to find V, I, and P are the same as they are in a dc circuit. Generally, effective (rms) values are used in these computations, unless otherwise indicated.

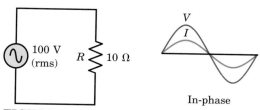

FIGURE 12–16   Basic ac resistive circuit

If you want to determine the instantaneous voltage at the 30° point of the sine wave, the value would equal the sine of 30° times the maximum value. In our sample case, the sine of 30° = 0.5. Therefore, if the maximum voltage (V_{max} or V_p) is 141.4 volts, the instantaneous value at 30° = 0.5 × 141.4 = 70.7 V.

Formula 12–5	$v = V_p \sin \theta$

where v = instantaneous voltage value
V_p = maximum voltage value
$\sin \theta$ = sine of the angle value

 V peak value, $\boxed{\times}$, angle value, θ $\boxed{\sin}$, $\boxed{=}$

The analyses you have been using for dc circuits also applies to purely resistive circuits with more than one resistor. The only difference is that you typically will deal with ac effective values (rms). When you know only peak or peak-to-peak values, you have to use the factors you have been studying to find the rms value(s). In later chapters, you will study what happens when out-of-phase quantities are introduced into circuits via "reactive" components.

Try to apply your new knowledge of ac quantities and relationships to the following practice problems.

PRACTICE PROBLEMS IV

1. Refer to the series circuit in Figure 12–17 and determine the following parameters:
   a. Effective value of V_1
   b. Effective value of V_2
   c. I_p
   d. Instantaneous peak power
   e. P_1

2. Refer to the parallel circuit in Figure 12–18 and determine the following parameters:
   a. Effective value of I_1
   b. Effective value of I_2
   c. I_p
   d. $V_{p\text{-}p}$

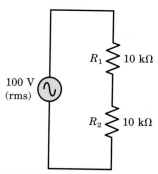

FIGURE 12–17   Purely resistive series ac circuit

FIGURE 12–18   Purely resistive parallel ac circuit

FIGURE 12–19   Purely resistive series-parallel ac circuit

3. Refer to the series-parallel circuit in Figure 12–19 and determine the following parameters:
   a. Effective value of V_1
   b. Effective value of V_2
   c. Effective value of V_3
   d. P_1 (NOTE: Use appropriate rms values.)

OTHER PERIODIC WAVEFORMS

Important Waveshapes

Although the sine wave is a fundamental ac waveform, there are nonsinusoidal waveforms that are important for the technician to know and understand. Modern digital electronics systems, such as computers, data communications, radar, and pulse systems, and circuits requiring ramp waveforms, require that the technician become familiar with their features. It is not necessary in this chapter to deal in-depth with these various waveforms, but it is important to at least introduce you to them.

The most important waveforms include the **square wave,** the **rectangular wave,** and the **sawtooth** (ramp-shaped) **wave.** Refer to Figure 12–20 and note the comparative characteristic of each waveform with the sine wave in terms of period, wave shape, and peak-to-peak values.

How Nonsinusoidal Waveshapes are Formed

It is interesting to note that a pure ac sine wave is comprised of one single frequency. Nonsinusoidal signal waveforms can be mathematically shown

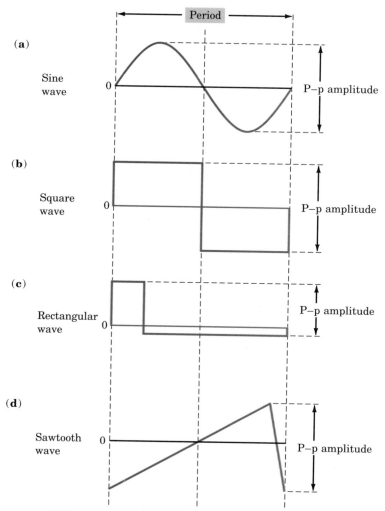

FIGURE 12–20   Comparison of sinusoidal and nonsinusoidal waveforms in terms of waveshape and peak-to-peak values

to be composed of a "fundamental" frequency sine wave *plus* a number of multiples of that frequency, called "harmonics." In other words, if the waveform is not sinusoidal, it indicates the presence of harmonics. For example, the square wave is composed of a fundamental sine-wave frequency and a large number of odd harmonic signals. The greater the number of odd harmonic signals that are added, or present, the closer to an ideal square wave the resultant complex waveform becomes. See Figure 12–21. Odd harmonics are those that represent frequencies of 3 times, 5 times, 7 times, and so on, the fundamental frequency. For example, if the fundamental frequency is 100 Hz, then the 3rd harmonic is 300 Hz, the 5th harmonic is 500 Hz, and so on.

Other nonsinusoidal waveforms may be created by the addition of even harmonic signals, or a mixture of even and odd harmonics, Figure 12–22. Even harmonics are even multiples of the fundamental frequency involved.

FIGURE 12–21   Graphic illustration of forming of a square wave

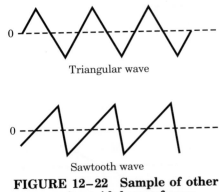

FIGURE 12–22   Sample of other nonsinusoidal waveforms

That is, 200 Hz is the 2nd harmonic of a fundamental frequency of 100 Hz; 400 Hz is the 4th harmonic frequency, and so on.

Important Parameters for Nonsinusoidal Waveforms

As you have already seen, the period of a nonsinusoidal waveform is determined in the same way the period is determined for a sine-wave signal; that is, the time it takes for one complete cycle of the signal, Figure 12–20. You have also seen that the peak-to-peak amplitude of nonsinusoidal signals is determined in similar fashion as for the sine-wave signal. That is, from the top-most point of the waveform to the bottom-most point on the waveform, Figure 12–20. Let's look at some more parameters of interest for these nonsinusoidal waveforms.

Pulse-Width, Duty Cycle, and Average Values

Notice how the **pulse width** is identified in the series of cycles of rectangular waves shown in Figure 12–23. By comparing the pulse width to the time for one period, we can determine the **duty cycle.** That is, the *duty cycle equals the pulse width divided by the time for one period.* Since duty cycle is

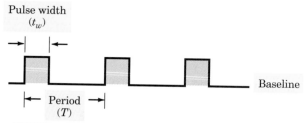

FIGURE 12–23 Pulse width and period of a pulse-type waveform

generally expressed in percentage, we then take this decimal number times 100% to find the percentage duty cycle:

Formula 12–6	$\% \text{ duty cycle} = \dfrac{t_w \text{ (Pulse width)}}{T \text{ (Time for 1 period)}} \times 100\%$

 pulse width time value, $\boxed{\div}$, time for one cycle, $\boxed{\times}$, 100, $\boxed{=}$

EXAMPLE

If the time for one period of this repetitive pulse signal is 20 μs and the pulse width is 4 μs, then the percentage duty cycle is:

$$\% \text{ duty cycle} = \frac{4 \text{ μs}}{20 \text{ μs}} \times 100\% = 0.2 \times 100 = 20\% \text{ duty cycle}$$

PRACTICE PROBLEMS V

What is the duty cycle of a repetitive pulse waveform if $T = 15$ μs and the $t_w = 1$ μs?

Finding Average Value for a Nonsinusoidal Waveform

In a number of situations in electronic circuits, these nonsinusoidal waveforms are not symmetrically balanced above and below the zero reference line. Many times a dc component is purposely added to "offset" the waveform, for operational purposes. To find the average value of voltage (V_{avg}) of such waveforms, it is necessary to understand where the "baseline" of the waveform is considered to be, and be able to determine the duty cycle and amplitude of the signal. The basic formula for finding V_{avg} is:

Formula 12–7	$V_{avg} = \text{Baseline value} + (\text{duty cycle} \times \text{p-p amplitude})$

 baseline value, $\boxed{+}$, $\boxed{(}$, duty cycle decimal value, $\boxed{\times}$, p-p value, $\boxed{)}$, $\boxed{=}$

EXAMPLES

In Figure 12–24a, you can see that the baseline has been defined as being at the zero volts level. Also, you can observe that the peak-to-peak amplitude of the waveform is 5 volts. The duty cycle can be calculated as being 0.1 (or 10%). Thus:

$$V_{avg} = 0 + (0.1 \times 5) = 0 + 0.5 = 0.5 \text{ V}$$

In Figure 12–24b, the conditions are changed so that the baseline is now at +2 V. What happens to V_{avg} in this case?

$$V_{avg} = 2 + (0.1 \times 5) = 2 + (0.1 \times 5) = 2 + 0.5 = 2.5 \text{ V}$$

PRACTICE PROBLEMS VI

1. Refer to Figure 12–24c and determine the V_{avg} for these conditions.

2. What would V_{avg} be if the waveform were symmetrical above and below the reference point (that is, if the baseline were −3 V, in this case)?

(a)

(b)

(c)

FIGURE 12–24   Finding V_{avg} for a
rectangular waveform

Later in your studies, you will be studying more details about pulse-type signals. Topics such as waveshaping and the waveform characteristics of ramp, slope, and so on will be studied at the time you can apply them to circuits you are studying. At this time, we simply wanted to introduce you to the fact that there are ac signals other than sine waves, and familiarize you with some of their basic characteristics.

SUMMARY

▶ Alternating current (ac) varies continually in magnitude and periodically reverses in direction (polarity).

▶ Whereas ac is bidirectional, dc is unidirectional. Also, ac alternates in polarity, but dc has only one polarity. Furthermore, ac can be transformed or stepped up or down via transformer action, whereas dc cannot.

▶ The sine wave is related to angular motion in degrees, and any given point's value throughout the sine wave relates to the trigonometric sine function. Maximum values occur at 90° and 270° (−90°) points of the sine wave. Zero values occur at 0°, 180°, and 360° points of each cycle. (Assuming the wave is started at the zero level.)

▶ A sine-wave voltage is generated by relative motion between conductor(s) and a magnetic field, assuming the conductor(s) links the flux lines via a rotating motion. The ac generator output is typically delivered to the load via slip rings and brushes.

▶ Quantities having magnitude and direction are frequently represented by lines or arrows called vectors. A vectors' length and relative direction illustrate the quantity's value and angle.

▶ When representing angles relative to ac quantities, the zero reference point is horizontal and to the right. Angles increase positively from that reference plane in a counterclockwise (CCW) direction.

▶ When representing angles relative to the coordinate system, angles between 0° and 90° are in quadrant 1; angles between 90° and 180° are in quadrant 2; angles between 180° and 270° are in quadrant 3; and angles between 270° and 360° are in quadrant 4. These positive angles apply with CCW rotation of the phasor.

▶ A cycle represents one complete set of recurring events or values. In ac, the completion of a positive alternation (half-cycle) and a negative alternation (or vice-versa) completes a cycle.

▶ A period is the time it takes for one cycle. It is calculated by the formula $T = 1/f$.

▶ Frequency is the number of cycles per second, and Hertz (Hz) indicates cycles per second. Frequency is calculated using the formula $f = 1/T$, if the time for one cycle (period) is known. Frequency and period are inversely related.

▶ Referring to a sine wave, rate of change indicates how much amplitude changes for a given amount of time. This rate of change is maximum near the zero level points and minimum near the maximum level points.

▶ In purely resistive ac circuits, voltage and current are in-phase with each other.

▶ When ac quantities are out-of-phase, they do not reach the same relative levels at the same time. A generalization is the waveform that reaches maximum positive level first is leading the other quantity in-phase.

▶ Phase angle is the difference in angle (time difference) between *two equal-frequency* sine waves. If sine waves A and B are in-phase, there is zero phase angle between them. If sine wave A reaches maximum positive level 1/4 cycle before sine wave B, sine wave A is leading sine wave B by 90°. Thus, the two signals are out-of-phase by that amount. If sine wave A reaches maximum positive level 1/8 cycle prior to sine wave B, sine wave A is

1/8 cycle prior to sine wave B, sine wave A is leading sine wave B in-phase by 45°.

▶ Average value of ac voltage or current is the average level under one alternation's waveform. Average value is equal to 0.637 times peak value. (NOTE: Average value of a symmetrical waveform over a complete cycle equals zero.)

▶ Effective value (rms) of ac voltage or current is the value that causes the same heating effect as an equal value of dc voltage or current. Effective value is computed as 0.707 times peak value, or 0.3535 times peak-to-peak value, or 1.11 times average value.

▶ Peak value (maximum positive or negative value) of ac voltage or current is 1.414 times the effective value. It can also be calculated as being half the peak-to-peak value.

▶ Peak-to-peak value is the total difference in voltage or current between the positive and negative maximum values. Peak-to-peak value is calculated as two times peak value, or 2.828 times effective value.

▶ In ac resistive circuits, average power is calculated with the effective values (rms) of I and V.

▶ Several nonsinusoidal waveforms are the square wave, rectangular wave, and sawtooth wave.

▶ The period, frequency, and peak-to-peak amplitude values of nonsinusoidal waveforms (square, triangular, rectangular, and sawtooth) are determined in the same fashion as the period, frequency, and peak-to-peak amplitude of a sine wave are identified.

▶ The duty cycle of repetitive pulse-type waveforms is found by dividing the pulse width by the time for one period. This is typically multiplied by 100 to find the percentage duty cycle.

▶ The average voltage value (V_{avg}) of pulse-type waveforms is determined by adding the baseline voltage to the product of the duty cycle and the peak-to-peak amplitude of the waveform.

FORMULAS AND CALCULATOR SEQUENCES SUMMARY

FORMULA 12-1 T (s) $= \dfrac{1}{f}$ (Hz)

frequency in hertz, $\boxed{1/x}$

FORMULA 12-2 f (Hz) $= \dfrac{1}{T}$ (s)

time for one cycle in seconds, $\boxed{1/x}$

FORMULA 12-3 Effective value (rms) $= 0.707 \times$ Peak value

0.707, $\boxed{\times}$, peak value of voltage, $\boxed{=}$

FORMULA 12-4 Average value $= 0.637 \times$ Peak value

0.637, $\boxed{\times}$, peak value of voltage, $\boxed{=}$

FORMULA 12–5 $v = V_p \sin \theta$

V peak value, $\boxed{\times}$, angle value, $\boxed{\text{SIN}}$, $\boxed{=}$

FORMULA 12–6 $\% \text{ duty cycle} = \dfrac{t_w \text{ (pulse width)}}{T \text{ (time for one period)}} \times 100$

pulse width time value, $\boxed{\div}$, time for one cycle, $\boxed{\times}$, 100, $\boxed{=}$

FORMULA 12–7 $V_{avg} = \text{Baseline value} + (\text{duty cycle} \times \text{p-p amplitude})$

baseline value, $\boxed{+}$, $\boxed{(}$, duty cycle decimal value, $\boxed{\times}$, p-p value, $\boxed{)}$, $\boxed{=}$

REVIEW QUESTIONS

1. What are two differences between ac and dc?

2. What is the period of a 2-kHz voltage?

3. What is the frequency of an ac having a period of 0.2 μsec?

4. How much time does one alternation of 60-Hertz ac require?

5. Draw an illustration showing one cycle of a 10-V (rms), 1 MHz ac signal. Identify and appropriately label the positive alternation; the period; the peak value; the peak-to-peak value; the effective value; and points on the sine wave representing angles of 30°, 90°, and 215°.

6. The peak value of a sine wave is 169.73 volts. What is the rms value?

7. What is the frequency of a signal having one-fourth the period of a signal whose T equals 0.01 seconds?

8. Draw a phasor diagram showing voltage A being approximately twice as great as voltage B and leading B by about 45°. (Use B as reference.)

9. What is the peak-to-peak voltage across a 10-kΩ resistor that dissipates 50 mW?

10. Draw an illustration showing one cycle of a 30 V p-p square wave. Assume the signal has a frequency of 100-Hz. Label the illustration indicating the period and the number of seconds in the positive alternation.

11. If the rms value of an ac voltage doubles, the peak-to-peak value must _____.

12. The ac effective voltage across a given resistor doubles. What happens to the power dissipated by the resistor?

13. The period of an ac signal triples. What must have happened to the frequency of the signal?

14. The rms value of a given ac voltage is 75 volts. What is the peak value? What is the peak-to-peak value?

15. What time is needed for one alternation of a 500-Hz ac signal?

16. Refer to Figure 12–25 and answer the following:
    a. Which waveform illustrates the signal with the longest period?
    b. What is the peak-to-peak voltage value of waveform B?
    c. What is the effective value of voltage waveform A?
    d. If the period of waveform A equals 0.002 milliseconds, what is the frequency represented by waveform B?
    e. If the peak-to-peak value of waveform A changes to 60 volts, what is the new effective voltage value?

FIGURE 12–25

CHALLENGE QUESTIONS

1. Explain the terms fundamental and harmonic.
2. Define the term periodic wave.
3. Draw a graphic illustration of one and one-half cycles of a rectangular wave. Assume the amplitude is 10 V, the baseline is +3 V, the period is 25 ms, and the duty cycle is 25%. Label all appropriate parameters on the drawing.
4. Determine the V_{avg} for the signal described in question 3.

TEST YOURSELF

1. Draw one cycle of a sine wave and a cosine wave on the same "baseline." Assume each maximum value is 10 units. Appropriately label the diagram to show the following information:
   a. Angles of 0, 45, 90, 135, 180, 225, 270, 315, and 360 degrees, marked on the x-axis.
   b. Amplitude values of the sine wave at 45 and 90 degrees, respectively. (Use a calculator to determine values.)
   c. Amplitude values of the cosine wave at 0, 45, and 90 degrees, respectively. (Use a calculator to determine values.)
   d. Indicate which wave is leading in phase and by how much.

2. One alternation of a signal takes 500 μs. What is its frequency? What is the time of two periods? (Use a calculator to find answers.)

3. Draw the diagram of a three-resistor series-parallel circuit having a 10-kΩ resistor (labeled R_1) in series with the source, and two 100-kΩ resistors (labeled R_2 and R_3) in parallel with each other and in series with R_1. Assume a 120-V ac source. Calculate the following parameters, and appropriately label them on the diagram.

   a. V_{R_1}
   b. I_{R_1}
   c. P_{R_1}
   d. I_{R_2}
   e. V_{R_3}
   f. θ between V applied and I total

The Oscilloscope

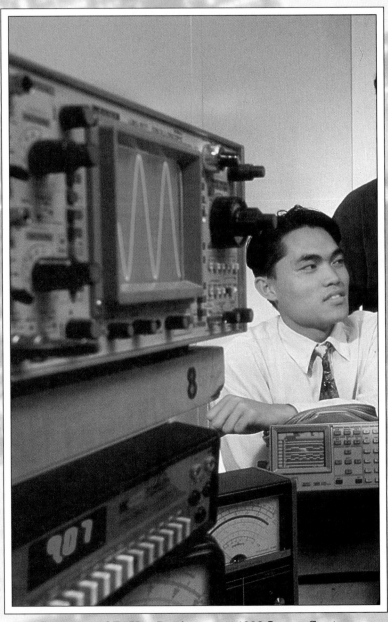

13

KEY TERMS

Attenuator

Cathode-ray tube (CRT)

Gain control(s)

Horizontal amplifier

Oscilloscope

Position control(s)

Sweep circuit(s)

Sweep frequency control(s)

Synchronized

Synchronization control(s)

Vertical volts/div control(s)

Vertical amplifier

OUTLINE

Background information

Key sections of the scope

The CRT

Combining horizontal and vertical signals to view a waveform

Measuring voltage and current with the scope

Using the scope for phase comparisons

Measuring frequency with the scope

CHAPTER PREVIEW

The **oscilloscope,** frequently called a scope, is without doubt one of the most versatile tools available to technicians and engineers. It can display waveforms, measure voltages and currents, measure and compare phase relationships, and measure and compare frequencies.

In your studies, you have been introduced to other measuring instruments and to some important ac quantities. This chapter will increase your knowledge about measuring techniques. This chapter will also reinforce your knowledge about the important ac quantities you will measure and analyze throughout your remaining studies and in your career as a technician.

The purpose of this chapter is not to discuss the details of circuitry involved in making a scope function, nor to instruct you on subjects that are best learned with actual experience. However, this chapter will, perhaps, inspire you to become familiar with this valuable tool and will be a good background for you to become proficient in using this device.

OBJECTIVES

After studying this chapter, you should be able to:

• List the key parts of the **oscilloscope**

• List precautions when using scopes

• List procedures when measuring voltage with a scope

• List procedures to display and interpret waveforms

• List procedures relating to phase measurement

• List procedures when measuring frequency with a scope

• Use the SIMPLER troubleshooting sequence to solve the Chapter Challenge problem

BACKGROUND INFORMATION

Scopes come in a variety of brands and complexity. Two common categories of scopes include those used for general-purpose applications, and the more complex (and expensive) laboratory-quality scopes, Figure 13–1. Also, the "digital storage" oscilloscope (DSO) is now fairly commonplace, Figure 13–2. These scopes have the ability to store signal waveforms in memory for either immediate or later retrieval and display. This type scope makes possible such applications as waveform comparisons at different times or locations, complex, long-duration and/or short-duration signal-change analysis, and other unique types of signal comparisons and analyses. These are analyses that would be difficult or impossible to perform using standard analog "real time" (what you see is happening right now) scopes. In order to store signals, this scope "digitizes" the analog signals by sampling signal levels

(a)

(b)

FIGURE 13–1   (a) General-purpose oscilloscope, (b) Lab-quality
oscilloscope (Photos courtesy of Tektronix, Inc.)

FIGURE 13–2   A digital storage oscilloscope (DSO)
(Courtesy of B & K Precision)

many times over the duration of the signal waveform. The digital data is then placed in memory for retrieval at the desired time.

Two general areas of commonality between scopes include:

1. All scopes have a cathode-ray tube (CRT) where the displays are viewed.

2. All scopes have controls and circuits that adjust the display to help you analyze voltage, time, and frequency parameters of the signal(s) under test.

Let's take a brief look at these two critical areas related to scopes.

KEY SECTIONS OF THE SCOPE

The key sections of a scope that will be discussed are:

1. Cathode-ray tube

2. Intensity and focus controls

3. Position controls

4. Sweep frequency controls

5. Vertical section

6. Horizontal section

7. Synchronization controls

A typical scope front panel layout is shown in Figure 13–3. Use this figure while you read the following descriptions of the various controls and circuits.

THE CRT

All oscilloscopes have a **cathode-ray tube (CRT),** which has a "screen." The screen end of the CRT is the part that displays visual information that is interpreted by the technician or engineer when using the scope. An "electron gun," which we will explain shortly, shoots an electron stream, or beam, at the inside coating of the CRT screen. The screen, in turn, lights up wherever the electron beam is striking the CRT screen coating.

The Screen

As indicated, the purpose of the screen is to convert the electron beam energy into light. This conversion is accomplished by the impact of the electron beam on the fluorescent screen material. The screen material has two characteristics, which are defined as follows:

1. Luminescence — the property that converts the energy of the electron beam striking the screen's surface into light.

2. Phosphorescence — the property that causes the screen to continue to glow for a short time after the actual impact of the electron beam.

FIGURE 13–3   Typical controls on a dual-trace scope (simplified)

Without going into too much detail, it is interesting to know that the type and nature of the coating material used can provide long persistence time (such as used in radar screens), medium persistence time (such as used in most oscilloscopes), and short persistence time (such as used in televisions).

The Electron Gun

The electron beam, about which we have been talking, is created and controlled by various elements along the length of the cathode-ray tube. Figure 13–4 illustrates these electrodes within the CRT. Refer to the figure as you study the following brief discussion of the function of each of these electrodes.

THE HEATER.   Functions to heat the cathode to a hot enough temperature so that the cathode will emit some electrons from its surface.

THE CATHODE.   A cylindrically shaped element that has a special coating on the side toward the CRT screen. When heated, it emits electrons, or supplies the electrons to be formed into and controlled as an electron stream, or beam.

FIGURE 13–4 Elements in a typical cathode-ray tube (CRT)

THE CONTROL ELECTRODE.   A metal sleeve, open at the cathode end and closed, except for a small aperature, at the end facing the CRT screen. It concentrates the electrons into the shape of a beam, and controls the number of electrons moving toward the screen.

THE ACCELERATING ANODES.   The first (pre-accelerating) anode and second (accelerating) anode work together to perform two functions. One function is to give high velocity to the electron beam as it travels toward the screen. The second function is to combine with the focusing anode to determine the focus characteristics, or fineness of definition, of the beam. NOTE: There is a conductive coating on the inside of the glass CRT tube that extends almost to the screen. This aquadag (coating) is electrically connected to the 2nd accelerating anode, and aids in the acceleration process.

THE FOCUSING ANODE.   A cylindrically shaped electrode with a small opening at either end. It is used to concentrate, or focus, the electron stream so that when it strikes the screen, it is a fine point.

THE DEFLECTION PLATES.   Function to deflect the electron stream in accordance with the signal voltage being observed. Obviously, the vertical deflection plates control the upward and downward movement of the electron beam, and the horizontal deflection plates control the left-to-right, or horizontal, movement of the electron beam.

Now, let's see how these various control electrodes translate to some of the front-panel controls on an oscilloscope.

PRACTICAL NOTES

Caution! One thing you should learn when operating a scope is that it is not good to leave a bright spot at one position on the CRT. The CRT screen material can be burned or damaged if this occurs for any great length of time. Never have the trace bright enough to cause a "halo" effect on the screen.

Position Controls

In Figure 13–4 you can see this CRT has "deflection plates" that control the position and movement of the electron beam coming from the opposite end of the tube and striking the back of the screen. The position where this electron beam strikes the back of the screen (CRT) is controlled by the electrostatic field set up between the plates when there is a difference of potential. (It is worth mentioning that positional control of the electron beam [or stream] in televisions is normally accomplished through *electro-magnetic* fields, rather than *electrostatic* fields used by most oscilloscopes.)

The reactions of the electron beam to these fields are illustrated in Figure 13–5. Notice the electron beam is attracted to the plate(s) having a positive charge and repelled from any deflection plates having a negative charge, or potential. Therefore, the position where the electron beam strikes the screen can be controlled by the voltages present at the deflection plates. Knowing that deflection of the electron beam is toward positive deflection plates and away from negative deflection plates, you can understand the

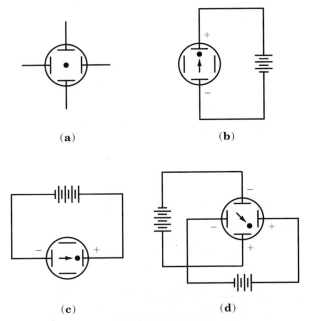

(a) (b)

(c) (d)

FIGURE 13–5   Electron-beam movement caused by dc voltages on deflection plates

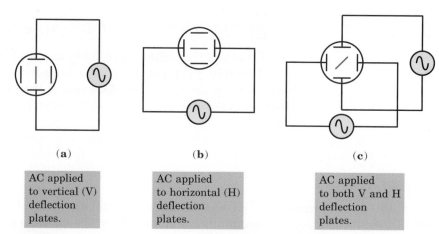

(a)

AC applied to vertical (V) deflection plates.

(b)

AC applied to horizontal (H) deflection plates.

(c)

AC applied to both V and H deflection plates.

FIGURE 13–6   Electron-beam movement caused by ac voltages on deflection plates

position controls simply make the appropriate deflection plates either more positive or more negative, depending on which way we want the electron beam to move on the CRT face.

Look again at Figure 13–3. You can see these front-panel controls that influence both the vertical and horizontal position of the trace. Notice what happens when we apply ac to the deflection plates, Figures 13–6a, b, and c.

Horizontal Sweep Frequency Control(s)

A sawtooth (or ramp-type) waveform applied to the horizontal plates, creates a horizontal trace on the screen, Figure 13–7. The voltage applied to

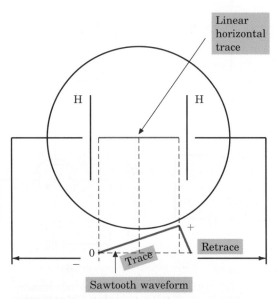

Linear horizontal trace

H   H

0

Trace

Retrace

Sawtooth waveform

FIGURE 13–7   Linear trace caused by sawtooth voltage on horizontal (H) plates

the horizontal plates is sometimes called the horizontal "sweep voltage," because the voltage causes the electron beam to horizontally sweep across the screen, creating a horizontal trace, or line.

The electron beam moves, or traces, at a constant rate (linear speed) across the screen from left to right, then quickly retraces, or "flies back," to its starting point on the left. During the very short retrace (or flyback) time, a blanking signal is applied to the system so that the scope screen will not display the electron beam's movement across the screen (from right to left) during retrace time.

Because the electron beam is moving at a constant speed across the screen during the left-to-right trace time, a horizontal "time base" has been established. In other words, we know how much time it takes for the beam to move from one horizontal point to another on the screen. Because of this, we can measure times and calculate frequencies of signal waveforms that are displayed, as you will see later.

The number of times the beam traces across the screen in a second is determined by the frequency of the sweep (sawtooth) voltage. You see a horizontal line on the screen when this happens at a rapid rate. Two reasons for this are:

1. The CRT screen material has a persistence that causes continual emission of light from the screen for a short time after the electron beam has stopped striking that area.

2. The retinas in our eyes also have a persistence feature. This is why you don't see flicker between frames as a movie film is projected, or flicker between frames in the TV "scanning" process.

The **sweep frequency controls** (horizontal time variable, "VAR", and horizontal sec/div) adjust the frequency output of the horizontal **sweep circuitry** in the oscilloscope. Naturally, this determines the number of times per second the electron beam is traced horizontally across the CRT screen. The horizontal frequency control allows us to view signals of different frequencies.

Having accumulated this information, let's now see how the electron beam is further controlled, amplified, or attenuated to provide meaningful displays on the CRT screen.

Vertical Section

Key elements in the vertical section are the vertical input jack(s) (sometimes called Y input), and the **vertical attenuator** and **vertical amplifier,** with related controls. When the scope is a single-trace scope, there is only one vertical input jack. When the scope is a dual-trace scope, there are two vertical input jacks. (See the jacks in Figure 13–3.)

The vertical attenuator and amplifier circuitry, and associated controls, decrease or increase the amplitude of the signals to be viewed; that is, the signals that are applied to the vertical section via the vertical input jack(s). View Figure 13–3 and note the calibrated **vertical volts/div** and the **variable control(s),** which are used for signal level control. Using these controls allows you to adjust for larger or smaller vertical deflection of the CRT trace with a given vertical input signal amplitude.

Horizontal Section

We have briefly discussed some of the important horizontal deflection elements and controls. One of these controls adjusts the frequency of the internally generated **horizontal** sweep signal.

Other possible elements related to the horizontal trace may include a horizontal **gain control** that changes the length of the horizontal trace line; and a jack (often marked "Ext X" input) that enables inputting a signal from an external source to the horizontal deflection system in lieu of using the internally generated sweep signal.

Synchronization Controls

The **synchronization controls** synchronize the observed signal with the horizontal trace; thus, the waveform appears stationary (assuming the signal has a periodic waveform). The effect is much like a strobe light that provides "stop-action" when timing an automobile, or the effect of making spinning fan blades appear to be standing still, using a light that is blinking at the appropriate rate. Without going into the technical details, suffice it to say that by starting or "triggering" the trace (left-to-right horizontal sweep) of the electron beam in proper time relationship with the signal to be observed on the scope (the signal fed to the vertical deflection plates), a stable waveform is displayed. If the horizontal trace and the vertical system signals are not **synchronized** (i.e., do not start from a given waveform point at a specific time), the waveform is a "moving target" and hard to view.

Some of the jacks and controls typically associated with this synchronizing function are also shown in Figure 13–3, and include:

1. trigger source selector switches (for selecting the synchronizing signal source),

2. "trigger holdoff and level" controls, which help set the appropriate level for the triggering signal to work best.

IN-PROCESS LEARNING CHECK I

1. The part of the oscilloscope producing the visual display is the _____ _____ _____.

2. The scope control that influences the brightness of the display is the _____ control.

3. A waveform is moved up or down on the screen by using the _____ _____ control.

4. A waveform is moved left or right on the screen by using the _____ _____ control.

5. For a signal fed to the scope's vertical input, the controls that adjust the number of cycles seen on the screen are the horizontal _____ and the horizontal _____ control.

6. The control(s) that help keep the waveform from moving or jiggling on the display are associated with the _____ circuitry.

COMBINING HORIZONTAL AND VERTICAL SIGNALS TO VIEW A WAVEFORM

Having an electron beam cause a horizontal line across the CRT screen by applying voltage(s) to the horizontal deflection plates alone is not useful for displaying and analyzing waveforms. Causing an electron beam to vertically move by applying voltage(s) to the vertical deflection plates alone also is not useful for displaying and analyzing waveforms. However, useful waveforms appear when appropriate voltages are simultaneously applied to both the horizontal and vertical deflection plates.

Look at Figure 13–8. Notice how a sinusoidal waveform is displayed when a sine-wave voltage is applied to the vertical deflection plates, and when a sawtooth voltage of the same frequency (from internal sweep circuitry) is simultaneously applied to the horizontal deflection plates of the CRT. In this case, the frequency is 60 Hertz.

Again refer to Figure 13–8 and study the following discussion. Look at the conditions at times t_0, t_1, t_2, t_3, and t_4.

1. The sawtooth voltage is applied to the horizontal deflection plates so that at the beginning of the sawtooth, the electron beam is at the left side of the screen (viewed by the person looking at the front of the scope). Time is t_0.

2. As the sawtooth voltage increases in a positive direction, the right-hand deflection plate becomes increasingly positive. Thus, the beam moves from the screen's left side toward its right side. Time is t_1.

3. The 60-Hz sine-wave voltage is applied to the vertical deflection plates. At time zero (t_0) the beam is horizontally on the screen's left side. Since no vertical voltage is at t_0 of the sine wave, a spot appears on the screen's left side.

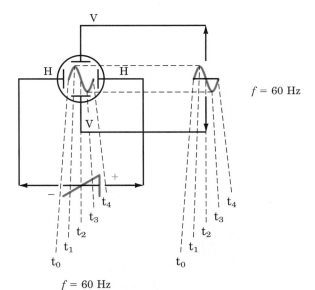

$f = 60$ Hz

FIGURE 13–8   Waveform with sawtooth on horizontal (H) plates and sine wave on vertical (V) plates

4. As the horizontal sawtooth signal causes the beam to begin its horizontal trace to the right, the voltage on the vertical deflection plates causes the beam to simultaneously move to its maximum upward position. Time is t_1.

5. By the time the vertical signal has decreased from maximum positive value back to zero (180° into the sine wave), the horizontal signal has caused the beam to arrive at the center of the screen. In other words, the positive alternation of the sine wave has been traced and displayed on the screen at t_2.

6. The same rationale is used when tracing the negative alternation's waveform. That is, as the beam continues its horizontal trek across the screen, the signal applied to the vertical deflection plates is causing the beam to be sequentially moved to the maximum downward position, then back to the zero voltage position, vertically (center of the screen vertically). During this sequence, the negative alternation of the sine wave has been displayed.

7. Throughout the cycle, one cycle of the sawtooth waveform applied to the horizontal plates has caused one horizontal trace. At the same time, the electron beam has been moved up and down, corresponding to the voltage throughout one sine-wave signal cycle applied to the vertical plates. The result is a sine-wave waveform viewed on the CRT.

8. Because these events have occurred 60 times in one second, our eyes perceive a single sine-wave waveform that traces over itself 60 times in one second.

EXAMPLE

A horizontal sweep frequency of 60 Hz and a vertical signal of 120 Hz result in two cycles of vertical deflections occurring at the same period that one sweep occurred. The display shows two full cycles of a sine wave, Figure 13−9.

FIGURE 13−9   Result of two cycles of sine-wave signal per one horizontal sweep

PRACTICE PROBLEMS I

Assume the vertical frequency is four times the horizontal frequency. What does the waveform look like?

By looking at the electron beam position at a number of instantaneous values of the applied horizontal and vertical signals, you can determine what type of waveform will appear on the scope screen, Figure 13–10. These concepts help interpret what we see on the CRT screen. Later you will see how these concepts help us measure frequency and phase of ac signals.

MEASURING VOLTAGE AND CURRENT WITH THE SCOPE

General Information

Both dc and ac voltage measurements can be made with oscilloscopes. It is useful to know how to make these measurements and to interpret waveforms in terms of voltage amplitudes. We will discuss two common methods used to measure voltages with the scope. One method is to use a "built-in" calibration voltage to set up the scope for measuring voltages. This calibrating voltage method is generally found on older oscilloscopes; however, it is still used enough for us to take time to briefly discuss and illustrate the basic technique.

A second method is to interpret the scope display in reference to the "volts/division" control setting when the "V VAR" (variable) control is set at the "Cal" position. This technique is presently the more commonly used method.

As indicated earlier, learning to use the scope in a practical way *only comes through "hands-on" experience*. The descriptions in this chapter provide only conceptual knowledge. Therefore, you will have to expand on these concepts by actual laboratory experiences—be sure you do!

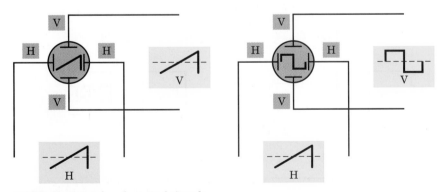

NOTE: Horizontal and vertical signals
are the same frequency.

FIGURE 13–10   Illustration of how different waveforms are traced

Scope screen

**FIGURE 13–11   Typical square-wave
calibration voltage**

Calibration Voltage Approach

Many scopes have built-in calibrating voltage sources used to set up the scope for interpretation of voltage levels. Typically, the calibrating voltage output is 0.5 V or 1 V peak-to-peak value. Also, this calibrating voltage is often a square wave, Figure 13– 11.

   Let's assume a 1-volt peak-to-peak calibrating voltage and see how calibration enables us to measure voltages. Refer to Figure 13–12, as you read the following discussion.

1. With *no* signals connected to the vertical input jack, the horizontal sweep and gain controls, vertical and horizontal positioning controls, and "sync" controls are adjusted to produce a horizontal line trace across the screen, that is centered both vertically and horizontally on the screen, Figure 13–12a.

2. A "jumper" wire connects the calibrating voltage output terminal on the scope to the vertical input jack, Figure 13–12b.

3. Horizontal frequency controls are adjusted so one cycle of the calibration square wave is displayed on the screen, Figure 13–12c.

4. Vertical "variable" level control is set to the "Cal" position. The volts/ div selector switch is then adjusted so the waveform is displayed with

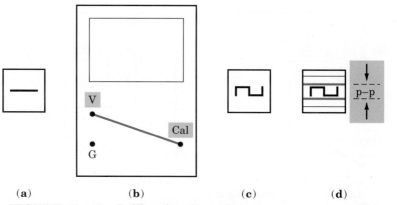

(a)   (b)   (c)   (d)

**FIGURE 13–12   Calibrating for voltage measurement with
calibration signal**

an "easy-to-read" amount of deflection on the screen calibrations (graticule). For example, one major division of vertical deflection appears, Figure 13–12d.

5. When this is done, each vertical square, or division of deflection caused by a voltage applied to the vertical input jack represents one volt (1 V) of input voltage.

6. The scope settings are now adjusted for one division to equal 1 V (peak-to-peak). Let's see how measurements are made.

AC Voltage Measurement

After removing the jumper wire from the vertical input jack (i.e., removing the calibrating voltage signal from the vertical input), there is a vertically and horizontally centered trace across the scope.

Let's now assume we want to measure a small ac (sine-wave) voltage that is connected between the vertical input jack and ground of the scope. Let's further assume a sine wave appears on the face of the scope. The sine wave is four divisions high, from top to bottom (peak-to-peak). What is the peak-to-peak value of the voltage being measured?

If you said four volts peak-to-peak, you are correct. Since each division in height represents one volt (peak-to-peak), then a four-divisions-high voltage must be four volts peak-to-peak.

EXAMPLE

What is the rms value of the voltage just measured? (Here is where you apply your knowledge gained from the last chapter!)

1. If the voltage is 4-V peak-to-peak, then it must be 2-V peak.

2. If it is 2-V peak, then its rms value must be 0.707 × 2 V, or 1.414 volts. You can also use the peak-to-peak value × 0.3535 and get the same results.

PRACTICE PROBLEMS II

1. Suppose the voltage in our example had measured 10 volts peak-to-peak on the scope. What would the rms value be? What would be the peak value of this voltage?

2. Assume the scope has been calibrated or set up so that a 1-volt peak-to-peak signal causes *one-half division vertical deflection*. Refer to the scope pattern in Figure 13–13 and determine the following parameters:
   a. Peak-to-peak value of the ac signal displayed
   b. RMS value of the ac signal displayed

FIGURE 13–13

FIGURE 13–14   Direct current (dc) voltage measurement

In summary, when measuring ac voltage with a scope, what you see on the screen is deflection caused by the peak-to-peak value of the voltage being measured. You must then use the appropriate calculation factors to interpret the waveform and determine the effective rms or other values of interest.

Refer to Figure 13–14 as you read the following discussion on dc voltage measurement.

DC Voltage Measurement

First, the "dc-gnd-ac" input mode switch is set to dc. For simplicity, assume the scope is set up so that a 1-V peak-to-peak signal causes a one-division-high waveform on the screen. With no voltage applied between the vertical input jack and ground, the scope shows a centered horizontal trace.

Assume when you connect the dc voltage to be measured to the vertical input, the trace jumps up one and one-half divisions from its starting point. What is the dc voltage being measured? If you said 1.5 volts dc, you are right! Furthermore, you might have indicated the end of the source connected to the vertical input jack was positive at the jack side and negative at the ground connection side, because the trace moved up. Obviously, if you reversed the connections, the trace would move down from center by one and one-half divisions.

PRACTICE PROBLEMS III

1. Assume the scope is set up so that a 1-V peak-to-peak signal voltage causes two vertical divisions of deflection. Furthermore, assume that with 0 V applied to the vertical input jack, the horizontal trace line is centered on the screen. Refer to Figure 13–15 and determine the amount and polarity (with respect to ground) of the dc voltage being measured.

2. If 1-V peak-to-peak voltage causes 2.5 divisions of deflection, and the scope pattern looks like that in Figure 13–15, what is the value of the dc voltage being measured, and what is its polarity with respect to ground?

Display with input voltage

0 V Display

FIGURE 13–15

Volts-per-Division Control Setting Approach

By making sure that the "V-Variable" control is set in the "Cal" position, you can directly interpret voltage readings of scope patterns by reference to the vertical volts/div control setting. In Figure 13–16, you can see that our example control can be set at eight different positions. The beauty in this is that you can select the position that makes the signal display the most readable amplitude for interpreting the signal being measured. The signal voltage value equals the volts/div setting times the number of divisions of deflection.

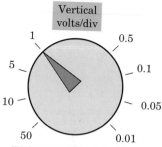

FIGURE 13–16   Using the calibrated volts/div switch setting to determine voltage

EXAMPLE

Observe in Figure 13–17 that the volts/div control is set at the 0.1 position. With the "V-Variable" control in the "Cal" position, each tenth-volt amplitude of the signal will cause one division displacement of the trace on the scope display. In our sample then, there are six divisions of deflection caused by the peak-to-peak amplitude of the signal. The interpretation is that the signal being measured must be 0.6 volts peak-to-peak. The rms value is 0.3535×0.6, or 0.21 volts. If the tested voltage were a dc voltage of 0.6 volts, the trace would have jumped upward or downward by six divisions, depending on the polarity applied to the scope input jacks. This would cause the trace to be "off-screen." You would have to move the zero-volts reference trace off center by two or more vertical divisions in order to see the 0.6-volt dc deflection.

 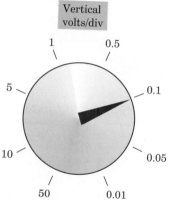

FIGURE 13–17   Example of volts/div. voltage measurement approach

An alternative approach would be to set the **volts/div control** to the next higher setting (0.5 V/div). Then, the display would only jump one and one-fifth squares, which would be viewable. Of course, you would interpret the value based on the new volts/div setting.

PRACTICAL NOTES

Once you have calculated the rms value for a given measurement, you don't need to recalculate to find the rms value for other measurements if you DO NOT change the vertical control settings. In the preceding example, the scope was set up so that 0.6 volts peak-to-peak, or 0.21 volts rms, caused six divisions of trace deflection. If the settings are not changed, you know that if you had three divisions of deflection, the rms must be one-half of 0.21 volts rms, or 0.105 volts rms. In other words, you would know that since 0.21 volts rms caused six divisions of deflection, then each division must equal 0.21/6 volts rms, 0.035 volts rms per division.

PRACTICE PROBLEMS IV

1. If the vertical volts/div control in Figure 13–17 were set on the "5" position and the scope display were the same as shown, what would be the peak-to-peak and rms values of the signal being checked?

2. What is the rms volts-per-division sensitivity setup used in question 1?

3. If you were going to measure a signal which you knew was about twice as great in amplitude as the one measured in question 1, which setting of the vertical volts/div control would you use?

FIGURE 13–18   An example of a "times ten" test probe (Courtesy of B & K Precision)

Attenuator Test Probes

Another technique often used with many scopes is a test probe that attenuates the signal by a given factor, for example ten times attenuation, Figure
13–18. Thus, you multiply by ten times what you are reading on the scope
to find the actual value of voltage applied to the probe. This probe often has
another purpose—it prevents the circuit being tested from being disturbed
by the scope's input circuitry, or test cable. Also, the cable is a shielded
cable, which minimizes stray signal pickup.

PRACTICAL NOTES

There are numerous types of test probes and associated cables available
for use with oscilloscopes. Some are switchable and have one switch position
for measuring low frequencies and another switch position for measuring
high frequencies. Many have switchable attenuation; for example, X1
(times 1) and X10 (times 10). Sometimes you will see these written as 1:1
and 10:1 modes. As indicated earlier, the X10 mode indicates that the
signal has been attenuated ten times; therefore, you would multiply your
scope reading by 10 to find the actual signal level input to the test probe.
There are also "current probes" available for use with scopes; however, the
voltage probes are the more common types found in practice. As you proceed
in your studies, you will be learning other parameters related to scope input
circuit characteristics and matching of test probes/cables to the circuit conditions.

PRACTICE PROBLEMS V

1. Refer to the waveform shown in Figure 13–19 and determine the
   peak-to-peak and rms values of the signal. Assume that the volts/
   div setting of the vertical input signal controls is set at 0.5 V/div
   and that you are using a 1:1-type test probe.

2. Assume that the same waveform pattern is showing on the scope
   as shown in Figure 13–19; however, this time the volts/div setting is 100 mV and you are using a 10:1 test probe. What are the
   signal's peak-to-peak and rms voltage values?

FIGURE 13–19

Summary of Voltage Measurement

Two basic techniques can be used. One method is to set up the scope controls
using a calibrating voltage (either from an internal scope source or from a
known-value external source). A second, more common technique is to use
the scope's designed "volts/div" controls to obtain a known sensitivity to
signal inputs.

Measuring Voltage to Find Current

Current can be calculated by the Ohm's law technique you learned in previous chapters. If you use the scope to determine the rms voltage across a 1-kΩ resistor, for example, you automatically know the number of volts measured equals the number of milliamperes through the 1-kΩ resistor. That is, $I = V/R$, and V divided by kΩ equals mA. Obviously, Ohm's law applies to any value resistor. The 1-kΩ value is merely convenient to use, where possible.

USING THE SCOPE FOR PHASE COMPARISONS

In addition to exhibiting waveforms and measuring voltage (or current), the scope can compare the phase of two signals. Two methods for doing this are:

1. An overlaying technique where the two waveforms are visually superimposed on each other. This is accomplished by a dual-trace scope, or by an electronic switching device. Therefore, the displacement of the two signals' positive peaks, or the points where they cross the zero reference axis, determines the phase difference.

EXAMPLE

Adjust the horizontal sweep frequency so exactly one cycle of the signal applied to the A (or #1) input covers the screen. (NOTE: ten divisions are typical).

The second signal is applied to the scope B (or #2) input. Since the screen is divided into ten divisions, and ten divisions represent 360 degrees, then each division represents 36 degrees. By noting the amount of offset between the two waveform traces and knowing that each division difference equals 36 degrees, you can easily interpret the phase difference between the two signals.

PRACTICE PROBLEMS VI

Suppose the horizontal frequency in our example were adjusted so there were exactly 2 cycles across the 10 divisions of the screen. How many degrees would each major division on the screen represent?

2. Another method used to compare the phase of two signals is the **Lissajous** (pronounced liz-a-ju) pattern(s) technique. With this method, one signal is applied to the horizontal plates and the other signal is applied to the vertical plates. The resulting waveform (known as a Lissajous pattern) enables the technician to determine the phase differential between the two signals. (NOTE: This technique is not a frequently used method for phase measurement.) For more information regarding Lissajous patterns, see Appendix G.

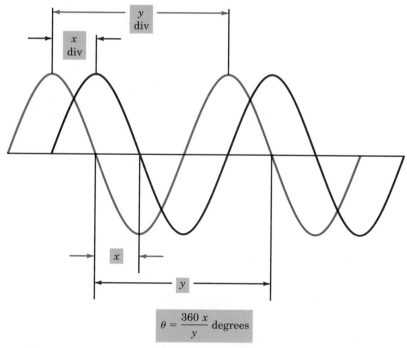

$$\theta = \frac{360\,x}{y} \text{ degrees}$$

x and y are measured in divisions from the scope graticule.

FIGURE 13–20   Overlay technique for phase measurement

The Overlaying Technique

If a dual-trace scope is used:

1. Two horizontal traces are obtained on the scope. They are then super-imposed at the center of the screen by using the positioning controls, as appropriate.

2. Signal #1 is applied to one channel's vertical input jack. Scope sweep frequency is adjusted for one sine wave. Vertical level controls are adjusted to obtain a convenient size waveform for viewing purposes.

3. Signal #2 is connected to the other channel's vertical input jack. Again, appropriate adjustments are made to appropriately display signal #2. The signals should be adjusted to the same vertical size and should be the same frequency.

4. Look at Figure 13–20 to see one method to interpret the display in terms of phase difference between the signals.

MEASURING FREQUENCY WITH THE SCOPE

Today's technology has given us devices called digital frequency counters that are excellent for making frequency measurements. However, there are still many times when you have a scope connected to the circuit under test, and you do not want or need to pull out another piece of test equipment when you already have a pattern on the scope that you can interpret for frequency measurement.

Let's take a quick look at how this is done. You have already learned that the sweep frequency controls allow us to have a "linear horizontal time base sweep" whose time per division we can easily determine. You have learned in an earlier chapter that a signal's period (time for one cycle) and its frequency are inversely proportional. That is $f = 1/T$ and $T = 1/f$. Add your calculator's 1/x (reciprocal) function and you have all the tools you need to find frequency quickly from a scope pattern.

EXAMPLE

Observe Figure 13–21. The sweep frequency control is set on 5 ms/div. This means the horizontal trace takes 5 milliseconds for each division of horizontal movement across the scope screen. With this knowledge, you can see that the signal being looked at is taking about 50 milliseconds for one cycle, or the period equals 50 ms. Using your calculator $\boxed{1/x}$ function and inputting 50, $\boxed{\text{EE}}$ (or $\boxed{\text{EXP}}$), $\boxed{+/-}$, 3, then using the $\boxed{1/x}$ function key, the answer is shown as 20. This means the frequency is 20 Hz. Of course, you probably could have easily interpreted in your head that 50 milliseconds = 0.05 seconds and simply used the reciprocal function on your calculator for 0.05 to get to the answer of 20.

PRACTICE PROBLEMS VII

1. Assume the s/div control is set on the 0.1 s/div setting. Also assume that the scope pattern shows exactly 4 cycles of the signal to be measured across its 10 horizontal divisions.
   a. What is the period of the signal being measured?
   b. What is the frequency of the signal being measured?

2. Assume the s/div control is set on the 50 μs/div setting. This time, the pattern shows exactly 10 cycles of the signal under test.
   a. What is the signal's period?
   b. What is the signal's frequency?

(Horizontal seconds/divison set at 5ms)

FIGURE 13–21 Example of frequency measurement

Final Comments About the Oscilloscope

Certainly, reading this chapter has not made you a professional oscilloscope user! We only hope that by briefly discussing the remarkable versatility of the oscilloscope, you will desire to become knowledgeable and practiced in applying this instrument throughout your studies and career. The ability of the scope to display waveforms so that increments of time are measured and interpreted, and its ability to measure relative amplitudes makes this instrument one of the technician's most valuable assets. Learn to use it!

Figure 13–22 shows a very simplified block diagram of the elemental parts of a scope. As you can see, the vertical input signal is processed through the vertical amplifier/attenuator system prior to being applied to the vertical deflection plates of the CRT. The processing can either be in the form of amplification or attenuation of the signal to be displayed by the scope. Generally, the signal applied to the vertical input system is the signal you are trying to analyze or check.

The horizontal amplifier system has two possible inputs. The one most-often used is the input coming from the horizontal timebase generator (called the sweep system in our diagram). The output of this horizontal amplifier system is then fed to the horizontal deflection plates of the CRT. This produces the horizontal trace on the scope. Sometimes, it is desired to use a horizontal signal source other than the internal sweep generator circuitry. In that case, the horizontal signal is input to the horizontal amplifier system via the H input jack (sometimes labeled the "x" input jack).

Of course, the dc power supply shown in our diagram supplies the appropriate dc voltages to operate the CRT, scope amplifier circuits, timebase generator circuits, and so on throughout the oscilloscope circuitry.

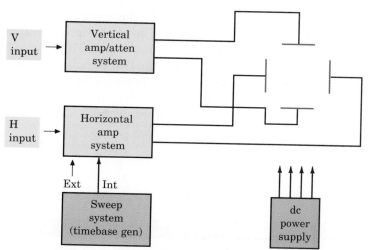

FIGURE 13–22 Simple oscilloscope functional block diagram

PRACTICAL NOTES

Some precautions: When using a scope, the technician should use common sense!

1. Do not leave a "bright spot" on the screen for any length of time.

2. Do not apply signals that exceed the scope's voltage rating.

3. Do not try to make accurate measurements on signals whose frequency is outside the scope's frequency specifications.

4. Be aware that the scope's input circuitry (including the test probe) can cause loading effects on the circuitry under test—use the correct probe for the job!

5. Check the attenuator probe setting (1:1 or 10:1) in order to interpret the scope display properly.

SUMMARY

► The oscilloscope (scope) is an instrument providing a visual presentation of electrical parameters with respect to time and amplitude.

► Intensity and focus controls adjust brightness and clarity of the visual presentation.

► Horizontal controls adjust the frequency of the internally generated horizontal sweep signal; select outside or inside sources for this sweep signal; horizontally position the visual presentation; and adjust the horizontal trace length.

► Vertical controls adjust the vertical amplitude and the vertical position of signals displayed on the screen.

► Synchronization control(s) adjust the start time of the horizontal sweep signal so the visual presentation of a signal is stable, or standing still.

► Scope presentations display the peak-to-peak values of the waveform. Other significant values are interpreted from the waveform by appropriate interpretation factors. For example, rms value = 0.707 × peak, or 0.3535 × peak-to-peak value.

► To measure voltage with an oscilloscope, the appropriate vertical controls are adjusted so the user knows the vertical sensitivity of the scope. That is, how many volts are represented by each division of vertical deflection on the calibrated screen.

► To measure the phase difference between two signals with an oscilloscope, a dual-trace scope is used so each signal is applied to a separate vertical input. Then, by appropriate adjustment, the two signals are superimposed on the visual presentation. Thus, the technician can interpret the phase differential in terms of the two signals.

REVIEW QUESTIONS

1. What is the name and abbreviation of the oscilloscope component that displays signals on its screen?

2. What are four practical uses of the oscilloscope?

3. Why should the intensity control be set to provide the minimum brightness, allowing good readability of the presentation?

4. What control(s) set the sweep frequency, therefore determining the number of cycles of a given signal viewed?

5. To what input terminal(s) is the signal to be viewed normally applied?

6. To what terminal(s) is an external sweep voltage applied?

7. To what terminal(s) is an external synchronizing signal applied?

8. The time/div control and horizontal gain control settings are set so the horizontal sweep line represents 0.1 ms per division. What is the frequency of a signal if one displayed cycle uses four horizontal divisions on the horizontal sweep line?

9. Along what axis is deflection caused by a signal applied to the vertical deflection system?

10. Why should the setting of the scope vertical gain control and vertical input switch remain unchanged after initial calibration, if using the scope for voltage measurement?

11. Refer to Figure 13–23 and answer the following questions:

a. What is the peak-to-peak value of voltage indicated by the waveform display?

b. What is the value of T for the signal displayed?

c. If you want the display to show only one cycle of the signal, which control(s) do you adjust on the scope?

d. What is the time-per-division setting to accomplish a one-cycle display?

e. The vertical sensitivity is set for 5-volts per division (calibrated). What are the rms, peak, and peak-to-peak values of voltage of the signal?

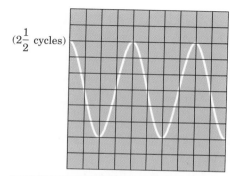

$(2\frac{1}{2}$ cycles$)$

Vertical sensitivity is set at 5 V/div
Horizontal time/cm is set at 0.001 s/div

FIGURE 13–23

CHALLENGE QUESTIONS

Since the best method to familiarize yourself with an oscilloscope is by seeing one, the following questions pertain to observing the oscilloscope you may have available in your training environment. Since the best method to learn how to use an oscilloscope is hands-on experience, we further recommend that you also perform "hands-on" laboratory projects(s) related to using the scope. (See the end-of-chapter "Performance Projects Correlation Chart" for suggestions along this line.)

1. List the precise names of the scope controls observed on your training program's oscilloscope that:
   a. Control the trace focus.
   b. Control the trace intensity.
   c. Control the vertical position of the trace(s).
   d. Control the horizontal position of the trace(s).
   e. Allow for scope adjustment and measurement of vertical input signal amplitude(s).
   f. Provide for measuring vertical input signal frequency(ies).

2. Observe the switches and controls on your training program's oscilloscope, and list the following data:
   a. Maximum possible vertical input signal sensitivity setting. (The smallest volts/div setting)
   b. Minimum possible vertical input signal sensitivity

TEST YOURSELF

1. Refer to Figure 13–24 and for each pattern shown (except for the model waveform pattern) list the oscilloscope controls used to cause the waveform shown to match the model waveform. (NOTE: a = model waveform.)

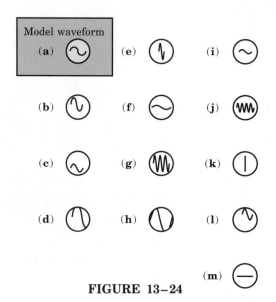

FIGURE 13–24

PERFORMANCE PROJECTS CORRELATION CHART

Suggested performance projects that correlate with topics in this chapter are:

CHAPTER TOPIC	PERFORMANCE PROJECT	PROJECT NUMBER
Key Sections of the Scope	Basic Operation	36
Measuring Voltage and Current with the Scope	Voltage Measurements	37
Using the Scope for Phase Comparisons	Phase Comparisons (Part A)	38
Measuring Frequency with the Scope	Frequency Measurement	39

CHAPTER TROUBLESHOOTING CHALLENGE

Following the SIMPLER troubleshooting sequence, find the problem with this circuit. As you follow the sequence, record your responses to each step on a separate sheet of paper.

SYMPTOMS Gather, verify, and analyze symptom information. (Look at the "Starting Point Information.")

IDENTIFY Identify initial suspect area for the location of the trouble.

MAKE Make a decision about "What type of test to make" and "Where to make it." (To simulate making a decision about each test you want to make, select the desired test from the "TEST" column listing below.)

PERFORM Perform the test. (Look up the result of the test in Appendix C. Use the number in the "RESULTS" column to find the test result in the Appendix.)

LOCATE Locate and define a new "narrower" area in which to continue troubleshooting.

EXAMINE Examine available information and again determine "what type of test" and "where."

REPEAT Repeat the preceding analysis and testing steps until the trouble is found. What would you do to restore this circuit to normal operation? When you have solved the problem, compare your results with those shown on the following pages.

CHALLENGE CIRCUIT 6

STARTING POINT INFORMATION

1. Simplified block diagram
2. Vertical input signal is 300-Hz ac sine-wave
3. Output of sweep signal circuitry is a 150-Hz sawtooth waveform
4. Horizontal sweep = 150 Hz
5. Trace seen on CRT is a horizontal line from left to right

TEST — Results in Appendix C

Signal/Voltage Checks
Point A to Ground ....(70)
Point B to Ground..........................(2)
Point C to Ground ....(113)
Point D to Ground ....(57)
Point E to Ground..........................(23)
Point F to Ground..........................(106)

CIRCUIT

6

Symptoms

CHALLENGE CIRCUIT 6

Symptoms Although there should be a waveform on the CRT screen, there is only a straight horizontal line. This implies there is a horizontal signal being fed to the CRT horizontal plates, but no vertical signal is reaching the CRT vertical plates.

Identify initial suspect area: The initial suspect area includes all elements involved in the vertical signal path to the CRT vertical plates. That is, the signal input to the vertical amplifier, the vertical amplifier, and the path from the vertical amplifier output to the CRT vertical plates.

Make test decision: We'll use the "inputs-outputs" technique and first check the input of the vertical amplifier to see if an input signal is present.

Perform 1st Test: Check the presence of an input signal for the vertical amplifier at point D. The input signal is present.

Locate new suspect area: The new suspect area is the remainder of the vertical signal path to the CRT plates.

Examine available data.

Repeat analysis and testing:

Check the signal "out of" the vertical amplifier at point E to ground. No signal is present, indicating a problem in the vertical amplifier circuitry. We have performed block-level troubleshooting. At this point, the vertical amplifier requires component-level troubleshooting.

After component-level troubleshooting has been completed and the needed repairs performed, test the system with the same signals. The result is that two sine-wave cycles are seen on the CRT screen.

1st Test

2nd Test

3rd Test

THE SIMPLER SEQUENCE
FOR TROUBLESHOOTING

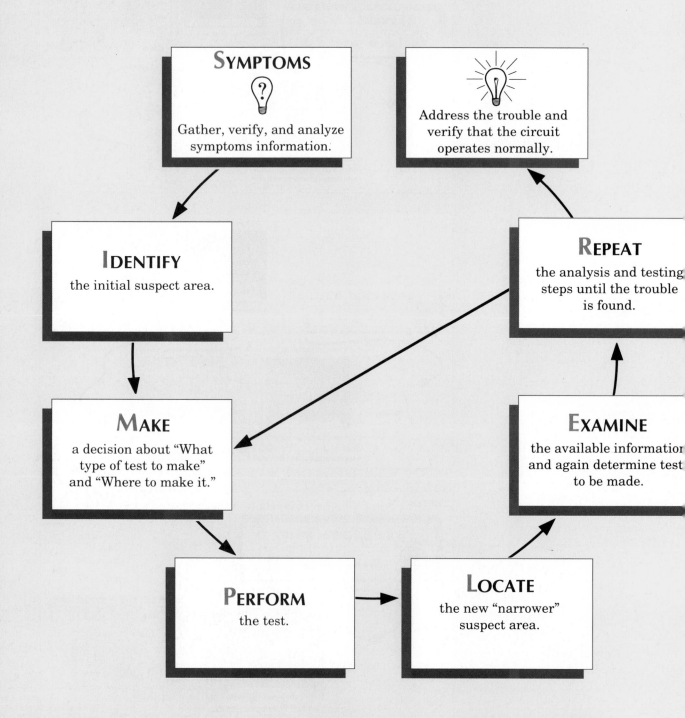

SYMPTOMS

Gather, verify, and analyze symptoms information.

Address the trouble and verify that the circuit operates normally.

IDENTIFY

the initial suspect area.

REPEAT

the analysis and testing steps until the trouble is found.

MAKE

a decision about "What type of test to make" and "Where to make it."

EXAMINE

the available information and again determine test to be made.

PERFORM

the test.

LOCATE

the new "narrower" suspect area.

PART IV

REACTIVE

COMPONENTS

Inductance

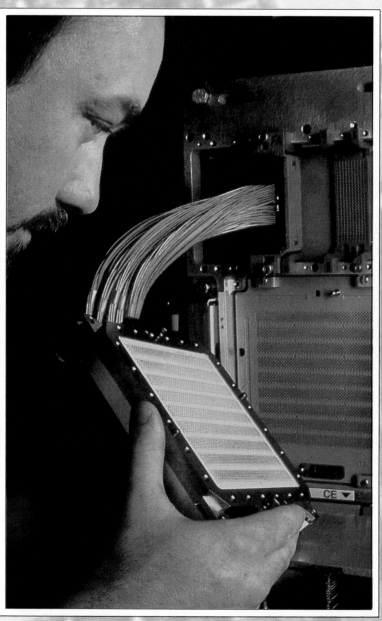

Courtesy of IBM

14

KEY TERMS

cemf

ϵ

Faraday's law

H

Henry

Inductance
(self-inductance)

Joule

Lenz's law

L, L_T

L/R time constant

τ

OUTLINE

Background information

Review of Faraday's and Lenz's laws

Self-inductance

Factors that determine inductance

Inductors in series and parallel

Energy stored in the inductor's magnetic
field

The L/R time constant

Using epsilon (ϵ) to help calculate
circuit parameters

Summary comments about inductors

Troubleshooting hints

CHAPTER PREVIEW

The electrical property of resistance and its component, a resistor, are important elements in your understanding of electronics. Likewise, the property of inductance and its related component, an inductor, are equally important. A significant contrast between resistors and inductors is that resistors dissipate electrical energy and inductors have the ability to temporarily store electrical energy.

Inductors have many uses in electronics. For example, they are used in "tuned" circuits, which enable selection of a desired radio or television station. They are used as filter elements to smooth waveforms or to suppress unwanted frequencies. Inductance is also used in transformers, which are devices that increase or decrease ac voltages, currents, and impedances. The inductor (or coil) in automobiles has been used for years to help develop high voltages, which create the spark across the spark plug gaps. These applications and many others use this unique electrical property of inductance.

In this chapter, we will discuss the property of inductance as it relates to dc circuit action. In the following chapters, you will study the action of this electrical property under ac conditions. An understanding of both will be very useful to you!

OBJECTIVES

After studying this chapter, you should be able to:

- Define **inductance** and **self-inductance**

- Explain **Faraday's** and **Lenz's laws**

- Calculate induced **cemf** values for specified circuit conditions

- Calculate inductance values from specified parameters

- Calculate inductance in series and parallel

- Determine energy stored in a magnetic field

- Draw and explain time-constant graphs

- Calculate time constants for specified circuit conditions

- Use ϵ to aid in finding circuit quantities in dc RL circuits

- List common problems of inductors

BACKGROUND INFORMATION

From previous chapters, you know that electrical resistance (R) is that property in a circuit that *opposes current flow*. It is also important for you to know that **inductance** (L) is the property in an electrical circuit that *opposes a change in current*.

Inductors come in a variety of physical and electrical sizes, Figure 14–1. The forms you see here are components that have been designed to oppose changes in current. It should be pointed out that even a straight piece of

FIGURE 14-1   Examples of various inductors and symbols (Photos by Michael A. Gallitelli)

wire exhibits a small amount of inductance. Sometimes this "stray" (many times undesired) inductance presents conditions that must be contended with in circuit designs and layouts. This is particularly true in high-frequency circuits, which you will study later in your training. Again, remember that inductance does not oppose current flow, as such, but opposes a *change* in current flow.

Another important characteristic of inductance and inductors is they store electrical energy in the form of the magnetic field surrounding them. As current increases through the inductor, a resulting magnetic field ex-

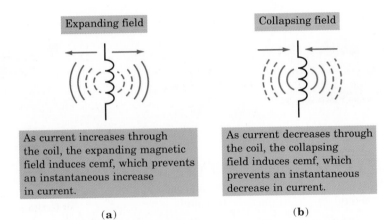

(a)   (b)

FIGURE 14–2   Expanding and collapsing fields

pands, surrounding the coil, Figure 14–2a. When current tries to decrease, the expanded field collapses and tries to prevent the current from decreasing, Figure 14–2b. (Recall Lenz's law!) As the field expands, it absorbs and stores energy. When the field collapses, it effectively returns that stored energy to the circuit.

Another feature of inductance and inductors is that an emf (electromotive force or voltage) is induced when a current *change* occurs through them. This induced emf is due to the expanding or collapsing magnetic field as it cuts the conductors of the coil. This self-induced emf is sometimes called **counter-emf,** or **cemf,** because its polarity opposes the change that induced it, Figure 14–3. The term **self-inductance** is frequently used to refer to inductors that create this situation.

In a previous chapter, we discussed magnetism and electromagnetism. Additionally, you learned the factors determining the strength of this electromagnetic field around the coil—number of turns; coil length (number of turns per unit length); cross-sectional area of coil; amount of current through the coil; and the type of core material (characteristics involved in the flux path). Keep these factors in mind, as there is a relationship

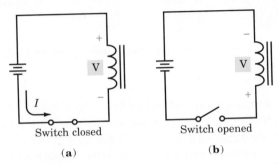

Switch closed   Switch opened

(a)   (b)

The polarity of the induced cemf is opposite to that of the source as the magnetic field expands.

(b) The polarity of the induced voltage reverses as the field collapses.

FIGURE 14–3   The concept of cemf

between these items and the property of inductance, which is addressed in this chapter.

In the earlier discussion of electromagnetism, we examined a constant level dc current through the coil. In this chapter, we will begin to examine what happens during the switch-on/switch-off "transient" (temporary condition) periods when dc is applied or removed from a circuit containing inductance.

REVIEW OF FARADAY'S AND LENZ'S LAWS

Faraday's Law

Recall **Faraday's law:** The emf induced in a circuit is proportional to the time-rate of change in the magnetic flux linking the circuit. In other words, the amount of induced emf depends on the rate of cutting the flux (ϕ).

Also, recall that in the SI system of units, a weber is 10^8 lines of flux. The formula to determine the induced voltage due to cutting flux lines (or flux linkages) is defined as:

$$V_{ind} = \frac{d\phi}{dt}$$

where $\dfrac{d\phi}{dt}$ indicates rate of cutting or linking flux lines. For a coil then:

Formula 14–1	$V_{ind} = N \text{ (number of turns)} \times \dfrac{d\phi \text{ (Wb)}}{dt \text{ (s)}}$

 number of turns, $\boxed{\times}$, $\boxed{(}$, number of webers cut, $\boxed{\div}$, amount of time, $\boxed{)}$, $\boxed{=}$

EXAMPLE

When a coil has 500 turns and the rate of flux change is 2 webers per second,

$$V_{ind} = 500 \times \frac{2}{1} = 1{,}000 \text{ V}$$

PRACTICE PROBLEMS I

What is the induced voltage (per turn) when 2×10^8 lines of flux are cut in 0.5 second?

Lenz's Law

Recall **Lenz's law:** The direction of an induced voltage (or current) opposes the change that caused it. That is, when the amount of magnetic flux linking an electric circuit changes, an emf is induced that produces a current opposing this flux change.

To summarize, Faraday's law describes how much voltage is induced by the parameters affecting it, and Lenz's law describes the polarity and nature of this induced voltage and the current produced. Let's now begin to correlate these laws with some new information as we examine inductance and its dc characteristics.

SELF-INDUCTANCE

In addition to inductance being defined as the property of a circuit that opposes a change in current, another definition for self-inductance (or inductance) is *the property of an electric circuit where a voltage is induced when the current flowing in the circuit changes*. You have learned the symbol for inductance, **L**. The **unit of inductance** is the **henry**. Named in honor of Joseph Henry, this unit of inductance is abbreviated **H**.

As you might suspect, the amount or magnitude of induced emf produced is related to the amount of inductance *(L)* in henrys (H) and the rate of current change. A formula expressing this is:

Formula 14–2	$V_L = L\dfrac{di \text{ (change in current, in amperes)}}{dt \text{ (change in time, in seconds)}}$

 inductance value, $\boxed{\times}$, $\boxed{(}$, change in current, $\boxed{\div}$, amount of time, $\boxed{)}$, $\boxed{=}$

where L = inductance in henrys

$$\frac{di}{dt} = \text{rate of change of current} \left(\frac{A}{s}\right)$$

It should be noted that the polarity of the induced voltage is opposite to the source voltage that caused the current through the coil.

From the above, a definition for a henry of inductance states that *a circuit has an inductance of one henry when a rate of change of one ampere per second causes an induced voltage of one volt*. The induced voltage is in the opposite direction to the source voltage; therefore, it is called **counter-emf** or **back-emf**. Again see Figure 14–3.

EXAMPLE

When the rate of current change is 5 amperes per second through an inductance of 10 henrys, what is the magnitude of induced emf?

$$V_L \div L\frac{di \text{ (change in current, in amperes)}}{dt \text{ (change in time, in seconds)}}$$

$$V_L = 10 \times 5 = 50 \text{ V (Or, induced cemf} = 50 \text{ V.)}$$

PRACTICE PROBLEMS II

What is the counter-emf produced by an inductance of 5 henrys if the rate of current change is 3 amperes per second?

From these definitions and formulas, a useful formula to calculate inductance value is:

Formula 14–3	$L = \dfrac{V_L}{\dfrac{di}{dt}}$

where L = inductance in henrys

$\dfrac{di}{dt}$ = rate of current change $\left(\dfrac{A}{s}\right)$

 V_L value, $\boxed{\div}$, $\boxed{(}$, change in current, $\boxed{\div}$, amount of time, $\boxed{)}$, $\boxed{=}$

EXAMPLE

What is the circuit inductance if a current change of 12 amperes in a period of 6 seconds causes an induced voltage of 5 volts?

$$L = \frac{V_L}{\dfrac{di}{dt}} = \frac{V_L}{1} \times \frac{dt}{di} = \frac{5}{1} \times \frac{6}{12} = 5 \times 0.5 = 2.5 \text{ H}$$

FACTORS THAT DETERMINE INDUCTANCE

As implied earlier, some of the same factors that affect the strength of a coil's electromagnetic field are also related to the amount of coil inductance.

Look at Figure 14–4. The physical properties of an inductor that affect its inductance include:

1. *Number of turns.* The greater the number of turns, the greater the inductance value. In fact, inductance is proportional to the square of the turns. That is, when you double the number of turns in a coil of a given length and diameter, the inductance quadruples, assuming the core material is unchanged.

2. *Cross-sectional area (A) of the coil.* This feature relates to the square of the coil diameter. The greater the cross-sectional area, the greater the inductance value. When the cross-sectional area doubles, the inductance (L) doubles.

Inductance:

| *Directly* related to relative permeability of core | | *Directly* related to cross-sectional area |

| *Inversely* related to coil's length | | *Directly* related to the square of the number of turns |

FIGURE 14–4   Factors that affect the amount of inductance

3. *Length of the coil.* The longer the coil for a given diameter and number of turns, the *less* the inductance value. This is due to a less concentrated flux, thus, less flux cut per unit time and the less the induced voltage produced.

4. *Relative permeability of the core.* For air-core inductors, the relative permeability (μ_r) is virtually one. The higher the permeability of the inductor's core material, the greater the inductance value. Recall that permeability relates to the magnetic path's ability to concentrate magnetic flux lines.

As a technician you probably will not design inductors. However, Formula 14–4 illustrates the relationship of inductance to the various factors we have been discussing. In SI units, the formula is:

Formula 14–4	$L = 12.57 \times 10^{-7} \times \dfrac{\mu_r N^2 A}{l}$

12.57, $\boxed{\text{EE}}$ or $\boxed{\text{EXP}}$, $\boxed{+/-}$, 7, $\boxed{\times}$, $\boxed{(}$, μ_r value, $\boxed{\times}$, number of turns, $\boxed{x^2}$, $\boxed{\times}$, area of coil in square meters, $\boxed{)}$, $\boxed{\div}$, length of coil in meters, $\boxed{=}$

where L = inductance in henrys

μ_r = relative permeability

N = number of turns

A = cross-sectional area in square meters

l = length of coil in meters

12.57×10^{-7} = the absolute permeability of air

IN-PROCESS LEARNING CHECK I

Fill in the blanks as appropriate.

1. All other factors remaining the same, when an inductor's number of turns increases four times, the inductance value _____ by a factor of _____.

2. All other factors remaining the same, when an inductor's diameter triples, the inductance value _____ by a factor of _____.

3. All other factors remaining the same, when an inductor's length increases, the inductance value _____.

4. When the core of an air-core inductor is replaced by material having a permeability of ten, the inductance value _____ by a factor of _____.

EXAMPLE

An air-core coil that is 0.01 meters in length with a cross-sectional area of 0.001 square meters has 2,000 turns. What is its inductance in henrys?

$$L = 12.57 \times 10^{-7} \times \left(\frac{1 \times (2{,}000)^2 \times 0.001}{0.01} \right)$$

$$L = 12.57 \times 10^{-7} \times \left(\frac{(40 \times 10^5) \times 1 \times 10^{-3}}{10 \times 10^{-3}} \right)$$

$$L = 12.57 \times 10^{-7} \times 4 \times 10^5 = \text{about } 0.5 \text{ H, or } 500 \text{ mH}$$

Since the μ_r of air is one, the formula is modified for *air-core* inductors and stated as:

Formula 14–5	$L = 12.57 \times 10^{-7} \times \dfrac{N^2 A}{l}$

  12.57 , $\boxed{\text{EE}}$ or $\boxed{\text{EXP}}$, $\boxed{+/-}$, 7, $\boxed{\times}$, $\boxed{(}$, number of turns, $\boxed{x^2}$, $\boxed{\times}$, area of coil in square meters, $\boxed{)}$, $\boxed{\div}$, length of coil in meters, $\boxed{=}$

INDUCTORS IN SERIES AND PARALLEL

For our purposes, we will assume the inductors in this section do not have any *coupling* between them. That is, there are no flux lines from one inductor linking the turns of another inductor. In technical terms, no *mutual inductance* exists between coils.

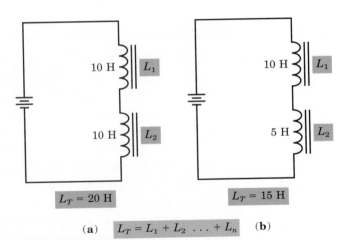

$$L_T = 20 \text{ H} \qquad\qquad L_T = 15 \text{ H}$$

(a) $L_T = L_1 + L_2 \ldots + L_n$ (b)

FIGURE 14–5 Total inductance of series inductor⸱

Inductance in Series

Finding the total inductance of inductors in *series,* Figure 14–5, is quite simple. Since they are in series, the current change through each inductor is the same. Thus, when they are equal-value inductors, the induced voltage in each is equal, and the total inductance is two times the value of either inductor. This implies that when the inductors are not equal in value, the total inductance (L_T) is the sum of the inductance values in series. That is, $L_T = L_1 + L_2$.

Stated another way: *Inductors in series add inductances in the same way that resistors in series add resistances!*

Formula 14–6	$L_T = L_1 + L_2 \ldots + L_n$

L_1 value, $\boxed{+}$, L_2 value, . . . , $\boxed{=}$

REMINDER: This formula holds true as long as there is no coupling of flux lines from one inductor to another; that is, there is no coupling between inductors.

EXAMPLE

When the inductors in Figure 14–5a are tripled in value, the total inductance of the circuit on the left is 60 H, rather than 20 H, and the total inductance of Figure 14–5b is 45 H, rather than 15 H.

PRACTICE PROBLEMS III

Determine the total inductance of the circuit shown in Figure 14–6.

10 mH

6 mH

100 μH

FIGURE 14–6

Inductance in Parallel

As you might expect, the total inductance of inductors in parallel is less than the least value inductor in parallel. This is analogous to resistors in parallel. The general formulas used to compute total inductance for inductors in parallel, Figure 14–7, are the reciprocal formulas:

Formula 14–7	$\dfrac{1}{L_T} = \dfrac{1}{L_1} + \dfrac{1}{L_2} + \dfrac{1}{L_3} \cdots + \dfrac{1}{L_n}$

 L_1 value, $\boxed{1/x}$, $\boxed{+}$, L_2 value, $\boxed{1/x}$, $\boxed{+}$, L_3 value, $\boxed{1/x}$, . . . , $\boxed{=}$

Formula 14–8	$L_T = \dfrac{1}{\dfrac{1}{L_1} + \dfrac{1}{L_2} + \dfrac{1}{L_3} \cdots + \dfrac{1}{L_n}}$

L_1 value, $\boxed{1/x}$, $\boxed{+}$, L_2 value, $\boxed{1/x}$, $\boxed{+}$, L_3 value, $\boxed{1/x}$, $\boxed{=}$, $\boxed{1/x}$

For the problem illustrated in Figure 14–7, you can use the reciprocal key of your calculator $\boxed{1/x}$ to quickly find the L_T. The keying sequence would be:

10, $\boxed{1/x}$, $\boxed{+}$, 5, $\boxed{1/x}$, $\boxed{+}$, 10, $\boxed{1/x}$, $\boxed{=}$, $\boxed{1/x}$

(NOTE: If all units are in henrys, the answer is in henrys.) Also, the product-over-the-sum approach, used in Figure 14–8, works for two inductors in parallel:

Formula 14–9	$L_T = \dfrac{L_1 \times L_2}{L_1 + L_2}$

L_1 value, $\boxed{\times}$, L_2 value, $\boxed{\div}$, $\boxed{(}$, L_1 value, $\boxed{+}$, L_2 value, $\boxed{)}$, $\boxed{=}$

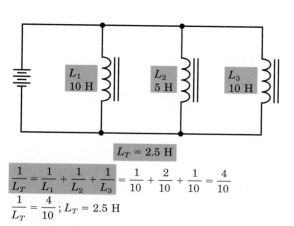

$L_T = 2.5 \text{ H}$

$$\frac{1}{L_T} = \frac{1}{L_1} + \frac{1}{L_2} + \frac{1}{L_3} = \frac{1}{10} + \frac{2}{10} + \frac{1}{10} = \frac{4}{10}$$

$$\frac{1}{L_T} = \frac{4}{10}; L_T = 2.5 \text{ H}$$

FIGURE 14–7   Finding L_T using the reciprocal formula

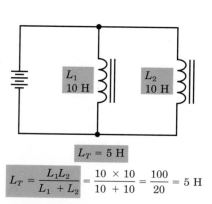

$L_T = 5 \text{ H}$

$$L_T = \frac{L_1 L_2}{L_1 + L_2} = \frac{10 \times 10}{10 + 10} = \frac{100}{20} = 5 \text{ H}$$

FIGURE 14–8   Finding L_T using the product-over-the-sum formula

EXAMPLE

You have probably already surmised that the same time-saving techniques can be used for inductance calculations as you used in resistance calculations. For example, if the two inductors in parallel are of equal value, the total inductance equals half of one inductor's value. This is similar to two 10-Ω resistors in parallel having an equivalent total resistance of 5 Ω.

PRACTICE PROBLEMS IV

1. What is the total inductance of three 15-H inductors that are connected in parallel and have no mutual inductance?

2. Use your calculator and the reciprocal method and find the total inductance of four inductors in parallel. Their individual values are: 10 mH, 12 mH, 15 mH, and 20 mH. Assume no coupling between them.

ENERGY STORED IN THE INDUCTOR'S MAGNETIC FIELD

The magnetic field surrounding an inductor when current passes through the inductor is a form of stored energy, Figure 14–9. The current producing the electromagnetic field is provided by the circuit's source. Hence, this stored energy is supplied by the source to the magnetic field.

When the source is removed from the circuit, the field collapses and returns the stored energy to the circuit in the form of induced emf. This induced emf tries to keep the current through the coil from decaying.

If the inductor were a perfect inductor with no resistance, no I^2R loss would occur; therefore, no power would be dissipated in this energy storing-and-returning process. NOTE: In reality, inductors have some resistance (the resistance of the wire), and they dissipate power; however, for the present discussion, we won't analyze the details of this fact. Note also that in dc circuits, the inductor's resistance (plus any other resistance in series with the inductor) and the amount of voltage applied determine the amount of current through the coil (Ohm's law).

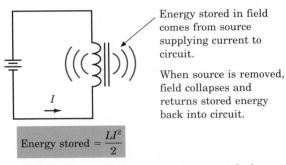

Energy stored in field comes from source supplying current to circuit.

When source is removed, field collapses and returns stored energy back into circuit.

$$\text{Energy stored} = \frac{LI^2}{2}$$

FIGURE 14–9   Concept of energy being stored in inductor's magnetic field

$$I = \frac{V}{R} = \frac{20 \text{ V}}{10 \text{ }\Omega} = 2 \text{ A}$$

FIGURE 14–10 Direct current (dc) through inductor found by using Ohm's law

Notice in Figure 14–10, if a coil has 10 ohms of resistance, and a dc voltage of 20 volts is applied, the dc current through the coil will be 20 V/10 Ω, or 2 amperes, after the field has finished building up.

Computing the energy stored in an inductor's field is not a daily calculation for most technicians. Many technicians, however, take FCC examinations to obtain commercial radio licenses. Since this computation is required for these tests, we will show you how this calculation is performed.

In an earlier chapter on Ohm's law, we discussed the unit of electrical energy, the **joule.** A joule is sometimes defined as the amount of electrical energy required to move one coulomb of electrical charge between two points having a potential difference of one volt. Also remember the watt of power relates to the joule since the performance of electrical work at a rate of one joule per second equals one watt.

Relating all of these factors, it can be shown that the stored energy correlates to the average power supplied to the inductor circuit and the time involved. The amount of energy stored, in turn, relates to the amount of inductance and current involved.

Accumulating these facts, then:

Formula 14–10	Energy = 0.5 LI^2

where L = inductance in henrys

I = current in amperes

Energy = joules (or watt-seconds)

 0.5, $\boxed{\times}$, $\boxed{(}$, L value, $\boxed{\times}$, I value, $\boxed{x^2}$, $\boxed{)}$, $\boxed{=}$

EXAMPLE

What is the energy stored by an inductor of 20 henrys that has a current of 5 amperes passing through it?
Substituting into the formula:

$$\text{Energy} = 0.5 \, LI^2$$

$$\text{Energy} = \frac{20 \times (5)^2}{2} = \frac{20 \times 25}{2} = \frac{500}{2} = 250 \text{ J}$$

PRACTICE PROBLEMS V

A dc voltage of 100 volts is applied to a 1-H inductor that has 20 Ω of dc resistance. (a) What is the dc current through the coil? (b) How many joules of energy are stored in the inductor's magnetic field?

THE *L/R* TIME CONSTANT

We earlier established that an inductor opposes a change in current flow due to the expanding or contracting magnetic field as it cuts the turns of the coil. The induced voltage opposes the change that originally induced it (Lenz's law). This implies that even when dc voltage is first applied to a coil, the current cannot instantly rise to its maximum value (the V/R value), which is true!

Let's examine the transient conditions, or the temporary state caused by a sudden change in circuit conditions in a simple inductive circuit when dc is first applied. Refer to Figures 14–11 and 14–12 as you study the following discussion.

1. In Figure 14–11, you see the conditions that exist in a purely resistive circuit. When the switch is closed, the current immediately reaches its dc (Ohm's law) value determined by the voltage applied and the resistance involved.

2. In Figure 14–12a, the inductor in the resistor-inductor *(RL)* circuit prevents current from rising to its maximum value during the first moments after the switch is closed. This is because the current increases from zero to some value, causing an expansion of the magnetic field in the inductor. The expanding field cuts the turns of the coil, inducing a back-emf (cemf) that opposes the current change. Therefore, the current does not abruptly rise from zero to the final (stable) dc value.

3. In Figure 14–12b, at time t_1, the current has arrived at only 63.2% of its final value. The time from t_0 to t_1 is equal to the value of L/R, where L is in henrys and R is the circuit resistance value in ohms.

When switch is closed, I immediately rises to $\dfrac{V}{R}$ value.

FIGURE 14–11   Direct current (dc) change in purely resistive circuit

When switch is closed, current rises "exponentially" rather than instantly to a value = V/R.

It takes five L/R time constants for current to complete the change.

(a)   (b)

FIGURE 14–12   Current change in circuit containing inductance

This value *(L/R)* is known as one **time constant.** The symbol for the time constant is the Greek letter tau (τ). In a circuit containing L and R, the formula for time constant is:

Formula 14–11	$\tau = \dfrac{L}{R}$

inductance value (henries), $\boxed{\div}$, resistance value (ohms), $\boxed{=}$

Let's think about Formula 14–11 for a moment. One time constant (τ) equals the value of the inductance *(L)* in henries divided by the resistance *(R)* in ohms. The amount of time for one time constant is *directly* related to the value of inductance. The time for one time constant is inversely related to the resistance value. Why might this be true?

You have already learned that the greater the inductance value, the greater will be the amount of counter-emf generated for a given amount of current change ($V_L = L \times d\phi/dt$). This indicates that the larger the L value, the greater the opposition to current change. This means it will take more time for the current (and field) to build up to their maximum, or "steady-state" level, which occurs after five time constants. In other words: As L increases, the time for one time constant (τ) increases. Conversely, as L decreases, τ decreases.

In accordance with Ohm's law, the smaller the resistance value, the higher the final current will be ($I = V/R$); thus, the greater the required change in current to arrive at the higher current value, and the longer the time involved for one time constant. In other words: As R decreases, the time for τ increases. Conversely, as R increases, τ decreases.

With these thoughts in mind, continue examining the graphing of the rise of current through the RL circuit, with dc applied. Incidentally, this type graph is often referred to as the "Universal Time-Constant Graph" or chart.

4. In Figure 14–12b, at time t_2 (where the time from t_1 to t_2 represents a second time constant), the current has risen 63.2% of the remaining distance to the final dc current level. That is, current rises 63.2% of the remaining 36.8% to be achieved, or 63.2% of 36.8% equals an additional 23.3%. Thus, in two time constants the current achieves 63.2% + 23.3%, or 86.5% of its final maximum value.

5. As time continues, the current rises another 63.2% of the remaining value for each time constant's worth of time. After *five time constants*, the current has virtually arrived at its final dc value. In this case,

$$\frac{5 \text{ V}}{5 \text{ }\Omega} = 1 \text{ A}$$

Therefore, the amount of time required for current to rise to maximum *V/R* value is directly related to the value of inductance *(L)* and inversely related to the circuit *R*, as indicated by the *L/R* relationship.

The inductor's expanding magnetic field and the resulting induced voltage cause current to exponentially increase (in a nonlinear fashion), rather than instantly as it would in a purely resistive circuit.

What happens when the inductor's field collapses, that is, when the source is removed after the field has fully expanded? In Figure 14–13, when the switch is in position A, the source is connected to the *RL* circuit.

When the switch is in position B, the source is disconnected from the *RL* circuit. However, a current path is provided for the *RL* circuit, even when the source is disconnected.

Assume the switch was in position A for at least five time constants and the current has reached its maximum level. Therefore, the magnetic field is fully expanded. When the switch is moved to position B, the expanded field collapses, inducing an emf into the coil, which attempts to keep the current

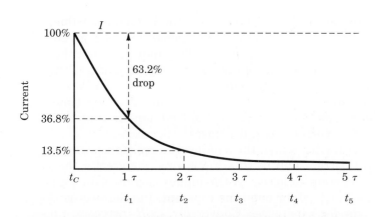

It takes five *L/R* time constants for current to complete the change.

FIGURE 14–13   Circuit with same *L/R* time constant for current decay as for current buildup

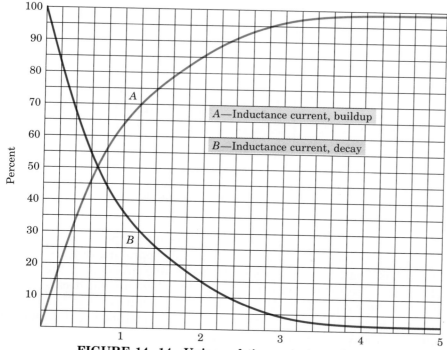

Percent

1   2   3   4   5

A—Inductance current, buildup

B—Inductance current, decay

FIGURE 14–14   Universal time-constant chart

from changing. The inductor's collapsing field and induced emf have now, in effect, become the source. Obviously, as the field fully collapses, current decays and eventually becomes zero. The following comments describe what happens during this declining current time. Figure 14–14 shows a graphic display of both the current build-up and decay. Refer to it while reading the following discussion.

1. The current decays 63.2% from maximum current value toward zero in the first time constant. In other words, it falls to 36.8% of maximum value.

2. During the second τ, it falls 63.2% of the remaining distance, or another 23.3% for a total of 86.5% decay. In other words, the current is now at 13.5% of its maximum value.

3. This continues, until at the end of the five L/R time constants, the current is virtually back at the zero level.

4. Given the same L/R conditions, the current decay is the mirror image (or the inverse) of the current-rise condition, described earlier.

Let's summarize. In a dc circuit comprised of an inductor and a series resistance (RL circuit), it takes five L/R time constants for current to change *from one static level to a "changed" new level,* whether it be an increase or a decrease in current levels. (NOTE: This is true whether the change is from 0 to 1 ampere, from 1 to 5 amperes, or from 4 amperes to 2 amperes, and so forth.) This is an important concept, so remember it!

Observe Figure 14–15 and notice that in a *dc RL* circuit, the sum of V_L and V_R must equal V applied. This means as current increases through the

FIGURE 14–15   Simple dc,
RL circuit

RL circuit, V_R is increasing and V_L is decreasing proportionately until V_R equals approximately V applied and V_L equals approximately 0 V when the current has finally reached a steady-state value. (Assuming the R of the inductor is close to zero ohms). Conversely, when current decays, V_R also decays at the same rate to the steady-state minimum value the circuit is moving toward. The τ curves are one convenient method to determine what values V_R and V_L have after specified times.

EXAMPLE

Refer to Figure 14–15. What is the value of V_R one second after the switch closed?

Answer: First, compute how many time constants are represented by one second of time:

$$\tau = \frac{L}{R} = \frac{10 \text{ H}}{12 \ \Omega} = 0.833 \text{ seconds}$$

Therefore, 1 second = 1/0.833 time constants, or 1 second = 1.2 τ.

Next, refer to the τ curves and note that at 1.2 τ, the current rises to 70% of maximum value. Therefore, $V_R = 0.7 \times V$ applied. Thus, V_R at that instant equals $0.7 \times 10 \text{ V} = 7$ V and V_L is 3 V.

PRACTICE PROBLEMS VI

Using the same circuit as in Figure 14–15, determine the V_R and V_L values two seconds after the switch is closed.

USING EPSILON (ε) TO HELP CALCULATE CIRCUIT PARAMETERS

As you have seen, you can come fairly close in approximating circuit conditions and values from the exponential Universal Time Constant Chart.

Another more precise way to calculate circuit parameters at any given instant is to use natural logarithms and epsilon, rather than graphical means.

The curves represented on the time-constant graph are sometimes called exponential curves. Many things in nature follow this exponential curve. The natural logarithms are mathematical values related to exponential changes. The base number of the natural logarithms is 2.71828. This number is often referred to by the Greek letter epsilon (ϵ). That is, ϵ equals 2.71828.

Graph Based on ϵ Raised to Negative Powers

Observe in Figure 14–16 that the descending curve on the graph is actually a plot that can be produced by raising ϵ (2.71828) to negative powers. That is, if we find ϵ^{-1}, it has a value of 0.3679, ϵ^{-2} equals 0.1353, and so on. The ascending curve on the graph represents 1 minus the other curve's value at each point along the curve. For example, when the descending curve is at 0.3679, the ascending curve is at $1 - 0.3679$, or 0.6321, and so on.

Relationship of Curves to Circuit Parameters

For our purposes here, the ascending curve represents the increase in current over five time constants. It can also represent the increasing voltage across the resistor in the RL circuit as current increases to maximum, or "full," value (I_f) over five time constants. As I increases, $I \times R$ increases, in step. The descending curve represents the current decay when the external voltage source is removed from the circuit and the inductor's field is collapsing (assuming there is a current path when the external source is removed, Figure 14–13). Also, the descending curve can represent the decay in voltage drop across the resistor, since its $I \times R$ drop is going to be in step with the current conditions.

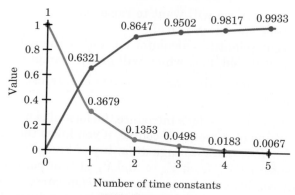

Plots of Epsilon to Negative Powers
and of 1 minus Epsilon to Negative Powers

FIGURE 14–16

As you know, in the dc circuit, the $V_R + V_L$ must add up to equal V_{applied}. At any given point along these curves, the decimal values of the descending and ascending curves must add up to 1, representing 100% of the maximum, or final, state value.

Relationship of Negative Powers for Epsilon and L/R Time Constant

As stated, the negative power to which epsilon (2.71828) is raised, determines where we are on the descending exponential curve. How does this curve relate to time constants?

You have already learned that after five L/R time constants, the circuit parameters will have changed from wherever they are to the new "stable state" conditions. The relationship of importance to us is: *how much time is being allowed for the change versus the time for one time constant?*

If the time being allowed for the change is five time constants, or greater: 1) current will be at its final value, if it is building up, or at zero value, if decaying; 2) V_R will be equal to V_{applied} if current is building up, or at zero value, if current is decaying; 3) V_L will be at zero value in either case.

If the time being allowed for the change is only one time constant, then current will be at 63.2% of its new maximum, if current is building up, or at 36.8% of maximum, if current is decaying.

The bottom line is, if we know the amount of time being allowed and know, or can calculate, the L/R time constant, we know how many time constants are being allowed for the change. *This ratio of time allowed to time for one time constant is the number we want to use as the negative power of epsilon (ϵ) in calculating circuit parameters!*

The method of finding the number of time constants allowed is

$$\text{Number of time constants} = \frac{\text{Time allowed for change } (t)}{\text{Time for one } L/R \text{ time constant}}$$

$$\text{Number of } \tau = \frac{t}{\dfrac{L}{R}} = \frac{t}{1} \times \frac{R}{L} = \frac{Rt}{L}$$

So the negative power we will want to raise epsilon to in our calculations related to L/R time constants is Rt/L.

Of course, your scientific calculator has either the natural log function key $\boxed{e^x}$, or a key labeled $\boxed{\ln_x}$, which will greatly simplify calculations involving raising epsilon to the various powers of interest. In our examples, we will use the $\boxed{e^x}$ key.

Armed with these facts, let's introduce the formulas used to solve for the instantaneous value of current. Of course, if you know the current, you can calculate the $I \times R$ resistor voltage drop. If you know the applied voltage, you can then subtract V_R from V_S to find V_L at that point in time.

The circuit conditions assumed here are: On current buildup, we are increasing from zero to maximum or final value. On current decay, we are decreasing from the present level to zero current.

Formula 14–12 (Increasing I)	$i = \dfrac{V_S}{R}(1 - \epsilon^{-Rt/L})$

where i = instantaneous current value

 R value, $\boxed{\times}$, time value, $\boxed{\div}$, inductance value, $\boxed{=}$, $\boxed{+/-}$, $\boxed{\text{2nd}}$, $\boxed{e^x}$,

$\boxed{+/-}$, $\boxed{+}$, 1, $\boxed{=}$, $\boxed{\times}$, V_S value, $\boxed{\div}$, R value, $\boxed{=}$

EXAMPLES

Refer to Figure 14–17 and determine the circuit current 1 millisecond after the switch is closed.

$$i = \frac{V_S}{R}(1 - \epsilon^{-Rt/L})$$

$$i = \frac{10\text{ V}}{10\text{ k}\Omega}(1 - \epsilon^{-10\text{ k}\Omega \times 1\text{ ms}/10\text{ H}})$$

$$i = 1\text{ mA }(1 - \epsilon^{-1}) = 1\text{ mA }(1 - 0.3678) = 1\text{ mA} \times 0.6322$$

$$i = 0.0006322\text{ A or }0.6322\text{ mA}$$

This checks out against our exponential graph. The time allowed for change was 1 time constant and the current increased from zero to about 63% of its maximum possible value of 1 mA.

$$V_R = 0.63\text{ mA} \times 10\text{ k}\Omega,\text{ or }6.3\text{ V};\ V_L = 10\text{ V} - 6.3\text{ V} = 3.7\text{ V}$$

The calculator keystrokes for finding $\epsilon^{-Rt/L}$ for our example are: 10, $\boxed{\text{EE}}$, 3, $\boxed{\times}$, 1, $\boxed{\text{EE}}$, $\boxed{+/-}$, 3, $\boxed{\div}$, 10, $\boxed{=}$, $\boxed{+/-}$, $\boxed{\text{2nd}}$, $\boxed{e^x}$. This yields the 0.3678, which easily plugs into the formula in place of the ϵ^{-1} from our example.

The complete calculator sequence for this problem would be: 10, $\boxed{\text{EE}}$, 3, $\boxed{\times}$, 1, $\boxed{\text{EE}}$, $\boxed{+/-}$, 3, $\boxed{\div}$, 10, $\boxed{=}$, $\boxed{+/-}$, $\boxed{\text{2nd}}$, $\boxed{e^x}$, $\boxed{+/-}$, $\boxed{+}$, 1, $\boxed{=}$, $\boxed{\times}$, 10, $\boxed{\div}$, 10, $\boxed{\text{EE}}$, 3, $\boxed{=}$

FIGURE 14–17

Observe that this sequence indicates that first you find the product of $R \times t$, then divide it by the L value, yielding the value of Rt/L (the number of time constants being allowed for the change). Then use that result as the negative power to raise epsilon to in order to find out where we are on the exponential curve. Then, make that result negative with the $\boxed{+/-}$ button in order to subtract it from a $+1$. The resultant decimal value is then multiplied times the value of maximum current found by V/R. The result, for our example, is the circuit current present after the one time constant allowed for the current change from zero.

PRACTICE PROBLEMS VII

Suppose for the example circuit in Figure 14–17, we want to compute the current 2.5 milliseconds after the switch was closed. What would be the value of current? (Use the calculator sequence, then see if the result makes sense by checking against the τ chart).

To find the value of decaying (decreasing) current, it is not necessary to subtract from 1, since we are not trying to find the value of the ascending curve, but rather we want the descending curve value. The formula for instantaneous value of current during decreasing current is:

$$\text{Formula 14–13 (Decreasing } I) \qquad i = \frac{V_S}{R}\,\epsilon^{-Rt/L}$$

 R value, $\boxed{\times}$, time value, $\boxed{\div}$, inductance value, $\boxed{=}$, $\boxed{+/-}$, $\boxed{\text{2nd}}$, $\boxed{e^x}$, $\boxed{\times}$, V_S value, $\boxed{\div}$, R value, $\boxed{=}$

EXAMPLE

Refer to the circuit shown in Figure 14–18. Assume that the current is at maximum, final-state value, then the switch is moved from point A to point B. What is the value of circuit current after 68 microseconds?

$$i = \frac{V_S}{R} \times \epsilon^{-Rt/L}$$

$$i = \frac{6\text{ V}}{2.2\text{ k}\Omega} \times \epsilon^{-(2.2\text{ k}\Omega \times 68\text{ }\mu\text{s})/50\text{ mH}} = 2.72\text{ mA} \times \epsilon^{-2.992} = 2.72\text{ mA} \times 0.05$$

$$= 0.136\text{ mA}$$

This checks out against our τ chart. After about 3 time constants, the current should have decayed to approximately 0.0498 times maximum current. Our calculations indicated that after 2.992 time constants, the current is at 0.05 times its original value. $V_R = 0.136\text{ mA} \times 2.2\text{ k}\Omega = 0.299$ volts at this point in time.

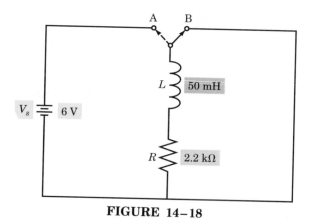

FIGURE 14–18

The complete calculator sequence for this problem would be:

2.2, [EE] , 3, [×] , 68, [EE] , [+/−] , 6, [÷] , 50, [EE] , [+/−] , 3, [=] , [+/−] , [2nd] ,
[eˣ] , [×] , 6, [÷] , 2.2, [EE] , 3, [=]

![pencil] **PRACTICAL NOTES**

Adaptations of the current formulas can be made to find resistor and inductor voltage at any instant during current rise or at any instant during current decay. The formulas that may be used are as follows:

Voltage across resistor during current buildup:

$$V_R = V_S \times (1 - \epsilon^{-Rt/L})$$

Voltage across resistor during current decay:

$$V_R = V_S \times (\epsilon^{-Rt/L})$$

Voltage across inductor during current buildup or decay:

$$V_L = V_S \times (\epsilon^{-Rt/L})$$

A Rapidly Collapsing Field Induces Very High Voltage(s)

In a circuit, such as in Figure 14–15, when the source voltage is disconnected, there is no low-resistance current path, as was shown earlier when explaining the current decay events. For purposes of explanation, let's assume the resistance across the open-switch contacts is extremely high. What happens to the total R in the current path when the switch opens? The R drastically increases (approaching infinity). What impact does this have on the L/R time constant? Since R drastically increases and L remains the same, the time constant must greatly decrease!

The result of all this activity is the expanded field collapses almost instantly. This means many flux lines cut the turns of the inductor in a very short period of time; thus, a very high voltage is induced in the coil. (NOTE: This phenomenon produces high voltage and high-voltage "arcs" jumping the gaps of spark plugs in your automobile.) In Figure 14–15, the induced voltage could be high enough to cause an arc across the open-switch contacts. The stored magnetic energy represented by the collapsing field is thus dissipated by the energy used in producing the arc.

SUMMARY COMMENTS ABOUT INDUCTORS

You have studied that inductors have properties of inductance (which oppose a *change* in current) and resistance in the wire from which the inductor is wound (which opposes current flow). The amount of inductance, L (measured in henrys), relates to the physical properties of the coil, including the length, cross-sectional area, number of turns, and type of core. Also, there is a relationship between the amounts of inductance and induced cemf that develops for any given rate of current change through the coil.

Also, you know when a dc voltage is first applied to a circuit with an inductor present, current does not instantly rise to the value dictated by the dc resistance of the circuit divided by the applied voltage. This delay is due to the inductor's opposing any current change. It takes five L/R time constants for the complete change from one current level (even if it is zero amperes) to a new current level. Under opposite conditions (i.e., if the dc source is removed from the inductor circuit) current will not abruptly decrease to zero, since the inductor opposes the current change and the collapsing field induces an emf in the coil to attempt to prevent the current from changing. When the time constant is very small during the decay sequence, the field quickly collapses, inducing a very high voltage in the coil. Wise circuit designers, therefore, make sure that current is not abruptly interrupted through the inductors in general dc circuits. When it must be abruptly interrupted, preventive measures are generally used to avoid unwanted voltage spikes. (NOTE: There are some special applications where high voltage is desired, and the component ratings and circuit design are appropriately chosen for this purpose.)

Laminated iron-core inductors, such as those used in power supplies and audio circuitry, are often in henrys of inductance. Powdered iron-core and variable ferrite-core inductors can have inductance ranges in millihenrys (mH). Air-core inductors are typically in the microhenrys (μH) range.

Also, inductors are wound with various sizes of wire and in varying numbers of turns which affect the dc resistance of the coil. Later, you will study how the resistance to inductance ratios affects inductors' operation in certain ac circuit applications.

Excellent information is available about the details of winding coils to desired specifications in *The Radio Amateur's Handbook,* published by the American Radio Relay League (ARRL). Also, there are numerous inexpensive, special slide rules that allow you to establish required parameters and to read the number of turns, coil length, coil diameter, and so on, needed for the desired results.

TROUBLESHOOTING HINTS

Typical Troubles

Troubles in inductors are usually associated with either opens or shorts. For this reason, the ohmmeter is a useful instrument to check the condition of an inductor—*provided* you know the normal resistance for the coil being checked.

Usual values range from 1 Ω to several hundred ohms for iron-core "chokes" and inductors with many turns and inductance values in the henrys range. Inductors in the millihenrys and microhenrys ranges typically have dc resistances from less than 1 Ω to something less than 100 Ω.

CAUTION! Care should be taken when removing ohmmeter test leads from large inductances. The "inductive kick" caused by the quick collapse of the inductor's field when the ohmmeter is removed can give you a nasty shock! This field is created by current produced through the inductor by the ohmmeter's dc source during testing. Be sure your hands or fingers are not touching or near the metallic tips of the test probes when removing the test leads from the inductor connections.

OPEN COIL.   If an inductor is open, it is obvious with an ohmmeter check that it measures infinite ohms. *CAUTION!* Be sure there are no sneak paths when you measure. If it is easy to do, it is good to isolate the component being measured from the rest of the circuitry when checking it. Probably the most common trouble is this open condition. Either there is a broken turn in the coil, or a bad connection to the coil, and so forth.

SHORTED TURNS.   Because each coil turn must be electrically insulated from all other coil turns for the inductor to operate properly, when the varnish or insulation on adjacent turns breaks down, these turns can short together. This is more difficult to detect with an ohmmeter test. The key is to know what R value the given coil should read. If there are shorted turns, the R value will read lower than normal. (NOTE: If only one or two turns are shorted, it may be impossible to detect by simple ohmmeter measurements!)

Obviously, the more turns that are shorted, the greater is the difference between the normal value and that which is measured. It is also possible in iron-core inductors to have turns shorting to the core, or being shorted by the core.

Another possibility is the whole inductor is being shorted from one end to the other. The reading will then be zero Ω. This condition, however, is less apt to happen than simply having shorted turns. Many times, an inductor with shorted turns will give the technician a clue by the burned insulation smell or visual effects shown by an extremely overheated component.

PRACTICAL NOTES

When selecting inductors for replacement or new designs, be sure to notice the important parameter ratings. Four important ratings you should note are:

1. the inductance value

2. the current rating

3. the dc resistance of the inductor

4. any voltage and insulation ratings or limitations related to the inductor's usage.

SUMMARY

▶ Inductance or self-inductance is the property of a circuit that opposes any *change* in current flow. Any change in current creates moving magnetic flux lines that, when linked (or cutting conductor[s]), induces an emf opposing the current change that initially produced it (Lenz's law).

▶ An inductor is a component that has the property of inductance. Any current-carrying conductor possesses some amount of inductance. However, most inductors are created by forming a conductor into loops of wire wound around some sort of structure. The amount of inductance created depends on the number of turns, the cross-sectional area, the length of the coil, and the permeability (magnetic characteristics) of the core material.

▶ Inductors come in a variety of physical and electrical sizes. Iron-core inductors typically have inductances in the henrys ranges. Powdered iron-core and variable ferrite-core inductors have values in the millihenrys range. Air-core inductors are typically in the microhenrys range.

▶ The amount of back-emf induced in an inductor relates to the amount of inductance involved and the rate of current change. That is,

$$V_L = L \frac{di}{dt}$$

where L = inductance in henrys

$\dfrac{di}{dt}$ = the rate of current change in amperes per second

▶ Inductance value is calculated when the inductor's physical characteristics and the core's magnetic properties (permeability) are known. That is,

$$L = (12.57 \times 10^{-7}) \times \frac{(\mu_r N^2 A)}{\text{length}}$$

where 12.57×10^{-7} is the absolute permeability of air, μ_r is the relative permeability of the core, N is the number of turns, and A is the core's cross-sectional area in square meters.

▶ Total inductance of inductors in series (with no mutual coupling) is calculated by adding the individual inductance values. That is, $L_T = L_1 + L_2 \ldots + L_n$ This method is similar to finding total resistance of series resistors.

▶ Total inductance of inductors in parallel (with no mutual coupling) is calculated in a similar fashion to that used when calculating equivalent resistance of parallel resistors. That is, either the reciprocal formula or the product-over-the-sum formula may be used. These methods are

$$\frac{1}{L_T} = \frac{1}{L_1} + \frac{1}{L_2} \cdots + \frac{1}{L_n}$$

$$\text{and } L_T = \frac{L_1 \times L_2}{L_1 + L_2}$$

▶ Energy is stored in the inductor magnetic field when there is current flowing through the inductor. The amount of energy stored is calculated in joules. That is,

$$E = \frac{LI^2}{2}$$

where E is energy in joules, L is inductance in henrys, and I is current in amperes.

▶ Current through an inductor does not instantly rise to its maximum value. Neither does it instantly fall from the value it is at to the minimum value where it decays. Current takes five time constants to change from the level it is at to a new stable level. A time constant (τ) is equal to the inductance in henrys, divided by the R in ohms, or $\tau = L/R$. (NOTE: The L/R time for the increasing current condition may be different than the L/R time for the decaying current condition depending on the circuitry involved).

▶ The universal time-constant chart is a graph of the exponential growth and decay of many things in nature. Looking at the curve of increasing values, after one time constant (τ), the quantity reaches 63.2% of maximum value. After two time constants, the quantity has reached 86.5% of maximum value and so on. After five time constants, the quantity has virtually reached 100% of the maximum value to be achieved. Looking at the decay curve, after one time constant, the value has fallen from 100% of maximum to 36.8% of maximum (or a decay of 63.2%). After two time constants, the quantity has fallen to 13.5% of the original maximum value, and so forth. After five time constants, the quantity has fallen to the minimum value involved.

▶ Natural logarithms and epsilon can be used to mathematically determine instantaneous values of dc transient circuit quantities, such as instantaneous current (i) in dc RL circuits.

For increasing current: $i = \dfrac{V_S}{R} \times (1 - \epsilon^{-Rt/L})$

For decaying current: $i = \dfrac{V_S}{R} \times \epsilon^{-Rt/L}$

▶ When an inductor's magnetic field rapidly collapses, a high voltage is induced. Therefore, in dc circuits, precautions should be taken to prevent abrupt interruptions of current through large inductances. Also, circuit design should minimize the effects of such changes.

▶ Typical problems with inductors are opens or shorts. The ohmmeter is useful to check for these problems, provided the technician has knowledge of normal resistance values for the component being tested. Shorted turns are more difficult to find with ohmmeter checks.

FORMULAS AND CALCULATOR SEQUENCES SUMMARY

FORMULA 14–1 $V_{ind} = N$ (number of turns) $\times \dfrac{d\phi(\text{Wb})}{dt(\text{s})}$

number of turns, $\boxed{\times}$, $\boxed{(}$, number of webers cut, $\boxed{\div}$, amount of time, $\boxed{)}$, $\boxed{=}$

FORMULA 14–2 $V_L = L\dfrac{di}{dt}$

inductance value, $\boxed{\times}$, $\boxed{(}$, change in current, $\boxed{\div}$, amount of time, $\boxed{)}$, $\boxed{=}$

FORMULA 14–3 $L = \dfrac{V_L}{\dfrac{di}{dt}}$

V_L value, $\boxed{\div}$, $\boxed{(}$, change in current, $\boxed{\div}$, amount of time, $\boxed{)}$, $\boxed{=}$

FORMULA 14–4 $L = 12.57 \times 10^{-7} \times \dfrac{\mu_r N^2 A}{l}$

12.57, $\boxed{\text{EE}}$ or $\boxed{\text{EXP}}$, $\boxed{+/-}$, 7, $\boxed{\times}$, $\boxed{(}$, μ_r value, $\boxed{\times}$, number of turns, $\boxed{x^2}$, $\boxed{\times}$, area of coil in square meters, $\boxed{)}$, $\boxed{\div}$, length of coil in meters, $\boxed{=}$

FORMULA 14–5 $L = 12.57 \times 10^{-7} \times \dfrac{N^2 A}{l}$

12.57, $\boxed{\text{EE}}$ or $\boxed{\text{EXP}}$, $\boxed{+/-}$, 7, $\boxed{\times}$, $\boxed{(}$, number of turns, $\boxed{x^2}$, $\boxed{\times}$, area of coil in square meters, $\boxed{)}$, $\boxed{\div}$, length of coil in meters, $\boxed{=}$

FORMULA 14–6 $L_T = L_1 + L_2 \ldots + L_n$

L_1 value, $\boxed{+}$, L_2 value, $\ldots$, $\boxed{=}$

FORMULA 14–7 $\dfrac{1}{L_T} = \dfrac{1}{L_1} + \dfrac{1}{L_2} + \dfrac{1}{L_3} \ldots + \dfrac{1}{L_n}$

L_1 value, $\boxed{1/x}$, $\boxed{+}$, L_2 value, $\boxed{1/x}$, $\boxed{+}$, L_3 value, $\boxed{1/x}$, $\ldots$, $\boxed{=}$

FORMULA 14–8 $L_T = \dfrac{1}{\dfrac{1}{L_1} + \dfrac{1}{L_2} + \dfrac{1}{L_3} \ldots + \dfrac{1}{L_n}}$

L_1 value, $\boxed{1/x}$, $\boxed{+}$, L_2 value, $\boxed{1/x}$, $\boxed{+}$, L_3 value, $\boxed{1/x}$, $\boxed{=}$, $\boxed{1/x}$

FORMULA 14–9 $L_T = \dfrac{L_1 \times L_2}{L_1 + L_2}$

L_1 value, $\boxed{\times}$, L_2 value, $\boxed{\div}$, $\boxed{(}$, L_1 value, $\boxed{+}$, L_2 value, $\boxed{)}$, $\boxed{=}$

FORMULA 14–10   Energy $= 0.5\, LI^2$

0.5, $\boxed{\times}$, $\boxed{(}$, L value, $\boxed{\times}$, I value, $\boxed{x^2}$, $\boxed{)}$, $\boxed{=}$

FORMULA 14–11 $\tau = \dfrac{L}{R}$

> inductance value (henries), $\boxed{\div}$, resistance value (ohms), $\boxed{=}$

FORMULA 14–12 $i = \dfrac{V_S}{R} \times (1 - \epsilon^{Rt/L})$ (NOTE: This is formula for increasing current)

> R value, $\boxed{\times}$, time value, $\boxed{\div}$, inductance value, $\boxed{=}$, $\boxed{+/-}$, $\boxed{\text{2nd}}$,
> $\boxed{e^x}$, $\boxed{+/-}$, $\boxed{+}$, 1, $\boxed{=}$, $\boxed{\times}$, V_S value, $\boxed{\div}$, R value, $\boxed{=}$

FORMULA 14–13 $i = \dfrac{V_S}{R} \times \epsilon^{-Rt/L}$ (NOTE: This is formula for decreasing current)

> R value, $\boxed{\times}$, time value, $\boxed{\div}$, inductance value, $\boxed{=}$, $\boxed{+/-}$, $\boxed{\text{2nd}}$,
> $\boxed{e^x}$, $\boxed{\times}$, V_s value, $\boxed{\div}$, R value, $\boxed{=}$

REVIEW QUESTIONS

1. Define inductance.
2. In what form is electrical energy stored by an inductor?
3. Define the four important physical elements that affect the amount of inductance.
4. Explain Lenz's law.
5. Explain Faraday's law.
6. State the formula for one time constant. Define terms.
7. If the iron core is removed from an inductor so that it becomes an air-core inductor, will the coil's inductance increase, decrease, or remain the same?
8. When the current through an inductor has reached its steady-state condition and the current through the coil remains the same, is there an induced back-emf?
9. When additional inductors are connected in series with a circuit's existing inductors, will the total inductance of the circuit increase, decrease, or remain the same? (Assume no mutual coupling exists between inductors.)
10. When additional inductors are connected in parallel with a circuit's existing inductors,

will the total inductance of the circuit increase, decrease, or remain the same? (Assume no mutual coupling exists between inductors.)
11. When the number of turns on an inductor that has an inductance of 100 mH doubles, while keeping its length, cross-sectional area, and core material constant, what is its new inductance value? (Express your answer in three ways: μH, mH, and H.)
12. Current through an inductor is changing at a rate of 100 mA/s and the induced voltage is 30 mV. What is the value of inductance?
13. What is the new inductance for a 100-μH air-core coil whose core is replaced with one having a relative permeability of 500?
14. How much energy is stored in the magnetic field of a 15-H inductor that is carrying current of 2 A?
15. A 100 mH, a 0.2 H, and a 1,000-μH inductor are connected in series. Assuming no mutual coupling exists between them, what is the circuit's total inductance? (Express your answer in mH.)
16. A 10-H and a 15-H inductor are connected in parallel. Assuming no mutual coupling

exists between them, what is the resultant total inductance?

17. A series RL circuit consists of a 250-mH inductor and a resistor having a resistance of 100 Ω. What is the value of one time constant? How long will it take for current to change completely from one level to another?

18. Refer to the Universal Time Constant (UTC) chart, Figure 14–14, and determine what percentage of applied voltage appears across the resistor in a series RL circuit af-

ter four time constants. What percentage of the applied voltage is across the inductor at this same time?

19. A 500-turn inductor has a total inductance of 0.4 H. What is the inductance when connections are made at one end of the coil and at a tap point connection at the 250th turn?

20. How long will it take the voltage across the resistance in a series RL circuit consisting of 5 H of inductance and 10 Ω of resistance to achieve 86.5% of V applied? (Assume the inductor has negligible resistance.)

CHALLENGE QUESTIONS

Perform the calculations for questions 1 and 2 using your calculator and the $\boxed{e^x}$ function, as appropriate.

1. Refer to Figure 14–19. What is the voltage across the inductor after the switch has been closed for 0.6 microseconds?

2. Refer to Figure 14–20. How much time will it take for the voltage across the resistor to reach 38 volts? (Round off, as appropriate.)

FIGURE 14–19

FIGURE 14–20

TEST YOURSELF

1. If a 1,000-turn inductor has an inductance of 4 H, how many additional turns does it take to increase the inductance to 6 H?

2. Draw a graph of the V_R and V_L voltages with respect to V applied for an inductor circuit consisting of a 10-H inductor, a series 10-Ω resistance, and a source voltage of 20 V. Identify the specific voltages across L and R at the 1τ, 2τ, 3τ, and 4τ points on your graph.

3. Explain why a 1,000-turn coil, consisting of wire that has a varnish insulation with a voltage breakdown rating of only 100 volts, can safely handle an applied voltage across the coil at 2,000 volts.

PERFORMANCE PROJECTS CORRELATION CHART

Suggested performance projects in the Laboratory Manual that correlate with topics in this chapter are:

CHAPTER TOPIC	PERFORMANCE PROJECT	PROJECT NUMBER
Self-Inductance	L Opposing Change of Current	40
L/R Time Constant		
Inductors in Series and Parallel	Total Inductance in Series and Parallel	41

After completing the above projects, perform "Summary" checkout at the end of the performance projects section in the Laboratory Manual.

Inductive Reactance
in AC

15

KEY TERMS

Inductive reactance

Leading in phase

Phase angle

Q (figure of merit)

OUTLINE

V and I relationships in a purely resistive ac circuit

V and I relationships in a purely inductive ac circuit

Concept of inductive reactance

Relationship of X_L to inductance value

Relationship of X_L to frequency of AC

Methods to calculate X_L

Inductive reactances in series and parallel

Final comments about inductors and inductive reactance

Inductor Q factor

CHAPTER PREVIEW

In the previous chapter, we learned that an inductor opposes a current change but provides very little opposition to a steady (dc) current. The dc winding resistance is the only opposition to dc current; whereas the opposition to the current change is because an inductor reacts to changing current by producing a back-emf. You saw that energy is stored in the inductor's magnetic field during current buildup and returned to the circuit during current decay. If the inductor were "perfect" (without dc wire resistance or other losses), it would return all the stored energy to the circuit, dissipate no power, and have no effect on dc circuit parameters. Practical inductors have some resistance and dissipate some power, even in dc circuits. However, these losses are not due to the property of inductance.

Inductors react to ac current, which is constantly varying. Also, they produce an opposition to ac current known as inductive reactance, the reactance that will be studied in this chapter.

Important concepts discussed earlier were the inductance value and rate of current change relationships to the induced voltage appearing across the coil. In this chapter, we will relate these concepts to the ac circuit and explore what happens when inductance is present in a sine-wave ac environment. We will show how these characteristics translate into how inductance influences parameters in ac circuits.

OBJECTIVES

After studying this chapter, you should be able to:

- Illustrate V-I relationships for a purely resistive ac circuit
- Illustrate V-I relationships for a purely inductive ac circuit
- Explain the concept of **inductive reactance**
- Write and explain the formula for inductive reactance
- Use Ohm's law to solve for X_L
- Use the X_L formula to solve for inductive reactance at different frequencies and with various inductance values
- Use the X_L formula to solve for unknown L or f values
- Determine X_L, I_L, and V_L values for series and parallel-connected inductances
- Use the SIMPLER troubleshooting sequence to solve the Chapter Challenge problem

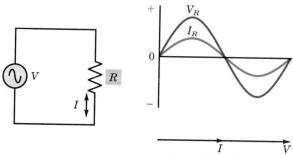

**FIGURE 15–1   *V* and *I* relationships in a
purely resistive ac circuit**

V AND *I* RELATIONSHIPS IN A PURELY RESISTIVE AC CIRCUIT

For a quick review, observe Figure 15–1. Note in a purely resistive circuit, the voltage across a resistor is in-phase with the current through it. Both the voltage and current pass through the zero points and reach the maximum points of the same polarity at the same time.

V AND *I* RELATIONSHIPS IN A PURELY INDUCTIVE AC CIRCUIT

For purposes of explanation, let's assume the inductor is an ideal inductor, having only inductance and no resistance. What phase relationship will the current through the coil and the voltage across the coil have with respect to each other? Examine Figure 15–2 as you study the following explanations and comments.

1. In the circuit shown, the inductor is connected across the source output terminals. Therefore, recalling the rule about voltage in parallel circuits, you know the impressed voltage across the coil and the source applied voltage are the same.

2. In the circuit shown, there is only one current path. Therefore, recalling the rule about current in series circuits, you know the current through all circuit parts is the same. That is, the circuit current (I) and the current through the inductor (I_L) are the same.

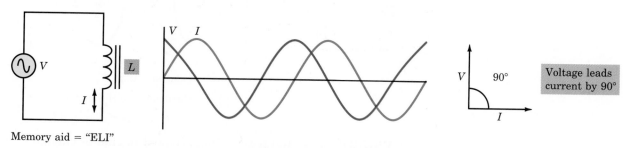

Memory aid = "ELI"

FIGURE 15–2   *V* and *I* relationships in a purely inductive ac circuit

3. Recall the induced voltage (cemf) across the inductor is maximum when the rate of current change is maximum and minimum when the rate of current change is minimum, Figure 15–3.

4. Also, remember for a sine wave the maximum rate of change occurs as the waveform crosses the zero points. The minimum rate of change occurs at the maximum positive and negative peaks of the sine wave, Figure 15–3.

5. The implication is that maximum induced V occurs across the inductor when the current sine wave is either increasing from (or through) zero in the positive direction or decreasing from (or through) zero in the negative direction. See Figure 15–3 again.

6. The *amplitude* of the induced voltage is determined by the rate of current change and the value of L.

7. Two factors determine the *polarity* of the induced voltage:
   a. the direction of the current flow through the coil, and
   b. whether the current is trying to increase or decrease.

8. Since the induced voltage opposes any current change, when the current tries to increase, the cemf will oppose it. When the current tries to decrease, the cemf will again oppose it (Lenz's law).

9. Figure 15–2 and 15–3 illustrate the results of these actions and reactions. In essence, because the induced back-emf is maximum when the rate of current change is maximum and minimum when the rate of current change is minimum, the current lags behind the impressed voltage. The time-lag, or delay of current behind voltage, has obviously been caused by the inductance's opposition to changes in current flow. The current and voltage are out of phase. The 90° **phase angle** is due to the fact that at the moment when the current increases or decreases, the inductor instantly develops the cemf, which opposes the change in current.

10. Since the V and I are one-fourth cycle apart, they are 90° out of phase with each other. A complete cycle represents 360° or 2π radi-

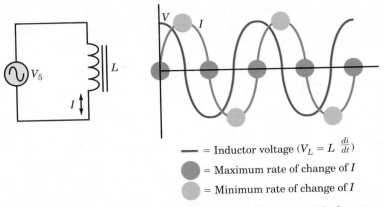

$\underline{\qquad}$ = Inductor voltage ($V_L = L\,\frac{di}{dt}$)

⬤ = Maximum rate of change of I

◯ = Minimum rate of change of I

FIGURE 15–3   Inductor voltage related to rate of change of current

ans, where a radian equals 57.3°, or the angle described by an arc on the circle's circumference that is equal in length to the radius.

11. Using our rule of thumb from a previous chapter, you can see the voltage is at maximum positive value prior to the current reaching maximum positive value. Hence, the voltage is **leading** the current by 90°. Incidentally, it is also accurate to say current is lagging voltage by 90°. Recall the reference point concept—John is taller than Bill, or Bill is shorter than John.

At this point we will introduce a helpful phrase that will serve as a memory aid and that will help throughout your studies in ac and in your career. The understanding of the complete expression will come only after you have studied capacitors and inductors. However, the phrase can also be helpful to you now.

Here's the phrase: **Eli the ice man.** It's a short saying, so you should not have trouble memorizing it, particularly after you understand the implications!

You have just learned in a sine-wave ac circuit environment, voltage across an inductor leads the current through the inductor by 90° (in a pure inductance). The word "Eli" helps you remember that voltage leads current since the E (emf or voltage) is before the I (current). Obviously the "L" indicates we are talking about an inductor. (You have probably already surmised that "ice" in this special phrase refers to current (I) leading voltage (E) in a capacitor (C), but we'll discuss this in a later chapter.)

PRACTICAL NOTES

Although the memory aid can be helpful to some, it is more important that you learn and understand the reasons for the electrical quantity relationships. As you are learning in this chapter, current lags voltage in an inductor due to the opposition an inductor shows to a change in current. You will be learning in an upcoming chapter that the reason current leads voltage across a capacitor is that a capacitor opposes a change in voltage. The details of why this is true will become apparent as you continue studying these subjects. We simply want to indicate that the memory aid should not replace your understanding of these matters.

CONCEPT OF INDUCTIVE REACTANCE

Since an inductor reacts to the constantly changing current in an ac circuit by producing an emf that opposes current changes, inductance displays opposition to ac current, Figure 15–4. Recall the inductor did not show opposition to dc current, except for the small resistance of the wire. This lack of opposition is because current must be changing to cause a changing flux that induces the back-emf we've been discussing. In dc the current is not changing, except for the temporary switch-on and switch-off conditions discussed earlier.

With ac applied, current is much less than with dc applied to same circuit. The additional opposition to current in the ac circuit is due to X_L of the inductor.

FIGURE 15–4 Concept of inductive reactance to ac

This special opposition an inductor displays to ac current is **inductive reactance**. The symbol "X" denotes reactance. Since it is inductive reactance, the correct abbreviation or symbol is $\mathbf{X_L}$. Because X_L is an opposition to current, the *unit of measure for inductive reactance (X_L) is the ohm!*

RELATIONSHIP OF X_L TO INDUCTANCE VALUE

We have already stated that the amount of opposition inductance shows to ac current is directly related to the amount of back-emf induced. For a given rate of current change, the amount of back-emf induced is directly related to the value of coil inductance (i.e., the greater the L, the greater the back-emf produced). Therefore, the higher the value of L, the higher will be the inductive reactance (X_L) value. In other words, if the L value doubles, the number of ohms of X_L doubles; if the L in the circuit decreases to half its original value, X_L decreases to half, and so on, Figure 15–5.

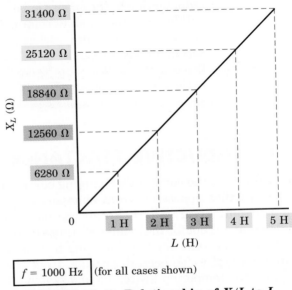

$f = 1000$ Hz (for all cases shown)

FIGURE 15–5 Relationship of X/L to L

EXAMPLE

If a purely inductive circuit has a 5-H inductor with a second 5-H inductor connected in series, what happens to the circuit current? (Assume all other circuit parameters remain the same.)

Answer: Current decreases to half the original value, since the total opposition to current (X_{L_T}) has doubled when inductance was doubled.

PRACTICE PROBLEMS I

1. For the circuit in Figure 15–6, what is the value of total X_L when two more equal value inductors are added in series with the existing inductor, thus making $L_T = 30$ H? (Assume all other circuit conditions remain the same.)

2. If the three 10-H inductors in question 1 are parallel-connected rather than series-connected, what is the total X_L?

FIGURE 15–6

RELATIONSHIP OF X_L TO FREQUENCY OF AC

Rate of Change Related to Frequency

It has been previously established that the amount of back-emf produced in an inductor relates to both the value of inductance and to the rate of current change. As you know, the faster the rate of change, the greater the induced emf produced. And the greater the back-emf, the higher the value of opposition or X_L.

Let's take a look at how changing the frequency of an ac signal affects this rate of change. Refer to Figure 15–7 as you study the following discussion.

1. Only one cycle is shown for the lower of the two frequency signals displayed. Two cycles of the higher frequency signal are shown.

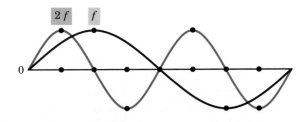

Rate of change = higher for higher frequency

FIGURE 15–7   Comparison of rate of change for two frequencies

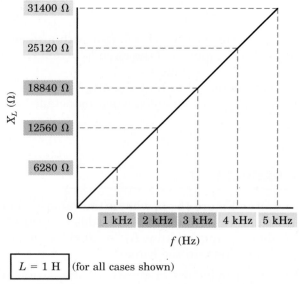

FIGURE 15–8   Relationship of X/L to f

2. The two signals have the same maximum (peak) amplitude. You can see the higher frequency signal must have a higher rate of change of current to reach its maximum amplitude level in half the time. Carrying this thought throughout the whole cycle, it is apparent that the higher frequency signal has twice the rate of change compared to the lower frequency signal.

3. Since the back-emf is directly related to the rate of change, when frequency doubles (which doubles the rate of change), the inductive reactance also doubles. When frequency decreases, rate of current change decreases and, consequently, the inductive reactance decreases proportionately, Figure 15–8.

Rate of Change Related to Angular Velocity

In an earlier chapter, a rotating vector to describe a sine wave was introduced. At this point, we want to indicate a relationship exists between frequency and the angular velocity of the quantity represented by a rotating vector.

Angular velocity is the speed that a vector rotates about its axis. The symbol representing angular velocity is the Greek letter omega, ω. Thus, the relationship of angular velocity to frequency is ω equals $2\pi f$ (radians per second). (Recall that 2π radians equal 360 degrees, or one complete cycle). We introduce angular velocity and its mathematical representation at this point to acquaint you with the expression "$2\pi f$", which is used in formulas dealing with ac quantities. You will see this expression as you continue in your studies and in your work as a technician.

IN-PROCESS LEARNING CHECK I

Fill in the blanks as appropriate.

1. When current increases through an inductor, the cemf (helps, hinders) _____ the current increase.

2. When current decreases through an inductor, the cemf (helps, hinders) _____ the current decrease.

3. In a pure inductor, the _____ leads the _____ by 90 degrees.

4. What memory aid helps to remember the relationship described in question 3? _____

5. The opposition that an inductor shows to ac is termed _____.

6. The opposition that an inductor shows to ac (increases, decreases, remains the same) _____ as inductance increases.

7. The opposition that an inductor shows to ac (increases, decreases, remains the same) _____ as frequency decreases.

8. X_L is (directly, inversely) _____ related to inductance value.

9. X_L is (directly, inversely) _____ related to frequency.

METHODS TO CALCULATE X_L

The X_L Formula

Two common methods are used to calculate the value of X_L. One method is to use Ohm's law. The other method is the inductive reactance formula,

Formula 15–1	$X_L = 2\pi fL$

where X_L = inductive reactance in ohms

2π = 6.28 (approximately)

f = frequency in hertz

L = inductance in henrys

 6.28, $\boxed{\times}$, frequency value, $\boxed{\times}$, inductance value, $\boxed{=}$

You can see from the formula and previous discussions that inductive reactance is directly related to both frequency and inductance. That is, when *either* frequency *or* inductance increases, inductive reactance increases. When *either* frequency *or* inductance decreases, X_L decreases. That is, when one factor doubles while the other one halves, the reactance stays the same, and so on. You get the idea!

The X_L formula can also be transposed to solve for L or f, if the other two factors are known.

Formula 15–2	$L = \dfrac{X_L}{2\pi f}$

 X_L value, $\boxed{\div}$, $\boxed{(}$, 6.28, $\boxed{\times}$, frequency value, $\boxed{)}$, $\boxed{=}$

That is, $L = \dfrac{X_L}{6.28\,f}$ henrys

and

Formula 15–3	$f = \dfrac{X_L}{2\pi L}$

X_L value, $\boxed{\div}$, $\boxed{(}$, 6.28, $\boxed{\times}$, inductance value, $\boxed{)}$, $\boxed{=}$

That is, $f = \dfrac{X_L}{6.28\,L}$

EXAMPLE

What is the inductive reactance of an iron-core choke having 5 H of inductance that is in a 60-Hz ac circuit?

$$X_L = 2\pi fL = 6.28 \times 60 \times 5 = 1{,}884\ \Omega$$

EXAMPLE

What is the inductance of an inductor that offers 2 kΩ of inductive reactance at a frequency of 1.5 MHz?

$$L = \frac{X_L}{2\pi f} = \frac{2 \times 10^3}{6.28 \times (1.5 \times 10^6)} = 212\ \mu\text{H}$$

EXAMPLE

What frequency of operation would cause a 30-μH inductor to produce 4 kΩ of inductive reactance?

$$f = \frac{X_L}{2\pi L} = \frac{4 \times 10^3}{6.28 \times (30 \times 10^{-6})} = 21.23\ \text{MHz}$$

PRACTICE PROBLEMS II

1. What is the X_L of a 10-mH inductor that is operating at a frequency of 100 kHz?

2. What is the inductance in a purely inductive circuit, if the frequency of the applied voltage is 400 Hz and the inductive reactance is 200 Ω?

3. What frequency causes a 100-mH inductor to exhibit 150 Ω of inductive reactance?

Using the Ohm's Law Approach

The ac current through an inductor is also calculated by using the Ohm's law expression: $I_L = V_L/X_L$. We can transpose this formula to find any one of the three factors, provided we know the other two factors. (We have previously done this using Ohm's law expressions.) This means that:

$$X_L = \frac{V_L}{I_L}$$

$$V_L = I_L X_L$$

$$I_L = \frac{V_L}{X_L}$$

PRACTICAL NOTES

When using Ohm's law, as indicated, it is common to use the (rms) or effective values of voltage and current in your calculations, unless otherwise indicated!

PRACTICE PROBLEMS III

Use Ohm's law and answer the following.

1. What is the inductive reactance of an inductor whose voltage measures 25 V and through which an ac current of 5 mA is passing?

2. What is the voltage across a 2-H inductor passing 3 mA of current? Assume the frequency of operation is 400 Hz. (Hint: Calculate X_L using the reactance formula first.)

3. What is the current through an inductor having 250 Ω of inductive reactance if the voltage across the inductor measures 10 V?

INDUCTIVE REACTANCES IN SERIES AND PARALLEL

You are aware that inductances in series are additive just like series resistances are additive. Likewise, you have already learned that inductive re-

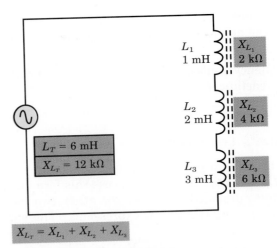

**FIGURE 15-9 Inductive reactances
in series**

actance, or X_L, is directly related to the amount of inductance. Therefore, the ohmic values of inductive reactances of series inductors add like the ohmic values of series resistances add. In other words, $X_{L_T} = X_{L_1} + X_{L_2} \cdots + X_{L_n}$, Figure 15-9.

Inductance values in parallel are also treated in the same manner that parallel resistance values are treated. That is, the reciprocal or product-over-the-sum approaches are used to find the total or equivalent value(s). Since inductive reactance is directly related to inductance value, total X_L is again correlated to the total inductance of parallel-connected inductors like it was to the total inductance of series-connected inductors, Figure 15-10.

Useful formulas to find the total inductive reactance of parallel-connected inductors are:

Formula 15-4	$\dfrac{1}{X_{L_T}} = \dfrac{1}{X_{L_1}} + \dfrac{1}{X_{L_2}} \cdots + \dfrac{1}{X_{L_n}}$

X_{L_1} value, $\boxed{1/x}$, $\boxed{+}$, X_{L_2} value, $\boxed{1/x}$, $\cdots$, $\boxed{=}$

FIGURE 15-10   Inductive reactances in parallel

or

Formula 15–5	$X_{L_T} = \dfrac{1}{\dfrac{1}{X_{L_1}} + \dfrac{1}{X_{L_2}}, \text{etc.}}$

 X_{L_1} value, $\boxed{1/x}$, $\boxed{+}$, X_{L_2} value, $\boxed{1/x}$, $\boxed{+}$, ... , $\boxed{=}$, $\boxed{1/x}$

PRACTICAL NOTES

   Once again, the reciprocal ($\boxed{1/x}$) function on your calculator is found to be very useful. You can solve for total inductive reactance of parallel-connected inductive reactances using the same techniques you have previously used to solve for total resistance and total inductance of parallel-connected inductances. (NOTE: Assume no coupling between inductances.)

For two inductors in parallel, you can use:

Formula 15–6	$X_{L_T} = \dfrac{X_{L_1} \times X_{L_2}}{X_{L_1} + X_{L_2}}$

 X_{L_1} value, $\boxed{\times}$, X_{L_2} value, $\boxed{\div}$, $\boxed{(}$, X_{L_1} value, $\boxed{+}$, X_{L_2} value, $\boxed{)}$, $\boxed{=}$

Where appropriate, you can use all the "shortcuts" you have previously learned about parallel circuits—two 25-ohm reactances in parallel yield a total equivalent reactance of 12.5 ohms, or half of either reactor's value, and so forth.

FINAL COMMENTS ABOUT INDUCTORS AND INDUCTIVE REACTANCE

Collecting what you now know about inductance and reactance, the following comments can be made:

1. The inductive reactance of an inductor limits or opposes the ac current flow through the inductor. That is, given an inductor in a circuit with a specified value of dc applied voltage, if that same inductor is put into a circuit with the same ac voltage (rms value), the current will be less in the ac circuit than in the dc circuit. This situation is due to the opposition of the inductive reactance to ac current, Figure 15–4.

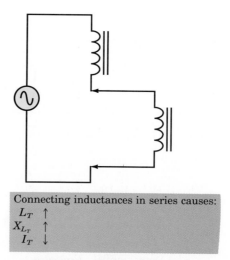

Connecting inductances in series causes:
L_T ↑
X_{L_T} ↑
I_T ↓

**FIGURE 15–11   Effects of
connecting inductances in series**

2. Connecting inductances in series causes total inductance to increase, total inductive reactance to increase, and current to decrease (assuming a constant V applied), Figure 15–11.

3. Series-connected inductors may act as a series, inductive ac voltage divider with the greatest voltage across the largest inductance and the smallest voltage across the smallest inductance, and so on. NOTE: In a series circuit containing *nothing but inductors,* the total voltage equals the sum of the individual voltages across the inductors. The current through the inductors is 90° out-of-phase with the voltage across each inductor. Each of the inductor's voltages is in-phase with the other inductors' voltages, thus are added to obtain the total voltage, Figure 15–12.

4. Connecting inductances in parallel causes total inductance to decrease, total inductive reactance to decrease, and current to increase (assuming a constant V applied), Figure 15–13.

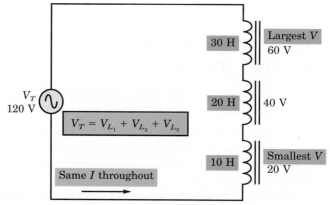

30 H   Largest V
        60 V

V_T
120 V   20 H   40 V

$V_T = V_{L_1} + V_{L_2} + V_{L_3}$

10 H   Smallest V
        20 V

Same I throughout

**FIGURE 15–12   Voltage division across
series-connected inductors**

Connecting inductances in parallel causes:

L_T ↓
X_{L_T} ↓
I_T ↑

**FIGURE 15–13   Effects of
connecting inductors in parallel**

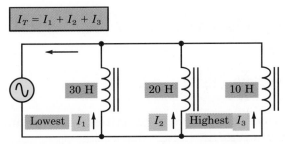

$I_T = I_1 + I_2 + I_3$

30 H   20 H   10 H

Lowest I_1 ↑ I_2 ↑   Highest I_3 ↑

**FIGURE 15–14   Current division through
parallel-connected inductors**

5. Parallel-connected inductors can act as a parallel inductive ac current divider with the highest current through the lowest inductance and the lowest current through the highest inductance and so on. NOTE: In a parallel circuit containing *nothing but inductors,* the total current equals the summation of the individual inductor branch currents. Again, there is a 90° phase difference between the current through each inductor and the voltage across that inductor. But since all branch currents are lagging the circuit voltage by 90°, the branch currents are in-phase with each other, thus are added to find total current, Figure 15–14.

INDUCTOR Q FACTOR

We have already implied that a perfect inductor would have zero resistance, and that the only opposition this inductor would show to ac current flow would be the inductive reactance. It has also been stated that every coil has some resistance in the wire making up the coil. In many cases, this resistance is so small that it has negligible effect on the circuit operation. In other cases, this is not true. In fact, under certain conditions it may have considerable effect. For this reason, inductors sometimes have a **Q** rating, which is a ratio describing the energy the inductor stores in its field compared to the energy it dissipates due to I^2R losses.

In essence, the ratio of the X_L the inductor shows (at a given frequency) to the R it has, depicts this ratio. That is:

| Formula 15–7 | $Q = \dfrac{X_L}{R}$ |

X_L value, $\boxed{\div}$, R value, $\boxed{=}$

EXAMPLE

The Q of a coil having 100-mH inductance and 50-Ω resistance that is operating at a frequency of 5 kHz would be:

$$Q = \frac{X_L}{R}$$

$$X_L = 6.28 \times 5 \times 10^3 \times 100 \times 10^{-3} = 3{,}140 \ \Omega$$

$$R = 50 \ \Omega$$

$$Q = \frac{X_L}{R} = \frac{3{,}140}{50} = 62.8$$

(Q has no units since it is a ratio of quantities with like units.)

The preceding discussion indicates that the lower the R that a given inductor (L) has, the higher is its **figure of merit** or **Q**. (Also, the higher the frequency of operation, the greater the X_L and the higher the Q for a given inductor.)

It is generally desirable to use high Q coils, particularly at radio frequencies. A coil is considered to be high Q if its Q value is greater than approximately 20–25. Of course, this is an arbitrary value. It does, however, give some indication of the inductance-to-resistance ratio of the inductor. Therefore, the value is the ratio of energy stored to energy dissipated or the X_L/R ratio we have been examining.

SUMMARY

▶ Inductive reactance is the opposition that inductance displays to ac (or pulsating) current.

▶ Inductive reactance is associated with the amount of back-emf generated by the inductor, which is associated with the value of inductance (L) and the rate of change of current.

▶ Inductive reactance is measured in ohms. The symbol for inductive reactance is X_L. X_L is calculated by the formula $X_L = 2\pi fL$, where

X_L is in ohms, f is in hertz, and L is in henrys. The 2π is the constant equal to two times pi or approximately 6.28.

▶ Voltage leads current by 90° in a purely inductive circuit. (Voltage and current are in-phase for a purely resistive circuit).

▶ The first part of the memory-aid phrase helps you remember the phase relationship of voltage and current relative to inductors. The phrase is *Eli the ice man*. The word *"Eli"* indicates that E comes before I in an inductor (L).

▶ Inductive reactance is *directly proportional* to both inductance (L) *and* to frequency (f).

▶ Quality (or the figure of merit) of an inductor is sometimes described as its Q. This factor equates the ratio of the inductor's X_L to its R. Thus, $Q = X_L/R$.

▶ Inductive reactances in series add just as resistances in series add. That is, $X_{L_T} = X_{L_1} + X_{L_2} \ldots + X_{L_n}$.

▶ Total (equivalent) inductive reactance of inductances in parallel is computed using similar methods to those used to find total or equivalent resistance of resistors in parallel.

That is, the reciprocal formula or the product-over-the-sum formula is used.

The reciprocal formula(s) include:

$$\frac{1}{X_{L_T}} = \frac{1}{X_{L_1}} + \frac{1}{X_{L_2}} + \ldots \cdot \frac{1}{X_{L_n}} \text{ or}$$

$$X_{L_T} = \frac{1}{\dfrac{1}{X_{L_1}} + \dfrac{1}{X_{L_2}}}$$

The product-over-the-sum formula is:

$$X_{L_T} = \frac{X_{L_1} \times X_{L_2}}{X_{L_1} + X_{L_2}}$$

 FORMULAS AND CALCULATOR SEQUENCES SUMMARY

FORMULA 15–1 $X_L = 2\pi f L$

6.28, $\boxed{\times}$, frequency value, $\boxed{\times}$, inductance value, $\boxed{=}$

FORMULA 15–2 $L = \dfrac{X_L}{2\pi f}$

X_L value, $\boxed{\div}$, $\boxed{(}$, 6.28, $\boxed{\times}$, frequency value, $\boxed{)}$, $\boxed{=}$

FORMULA 15–3 $f = \dfrac{X_L}{2\pi L}$

X_L value, $\boxed{\div}$, $\boxed{(}$, 6.28, $\boxed{\times}$, inductance value, $\boxed{)}$, $\boxed{=}$

FORMULA 15–4 $\dfrac{1}{X_{L_T}} = \dfrac{1}{X_{L_1}} + \dfrac{1}{X_{L_2}} \ldots + \dfrac{1}{X_{L_n}}$

X_{L_1} value, $\boxed{1/x}$, $\boxed{+}$, X_{L_2} value, $\boxed{1/x}$, $\ldots$, $\boxed{=}$

FORMULA 15–5 $X_{L_T} = \dfrac{1}{\dfrac{1}{X_{L_1}} + \dfrac{1}{X_{L_2}}}$, etc.

X_{L_1} value, $\boxed{1/x}$, $\boxed{+}$, X_{L_2} value, $\boxed{1/x}$, $\boxed{+}$, $\ldots$, $\boxed{=}$, $\boxed{1/x}$

FORMULA 15–6 $X_{L_T} = \dfrac{X_{L_1} \times X_{L_2}}{X_{L_1} + X_{L_2}}$

X_{L_1} value, $\boxed{\times}$, X_{L_2} value, $\boxed{\div}$, $\boxed{(}$, X_{L_1} value, $\boxed{+}$, X_{L_2} value, $\boxed{)}$, $\boxed{=}$

FORMULA 15–7 $Q = \dfrac{X_L}{R}$

X_L value, $\boxed{\div}$, R value, $\boxed{=}$

REVIEW QUESTIONS

1. What is the inductance of an inductor that exhibits 376.8 Ω reactance at a frequency of 100 Hz? What is inductive reactance at a frequency of 450 Hz?

2. What is the X_L of a 5-H inductor at 120 Hz? What is the value of X_L at a frequency of 600 Hz?

3. If circuit conditions are changed so that frequency triples and the amount of inductance doubles, how many times greater will be the new X_L compared to the original value?

4. Assume circuit conditions change for an inductive circuit so that frequency is halved and inductance is tripled. What is the relationship of the new X_L to the original value?

5. In a circuit containing inductance, when frequency decreases, will circuit current increase, decrease, or remain the same?

6. When the frequency of applied voltage for a given inductive circuit doubles and the circuit voltage increases by three times, will the circuit current increase, decrease, or remain the same?

7. When the frequency applied to a given inductor decreases, will the Q of the coil change? If so, in what way?

8. Does the R of a given inductor change with frequency?

9. What is the total inductance of the circuit in Figure 15–15? When the voltage applied is 100 V and the current through L_1 is 100 mA,

what is the frequency of the applied voltage? (Use one decimal place in your calculations.)

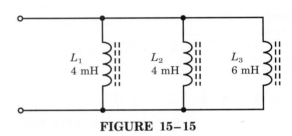

FIGURE 15–15

10. Find L_T for the circuit in Figure 15–16. At approximately what frequency does X_{L_T} equal 5,000 ohms?

FIGURE 15–16

11. Find the Q of the coil in the circuit shown in Figure 15–17. When the frequency triples and the R of the inductor doubles, does Q increase, decrease, or remain the same? If the applied voltage were dc instead of ac, would Q change? Explain.

FIGURE 15–17

12. Draw a waveform showing two cycles that illustrates the circuit V and I for the circuit in Figure 15–18. Appropriately label the peak values of V and I, assuming the source voltage is shown in rms value.

FIGURE 15–18

13. Find V_T, V_{L_1}, and I_T for the circuit in Figure 15–19.

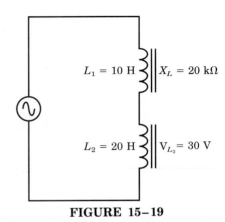

FIGURE 15–19

14. What is the branch current through L_2 for the circuit in Figure 15–20?

15. What is the value of the source voltage in the circuit of Figure 15–20?

FIGURE 15–20

CHALLENGE QUESTIONS

1. In your own words, explain why inductive reactance is directly related to the value of inductance.

2. In your own words, explain why inductive reactance is directly related to the value of frequency.

3. Explain why you think, in most cases, it is best to have an inductor with a high Q factor.

TEST YOURSELF

1. Determine the total inductance of the circuit in Figure 15–21.

FIGURE 15–21

2. Refer again to Figure 15–21 and answer the following with "I" for increase, "D" for decrease, and "RTS" for remain the same.
   a. If frequency doubles and V applied stays the same,
      L_T will _____;
      X_{L_T} will _____;
      I_T will _____;
      V_{L_4} will _____; and
      I_1 will _____.
   b. If L_2 changes to a 1-H inductor,
      L_T will _____;
      X_{L_1} will _____;
      I_T will _____;
      V_{L_1} will _____; and
      f will _____.

PERFORMANCE PROJECTS CORRELATION CHART

Suggested performance projects in the Laboratory Manual that correlate with topics in this chapter are:

CHAPTER TOPIC	PERFORMANCE PROJECT	PROJECT NUMBER
Concept of Inductive Reactance	Induced Voltage	42
Relationship of X_L to L	Relationship of X_L to L and f	43
Relationship of X_L to f		
Methods to Calculate X_L	The X_L Formula	44

After completing the above projects, perform "Summary" checkout at the end of the performance projects section in the Laboratory Manual.

Following the SIMPLER troubleshooting sequence, find the problem with this circuit. As you follow the sequence, record your responses to each step on a separate sheet of paper.

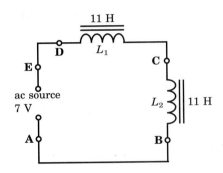

CHALLENGE CIRCUIT 7

SYMPTOMS Gather, verify, and analyze symptom information. (Look at the "Starting Point Information.")

IDENTIFY Identify initial suspect area for the location of the trouble.

MAKE Make a decision about "What type of test to make" and "Where to make it." (To simulate making a decision about each test you want to make, select the desired test from the "TEST" column listing below.)

PERFORM Perform the test. (Look up the result of the test in Appendix C. Use the number in the "RESULTS" column to find the test result in the Appendix.)

LOCATE Locate and define a new "narrower" area in which to continue troubleshooting.

EXAMINE Examine available information and again determine "what type of test" and "where."

REPEAT Repeat the preceding analysis and testing steps until the trouble is found. What would you do to restore this circuit to normal operation? When you have solved the problem, compare your results with those shown on the following pages.

STARTING POINT INFORMATION

1. Circuit diagram
2. Each inductor is rated as 11 H, 400 Ω dc resistance
3. The ac voltage measured across L_1 is significantly lower than the voltage measured across L_2

TEST   Results in Appendix C

TEST	Results in Appendix C
V_{A-B}	(66)
V_{B-C}	(31)
V_{C-D}	(4)
V_{D-E}	(82)
V_{E-A}	(43)
R_{B-C}	(17)
R_{C-D}	(114)

CIRCUIT

7

Symptoms The series circuit inductors are rated at the same value, and in a series circuit the current through both inductors should be the same. However, there is a significant difference in the $I \times X_L$ drops across the two inductors. This implies one inductor has a problem.

Identify initial suspect area: Both inductors until further checks are made.

Make test decision: It is less likely that an inductor would increase rather than decrease in inductance (due to shorted turns or change in magnetic path). So let's start by examining the inductor with the lowest voltage drop (L_1). Our assumption is L_2 has remained at the rated value of L, while L_1 may have decreased. This would fit the symptom information regarding L_1's voltage drop being low.

Perform 1st Test: Check the dc resistance of L_1 versus its rated value to see if there might be a hint of shorted turns. (NOTE: Source is removed from the circuit.) R is 297 Ω and the rated value is 400 Ω, indicating the possibility of some shorted turns.

Locate new suspect area: L_1 is now a strong suspect area. But we still need to check L_2's parameters to compare and verify our suspicions about L_1.

Examine available data.

Repeat analysis and testing: A resistance check of L_2 might be necessary.

CHALLENGE CIRCUIT 7

Check $R_{\text{B-C}}$. $R_{\text{B-C}}$ is 401 Ω, which agrees with the rating.

Replace L_1 with the appropriately rated inductor. Then check the voltage across L_1. $V_{\text{C-D}}$ is now close to 3.5 V, which is half of the source voltage. This is normal for this circuit with equal inductor values.

Check the voltage across L_2 to verify that it is approximately equal to the voltage across L_1.

When checked, the result is also about 3.5 V. The trouble is L_1 has shorted turns. Replacing L_1 restores the circuit to normal.

1st Test

2nd Test

3rd Test

4th Test

RL Circuits in AC

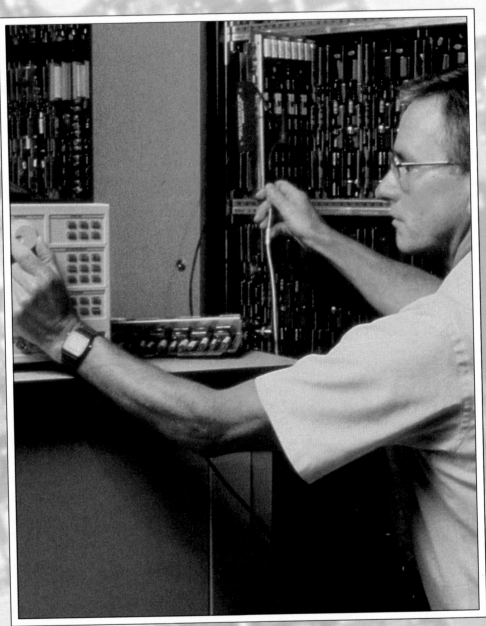

Courtesy of Hewlett-Packard Company

16

KEY TERMS

Cosine function

Impedance

Phase angle (θ)

Pythagorean theorem

Scalar

Sine function

Tangent function

Vector

OUTLINE

Review of simple R and L circuits

Using vectors to describe and determine magnitude and direction

Introduction to basic AC circuit analysis techniques

Fundamental analysis of series RL circuits

Fundamental analysis of parallel RL circuits

Examples of practical applications for inductors and RL circuits

CHAPTER PREVIEW

In the preceding chapters, you have studied the circuit characteristics of both dc and ac circuits containing only inductance. You have learned basic concepts related to inductors, such as inherent opposition to current changes, the self-induced back-emf, phase relationships of V and I, and other characteristics of inductance.

The purpose of this chapter is to introduce you to basic facts and circuit analysis techniques used to examine ac circuits containing both resistance and inductance (RL circuits). Impedance, a combination of resistance and reactance, will be discussed and analyzed. Further considerations about **phase angle** will be given. Primary methods of using **vectors** and other simple math techniques to analyze RL circuit parameters will also be introduced. Later in the text, you will study vector analysis techniques that involve polar and rectangular coordinate systems, and methods of expressing electrical quantities using these systems. For this chapter, however, we simply want to introduce you to basic concepts related to analyzing RL circuits in ac.

The concepts discussed in this chapter are important for you to learn, because they provide valuable knowledge for all your future ac circuit analyses endeavors. This chapter begins a series of chapters in which you will be performing calculator operations such as square roots, squares, trigonometric functions, and so on. We want to remind you that different models of scientific calculators may use slightly different keystrokes for certain operations. As you get involved in this chapter, study and/or refer back to the "Special Notations Regarding Scientific Calculators" chart, as required.

OBJECTIVES

After studying this chapter, you should be able to:

• Use **vectors** to determine magnitude and direction

• Determine circuit impedance using the **Pythagorean theorem**

• Determine V_T and I_T using the Pythagorean theorem

• Determine ac circuit parameters using trigonometry

• Calculate ac electrical parameters for series RL circuits

• Calculate ac electrical parameters for parallel RL circuits

• List at least three practical applications of inductive circuits

• Use the SIMPLER troubleshooting sequence to solve the Chapter Challenge problem

SPECIAL NOTATIONS REGARDING SCIENTIFIC CALCULATORS

There are numerous brands and types of scientific calculators on the market. Functions such as squaring a number, finding square roots, finding the arc tangent, finding antilogarithms, etc. may differ slightly from calculator to calculator. For the types of calculations we perform throughout this text, we want to mention some common variations you may see or use. If you are using a different type scientific calculator than the one which was used in creating calculator sequences in our text ("CALCULATOR #1") . . . the following equivalent keystrokes chart may be helpful to you.

TASK DESCRIPTION	CALCULATOR #1 KEYSTROKE(S) TI 36X SOLAR	CALCULATOR #2 KEYSTROKE(S) HP 20S	CALCULATOR #3 KEYSTROKE(S) (OTHERS)
Finding the reciprocal	1/x	1/x	2nd, 1/x
Squaring a number	x^2	↰, x^2	2nd, x^2
Taking a square root	$\sqrt{x}$	$\sqrt{x}$	$\sqrt{x}$
Finding the arctan	2nd, tan	↰, ATAN	INV, tan
Finding the antilog of a natural logarithm (ϵ)	2nd, e^x	e^x	e^x
Converting from polar to rectangular form: Finding the x-axis and y-axis (j) values:	phasor value, X ⋛ Y, angle value, 2nd, (P ▷ R) (Displays x-axis value) Follow above sequence with x ⋛ y key again (Displays y-axis (j) value)	phasor value, INPUT, angle value, ↗,→R (Displays y-axis (j) value) Follow above sequence with ↰, SWAP (Displays x-axis value)	phasor value, P-R key, angle value, = key (Displays x-axis value) Follow above sequence with X-Y key (Displays y-axis (j) value)
Converting from rectangular to polar form: Finding the phasor value and the angle	x-axis value, x ⋛ y, y-axis (j) value, "b", 3rd, (R ▷ P) Displays the phasor value Press the x ⋛ y key again (Displays the angle value)	x-axis value, INPUT, y-axis (j) value, ↰, →P, (Displays angle value) Follow above sequence with ↰, SWAP (Displays the phasor value)	x-axis value, R-P key, y-axis (j) value, =, (Displays phasor value) Follow the above sequence with X-Y key, (Displays the angle value)

REVIEW OF SIMPLE *R* AND *L* CIRCUITS

Simple *R* Circuit

The ac current through a resistor and the ac voltage across a resistor are in-phase, as you have already learned. You know this may be shown by a waveform display or phasor diagram, Figure 16–1.

In the purely resistive *series* ac circuit, Figure 16–2a, you simply add the individual series resistances to find total circuit opposition to current. Or you add voltage drops across resistors to find total circuit voltage, since all voltage drops are in-phase with each other.

In purely resistive *parallel* circuits, Figure 16–2b, total resistance is found by using the product-over-the-sum or reciprocal methods. The total

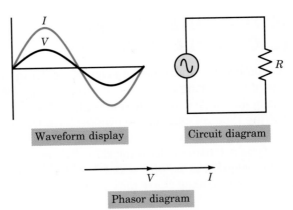

FIGURE 16−1   *V* and *I* in a simple *R* circuit

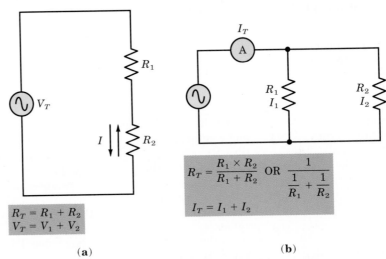

FIGURE 16−2   **Purely resistive circuits**

circuit current is found by adding the branch currents, since all branch currents are in-phase with each other.

Simple *L* Circuit

In a purely inductive ac circuit, you know current through the inductor and voltage across the inductor are 90° out-of-phase with each other. Again, this is illustrated by a waveform display, Figure 16−3a or a phasor diagram, Figure 16−3b.

You know that in purely inductive series circuits, total inductance is the sum of the individual inductances; total inductive reactance (X_{L_T}) is the sum of all X_Ls; and total voltage equals the sum of all V_Ls, since each V_L is in-phase with the other V_Ls although they are not in-phase with the current, Figure 16−4a. In purely inductive parallel circuits, total circuit opposition to current (X_{L_T}) is found by using reciprocal or product-over-the-sum approaches. Also, total circuit current equals the sum of the branch currents, since branch currents are in-phase with each other although not in-phase with the circuit voltage, Figure 16−4b.

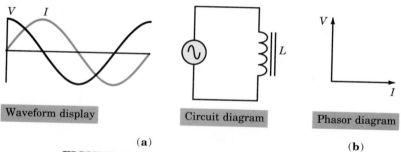

Waveform display   Circuit diagram   Phasor diagram

(a)   (b)

FIGURE 16–3   *V* and *I* in a simple *L* circuit

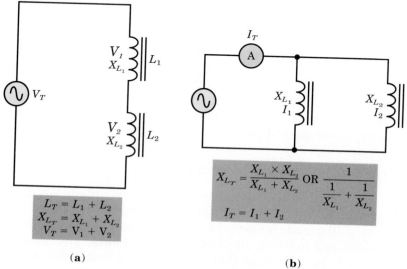

$$L_T = L_1 + L_2$$
$$X_{L_T} = X_{L_1} + X_{L_2}$$
$$V_T = V_1 + V_2$$

$$X_{L_T} = \frac{X_{L_1} \times X_{L_2}}{X_{L_1} + X_{L_2}} \text{ OR } \frac{1}{\dfrac{1}{X_{L_1}} + \dfrac{1}{X_{L_2}}}$$

$$I_T = I_1 + I_2$$

(a)   (b)

FIGURE 16–4   Purely inductive circuits

USING VECTORS TO DESCRIBE AND DETERMINE MAGNITUDE AND DIRECTION

Background

When ac circuits contain reactances and resistances, their voltage(s) and current(s) are not in-phase with each other. In this case, we cannot add these quantities together to find resultant totals as we did with dc circuits or ac circuits that are purely resistive or reactive. Thus, other methods are used to analyze ac circuit parameters.

One of these methods is to represent circuit parameters using **vectors.** A *vector quantity expresses both magnitude and direction.* Examples of quantities having both magnitude and direction that are illustrated using vectors include mechanical forces, the force of gravity, wind velocity and direction, and magnitudes and relative angles of electrical voltages and currents in ac.

Scalars, on the other hand, define quantities that exhibit only magnitude. Units of measure, such as the henry or the ohm, are scalar quantities. Even dc circuit parameters, or instantaneous values of ac electrical quan-

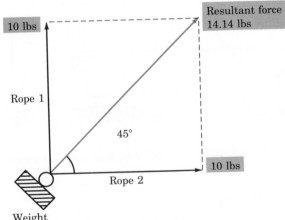

FIGURE 16–5　Vectors show both magnitude and direction.

tities are treated as scalar quantities, since there are no time differences or angles involved. Simple addition is used to add scalar quantities, but this is not true for vector quantities.

Plotting and Measuring Vectors

Vectors can be plotted to scale to determine *both* magnitude and direction of given quantities, Figure 16–5. If a force of 10 pounds is exerted on Rope 1 and a force of 10 pounds is exerted on Rope 2, what is the total force exerted on the weight, and in which direction is this total force? The resultant force is 14.14 pounds in a direction that is 45° removed from either rope's direction.

Look at Figure 16–6. Notice these vectors are plotted to scale so you can then *measure* both the magnitude and direction of the resultant force. *Measure the length* of the resultant vector *to find magnitude,* and *measure the*

FIGURE 16–6　Plotting vectors to scale

FIGURE 16–7   Methods of formatting vectors

angle between the original forces and the resultant force by a protractor *to find direction.*

Observe Figure 16–7a, which illustrates how vectors are laid end-to-end or head-to-tail to determine resultant values. Also, notice in Figure 16–7b that a parallelogram is drawn that projects the same magnitudes as the end-to-end approach. Although either method can be used to find resultant values, we will use the parallelogram approach most frequently.

Later in the chapter you will see how this plotting and measuring technique is used to analyze *RL* circuit parameters. Although it is not always convenient to have graph paper and protractor handy, or to draw circuit parameters to scale when analyzing a circuit, the plotting approach can be used. Also, it is a useful tool to help introduce you to vectors.

Even when not plotted to scale for the direct measurement method, vectors are useful to illustrate ac circuit parameters and show relative magnitudes and angles. Sketching this diagram helps you understand the situation and verify your computations.

PRACTICAL NOTES

Even when not precisely measuring vectors to determine resultant values, get in the habit of approximating the *relative* lengths and directions of vectors when you draw them to solve ac problems. This allows you to visually determine if your computations are making sense! A good practice is to draw simple vector diagrams whenever you are analyzing ac circuits—they help visualize the situation.

PRACTICE PROBLEMS I

Assume the forces on the ropes in Figure 16–5 are 3 pounds for Rope 1 and 4 pounds for Rope 2. What are the magnitude and direction of the resultant force on the weight? (NOTE: Use the plotting-to-scale approach and remember to keep things relative to the scale you use.)

✔ IN-PROCESS LEARNING CHECK I

1. In a purely resistive ac circuit, the circuit voltage and current are _____-phase.

2. In a purely inductive ac circuit, the current through the inductance and the voltage across the inductance are _____ degrees out-of-phase. In this case, the _____ leads the _____.

3. A quantity expressing both magnitude and direction is a _____ quantity.

4. The length of the vector expresses the _____ of a quantity.

INTRODUCTION TO BASIC AC CIRCUIT ANALYSIS TECHNIQUES

Most ac circuit analyses solve for quantities that are illustrated with vectors in right-triangle form. Recall from our previous discussions that a right triangle has one angle forming 90° and two angles that total another 90°, Figure 16–8. Many electrical parameters have this 90° phase difference between quantities. For example, in perfect inductors and capacitors, the voltages and currents are 90° out-of-phase. Also, ac circuit resistive and reactive voltage drops or currents are often represented as 90° out-of-phase with each other.

We will discuss two basic techniques used to analyze these right-triangle problems. The first method is the **Pythagorean theorem,** which is useful to determine the *magnitude* of quantities. The other method briefly addressed in this chapter is trigonometric functions that determine the circuit phase angles. Also, trigonometry can be used to find the magnitude of various electrical quantities, if the phase angle is known.

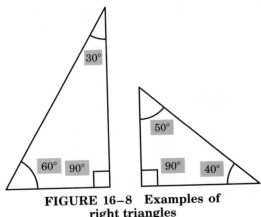

FIGURE 16–8   Examples of right triangles

PRACTICAL NOTES

NOTE: The calculations are easy if you have a commonly used scientific calculator with square, square root, and "trig" function keys. You probably own or presently have access to such a calculator. If not, you might want to gain access to one and learn to use these functions as you proceed through the remainder of the chapter. If this device is not available, paper and pencil will work just fine! NOTE: *Be aware* that some calculators only accept radian measure! Therefore, to convert degrees to number of radians, divide the number of degrees by 57.3; and to convert radians to number of degrees, multiply the number of radians by 57.3. Recall that 2π radians equal 360°; that is, 6.28 times 57.3° equals 360°.

Pythagorean Theorem

The Pythagorean theorem states the square of the hypotenuse of a right triangle equals the sum of the squares of the other two sides.

Formula 16–1	$c^2 = a^2 + b^2$

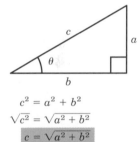

a value, $\boxed{x^2}$, $\boxed{+}$, b value, $\boxed{x^2}$, $\boxed{=}$

$$c^2 = a^2 + b^2$$
$$\sqrt{c^2} = \sqrt{a^2 + b^2}$$
$$c = \sqrt{a^2 + b^2}$$

FIGURE 16–9
Pythagorean theorem

Refer to Figure 16–9 and review the equations related to the right triangle and Pythagorean theorem.

Now, let's transfer this information into finding the resultant value, or magnitude, of two voltages that are displaced from each other by 90 degrees and vectorially added.

EXAMPLE

Refer to Figure 16–10 and observe the following steps: In the diagram shown,

1. V_S represents the hypotenuse of the right triangle.

2. V_1 represents one of the other two sides of the right triangle.

3. V_2 represents the remaining side of the right triangle.

4. Transferring this information into the Pythagorean formula:

$$V_S = \sqrt{V_1{}^2 + V_2{}^2}$$
$$V_S = \sqrt{25^2 + 25^2}$$
$$V_S = \sqrt{1{,}250} = 35.35$$

FIGURE 16–10

Using the calculator, the sequence of keystrokes is:

25, $\boxed{x^2}$, $\boxed{+}$, 25, $\boxed{x^2}$, $\boxed{=}$, $\boxed{\sqrt{x}}$ (or, $\boxed{\text{INV}}$, $\boxed{x^2}$, for some calculators)

NOTE: Some calculators require preceding the $\boxed{x^2}$ or the $\boxed{\sqrt{x}}$ key with the "second function" key.

FIGURE 16–11

PRACTICE PROBLEMS II

Refer to Figure 16–11. Use the Pythagorean theorem and solve for the magnitude of side V_S.

Trigonometric Functions

Trigonometric functions result from special relationships existing between the sides and angles of right triangles. "Trig" is a common shortened term for trigonometry or trigonometric. We will only discuss three basic trig functions used for your study of ac circuits—sine, cosine, and tangent functions.

Notice the side opposite (opp), the side adjacent (adj), and the hypotenuse (hyp) in Figure 16–12. As you can see, the relationship of each side to the "angle of interest," θ, determines whether it is called the opposite or the adjacent side. The hypotenuse is *always* the longest side of the triangle.

Refer to Figure 16–12 while you study the following.

1. **Sine** of the angle (sin θ) is the ratio of the side opposite to the hypotenuse.

Formula 16–2	$\sin \theta = \dfrac{\text{opposite}}{\text{hypotenuse}}$

opposite side value, $\boxed{\div}$, hypotenuse value, $\boxed{=}$

NOTE: In Figure 16–12, we have labeled the opposite side to the angle of interest as "a" and the hypotenuse as "c." Thus, $\sin \theta = a/c$.

2. **Cosine** of the angle (cos θ) is the ratio of the side adjacent to the hypotenuse.

Formula 16–3	$\cos \theta = \dfrac{\text{adjacent}}{\text{hypotenuse}}$

adjacent side value, $\boxed{\div}$, hypotenuse value, $\boxed{=}$

NOTE: In Figure 16–12, $\cos \theta = b/c$.

3. **Tangent** of the angle (tan θ) is the ratio of the side opposite to the side adjacent.

Formula 16–4	$\tan \theta = \dfrac{\text{opposite}}{\text{adjacent}}$

opposite side value, $\boxed{\div}$, adjacent side value, $\boxed{=}$

NOTE: In Figure 16–12, $\tan \theta = a/b$.

Take time to learn and understand the formula for each of these three functions now! It will be worth your time!

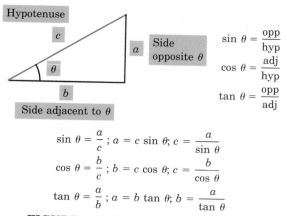

$$\sin \theta = \frac{a}{c} \; ; a = c \sin \theta; c = \frac{a}{\sin \theta}$$

$$\cos \theta = \frac{b}{c} \; ; b = c \cos \theta; c = \frac{b}{\cos \theta}$$

$$\tan \theta = \frac{a}{b} \; ; a = b \tan \theta; b = \frac{a}{\tan \theta}$$

FIGURE 16–12 Labeling sides of a right triangle

If we assume a rotating vector with a unit length of one starts at 0° and rotates counterclockwise to the 90° position, the values of sine, cosine, and tangent vary within limits. That is:

sin θ increases from 0 to 1.0 (with angles from 0° to 90°)
cos θ decreases from 1.0 to 0 (with angles from 0° to 90°)
tan θ increases from 0 to ∞ (with angles from 0° to 90°)

See if you can relate this to the sine and cosine waves in Figure 16–13. Assume the maximum (peak) value is one volt for each waveform. Relate the zero points and the maximum points to the angles. Do they compare?

Figure 16–14 carries this idea one step further. Notice how these functions vary in the four quadrants of the coordinate system we discussed in an earlier chapter. Again, try relating the sine and cosine waves in Figure 16–13 to the statements in Figure 16–14.

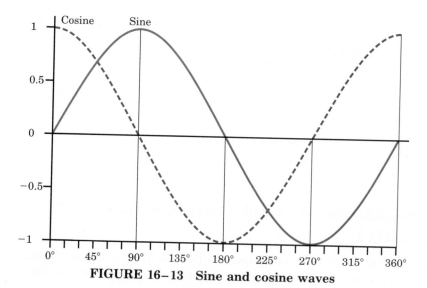

FIGURE 16–13 Sine and cosine waves

| | SIN θ | | COS θ | | TAN θ | |
QUADRANT	FROM	TO	FROM	TO	FROM	TO
I (0° to 90°)	0	1.0	1.0	0	0	∞
II (90° to 180°)	1.0	0	0	−1.0	−∞	0
III (180° to 270°)	0	−1.0	−1.0	0	0	∞
IV (270° to 360°)	−1.0	0	0	1.0	−∞	0

FIGURE 16–14   Variations in values of trig functions

EXAMPLE

What is the phase angle if the side opposite is 2 units and the hypotenuse is 2.828 units?

Referring to the trig formulas for sine, cosine, and tangent, the trig formula containing the side opposite and the hypotenuse is the formula for the sine of the angle. Therefore,

$$\sin \theta = \frac{2}{2.828} = 0.707$$

Using a calculator to find what angle has a sine equal to 0.707, we use the following keystrokes:

0.707, 2nd , SIN⁻¹ , and the answer is 45 degrees. (To check this out, try entering 45 into the calculator keypad, then press the SIN function key. Does the readout show 0.707?) (NOTE: If you do not have a scientific calculator available, you can look up this data on a "trig table," which can be found in numerous reference books and booklets.)

Now try using your knowledge and a calculator (or trig table) to solve the following practice problems.

PRACTICE PROBLEMS III

1. What is the phase angle if the side adjacent has a magnitude of 5 units and the hypotenuse has a magnitude of 10 units?

2. What is the tangent of 45°?

3. What is the sine of 30°?

FUNDAMENTAL ANALYSIS OF SERIES *RL* CIRCUITS

Now we will take the knowledge you have acquired and apply it to some practical situations and problems, so you can see the benefit of this knowledge!

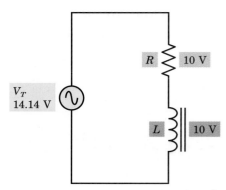

FIGURE 16–15   Series *RL* circuit voltages

Refer to Figure 16–15. Notice if you added the voltages across the series resistor and the inductor to find total applied voltage, you would get 20 volts. This answer is wrong! The actual applied voltage is 14.14 volts.

Let's see why 20 volts is the wrong value! Recall in series circuits, the current is the same through all components. Again referring to Figure 16–15 and from our previous discussions, we know the voltage across the resistor is in-phase with the current. We also know the voltage across the inductor leads the current by about 90°, which means the voltages across the resistor and the inductor are *not* in-phase with each other. Since the current through both is the same current, one voltage is in-phase with the current (V_R), and one voltage is not in-phase with the current. Therefore, we cannot add the voltages across R and L to find total voltage, as we would do in a simple dc circuit. Rather, we must use a special method of addition called vector addition to find the correct answer. The voltages must be added "vectorially" not "algebraically."

Let's try the two methods previously discussed to find total voltage in series *RL* circuits.

Using the Pythagorean Theorem to Analyze Voltage in Series *RL* Circuits

Figure 16–16 illustrates how V_R, V_L, and V_T are typically represented in a right triangle by phasors or vectors.

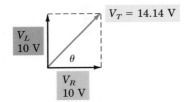

FIGURE 16–16   Voltage vector diagram

EXAMPLE

Let's transfer the Pythagorean theorem formula in Figure 16–9, $c = \sqrt{a^2 + b^2}$, to this situation. As you can see, the longest side is V_T. This is equivalent to the c in the original Pythagorean formula. The other two quantities (V_R and V_L) are the other quantities in the formula. Thus, the formula to solve for V_T becomes:

Formula 16–5	$V_T = \sqrt{V_R^2 + V_L^2}$

V_R value, $\boxed{x^2}$, $\boxed{+}$, V_L value, $\boxed{x^2}$, $\boxed{=}$, $\boxed{\sqrt{x}}$

Therefore, $V_T = \sqrt{10^2 + 10^2} = \sqrt{200} = 14.14$ volts.

(NOTE: V_T is *not* equal to the sum of V_R and V_L. It is greater than either one alone, but not equal to their sum.)

FIGURE 16–17

EXAMPLE

Assume the voltage values in Figure 16–15 change so that V_R is 30 volts and V_L is 40 volts. What is the new V applied value?

Answer: The V_T is 50 volts, Figure 16–17.

PRACTICE PROBLEMS IV

In a simple series RL circuit, V_R is 120 V and V_L is 90 V. What is the V applied value? SUGGESTION: Use your calculator.

You can see that using the Pythagorean theorem, it is easy to find the magnitude of total voltage in a series RL circuit— simply substitute the appropriate values in the formula. We are not able to find the resultant angular direction by this method. However, the trig functions help us.

Using Trig to Solve for Angle Information in Series *RL* Circuits

Look at Figure 16–18 as you study the following statements.

1. To find the desired angle information using trig functions, identify the sides of the right triangle in terms of the circuit parameters of interest. Note in Figure 16–18 that V_L is in the upward (+90°) direction, and V_R is at the 0° reference direction. In *series* circuits, the current is the same throughout the circuit and is in-phase with the resistive voltage. Therefore, I is also on the 0° axis.

PRACTICAL NOTES

For series circuits since current is common to all parts, *current is commonly the reference phasor* and is plotted on the reference phasor 0-degree axis.

$$\text{Sin } \theta = \frac{\text{opp}}{\text{hyp}} \text{ OR } \frac{V_L}{V_T}$$

$$\text{Sin } \theta = \frac{30}{50} \text{ OR } 0.60$$

Angle whose sine equals
0.60 = 36.9°
$$\theta = 36.9°$$

NOTE: Phase angle is considered the angular difference in phase between circuit total voltage and circuit total current. Get in the habit of looking at the sketch to see if the answer is logical. It is obvious that the angle in this case must be less than 45° because the vertical leg is shorter than the horizontal leg. The answer looks and seems logical!

FIGURE 16–18   Using trig to find angle information

Since these directions are important, let's restate them! They are general rules of thumb that you will use from now on! V_L is vertical (up), V_R is horizontal to the right, and I is in-phase with V_R.

2. Referring to the values in Figure 16–18, let's find the *circuit phase angle* that is defined as *the phase angle between circuit applied voltage and circuit total current.*

 a. Consider the angle between V_T and I_T the angle of interest. Thus, the side opposite to this angle is the V_L value and the hypotenuse is V_T.

 b. Use the information relating these two sides in the sine function formula to find the angle. (Recall that sin θ = opp/hyp.) Therefore, in our example, sin θ = 30/50 = 0.60.

 c. Using a calculator, or computer, (or trig table) to find the angle having a sine function value of 0.60, we find that it is 36.9°.

PRACTICAL NOTES

   In ac circuits containing both resistance and reactance, the phase difference in degrees between applied voltage and total current must fall between 0° (indicating a purely resistive circuit) and 90° (indicating a purely reactive circuit). Since both resistance and reactance are present, the circuit cannot act purely resistive or purely reactive.

 d. In this example, since we know all sides, we can also use the other two trig functions. That is,

$$\cos \theta = \frac{\text{adj}}{\text{hyp}} = \frac{V_R}{V_T} = \frac{40}{50} = 0.8$$

The angle with a cosine value of 0.8 is 36.9°.

$$\tan \theta = \frac{\text{opp}}{\text{adj}} = \frac{V_L}{V_R} = \frac{30}{40} = 0.75$$

The angle with a tangent value of 0.75 is 36.9°.

PRACTICAL NOTES

Rather than having to say in words "the angle having a tangent value equal to . . . ," it is common to use the mathematical term "arc tangent," or "arctan." In other words, if I find the arctan of 0.75, it is an angle of 36.9 degrees. You will also see arctan written as "$\tan^{-1}$." To find the arctan on a calculator, input the ratio of the opposite side to the adjacent side. That is: (opposite side value, $\boxed{\div}$, adjacent side value, $\boxed{=}$), then press the $\boxed{\text{2nd}}$ and $\boxed{\text{TAN}^{-1}}$ function keys (in that order). You will then see the $\tan^{-1}$ (arctan, or angle value) displayed in the readout.

Naturally, you can find the arc sine or arc cosine using the same technique. For the sine, the values to input are the opposite side divided by the hypotenuse. For the cosine, the values to input are the adjacent side divided by the hypotenuse. Once you have the sine or cosine value, use the $\boxed{\text{2nd}}$ and $\boxed{\text{SIN}^{-1}}$ function keys or the $\boxed{\text{2nd}}$ and $\boxed{\text{COS}^{-1}}$ function keys as appropriate to find the arc sine or arc cosine, respectively. (Note: This gives you the angle value.)

e. Use the sine, cosine, or tangent functions to solve for the unknown angle depending on the information available. Obviously, when you know the values for the opposite side and the hypotenuse, use the sine function. When you know the values for side adjacent and the hypotenuse, use the cosine function. When you know only the values for side opposite the angle of interest and the side adjacent to the angle of interest, use the tangent function.

Stated another way, select and use the formula containing the two known values and the value you are solving for.

PRACTICE PROBLEMS V

1. Draw the vector diagram for the voltages in Figure 16–19.

2. Use trig and solve for the phase angle of the circuit in Figure 16–19.

3. Which trig function did you use in this case?

4. If the voltages had been reversed; i.e., $V_R = 20$ volts and $V_L = 30$ volts, would the phase angle have been greater, smaller, or the same?

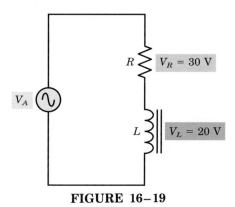

FIGURE 16–19

Using Trig to Solve for Voltages in Series *RL* Circuits

You have seen how trig helps find angle information. You have also learned the Pythagorean theorem is useful to find magnitude or voltage value in series *RL* circuits. Can trig also be used to find magnitude? Yes, it can! Let's see how!

EXAMPLE

If the angle information is known, determine the sine, cosine, or tangent values for that angle. Next, use the appropriate, available sides information, and find the value of the unknown side. Look at Figure 16–20 and let's see how to do this.

1. Start by making sure the formula for the trig function used includes the unknown side of interest, plus a known side. This allows us to solve for the unknown by using equation techniques.

2. To solve for V_L, use the sine function since the sin θ uses the opposite side divided by the hypotenuse. V_L is the unknown side of interest and is the side opposite the known angle. Since we know the hypotenuse value, we have our starting point.

$$\text{Sin } \theta = \frac{V_L}{V_T}$$

$$\text{Sin } \theta = \frac{V_L}{100}$$

$$\text{Sin } 60° = \frac{V_L}{100}$$

$$0.866 = \frac{V_L}{100}$$

$$86.6 \text{ V} = V_L$$

$$\text{Cos } \theta = \frac{V_R}{V_T}$$

$$0.5 = \frac{V_R}{100}$$

$$50 \text{ V} = V_R$$

FIGURE 16–20

3. Sin θ = opp/hyp, thus sin 60° = $V_L/100$. Now we determine the sine of 60° and substitute in the equation. Use the trig table, calculator, or computer and find the sine of 60° is 0.866. (NOTE: Round-off to 0.87.) Put this in the formula, thus 0.87 = $V_L/100$. Transpose and solve: V_L = 100 × 0.87 = 87 volts.

4. To solve for V_R, use either the cosine or tangent functions with the information now known. Cosine uses the adjacent divided by the hypotenuse and tangent uses the opposite divided by the adjacent. In either case, a known and the desired unknown (the adjacent) values appear in the formulas.

5. Assuming we do not know the V_L value, we use the cosine function. Cos θ = adj/hyp, thus cos 60° = $V_R/100$. Next, determine the cosine of 60°, which equals 0.5. Thus, substitute in the formula, 0.5 = $V_R/100$. Transpose and solve: V_R = 100 × 0.5 = 50 volts.

Now, let's see if these results make sense with respect to our vector diagram in Figure 16–20. Since the angle is greater than 45°, we know V_L must be greater than V_R and it is!

Now we can use the tangent function to check our results, since it uses the opposite divided by the adjacent sides. Insert the values found for V_L and V_R into the tan θ = opp/adj formula to find that 86.6/50 = 1.732. The angle whose tangent is 1.732 is 60°. It checks out! Now, you try some problems.

PRACTICE PROBLEMS VI

---

1. A series RL circuit has one resistor and one inductor. What is the V_R value if the applied voltage is 50 volts and the circuit phase angle is 45°?

2. A series RL circuit comprised of a single resistor and inductor has an applied voltage value of 65 volts and the circuit phase angle is 60 degrees.
   a. What is the value of V_R?
   b. What is the value of V_L?

3. A series RL circuit comprised of a single resistor and inductor has values where V_L = 36 volts and V_R = 24 volts.
   a. What is the value of the angle between V_R and V_T?
   b. What is the value of V_T?

Analyzing Impedance in Series *RL* Circuits

You have learned how to analyze various voltages in series RL circuits by measuring vectors, using the Pythagorean theorem, and applying basic trig functions. Now, let's look at another important ac circuit consideration, impedance.

Impedance is the total opposition a circuit offers to ac current flow at a given frequency. This means impedance is the combined opposition of all

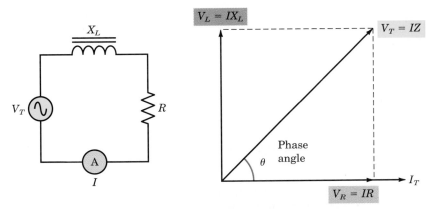

FIGURE 16-21   Series *RL* circuit quantities

resistances and reactances in the circuit. Since impedance is sensitive to frequency, there must be some reactance—*Pure* resistance is not sensitive to frequency. The symbol for impedance is "Z", and the unit of impedance is the ohm.

Before discussing impedance, let's review some important facts you already know, Figure 16-21.

1. The voltage drop across the resistor is in-phase with the circuit current through the resistor, and V_R equals IR. Note that this is a series circuit. Since current is the common factor in series circuits, the current is the reference vector and is at the 0° position of the coordinate system. Also V_R is at 0° since it is in-phase with I.

2. The voltage drop across the inductor leads the circuit current by 90° (assuming a perfect inductor with no resistance), and V_L equals IX_L. Since V_L leads I by 90°, the IX_L vector is leading the I vector by 90°.

3. As you know, total circuit voltage is the *vector resultant* of the individual series voltages. Since the circuit is not purely resistive, V_T cannot be at 0°. Since the circuit is not purely inductive, V_T cannot be at 90°. Remember in circuits containing both resistance and reactance, the resultant circuit phase angle between circuit V and circuit I is between 0° and 90°. As you can see from the diagram, total circuit voltage equals IZ, where Z is the total opposition to ac current from the combination of resistance and reactance. Since I is at 0° and IZ is at an angle between 0° and 90°, when we plot Z, it also will be between 0° and 90°.

Let's pursue these thoughts. Refer to Figure 16-22 as you study the following discussion about how to plot a diagram of R, X_L, and Z (impedance diagram) for series *RL* circuits.

1. The V_R equals IR, and I is plotted at the 0° reference vector position. Also, since the resistor voltage is in-phase with the current through the resistor, R is plotted at the 0° reference position when plotting an impedance diagram of R, X_L, and Z.

2. Since V_L equals IX_L and V_L is plotted at 90° from the reference vector position, it is logical that X_L is plotted at the 90° position when plot-

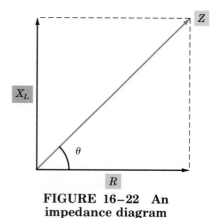

FIGURE 16–22   An impedance diagram

ting R, X_L, and Z on an impedance diagram. Thus, the result of I times X_L is plotted at 90°.

3. The V_T equals IZ, and V_T is plotted at an angle (depending on the vector resultant of V_R and V_L) between 0° and 90°. Then it is logical to plot Z at the same angle in an impedance diagram where V_T is plotted in the voltage diagram of the same series RL circuit. In fact, the angle between Z and R is the same as the circuit phase angle between V_T and I_T.

Knowing all this, how can the techniques to analyze voltages in series RL circuits be applied to impedance? As you can see by comparing the diagrams of Figures 16–21 and 16–22, the layout of the impedance diagram looks just like the diagram used in the voltage vector diagram analysis. This implies the techniques used to find impedance (Z) are those techniques used to find V_T. Recall these methods include:

1. Draw to scale and measure to find magnitude and direction.

2. Apply the Pythagorean theorem to find magnitudes.

3. Apply trig functions to find angles or magnitudes.

EXAMPLE

Figure 16–23 illustrates the drawing-to-scale approach. Figure 16–24 shows the Pythagorean theorem version used for impedance solutions. That is:

| Formula 16–6 | $Z \ = \ \sqrt{R^2 + X_L{}^2}$ |

R value, $\boxed{x^2}$, $\boxed{+}$, X_L value, $\boxed{x^2}$, $\boxed{=}$, $\boxed{\sqrt{x}}$

Finally, Figure 16–25 solves for the angle using the simple trig technique. Now you try applying these techniques in the following situation.

FIGURE 16–23

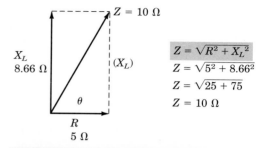

$$Z = \sqrt{R^2 + X_L^2}$$
$$Z = \sqrt{5^2 + 8.66^2}$$
$$Z = \sqrt{25 + 75}$$
$$Z = 10 \ \Omega$$

(NOTE: Z is not equal to the sum of R and X_L. It is greater than either one alone but not equal to their sum.)

FIGURE 16–24

PRACTICE PROBLEMS VII

1. Draw the schematic diagram of a series RL circuit containing 15 Ω of resistance and 20 Ω of inductive reactance.

2. Draw an impedance diagram for the circuit described in question 1 (to scale). Measure and record the impedance value. Use a protractor and measure the angle between R and Z. Label this angle on the diagram.

3. Use the Pythagorean theorem formula where $Z = \sqrt{R^2 + X_L^2}$ and calculate and record the Z value. Does the answer agree with your measurement results?

4. Use appropriate trig formula(s) and solve for the angle where Z should be plotted. Show your work. Does this answer agree with the angle you measured using the protractor in Question 2?

Summary of Analyses of Series *RL* Circuits

1. The larger the R value compared to the X_L value, the more resistive the circuit will act. Conversely, the larger the X_L compared to the R, the more inductive the circuit will act.

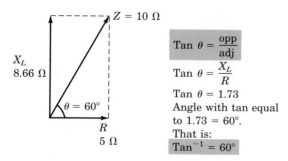

$$\text{Tan } \theta = \frac{\text{opp}}{\text{adj}}$$
$$\text{Tan } \theta = \frac{X_L}{R}$$
Tan θ = 1.73
Angle with tan equal to 1.73 = 60°.
That is:
$$\text{Tan}^{-1} = 60°$$

FIGURE 16–25

2. The more resistive the circuit, the closer the circuit phase angle is to 0°. The more inductive the circuit, the closer the circuit phase angle is to 90° because circuit voltage is leading circuit current, which is the 0° reference vector in series circuits. Recall in a purely resistive series circuit, phase angle equals 0°. In a purely inductive series circuit, phase angle equals 90°. Refer to Figure 16–26 for some examples of R, X_L, Z, V_R, V_L, V_T, and phase angle (θ) relationships.

3. Since current (I) is the same through all parts of a series circuit, the ratio of R and X_L is the same as the relationship (or ratio) of V_R and V_L. That is, $V_R = IR$ and $V_L = IX_L$. For example, when the R value is two times the X_L value, the V_R value is twice the V_L value.

4. Total circuit impedance is the result of both factors, resistance and reactance. Impedance is greater than each factor but *not* equal to their sum. Right-triangle methods are used to find its value. Refer again to Figure 16–24.

5. Total circuit voltage is the vector resultant of the resistive and reactive voltages in series. This voltage is greater than each factor but *not* equal to their sum. Right-triangle methods are used. Refer again to Figure 16–17.

6. Phase angle for series RL circuits is found using sine, cosine, or tangent functions. For example:

$$\sin \theta = \frac{V_L}{V_T}$$

$$\cos \theta = \frac{V_R}{V_T}$$

$$\tan \theta = \frac{V_L}{V_R}$$

NOTE: The tangent function is useful for many ac circuit problems because it does not require knowing the hypotenuse value.

R Ω	V_R (IR)	X_L Ω	V_L (IX_L)	$Z\Omega$ ($\sqrt{R^2 + X_L^2}$)	V_T (IZ) VOLTS	θ (PHASE ANGLE) $\left(\text{Tan} = \dfrac{X_L}{R}\right)$ $\left(\text{or } \dfrac{V_L}{V_R}\right)$
10	10 V	5	5 V	11.18	11.18	26.56°
10	10 V	10	10 V	14.14	14.14	45°
5	5 V	10	10 V	11.18	11.18	63.43°
100	100 V	10	10 V	100.5	100.5	5.71°
10	10 V	100	100 V	100.5	100.5	84.29°

Assume an I_T value of one ampere in each case. (This means V applied is changing.)

FIGURE 16–26   Examples of parameter relationships in series RL circuits

For the impedance diagram:

$$\sin \theta = \frac{X_L}{Z}$$

$$\cos \theta = \frac{R}{Z}$$

$$\tan \theta = \frac{X_L}{R}$$

7. A greater R or X_L value results in greater circuit impedance and lower current value for a given applied voltage.

8. The circuit current value is found by using Ohm's law, where

$$I_T = \frac{V_T}{Z_T}$$

9. *Impedance diagrams can only be drawn and used to analyze series ac circuits!* Also, they can be used to analyze series portions of more complex circuits but *should not* be drawn to analyze parallel circuits.

10. An impedance diagram is not a phasor diagram, since the quantities shown are not sinusoidally varying quantities. However, because Z is the vector resulting from R and X, it can be graphed.

FUNDAMENTAL ANALYSIS OF PARALLEL *RL* CIRCUITS

Basically, the same strategies we have been using in the series circuit solutions can be applied to parallel circuits. That is, we can use vectors, the Pythagorean theorem, and trig functions. The differences are in *which* parameters are used on the diagrams and in the formulas.

Because *voltage* is the common factor in parallel circuits, it becomes the reference vector at the 0° position on our diagrams. Since I_T equals the vector sum of the branch currents in parallel circuits, our diagrams illustrate the branch currents in appropriate right-triangle form. (Recall that in series circuits, current was the reference and voltages are plotted in right-triangle format.)

Another contrast between parallel RL circuit analysis and series RL circuit analysis is that we *do not* use impedance diagrams to analyze parallel RL circuits.

With these facts in mind, let's apply the principles you have learned to some parallel RL circuit analysis. Refer to Figure 16–27 as you study the following comments.

1. V_T is at the reference vector position, because it is common to both the resistive and the reactive branches.

2. Since I_R (resistive branch current) is in-phase with V_R (or V_T), it is also at the 0° position.

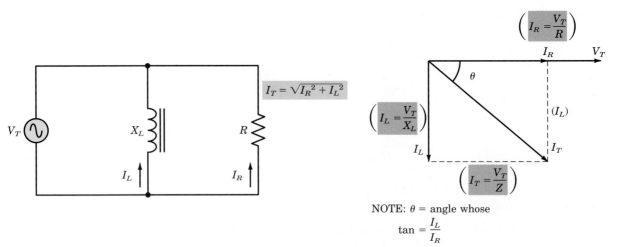

FIGURE 16–27   Parallel *RL* circuit analysis

3. We know the current through a coil lags the voltage across the coil by 90°. Thus, I_L is lagging V_T (or V_L) by 90°. (Remember in the four-quadrant system that $-90°$ is shown as straight down.)

4. The total current is the vector resultant of the resistive and the reactive branch currents. Thus, it is between 0° and $-90°$.

5. The higher the resistive branch current relative to the reactive branch current, the more resistive the circuit will act. This means the smaller phase angle is between V_T and I_T. Also, the greater the inductive branch current relative to the resistive branch current, the more inductive the circuit will act. This means the phase angle is greater between V_T and I_T.

6. In a parallel *RL* circuit, total impedance is the resultant opposition offered to current by the combined circuit resistance and reactance. That is,

$$Z = \frac{V_T}{I_T}$$

Because of the phase relationships in a circuit with both resistance and reactance, the impedance is less than the impedance of any one branch. However, the shortcut methods used in purely resistive or reactive circuits will not work. For example, if there are 10 Ω of resistance in parallel with 10 Ω of reactance, the impedance will *not* be 5 Ω but 7.07 Ω. This is because the total current is not the simple arithmetic sum of the branch currents, as it is in purely resistive or reactive circuits. Rather, total current is the vector sum.

7. Branch currents are found using Ohm's law, where

$$I_R = \frac{V_R}{R} \text{ and } I_L = \frac{V_L}{X_L}$$

8. Total current is found using the Pythagorean formula, where:

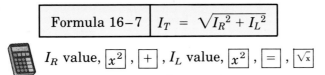

$$\boxed{\text{Formula 16-7} \quad I_T = \sqrt{I_R^2 + I_L^2}}$$

I_R value, $\boxed{x^2}$, $\boxed{+}$, I_L value, $\boxed{x^2}$, $\boxed{=}$, $\boxed{\sqrt{x}}$

EXAMPLE

Parameters for the circuit in Figure 16–28 are $V_T = 150$ V, $R = 50\ \Omega$, and $X_L = 30\ \Omega$. Find I_T, Z, and the circuit phase angle (θ).

1. Label the known values on the diagram, Figure 16–29.

2. Determine the branch currents:

$$I_R = \frac{V_T}{R} = \frac{150\text{ V}}{50\ \Omega} = 3\text{ A}$$

$$I_L = \frac{V_T}{X_L} = \frac{150\text{ V}}{30\ \Omega} = 5\text{ A}$$

3. Draw a current vector diagram and label knowns, Figure 16–30.

4. Use Pythagorean theorem and solve for I_T:

$$I_T = \sqrt{I_R^2 + I_L^2}$$
$$I_T = \sqrt{9 + 25}$$
$$I_T = \sqrt{34}$$
$$I_T = 5.83\text{ A}$$

5. Use Ohm's law and solve for Z:

$$Z = \frac{V_T}{I_T}$$

$$Z = \frac{150\text{ V}}{5.83\text{ A}}$$

$$Z = 25.73\ \Omega$$

FIGURE 16–28

FIGURE 16–29

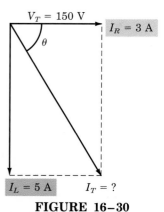

FIGURE 16–30

6. Use the trig formula and solve for phase angle:

$$\text{Tan } \theta = \frac{I_L}{I_R}$$

$$\text{Tan } \theta = 1.666$$

Angle whose tangent = 1.666 is 59° (i.e., $\text{Tan}^{-1}\,1.666 = 59°$)

Answers: $I_T = 5.83$ A, $Z = 25.73$ Ω, and $\theta = -59°$

(NOTE: The angle is considered a negative angle because the circuit current is lagging the reference circuit voltage by 59°.)

Now you try the following practice problem.

PRACTICE PROBLEMS VIII

A circuit similar to the one in Figure 16–28 has the following parameters: $V_T = 300$ V, $R = 60$ Ω, and $X_L = 100$ Ω. Perform the same steps previously shown and find 1) I_T, 2) Z, and 3) the phase angle (θ).

A Formula to Find Z Without Knowing I

To find the impedance (Z) of a simple parallel RL circuit, you can use the following formula:

Formula 16–8	$Z = \dfrac{RX_L}{\sqrt{R^2 + X_L^2}}$

 R value, $\boxed{\times}$, X_L value, $\boxed{\div}$, $\boxed{(}$, R value, $\boxed{x^2}$, $\boxed{+}$, X_L value, $\boxed{x^2}$, $\boxed{)}$, $\boxed{\sqrt{x}}$, $\boxed{=}$

EXAMPLE

Using the same values of R and X_L just used in our preceding example:

$$Z = \frac{RX_L}{\sqrt{R^2 + X_L^2}} = \frac{50 \times 30}{\sqrt{50^2 + 30^2}} = \frac{1,500}{\sqrt{3,400}} = 25.72 \text{ Ω}$$

As you can see, it checks out with our previous solution in which we used Ohm's law to solve for Z. Now you try one using this Z formula technique.

PRACTICE PROBLEMS IX

Assume a simple parellel RL circuit in which the resistance equals 47 Ω and the inductive reactance equals 100 Ω. What is the impedance value of this circuit?

Summary of Analysis of Parallel *RL* Circuits

1. The greater the resistive branch *current* compared to the inductive branch *current,* the more resistive the circuit acts. Conversely, the greater the inductive branch *current* compared to the resistive branch current, the more inductive the circuit acts.

2. The more resistive the circuit is, the closer the circuit phase angle is to 0°. The more inductive the circuit is, the closer the circuit phase angle is to −90°. These situations result from circuit current lagging behind the circuit voltage, which is the 0° reference vector in parallel circuits. Recall in a purely *resistive* parallel circuit $\theta = 0°$, but in a purely inductive *parallel* circuit $\theta = -90°$. Refer to Figure 16–31 to see some examples of the relationships of R, X_L, Z, I_R, I_L, I_T, and phase angle (θ) with selected circuit parameters purposely varied.

3. Branch current is *inverse* to the resistance or reactance of a given branch. For example, when R is two times X_L, current through the *resistive* branch is half the current through the *reactive* branch.

4. Total circuit impedance of a given parallel RL circuit is less than the opposition of any one branch. However, impedance cannot be found with product-over-the-sum or reciprocal formulas used in dc circuits. Neither can impedance be found using an impedance diagram approach. Ohm's law or other methods are used.

5. Total circuit current is the vector resultant of the resistive and reactive branch currents. Current is greater than any one branch current but *not* equal to their sum. Right-triangle methods are used. Refer again to Figure 16–27.

6. Phase angle for parallel RL circuits is found using sine, cosine, or tangent trig functions. For example:

$$\sin \theta = \frac{I_L}{I_T}$$

$$\cos \theta = \frac{I_R}{I_T}$$

$$\tan \theta = \frac{I_L}{I_R}$$

NOTE: The tangent function is useful for many ac circuit problems because it does not require knowing the hypotenuse value.

7. A greater R or X_L value results in greater circuit impedance and lower circuit current value for a given applied voltage.

8. The circuit impedance value is found by using Ohm's law, where

$$Z_T = \frac{V_T}{I_T}.$$

9. *Impedance diagrams can only be drawn and used to analyze series ac circuits!* Also, they can be used to analyze series portions of more complex circuits but *should not* be drawn to analyze parallel circuits. Ob-

R Ω	I_R $\left(\dfrac{V_T}{R}\right)$	X_L Ω	I_L $\left(\dfrac{V_T}{X_L}\right)$	I_T $\left(\sqrt{I_R^2 + I_L^2}\right)$	Z $\left(\dfrac{V_T}{I_T}\right)$	Θ (PHASE ANGLE) $\left(\dfrac{\text{Tan } I_L}{I_R}\right)$
10	5 A	5	10 A	11.18 A	4.47 Ω	$-63.43°$
10	5 A	10	5 A	7.07 A	7.07 Ω	$-45°$
5	10 A	10	5 A	11.18 A	4.47 Ω	$-26.56°$
100	0.5 A	10	5 A	5.02 A	9.96 Ω	$-84.29°$
10	5 A	100	0.5 A	5.02 A	9.96 Ω	$-5.7°$
						NOTE: The larger the ($-$) phase angle, the more inductive the circuit is acting.

Assume a voltage applied (V_T) of 50 V for each case. (This means I_T will be changing.)

FIGURE 16–31   Examples of parameter relationships in parallel *RL* circuits

viously, since the impedance of a parallel *RL* circuit is smaller than any one branch, the hypotenuse cannot be shorter than one of the sides when using the right-triangle techniques. Therefore, an impedance diagram cannot be drawn for parallel *RL* circuits.

EXAMPLES OF PRACTICAL APPLICATIONS FOR INDUCTORS AND *RL* CIRCUITS

Inductors and inductances have a variety of applications. Some of these applications use the property of inductance or mutual inductance. Other applications involve using inductance with resistance or capacitance, or both. Inductance can be used in circuits operating at power-line (low) frequencies; circuits operating in the audio-frequency range; and circuits operating at higher frequencies, known as radio frequencies.

The following is a brief list of applications.

1. Power-frequency applications, Figure 16–32.
   a. Power transformers
   b. Filter circuits (both power line and power supply filters)
   c. Solenoids, electromagnets, relays

2. Audio-frequency applications, Figure 16–33.
   a. Audio transformers (used for coupling and Z-matching)
   b. Audio chokes (as loads for amplifiers)
   c. Filter circuits

3. Radio-frequency applications, Figure 16–34.
   a. Tuning circuit applications
   b. Radio chokes/filter
   c. Waveshaping applications (using L/R time constant characteristics)

Typically, inductance values used at power-line frequencies are in the henrys range. For audio frequencies, values are often in the tenths of a henry, or millihenry ranges. For radio frequency *(RF)* applications, values are typically in the microhenrys range.

Primary   Secondary

Power transformers
(step voltages up or down)

Filter circuits
help smooth pulsating
voltages and line surges

Electromagnets in meters

(a)

Solenoids

(b)

Relays
(often used to cause switching actions)

(c)

**FIGURE 16–32   Examples of applications at low ac (or pulsating dc)
frequencies (Photo a courtesy of Simpson Electric Company;
photos b and c courtesy of Guardian Electric Mfg. Co.)**

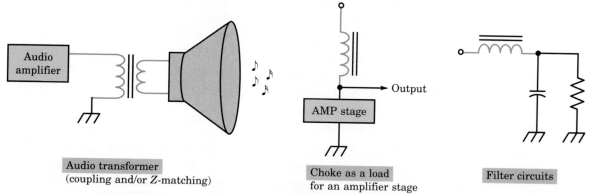

Audio
amplifier

Output

AMP stage

Audio transformer
(coupling and/or *Z*-matching)

Choke as a load
for an amplifier stage

Filter circuits

FIGURE 16–33   Examples of applications at audio frequencies

FIGURE 16–34   Examples of radio-frequency (RF) and other applications

One key characteristic of inductors allowing them to be used in these applications is their sensitivity to frequency. (Recall, X_L varies with frequency.) Their relationship to creating and reacting to magnetic fields is another very useful characteristic.

You have already studied the basic principles of inductors, which make the varied applications shown in Figure 16–33 possible. For example, you know inductive reactance varies with frequency. The application of inductors in filter and tuning circuits depends on this principle. Also, you studied the principle for waveshaping when you studied L/R time constants.

As you can see, inductors, the property of inductance, and inductive reactance are valuable tools in the field of electronics. Your understanding of inductors will be of great help in your career.

SUMMARY

▶ Phase angle is the difference in phase between the circuit applied voltage (V_T) and the circuit current (I_T).

▶ Circuit current and voltage are in-phase in purely resistive ac circuits; are 90° out-of-phase in purely inductive circuits; and are between 0° and 90° out-of-phase in circuits containing both resistance and inductance.

▶ For series ac circuits containing both resistance and inductive reactance, total circuit voltage and impedance cannot be found by adding voltage values or oppositions to current, as is done in dc circuits. Rather, values must be determined by vectors or right-triangle techniques.

▶ For voltage-current vector diagrams relating to series RL circuits, current is the reference vector positioned at the 0° location since it is common throughout the circuit. Resistive voltage drop(s) are also at the 0° position, since they are in-phase with current. Inductive voltages are shown at 90° (since V_L leads I by 90°). Total circuit voltage is plotted as the vector resultant of the resistive and inductive voltages.

▶ For impedance diagrams (used *only* for *series* circuits), the R values are plotted at the 0° reference position. The X_L value is plotted vertical (up) at 90° (assuming a perfect inductor). Impedance is plotted at a position that is the vector resultant of the R and X_L vectors.

▶ The Pythagorean theorem applied to *series RL* circuits is used to calculate the total voltage value where $V_T = \sqrt{V_R^2 + V_L^2}$. It is also useful to determine the impedance *(Z)* value where $Z = \sqrt{R^2 + X_L^2}$.

▶ The trig functions of sine, cosine, and tangent are used to determine phase angle, if two sides are known. Also, these functions can be used to determine voltage and impedance values of *series RL* circuits, if the phase angle and one side are known.

▶ For *parallel* ac circuits containing both resistance and inductive reactance, total circuit current cannot be found by adding branch currents, as is done in dc circuits. Right-triangle methods are again used.

▶ For voltage-current vector diagrams relating to *parallel RL* circuits, *voltage* is the *reference vector* at the 0° location since it is common throughout the circuit. Resistive branch current(s) are also at the 0° position, since they are in-phase with the voltage. Inductive branch currents are at −90° since I_L lags V by 90°. Total circuit current is plotted as the vector resultant of the resistive and inductive branch currents.

▶ The Pythagorean theorem applied to *parallel RL* circuits is used to calculate the total current value where $I_T = \sqrt{I_R^2 + I_L^2}$.

▶ A formula that may be used to find Z in a simple parallel RL circuit is:

$$Z = \frac{RX_L}{\sqrt{R^2 + X_L^2}}$$

▶ The trig functions (sine, cosine, and tangent) are also used to determine phase angle in parallel circuits, as well as series circuits, if two sides are known. They can be used to determine current values in *parallel RL* circuits, if the phase angle and one side are known.

▶ Plotting vectors to scale and then measuring lengths and angles is one method to determine parameters of interest in RL circuits. For series RL circuits, the resistive and inductive voltage values are plotted to find V_T and θ. For parallel RL circuits, the resistive and inductive branch current values are plotted to find I_T and θ.

▶ Impedance is the total opposition offered to ac current by a combination of resistance and reactance. The symbol for impedance is "Z", and the unit of impedance is the ohm.

▶ Inductors, either alone or in conjunction with other components (R or C), are used in many applications. General uses include transformers, tuning circuits, filter circuits, and wave-shaping circuits.

FORMULAS AND CALCULATOR SEQUENCES SUMMARY

FORMULA 16–1 $c^2 = a^2 + b^2$

a value, $\boxed{x^2}$, $\boxed{+}$, b value, $\boxed{x^2}$, $\boxed{=}$

FORMULA 16–2 $\sin \theta = \dfrac{\text{opposite}}{\text{hypotenuse}}$

opposite side value, $\boxed{\div}$, hypotenuse value, $\boxed{=}$

FORMULA 16–3 $\cos \theta = \dfrac{\text{adjacent}}{\text{hypotenuse}}$

adjacent side value, $\boxed{\div}$, hypotenuse value, $\boxed{=}$

FORMULA 16–4 $\tan\theta = \dfrac{\text{opposite}}{\text{adjacent}}$

opposite side value, $\boxed{\div}$, adjacent side value, $\boxed{=}$

FORMULA 16–5 $V_T = \sqrt{V_R{}^2 + V_L{}^2}$

V_R value, $\boxed{x^2}$, $\boxed{+}$, V_L value, $\boxed{x^2}$, $\boxed{=}$, $\boxed{\sqrt{x}}$

FORMULA 16–6 $Z = \sqrt{R^2 + X_L{}^2}$

R value, $\boxed{x^2}$, $\boxed{+}$, X_L value, $\boxed{x^2}$, $\boxed{=}$, $\boxed{\sqrt{x}}$

FORMULA 16–7 $I_T = \sqrt{I_R{}^2 + I_L{}^2}$

I_R value, $\boxed{x^2}$, $\boxed{+}$, I_L value, $\boxed{x^2}$, $\boxed{=}$, $\boxed{\sqrt{x}}$

FORMULA 16–8 $Z = \dfrac{RX_L}{R^2 + X_L{}^2}$

R value, $\boxed{\times}$, X_L value, $\boxed{\div}$, $\boxed{(}$, R value, $\boxed{x^2}$, $\boxed{+}$, X_L value, $\boxed{x^2}$, $\boxed{)}$,
$\boxed{\sqrt{x}}$, $\boxed{=}$

REVIEW QUESTIONS

1. A series RL circuit contains a resistance of 20 Ω and an inductive reactance of 30 Ω. Find the following parameters using the specified techniques.
   a. Use the plot-and-measure technique to find Z. (Round-off to the nearest whole number.) Measure the θ with a protractor. (Round-off to the nearest degree.)
   b. Use Pythagorean theorem to find Z.
   c. Use trig to find the angle between R and Z.

2. A parallel RL circuit has one branch consisting of a 200-Ω resistance, and another branch consisting of an inductor with X_L equal to 150 Ω. The applied voltage is 300 V. Find the values for the following parameters. Draw and label the schematic and vector diagrams.
   a. I_R
   b. I_L
   c. I_T
   d. θ
   e. Z

3. Draw the circuit diagram, voltage-current vector diagram and impedance diagram for a series RL circuit having one R and one L. Voltage drop across the 10-kΩ resistor is 5 V, and circuit phase angle is 45°. Label all components and vectors, and show all calculations.

4. Draw the circuit diagram and voltage-current vector diagram of a parallel RL circuit having one R and one L. Assume a total current of 12 A and a phase angle of $-75°$. Label all components and vectors in the diagrams, and show all calculations. (Round answers to the nearest whole number and the nearest degree values.)

Answer the following with "I" for increase, "D" for decrease, and "RTS" for remain the same.

5. If frequency of the applied voltage in Question 2 doubles,
   a. R will _____
   b. L will _____
   c. X_L will _____

d. Z will _____
e. θ will _____
f. I_T will _____
g. V_T will _____

6. If the inductance in Question 3 doubles while R remains the same,
   a. R will _____
   b. L will _____
   c. X_L will _____
   d. Z will _____
   e. θ will _____
   f. I_T will _____
   g. V_T will _____

7. If the R value in Question 3 doubles while X_L remains the same,
   a. R will _____
   b. L will _____
   c. X_L will _____
   d. Z will _____
   e. θ will _____
   f. I_T will _____
   g. V_T will _____

8. If the inductance in Question 4 doubles while R remains the same,
   a. R will _____
   b. L will _____
   c. X_L will _____
   d. Z will _____
   e. θ will _____
   f. I_T will _____
   g. V_T will _____

9. If the R value in Question 4 doubles while X_L remains the same,
   a. R will _____
   b. L will _____
   c. X_L will _____
   d. Z will _____
   e. θ will _____
   f. I_T will _____
   g. V_T will _____

10. If both R and X_L values in Question 4 are halved:
    a. R will _____
    b. L will _____
    c. X_L will _____
    d. Z will _____
    e. θ will _____
    f. I_T will _____
    g. V_T will _____

11. Refer to Figure 16–35 and solve for the following:
    a. Z
    b. V_R
    c. θ

67 V
1060 Hz

10 kΩ

3 H

FIGURE 16–35

12. Refer to Figure 16–36 and solve for the following:
    a. X_L
    b. V_L
    c. V_R
    d. θ

200 V
60 Hz

30 H

R

10 mA

A

FIGURE 16–36

13. Refer to Figure 16–37 and solve for the following:
    a. I
    b. Z
    c. V_L
    d. X_L
    e. L

200 V
60 Hz

L

170 V

R
34 kΩ

FIGURE 16–37

CHALLENGE QUESTIONS

1. What calculator keystrokes would you use to find the sine of 57 degrees?

2. What calculator keystrokes would you use to find the $\tan^{-1}$ related to a ratio of 0.80?

3. What calculator keystrokes would you use to find the cosine of 34 degrees?

4. What calculator keystrokes would you use to find the arcsine for a ratio of 0.75?

5. What, if any, are the advantages of using the tangent function for ac circuit analysis?

6. How many radians are represented by 75 degrees?

TEST YOURSELF

1. Draw the circuit diagram and all vector diagrams of the series RL circuit described below to find and illustrate all voltage, current, impedance, and phase-angle parameters. Show all calculations.
   Given: Inductance = 10 H
   Voltage leads current by 60°
   Frequency = 1 kHz
   Total voltage (V_T) = 72.5 V

2. Draw the circuit diagram and all vector diagrams for the parallel RL circuit described below to find and illustrate all voltage, current, and phase-angle parameters. Show all calculations.
   (NOTE: Rounding-off is acceptable for this problem.)
   Given: Inductance = 1.99 H
   Frequency = 400 Hz
   Resistance = 7 kΩ
   V_T = 21 V

3. Draw the circuit diagram for a series RL circuit containing two 4-mH inductances and two 5-kΩ resistors. The circuit has a frequency of 200 kHz. Find I_T if the applied voltage is 140 V. (Round-off to the nearest kilohm, volt, and milliampere, as appropriate.)

PERFORMANCE PROJECTS CORRELATION CHART

Suggested performance projects in the Laboratory Manual that correlate with topics in this chapter are:

CHAPTER TOPIC	PERFORMANCE PROJECT	PROJECT NUMBER
Fundamental Analysis of Series RL Circuits	Review of V, I, R, Z, and θ Relationships in a Series RL Circuit	45
	V, I, R, Z, and θ Relationships in a Series RL Circuit	46
Fundamental Analysis of Parallel RL Circuits	Review of V, I, R, Z, and θ Relationships in a Parallel RL Circuit	47
	V, I, R, Z, and θ Relationships in a Parallel RL Circuit	48

After completing the above projects, perform "Summary" checkout at the end of the performance projects section in the Laboratory Manual.

THE SIMPLER SEQUENCE FOR TROUBLESHOOTING

Following the SIMPLER troubleshooting sequence, find the problem with this circuit. As you follow the sequence, record your responses to each step on a separate sheet of paper.

CHALLENGE CIRCUIT 8

SYMPTOMS Gather, verify, and analyze symptom information. (Look at the "Starting Point Information.")

IDENTIFY Identify initial suspect area for the location of the trouble.

MAKE Make a decision about "What type of test to make" and "Where to make it." (To simulate making a decision about each test you want to make, select the desired test from the "TEST" column listing below.)

PERFORM Peform the test. (Look up the result of the test in Appendix C. Use the number in the "RESULTS" column to find the test result in the Appendix.)

LOCATE Locate and define a new "narrower" area in which to continue troubleshooting.

EXAMINE Examine available information and again determine "what type of test" and "where."

REPEAT Repeat the preceding analysis and testing steps until the trouble is found. What would you do to restore this circuit to normal operation? When you have solved the problem, compare your results with those shown on the following pages.

STARTING POINT INFORMATION

1. Circuit diagram
2. V_{R_1} is higher than it should be
3. V source checked as being 7.8 V

TEST — Results in Appendix C

V_{R_1} . . . .(34)
V_{L_1} . . . .(89)
V_{L_2} . . . .(48)
R_1 . . . .(109)

Meter lead
Meter lead

Symptoms The voltage drop across the resistor is slightly higher than it should be. This could be due to an inductor that has decreased in value due to shorted turns or a resistor that has increased or opened since, at this point, we don't know how high the voltage is across the resistor.

Identify initial suspect area: All components (resistor and two inductors) are in the initial suspect area.

Make test decision: Since inductors typically open more frequently than they short and since we don't know how high the voltage is across the resistor, it makes sense to check the voltage across R_1.

Perform 1st Test: V_{R_1} is 6.4 V. If everything were normal, the resistor would drop 5.9 V, and each inductor would drop about 2.6 V, since their inductive reactances at this frequency would be close to 4.35 kΩ each. The total X_L is about 8.7 kΩ and the rated R is 10 kΩ. Z would then equal about 13.3 kΩ, and circuit current would be approximately 0.59 mA. Thus, the vector sums of the resistive and inductive voltage drops would equal approximately 7.8 V.

Locate new suspect area: The new suspect area stays the same as the initial suspect area since we do not have enough information to eliminate the inductors or the resistor.

Examine available data.

Repeat analysis and testing: It would be good to look at the inductor voltage drops so we can draw some conclusions.

CHALLENGE CIRCUIT 8

Measure V_{L_1}. V_{L_1} is close to 2.3 volts, which is slightly lower than normal.

Measure V_{L_2}. V_{L_2} is close to 2.36 volts, which is slightly lower than normal.

Measure R_1. R_1 is measured with power removed and equals 12 kΩ, which is higher than the 10 kΩ it should be.

Since R_1 resistance is high, replace R_1 with a new 10-kΩ resistor. After replacing the resistor, all the circuit parameters become normal.

2nd Test

3rd Test

4th Test

5th Test

Basic Transformer Characteristics

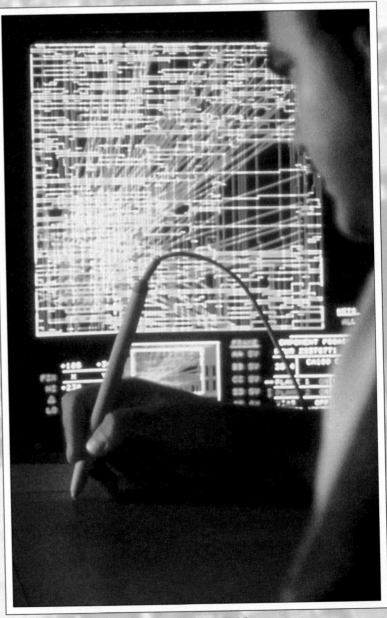

Courtesy of IBM

17

KEY TERMS

Autotransformer

Coefficient of coupling
(k)

Copper loss

Core loss

Current ratio

Impedance ratio

Isolation transformer

Mutual inductance
(M or L_M)

Transformers

Turns ratio

Voltage ratio

OUTLINE

Background information

Coefficient of coupling

Mutual inductance and transformer
action

Mutual inductance between coils other
than transformers

Important transformer ratios

Transformer losses

Characteristics of selected transformers

Troubleshooting hints

584

CHAPTER PREVIEW

We stated earlier that one application of electromagnetic induction is the transformer. Furthermore, you know these devices have application in different frequency ranges and for various purposes. That is, there are power transformers, audio transformers, and radio-frequency (rf) transformers, which all find various applications in many types of electronic circuits and systems. Power supplies, audio amplifiers, radio and TV receivers, transmitters, and other systems and subsystems use the unique characteristics of these devices.

In this chapter, you will study the basic features and types of transformers that a technician needs to understand. Practical information, such as schematic symbols and color codes, will be provided. Turns ratios, voltage ratios, current ratios, impedance ratios, and other major aspects about transformers will be studied. Finally, typical malfunctions and troubleshooting techniques will be examined.

OBJECTIVES

After studying this chapter, you should be able to:

- Define **mutual inductance**
- Calculate mutual inductance values
- Calculate **coefficient of coupling** values
- Calculate **turns, voltage, current,** and **impedance ratios**
- List, draw, or explain physical, magnetic, electrical, and schematic characteristics of various **transformers**
- List common transformer color codes
- Define at least two types of **core losses**
- List common problems found in transformers
- List troubleshooting procedures

BACKGROUND INFORMATION

You know when a changing current passes through a conductor wire or a coil, an expanding or collapsing magnetic field is produced. This expanding or collapsing field, in turn, induces a counter-emf (cemf). This process of producing voltage through a changing magnetic field is known as electromagnetic induction. When the induced voltage is created across the current-carrying conductor or coil itself, it is called self-induction. As you know, the symbol for this self-inductance is L.

Another important kind of induction is **mutual induction.** The unit of mutual inductance is the henry and the symbol is M. (NOTE: Sometimes you will also see the abbreviation L_M.)

One definition for mutual inductance is the property of inducing voltage in one circuit by varying the current in another circuit. Refer to

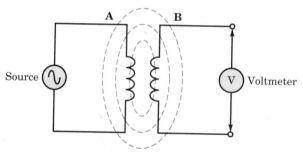

FIGURE 17–1 **Mutal induction by
flux linkage**

Figure 17–1 and note that Coil A is connected to an ac source. Coil B is not
connected to any source. However, it is located close to coil A so that most
of the flux produced in Coil A by current from the source cuts or "links" Coil
B. Since a *conductor, magnetic field,* and *relative motion* will induce voltage,
a voltage is induced in Coil B because of the changing current in Coil A. By
definition, there must be mutual inductance. These two circuits (Coil A and
Coil B) are positioned so that energy is transferred by *magnetic linkage.*
This coupling between circuits is also sometimes called *inductive coupling.*
Inductive coupling is due to flux linkages.

COEFFICIENT OF COUPLING

In Figure 17–2 you can see smaller and greater amounts of flux linkage
between coils. A term expressing this relationship by the fractional amount
of total flux linking the two coils is **coefficient of coupling,** represented by
the letter **k.** Coefficient of coupling and mutual inductance are *not* the same.
However, they are related. Figure 17–2b shows the coils with the higher
coefficient of coupling. The coefficient of coupling is therefore computed as
the fractional amount of the total flux linking the two circuits. When *all* the
flux from one circuit links the other, coefficient of coupling *(k)* equals one.
When two-thirds of the flux links the other circuit, *k* equals 0.66, and so
forth.

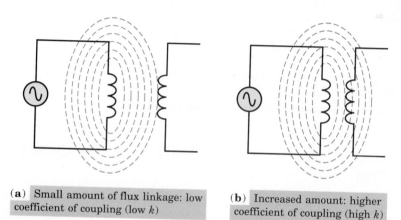

(**a**) Small amount of flux linkage: low
coefficient of coupling (low *k*)

(**b**) Increased amount: higher
coefficient of coupling (high *k*)

FIGURE 17–2   **Coefficient of coupling related to degree of
flux linkage**

| Close proximity means more coupling | Greater distance (less proximity) means less coupling | Aligned windings means more coupling | Perpendicular coils have small amount of coupling | Windings on same iron core have high degree of coupling | Air core separate windings have small amount of coupling |

Proximity   Relative position   Core material and how it is wound

FIGURE 17–3   Factors influencing coefficient of coupling (k)

See Figure 17–3 for examples of factors that influence the coefficient of coupling. These factors include the proximity of coils (the closer they are the higher the k); the relative positions of coils with respect to each other (parallel coils at a given distance have greater k than perpendicular coils); and other factors, such as when the coils are wound on the same core. Coils wound on the same iron core have a coefficient of coupling almost equal to unity (1). Virtually all the flux produced by one coil links the other coil, with little leakage flux occurring. Air-core coils (often used in rf circuits) have a k that indicates a percentage of coupling from less than 5% ($k = 0.05$) to about 35% ($k = 0.35$). Obviously, air-core coils have more leakage flux.

✔ **IN-PROCESS LEARNING CHECK I**

Fill in the blanks as appropriate.

1. Producing voltage via a changing magnetic field is called electromagnetic _____.

2. Inducing voltage in one circuit by varying current in another circuit is called _____ .

3. The fractional amount of the total flux that links two circuits is called the _____ of coupling, which is represented by the letter _____. When 100% of the flux links the two circuits, the _____ of coupling has a value of _____.

4. The closer coils are, the _____ the coupling factor produced. Compared to parallel coils, perpendicular coils have a _____ degree of coupling.

MUTUAL INDUCTANCE AND TRANSFORMER ACTION

As stated, when two circuits are coupled magnetically, they have mutual inductance. Such is the case for transformers. **Transformers** are devices that transfer energy from one circuit to another by electromagnetic induction (mutual induction). Typically, transformers have their *primary* and *secondary* windings wound on a single core.

Refer to Figure 17–4. Notice the winding connected to the source is the primary winding. The winding connected to the load is the secondary winding. Some transformers can have more than one secondary winding. Also, observe the schematic symbol(s) used for various transformers.

Another feature shown on some diagrams, Figure 17–4, are sense dots. Sense dots indicate the ends of the windings that have the same polarity at any given moment. Typically, when sense dots are *not* shown, there is 180-degree difference between the primary and secondary voltage polarity. That is, if the top of the primary is positive with respect to a given reference point, the top of the secondary is negative, at that same moment.

Two circuits have a mutual inductance of 1 henry when a rate of current change of one ampere per second in the first circuit induces a voltage of one volt in the second circuit.

Mutual inductance between windings is directly related to the inductance of each winding and the coefficient of coupling. The formula is:

$$\boxed{\text{Formula } 17-1 \quad \bigg| \quad M = k\sqrt{L_1 \times L_2}}$$

k value, $\boxed{\times}$, $\boxed{(}$, L_1 value, $\boxed{\times}$, L_2 value, $\boxed{)}$, $\boxed{\sqrt{x}}$, $\boxed{=}$

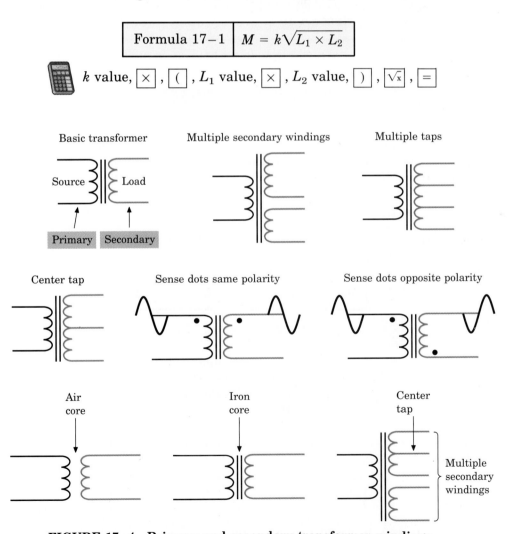

FIGURE 17–4   Primary and secondary transformer windings

where M = mutual inductance in henrys
L_1 = self-inductance of the primary and L_2 = self-inductance of
the secondary (for transformers) where Ls are in henrys

EXAMPLE

A transformer has a k of 1 (100% coupling of flux lines), and the primary and secondary each have inductances of 3 H. The mutual inductance (M or L_M) is computed as $M = 1\sqrt{3 \times 3} = 1\sqrt{9} = 3$ H. From our definition, this indicates a current change of one ampere per second in the primary induces 3 volts in the secondary.

PRACTICE PROBLEMS I

What is the mutual inductance of a transformer where the coefficient of coupling is 0.95, the primary inductance is 10 H, and the secondary inductance is 15 H?

MUTUAL INDUCTANCE BETWEEN COILS OTHER THAN TRANSFORMERS

We have been discussing mutual inductance in transformers. How does mutual inductance (coupled inductance) affect the total inductance of coils other than transformers that are located where their fields can interact?

Series Coils

When two coils are connected in *series* so their fields *aid* each other, total inductance is calculated from the formula:

$$\boxed{\text{Formula 17-2} \quad L_T = L_1 + L_2 + 2M}$$

 L_1 value, $\boxed{+}$, L_2 value, $\boxed{+}$, 2, $\boxed{\times}$, M value, $\boxed{=}$

where L_T = total inductance (H)
L_1 = inductance of one coil (H)
L_2 = inductance of other coil (H)
M = mutual inductance (H)

When two coils are connected in *series* so their fields oppose each other, the formula becomes:

$$\boxed{\text{Formula 17-3} \quad L_T = L_1 + L_2 - 2M}$$

L_1 value, $\boxed{+}$, L_2 value, $\boxed{-}$, 2, $\boxed{\times}$, M value, $\boxed{=}$

Therefore, the general formula for series inductors is:

| Formula 17–4 | $L_T = L_1 + L_2 \pm 2M$ |

 L_1 value, $\boxed{+}$, L_2 value, $\boxed{+}$ or $\boxed{-}$ as appropriate, 2, $\boxed{\times}$,

M value, $\boxed{=}$

EXAMPLE

Two inductors are series connected so their fields aid one another. Inductor 1 has an inductance of 8 H. Inductor 2 also has an inductance of 8 H. Their mutual inductance is 4 H. What is the total inductance of the circuit?

$$L_T = L_1 + L_2 + 2M = 8 + 8 + 8 = 24 \text{ H}$$

PRACTICE PROBLEMS II

What is the total inductance of two series-connected 10-H inductors whose mutual inductance is 5 H and whose magnetic fields are opposing?

Parallel Coils

When two inductors are in *parallel* with their fields *aiding* each other, the formula is:

| Formula 17–5 | $L_T = \dfrac{1}{\dfrac{1}{L_1 + M} + \dfrac{1}{L_2 + M}}$ |

 L_1 value, $\boxed{+}$, M value, $\boxed{=}$, $\boxed{1/x}$, $\boxed{+}$, $\boxed{(}$, L_2 value, $\boxed{+}$,

M value, $\boxed{)}$, $\boxed{1/x}$, $\boxed{=}$, $\boxed{1/x}$

When two inductors are in *parallel* with their fields *opposing* each other, the formula becomes:

| Formula 17–6 | $L_T = \dfrac{1}{\dfrac{1}{L_1 - M} + \dfrac{1}{L_2 - M}}$ |

L_1 value, $\boxed{-}$, M value, $\boxed{=}$, $\boxed{1/x}$, $\boxed{+}$, $\boxed{(}$, L_2 value, $\boxed{-}$,

M value, $\boxed{)}$, $\boxed{1/x}$, $\boxed{=}$, $\boxed{1/x}$

PRACTICE PROBLEMS III

1. What is the total inductance of two parallel coils whose fields are aiding, if $L_1 = 8$ H, $L_2 = 8$ H, and $M = 4$ H?

2. What is the total inductance of two parallel coils whose fields are opposing, if $L_1 = 8$ H, $L_2 = 8$ H, and $M = 4$ H?

Relationship of *k* and *M*

The coefficient of coupling and mutual inductance are related by the following formulas for k and M.

Formula 17–7	$k = \dfrac{M}{\sqrt{L_1 \times L_2}}$

 M value, $\boxed{\div}$, $\boxed{(}$, L_1 value, $\boxed{\times}$, L_2 value, $\boxed{)}$, $\boxed{\sqrt{x}}$, $\boxed{=}$

Recall Formula 17–1 is $M = k\sqrt{L_1 \times L_2}$. As you can see from these two formulas, mutual inductance and coefficient of coupling are directly related.

IMPORTANT TRANSFORMER RATIOS

We will begin this section by making the following assumptions.

1. With *no load* connected to the secondary winding, the primary winding acts like a simple inductor.

2. Current flowing in the primary depends on applied voltage and inductive reactance of the primary.

$$I_P = \frac{V}{X_P}$$

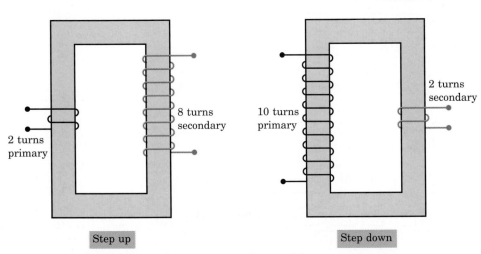

FIGURE 17–5   Step-up and step-down transformers

3. The primary current produces a magnetic field that cuts both the primary and secondary turns.

4. A cemf is produced in the primary. This cemf almost equals the applied voltage due to self-inductance.

5. A cemf is produced in *each turn* of the secondary, which equals the voltage induced in *each turn* of the primary due to mutual inductance. This means when there are equal turns on the secondary and the primary, their voltages are equal.

Using these assumptions, let's look at the transformation ratio. That is, the ratio between electrical output and input parameters under specified conditions. When 100% of the primary flux links the secondary winding (when $k = 1$), the ratio of the induced voltages (i.e., the ratio of secondary voltage to the primary voltage) is the same as the ratio of turns on the secondary compared to the turns on the primary. The symbol for transformation ratio is the Greek letter alpha (α). Thus,

$$\alpha = \frac{V_S}{V_P} = \frac{N_S}{N_P}$$

where V_S is secondary induced voltage; V_P is primary induced voltage; N_S is the number of turns on the secondary winding; and N_P is the number of turns on the primary winding.

Observe Figure 17–5 and notice how the terms step-up and step-down relate to transformers. When the secondary (output) voltage is higher than the primary (input) voltage, the transformer is a step-up transformer. Conversely, when the secondary voltage is lower than the primary voltage, the transformer is a step-down transformer.

Turns Ratio

It is often stated the ratio of secondary turns to primary turns (N_S/N_P) is the **turns ratio** (TR). For example, when the secondary has 1,000 turns and the primary has 250 turns, the turns ratio is

$$\frac{1,000}{250} = 4{:}1$$

This is called the secondary-to-primary (s-p) turns ratio, Figure 17– 6a.

PRACTICAL NOTES

A number of texts use the opposite approach; i.e., primary-to-secondary ratio. Either approach works as long as formulas are consistently set up using the same approach. In any case, the technician should be careful in making transformer calculations. Be *consistent* in the way you set up the various ratios. We will use both ratios, as appropriate to promote familiarity.

FIGURE 17–6 Turns and voltage ratio (Adapted with permission from Duff and Herman, *Alternating Current Fundamentals,* 3rd edition, © 1986 by Delmar Publishers Inc., Albany, New York)

Voltage Ratio

As you learned with the ideal transformer, the **voltage ratio** has the same ratio as the turns ratio, Figure 17–6b. That is:

$$\text{Formula 17–8} \quad \frac{V_S}{V_P} = \frac{N_S}{N_P}$$

 V_S value, $\boxed{\div}$, V_P value, $\boxed{=}$, (should equal) N_S value, $\boxed{\div}$, N_P value, $\boxed{=}$

EXAMPLE

1. What is the secondary-to-primary turns ratio of a transformer having eight times as many turns on the primary as on the secondary?

$$TR = \frac{N_S}{N_P} = \frac{1}{8} \text{ or } 1{:}8$$

2. What is the secondary voltage on a transformer having a turns ratio of 14:1 (s-p), if the applied voltage is 20 V?

$$\frac{V_S}{V_P} = \frac{N_S}{N_P}; \frac{V_S}{20} = \frac{14}{1}; V_S = 20 \times 14 = 280 \text{ V}$$

PRACTICE PROBLEMS IV

1. What is the s-p turns ratio of an ideal transformer whose secondary voltage is 300 V when 100 V is connected to the primary?

2. What is the s-p turns ratio of an ideal transformer that steps-up the input voltage six times?

3. What is the secondary voltage of an ideal transformer whose s-p turns ratio is 5:1 and whose primary voltage is 50 V?

4. What is the voltage ratio of an ideal transformer whose s-p turns ratio is 1:4? Is this a step-up or a step-down transformer?

Current Ratio

Up to now, our discussions have been about transformers with no load connected to the secondary(ies). Notice in Figure 17–7 that when a load is connected to the secondary, current flows through the load and the secondary. This current produces a magnetic field related to the secondary current. The polarity produced by the secondary current cancels some of the primary field. Hence, the cemf is reduced in the primary, which reduces the impedance of the primary to current flow. This reduced impedance causes primary current to increase. The interaction between the secondary and the primary is termed reflected impedance. In other words, the impedance reflected across the input or primary results from current in the output or secondary.

In Figure 17–7, the power available to the load from the transformer secondary must come from the source that supplies the primary, since a transformer does not *create* power, but *transfers* power from the primary circuit to the secondary circuit. Transformers are not 100% efficient. Therefore, the power delivered to the primary by its source is slightly higher than

FIGURE 17–7   Current ratio

the power delivered by the secondary to its load. Since transformers are highly efficient (generally ranging from 90-98% efficient), Figure 17−7 has assumed 100% efficiency to illustrate this principle. Later in the chapter, we will look at some losses that prevent actual 100% efficiency.

Also notice in Figure 17−7 the power delivered to the load is 40 W, where $V_S = 20$ V, $R_S = 10$ Ω, and $I_S = 2$ A. The power delivered to the primary is 40 W, where $V_P = 100$ V and $I_P = 0.4$ A. This is a step-down transformer where the voltage has been stepped down five times. That is,

$$\frac{N_S}{N_P} = 1:5 \text{ and } \frac{V_S}{V_P} = 1:5$$

What about the **current ratio?** Observe the current in the secondary is five times higher than the current in the primary. Using our secondary-to-primary principle, the current ratio is 5:1, or the inverse of the voltage ratio.

This makes sense, if we assume the power in the secondary is the same as the power in the primary. In other words, $V_S I_S = V_P I_P$. If voltage were decreased by five times, the current would have to be increased by five times so the V times I (or P) would be the same in both primary and secondary.

(NOTE: Power supplied to the primary must equal power delivered to the secondary plus power lost.)

Now let's relate current ratio to the turns and voltage ratios. Thus,

$$\frac{I_P}{I_S} = \frac{N_S}{N_P} = \frac{V_S}{V_P}$$

Note the inversion in the current ratio. Since quantities equaling the same thing are equal to each other, we can list three useful equations from the preceding information.

Formula 17−9	$\dfrac{N_S}{N_P} = \dfrac{V_S}{V_P}$

N_S value, $\boxed{\div}$, N_P value, $\boxed{=}$, (should equal) V_S value, $\boxed{\div}$, V_P value, $\boxed{=}$

Formula 17−10	$\dfrac{N_S}{N_P} = \dfrac{I_P}{I_S}$

N_S value, $\boxed{\div}$, N_P value, $\boxed{=}$ (should equal) I_P value, $\boxed{\div}$, I_S value, $\boxed{=}$

Formula 17−11	$\dfrac{V_S}{V_P} = \dfrac{I_P}{I_S}$

V_S value, $\boxed{\div}$, V_P value, $\boxed{=}$, (should equal) I_P value, $\boxed{\div}$, I_S value, $\boxed{=}$

If you know any three of the quantities in any of these equations, you can solve for the unknown quantity, assuming 100% efficiency.

EXAMPLES

1. A transformer has 1,000 turns on the secondary, 250 turns on the primary, and the primary voltage is 120 V. What is the secondary voltage value?

$$\frac{N_S}{N_P} = \frac{V_S}{V_P}$$

$$\frac{1,000}{250} = \frac{V_S}{120}$$

$$250V_S = 120,000$$

$$V_S = \frac{120,000}{250} = 480 \text{ V}$$

2. A transformer has a secondary-to-primary turns ratio of 1:5. What is the primary current when the secondary load current is 0.5 A?

$$\frac{N_S}{N_P} = \frac{I_P}{I_S}$$

$$\frac{1}{5} = \frac{I_P}{0.5}$$

$$5I_P = 1 \times 0.5$$

$$I_P = \frac{0.5}{5} = 0.1 \text{ A}$$

3. The primary-to-secondary current ratio of a transformer is 4:1, and the secondary voltage is 60 V. What is the voltage on the primary?

$$\frac{V_S}{V_P} = \frac{I_P}{I_S}$$

$$\frac{60}{V_P} = \frac{4}{1}$$

$$4V_P = 60 \times 1$$

$$V_P = \frac{60}{4} = 15 \text{ V}$$

Impedance Ratio

The **impedance ratio** is related to the turns ratio, Figure 17–8. Also, since inductance and thus the inductive reactance of a coil, relates to the square of the turns, the impedance also has a similar relationship. A load of a given

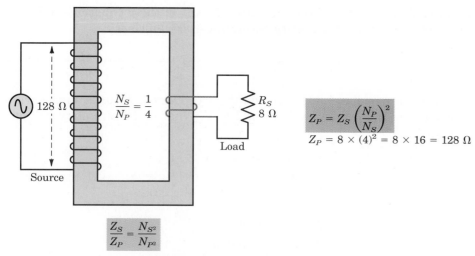

$$Z_P = Z_S \left(\frac{N_P}{N_S}\right)^2$$
$$Z_P = 8 \times (4)^2 = 8 \times 16 = 128 \ \Omega$$

$$\frac{Z_S}{Z_P} = \frac{N_{S^2}}{N_{P^2}}$$

FIGURE 17–8   Impedance ratio

impedance connected to the transformer secondary is transformed to a different value by what the source sees "looking into" the primary (unless the turns ratio is 1:1). In fact, for an ideal (zero loss) transformer, the impedance transformation (secondary-to-primary) is proportional to the square of the secondary-to-primary turns ratio. That is,

Formula 17–12 $$\left(\frac{N_S}{N_P}\right)^2 = \frac{Z_S}{Z_P}$$

where Z_S = the impedance of the load connected to the secondary
Z_P = the impedance looking into the primary from source
$\dfrac{N_S}{N_P}$ = the secondary-to-primary turns ratio

  N_S value, $\boxed{\div}$, N_P value, $\boxed{=}$, $\boxed{x^2}$, (should equal) Z_S value,
$\boxed{\div}$, Z_P value, $\boxed{=}$

Another way of thinking about impedance ratios and their relationship to turns ratios is to consider that impedance can be regarded as the ratio of V to I. That is $Z = V/I$. For example, if we consider a step-up transformer that steps up voltage by two times, we understand the current is "stepped-down" by two times (or is one-half). Then our ratio of V/I has become $2V/0.5I$, which translates to a ratio of 4. Thus, the impedance of the winding with the higher turns is four times that of the winding with the lesser turns. Of course, the converse is true if the transformer is a step-down transformer. The winding with the lesser turns would have half the voltage and twice the current; thus, the $0.5V/2I$ ratio would indicate that the winding with the lower number of turns has an impedance that is one-fourth that of the winding with the greater number of turns. Therefore, we see again, that the impedance ratio is related to the square of the turns ratio.

EXAMPLE

When the N_S/N_P turns ratio of a transformer is 1:4 and the impedance connected to the secondary is 8 Ω, the primary impedance is related to the square of the turns ratio. Hence, Z_P is 16 times Z_S. That is $Z_P = 8 \times 16 = 128$ Ω, Figure 17–8. A practical formula expressing this in terms of *primary-to-secondary* turns ratio is:

Formula 17–13	$Z_P = Z_S \left(\dfrac{N_P}{N_S}\right)^2$

 Z_S value, $\boxed{\times}$, $\boxed{(}$, N_P value, $\boxed{\div}$, N_S value, $\boxed{)}$, $\boxed{x^2}$, $\boxed{=}$

EXAMPLE

A transformer has a secondary-to-primary turns ratio of 0.5, and a load resistance of 2,000 ohms is connected to the secondary. What is the apparent primary impedance?

Since the secondary-to-primary turns ratio is 0.5, the secondary has half as many turns as the primary, or the primary has twice as many turns as the secondary. Appropriately substitute in the formula:

$$Z_P = Z_S \left(\frac{N_P}{N_S}\right)^2$$

$$Z_P = 2{,}000 \ \Omega \times (2)^2$$

$$Z_P = 2{,}000 \ \Omega \times 4 = 8{,}000 \ \Omega$$

PRACTICE PROBLEMS V

In the prior example, suppose the turns ratio reverses. In other words, the secondary has twice as many turns as the primary. What is the apparent primary impedance?

These discussions indicate the primary impedance of a very efficient transformer (as it looks to the power source feeding the primary) is determined by the load connected to the secondary and the turns ratio. Because this is true, transformers are often used to match impedances. That is, a transformer is used so that a secondary load of given value is transformed to the optimum impedance value into which the primary source should be looking. (The source may be an amplifier stage, another ac signal, or power source.) Therefore, a useful formula is:

$$\boxed{\text{Formula 17--14} \quad \frac{N_P}{N_S} = \sqrt{\frac{Z_P}{Z_S}}}$$

where $\dfrac{N_P}{N_S}$ = the required *primary-to-secondary* turns ratio
Z_P = the primary impedance desired (required)
Z_S = the impedance of the load connected to the secondary

 N_P value, $\boxed{\div}$, N_S value, $\boxed{=}$, (should equal) $\boxed{(}$, Z_P value, $\boxed{\div}$, Z_S value, $\boxed{)}$, $\boxed{\sqrt{x}}$

EXAMPLE

An amplifier requires a load of 1,000 ohms for best performance. The amplifier output is to be connected to a loudspeaker having an impedance of 10 ohms. What must the turns ratio *(primary-to-secondary)* be for a transformer used for impedance matching?

$$\frac{N_P}{N_S} = \sqrt{\frac{Z_P}{Z_S}}$$

$$\frac{N_P}{N_S} = \sqrt{\frac{1,000}{10}} = \sqrt{100} = 10$$

This indicates the primary must have 10 times as many turns as the secondary.

PRACTICE PROBLEMS VI

1. What is the primary-to-secondary impedance ratio of a transformer that has 1,000 turns on its primary and 200 turns on its secondary?

2. What impedance would the primary source see looking into a transformer where the primary-to-secondary turns ratio is 16 and the load resistance connected to the secondary is 4 Ω?

TRANSFORMER LOSSES

Because of energy losses, practical transformers do not meet all the criteria for the ideal, 100% efficient devices we have been discussing. These transformer losses consist primarily of two categories. One category is **copper loss,** due to the I^2R copper wire losses that make up the primary and secondary windings. Current passing through the resistance of the wire produces energy loss in the form of heat.

The other category of transformer losses is termed **core loss.** Core losses result from two factors. One factor is called *eddy current* losses. Due to the changing magnetic fields and the iron-core conductivity, currents are produced in the iron core that do not aid transformer output. These eddy currents produce an I^2R core loss related to the core material resistance and the current amount. To reduce the current involved, the iron core is usually constructed of thin sheets of metal called *laminations*. These laminations are electrically insulated from each other by varnish or shellac. Reducing the cross-sectional area of the conductor by laminating significantly increases the core resistance, which decreases the current, and greatly reduces the I^2R loss.

The other core loss is called *hysteresis loss*. Recall magnetic domains in the magnetic core material require energy to rearrange. (Recall in an earlier chapter, the lagging effect of the core magnetization behind the magnetizing force was illustrated by the hysteresis loop). Since ac current constantly changes in magnitude and direction, the tiny molecular magnets within the core are constantly being rearranged. This process requires energy that causes some loss.

Hysteresis loss increases as the frequency of operation is increased. For this reason, iron-core transformers, such as those used at power frequencies, are not used in radio frequency applications.

Copper losses, core losses, leakage flux, and other factors prevent a transformer from being 100% efficient. Efficiency percentage is calculated as output power divided by input power times 100.

Formula 17–15	Efficiency % $= \dfrac{P_{out}}{P_{in}} \times 100$

P_{out} value, $\boxed{\div}$, P_{in} value, $\boxed{\times}$, 100, $\boxed{=}$

To minimize these various losses, the physical features of coils and transformers used for applications in power, audio, and rf frequency ranges slightly differ. We will investigate some of these characteristics in the next section.

CHARACTERISTICS OF SELECTED TRANSFORMERS

Power Transformers

Power transformers, Figure 17–9, typically have the following features:

1. Heavy (due to iron core) and can be very large, like the transformers on telephone poles. However, power transformers in electronic devices are much smaller and vary considerably in size.

2. Laminated iron core to reduce eddy-current losses.

3. As short a flux path as possible to reduce leakage flux and minimize magnetizing energy needed.

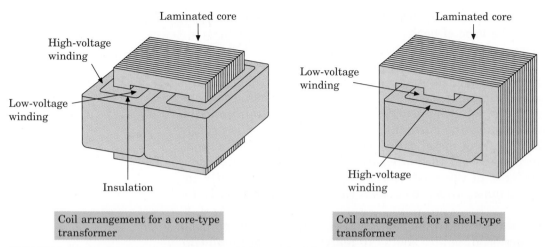

FIGURE 17–9   Power transformers (Adapted with permission from Perozzo, *Practical Electronics Troubleshooting,* by Delmar Publishers Inc., Albany, New York)

4. Use one of two core shapes—core type (primary and secondary windings on separate legs of the core) and shell type (both primary and secondary windings on the center leg), again see Figure 17–9.

5. One or more primary and secondary windings.

6. May have *one* tapped winding, such as the **autotransformer,** Figure 17–10. The autotransformer is a special single-winding transformer.

When the input source is connected from one end to the tap while the output is across the whole coil, the voltage is stepped up. If the input voltage is applied across the whole coil and the output taken from the tap to one end, the voltage is stepped down.

Variacs or *Powerstats* are trade names for autotransformers having a movable wiper-arm tap that allows varying voltage from zero volts to the maximum level. They are often used as a variable ac source in experimental setups.

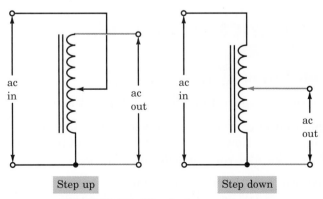

FIGURE 17–10   Autotransformer

> ### ⚠ SAFETY HINTS
>
> *Caution:* Bad shock can occur if no **isolation transformer** is used on the autotransformer input side, because there is no isolation from the raw line voltage and ground connections in the autotransformer circuit. For this reason, do *not* use autotransformers with a movable wiper-arm tap unless you plug the autotransformer into an isolation transformer output, which isolates it from the raw ac power source.

7. Isolates the circuitry connected to its secondary from the primary ac source (e.g., the ac powerline). A 1:1 turns ratio transformer is specifically designed for isolation applications, Figure 17–11. (*Note:* all standard transformers isolate secondary from primary source regardless of turns ratios, except for autotransformers.)

   When using this special 1:1 turns-ratio isolation transformer, the load on the secondary has no *direct* connection to the primary source. This provides a safety measure for people working with the circuitry connected to the isolation transformer secondary.

8. Cannot operate at high frequency due to excessive losses that would be created.

9. Color-coded leads, typically color-coded as shown in Figure 17–12a.

Audio Transformers

Audio transformers are similar to power transformers in their construction. Some typical features include:

1. Smaller and lighter than most power transformers.

2. Color-coded as shown in Figure 17–12b.

FIGURE 17–11   Using an isolation transformer for safety when using a powerstat or variac-type variable voltage autotransformer

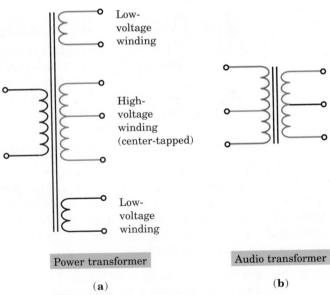

FIGURE 17–12   Typical power and audio transformer color-coding systems

RF Transformers and Coils

Because the flux is changing *very rapidly,* rf transformers or coils, Figures 17–13 and 17–14, have different features compared to power or audio transformers. Some features include:

1. Cores are air core, or powdered iron core, depending on frequency and application. Air-core coils have virtually zero hysteresis or eddy-current losses. Powdered iron cores are made by grinding metal materials and nonconductive materials into granules, which are pressed into a cylindrical-shaped "slug." Each metal particle is therefore insulated, greatly reducing losses such as eddy currents. See Figure 17–13a for examples of this coil.

   Also, nonconductive ferrite materials can be used in rf coils or transformers. As you recall, ferrite materials are special magnetic materials that provide easy paths for magnetic flux but are nonconductive. Often, toroid-type cores are used in radio-frequency circuits and lower-frequency applications, Figure 17– 13b.

2. Rf coils and transformers typically have very low inductance values. Generally, these values are in the microhenry or millihenry ranges.

3. Figure 17–14 shows examples of air-core coils found in transmitter applications.

TROUBLESHOOTING HINTS

Transformer problems are virtually the same problems as those described in the chapter on inductance. Two basic problems are opens or shorts. The primary test to detect these defects is the ohmmeter.

(a)

Open lead style, clamp fixing

Printed circuit style

Terminal pins
0.08 dia. × 0.16 long
located on 0.1 in. grid

A

B

Threaded
bushes

(b)

FIGURE 17–13   Examples of various types coils and cores: (a) Powdered iron-core type coils; (b) Toroidal cores (Photo courtesy of Miller Division of Bell Industries)

(a)

(b)

**FIGURE 17–14   Examples of air-core rf coils and transformers
(Courtesy of James Miller Electronics)**

⚠️ **SAFETY HINTS**

Once again, we want to remind you that when checking large
inductances (including transformer windings) with an ohmmeter, take precautions not to get shocked by the "inductive kick" of the winding's collapsing field when removing the ohmmeter leads from the winding under test.

Open Winding(s)

With an open winding, the obvious result is zero secondary output voltage,
even when source voltage is applied to the primary. When the primary is *not*
open and only one secondary of a multiple-secondary transformer is open,
only the open winding shows zero output. To check these conditions, remove
the power and use the ohmmeter.

Refer to Figure 17–15a to see an open winding. In this case, the ohmmeter registers an infinite resistance reading, indicating the winding under
test has an open. Both the primary and secondary windings should be
checked! Normal readings are the specified dc resistance values for each
winding.

Shorted Winding(s)

Shorted turns or a shorted winding can cause a fuse in the primary circuit
to blow (open) due to excessive current. Another possible clue to this condition is if the transformer operates at a hotter than normal temperature.
Also, another hint of shorted turns is lower than normal output voltage.
Actually, partially shorted windings or windings with very low *normal* resistances are hard to check only using an ohmmeter. Therefore, watch for
these clues when troubleshooting transformer problems.

FIGURE 17–15 Typical transformer troubles

However, the ohmmeter is useful to detect certain types of transformer short conditions. Figure 17–15b shows a total-winding short condition. In this case, the ohmmeter reads virtually zero ohms when it should be reading the normal winding resistance value, if there were no short.

Figure 17–15c illustrates the condition when the primary and secondary windings are shorted together. In this case, each winding resistance might be normal; however, there would be a low resistance reading between windings. When conditions are normal, "infinite ohms" is measured between primary and secondary windings.

Another possible short condition is when one or both windings are shorted to the metal core, Figure 17–15d. Again, if things are normal, an infinite ohms reading will be present between either winding and the core material.

PRACTICAL NOTES

The following discussion describes one practical approach used to determine whether a short in a *power transformer* exists. This approach is a simple test involving the transformer in question and a light bulb. In Figure 17–16, you can see a light bulb is placed in series with the transformer primary winding. When a short is present in *either* the primary or secondary, the lamp glows brightly due to the higher than normal primary current. When no short exists, the lamp does *not* glow or may glow dimly.

If lamp glows dimly or not at all = no short(s)
If lamp glows brightly = short(s); (throw transformer away!)

NOTE: Use small bulb rating for small transformers;
use larger bulb rating for large transformers.

FIGURE 17–16   One way to find shorted turns

The requirements for bulb size and its electrical ratings are:

1. Voltage rating should equal the same voltage rating as the transformer primary.

2. For small transformers, use a small wattage lamp.

3. For large transformers, use a large wattage lamp.

Be aware that choosing the correct bulb size and interpreting the results come only with experience. This test is simply mentioned to show you some innovative ways technicians find to learn *what they need to know*, even in difficult situations.

PRACTICAL TRANSFORMER RATINGS

POWER TRANSFORMERS	AUDIO TRANSFORMERS
Input volts	Primary impedance
Output V, I, and kVA ratings	Secondary impedance
Insulation ratings	Maximum V, I, and P
Size	Size
Mounting dimensions	Mounting dimensions
Weight	

SUMMARY

▶ When the magnetic flux of two coils link and interact, mutual inductance and inductive coupling occur between the two circuits.

▶ When mutual inductance is present, current changing in one circuit causes an induced voltage in the other circuit. The symbol for mutual inductance is M or L_M, and the unit of measure is the henry.

▶ One henry of mutual inductance is present when a change of one ampere per second in one circuit induces one volt in the mutually coupled circuit.

▶Expressing the fraction of total flux linking one circuit to another circuit is called coefficient of coupling. The abbreviation for coefficient of coupling is k. NOTE: Multiplying k by 100 yields the percentage of total flux lines linking the circuits.

▶The relationship between mutual inductance and coefficient of coupling is shown by the formulas:

$$M = k\sqrt{L_1 \times L_2} \text{ and } k = \frac{M}{\sqrt{L_1 \times L_2}}$$

where M = mutual inductance in henrys
$\quad k$ = coefficient of coupling
$\quad L_1$ = self-inductance of coil or transformer winding 1
$\quad L_2$ = self-inductance of coil or transformer winding 2

▶Mutual inductance can exist between coils that are not transformer windings but that are close enough for flux linkage to occur. This can affect total inductance for the circuit.

▶Transformers are comprised of separate coils or windings, having inductive coupling. The primary winding is connected to an ac source. The load(s) are connected to the secondary winding(s).

▶One feature of a two-winding transformer is that the primary and secondary are magnetically coupled, yet electrically isolated from each other. This feature can keep circuits and loads from being directly connected to the power lines, which increases safety.

▶One key transformer parameter is the turns ratio. Turns ratio is expressed either by the primary-to-secondary ratio N_P/N_S, or secondary-to-primary ratio N_S/N_P. The latter approach is used frequently.

▶The ratio of voltages between primary and secondary is the same as the ratio of turns. That is,

$$\frac{N_S}{N_P} = \frac{V_S}{V_P}$$

▶The current ratio of a transformer is the inverse of the voltage ratio. When voltage is stepped up from primary to secondary, current is stepped down by a proportionate factor:

$$\frac{V_S}{V_P} = \frac{I_P}{I_S}$$

▶The impedance ratio of a transformer is related to the square of the turns ratio. That is,

$$\frac{Z_P}{Z_S} = \frac{N_P^2}{N_S^2}$$

▶When a transformer has 100% efficiency, the power delivered to the load by the secondary equals the power supplied to the primary by the ac source. The efficiency of iron-core transformers is usually in the 90–98% range.

▶Two categories of transformer losses are copper losses and core losses. Copper losses relate to the I^2R losses of the copper wire in the primary and secondary windings. Core losses result from eddy currents being induced in the core by the changing magnetic flux, and the energy wasted due to hysteresis losses. Eddy currents in the core material cause I^2R losses.

▶Transformers have various applications at different frequencies. Three transformer categories include power transformers, audio transformers, and rf transformers.

▶Power and audio transformers generally have laminated iron-cores, for efficiency's sake. Rf transformers usually use air cores, special powdered iron, or ferrite in their core material.

▶Transformers can develop abnormal operations because of opens or shorts. Opens are evident by lack of output. Shorts between turns, from winding-to-winding or from winding-to-core, can be seen by overheating, blown fuses, and so forth.

FORMULAS AND CALCULATOR SEQUENCES SUMMARY

FORMULA 17–1 $M = k\sqrt{L_1 \times L_2}$

k value, $\boxed{\times}$, $\boxed{(}$, L_1 value, $\boxed{\times}$, L_2 value, $\boxed{)}$, $\boxed{\sqrt{x}}$, $\boxed{=}$

FORMULA 17–2 $L_T = L_1 + L_2 + 2M$

L_1 value, $\boxed{+}$, L_2 value, $\boxed{+}$, 2, $\boxed{\times}$, M value, $\boxed{=}$

FORMULA 17–3 $L_T = L_1 + L_2 - 2M$

L_1 value, $\boxed{+}$, L_2 value, $\boxed{-}$, 2, $\boxed{\times}$, M value, $\boxed{=}$

FORMULA 17–4 $L_T = L_1 + L_2 \pm 2M$

L_1 value, $\boxed{+}$, L_2 value, $\boxed{+}$ or $\boxed{-}$ as appropriate, 2, $\boxed{\times}$, M value, $\boxed{=}$

FORMULA 17–5 $L_T = \dfrac{1}{\dfrac{1}{L_1 + M} + \dfrac{1}{L_2 + M}}$

L_1 value, $\boxed{+}$, M value, $\boxed{=}$, $\boxed{1/x}$, $\boxed{+}$, $\boxed{(}$, L_2 value, $\boxed{+}$, M value, $\boxed{)}$, $\boxed{1/x}$, $\boxed{=}$, $\boxed{1/x}$

FORMULA 17–6 $L_T = \dfrac{1}{\dfrac{1}{L_1 - M} + \dfrac{1}{L_2 - M}}$

L_1 value, $\boxed{-}$, M value, $\boxed{=}$, $\boxed{1/x}$, $\boxed{+}$, $\boxed{(}$, L_2 value, $\boxed{-}$, M value, $\boxed{)}$, $\boxed{1/x}$, $\boxed{=}$, $\boxed{1/x}$

FORMULA 17–7 $k = \dfrac{M}{\sqrt{L_1 \times L_2}}$

M value, $\boxed{\div}$, $\boxed{(}$, L_1 value, $\boxed{\times}$, L_2 value, $\boxed{)}$, $\boxed{\sqrt{x}}$, $\boxed{=}$

FORMULA 17–8 $\dfrac{V_S}{V_P} = \dfrac{N_S}{N_P}$

V_S value, $\boxed{\div}$, V_P value, $\boxed{=}$, (should equal) N_S value, $\boxed{\div}$, N_P value, $\boxed{=}$

FORMULA 17–9 $\dfrac{N_S}{N_P} = \dfrac{V_S}{V_P}$

N_S value, $\boxed{\div}$, N_P value, $\boxed{=}$, (should equal) V_S value, $\boxed{\div}$, V_P value, $\boxed{=}$

FORMULA 17–10 $\dfrac{N_S}{N_P} = \dfrac{I_P}{I_S}$

N_S value, $\boxed{\div}$, N_P value, $\boxed{=}$ (should equal) I_P value, $\boxed{\div}$, I_S value, $\boxed{=}$

FORMULA 17–11 $\dfrac{V_S}{V_P} = \dfrac{I_P}{I_S}$

V_S value, $\boxed{\div}$, V_P value, $\boxed{=}$, (should equal) I_P value, $\boxed{\div}$, I_S value, $\boxed{=}$

FORMULA 17–12 $\left(\dfrac{N_S}{N_P}\right)^2 = \dfrac{Z_S}{Z_P}$

N_S value, $\boxed{\div}$, N_P value, $\boxed{=}$, $\boxed{x^2}$ (should equal) Z_S value, $\boxed{\div}$, Z_P value, $\boxed{=}$

FORMULA 17–13 $Z_P = Z_S \left(\dfrac{N_P}{N_S}\right)^2$

Z_S value, $\boxed{\times}$, $\boxed{(}$, N_P value, $\boxed{\div}$, N_S value, $\boxed{)}$, $\boxed{x^2}$, $\boxed{=}$

FORMULA 17–14 $\dfrac{N_P}{N_S} = \sqrt{\dfrac{Z_P}{Z_S}}$

N_P value, $\boxed{\div}$, N_S value, $\boxed{=}$, (should equal) $\boxed{(}$, Z_P value, $\boxed{\div}$, Z_S value, $\boxed{)}$, $\boxed{\sqrt{x}}$

FORMULA 17–15 Efficiency % $= \dfrac{P_{out}}{P_{in}} \times 100$

P_{out} value, $\boxed{\div}$, P_{in} value, $\boxed{\times}$, 100, $\boxed{=}$

REVIEW QUESTIONS

1. What is the primary current of a transformer whose source is 120 V and secondary is supplying 240 V at 100 mA?

2. What is the secondary voltage of a transformer having a secondary-to-primary turns ratio of 5.5:1, if the source voltage is 100 V?

3. What is the primary-to-secondary turns ratio of a transformer whose primary voltage is 120 V and secondary voltage is 6 V?

4. If a transformer has a primary-to-secondary turns ratio of 5:1 and the load resistance connected to the secondary is 8 Ω, what impedance is seen looking into the primary?

5. What is the mutual inductance of two coils having a coefficient of coupling of 0.6, if each coil has a self-inductance of 2 H?

6. What is the total inductance of two coils in series having 0.3 k, if each of the coils has a self-inductance of 10 mH? (Assume their fields are aiding.)

7. What is the total inductance of two coils in parallel having 0.5 k, if each coil has a self-

inductance of 10 H. (Assume their fields are opposing.)

8. What is the secondary voltage of an ideal transformer whose secondary-to-primary turns ratio is 15 and primary voltage is 12 V?

9. How many volts-per-turn are in a transformer having a 500-turn primary and a 100-turn secondary, if the primary voltage is 50 V?

10. A 6:1 step-down transformer has a secondary current of 0.1 A. Assuming an ideal transformer, what is the primary current?

11. What is the number of turns in the primary winding of the transformer shown in Figure 17–17?

FIGURE 17–17

12. Assuming an ideal transformer, what is the power delivered to R_L in Figure 17–18?

FIGURE 17–18

13. What is the primary-to-secondary turns ratio in Figure 17–19?

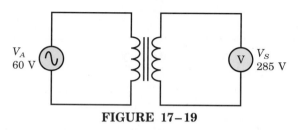

FIGURE 17–19

14. What is the primary current in the transformer circuit in Figure 17–20? (Assume an ideal transformer.)

FIGURE 17–20

15. What is the voltage applied to the primary of the circuit in Figure 17–21?

FIGURE 17–21

CHALLENGE QUESTIONS

1. Research the term "flyback transformer." Indicate where it is used, and whether its output is a high or low voltage. Also, list the frequency at which a TV flyback transformer is operating.

2. Perform research to find out what is meant by a "ringing test" in relation to checking for shorted turns in a transformer or inductor.

TEST YOURSELF

1. An amplifier having an output impedance of 5,000 Ω is feeding its output signal to a matching transformer. If the speaker connected to the secondary is a 16-Ω speaker, and the number of turns on the secondary is 200 turns, what number of turns should the primary winding have to provide the best impedance match? (Assume a perfect transformer.)

2. The secondary-to-primary turns ratio of a transformer is 1:8. What is the primary current, if the power delivered to a 200-Ω load resistance is 405 mW?

3. If the impedance ratio (secondary-to-primary) of an ideal transformer is 1:36 and the number of turns on the primary is 1,200, how many turns are on the secondary winding?

4. A TV receiver contains a transformer in its power supply. If you unplug the receiver from its 120-V, 60-Hz wall plug (ac source), turn the on/off switch to the on position, and connect an ohmmeter to each lead of the ac cord's plug, the ohmmeter reads 10 Ω. Using Ohm's law, you think the current supplied to the transformer is 12 A, since $I = V/R$, or 120 V/10 Ω. Explain why the circuit is fused with only a 5-A fuse, yet, when the TV receiver is turned on, the fuse does not blow!

PERFORMANCE PROJECT CORRELATION CHART

Suggested performance projects that correlate with topics in this chapter are:

CHAPTER TOPIC	PERFORMANCE PROJECTS	PROJECT NUMBER
Mutual Inductance and Transformer Action	Turns, Voltage, and Current Ratios	49
and	and	
Important Transformer Ratios	Turns Ratio(s) versus Impedance Ratios	50

Capacitance

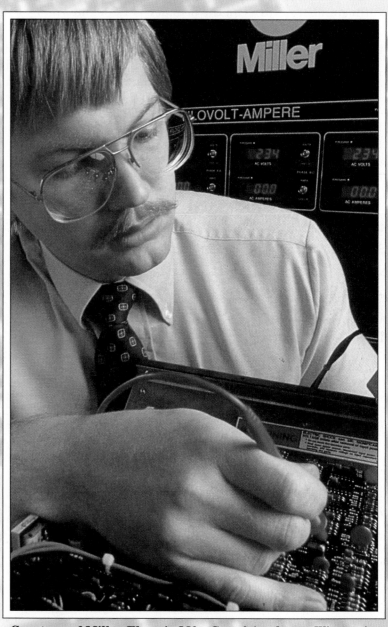

Courtesy of Miller Electric Mfg. Co. of Appleton, Wisconsin

18

KEY TERMS

Capacitance

Capacitor

Charging a capacitor

Dielectric

Dielectric constant

Dielectric strength

Discharging a capacitor

Electrostatic (electric) field

RC time constant

OUTLINE

Definition and description of a capacitor

The electrostatic field

Charging and discharging action

The unit of capacitance

Energy stored in capacitor's electrostatic field

Factors affecting capacitance value

Capacitance formulas

Total capacitance in series and parallel

Finding voltage when three or more capacitors are in series

The *RC* time constant

Types of capacitors

Typical color codes

Typical problems and troubleshooting techniques

CHAPTER PREVIEW

Three basic properties of electrical circuits are resistance, inductance, and capacitance. You have studied resistance (opposition to current flow) and inductance (property opposing a *change in current*). In this chapter, you will study capacitance (property opposing a *change in voltage*).

You will learn that capacitors have the unique ability to store electrical energy because of an electric field. Even as current produces a magnetic field, voltage establishes an electric field. This phenomenon and other important dc characteristics of capacitors will be analyzed.

Also, you will study how capacitors charge and discharge; the physical factors affecting capacitance values; some examples of capacitors; how to calculate total capacitance for circuits containing more than one capacitor; examples of capacitor color codes; and some common capacitor problems and troubleshooting techniques.

OBJECTIVES

After studying this chapter, you should be able to:

• Define **capacitor, capacitance, dielectric, dielectric constant, electric field,** farad, *RC* **time constant,** and leakage resistance

• Describe **capacitor charging** action and **discharging** action

• Calculate charge, voltage, **capacitance,** and stored energy, using the appropriate formulas

• Determine total **capacitance** in circuits with more than one **capacitor** (series and parallel)

• Calculate circuit voltages using appropriate *RC* **time constant** formulas

• List and describe the physical and electrical features of at least four types of **capacitors**

• List typical **capacitor** problems and describe troubleshooting techniques

• Use the SIMPLER troubleshooting sequence to solve the Chapter Challenge problem

DEFINITION AND DESCRIPTION OF A CAPACITOR

In *physical* terms, a **capacitor** is an electrical component, generally consisting of two conducting surfaces (often called capacitor plates) that are separated by a nonconductor (called the dielectric), Figure 18−1.

The nonconducting material, or **dielectric,** can be any number of materials. Some examples of dielectric materials are vacuum, air, waxed paper, plastic, glass, ceramic material, aluminum oxide, and tantalum oxide.

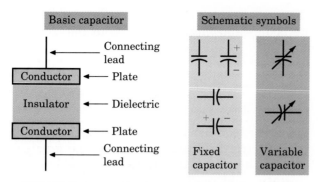

FIGURE 18–1   Basic capacitor and symbols

The schematic symbol for the capacitor plates is one straight line and one curved line, each perpendicular to the lines representing the wires or circuit conductors connected to the two plates, again see Figure 18–1.

In *electrical* terms, a capacitor is an electrical component that stores electrical charge when voltage is applied.

THE ELECTROSTATIC FIELD

You know that unlike charges attract and like charges repel. Refer to Figure 18–2 as you study the following discussion.

Lines are used to illustrate **electrostatic fields** between charged bodies, much like lines illustrate magnetic flux between magnetic poles. Direction of the electrostatic field lines is usually shown from the positive charged body to the negative charged body. (NOTE: An electron placed in this field travels in the opposite direction from the electrostatic field. In other words, the electron is repelled from the negative body and attracted to the positive body.)

$$F = \frac{kQ_1Q_2}{d^2}$$

The direction an electron, or negative charge would move *if* placed in this field is in the opposite direction from that used to illustrate the electrostatic field. (NOTE: There is not an actual electron flow through a good dielectric). A positive charge, placed in the field, would move in the direction used to represent the field lines.

FIGURE 18–2   The electrostatic field

The force between charged bodies is *directly* related to the product of the charges on the bodies and *inversely* related to the square of the distance between them.

$$F = \frac{kQ_1 \times Q_2}{d^2} \text{ (coulomb's law)}$$

This means the closer together a capacitor's plates are for a given charge, the stronger the electric field produced. Also, the greater the charge (in coulombs) stored on the plates, the stronger the field produced.

The electrostatic field represents the storage of electric energy. This energy originates with a source and can be returned to a circuit when the source is removed. (This is analogous to energy stored in the buildup of an inductor's magnetic field, which may be returned to the circuit when the magnetic field collapses.)

CHARGING AND DISCHARGING ACTION

Figure 18–3a illustrates a capacitor with no *charge,* that is, no electrostatic field exists between the plates. The plates and dielectric between them are electrically *neutral.*

Charging Action

In Figure 18–3b, the circuit switch is closed and electrons move for a short time, **charging the capacitor.** Let's look at this action.

As soon as the switch is closed, neither capacitor plate has an electron excess or deficiency. Also, there is no voltage across the plates or electric field across the dielectric to oppose electron movement to or from the plates.

Electrons leave the negative terminal of the battery and travel to the capacitor plate, which is electrically connected to it. The electrons do not travel through the nonconductive dielectric material between the capacitor plates. Rather, the electrons accumulate on that side of the capacitor, establishing an electron excess and a negative charge at that location. Incidentally, the electron orbital paths in the atoms of the dielectric material

FIGURE 18–3   Capacitor charging action

(unless it is vacuum) are distorted, since they are repelled by the negative charged plate of the capacitor, Figure 18–4. (NOTE: The distorted orbits represent storage of some electrical energy.)

As electrons accumulate on one side of the capacitor, making that side negative, electrons are being simultaneously attracted to the source's positive terminal from the capacitor's opposite side. Recall in a series circuit, the current is the same through all circuit parts.

Obviously, as an electron surplus builds on one plate and an equal electron deficiency occurs on the other plate, a potential (voltage) difference is established between the capacitor plates.

Notice in Figure 18–3c the voltage polarity established across the capacitor plates is one that *series-opposes* the source voltage (V_S). When sufficient *charging current* has flowed to cause the capacitor voltage to equal the source voltage, no more current can flow. In other words, the capacitor has *charged* to the source voltage.

Current flowing from the negative source terminal to one capacitor plate and to the positive source terminal from the other capacitor plate is maximum the moment the switch is closed. For a moment, the capacitor acts almost like a short. NOTE: Current does not flow *through* the capacitor via the nonconductive dielectric. Rather, current acts like it has flowed because of the electron movement on either side of the capacitor. Charging current is maximum the first instant the switch is closed and decreases to zero as the capacitor becomes *charged* to the source voltage value.

When the source voltage increases to a higher voltage, current again flows until the capacitor becomes charged to this new voltage level. Once the capacitor has charged to the source voltage, it acts like an open and blocks dc, again see Figure 18– 3c.

In summary, maximum charging current flows when voltage is first applied to the capacitor. As the capacitor charges, voltage across the capacitor increases to the source voltage value and charging current decreases. When the capacitor is charged to V_S, no further current flows and the capacitor is holding a charge, which causes a voltage equal to V_S across its plates. The electrical energy has been stored in the electrostatic (dielectric) field of the capacitor.

When we open the switch in the charged capacitor circuit, Figure 18–5, and we assume the dielectric has infinite resistance to the current, (i.e., no leakage current flows between the plates through the dielectric), the capacitor remains charged *as long* as no external *discharge* path for current is provided.

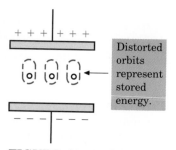

FIGURE 18–4   Distorted orbits represent stored electrical energy.

FIGURE 18–5   Capacitors retain charge until it bleeds off either through leakage R of dielectric or through some external current path.

⚠ SAFETY HINTS

Caution: **Always discharge circuit capacitors after power has been removed and *prior* to working on circuits containing them!** Technicians have been killed by accidentally becoming the discharge path! By providing an appropriate discharge path to bleed the charge from circuit capacitors prior to working on the circuit, this danger is avoided.

Many power supplies have a "bleeder resistor," which is intended to bleed off the charge from power supply capacitors when the power is turned off. IT IS NOT SAFE to depend on the bleeder working, as they occasionally become open. A number of technicians have been killed, assuming the bleeder resistor would protect them!

Some dangerous equipment has a "shorting stick" mounted inside the cabinet of the equipment. This is a long, nonconductive, wooden or plastic stick with a large braided conductor attached at one end. The other end of the braided conductor is permanently connected to the system ground. The technician holds the end of the stick at the opposite end from which the conductor is attached to the stick, then positions the conductor end of the stick to touch the capacitor terminals to be discharged. This of course discharges the capacitor through the braid to ground. Caution must be used in doing this due to the large arc that can result. All operator safety precautions recommended by the equipment manufacturer and common sense MUST be followed when using this procedure.

The point to be made here is: NEVER ASSUME LARGE CAPACITORS IN SYSTEMS ARE DISCHARGED JUST BECAUSE THE POWER HAS BEEN TURNED OFF!

Discharging Action

Figure 18–6 illustrates **discharging a capacitor.** An external path for electron movement has been supplied across the capacitor plates. The excess electrons on the bottom plate quickly move through this path to the top plate, which has an electron deficiency. The charge difference between the plates is neutralized, and the potential difference between the plates falls to zero. When full neutralization occurs and the potential difference between plates is zero, we have completely discharged the capacitor.

On discharge, V_C decreases to zero volts.

FIGURE 18–6   Discharging a capacitor

✔ IN-PROCESS LEARNING CHECK I

Fill in the blanks as appropriate.

1. A capacitor is an electrical component consisting of two conducting surfaces called _____ that are separated by a nonconductor called the _____.

2. A capacitor is a device that stores electrical _____ when voltage is applied.

3. Electrons _____ travel through the capacitor dielectric.

4. During charging action, one capacitor plate collects _____, making that plate _____. At the same time, the other plate is losing _____ to become _____.

5. Once the capacitor has charged to the voltage applied, it acts like an _____ circuit to dc. When the level of dc applied increases, the capacitor _____ to reach the new level. When the level of dc applied decreases, the capacitor _____ to reach the new level.

6. When voltage is first applied to a capacitor, _____ charge current occurs.

7. As the capacitor becomes charged, current _____ through the circuit in series with the capacitor.

THE UNIT OF CAPACITANCE

You have seen that **capacitance** relates to the *capacity* of a *capacitor* to store electrical charge. As you might expect, the more charge a capacitor stores for a given voltage, the larger its capacitance value must be. The unit of capacitance is the farad, named in honor of Michael Faraday. The farad is that amount of capacitance where a charge of one coulomb develops a potential difference of one volt across the capacitor plates, or terminals. The farad is abbreviated as capital F.

From this definition, relationships between capacitance *(C)*, voltage *(V)*, and charge *(Q)* are:

$$\boxed{\text{Formula 18–1} \quad C = \frac{Q}{V}}$$

 charge value in coulombs, $\boxed{\div}$, voltage value, $\boxed{=}$

$$\boxed{\text{Formula 18–2} \quad Q = CV}$$

 capacitance value in farads, $\boxed{\times}$, voltage value, $\boxed{=}$

$$\boxed{\text{Formula 18–3} \quad V = \frac{Q}{C}}$$

where V = volts
Q = coulombs
C = capacitance in farads

 charge value in coulombs, $\boxed{\div}$, capacitance value in farads, $\boxed{=}$

As you can see from Formula 18–2, the amount of charge a capacitor stores is *directly related to* the amount of *capacitance and* the *voltage* across the capacitor.

EXAMPLE

When 10 coulombs of charge are stored on a capacitor charged to 2 volts, the capacitance (in farads) is calculated as:

$$C \text{ (farads)} = \frac{Q \text{ (coulombs)}}{V \text{ (volts)}}$$

$$C(F) = \frac{10 \text{ C}}{2 \text{ V}}$$

$$C = 5 \text{ F}$$

NOTE: The unit of capacitance *(C)* is the farad (F). The unit of charge *(Q)* is the coulomb, also abbreviated as "C."

It should be noted that the farad is a huge amount of capacitance. In fact, it would take a room-sized capacitor to have that much capacitance. Practical capacitors are usually in the range of millionths of a farad (μF) or millionths of a millionth of a farad, formerly called micro-micro farad (μμF). This value is now called picofarad, pF. Using powers of ten, these values are expressed as:

One microfarad = 1 μF = 1×10^{-6} F
One picofarad = 1 pF = 1×10^{-12} F

PRACTICE PROBLEMS I

1. What is the capacitance of a capacitor that develops 25 V across its plates when storing a 100-μC charge? (Express answer in microfarads.)

2. What is the charge on a 10-μF capacitor when the capacitor is charged to 250 V? (Express answer in μC.)

3. What voltage develops across the plates of a 2-μF capacitor with a charge of 50 μC?

ENERGY STORED IN CAPACITOR'S ELECTROSTATIC FIELD

It was stated earlier that electrical energy supplied by the source during the capacitor charge is stored in the electrostatic field of the charged capacitor. This energy is returned to the circuit when the capacitor discharges.

For inductors, the energy (joules) stored in the magnetic field is related to the inductance and the current creating that field

$$W = \frac{1}{2}LI^2$$

However, for capacitors, the joules of stored energy are related to the capacitance and the voltage, or potential difference, between capacitor plates created by the electrostatic field. The formula for energy (in joules) is:

Formula 18–4	Energy (J) $= \dfrac{1}{2}CV^2$

 capacitance value in farads, $\boxed{\times}$, voltage value, $\boxed{x^2}$, $\boxed{\div}$, 2, $\boxed{=}$

EXAMPLE

How many joules are stored in a 5-μF capacitor that is charged to 250 V?

$$\text{Energy } (J) = \frac{1}{2}CV^2$$

$$\text{Energy} = \frac{5 \times 10^{-6} \times (250)^2}{2} = \frac{0.3125}{2} = 0.15625 \text{ J}$$

PRACTICE PROBLEMS II

How much electrical energy is stored in a 10-μF capacitor that is charged to 100 V?

FACTORS AFFECTING CAPACITANCE VALUE

Capacitance value is related to how much charge is stored per unit voltage. The physical factors influencing the capacitance of parallel-plate capacitors are size of plate area; thickness of dielectric material (spacing between plates); and type of dielectric material, Figure 18–7. Let's examine these factors.

• **Plate area** Obviously, the larger the area of the plates facing each other, the greater the number of electrons (and charge) that can be stored. Therefore in Figure 18–7, C is directly related to plate area. For example, when the plate area doubles, there is room for twice as many

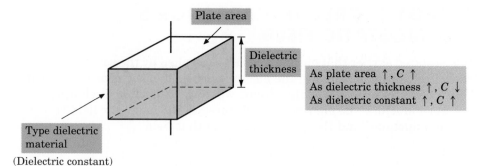

FIGURE 18–7   Physical factors affecting capacitance value

electric lines of force with the same potential difference between plates. Thus, the amount of charge the capacitor holds per given voltage has doubled and the capacitance has doubled. The area of the plates directly facing each other equals the area of one plate. Using SI units, this area is expressed in square meters.

- **Distance between plates (dielectric thickness)** From previous discussions, it is logical to assume the further apart the plates, the less the electric field strength established for a given charge and potential difference between plates. Thus, the less the electrical energy or charge stored in this electric field. In other words, the *capacitance value is inverse to the spacing between plates.* Conversely, the closer the plates (the thinner the dielectric material) are, the stronger the electric field intensity per given voltage and the higher the capacitance value. In SI units, this distance (thickness of dielectric) is expressed in meters.

- **Type of Dielectric Material** Different dielectric materials exhibit different abilities to concentrate electric flux lines. This property is sometimes called *permittivity,* which is analogous to conductivity of conductors. Vacuum (or air) has a permittivity of one. The relative permittivity of a given dielectric material is the ratio of the absolute permittivity (ϵ_o) of the given dielectric to the absolute permittivity of free space (vacuum) (ϵ_v). This relative permittivity is called the **dielectric constant** *(k)* of the material.

Dielectric Constants

Refer to the chart in Figure 18–8 for examples of the average dielectric constant of various materials. The ratio reflecting the dielectric constant is shown as:

Formula 18–5	$k = \dfrac{\epsilon_o \text{ (absolute permittivity of dielectric material)}}{\epsilon_v \text{ (absolute permittivity of vacuum)}}$

  ϵ_o value, $\boxed{\div}$, ϵ_v value, $\boxed{=}$

(NOTE: Sometimes the symbol K_ϵ is used in lieu of k to represent dielectric constant.)

KIND OF MATERIAL	APPROXIMATE K
Air	1.0
Glass	8.0
Waxed paper	3.5
Mica	6.0
Ceramic	100.0 +

NOTE: Values vary depending on quality, grade, environmental conditions, and frequency.

FIGURE 18–8   Sample dielectric constants

The higher the dielectric constant is, the greater the density of electric flux lines established for any given plate area and spacing between plates. Therefore, capacitance is directly related to the dielectric constant. For example, when the average dielectric constant (k or K_ϵ) of a given material is five, the capacitance is five times as great with this dielectric material compared to a dielectric of air or vacuum.

Dielectric Strength

The electric field intensity is increased by increasing the voltage across a dielectric. As we continue increasing the potential difference across the capacitor plates, eventually a point is reached where the electron orbits in the dielectric material are stressed (distorted in their orbital paths), causing electrons to be torn from their orbits. At this time, the dielectric material breaks down and typically punctures or arcs, becoming a conductor. The material is no longer functioning as a nonconducting dielectric material (unless it is air or vacuum, which are self-healing). The breakdown value relates to the **dielectric strength** of the material. The practical measure of dielectric strength is called the *breakdown voltage*. The breakdown voltage depends on the kind of material and its thickness. Figure 18–9 lists some materials and their approximate average dielectric strength ratings in terms of volts per mil (thousandths of an inch) breakdown voltage ratings. Be aware these numbers will vary, depending on manufacture, frequency of operation, and so on.

KIND OF MATERIAL	APPROXIMATE DIELECTRIC STRENGTH (VOLTS PER 0.001 INCH)
Air	80
Glass	200 + (up to 2kV)
Waxed paper	1,200
Mica	2,000
Ceramic	500 + (up to $\approx$ 1kV)

FIGURE 18–9   Examples of dielectric strengths

CAPACITANCE FORMULAS

As you have learned, capacitance depends on the surface dimensions of plates facing each other, and the spacing and the dielectric constant of the material between plates. Using the appropriate factors to compute capacitance in farads using the SI units, the formula for capacitance of a parallel-plate capacitor with air (or vacuum) as the dielectric is:

Formula 18–6	$C = \dfrac{8.85A}{10^{12}s}$

where C = capacitance in farads
A = area of the plates facing each other, in square meters
s = spacing between plates, in meters

 8.85, $\boxed{\times}$, area of plates, $\boxed{\div}$, $\boxed{(}$, 1, $\boxed{\text{EE}}$, 12 , $\boxed{\times}$, spacing, $\boxed{)}$, $\boxed{=}$

NOTE: If the plates have different dimensions, *use the smaller plate area for "A"* in the formula.

To calculate capacitance when dielectric materials other than air or vacuum are used, use this formula and insert the dielectric constant *(k)* in the formula. Thus:

Formula 18–7	$C = \dfrac{8.85kA}{10^{12}s}$

Again, C = capacitance in farads
A = area of the plates facing each other in square meters
k = the dielectric constant
s = the spacing between the parallel plates in meters

8.85, $\boxed{\times}$, k value, $\boxed{\times}$, area of plates, $\boxed{\div}$, $\boxed{(}$, 1, $\boxed{\text{EE}}$, 12 , $\boxed{\times}$, spacing, $\boxed{)}$, $\boxed{=}$

For your information, a practical formula used to determine the capacitance in *microfarads* (μF) of multiplate capacitors is in Figure 18–10. NOTE: This

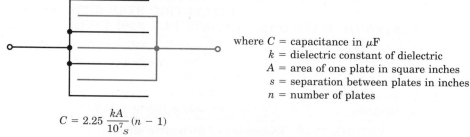

where C = capacitance in μF
k = dielectric constant of dielectric
A = area of one plate in square inches
s = separation between plates in inches
n = number of plates

$$C = 2.25 \frac{kA}{10^7 s} (n - 1)$$

FIGURE 18–10   Multiplate capacitor

formula uses inches and square inches, rather than meters and square meters.

Formula 18–8	$C = 2.25\dfrac{kA}{10^7 s}\,(n \; - \; 1)$

where C = capacitance in μF
$\quad\; k$ = dielectric constant in material between plates
$\quad\; A$ = area of one plate in square inches
$\quad\; s$ = separation between plate surfaces in inches
$\quad\; n$ = number of plates

 2.25, $\boxed{\times}$, k value, $\boxed{\times}$, A value, $\boxed{\div}$, $\boxed{(}$, 1, $\boxed{\text{EE}}$, 7, $\boxed{\times}$, s value, $\boxed{)}$, $\boxed{\times}$, $\boxed{(}$, n value, $\boxed{-}$, 1 , $\boxed{)}$, $\boxed{=}$

EXAMPLES

1. A two-plate capacitor has a dielectric with a dielectric constant of 3. The plates are each 0.01 square meters in area and spaced 0.001 meter apart. What is the capacitance?

$$C = \frac{8.85kA}{10^{12}s} = \frac{8.85 \times 3 \times 0.01}{10^{12} \times 0.001}$$

$$C = \frac{0.2655}{1 \times 10^9} = 0.0000000002655 \text{ F}$$

$$C = 265.5 \text{ pF}$$

2. What is the capacitance in μF of a capacitor if the plate area of one plate is 2 square inches, and the dielectric has a k of 100 and is 0.005 inches thick? (Assume a two-plate capacitor.)

$$C = 2.25\,\frac{kA}{10^7 s}\,(n - 1)$$

$$C = \frac{2.25 \times 100 \times 2}{10^7 \times 0.005} \times (2 - 1) = \frac{450}{50,000} \times 1 = 0.009 \text{ μF}$$

PRACTICE PROBLEMS III

1. A capacitor has four plates of equal size. Each plate has an area of 1 square inch. This capacitor uses a dielectric having a k of 10. The dielectric thickness between each pair of plates is 0.1 inch. What is the capacitance value in μF?

2. Each plate of a two-plate capacitor has an area of 0.2 square meters. The plates are uniformly spaced 0.005 meters apart and the dielectric is air. What is the capacitance?

PRACTICAL NOTES

As a technician, you will rarely, if ever, have to actually measure capacitor plate dimensions, plate-spacing, and so on, and use the formulas we have shown you to calculate capacitance value. These formulas were introduced as a reinforcement tool to help you understand the important physical factors that influence the value of capacitance.

Technicians do, however, have to check and measure capacitances quite routinely. There are several pieces of test equipment that can be used for doing this (see Figure 18–11):

1. DMMs that have capacity-checking range(s)

2. "Dedicated" digital capacitance meters or checkers

3. LCR meters that can check inductance, resistance, and/or capacitance

4. Other (such as bridges, dip-meters, and so on)

TOTAL CAPACITANCE IN SERIES AND PARALLEL

As a technician, you may have to determine circuit parameters with capacitors connected in series or parallel. The following information will help you learn to perform such required circuit analysis.

(a)

(b)

(c)

FIGURE 18–11   Test instruments used to measure capacitance (Photo a courtesy of John Fluke Mfg. Co., Inc.; photos b and c courtesy of B & K Precision)

Capacitance of Series Capacitors

Refer to Figure 18–12 as you study the following comments.

1. Connecting two capacitors in series increases the dielectric thickness, while the plate areas connected to the source remain the same, Figure 18–12. From our previous discussions, you can surmise that connecting capacitors in series *decreases* the resultant total capacitance.

2. Finding total capacitance of *capacitors in series* is analogous to finding total resistance of *resistors in parallel*.

3. The formulas to determine total capacitance (C_T) of series-connected capacitors are as follows:
   a. For two capacitors in series:

$$\text{Formula 18–9} \quad C_T = \frac{C_1 \times C_2}{C_1 + C_2}$$

 C_1 value, $\boxed{\times}$, C_2 value, $\boxed{\div}$, $\boxed{(}$, C_1 value, $\boxed{+}$, C_2 value, $\boxed{)}$, $\boxed{=}$

   b. For equal-value capacitors in series:

$$\text{Formula 18–10} \quad C_T = \frac{C}{N}$$

where C is the value of one of the equal-value capacitors and N is the number of capacitors in series.

 capacitor value of one of the equal-value capacitors, $\boxed{\div}$, number of capacitors, $\boxed{=}$

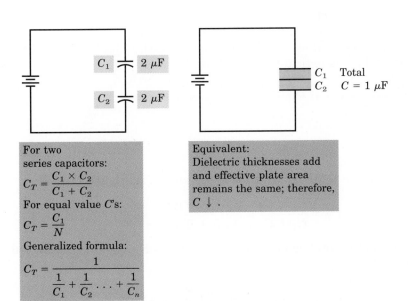

FIGURE 18–12   Total capacitance of series capacitors

c. The general formula for any number and value of capacitors in series:

| Formula 18–11 | $C_T = \dfrac{1}{\dfrac{1}{C_1} + \dfrac{1}{C_2} + \dfrac{1}{C_3} \cdots + \dfrac{1}{C_n}}$ |

C_1 value, $\boxed{1/x}$, $\boxed{+}$, C_2 value, $\boxed{1/x}$, . . . , $\boxed{=}$, $\boxed{1/x}$

Voltage Distribution Related to Series Capacitors

Because the current throughout a series circuit is the same at any point, capacitors connected in series have the same number of coulombs (C) of charge (Q). The potential difference between plates of equal-value capacitors for a given charge is equal. But for different-value capacitors, potential difference between plates is not equal. Recall the $V = Q/C$ formula, where V is directly related to Q and inversely related to C. This indicates in a series circuit (since the Q is the same), the voltage across any given capacitor is *inversely* proportional to its C value. For example, if we have a 5-μF capacitor and a 10-μF capacitor in series, the voltage across the 5-μF capacitor is two times the voltage across the 10-μF capacitor, Figure 18–13. To obtain capacitances in the ranges described (i.e., multi-microfarad), the electrolytic capacitor is typically used. You will study about this capacitor later in the chapter. However, you should take note of the preceding comments and following examples about voltage distribution.

FIGURE 18–13   Voltage distribution across series capacitors

PRACTICAL NOTES

Due to the large tolerances typical of electrolytics, the voltage distribution may not "track" the theoretical values computed by using the "marked values" on the capacitors. If the capacitors' values are precisely as marked, the resulting voltage distribution will be as indicated in the preceding discussion.

EXAMPLE

1. What is the total capacitance of the circuit in Figure 18–14?

$$C_T = \frac{C_1 \times C_2}{C_1 + C_2} = \frac{3 \ \mu\text{F} \times 6 \ \mu\text{F}}{3 \ \mu\text{F} + 6 \ \mu\text{F}} = \frac{18 \ \mu\text{F}}{9} = 2 \ \mu\text{F}$$

2. What is the total capacitance of three 10-μF capacitors connected in series?

$$C_T = \frac{C}{N} = \frac{10 \ \mu\text{F}}{3} = 3.33 \ \mu\text{F}$$

3. When the voltage applied to the circuit in Figure 18–14 is 90 V, what are the V_{C_1} and V_{C_2} values?

Since the voltage distribution is *inverse* to the capacitance values, V_{C_1} equals twice the value of V_{C_2}, or two-thirds of the applied voltage. Thus,

$$V_{C_1} = \frac{2}{3} \times 90 = 60 \text{ V, and } V_{C_2} = \frac{1}{3} \times 90 = 30 \text{ V}$$

An alternate solution is to find Q by using the total capacitance value in the formula $Q = C_T \times V_T$. Then solve for each capacitor voltage value using $V = Q/C$. Therefore, since the total capacitance is 2 µF:

$$Q = 2 \times 10^{-6} \times 90 = 180 \text{ µC}$$

$$V_{C_1} = \frac{180 \text{ µC}}{3 \text{ µF}} = 60 \text{ V}$$

$$V_{C_2} = \frac{180 \text{ µC}}{6 \text{ µF}} = 30 \text{ V}$$

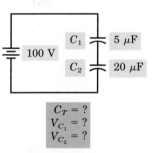

$$C_T = \frac{(3 \times 10^{-6}) \times (6 \times 10^{-6})}{(3 \times 10^{-6}) + (6 \times 10^{-6})}$$

$$C_T = \frac{18}{9} = 2 \text{ µF}$$

$$V_{C_1} = \frac{2}{3} \times V_T$$

$$V_{C_1} = 60 \text{ V}$$

$$V_{C_2} = \frac{1}{3} \times V_T$$

$$V_{C_2} = 30 \text{ V}$$

$$Q_T = V_T \times C_T$$

$$Q_T = 90 \text{ V} \times 2 \text{ µF} = 180 \text{ µC}$$

FIGURE 18–14

PRACTICE PROBLEMS IV

1. What is the total capacitance of the circuit in Figure 18–15?
2. What is the total capacitance of four 20-µF capacitors connected in series?
3. What are the voltages across each capacitor in Figure 18–15?

$$C_T = ?$$
$$V_{C_1} = ?$$
$$V_{C_2} = ?$$

FIGURE 18–15

FINDING VOLTAGE WHEN THREE OR MORE CAPACITORS ARE IN SERIES

A handy formula used to find voltage across a given capacitor when three or more capacitors are in series is (see Figure 18–16):

Formula 18–12	$V_X = V_S \dfrac{C_T}{C_X}$

where V_X = voltage across capacitor "x"
   V_S = the dc source voltage
   C_T = total series capacitance
   C_X = value of capacitor "x"

 V_S value, $\boxed{\times}$, $\boxed{(}$, C_T value, $\boxed{\div}$, C_X value, $\boxed{)}$, $\boxed{=}$

$$V_X = V_S \left(\frac{C_T}{C_X}\right)$$

$$V_X = 60 \left(\frac{7.5}{15}\right)$$

$$V_X = 60 \times 0.5$$

$$V_X = 30 \text{ V}$$

FIGURE 18–16   Series capacitors voltage formula

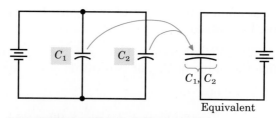

When plate areas are combined,
dielectric thickness remains unchanged.

**FIGURE 18–17   Capacitance of
parallel capacitors**

Capacitance of Parallel Capacitors

Refer to Figure 18–17 as you study the following information. Since the top and bottom plates of all the parallel capacitors are electrically connected to each other, the combined plate area equals the sum of the individual capacitor plate areas. However, the dielectric thicknesses do not change and are not additive.

Increasing the effective plate area while keeping the dielectric thickness(es) the same results in the total capacitance of the parallel-connected capacitors equaling the sum of the individual capacitances. In other words, *capacitances in parallel add like resistances add in series.* The formula for total capacitance (C_T) of parallel-connected capacitors is:

| Formula 18–13 | $C_T = C_1 + C_2 \ldots + C_n$ |

C_1 value, $\boxed{+}$, C_2 value, $\ldots$, $\boxed{=}$

Charge Distribution Related to Parallel Capacitors

Because the voltage across components connected in parallel is the same, the charge (in coulombs) contained by each capacitor in a parallel-capacitor circuit is *directly* related to each capacitance value. That is, the larger the capacitance value is, the greater is the charge for a given voltage, $Q = CV$.

300 V ⎓ C_1 10 μF C_2 20 μF C_3 30 μF

$$V_{C^1} = V_{C^2} = V_{C^3}$$

$Q_{C_1} = C_1 \times V = 10\ \mu\text{F} \times 300\ \text{V} = 3{,}000\ \mu C$
$Q_{C_2} = C_2 \times V = 20\ \mu\text{F} \times 300\ \text{V} = 6{,}000\ \mu C$
$Q_{C_3} = C_3 \times V = 30\ \mu\text{F} \times 300\ \text{V} = 9{,}000\ \mu C$

Charge on parallel capacitors is directly related
to each one's capacitance value.

**FIGURE 18–18   Charge distribution on
parallel capacitors**

For example, a 10-μF, 20-μF, and 30-μF capacitor are in parallel, and the voltage applied to this circuit is 300 V. Thus, the charge *(Q)* on the 10-μF capacitor is half that on the 20-μF capacitor and one-third that on the 30-μF capacitor, *(Q = CV)*. The *Q* on the 10-μF capacitor equals 10 μF × 300 V = 3,000 μC. The charge on the 20-μF capacitor equals 20 μF × 300 V = 6,000 μC. The charge on the 30-μF capacitor equals 30 μF × 300 V = 9,000 μC.

NOTE: Since total charge equals the sum of the charges, it can also be computed from the $Q_T = C_T \times V_T$ formula. Therefore, total charge = 60 μF × 300 V, so $Q_T = 18,000$ μC, Figure 18–18.

EXAMPLES

1. What is the total capacitance of a parallel circuit containing one 15-μF capacitor, two 10-μF capacitors, and one 3-μF capacitor?

$$C_T = C_1 + C_2 + C_3 + C_4 = 15 + 10 + 10 + 3 \text{ μF} = 38 \text{ μF}$$

2. What is the charge on the 3-μF capacitor, if the voltage applied to the circuit in Question 1 is 50 V?

$$Q = CV = 3 \times 10^{-6} \times 50 = 150 \text{ μC}$$

PRACTICE PROBLEMS V

1. What is the total capacitance of five 12-μF capacitors connected in parallel?

2. If V_T equals 60 V, what is the charge on each capacitor in Problem 1? What is the total charge on all capacitors?

THE *RC* TIME CONSTANT

At the beginning of the chapter, we said a capacitor opposes a *change* in voltage. You have seen how charge current is required to establish a potential difference between the capacitor plates. Obviously, the rate of charge for a given capacitor is directly related to the charge current. That is, the higher the current is, the quicker the capacitor is charged to V_S.

You also know current allowed to flow is *inversely* related to the resistance in the current path. Therefore, the larger the resistance value in the charge (or discharge) path of the capacitor is, the longer the time required for voltage to reach V_S across the capacitor during charge, or to reach zero volts during discharge.

The larger the capacitance *(C)* value is, the more charge required to develop a given potential difference between its plates, $V = Q/C$. Because a capacitor prevents an instantaneous voltage change across its plates (if there is any resistance in the charge path or discharge path), a capacitor is said to oppose a *change* in voltage across itself.

Recall when inductors were discussed in a previous chapter, there was an *L/R* time constant relating the time required to change current through the coil. You learned that *five L/R* time constants are needed for a complete change from one circuit current value to the new level.

In like manner, an **RC time constant** relating the time required to charge (or discharge) a capacitor exists. *To completely change from one voltage level to a "new level" of source voltage,* the capacitor needs *five RC time constants. One RC time constant (τ) = R (in ohms) × C (in farads).* That is:

Formula 18–14	$\tau = R \times C$

R value, $\boxed{\times}$, C value, $\boxed{=}$

Refer to Figure 18–19 as you study the following information about a typical series *RC* circuit.

The sum of the dc voltages across the capacitor and the resistor equals the source voltage (V_S) at all times.

As the capacitor is charged, the voltage across the capacitor is series-opposing the source voltage. Therefore, the voltage across the resistor at any given moment equals the source voltage minus the capacitor voltage. That is, $V_R = V_S - V_C$. Thus, as the capacitor charges to the V_S value, the resistor voltage decreases toward zero volts. When the capacitor is fully charged, V_C equals V_S and zero current flows, and no *IR* drop occurs across the resistor.

When the switch is initially closed, the capacitor has zero charge and zero voltage. Therefore, the source voltage is dropped by the resistor *(R)*. Since this is a series circuit, and the current is the same throughout a series circuit, Ohm's law reveals what the charging current is at that moment.

$$I = \frac{V_R}{R}$$

In other words, at that moment, the capacitor behaves like a short; therefore, only the resistor limits current.

As the charging current flows, a series-opposing potential difference across the capacitor develops, and the charging current diminishes at a nonlinear rate (toward zero). The *IR* drop across the resistor drops accordingly. For example, after a certain interval of time, the capacitor is charged

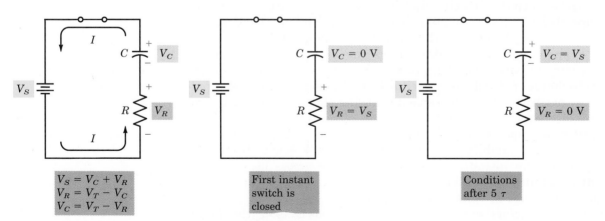

FIGURE 18–19   Voltages in an *RC* circuit

to half of V_S. At that time, one-half the source voltage is across the resistor, and one-half the source voltage is across the capacitor.

The percentage of voltage across the capacitor at any given moment relates to the RC time constant, where one time constant (*in seconds*) = R (Ω) $\times$ C (farads). Often the Greek lower case letter tau (τ) is used to represent one time constant. Note also that τ (sec) = R (megohms) $\times$ C (microfarads). Also, observe that τ *(μsec)* = R (Ω) $\times$ C (μF).

Refer to Figure 18–20 (curve A) and notice that after one time constant (τ), *the capacitor voltage (V_C)* equals 63.2% of V_S. After two τ, V_C equals 86.5% of V_S; after three τ, V_C equals 95%; after four τ, V_C equals 98.5%; and after *five* τ, V_C equals over 99% of V_S; therefore, it is considered fully charged. Are these percentages and corresponding numbers familiar? Of course! They are the same percentages and numbers used when the L/R time constant was discussed. In other words, the same exponential curves (or time constant chart curves) are applicable for analyzing voltages in a capacitor circuit as were used to analyze current in an inductor circuit.

Note in Figure 18–20 that curve B illustrates what happens to the resistor voltage as the capacitor is charged. That is, at the first instant V_R equals V_S. As the capacitor charges, V_R decreases toward zero. In effect, the resistor voltage and the circuit current are inverse to the capacitor voltage. As the capacitor voltage increases, the circuit current, and thus the voltage across the resistor, decrease.

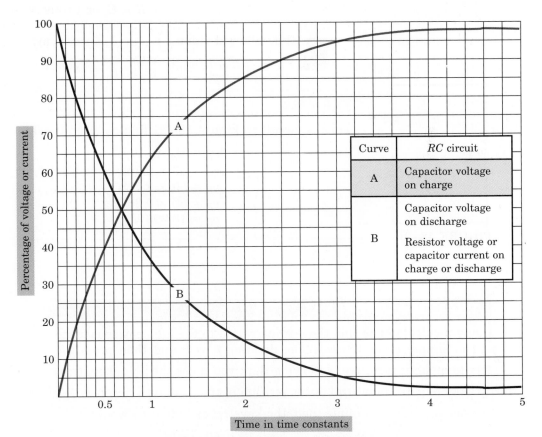

FIGURE 18–20   Time-constant chart

Using Exponential Formulas to Find Voltage(s)

Values along the exponential curves on the time constant chart in Figure 18–20 are more precisely found by using some simple formulas. The values along the descending exponential curve relate to the base of *natural logarithms,* called epsilon (ε), which has a value of 2.718. (Recall this number of natural growth was discussed in an earlier chapter).

When a capacitor is charging in a series *RC* circuit, the voltage across the resistor during the first instant is maximum, then decreases exponentially as the capacitor charges to the new voltage level. Thus, the resistor voltage follows the descending exponential curve shown in the time constant chart. The level of v_R for any given point of time relates to the length of time the capacitor has charged relative to the circuit's *RC* time constant. That is, the time allowed *(t)* divided by the time of one *RC* time constant (τ), or t/τ. Obviously, when the *t* (time allowed) equals five or more τ, then the capacitor has fully charged, no current is flowing, and v_R equals zero. When *t* is less than 5 τ, then v_R equals a value that is found by the formula:

$$\boxed{\text{Formula 18–15} \quad v_R = V_S \times \epsilon^{-t/\tau}}$$

where ε = epsilon (or 2.71828)
t = time allowed in seconds
τ = *R* × *C* (or 1 time constant)

 time allowed value, $\boxed{\div}$, $\boxed{(}$, *R* value, $\boxed{\times}$, *C* value, $\boxed{)}$, $\boxed{=}$, $\boxed{\pm}$, $\boxed{\text{2nd}}$, $\boxed{e^x}$, $\boxed{\times}$, V_S value, $\boxed{=}$

EXAMPLE

For the circuit in Figure 18–21 find v_R after 10,000 μs.

$$v_R = V_S \times \epsilon^{-t/\tau}$$
$$v_R = 50 \times 2.718^{-10,000(100,000 \times 0.1)}$$
$$v_R = 50 \times 2.718^{-1}$$
$$v_R = 50 \times 0.3679$$
$$v_R = 18.395 \text{ V}$$

FIGURE 18–21

PRACTICAL NOTES

Using a calculator to find the ε value raised to the −*t*/τ power is quite easy. You simply enter the result of dividing *t* by τ and then press the $\boxed{+/-}$ key to make it negative. Following that, you press the $\boxed{\text{2nd}}$ key, then the $\boxed{e^x}$ key to get the natural antilogarithm of the displayed number.

For our preceding example then, the keystrokes to find v_R would be:

 1, $\boxed{+/-}$, $\boxed{\text{2nd}}$, $\boxed{e^x}$, $\boxed{\times}$, 50 , $\boxed{=}$

EXAMPLE

For the circuit in Figure 18–21, what is the V_R after 1.8 time constants?

$$v_R = V_S = \epsilon^{-t/\tau}$$
$$v_R = 50 \times 2.718^{-1.8}$$
$$v_R = 50 \times 0.1653$$
$$v_R = 8.265 \text{ V}$$

The calculator keystrokes that can be used to perform the calculations in this example are:

 1.8 , $\boxed{+/-}$, $\boxed{\text{2nd}}$, $\boxed{e^x}$, $\boxed{\times}$, 50 , $\boxed{=}$

The 1.8, $\boxed{+/-}$, $\boxed{\text{2nd}}$, $\boxed{e^x}$, sequence gives the result of raising 2.718 to the negative 1.8th power, which is 0.1653. This result times the 50 V (V_A) provides the value of voltage across the resistor (8.265 V).

The capacitor voltage is the difference between the resistor voltage and the new voltage applied or source voltage for any given instant. A formula for v_C is:

| Formula 18–16 | $v_C = V_S (1 - \epsilon^{-t/\tau})$ |

 time allowed value, $\boxed{\div}$, $\boxed{(}$, R value, $\boxed{\times}$, C value, $\boxed{)}$, $\boxed{=}$, $\boxed{+/-}$, $\boxed{\text{2nd}}$, $\boxed{e^x}$, $\boxed{+/-}$, $\boxed{+}$, 1 , $\boxed{=}$, $\boxed{\times}$, V_A value, $\boxed{=}$

NOTE: The small letter v denotes an instantaneous voltage value for a given moment in time.

For the first example then:

$$v_C = 50 \times (1 - 0.3679)$$
$$v_C = 50 \times 0.6321$$
$$v_C = 31.605 \text{ V}$$

To check, add the V_R and the v_C (18.395 + 31.605); the result is 50 V (our applied voltage). It checks out!

For the second example (Figure 18–21):

$$v_C = 50 \times (1 - 0.1653)$$
$$v_C = 50 \times 0.8347$$
$$v_C = 41.735 \text{ V}$$

When we add v_R (8.265 V) + v_C (41.735 V), we get 50 V. It checks!

PRACTICE PROBLEMS VI

Use either the time-constant chart or the formulas and a scientific calculator to solve the following.

1. What is the voltage across the capacitor of the circuit in Figure 18–22, 20 μs after the switch is closed?

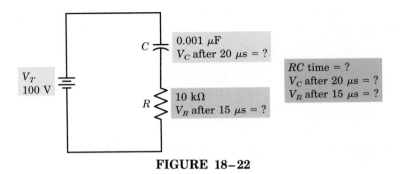

FIGURE 18–22

2. What is the voltage across the resistor in Figure 18–22, 15 μs after the switch is closed? What is the charging current value at that moment?

3. What is the value of v_C 30 μs after the switch is closed?

4. What is the value of v_R 25 μs after the switch is closed?

TYPES OF CAPACITORS

General Classification

There are two basic classifications of capacitors—*fixed and variable*. These capacitors come in various configurations, sizes, and characteristics. As the two names imply, fixed capacitors have one value of capacitance; variable capacitors have a range of capacitance values that may be adjusted, Figure 18–23. By changing the position of the *rotor* plates with respect to the fixed *stator* plates, Figure 18–23a, the plate areas facing each other are varied. This changes the capacitance value.

When the plates are fully meshed, capacitance is maximum. When the plates are fully unmeshed, capacitance value is minimum. The dielectric for this type of capacitor is air.

Another variable capacitor is shown in Figure 18–23b. This compression-type capacitor has only one movable plate. By turning the screw, the

(a)   (b)

FIGURE 18–23   Variable capacitors (Photo a courtesy of Amateur Radio Relay League; photo b courtesy of Cornell Dubilier Electronics and Sprague-Goodman Electronics)

spacing between the movable plate and the fixed plate is varied thus changing the capacitance value. When the movable plate is screwed tight, capacitance is maximum. As the screw is loosened, capacitance decreases.

Capacitors and Their Characteristics

Capacitors are generally referred to in terms of their dielectric material. The following discussion describes several common capacitors and their features.

1. *Paper and Plastic capacitors* use a variety of dielectric materials. Some of these include mylar, polystyrene and polyethylene for the plastic types, and waxed or oiled paper for the older, less expensive paper type. Typically, the plates are long strips of tinfoil separated by the dielectric material. The foil and dielectric material are frequently rolled into a cylindrical component, Figure 18–24. It is interesting to note that because of their construction, each plate has two active surfaces. This means to calculate the plate areas, use twice the area of one plate, rather than only the area of one plate. Typical capacitance values for this type capacitor range from about 0.001 μF to one or more μF. Paper and plastic capacitors are typically used in fairly low-frequency applications such as in audio amplifiers.

2. *Mica capacitors* use mica for the dielectric material. Because the mica dielectric has a high breakdown voltage, these low-capacitance, high-voltage capacitors are frequently found in high-voltage circuits. Often, their construction is alternate layers of foil with mica that is molded into a plastic case, Figure 18–25. They are compact, moisture-proof and durable. Voltage ratings are in thousands of volts. Typical capacitance values range from about 5 to 50,000 picofarads, depending on voltage ratings.

3. *Ceramic capacitors* are typified by their small size and high-dielectric strength. Ceramic capacitors generally come in the shape of a flat disk

Typical available capacitance values:
Approximately 0.001 to 1.0 μF

Typical working voltage dc (WVDC):
100 to 1,500 V

FIGURE 18–24   Paper-type capacitor (Adapted with permission from Duff and Herman, *Alternating Current Fundamentals,* 3rd editon, by Delmar Publishers Inc., Albany, New York)

FIGURE 18–25   Mica capacitors (Photo by Michael A. Gallitelli)

FIGURE 18–26   Ceramic capacitors (Photo by Michael A. Gallitelli)

(disk-ceramics), Figure 18–26, or in cylindrical shapes. These capacitors are also compact, moisture-proof, and durable. Typical available ranges having 1,000 volt ratings are from approximately 5 pF to about 5,000 pF. At lower voltage ratings, higher capacitance values are available. Due to good dielectric characteristics, both mica and ceramic capacitors can be used in applications from the audio-frequency range up to several hundred megahertz.

4. *Electrolytic capacitors* have several prominent characteristics, including:
   a. high capacitance-to-size ratio
   b. polarity sensitivity and terminals marked + and −
   c. allow more leakage current than other types
   d. low cost

Refer to Figure 18–27 for the typical construction of electrolytic can-type capacitors. The external aluminum can or housing is typically the

FIGURE 18–27   Aluminum-type electrolytic capacitors (Courtesy of Mallory North American Capacitor Co.)

negative plate (or electrode). The positive electrode external contact is
generally aluminum foil immersed or in contact with an electrolyte of
ammonium borate (or equivalent). If the electrolyte is a borax solution,
the capacitor is called a *wet electrolytic*. If the electrolyte is a gauze
material saturated with borax solution, the capacitor is called a *dry
electrolytic*. To create the dielectric, a dc current is passed through the
capacitor. This causes a very thin aluminum oxide film (about 10 mi-
crocentimeters thick) to form on the foil surface. This thin oxide film is
the dielectric. (NOTE: This type of capacitor is *not* named after its dielectric.) Because of the extremely thin dielectric, capacitance value
per given plate area is very high. Also, voltage breakdown levels are
relatively low. Furthermore, since the dielectric is very thin, some
leakage current occurs. Typically, this leakage is a fraction of a milliampere for each microfarad of capacitance.

SAFETY HINTS

A **very important** fact about electrolytics is you must *observe
polarity* when connecting them into a circuit. If current is passed in the
wrong direction through the capacitor, the chemical action that created the
dielectric layer will be reversed and will destroy the capacitor. A shorted
capacitor is actually created. Gas may build up in the component, causing
an explosion! Depending on the voltage ratings, common values of electrolytic capacitors range from 2 μF to several hundred μF. Usual voltage
ratings for this capacitor range from 10 volts to about 500 volts.

The main advantage of the electrolytic capacitor is the large capaci-
tance-per-size factor. Two obvious disadvantages are the polarity,
which must be observed, and the higher leakage current feature. Also,
in many capacitors of this type, the electrolyte can dry out with age
and depreciate the capacitor quality or render it useless.

PRACTICAL NOTES

There are some specially made "nonpolarized" electrolytic capacitors
that are used in ac applications. One application for this type capacitor is as
a motor-starting capacitor. This application requires a large capacitance
value that must operate in an ac environment. These capacitors are actually
constructed by putting two electrolytic capacitors "back-to-back" in one
package. In order to function properly, they are connected with series-opposing polarities.

Due to the losses of the dielectric at higher frequencies, electrolytic
capacitor applications are generally limited to power-supply circuits
and audio-frequency applications.

**FIGURE 18–28   Tantalum electrolytic capacitors
(Photo by Michael A. Gallitelli)**

5. *Tantalum capacitors,* Figure 18–28, are in the electrolytic capacitor family. They use tantalum instead of aluminum. One important quality of tantalum capacitors is they provide very high capacities in small-sized capacitors. They have lower leakage current than the older electrolytics. Also, they do not dry out as fast, thus have a longer shelf-life. Because they are generally only manufactured with low-voltage ratings, they are used in low-voltage semiconductor circuitry. Tantalum capacitors are expensive.

Ratings

There are several important factors that should be observed when selecting or replacing capacitors in various applications. Some of these include:

• Physical size and mounting characteristics

• Capacitance value (in μF or pF)

• Capacitance tolerance (in percentage of rated value)

• Voltage ratings (e.g., working volts dc or WVDC)

• Safe temperature operating range

• Temperature coefficient (for temperature-compensating ceramic capacitors used in tuned rf circuits)

• Power factor of the capacitor (expressing the loss characteristics of dielectric used in capacitor)

• Inductance characteristics (when used at high frequencies)

Taking into consideration these factors, Figure 18–29 shows examples of typical ratings for various types of capacitors.

TYPICAL RANGE OF RATINGS FOR VARIOUS CAPACITORS
(Shown in standard electronic parts catalogs)

CAPACITOR TYPE	CAPACITANCE RATINGS	VOLTAGE RATINGS	TOLERANCE RATINGS	OPER. TEMPS	TEMP. COEFF.
Paper	0.001–1.0 μF	100–1,500 V	±10% (common)	−40 to +85°C	N/A
Mica	5–50,000 pF	600 to several kV	±1–5%	−55 to +125°C	N/A
Ceramic	1–10,000 pF	1kV–6kV	±10–20%	−55 to +85°C	N or P 0–750
Electrolytic	10–1,000 μF	5–500 V	−10–+50%	−40 to +85°C	N/A

NOTE: Temperature coefficient ratings are generally given for ceramic-type capacitors that are designed to increase or decrease in capacitance with temperature change. The rating is normally given in "parts-per-million per degree Celsius." For example, if a capacitor decreases 500 ppm per degree Celsius, it would be labeled "N500."

FIGURE 18–29 Summary of typical ratings

TYPICAL COLOR CODES

Capacitors are color-coded with various systems. It is not critical to learn these systems at this time. However, you should be aware that they exist.

Appendix A displays some common methods of color-coding for mica, ceramic disc, ceramic tubular, and paper capacitors. As in other color-coding systems, the position and color of the dots or bands conveys the desired information. (NOTE: Electrolytic capacitors are generally large enough so that the information is printed on their body, rather than using color codes. Also, ceramic capacitors often have the critical information printed on them.)

TYPICAL PROBLEMS AND TROUBLESHOOTING TECHNIQUES

Capacitors develop problems more frequently than resistors or inductors. As with the other components, troubles may be related to a shorted capacitor, an open capacitor, or degenerated performance due to excessive leakage, age, and other causes.

Causes of Open Capacitors

One reason a capacitor acts similar to an open is that a broken connection occurs between an external lead and a capacitor plate. In fact, this situation can occur with any capacitor! In electrolytic capacitors, the electrolyte resistance can drastically increase due to drying out, age, or operation at higher than normal temperatures.

Causes of Shorted Capacitors

Shorts can occur due to external leads touching or plates shorted internally. Internal shorts are most common with paper capacitors and electrolytic capacitors. Common causes for the dielectric material breaking down are age and usage. Again, operating under high temperature conditions accelerates this aging process. In paper electrolytic capacitors, the dielectric material sometimes deteriorates with age, which decreases leakage resistance. The resulting increase in leakage current is equivalent to a *partial short* condition.

Troubleshooting Techniques

Two general methods for testing capacitors are *in-circuit* testing and *out-of-circuit* testing. There are capacitance testers that measure capacitance values, leakage characteristics, and so on; however, the ohmmeter is often useful for making out-of-circuit general-condition tests. Because of the capacitor charging action, there typically is a current surge when a voltage is first applied to the capacitor, and the current decreases to zero (or to leakage-current level) when the capacitor is fully charged. Often, this action is monitored with the ohmmeter mode on a VOM.

Using the Ohmmeter for General Tests

GENERAL PRECAUTIONS.   When using the ohmmeter to check electrolytic capacitors, you should *observe polarity*. First, determine the voltage polarity at the test leads of the ohmmeter when in the ohms function. Frequently, the red lead is positive and the black lead is negative *but not* for all brands and models of instruments—check with the instrument's literature or another voltmeter. Next, observe the polarity markings on the capacitor, as appropriate. Also, to protect the meter, *make sure the capacitor to be checked is discharged* prior to testing.

ACTION WITH GOOD CAPACITORS.   An out-of-circuit test is where one capacitor lead is disconnected from the circuit, or the capacitor is physically removed from the circuit. When checking a capacitor, use a high range (e.g., $R \times 1$ meg selector position) and connect the ohmmeter leads across the capacitor, Figure 18– 30. As shown, the ohmmeter pointer initially deflects close to the zero ohms position. As the capacitor charges to the voltage provided by the ohmmeter battery, the pointer moves across the scale toward the infinite resistance scale position, and eventually rests at a very high resistance. For small-value capacitors, this value can be close to infinite. For electrolytic capacitors, the leakage resistance by this method varies from about 0.25 to 1 megohm, depending on the capacitor value and condition. When the R reading is less than this, the capacitor probably needs to be replaced.

For small-value capacitors, the changing deflection action is very quick and little pointer movement is seen. For larger values of capacitors, the charging action and resultant deflection are slower due to the larger RC time constant involved.

OPEN CAPACITORS.   When an ohmmeter is connected across an open capacitor, the pointer does not move (or an infinite Ω reading is shown), and no charging action is observed. Of course, the technician should be aware of the capacitor type and value being checked. For example, small-value capacitors (such as mica capacitors) may not provide an observable deflection, while large-value capacitors should provide some discernable charging action.

FIGURE 18–30   Checking capacitors with an ohmmeter

SHORTED CAPACITORS.   When an ohmmeter is connected across a shorted capacitor, the pointer moves to zero ohms on the scale and remains there. Sometimes, a variable capacitor short is remedied by locating the plate(s) that touch. In this case, simply bend the plates apart. Fixed capacitors indicating a shorted condition must be replaced.

EXCESSIVE LEAKAGE INDICATION.   When an ohmmeter is connected across a capacitor having excessively high leakage current (excessively low leakage resistance), the obvious result is a lower-than-normal resistance reading when the pointer finally comes to rest. The capacitor shows a charging action; however, the leakage resistance is lower than it should be.

REPLACING ELECTROLYTIC CAPACITORS.   Electrolytic capacitors definitely deteriorate with age. Their electrolyte may dry out, causing an increase in resistance and degeneration of operation. Also, the dielectric resistance greatly decreases from constant use causing excessive leakage current and reduced effectiveness in operation. When selecting an electrolytic capacitor for replacement, pay attention to the age of the capacitor— shelf-life is an important consideration.

Closing Comments

Capacitors are tremendously useful components found in most electronic circuits. Their characteristics of opposing a change in voltage, of blocking dc current once charged, and of storing electrical energy make them useful in numerous applications.

You should be aware that resistance, inductance, and capacitance are present in each of the three basic components you have studied thus far. Also, there is stray, or "distributed," capacitance and inductance present in most circuits. For example, wires and conductors that carry current exhibit some inductance, even though it may be a small amount. The leads and foil of capacitors have some inductance, as well. There is capacitance between the turns of an inductor. Distributed capacitance is present between wires and chassis, or ground. Both inductance and capacitance are present along the length of antenna elements. Coaxial cables have capacitance between conductors and inductance along their length. A resistor has a small degree of both inductance and capacitance present, due to its structure.

These stray, or distributed, capacitances and/or inductances normally do not affect circuits operating at low frequencies. However, in your later studies of high-frequency components and circuits, you will see they can play a part in proper or improper operation of circuits and systems. You will learn that the effects of some of these stray electrical quantities can be minimized by using shielding and/or using proper location of components and wise routing of wires. In summary, then, the frequency where components are expected to operate determines how critical these stray or distributed *R, L,* and *C* values are to circuit operation.

SUMMARY

▶ Capacitance is the circuit property that opposes a voltage *change* due to the storing of electrical energy (charge) in an electrostatic field. The unit of capacitance is the farad (F). One farad of capacitance stores one coulomb of charge when one volt is applied or is present across its plates. Practical subunits of the farad are the microfarad (μF) and the picofarad (μμF or pF). The symbol for capacitance is *C*.

▶ A capacitor is a device that possesses the property of capacitance. Capacitance is present when there are two conductors separated by an insulator (nonconductor or dielectric). Capacitance can be lumped as in the capacitor component, or distributed as the capacitance between wires and a chassis.

▶ Capacitors come in various sizes and shapes and in various capacitance values. Capacitors are often named after the type of dielectric used in their construction; e.g., paper, mica, ceramic, and so on. Electrolytic capacitors are an exception and are *not* named after their aluminum oxide dielectric. This capacitor is also the only basic category that is polarity sensitive.

▶ Factors affecting the capacitance amount include conductive plate areas, spacing between the plates (dielectric thickness), and the type of dielectric material. The formula is

$$C = \frac{8.85kA}{10^{12}s}$$

The greater the plate areas, the greater the capacitance value. The greater the spacing between plates (the thicker the dielectric) is, the less the capacitance value. Also, the higher the dielectric constant *(k)* is, the higher the capacitance value.

▶ The capacitance, charge, and voltage relationships for a capacitor are expressed as:

$$Q = CV, C = \frac{Q}{V}, \text{ and } V = \frac{Q}{C}$$

where Q is charge in coulombs; C is capacitance in farads; and V is voltage across the capacitor in volts.

▶ Dielectric constant (k or K_ϵ) is the comparative charge that different materials store relative to that charge stored in vacuum or air. That is, vacuum and air have a reference k of one.

▶ When capacitors charge, they are accumulating electrons on their negative plate(s) and losing electrons from their positive plate(s). As this happens, the electron orbits in the dielectric material are stressed, or distorted. This stored energy is returned to the circuit when the capacitor discharges. When the capacitor discharges, the excess electrons on the negative plate(s) travel to the positive plate(s), eventually neutralizing the charge and the potential difference between plates.

▶ Leakage resistance is related to the resistance of the dielectric material to current flow. If the dielectric were a perfect nonconductor, the leakage resistance would be infinite. Most capacitors have very high leakage resistance, unless their voltage ratings are exceeded. Electrolytic capacitors have a lower leakage resistance than the other types due to the thinness and type of dielectric.

▶ When capacitors are connected in series, the total capacitance is less than the least capacitance in series. That is, capacitances in series add like resistances in parallel.

▶ When capacitors are in parallel, total capacitance equals the sum of all the individual capacitances. That is, capacitances in parallel add like resistances in series, where $C_T = C_1 + C_2 + \ldots C_n$.

▶ Capacitors take time to charge and discharge. The time depends on the amount of capacitance storing the charge and the amount of resistance limiting the charging or discharging current. One RC time constant equals R (in ohms) times C (in farads). Five RC time constants are required for the capacitor to completely charge or discharge to a new voltage level.

▶ Basic types of capacitors are paper, mica, ceramic, and electrolytic types that come in a variety of combinations. Some of these variations are epoxy-dipped tantalum capacitors, epoxy-coated polyester film capacitors, dipped-mica capacitors, polystyrene film capacitors, dipped-tubular capacitors, and so on.

▶ Capacitors have various color codes. Electronic Industries Association (EIA) and MIL (military) color codes are common. Tubular capacitors often use a banded color-code system (typically six bands). Older types of mica capacitors are frequently color coded with either a five-dot, or six-dot system. Ceramic capacitors may sometimes be color coded with MIL color codes, using three, four, or five color dots or bands.

▶ Typical problems of capacitors include opens, shorts, or degeneration in operation due to low leakage resistance, or high electrolyte resistance.

▶ An ohmmeter check of a good capacitor will indicate a charging action occurring when the ohmmeter is first connected to the capacitor. The capacitor is charging to the ohmmeter's internal voltage source value. This is displayed by the ohmmeter pointer initially moving to a lower resistance point on the scale then moving to a final resting point at very high resistance once the capacitor has charged.

▶ Typical R values after charging are 100 or more megohms. Electrolytic capacitors have a lower leakage resistance in the range of 1/2 to 1 megohm.

FORMULAS AND CALCULATOR SEQUENCES SUMMARY

FORMULA 18–1 $C = \dfrac{Q}{V}$

charge value in coulombs, $\boxed{\div}$, voltage value, $\boxed{=}$

FORMULA 18–2 $Q = CV$

capacitance value in farads, $\boxed{\times}$, voltage value, $\boxed{=}$

FORMULA 18–3 $V = \dfrac{Q}{C}$

charge value in coulombs, $\boxed{\div}$, capacitance value in farads, $\boxed{=}$

FORMULA 18–4   Energy (J) $= \dfrac{1}{2CV^2}$

capacitance value in farads, $\boxed{\times}$, voltage value, $\boxed{x^2}$, $\boxed{\div}$, 2, $\boxed{=}$

FORMULA 18–5 $k = \dfrac{\epsilon_0 \text{ (absolute permittivity of dielectric material)}}{\epsilon_V \text{ (absolute permittivity of vacuum)}}$

ϵ_o value, $\boxed{\div}$, ϵ_V value, $\boxed{=}$

FORMULA 18–6 $C = \dfrac{8.85A}{10^{12}s}$ (where A is in square meters and s is in meters)

8.85, $\boxed{\times}$ area of plates, $\boxed{\div}$, $\boxed{(}$, 1, $\boxed{\text{EE}}$, 12 , $\boxed{\times}$, spacing, $\boxed{)}$, $\boxed{=}$

FORMULA 18–7 $C = \dfrac{8.85kA}{10^{12}s}$

8.85, $\boxed{\times}$, k value, $\boxed{\times}$, area of plates, $\boxed{\div}$, $\boxed{(}$, 1, $\boxed{\text{EE}}$, 12 , $\boxed{\times}$, spacing, $\boxed{)}$, $\boxed{=}$

FORMULA 18–8 $C = 2.25\dfrac{kA}{10^7 s}(n - 1)$ (where A is in square inches and s is in inches)

2.25, $\boxed{\times}$, k value, $\boxed{\times}$, A value, $\boxed{\div}$, $\boxed{(}$, 1, $\boxed{\text{EE}}$, 7, $\boxed{\times}$, s value, $\boxed{)}$, $\boxed{\times}$, $\boxed{(}$, n value, $\boxed{-}$, 1 , $\boxed{)}$, $\boxed{=}$

FORMULA 18–9 $C_T = \dfrac{C_1 \times C_2}{C_1 + C_2}$ (series capacitors)

C_1 value, $\boxed{\times}$, C_2 value, $\boxed{\div}$, $\boxed{(}$, C_1 value, $\boxed{+}$, C_2 value, $\boxed{)}$, $\boxed{=}$

FORMULA 18–10 $C_T = \dfrac{C}{N}$ (series capacitors)

capacitor value of one of the equal-value capacitors, $\boxed{\div}$, number of capacitors, $\boxed{=}$

FORMULA 18–11 $C_T = \dfrac{1}{\dfrac{1}{C_1} + \dfrac{1}{C_2} + \dfrac{1}{C_3} \ldots + \dfrac{1}{C_n}}$ (series capacitors)

C_1 value, $\boxed{1/x}$, $\boxed{+}$, C_2 value, $\boxed{1/x}$, . . . , $\boxed{=}$, $\boxed{1/x}$

FORMULA 18–12 $V_X = V_S \dfrac{C_T}{C_X}$

V_S value, $\boxed{\times}$, $\boxed{(}$, C_T value, $\boxed{\div}$, C_X value, $\boxed{)}$, $\boxed{=}$

FORMULA 18–13 $C_T = C_1 + C_2 \ldots + C_n$ (parallel capacitors)

C_1 value, $\boxed{+}$, C_2 value, . . . , $\boxed{=}$

FORMULA 18–14 $\tau = R$ (ohms) $\times$ C (farads)

R value, $\boxed{\times}$, C value, $\boxed{=}$

FORMULA 18–15 $v_R = V_S \times \epsilon^{-t/\tau}$

time allowed value, $\boxed{\div}$, $\boxed{(}$, R value, $\boxed{\times}$, C value, $\boxed{)}$, $\boxed{=}$, $\boxed{+/-}$, $\boxed{\text{2nd}}$, $\boxed{e^x}$, $\boxed{\times}$, V_S value, $\boxed{=}$

FORMULA 18–16 $v_C = V_S(1 - \epsilon^{-t/\tau})$

time allowed value, $\boxed{\div}$, $\boxed{(}$, R value, $\boxed{\times}$, C value, $\boxed{)}$, $\boxed{=}$, $\boxed{+/-}$, $\boxed{\text{2nd}}$, $\boxed{e^x}$, $\boxed{+/-}$, $\boxed{+}$, 1 , $\boxed{=}$, $\boxed{\times}$, V_S value, $\boxed{=}$

REVIEW QUESTIONS

1. The capacitor plate area doubles while the dielectric thickness is halved. What is the relationship of the new capacitance value to the original capacitance value? (Assume the same dielectric material.)

2. What charge is a 100-pF capacitor storing, if it is charged to 100 V?

3. What is the capacitor value that stores 200 pC when charged to 50 V?

4. What voltage is present across a 10-μF capacitor having a charge of 2,000 μC?

5. What is the total capacitance of a circuit containing a 100-pF, 400-pF, and 1,000-pF capacitor connected in series? If the capacitors were connected in parallel, what would be the total capacitance?

6. A 10-μF, 20-μF, and 50-μF capacitor are connected in series across a 160-V source. What

voltages appear across each capacitor after it is fully charged? What value of charge is on each capacitor?

7. A 10-μF, 20-μF, and 50-μF capacitor are connected in parallel across a 100-V source. What is the charge on each capacitor? What is the total capacitance and total charge for the circuit?

8. Using

$$V_X = V_S \frac{C_T}{C_X}$$

determine the voltage across the 20-μF capacitor in a series circuit containing a 20-μF, 40-μF, and 60-μF capacitor across a dc source of 100 V. (Show all work.)

9. Use the appropriate exponential formula(s) and find the v_R value in a series RC circuit after 1.2 τ if V_S is 200 V. What is the v_C value? (Use a calculator with the e^x or l_{nx} key, if possible.)

CHALLENGE QUESTIONS

1. Resarch and describe:
   a. The procedure for using a DMM to measure capacitance values.
   b. For the particular meter you research, define the ranges of capacitance the meter can measure.

2. Research and describe:
   a. The procedure for using a dedicated capacitance meter to measure capacitance values.
   b. For the particular meter you research, define the ranges of capacitance the meter can measure.

3. Research and describe:
   a. The procedure for using an LCR meter to measure capacitance values.
   b. For the particular meter you research, define the ranges of capacitance the meter can measure.

4. Research and describe:
   a. Any other types of measurements, if any, that can be made relative to capacitors' characteristics or conditions.
   b. By what test equipment and/or means these other measurements might be made.

TEST YOURSELF

1. What is the total capacitance of the circuit in Figure 18–31?

2. What is the voltage across C_5 in Question 1?

3. What is the charge on C_6 in Question 1?

4. Assume each capacitor in Question 1 has a breakdown voltage rating of 300 V. What voltage should not be exceeded between points A and B?

5. For the voltage described between points A and B in Question 4, what is the V_S value that causes that condition to exist?

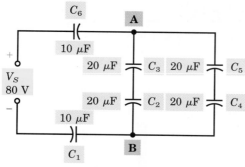

FIGURE 18–31

6. What is the total energy stored in the circuit of Question 1 with 80-volts V_S?

7. Assume a 15-μF capacitor is charged from a 150-V source and another 15-μF capacitor is charged from a separate 300-V source. If the capacitors are disconnected from their sources after being charged and connected in parallel with positive plate to positive plate and negative plate to negative plate, what is the resultant *voltage* and *charge* on each capacitor?

8. For the conditions described in Question 7, what is the total stored energy after the capacitors are connected?

PERFORMANCE PROJECT CORRELATION CHART

Suggested performance projects that correlate with topics in this chapter are:

CHAPTER TOPIC	PERFORMANCE PROJECT	PROJECT NUMBER
Charging and discharging action	Charge and Discharge Action and RC Time	51
The RC Time Constant		
Total Capacitance in Series and Parallel	Total Capacitance in Series and Parallel	52

THE SIMPLER SEQUENCE FOR TROUBLESHOOTING

CHAPTER TROUBLESHOOTING CHALLENGE

Following the SIMPLER troubleshooting sequence, find the problem with this circuit. As you follow the sequence, record your responses to each step on a separate sheet of paper.

CHALLENGE CIRCUIT 9

SYMPTOMS Gather, verify, and analyze symptom information. (Look at the "Starting Point Information.")

IDENTIFY Identify initial suspect area for the location of the trouble.

MAKE Make a decision about "What type of test to make" and "Where to make it." (To simulate making a decision about each test you want to make, select the desired test from the "TEST" column listing below.)

PERFORM Perform the test. (Look up the result of the test in Appendix C. Use the number in the "RESULTS" column to find the test result in the Appendix.)

LOCATE Locate and define a new "narrower" area in which to continue troubleshooting.

EXAMINE Examine available information and again determine "what type of test" and "where."

REPEAT Repeat the preceding analysis and testing steps until the trouble is found. What would you do to restore this circuit to normal operation? When you have solved the problem, compare your results with those shown on the following pages.

STARTING POINT INFORMATION

1. Circuit diagram
2. V_{C_2} is lower than it should be

TEST	Results in Appendix C
V_{C_1}	(116)
V_{C_2}	(15)
V_{C_3}	(94)

The circuit diagram labels: 1 μF (C_3) top, C_2 1 μF right, C_1 1 μF bottom, V_A 10 V (dc) left, nodes E, D, C, B, A.

(Circuit: V_A = 10 V (dc); C_1 = 1 μF, C_2 = 1 μF, C_3 = 1 μF; nodes A, B, C, D, E)

CIRCUIT

9

Source lead   Meter lead

Source lead   Meter lead

AUTO-RANGE DIGITAL MULTIMETER

From source +

From source –

Symptoms All we know is that the voltage across C_2 is lower than it should be. If things were normal, voltage across C_1 would be 3.3 V; the voltage across C_2 would be 3.3 V; and the voltage across C_3 would also be 3.3 V. Since Q is the same for all capacitors, the voltage distributions should be as described above. Also, we could predict that if C_2 were opened, its voltage would be 10 V (which does not agree with our symptom information); if C_2 were shorted or one of the other capacitors were opened, V_{C_2} would be 0 V; and if V_{C_2} were leaky, voltage would also be close to 0 V.

Identify initial suspect area: C_2 is the initial suspect area. However, we can't disregard the other capacitors since we have limited information.

Make test decision: Let's first check C_2's voltage to see where we are.

Perform 1st Test: Check V_2. V_{C_2} is close to 0 V, which is lower than normal.

Locate new suspect area: If either one of the other capacitors were open, the voltage for C_2 would still be 0 V. If either one of the other two capacitors were leaky or shorted, C_2's voltage would be higher than normal.

Examine available data.

Repeat analysis and testing: A check of V_{C_1} or V_{C_3} might be necessary.

CHALLENGE CIRCUIT 9

E   D C_3 C

1 μF

V_A
10 V (dc)

C_2 ⊣⊢ 1 μF

–

A C_1 B

1 μF

Check V_{C_3}. V_{C_3} is about 5 V, which is higher than normal, but C_3 is certainly not open or shorted. It looks like C_2 might be leaky. Let's take a look at C_2.

Lift one of C_2's leads and check the capacitor's dc resistance. The test reveals a lower than normal R.

Replace C_2 and check the circuit operation. The circuit checks out normal. C_2 was leaky.

2nd Test

3rd Test

4th Test

Capacitive Reactance
in AC

Courtesy of Hewlett-Packard Company

19

KEY TERMS

Capacitive reactance (X_C)

Rate of change of voltage

Reactance

OUTLINE

V and I relationships in a purely resistive AC circuit

V and I relationships in a purely capacitive AC circuit

Concept of capacitive reactance

Relationship of X_C to capacitance value

Relationship of X_C to frequency of AC

Methods to calculate X_C

Capacitive reactances in series and parallel

Voltages, currents, and capacitive reactances

Final comments about capacitors and capacitive reactance

CHAPTER PREVIEW

You have learned that a capacitor allows current flow, provided it is charging or discharging. However, a capacitor does not allow dc current to continue once charged to the source voltage. Also, you learned a capacitor does not instantly charge or discharge to a new voltage level; it requires current flow and time.

The unique characteristics of capacitors allow them to be used for a number of purposes in electronic circuits. Their ability to block dc current (once charged) makes them useful in coupling and decoupling applications for amplifiers, and so on. Their ability to store electrical energy provides great value when they are used for filtering circuits in power supplies and other circuits. Their charging and discharging characteristics allow them to be used in waveshaping applications and for timing circuits of all kinds.

This characteristic of allowing ac current due to the charging and discharging current will be studied in this chapter. You will learn there is an inverse relationship between the amount of *opposition* a capacitor gives to ac current and the circuit frequency and capacitance values involved. Capacitor opposition to ac current is called **capacitive reactance.** (Recall inductive reactance describes inductor opposition to ac current.)

In this chapter, you will examine capacitance as it relates to the capacitor action in a sine-wave ac circuit. You will study the phase relationship of voltage and current for capacitive components; learn to compute capacitive reactance; discover some applications of this characteristic; and learn other practical information about this important electrical parameter.

OBJECTIVES

After studying this chapter, you should be able to:

- Illustrate *V-I* relationships for purely resistive and purely capacitive circuits

- Explain **capacitive reactance**

- Use Ohm's law to solve for X_C value(s)

- Use capacitive reactance formula to solve for X_C value(s)

- Use X_C formula to solve for unknown C and f values

- Use Ohm's law and **reactance** formulas to determine circuit reactances, voltages, and currents for series- and parallel-connected capacitors

- List two practical applications for X_C

- Use the SIMPLER troubleshooting sequence to solve the Chapter Challenge problem

V AND I RELATIONSHIPS IN A PURELY RESISTIVE AC CIRCUIT

As a quick review, look at Figure 19–1. Note that V and I are in-phase in a purely resistive ac circuit.

V AND I RELATIONSHIPS IN A PURELY CAPACITIVE AC CIRCUIT

Observe Figure 19–2 as you study the following statements about the relationship of V and I in a purely capacitive ac circuit.

1. To develop a voltage (potential difference or p.d.) across capacitor plates, charge must move. That is, charging current must flow first.

2. When a sine-wave voltage is applied across a capacitor that is effectively in parallel with the source terminals, voltage across the capacitor must be the same as the applied voltage.

3. For the voltage across the capacitor to change, there must be alternate charging and discharging action. This implies charging and discharging current flow. The value of instantaneous capacitor current (i_C) directly relates to the amount of capacitance and the rate of change of voltage. (This principle fits well with the concept learned earlier where $Q = CV$.)

4. The formula that expresses the direct relationship of i_C to C and rate of change of voltage is:

Formula 19–1	$i_C = C\dfrac{dv}{dt}$

 Cap value, $\boxed{\times}$, voltage change value, $\boxed{\div}$, amount of time, $\boxed{=}$

where dv/dt indicates change in voltage divided by amount of time over which the change occurs, or the rate of change of voltage.

For example, if the voltage across a 100-pF capacitor changes by 100 V in 100 µs, then:

$$i_C = C\frac{dv}{dt}$$

$$i_C = 100 \times 10^{-12} \times \frac{100}{100 \times 10^{-6}}$$

$$i_C = \frac{10,000 \times 10^{-12}}{100 \times 10^{-6}}$$

$$i_C = 100 \times 10^{-6}, \text{ or } 100 \text{ µA}$$

V and I are in-phase

FIGURE 19–1 *V and I* phase relationships in a purely resistive circuit

FIGURE 19–2 *V and I* phase relationships in a purely capacitive circuit

keystrokes for 100 pF   keystrokes for 100 μs

(Answer is in amperes.)

5. Since the amount of charging (or discharging) current directly relates to the rate of change of voltage across the capacitor, capacitor current, i_C, is maximum when voltage rate of change is maximum and minimum when voltage rate of change is minimum.

6. Since **rate of change of voltage** is maximum near the zero points of the sine wave and minimum at the maximum points on the sine wave, current is maximum when voltage is minimum and vice-versa. In other words, I is 90° out-of-phase with V.

7. To illustrate these facts, Figure 19–2 shows that capacitor current *leads* the voltage across the capacitor by 90°. Remember our phrase mentioned in an earlier chapter—*"Eli, the ice man."* *Eli* represents voltage or emf *(E)* that comes before (leads) current *(I)* in an inductor *(L)* circuit. Obviously, the word "ice" represents that *current (I) leads voltage (E) in a capacitive (C) circuit.* If this phrase helps you remember phase relationships of voltage and current for inductors and capacitors, good! But if you understand the concepts related to inductors opposing a current change and capacitors opposing a voltage change, you will not need to remember this phrase!

✔ IN-PROCESS LEARNING CHECK I

Fill in the blanks as appropriate.

1. For a capacitor to develop a potential difference between its plates, there must be a _____ current.

2. For the potential difference between capacitor plates to decrease (once it is charged), there must be a _____ current.

3. When ac is applied to capacitor plates, the capacitor will alternately _____ and _____.

4. The value of instantaneous capacitor current directly relates to the value of _____ and rate of change of _____.

5. The amount of charging or discharging current is maximum when the rate of change of voltage is _____. For sine-wave voltage, this occurs when the sine wave is near _____ points.

6. Capacitor current _____ the voltage across the capacitor by _____ degrees. The expression that helps you remember this relationship is "Eli, the _____ man."

CONCEPT OF CAPACITIVE REACTANCE

As you have just seen, a sine-wave current develops in a capacitor circuit by the charge and discharge action of the capacitor when a sine-wave voltage is applied. This sine-wave capacitor current leads the sine-wave capacitor voltage by 90°.

In a purely capacitive circuit, the factor that limits the ac current is called capacitive reactance. The symbol used to represent this quantity is "X_C". It is measured in ohms. As you would anticipate, the higher the capacitive reactance, the lower the amount of current produced for a given voltage. That is,

$$I_C = \frac{V_C}{X_C}$$

which is simply an application of Ohm's law.

What factors affect this current-limiting trait called capacitive reactance? It is apparent that for a given voltage the larger the amount of charge moved per unit time, the higher the current and the lower the value of X_C produced and vice versa. Let's examine the relationship of X_C to the two factors that affect this movement of charge, or amount of current for a given voltage. These two factors are *capacitance* and *frequency*.

RELATIONSHIP OF X_C TO CAPACITANCE VALUE

The larger the capacitor, the greater the charge accumulated for a given V across its plates

$$V = \frac{Q}{C}$$

To match a given V applied, a higher charging current is required in a high C value circuit than is required with a smaller capacitance. This implies a lower opposition to ac current or a lower capacitive reactance (X_C). This means X_C *is inversely proportional to C.* In other words, the larger the C, the lower the X_C present; and the smaller the C, the higher the X_C present.

You will soon be introduced to a formula for calculating X_C; however, for now, notice in Figure 19–3a that for a given frequency of operation (100 Hz, in this case), the amount of capacitive reactance in ohms is *inversely related* to the capacitance value, which illustrates our previous statement. We will ask you to confirm the values shown in the figure after you study the X_C formula. For now, assume the values are valid. The point here is that the greater the C value, the less the X_C value, and vice-versa.

RELATIONSHIP OF X_C TO FREQUENCY OF AC

If the capacitor voltage is tracking the source V, as in the circuit of Figure 19–3b, then the higher the frequency and the less available time there is for the required amount of charge (Q) to move to and away from the respective

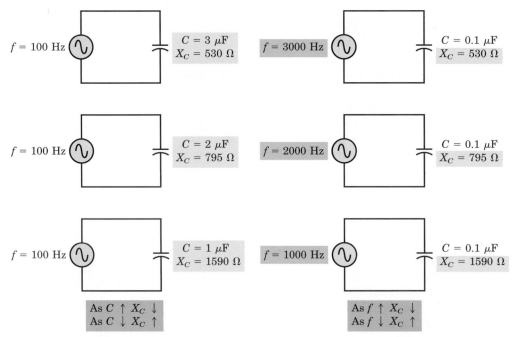

FIGURE 19–3   Relationship of X_C to capacitance and frequency

capacitor plates during charge and discharge action. As you know, it takes a given amount of Q to achieve a specific voltage across a given capacitor $(V = Q/C)$.

The higher the frequency of applied voltage, the faster is the rate of change of voltage for any given voltage level. As you learned earlier, the i_C directly relates to the voltage rate of change (dV/dt). Since higher frequencies produce a greater charge and discharge current for any given voltage level, as frequency increases opposition to current decreases. More current for a given V applied indicates there is *less* opposition, meaning capacitive reactance (X_C) decreases as frequency increases. That is, X_C *is inversely proportional to f*. In other words, the higher the frequency, the lower the X_C present; and the lower the frequency, the greater the X_C present. See Figure 19–3b.

METHODS TO CALCULATE X_C
Ohm's Law Method

Ohm's law is one practical method to calculate X_C, where

$$X_C = \frac{V_C}{I_C}$$

For example, if the effective ac voltage applied to a capacitive circuit is 100 volts, and the I_C is 1 ampere (rms), capacitive reactance (X_C) must equal 100 ohms. NOTE: Unless told otherwise, assume effective (rms) values of ac voltage and current in your measurements and computations relating to V_C,

I_C, and X_C. As an electronics student and technician, you will frequently make such measurements and calculations. Remember, the Ohm's law formula can also be transposed to yield

$$V_C = I_C \times X_C, \text{ and } I_C = \frac{V_C}{X_C}$$

The Basic X_C Formula

From the previous discussions, it is easy to understand the concepts illustrated by the formula to calculate X_C when only C and f are known. The formula states that:

Formula 19–2	$X_C \text{ (ohms)} = \dfrac{1}{2\pi fC}$

where f is frequency in hertz
    C is capacitance in farads

 6.28, $\boxed{\times}$, frequency value, $\boxed{\times}$, capacitance value, $\boxed{=}$, $\boxed{1/x}$

NOTE: If the capacitor value is in microfarads, follow the number for the capacitor value with the keystrokes: $\boxed{\text{EE}}$, $\boxed{+/-}$, 6 , to indicate the value times 10^{-6} power. If your calculator has the $\boxed{\text{EXP}}$ key, rather than the $\boxed{\text{EE}}$ key, use $\boxed{\text{EXP}}$, $\boxed{+/-}$, 6.

PRACTICAL NOTES

Some people find it covenient to alter the formula

$$X_C = \frac{1}{2\pi fC} \text{ to read } \frac{0.159}{fC}$$

This can be done because the value of 2π does not change. That is, it is a "constant." One divided by $2\pi = 0.159$; therefore, the formula can be simplified to read $0.159/(fC)$. When X_C calculations must be performed without a calculator, this modified formula may be easier to use. When a calculator is available, either formula format is easily used.

Notice that X_C is *inversely* proportional to both f and C. That is, if either f or C (or both) are increased, the resulting larger product in the denominator divided into one results in a smaller value of X_C. In fact, if either f or C is doubled, X_C is half. If either f or C is halved, X_C doubles, and so forth, Figure 19–4.

    Stated another way, larger C means more charging and discharging current flows to charge to and discharge from a given voltage. Thus, a larger C results in a lower X_C. Since a higher f means the voltage rate of change

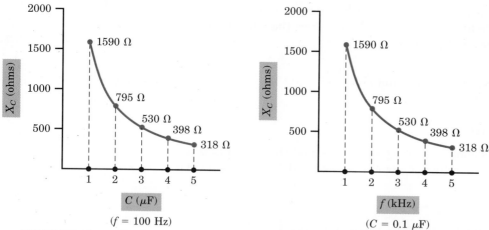

FIGURE 19–4   Examples of the inverse relationship of X_C to C and to f

is faster, more charging and discharging current must flow in the time allowed by the higher frequency; hence, a higher f results in a lower X_C.

Recall the "$2\pi f$" in the formula relates to angular velocity which relates to sine-wave frequency. The X_C formula is the second formula where you have found this angular velocity expression—the other formula was for inductive reactance where $X_L = 2\pi f L$.

EXAMPLE

1. What is the capacitive reactance of a 2-μF capacitor operating at a frequency of 120 Hz?

$$X_C = \frac{1}{2\pi f C} = \frac{1}{6.28 \times 120 \times (2 \times 10^{-6})} = \frac{1}{1{,}507.2 \times 10^{-6}} = 663.5\ \Omega$$

6.28, $\boxed{\times}$, 120, $\boxed{\times}$, 2, $\boxed{\text{EE}}$, $\boxed{+/-}$, 6, $\boxed{=}$, $\boxed{1/x}$

2. The frequency of applied voltage doubles while the C remains the same in a capacitive circuit. What happens to the circuit X_C?

$$X_C = \frac{1}{2\pi f C}, \text{ where } X_C \text{ for stated conditions} = \frac{1}{2\pi(2f)C}$$

Therefore, the denominator's product is twice as large as that divided into the unchanged numerator. That is, the X_C is half of the original frequency X_C.

3. In a capacitive circuit, if X_C quadruples while the frequency of the applied voltage remains the same, what must have happened to the C value in the circuit?

Since X_C is inversely proportional to f and C and f remains unchanged, C must decrease to one-fourth its original value. That is,

$$4 \times \text{original } X_C = \frac{1}{2\pi f(0.25C)}.$$

PRACTICE PROBLEMS I

1. Use the X_C formula and verify values of X_C shown in Figure 19–3. Do the formula results validate the values of X_C shown?

2. What is the capacitive reactance of a 5-μF capacitor operating at a frequency of 200 Hz?

3. What happens to the X_C in a circuit when f doubles and the capacitance value is halved at the same time?

Rearranging the X_C Formula to Find f and C

By rearranging the X_C equation, we can solve for frequency if both C and X_C are known, or solve for C if both f and X_C are known. When the equation is transposed, the new equations are as follows:

Formula 19–3	$f = \dfrac{1}{2\pi C X_C}$	and	Formula 19–4	$C = \dfrac{1}{2\pi f X_C}$

 6.28, $\boxed{\times}$, capacitance value, $\boxed{\times}$, X_C value, $\boxed{=}$, $\boxed{1/x}$ (to find f)

 6.28, $\boxed{\times}$, frequency value, $\boxed{\times}$, X_C value, $\boxed{=}$, $\boxed{1/x}$ (to find C)

Try these formulas on the following practice problems.

PRACTICE PROBLEMS II

1. What is the frequency of operation of a capacitive circuit where the total X_C is 1,000 ohms and the total capacitance is 500 pF?

2. What is the total circuit capacitance of a circuit exhibiting an X_C of 3,180 ohms at a frequency of 2 kHz?

CAPACITIVE REACTANCES IN SERIES AND PARALLEL

In the last chapter, you learned that capacitances in series add like resistances in parallel, and capacitances in parallel add like resistances in series. It is interesting that when we consider capacitive *reactances* in series and parallel, the converse is true.

Capacitive Reactances in Series

Since capacitive reactance (X_C) is an opposition to current flow, measured in ohms, capacitive reactances (current oppositions) in series add like resistances in series. That is, $X_{C_T} = X_{C_1} + X_{C_2} + X_{C_3} \ldots + X_{C_n}$, Figure 19–5.

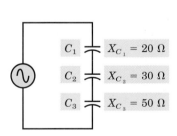

$X_{C_1} = 20\ \Omega$
$X_{C_2} = 30\ \Omega$
$X_{C_3} = 50\ \Omega$

$X_{C_T} = X_{C_1} + X_{C_2} + X_{C_3}$
$X_{C_T} = 100\ \Omega$

(X_C in series add like resistances in series)

FIGURE 19–5 X_C **in series**

EXAMPLE

What is the total X_C of a circuit containing three series-connected 3-μF capacitors operating at a frequency of 100 Hz?

Approach 1: Find total C, then use the X_C formula:

  C_T for three equal-value capacitors in series equals the C value of one capacitor divided by the number of equal-value capacitors ($C_T = C_1/N$).

$$C_T = \frac{3\ \mu F}{3} = 1\ \mu F$$

$$X_{C_T} = \frac{1}{2\pi f C_T} = \frac{1}{6.28 \times 100 \times 1 \times 10^{-6}} = 1{,}592\ \Omega$$

Approach 2: Find X_C of a 3-μF capacitor, then multiply by 3:

$$X_{C_1} = \frac{1}{2\pi f C_1} = \frac{1}{6.28 \times 100 \times 3 \times 10^{-6}} = 530.7\ \Omega$$

$$X_{C_T} = 3 \times 530.7\ \Omega = 1{,}592\ \Omega$$

PRACTICE PROBLEMS III

What is the total X_C of the circuit in Figure 19–6?

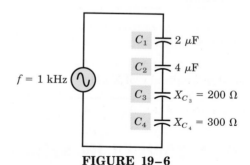

$f = 1\ \text{kHz}$

C_1 2 μF
C_2 4 μF
C_3 $X_{C_3} = 200\ \Omega$
C_4 $X_{C_4} = 300\ \Omega$

FIGURE 19–6

Capacitive Reactances in Parallel

Since capacitive reactance is an opposition to ac current flow, when we supply more current paths by connecting X_Cs in parallel, the resulting total opposition decreases just as it does when we parallel resistances.

Useful formulas to solve for total capacitive reactance of capacitors in parallel are as follows:

• General formula for X_Cs in parallel:

$$\text{Formula 19--5} \quad X_{C_T} = \cfrac{1}{\cfrac{1}{X_{C_1}} + \cfrac{1}{X_{C_2}} + \cfrac{1}{X_{C_3}} \cdots + \cfrac{1}{X_{C_n}}}$$

 X_{C_1} value, $\boxed{1/x}$, $\boxed{+}$, X_{C_2} value, $\boxed{1/x}$, $\boxed{+}$ X_{C_3} value, $\boxed{1/x}$, $\boxed{=}$, $\boxed{1/x}$

• Formula for two X_Cs in parallel:

$$\text{Formula 19--6} \quad X_{C_T} = \frac{X_{C_1} \times X_{C_2}}{X_{C_1} + X_{C_2}}$$

 X_{C_1} value, $\boxed{\times}$, X_{C_2} value, $\boxed{\div}$, $\boxed{(}$, X_{C_1} value, $\boxed{+}$, X_{C_2} value, $\boxed{)}$, $\boxed{=}$

• Formula for equal-value X_Cs in parallel:

$$\text{Formula 19--7} \quad X_{C_T} = \frac{X_{C_1}}{N}$$

where N = number of equal-value X_Cs in parallel

 X_{C_1} value, $\boxed{\div}$, number of equal value X_C's in parallel, $\boxed{=}$

EXAMPLE

What is the total X_C of a circuit containing three parallel-connected capacitor branches with values of 2 μF, 3 μF, and 5 μF, respectively? Assume the circuit is operating at a frequency of 400 Hz.

Two possible approaches are:

1. Find total C, then use X_C formula to find X_{C_T}.

2. Find X_C of each capacitor using X_C formula, then use the general formula to solve for X_{C_T}.

Of these two approaches, approach 1 is easiest for these circumstances. Therefore,

$$C_T = C_1 + C_2 + C_3$$
$$C_T = 2\ \mu F + 3\ \mu F + 5\ \mu F = 10\ \mu F$$

$$X_{C_T} = \frac{1}{2\pi f C_T} = \frac{1}{6.28 \times 400 \times 10 \times 10^{-6}} = \frac{1}{0.02512} = 39.8\ \Omega$$

PRACTICE PROBLEMS IV

1. What is the total capacitive reactance of a circuit containing three parallel-connected capacitors of 1 μF, 2 μF, and 3 μF, respectively. (Assume an operating frequency of 1 kHz.)

2. What is the total capacitive reactance of two parallel-connected capacitors where one capacitance is 5 μF and the other capacitance is 20 μF? (Circuit is operating at a frequency of 500 Hz.)

3. What is the total capacitive reactance of four 1,600-ohm capacitive reactances connected in parallel?

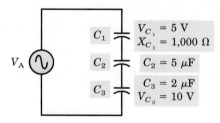

Find f; C_1; I; X_{C_2}; X_{C_3}; X_{C_T}; V_{C_2}; and V_A.

(Since we know two things about C_1, let's start there.)

$$I = \frac{V_{C_1}}{X_{C_1}} = \frac{5}{1,000} = 0.005 \text{ A, or } I = 5 \text{ mA.}$$

Since V_{C_3} is double V_{C_1}, its X_C must be double that of C_1 since they have the same I and $V = I \times X_C$. Thus, $X_{C_3} = 2,000$ Ω.

Since V_{C_1} is half V_{C_3}, C_1 must have twice the capacitance and half the X_C of C_3. Thus, $C_1 = 4$ μF.

Since C_2 is 2.5 times the C value of C_3 and X_C is inversely related to C, X_{C_2} must equal $\frac{X_{C_3}}{2.5}$ or $\frac{2,000}{2.5}$. Thus, $X_{C_2} = 800$ Ω.

$$V_{C_2} = I \times X_{C_2} = 5 \text{ mA} \times 800 \text{ Ω}$$
$$V_{C_2} = 4 \text{ V}$$

$$X_{C_T} = X_{C_1} + X_{C_2} + X_{C_3} = 1,000 + 800 + 2,000$$
$$X_{C_T} = 3,800 \text{ Ω}$$

$$C_T = \frac{1}{\left(\frac{1}{4} + \frac{1}{5} + \frac{1}{2}\right)}$$
$$C_T = 1.05 \text{ μF}$$

$$f = \frac{1}{2\pi C_T X_{C_T}} = \frac{1}{6.28 \times 1.05 \times 10^{-6} \times 3,800}$$
$$f = \frac{1}{0.02505} = 39.9 \text{ Hz}$$
$$f = 39.9 \text{ Hz or about 40 Hz}$$

$$V_A = I \times X_{C_T} = 5 \text{ mA} \times 3.8 \text{ kΩ}$$
$$V_A = 19 \text{ V}$$
$$\text{(Also, } V_1 + V_2 + V_3 = 19 \text{ V)}$$

FIGURE 19–7   Series capacitive circuit analysis

VOLTAGES, CURRENTS, AND CAPACITIVE REACTANCES

Our computations have concentrated on finding the capacitive reactances in capacitive circuits. Let's discuss voltage distribution throughout series capacitive circuits and current distribution throughout parallel capacitive circuits.

Capacitive reactances represent opposition to ac current (in ohms). When dealing with V, I, and X_C, we treat ohms of opposition like we treat resistances in resistive circuits. That is, voltage drops are calculated by using $I \times X_C$; and currents are calculated by using V/X_C, and so on.

Refer to Figure 19–7 to see series circuit parameters and computations illustrated. Then refer to Figure 19–8 and observe how parallel circuit parameters and computations are made. After studying these figures, try the following practice problems.

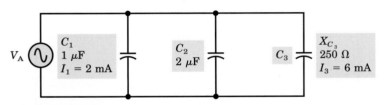

Find V_A; f; X_{C_2}; X_{C_1}; I_2; I_T.

Starting with information regarding C_3,

$$V_A = V_3 = I_3 \times X_{C_3}$$
$$V = 6 \text{ mA} \times 0.25 \text{ k}\Omega$$
$$V_A = 1.5\text{V}$$

Since I_3 is triple I_1, X_{C_3} must be one-third X_{C_1}. Therefore, C_3 must be triple C_1.

$$C_3 = 3 \ \mu\text{F}$$

$$X_{C_1} = 3 \times X_{C_3} = 3 \times 250 \ \Omega$$
$$X_{C_1} = 750 \ \Omega$$

$$X_{C_2} = \frac{1}{2} X_{C_1} = \frac{1}{2} \text{ of } 750 \ \Omega$$
$$X_{C_2} = 375 \ \Omega$$

$$I_2 = \frac{V_2}{X_{C_2}} = \frac{1.5 \text{ V}}{375 \ \Omega}$$
$$I_2 = 4 \text{ mA}$$

$$I_T = I_1 + I_2 + I_3 = 2 \text{ mA} + 4 \text{ mA} + 6 \text{ mA}$$
$$I_T = 12 \text{ mA}$$

f is found by using any one of the branches information or by using the total C and X_C parameters. That is,

$$f = \frac{1}{2\pi C X_C} = \frac{1}{6.28 \times 1 \times 10^{-6} \times 750}$$
$$f = \frac{1}{0.00471}$$
$$f = 212 \text{ Hz}$$

FIGURE 19–8   Parallel capacitive circuit analysis

PRACTICE PROBLEMS V

1. For the circuit in Figure 19–9, find X_{C_1}, X_{C_2}, X_{C_3}, X_{C_T}, C_2, C_3, V_2, and I_T.

2. For the circuit in Figure 19–10, find C_1, X_{C_1}, X_{C_2}, X_{C_3}, I_2, and I_T.

Find X_{C_1}; X_{C_2}; X_{C_3}; X_{C_T}; C_2; C_3; V_2; and I_T.

FIGURE 19–9

Find C_1; X_{C_1}; X_{C_2}; X_{C_3}; I_2; and I_T.

FIGURE 19–10

PARAMETER INVOLVED	CAPACITIVE CIRCUIT	INDUCTIVE CIRCUIT
Opposition to I	X_C in ohms	X_L in ohms
As f increases	X_C decreases	X_L increases
As f decreases	X_C increases	X_L decreases
As C increases	X_C decreases	not applicable
As C decreases	X_C increases	not applicable
As L increases	not applicable	X_L increases
As L decreases	not applicable	X_L decreases
Voltage across component	$I \times X_C$	$I \times X_L$
Instantaneous values	$i_C = C \times dv/dt$	$v_L = L \times di/dt$
Phase of V and I	I leads V by 90 degrees	V leads I by 90 degrees
To calculate X	$X_C = 1/2\pi fC$	$X_L = 2\pi fL$
Series-connected	$C_T < $ least C	$L_T = $ sum of L's
Parallel-connected	$C_T = $ sum of C's	$L_T < $ least L
Series-connected	$X_{C_T} = $ sum of X_C's	$X_{L_T} = $ sum of X_L's
Parallel-connected	$X_{C_T} < $ least X_C	$X_{L_T} < $ least X_L

FIGURE 19–11   Comparisons of C and L circuit characteristics

FINAL COMMENTS ABOUT CAPACITORS AND CAPACITIVE REACTANCE

Having studied capacitive reactance, you can more clearly see how capacitors are used. As you proceed in your studies and career, you should *keep in mind* the inverse relationship of capacitance and capacitive reactance. Also, you remember capacitances in series add like resistances in parallel. Keeping these ideas in mind will help prevent errors when analyzing circuits containing both inductances and capacitances.

A list of similarities and differences between capacitive and inductive circuits is shown in Figure 19–11. Review what you have learned about the operating features of these two electronic components. It is important that you thoroughly understand each component to analyze circuits that combine these components in various configurations.

SUMMARY

▶ Capacitive reactance is the opposition a capacitor displays to ac or pulsating dc current. The symbol is X_C and it is measured in ohms. Capacitive reactance is calculated by the formula

$$X_C = \frac{1}{2\pi fC}$$

where f is frequency in hertz and C is capacitance in farads.

▶ Factors influencing the amount of capacitive reactance (X_C) in a circuit are the amount of capacitance (C) and the frequency (f). Voltage does *not* affect capacitive reactance. As C *or* f increases, X_C decreases. As C *or* f decreases, X_C increases proportionately.

▶ In a purely capacitive circuit, current leads voltage by 90°. (Charging current must flow before a potential difference develops across the capacitor plates.)

▶ Capacitive reactance is calculated by using Ohm's law or the capacitive reactance formula, where

$$X_C = \frac{V_C}{I_C}, \text{ or } X_C = \frac{1}{2\pi fC}$$

▶ The capacitive reactance formula can be rearranged to find either f or C, if all other factors are known. For example,

$$f = \frac{1}{2\pi CX_C}, \text{ and } C = \frac{1}{2\pi fX_C}$$

▶ Capacitive reactances in series add like resistances in series. That is, $X_{C_T} = X_{C_1} + X_{C_2} + X_{C_3} \ldots + X_{C_n}$.

▶ Capacitive reactances in parallel add like resistances in parallel. The reciprocal formula, product-over-the-sum formula, or equal-value parallel formulas are used, as appropriate, to find total capacitive reactance of parallel-connected capacitors.

▶ All three variations of Ohm's law may be used with relation to capacitor parameters. That is,

$$I_C = \frac{V_C}{X_C}; V_C = IX_C; \text{ and } X_C = \frac{V_C}{I_C}$$

▶ Capacitance and capacitive reactance exhibit converse characteristics. In a *series* circuit, the *smallest* C has the largest voltage drop and the largest C the least voltage drop. In this same series circuit, the largest X_C drops the most voltage (because the smallest C has the highest X_C), and so forth. In *parallel* circuits, the largest C holds the largest amount of charge (Q) and the smallest C has the least Q.

FORMULAS AND CALCULATOR SEQUENCES SUMMARY

FORMULA 19–1 $i_C = C\dfrac{dv}{dt}$

Cap value, $\boxed{\times}$, voltage change value, $\boxed{\div}$, amount of time, $\boxed{=}$

FORMULA 19–2 $X_C \text{ (ohms)} = \dfrac{1}{2\pi fC}$

6.28, $\boxed{\times}$, frequency value, $\boxed{\times}$, capacitance value, $\boxed{=}$, $\boxed{1/x}$

FORMULA 19–3 $f = \dfrac{1}{2\pi C X_C}$

6.28, $\boxed{\times}$, capacitance value, $\boxed{\times}$, X_C value, $\boxed{=}$, $\boxed{1/x}$

FORMULA 19–4 $C = \dfrac{1}{2\pi f X_C}$

6.28, $\boxed{\times}$, frequency value, $\boxed{\times}$, X_C value, $\boxed{=}$, $\boxed{1/x}$

FORMULA 19–5 $X_{C_T} = \dfrac{1}{\dfrac{1}{X_{C_1}} + \dfrac{1}{X_{C_2}} + \dfrac{1}{X_{C_3}} \cdots \dfrac{1}{X_{C_n}}}$

X_{C_1} value, $\boxed{1/x}$, $\boxed{+}$, X_{C_2} value, $\boxed{1/x}$, $\boxed{+}$, X_{C_3} value, $\boxed{1/x}$, $\boxed{=}$, $\boxed{1/x}$

FORMULA 19–6 $X_{C_T} = \dfrac{X_{C_1} \times X_{C_2}}{X_{C_1} + X_{C_2}}$

X_{C_1} value, $\boxed{\times}$, X_{C_2} value, $\boxed{\div}$, $\boxed{(}$, X_{C_1} value, $\boxed{+}$, X_{C_2} value, $\boxed{)}$, $\boxed{=}$

FORMULA 19–7 $X_{C_T} = \dfrac{X_{C_1}}{N}$

X_{C_1} value, $\boxed{\div}$, number of equal value X_C's in parallel, $\boxed{=}$

REVIEW QUESTIONS

1. What is the capacitive reactance of a 0.00025-µF capacitor operating at a frequency of 5 kHz?

2. What capacitance value exhibits 250 Ω of reactance at a frequency of 1,500 Hz?

3. The frequency of applied voltage to a given circuit triples while the capacitance remains the same. What happens to the value of capacitive reactance?

4. Series-circuit capacitors are all exchanged for Cs, each having twice the capacitance of the capacitor it replaces. What happens to total circuit X_C and total C?

5. For the circuit in Figure 19–12 find C_T, X_{C_T}, V_A, C_1, C_2, and X_{C_2}.

$I = 2$ mA

FIGURE 19–12

6. For the circuit in Figure 19–12, if C_1 and C_2 were equal in value, what would be the new voltages on C_1 and C_2?

7. Assume the frequency doubles in the circuit in Figure 19–12. Answer the following statements with "I" for increase, "D" for decrease, and "RTS" for remain the same.
   a. C_1 will _____
   b. C_2 will _____
   c. C_T will _____
   d. X_{C_1} will _____
   e. X_{C_2} will _____
   f. X_{C_T} will _____
   g. Current will _____
   h. V_{C_1} will _____
   i. V_{C_2} will _____
   j. V_A *will* _____

8. For the circuit in Figure 19–13 find C_T, I_1, I_2, X_{C_T}, X_{C_1}, X_{C_2}, and I_T.

FIGURE 19–13

9. For the circuit in Figure 19–13, if C_2 is equal to C_1, what is the I_T?

10. Assume the frequency doubles in the circuit in Figure 19–13. Answer the following statements with "I" for increase, "D" for decrease, and "RTS" for remain the same.
    a. C_1 will _____
    b. C_2 will _____
    c. C_T will _____
    d. X_{C_1} will _____
    e. X_{C_2} will _____
    f. X_{C_T} will _____
    g. Current will _____
    h. V_{C_1} will _____
    i. V_{C_2} will _____
    j. V_A will _____

11. Refer to Figure 19–14 and answer questions 11–15:
    What is the X_C of C_1?

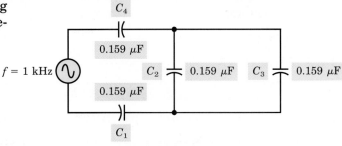

FIGURE 19–14

12. What is the X_{C_T}?

13. What is the value of C_T?

14. If f were changed to 5 kHz, what would be the new value of circuit X_{C_T}?

15. If C_2 were replaced by a larger capacitance value capacitor, would the voltage across C_4 increase, decrease, or remain the same?

CHALLENGE QUESTIONS

1. In your own words, explain when and why a capacitor might be said to be an "open circuit" with respect to dc, or that a capacitor "blocks" dc.

2. In your own words, explain how a capacitor can "pass" ac.

TEST YOURSELF

1. Draw a capacitive voltage divider circuit where the following conditions are found: (Label the diagram with all component and parameter values.)

   a. Three capacitors are used.
   b. The voltage across the middle capacitor will be double that found across the capacitors on either side of it.
   c. The frequency of operation is 2 kHz.
   d. The capacitor with maximum X_C has a reactance value of 79.5 Ω.
   e. $V_A = 159$ V

2. Draw a circuit configuration where the total capacitance is 0.66 μF, using six 1-μF capacitors.

PERFORMANCE PROJECTS CORRELATION CHART

Suggested performance projects that correlate with topics in this chapter are:

CHAPTER TOPIC	PERFORMANCE PROJECT	PROJECT NUMBER
V & I Relationships in a Purely Capacitive AC Circuit	Capacitance Opposing a Change in Voltage	53
Relationship of X_C to Capacitance Value and Relationship of X_C to Frequency of AC	X_C Related to Capacitance and Frequency	54
Methods to Calculate X_C	The X_C Formula	55

Following the SIMPLER troubleshooting sequence, find the problem with this circuit. As you follow the sequence, record your responses to each step on a separate sheet of paper.

SYMPTOMS Gather, verify, and analyze symptom information. (Looking at the "Starting Point Information.")

IDENTIFY Identify initial suspect area for the location of the trouble.

MAKE Make a decision about "What type of test to make" and "Where to make it." (To simulate making a decision about each test you want to make, select the desired test from the "TEST" column listing below.)

PERFORM Perform the test. (Look up the result of the test in Appendix C. Use the number in the "RESULTS" column to find the test result in the Appendix.)

LOCATE Locate and define a new "narrower" area in which to continue troubleshooting.

EXAMINE Examine available information and again determine "what type of test" and "where."

REPEAT Repeat the preceding analysis and testing steps until the trouble is found. What would you do to restore this circuit to normal operation? When you have solved the problem, compare your results with those shown on the following pages.

CHALLENGE CIRCUIT 10

STARTING POINT INFORMATION

1. Circuit diagram
2. When testing at test point "1," voltmeter reading is zero volts.
3. When testing at test point "2," voltmeter reading is 7.5 volts.
4. When testing at test point "3," voltmeter reading is 7.5 volts.

TEST   Results in Appendix C

V_{C_1} ....(28)
V_{C_2} ....(67)
V_{C_3} ....(102)
V_T ....(79)
"Bridging" with known good
0.001-μF capacitor:
Bridge C_1 ....(71)
Bridge C_2 ....(24)
Bridge C_3 ....(111)

CIRCUIT

10

Meter lead
Source lead
Meter lead
Source lead

Symptoms The capacitive voltage divider is not working correctly. The X_C's of the equal value capacitors should be equal. The $I \times X_C$ drops should all be approximately equal. At test point 1, the voltmeter should read about 2.5 V; at test point 2 about 5.0 V; and at test point 3 about 7.5 V.

Identify initial suspect area: If C_1 were shorted, the V at test point 1 would be 0 V. The V at test point 2 would be 3.75 V rather than 2.5 V. If C_3 were shorted, the V at switch position 1 would be 3.75 V not 0 V. If C_1 were open, the V at switch position 1 would be 7.5 V. If C_3 were open, the V at switch position 2 would be 0 V. The normal voltage at switch position 3 is 7.5 V (the source voltage), which is what is measured. Without making our first test, we have narrowed the initial suspect area to C_2. If it were open, the circuit would operate similar to the symptom descriptions.

Make test decision: Bridge the suspected open capacitor with a known good one of the same ratings and note any changes in circuit operation. We'll try the quick bridging technique. CAUTION: Make bridging connections with the power off. Then turn power on and make the necessary observations.

Perform 1st Test: Bridge C_2 while monitoring voltage at switch position 2. When this is done, V measures about 4.8 V, close to the nominal 5-V normal operation.

C_3 0.001 μF

C_2 0.001 μF

C_1 0.001 μF

100 Hz 7.5 V

2

3

1

V

CHALLENGE CIRCUIT 10

Locate new suspect area: This is not necessary since we've found our problem.

Examine available data.

Repeat analysis and testing: Verify that the circuit operates normally by checking the voltage at each switch position (**2nd, 3rd, and 4th Tests**). C_2 was open.

1st Test

2nd Test

3rd Test

4th Test

RC Circuits in AC

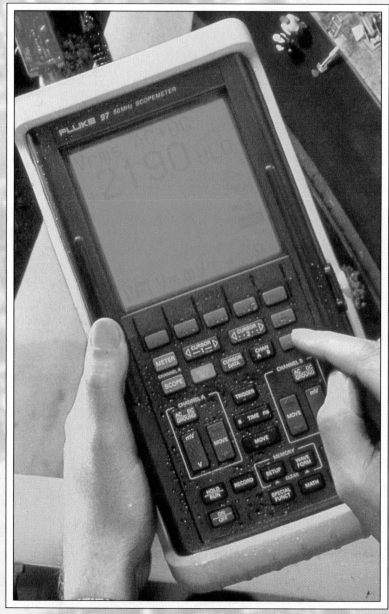

Courtesy of John Fluke Mfg. Co., Inc.

20

KEY TERMS

Current leading voltage

Long time-constant circuit

Short time-constant circuit

OUTLINE

Review of simple R and C circuits

Series RC circuit analysis

Phase-shift application of series RC network

Parallel RC circuit analysis

Similarities and differences between RC and RL circuits

Linear waveshaping and nonsinusoidal waveforms

Other applications of RC circuits

Troubleshooting hints and considerations for RC circuits

CHAPTER PREVIEW

Seldom does the technician find capacitors alone in a circuit. They are generally used with other components, such as resistors, inductors, transistors, and so on, to perform some useful function. In this chapter, we will investigate simple circuits containing both resistors and capacitors. Understanding *RC* circuits is foundational to dealing with practical circuits and systems, such as amplifiers, receivers, transmitters, computers, power supplies, and other systems and subsystems.

You will learn to analyze both series and parallel combinations of resistance and capacitance that operate under sine-wave ac conditions. An overview of how these components are used for waveshaping in nonsinusoidal applications will also be given. Also in this chapter, as in the earlier chapter "*RL* Circuits in AC," we will have you use the basic, or elemental, approaches of the Pythagorean theorem, and simple trig functions for circuit analysis. After completing this chapter, you will have had more practice in these techniques. Since after this chapter, you will have familiarity with basic vector notations for both *RL* and *RC* circuits, the next chapter will introduce additional powerful vector analysis and notation techniques. So be diligent in practicing the techniques and calculator usage afforded in this chapter. This will help you master the next chapter, which combines all your knowledge of *RL* and *RC* circuits, as well as teaching you some additional powerful analysis techniques. The concluding portion of this chapter will provide a summary and review of the similarities and differences between *RC* and *RL* circuits and supply some troubleshooting hints for *RC* circuits.

OBJECTIVES

After studying this chapter, you should be able to:

• Draw or describe operation of simple *R* and *C* circuits

• Analyze appropriate series and parallel *RC* circuit parameters using Pythagorean theorem

• Use vector analysis to analyze series and parallel *RC* circuit parameters

• List differences between *RC* and *RL* circuits

• Predict output waveform(s) from a waveshaping network

• List two practical applications for *RC* circuits

• Apply troubleshooting hints to help identify problems

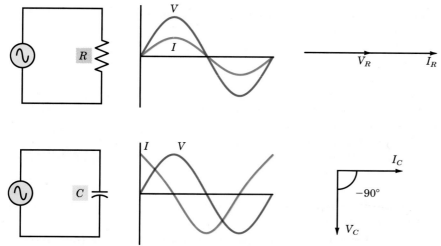

**FIGURE 20–1   Voltage-current relationships in purely *R*
and purely *C* circuits**

REVIEW OF SIMPLE *R* AND *C* CIRCUITS

Although you have studied purely resistive and capacitive circuits, review
Figure 20–1 as a quick refresher.

SERIES *RC* CIRCUIT ANALYSIS

As Figure 20–1 has just revealed, current and voltage are in-phase for
purely resistive ac circuits. Circuit **current leads voltage** by 90° in purely
capacitive circuits. Therefore, when resistance and capacitance are combined, the overall difference in angle between circuit voltage and current is
an angular difference between 0° and 90°, Figure 20–2. The actual angle

$$V_R = I \times R$$
$$V_C = I \times X_C$$
$$V_T = I \times Z \text{ (and vector result of } V_R \text{ and } V_C)$$
$$I = \frac{V_T}{Z}$$

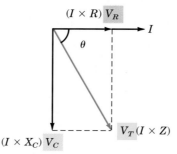

Angular difference between V_T
and I (θ) is greater than 0° but
less than 90°.

**FIGURE 20–2   Positional information for phasors in series
RC circuits**

depends on the relative values of the components' electrical parameters and the frequency of operation.

Positional Information for Phasors

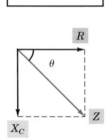

FIGURE 20–3
Impedance diagram for a series *RC* circuit

In series circuits, the current is used as the zero-degree reference vector for *V-I* vector diagrams because *current is common throughout a series circuit.* The *V-I* vector diagram, therefore, shows the resultant circuit voltage lagging behind the current by some angle between 0° and −90°, again see Figure 20–2. Remember, this is in the fourth quadrant of the coordinate system.

In *series RC* circuits (and *only* for *series* circuits), we can also develop an impedance diagram as well as a *V-I* vector diagram, Figure 20–3. Since the resistive voltage drop $(I \times R)$ vector must indicate the resistor voltage is in-phase with the circuit current, the impedance diagram line representing R is shown at the 0° position. Because the capacitor voltage $(I \times X_C)$ lags the capacitor (and circuit) current by 90°, the X_C on the impedance diagram is shown downward at the −90° position.

Summarizing for *series RC* circuits, I and V_R are shown at 0° and V_C is shown at −90° in *V-I* vector diagrams. The R is shown at 0° and X_C at − 90° for impedance diagrams. The angle between V_T and I (in *V-I* vector diagrams) and between Z and R (in impedance diagrams) is the value of the circuit phase angle. With this information, let's proceed with some analysis techniques.

Solving Magnitude Values by the Pythagorean Theorem

As in the chapter on *RL* circuits, we can use the Pythagorean theorem to solve for magnitudes of voltages, currents, and circuit impedance in series *RC* circuits.

Recall the Pythagorean theorem deals with right triangles and the relationship of triangle sides. The hypotenuse *(c)* relates to the other sides by the formula $c^2 = a^2 + b^2$, or $c = \sqrt{a^2 + b^2}$. Observe Figure 20–4 as you

FIGURE 20–4   Using the Pythagorean theorem approach

study the following statements to see how this theorem is applied in phasor addition for *V-I* vector and impedance diagrams for a series *RC* circuit.

Solving For V_T in the *V-I* Vector Diagram

1. The hypotenuse of the right triangle in the *V-I* vector diagram is V_T, or the vector resultant of combining V_R and V_C.

2. The side adjacent to the angle of interest (the phase angle) is V_R. The side opposite the angle is V_C.

3. Substituting into the Pythagorean formula, the formula to solve for the magnitude of the hypotenuse, or V_T, is:

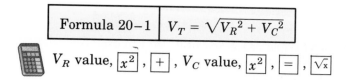

Formula 20–1	$V_T = \sqrt{V_R{}^2 + V_C{}^2}$

V_R value, $\boxed{x^2}$, $\boxed{+}$, V_C value, $\boxed{x^2}$, $\boxed{=}$, $\boxed{\sqrt{x}}$

EXAMPLE

Thus, in Figure 20–4, $V_T = \sqrt{50^2 + 50^2} = \sqrt{2{,}500 + 2{,}500} = \sqrt{5{,}000} = 70.7$ V.

50, $\boxed{x^2}$, $\boxed{+}$ 50, $\boxed{x^2}$, $\boxed{=}$, $\boxed{\sqrt{x}}$

Solving for Z in the Impedance Diagram

1. The hypotenuse of the right triangle in the impedance diagram is Z (impedance).

2. The side adjacent to the angle representing the phase angle is R. The side opposite the angle represents X_C.

3. Substituting into the Pythagorean formula, the formula to solve for the magnitude of the hypotenuse or Z, is:

Formula 20–2	$Z = \sqrt{R^2 + X_C{}^2}$

R value, $\boxed{x^2}$, $\boxed{+}$, X_C value, $\boxed{x^2}$, $\boxed{=}$, $\boxed{\sqrt{x}}$

EXAMPLE

Thus, in Figure 20–4, $Z = \sqrt{50^2 + 50^2} = \sqrt{2{,}500 + 2{,}500} = \sqrt{5{,}000} = 70.7$ Ω.

50, $\boxed{x^2}$, $\boxed{+}$, 50, $\boxed{x^2}$, $\boxed{=}$, $\boxed{\sqrt{x}}$

Try applying the Pythagorean formula to the following problems.

FIGURE 20–5

PRACTICE PROBLEMS I

1. For the circuit in Figure 20–5, use the Pythagorean approach and solve for V_T. Draw the V-I vector diagram and label all elements. (NOTE: You may want to try the drawing-to-scale approach learned in an earlier chapter to check your results.)

2. For the circuit in Figure 20–5, use the Pythagorean approach and solve for Z. Draw and label an appropriate impedance diagram.

PRACTICAL NOTES

Reminder: It is a good habit to sketch diagrams for any problem you are solving even if you don't draw them precisely. By sketching, you can approximate to see if the answers you get are logical.

Solving for Phase Angle Using Trigonometry

Recall from the RL circuits chapter that we can use simple trig functions to solve for the angles of a right triangle. Remember that

$$\sin \theta = \frac{\text{opp}}{\text{hyp}}, \cos \theta = \frac{\text{adj}}{\text{hyp}}, \text{ and } \tan \theta = \frac{\text{opp}}{\text{adj}}$$

Since we have already identified which sides are the side adjacent, side opposite, and the hypotenuse in the series RC circuit V-I vector and impedance diagrams, substitute into the trig formulas to solve for angle information.

Note the trig formula that does not require knowing the magnitude of the hypotenuse to solve for the angle information is the tangent function. Therefore, this function is often used.

PRACTICAL NOTES

With the accessibility of scientific calculators, finding phase angles (and other circuit parameters) in ac circuits has become increasingly simple. In this chapter, we are having you use, practice, and continue to learn the rudimentary concepts related to trig functions for phase-angle solutions.

In the next chapter, you will have the opportunity to learn rectangular and polar vector notation techniques and simple calculator keystrokes that allow conversion between these notation methods. As you will see, these notation methods and the calculator can take much of the "grief" out of ac circuit calculations.

> The foundations you have built in previous chapters, and are building in this chapter, will help you to better understand and use these techniques as you progress in your studies. We encourage you to try to understand concepts, rather than simply memorizing formulas!

Solving the Angle from the *V-I* Vector Diagram

EXAMPLE

Refer again to Figure 20–4. Since we know all three sides, we can use any of the three basic trig formulas. However, let's use the tangent function.

1. To find the phase angle, or the angle between V_T and I, the V_C and V_R values can be used.

2. Substitute these values into the tangent formula.

$$\tan \theta = \frac{\text{opp}}{\text{adj}} = \frac{V_C}{V_R} = \frac{50}{50} = 1$$

3. Using a calculator, we find the arctan θ, or the angle whose tangent equals 1 is 45°.

4. Therefore, the phase angle is 45°. This makes sense since both legs of the triangle are equal in length. If you draw this diagram to scale and use a protractor to measure the angle, the result should verify a 45° angle.

Solving the Angle from the Impedance Diagram

EXAMPLE

Refer to Figure 20–6 and use the tangent function to find the phase angle.

1. The angle between Z and R is the same value as the phase angle (the angle between the circuit voltage and circuit current).

FIGURE 20–6

2. Using the tangent function to find this angle requires the X_C and R values.
   a. Add the resistances to get the R value.

   $$R = 270\ \Omega + 270\ \Omega = 540\ \Omega$$

   b. Add the X_C's to get the X_C value.

   $$X_C = 300\ \Omega + 100\ \Omega = 400\ \Omega$$

3. Substitute appropriate values into the tangent formula.

   $$\tan \theta = \frac{\text{opp}}{\text{adj}} = \frac{X_C}{R} = \frac{400}{540} = 0.74$$

4. Find the $\tan^{-1}$ of 0.74 to get the angle.

   $$\tan^{-1} 0.74 = 36.5 \text{ degrees; thus, } \theta = 36.5 \text{ degrees}$$

   NOTE: When impedance for series RC circuits is plotted on an impedance vector diagram, the location of the Z vector is in the fourth quadrant, somewhere between 0 and -90 degrees removed from the reference vector (R). This means the phase angle is considered a negative angle. Thus, $\theta = -36.5$ degrees, in this case.

   Calculator keystrokes for this computation are:

$$400,\ \boxed{\div}\ ,\ 540,\ \boxed{=}\ ,\ \boxed{2^{\text{nd}}}\ ,\ \boxed{\text{TAN}^{-1}}$$

Apply these principles to solve the following problems.

FIGURE 20–7

PRACTICE PROBLEMS II

1. Refer to Figure 20–7. Draw a *V-I* vector diagram for the circuit shown, and find the phase angle using a basic trig function. (Round-off to the nearest volt.)

2. Draw an impedance diagram for the circuit in Figure 20–7. Then solve for the value of θ, using an appropriate trig function.

PHASE-SHIFT APPLICATION OF SERIES *RC* NETWORK

Background Information

Phase-shift networks are frequently needed and used in electronic circuits and electronically controlled systems. The series RC network is one of the simpler networks for achieving a desired phase shift (or phase difference). The network is used to create and/or control the phase difference between the voltage into the network, and the voltage being taken as output voltage from the network. These phase-shift networks are sometimes called RC "lead" or "lag" circuits.

The principle upon which these RC phase-shift networks operate is based upon the fact that a capacitor opposes a change in voltage. That is, it takes time for the capacitor to change from one voltage level to another (due to charging and discharging time required). You have already learned that in a series RC circuit, the circuit current may lead the circuit applied voltage (V_{in}) anywhere from zero degrees to 90 degrees, depending upon the relative values of R and X_C. Also, you have learned that the voltage across the resistor is in-phase with the circuit current.

We can control the relative R and X_C values by choosing or varying the component values. For example, we can use a variable resistor in the network to control the amount of phase shift. We can also select which component we will take the output voltage across. By selecting the appropriate component, we can choose whether the output voltage leads or lags the input voltage. A *"lead"* network is defined as one in which the *output voltage leads input voltage*. A *"lag"* network is one in which the *output voltage lags the input voltage*.

EXAMPLE (AN *RC* "LEAD" NETWORK)

Refer to Figure 20–8 as you study the following comments.

1. For the lead network, the output is taken across R.

2. For the lead network, then, V_R and V_{out} are the same voltage.

3. For our vector diagram, we will use V_{out} as the reference vector.

4. Since the capacitor causes the circuit current to "lead" the circuit voltage and the output voltage is taken across the resistor, the output voltage is leading the input voltage.

5. The amount of lead is controllable by the values of C and R, which control the relative values of X_C and R.

FIGURE 20–8   An *RC* "load" network

6. The circuit phase angle (θ) = arctan $\dfrac{X_C}{R}$

$$X_C = \frac{1}{2\pi f C} = \frac{1}{6.28 \times 400 \times 0.5 \times 10^{-6}} = 796 \ \Omega$$

$$\frac{X_C}{R} = \frac{796}{1,000} = 0.796$$

$$\text{arctan } 0.796 = 38.5 \text{ degrees}$$

7. For this network, the output voltage leads the input voltage by 38.5 degrees.

8. The value of V_{out} is determined by the voltage divider action of the RC network. In essence, the output voltage (in this case V_R) relates to the ratio of the R to the circuit impedance. That is, $V_{out} = V_{in} \times R/Z$. (Of course Z in this circuit equals $\sqrt{R^2 + X_C^2}$).

$$V_{out} = V_{in} (R/Z)$$

$$\text{where } Z = \sqrt{R^2 + X_C^2} = \sqrt{1,000^2 + 796^2} = 1,278 \ \Omega$$

$$V_{out} = 25 \text{ V} (1,000/1,278) = 25 \text{ V} \times 0.78 = 19.5 \text{ V}$$

PRACTICE PROBLEMS III

What would be the angle of "lead" should the variable resistor in Figure 20–8 be set at 500 ohms?

EXAMPLE (AN RC "LAG" NETWORK)

Refer to Figure 20–9 as you study the following information.

1. For the "lag" network, the output is taken across C.

2. For the lag network, then, V_C and V_{out} are the same voltage.

3. For our vector diagram, we will use V_R as the reference vector.

4. Since the capacitor causes the circuit current to lead the circuit voltage and the output voltage is taken across the capacitor, the output voltage is lagging the input voltage.

5. The circuit phase angle and the amount of lag are NOT the same thing, in this case. The circuit phase angle is the phase difference between V_R and V_{in}. Since the output is taken across the capacitor, the lag angle is the difference in degrees between V_C and V_{in}.

6. The circuit phase angle and the amount of lag are controllable by the values of C and R, which, in turn, control the relative values of X_C and R.

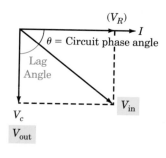

FIGURE 20–9

7. The circuit phase angle $(\theta) = \arctan \dfrac{X_C}{R}$

$$X_C = \frac{1}{2\pi fC} = \frac{1}{\sqrt{6.28 \times 400 \times 0.5 \times 10^{-6}}} = 796 \ \Omega$$

$$\frac{X_C}{R} = \frac{796}{1,000} = 0.796$$

$$\arctan 0.796 = 38.5°$$

8. For this network, V_{in} lags V_R by this 38.5 degrees (the circuit phase angle); however, the amount of lag between V_{out} compared to V_{in} equals 90 degrees minus 38.5 degrees or 51.5 degrees of lag.

   Lag angle = 90 degrees − circuit phase angle

9. To vary the amount of lag in this circuit, we can vary the R value. If the R value is decreased, the circuit becomes more capacitive and the phase angle will increase; however, the amount of lag will decrease. Remember that the lag angle is the angular difference between V_{in} and V_C (the output voltage). If the R value is increased, the circuit phase angle will decrease; however, the lag angle will increase.

10. The value of V_{out} is determined by the voltage-divider action of the RC network. In essence, the output voltage (in this case V_C) relates to the ratio of the X_C to the circuit impedance. That is, $V_{out} = V_{in} \times X_C/Z$. (Of course Z, in this circuit, equals $\sqrt{R^2 + X_C^2}$.

$$V_{out} = V_{in}\left(\frac{X_C}{Z}\right)$$

where $Z = \sqrt{R^2 + X_C^2} = \sqrt{1,000^2 + 796^2} = 1,278 \ \Omega$

$$V_{out} = 25 \text{ V} \left(\frac{796}{1,278}\right) = 25 \text{ V} \times 0.62 = 15.5 \text{ V}$$

PRACTICE PROBLEMS IV

What would be the angle of lag should the R be set at 750 ohms in the circuit of Figure 20–9?

Summary of Series RC Circuits

Figure 20–10 shows examples of series RC circuit parameters for various circuit conditions. These examples help reinforce the following generalizations about series RC circuits.

1. Ohm's law formulas can be used at any time to solve for individual component or total circuit voltages, currents, resistances, reactances, and impedances, as long as the appropriate parameter values are substituted into the formulas. That is,

$$V_C = I_C \times X_C, Z = \frac{V_T}{I_T}, \text{ and so forth}$$

CAPACITANCE CHANGED BY A FACTOR OF 2 (FROM 2 µF TO 1 µF)

f (Hz)	C (µF)	X_C (Ω)	R (Ω)	Z (Ω)	I (Amps)	V_C (IX_C)	V_R (IR)	V_T (IZ)	θ deg
100	2	795	7,950	7,989.65	0.025	19.90	199.01	200	−5.71
100	1	1,590	7,950	8,107.44	0.0246	39.22	196.12	200	−11.31

(C to half; X_C doubled; Z increased; I decreased; V_C increased; V_R decrased; and θ increased)

(a)

RESISTANCE CHANGED BY A FACTOR OF 2 (FROM 7,950 TO 3,975 Ω)

f (Hz)	C (µF)	X_C (Ω)	R (Ω)	Z (Ω)	I (Amps)	V_C (IX_C)	V_R (IR)	V_T (IZ)	θ deg
100	1	1,590	7,950	8,107.44	0.0246	39.22	196.12	200	−11.31
100	1	1,590	3,975	4,281.21	0.0467	74.28	185.70	200	−21.80

(R to half; Z decreased; I increased; V_C increased; V_R decreased; and θ increased)

(b)

FREQUENCY CHANGED BY A FACTOR OF 10 (FROM 100 Hz TO 10 Hz)

f (Hz)	C (µF)	X_C (Ω)	R (Ω)	Z (Ω)	I (Amps)	V_C (IX_C)	V_R (IR)	V_T (IZ)	θ deg
100	1	1,590	3,975	4,281.21	0.0467	74.28	185.7	200	−21.80
10	1	15,900	3,975	16,389.34	0.122	194.03	48.51	200	−75.96

(f to 1/10th; X_C increased ten times; Z increased; I decreased; V_R decreased; and θ increased greatly)

(c)

FIGURE 20–10   Sample series RC circuit parameters charts

2. The circuit voltage lags the circuit current by some angle between 0° and −90°. Conversely, the circuit current leads the circuit voltage by that same angular difference.

3. Since current is common throughout series circuits, current is the reference vector in phasor diagrams of V and I.

4. The larger the R value is with respect to the X_C value, the more resistive the circuit is. That is, the phase angle is smaller or closer to 0°. Conversely, the larger the X_C value is with respect to the circuit R value, the greater is the phase angle and the closer it is to −90°.

5. Impedance of a series RC circuit is greater than either the R value alone or the X_C value alone but *not* as great as their arithmetic sum. Impedance equals the vector sum of R and X_C.

6. When either R or X_C increases in value, circuit impedance increases and circuit current decreases.

7. Total voltage of a series RC circuit is greater than either the resistive voltage drops or the reactive voltages alone but *not* as great as their arithmetic sum. The voltages must be added vectorially to find the resulting V_T value.

8. For series RC circuits, both *V-I* vector and impedance diagrams are used. The quantities used in the *V-I* vector diagram are the circuit current as a reference vector at 0°; V_R (also at the 0° reference position); and V_C at −90°. V_T (or V_A) is plotted as the vector resulting from the resistive and reactive voltage vectors. For the impedance diagram, R is the reference vector and X_C is plotted downward at the −90° position. Circuit impedance (Z) is the vector sum resulting from R and X_C.

✔ IN-PROCESS LEARNING CHECK I

Fill in the blanks as appropriate.

1. In series RC circuits, the _____ is the reference vector, shown at 0°.

2. In the *V-I* vector diagram of a series RC circuit, the circuit voltage is shown _____ the circuit current.

3. In series RC circuits, an impedance diagram _____ be drawn.

4. Using the Pythagorean approach, the formula to find V_T in a series RC circuit is _____.

5. Using the Pythagorean approach, the formula to find Z in a series RC circuit is _____.

6. The trig function used to solve for phase angle when you do not know the hypotenuse value is the _____ function.

PARALLEL *RC* CIRCUIT ANALYSIS

In parallel *RC* circuits, the voltage is common to all parallel branches. Therefore, *voltage* is the reference vector for the *V-I* vector diagrams, Figure 20–11a. Again, you *will not* draw impedance diagrams for *parallel* circuits.

In drawing a *V-I* vector diagram, branch currents are plotted relative to the voltage reference vector. I_R is plotted at the zero-degree position and is the side adjacent to the angle of interest. I_C is plotted at the +90° position because it leads the voltage by 90° and is the side opposite the angle for computational purposes. Again, refer to Figure 20–11a.

Solving for I_T Using the Pythagorean Approach

For parallel *RC* circuits, the Pythagorean formula to solve for I_T becomes:

$$\boxed{\text{Formula 20–3} \quad I_T = \sqrt{I_R^2 + I_C^2}}$$

I_R value, $\boxed{x^2}$, $\boxed{+}$, I_C value, $\boxed{x^2}$, $\boxed{=}$, $\boxed{\sqrt{x}}$

(a) V_T 60 V

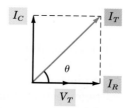

(b) V_T 60 V   15 A   20 A

$I_T = \sqrt{I_R^2 + I_C^2}$
$I_T = \sqrt{15^2 + 20^2}$
$I_T = \sqrt{625}$
$I_T = 25$ A

$\tan \theta = \dfrac{I_C}{I_R}$

$\tan \theta = \dfrac{20}{15}$

$\tan \theta = 1.33$
Angle whose tan = 1.33 is 53.1°

(c) V_T 60 V   15 A 4 Ω   20 A 3 Ω $Z = 2.4\ \Omega$

25 A

$Z = \dfrac{V_T}{I_T} = \dfrac{60}{25} = 2.4\ \Omega$

$R = \dfrac{V_R}{I_R} = \dfrac{60}{15} = 4\ \Omega$

$X_C = \dfrac{V_C}{I_C} = \dfrac{60}{20} = 3\ \Omega$

(NOTE: *Z* is less than either of the branch ohms.)

FIGURE 20–11   Parallel *RC* circuit analysis

EXAMPLE

Use the quantities in Figure 20–11b, and substitute into the preceding formula. Thus,

$$I_T = \sqrt{I_R^2 + I_C^2} = \sqrt{15^2 + 20^2} = \sqrt{625} = 25 \text{ A}$$

15, $\boxed{x^2}$, $\boxed{+}$, 20, $\boxed{x^2}$, $\boxed{=}$, $\boxed{\sqrt{x}}$

Solving for Phase Angle Using Trigonometry

The sine, cosine, and tangent functions, as related to a parallel *RC* circuit, are stated as:

$$\sin \theta = \frac{\text{opp}}{\text{hyp}} \text{ or } \sin \theta = \frac{I_C}{I_T}$$

$$\cos \theta = \frac{\text{adj}}{\text{hyp}} \text{ or } \cos \theta = \frac{I_R}{I_T}$$

$$\tan \theta = \frac{\text{opp}}{\text{adj}} \text{ or } \tan \theta = \frac{I_C}{I_R}$$

It should be noted that due to the fact that the tangent function does not require solving for the hypotenuse (I_T, in this case), it is often used to solve for phase angles.

EXAMPLES

Referring to the information in Figure 20–11b, trig functions can be used to solve for the phase angle (θ) as follows:

1. Using the tangent function:

$$\tan \theta = \frac{I_C}{I_R} = \frac{20}{15} = 1.33. \text{ The arctan } 1.33 = 53.1 \text{ degrees}$$

20, $\boxed{\div}$, 15, $\boxed{=}$, $\boxed{2^{\text{nd}}}$, $\boxed{\text{TAN}^{-1}}$

2. Using the sine function:

$$\sin \theta = \frac{I_C}{I_T} = \frac{20}{25} = 0.8. \text{ The arcsin } 0.8 = 53.1 \text{ degrees}$$

20, $\boxed{\div}$, 25, $\boxed{=}$, $\boxed{2^{\text{nd}}}$, $\boxed{\text{SIN}^{-1}}$

3. Using the cosine function:

$$\cos \theta = \frac{I_R}{I_T} = \frac{15}{25} = 0.6. \text{ The arccos } 0.6 = 53.1 \text{ degrees}$$

15, $\boxed{\div}$, 25, $\boxed{=}$, $\boxed{2^{\text{nd}}}$, $\boxed{\text{COS}^{-1}}$

Looking at the diagram, the angle makes sense because the current through the reactive branch is greater than the current through the resistive branch. The circuit is acting more reactive than resistive. In other words, the angle is greater than 45 degrees.

Solving for Impedance in a Parallel *RC* Circuit

An impedance diagram CANNOT be used for illustrating impedance in parallel circuits as it is used for series circuits. The circuit impedance for a parallel circuit is less than the impedance of any one branch; therefore, it cannot equal the hypotenuse, or vector sum of the oppositions, as it does in series circuits. (For example, if the resistive branch and the reactive branch had equal ohmic values, the circuit impedance would equal 70.7% of one of the branches). There are a couple of simple ways, however, to solve for the value of impedance in a basic parallel *RC* circuit.

1. Use Ohm's law, where $Z = V_T/I_T$.

2. Use the following formula:

$$\text{Formula 20--6} \quad \bigg| \quad Z = \frac{RX_C}{\sqrt{R^2 + X_C^2}}$$

 R value, $\boxed{\times}$, X_C value, $\boxed{\div}$, $\boxed{(}$, *R* value, $\boxed{x^2}$, $\boxed{+}$, X_C value, $\boxed{x^2}$, $\boxed{)}$, $\boxed{\sqrt{x}}$, $\boxed{=}$

Let's try these out!

EXAMPLES

Refer again to Figure 20–11b.

1. Using Ohm's law

$$Z = \frac{V_T}{I_T}$$

$$Z = \frac{60\text{ V}}{25\text{ A}}$$

$$Z = 2.4\ \Omega$$

2. Using the special formula:

$$Z = \frac{RX_C}{\sqrt{R^2 + X_C^2}}$$

To use this formula, we need to know the *R* and X_C values. Since the voltage is known, and each branch current is known, we can use Ohm's law to find these.

$$R = \frac{60\text{ V}}{15\text{ A}} = 4\ \Omega; \ X_C = \frac{60\text{ V}}{20\text{ A}} = 3\ \Omega$$

$$Z = \frac{4 \times 3}{\sqrt{4^2 + 3^2}} = \frac{12}{5} = 2.4\ \Omega$$

 4, $\boxed{\times}$, 3, $\boxed{\div}$, $\boxed{(}$, 4, $\boxed{x^2}$, $\boxed{+}$, 3, $\boxed{x^2}$, $\boxed{)}$, $\boxed{\sqrt{x}}$, $\boxed{=}$

FIGURE 20–12

PRACTICE PROBLEMS V

Refer to Figure 20–12 and solve the following problems.

1. Use the capacitive reactance formula, Ohm's law, and the Pythagorean approach, as appropriate, to solve for I_T.

2. Draw the V-I vector diagram for the circuit and determine the phase angle, using appropriate trig formula(s). Label all the elements in the diagram.

3. Determine the circuit impedance.

✔ IN-PROCESS LEARNING CHECK II

1. In parallel RC circuits, _____ is the reference vector in V-I vector diagram.

2. In parallel RC circuits, current through the resistor(s) is plotted _____ with the circuit current vector.

3. In parallel RC circuits, current through capacitor branch(es) is plotted at _____ with respect to circuit applied voltage.

4. The Pythagorean formula to find circuit current in parallel RC circuits is _____.

5. Is the Pythagorean formula used to find circuit voltage in a parallel RC circuit? _____

6. Can the tangent function be used to find phase angle in a parallel RC circuit? _____

7. Can an impedance diagram be used to find circuit impedance in a parallel RC circuit? _____

Summary of Parallel *RC* Circuits

Figure 20–13 provides several examples of parallel RC circuit parameters that illustrate the following facts. Study and understand these general principles before proceeding in the chapter.

1. Ohm's law formulas are used at any time to solve for individual component or overall circuit parameters, providing you substitute proper quantities into the formulas.

2. The applied voltage (V_T) is the reference vector in phasor diagrams of parallel RC circuits, since V is common to all branches.

CAPACITANCE CHANGED BY A FACTOR OF 2 (FROM 2 µF TO 1 µF)

f (Hz)	V_T (V)	C (µF)	X_C (Ω)	R (Ω)	I_C (Amps)	I_R (Amps)	I_T (Amps)	Z (Ohms)	θ deg
100	100	2	795	5,000	0.1258	0.200	0.1274	785.14	80.97
100	100	1	1,590	5,000	0.0629	0.200	0.0660	1,515.2	72.36

(C to half; X_C doubled; I_C halved; I_R = same; I_T decreased; and θ decreased)

(a)

RESISTANCE CHANGED BY A FACTOR OF 2 (FROM 5,000 Ω TO 2,500 Ω)

f (Hz)	V_T (V)	C (µF)	X_C (Ω)	R (Ω)	I_C (Amps)	I_R (Amps)	I_T (Amps)	Z (Ohms)	θ deg
100	100	1	1,590	5,000	0.0629	0.0200	0.0660	1,515.23	73.36
100	100	1	1,590	2,500	0.0629	0.0400	0.0745	1,341.64	57.54

(R to half; X_C = same; I_C = same; I_R doubled; I_T increased; and θ decreased)

(b)

FREQUENCY CHANGED BY A FACTOR OF 10 (FROM 100 Hz TO 10 Hz)

f (Hz)	V_T (V)	C (µF)	X_C (Ω)	R (Ω)	I_C (Amps)	I_R (Amps)	I_T (Amps)	Z (Ohms)	θ deg
100	100	1	1,590	2,500	0.0629	0.0400	0.0745	1,341.64	57.54
10	100	1	15,900	2,500	0.0063	0.0400	0.0405	2,469.66	8.94

(f to 1/10th; X_C increased 10x; I_C decreased to 1/10th; I_R = same; I_T decreased; and θ decreased greatly)

(c)

RESISTANCE CHANGED TO MAKE EQUAL WITH X_C

f (Hz)	V_T (V)	C (µF)	X_C (Ω)	R (Ω)	I_C (Amps)	I_R (Amps)	I_T (Amps)	Z (Ohms)	θ deg
10	100	1	15,900	2,500	0.0063	0.0400	0.0405	2,469.66	8.94
10	100	1	15,900	15,900	0.0063	0.0063	0.0089	11,243.00	45

(R made = X_C; X_C = same; I_C = same; I_R decreased greatly; I_T decreased greatly; and θ = 45 degrees (since I_C and I_R are equal)

(d)

V_T IS DOUBLED

f (Hz)	V_T (V)	C (µF)	X_C (Ω)	R (Ω)	I_C (Amps)	I_R (Amps)	I_T (Amps)	Z (Ohms)	θ deg
10	100	1	15,900	15,900	0.0063	0.0063	0.0089	11,243.00	45
10	200	1	15,900	15,900	0.0126	0.0126	0.0178	11,243.00	45

(V_T doubled; all currents doubled; X_C, R, Z, and θ did not change)

(e)

FIGURE 20–13   Sample parallel *RC* circuit parameters charts

3. The total circuit current leads the circuit voltage by some angle between $0°$ and $+90°$.

4. The *smaller* the R value with respect to the X_C value, the more resistive the circuit is. That is, the closer to $0°$ the phase angle is. Conversely, the larger the R value is with respect to the X_C value, the more capacitive the circuit is.

5. Impedance in a parallel RC circuit is less than the ohmic opposition of any one branch and must be found through Ohm's law techniques or a special formula. Impedance for a parallel RC circuit can *not* be found by drawing an impedance diagram.

6. When either R or X_C increases, impedance increases and circuit current decreases.

7. Total current in a parallel RC circuit is greater than any one branch current. However, total current is *not* as great as the sum of the branch currents. The branch currents must be added vectorially to find the resulting I_T value.

8. For a *V-I* vector diagram in a parallel RC circuit, voltage applied is the reference vector, I_R is plotted at $0°$, and I_C is plotted at $+90°$. I_T is plotted as the vector resulting from the resistive and reactive branch currents.

SIMILARITIES AND DIFFERENCES BETWEEN *RC* AND *RL* CIRCUITS

So that you can see and remember the key points about RL and RC circuits, Figure 20–14 provides an important list of similarities and differences. If

ITEM DESCRIPTION	SERIES *RC* CIRCUIT	PARALLEL *RC CIRCUIT*	SERIES *RL* CIRCUIT	PARALLEL *RL* CIRCUIT
Formulas	$V_T = \sqrt{V_R{}^2 + V_C{}^2}$ $I_T = I_R = I_C = \dfrac{V_T}{Z}$ $Z = \sqrt{R^2 + X_C{}^2}$ $\tan\theta = \dfrac{X_C}{R}$ or $\dfrac{V_C}{V_R}$ $X_C = \dfrac{0.159}{fC}$ or $\dfrac{V_C}{I_C}$	$V_T = I_T \times Z$ $I_T = \sqrt{I_R{}^2 + I_C{}^2}$ $Z = \dfrac{V_T}{I_T}$ $\tan\theta = \dfrac{I_C}{I_R}$ $X_C = \dfrac{0.159}{fC}$ or $\dfrac{V_C}{I_C}$	$V_T = \sqrt{V_R{}^2 + V_L{}^2}$ $I_T = I_R = I_L = \dfrac{V_T}{Z}$ $Z = \sqrt{R^2 + X_L{}^2}$ $\tan\theta = \dfrac{X_L}{R}$ or $\dfrac{V_L}{V_R}$ $X_L = 2\pi fL$ or $\dfrac{V_L}{I_L}$	$V_T = I_T \times Z$ $I_T = \sqrt{I_R{}^2 + I_L{}^2}$ $Z = \dfrac{V_T}{I_T}$ $\tan\theta = \dfrac{I_L}{I_R}$ $X_L = 2\pi fL$ or $\dfrac{V_L}{I_L}$
As $f\uparrow$	$X_C\downarrow Z\downarrow I\uparrow \theta\downarrow$	$X_C\downarrow Z\downarrow I\uparrow \theta\uparrow$	$X_L\uparrow Z\uparrow I\downarrow \theta\uparrow$	$X_L\uparrow Z\uparrow I\downarrow \theta\downarrow$
As $C\uparrow$	$X_C\downarrow Z\downarrow I\uparrow \theta\downarrow$	$X_C\downarrow Z\downarrow I\uparrow \theta\uparrow$	N/A	N/A
As $L\uparrow$	N/A	N/A	$X_L\uparrow Z\uparrow I\downarrow \theta\uparrow$	$X_L\uparrow Z\uparrow I\downarrow \theta\downarrow$
As $R\uparrow$	$Z\uparrow I\downarrow \theta\downarrow$	$Z\uparrow I\downarrow \theta\uparrow$	$Z\uparrow I\downarrow \theta\downarrow$	$Z\uparrow I\downarrow \theta\uparrow$

FIGURE 20–14   Similarities and differences between *RL* and *RC* circuits

you understand the facts listed, you will find the following chapters about circuits containing all three quantities of resistance, inductance, and capacitance easy to understand.

LINEAR WAVESHAPING AND NONSINUSOIDAL WAVEFORMS

Many electronic circuits depend on the ability to change nonsinusoidal circuit waveforms (such as square or rectangular waveforms) into different shapes for different purposes. For example, in radar systems it is not uncommon to need sharp or pointed waveforms to act as *trigger pulses* for circuits where timing action is performed.

FIGURE 20–15   Examples of *RC* circuit waveshaping

In Figure 20–15, a square-wave input fed to a series *RC* network is shaped to be somewhat sharp or peaked across the resistor. This output is called the *differentiated output.* By definition, a differentiating circuit is one whose output relates to the rate of change of voltage or current of the input.

On the other hand, across the capacitor an output called an *integrated output* is taken. In this case, the waveform is rounded-off or flattened-out. This seems logical since a capacitor opposes a voltage change. Therefore, the voltage can't make sudden or "sharp" changes across the capacitor. By definition, an integrating circuit output relates to the integral of the input signal.

The amount of change of the input signal waveshape at the output relates to the amount of time allowed for the capacitor to charge or discharge during one alternation of the input signal. If the capacitor has more than ample time (over five time constants) to make its voltage change, the circuit is a **short time-constant circuit.** By definition, a short time-constant circuit is one where the time allowed *(t)* is 10 or more times one *RC* time constant. That is, t/RC = 10 or more. *RC* time is short compared with *t*.

Conversely, if the *RC* time constant is such that the time allowed by the input signal alternation is short compared to the time needed for complete change across the *RC* network, it is a **long time-constant circuit.** That is, t/RC = 0.1 or less. *RC* time constant is long compared with the time allowed to charge or discharge.

Refer again to Figure 20–15. Notice both the differentiated and integrated waveforms present across the resistor and capacitor for long, medium, and short time-constant conditions. You will see practical applications of such circuits as you study a variety of waveshaping circuit applications.

In Figures 20–16 and 20–17, you can again see how the *RC* time of resistive-capacitive networks is useful in integrating or differentiating signals, causing desired change in the waveshape of square or rectangular wave signals.

OTHER APPLICATIONS OF *RC* CIRCUITS

In addition to waveshaping and timing applications, *RC* circuits have other valuable uses. One common function is that of coupling desired signals from one circuit point or one circuit stage to another. Often they are used to block dc while passing along the ac component of a signal, Figure 20–18 (page 700).

Capacitors are also used to bypass or filter undesired signals. You will see examples of power-supply filtering when you study rectifier and filter circuits.

Also, capacitor networks are used for ac and dc voltage dividers, Figure 20–19 (page 700). Recall, voltage divides inversely to the capacitance values when capacitors are connected in series. That is, the smallest *C* has the highest *V* and the largest *C* has the smallest *V*.

By now, it is obvious that capacitors are found in almost every electronic circuit and system you will study or work on as a technician. The importance of learning about them and their operation cannot be overemphasized.

These circuits will pass
part of any dc component
present in input.

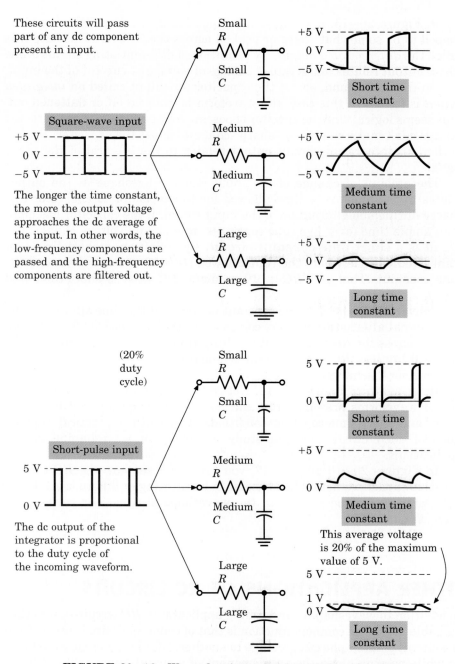

Square-wave input

The longer the time constant,
the more the output voltage
approaches the dc average of
the input. In other words, the
low-frequency components are
passed and the high-frequency
components are filtered out.

(20%
duty
cycle)

Short-pulse input

The dc output of the
integrator is proportional
to the duty cycle of
the incoming waveform.

This average voltage
is 20% of the maximum
value of 5 V.

FIGURE 20–16   Waveshaping—*RC* integrator circuits

These circuits block
any dc present in input.

Square-wave input

The longer the time constant,
the more the output waveform
approaches that of the input.
The high-frequency components
are all that are passed by a
short time constant.

Short-pulse input

Note here and in the above
example, if capacitor has
time to charge up, each change
of the input waveform
produces a separate pulse
output with the amplitude
of the change.

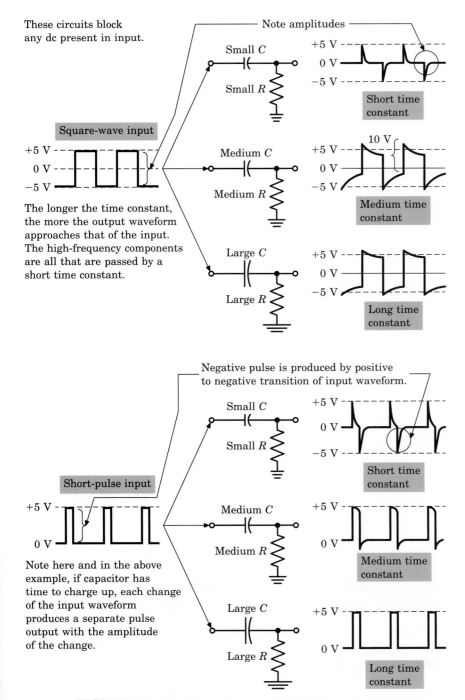

FIGURE 20–17   Waveshaping—RC differentiator circuits

Coupling capacitor:
−Blocks dc component of signal
−Passes along ac component of signal
−Has low X_C to signal frequencies;
  thus, most of signal appears across R.

FIGURE 20–18   Example of an *RC* coupling network

FIGURE 20–19   Capacitor voltage dividers

TROUBLESHOOTING HINTS AND CONSIDERATIONS FOR *RC* CIRCUITS

Based on information supplied earlier regarding typical troubles with capacitors, one can surmise some of the problems that can occur for the various applications where capacitors are used. Some examples follow.

When used as a filtering device—for example, in power supplies or in decoupling circuits—either an excessive leakage condition or excessive internal resistance condition will prevent proper filtering action. As you will learn, in such systems as amplifiers, radio or TV receivers a range from excessive hum to high-pitched squeals can be heard when such problems exist.

The only cure for this situation is to isolate the bad capacitor and *replace it!* Sometimes, bridging a degenerated capacitor with a good capacitor displays sufficient difference to help isolate the bad capacitor.

PRACTICAL NOTES

1. If a capacitor is used in a *timing circuit,* one symptom of a possible bad C (or R) is an improper time period.

2. If the capacitor is used in a waveshaping function, obvious clues are noted if an oscilloscope is used to analyze the normality or abnormality of the waveform(s).

3. If a capacitor is used in a *coupling circuit* where it is supposed to block dc and pass ac, the presence of measurable dc across the network resistor in series with the coupling capacitor shows the capacitor is probably leaking dc.

The primary tools used to help find defective capacitors in these various functions are the oscilloscope, digital multimeters, and capacitor testers. Another useful device is a component *substitution box,* which enables selection of an appropriate-value component. This component is temporarily substituted in the suspected area of trouble to see if the problem disappears with the substitute part.

As you study power supplies, amplifiers, oscillators, pulse and digital circuits, you will gain more familiarity with common symptoms and troubleshooting techniques associated with RC circuits. Perhaps these few comments will whet your appetite to learn more.

SAFETY HINTS

1. To avoid shock, don't make connections with the circuit on.

2. In a defective circuit, capacitors may not always be discharged. BE CAREFUL when inserting test equipment. Check that capacitors are discharged by making appropriate voltage measurements, and/or using safe discharge techniques.

SUMMARY

▶ Circuit current and circuit voltage are in-phase in purely resistive ac circuits. They are 90° out-of-phase in purely capacitive circuits. They are somewhere between 0° and 90° out-of-phase with each other in circuits containing both resistance and capacitance.

▶ To calculate total circuit voltage in a series RC circuit, vectorially add the resistive and capacitive voltages, as opposed to adding them with simple arithmetic.

▶ Since current is common throughout any series circuit, the reference vector for phasor diagrams of series RC circuits is the circuit current. The various circuit voltages (V_C and V_R) are then plotted with respect to the reference vector. V_C is shown lagging the current by 90°, and V_R is in-phase with the current.

▶ The reference vector for phasor diagrams of parallel RC circuits is the circuit voltage. The various parallel branch currents are plotted with respect to the reference vector. I_C is shown leading the voltage by 90°, and I_R is in-phase with the circuit voltage.

▶ Impedance is the opposition shown to ac current by the combination of resistance and reactance.

▶ Impedance diagrams can only be used for *series* circuits. They should not be used for analyzing parallel RC (or RL) circuits.

▶ The Pythagorean theorem approach is used to solve for the magnitudes of various phasors. In series RC circuits, use the formula to calculate Z and V_T. For parallel RC circuits, use the formula to calculate I_T.

▶ Trig functions are used to solve for both phase angles and magnitudes of various sides. One commonly used trig function is the tangent function, used for both series and parallel resistive and reactive circuits.

▶ In series RC circuits, the larger the C is, the smaller the X_C present. And the less the capacitance controls the circuit parameters. In series circuits, the smaller the X_C is with respect to the R, the less the $I \times X_C$ drop is compared to the $I \times R$ drop. Hence, the less capacitive the circuit acts, and the smaller the phase angle will be.

▶ In parallel RC circuits, the larger the C is, the smaller the X_C present. And the more the capacitance controls the circuit parameters. In parallel circuits, the smaller the X_C is with respect to the R, the higher is the current through the capacitive branch. Hence, the more capacitive the circuit acts, and the larger the phase angle will be.

▶ Any frequency change changes the X_C. Therefore, frequency change affects circuit current, component voltages, impedance, and phase angle for series RC circuits. Frequency changes also affect branch currents, impedance, and phase angle in parallel RC circuits.

▶ Capacitors and RC circuits have various applications, including filtering, coupling, voltage-dividing, waveshaping, and timing. These uses take advantage of the capacitor's sensitivity to frequency, ability to block dc while passing ac and to store charge.

▶ Because electrolytic capacitors age faster than other components, such as resistors or inductors, periodically replace them to assure optimum operation of the equipment where they are used.

FORMULAS AND CALCULATOR SEQUENCES SUMMARY

FORMULA 20–1 $V_T = \sqrt{V_R{}^2 + V_C{}^2}$ (Series RC circuits)

V_R value, $\boxed{x^2}$, $\boxed{+}$, V_C value, $\boxed{x^2}$, $\boxed{=}$, $\boxed{\sqrt{x}}$

FORMULA 20–2 $Z = \sqrt{R^2 + X_C{}^2}$ (Series RC circuits)

R value, $\boxed{x^2}$, $\boxed{+}$, X_C value, $\boxed{x^2}$, $\boxed{=}$, $\boxed{\sqrt{x}}$

FORMULA 20–3 $I_T = \sqrt{I_R{}^2 + I_C{}^2}$ (Parallel RC circuits)

I_R value, $\boxed{x^2}$, $\boxed{+}$, I_C value, $\boxed{x^2}$, $\boxed{=}$, $\boxed{\sqrt{x}}$

FORMULA 20–4 $X_C = \dfrac{1}{2\pi f C}$ (Any circuit) NOTE: Using $\pi = 6.28$, then,

6.28, $\boxed{\times}$, freq. value, $\boxed{\times}$, cap. value, $\boxed{=}$, $\boxed{1/x}$

NOTE: If the circuit capacitance value is in μF, use: Cap. value number, $\boxed{\text{EE}}$, $\boxed{+/-}$, 6 to express the "Cap. value" in microfarads for the calculator sequence.

TRIG FUNCTIONS:

$\sin \theta$ opp/hyp = V_C/V_T, or X_C/Z (series RC circuits)
$\sin \theta$ = opp/hyp = I_C/I_T (parallel RC circuits)

opp side value, $\boxed{\div}$, hyp value, $\boxed{=}$

$\cos \theta$ = adj/hyp = V_R/V_T, or R/Z (series RC circuits)
$\cos \theta$ = adj/hyp = I_R/I_T (parallel RC circuits)

adj side value, $\boxed{\div}$, hyp value, $\boxed{=}$

$\tan \theta$ = opp/adj = V_C/V_R, or X_C/R (series RC circuits)
$\tan \theta$ = opp/adj = I_C/I_R (parallel RC circuits)

opp side value, $\boxed{\div}$, adj side value, $\boxed{=}$

arctan θ = the angle represented by the tangent value

tan value, $\boxed{\text{2nd}}$, $\boxed{\text{TAN}^{-1}}$

IMPEDANCE OF PARALLEL RC CIRCUIT:

$$Z = \frac{RX_C}{\sqrt{R^2 + X_C^2}}$$

R value, $\boxed{\times}$, X_C value, $\boxed{\div}$, $\boxed{(}$, R value, $\boxed{x^2}$, $\boxed{+}$, X_C value, $\boxed{x^2}$, $\boxed{)}$, $\boxed{\sqrt{x}}$, $\boxed{=}$

REVIEW QUESTIONS

 Answer questions 1 through 5 in reference to Figure 20–20.

1. What is the circuit impedance?

2. What are the V_C, V_R, and V_T values?

3. What is the C value?

4. Is circuit voltage leading or lagging circuit current? By how much?

5. Answer the following with "I" for increase, "D" for decrease, and "RTS" for remain the same.
   If the frequency of the circuit in Figure 20–20 doubles and component values remain the same:
   a. R will
   b. X_C will
   c. Z will

d. I_T will
e. θ will
f. V_C will
g. V_R will
h. V_T will

FIGURE 20–20

704

PART IV   Reactive Components

Answer questions 6 through 10 in reference to Figure 20–21.

6. What are the I_C and I_T values?

7. What is the R value?

8. What is the Z value?

9. What is the frequency of operation?

10. Answer the following with "I" for increase, "D" for decrease, and "RTS" for remain the same.
    If the frequency of the circuit in Figure 20–21 doubles and component values remain the same:
    a. R will
    b. X_C will
    c. Z will
    d. I_T will
    e. θ will
    f. I_C will
    g. I_R will
    h. V_T will

FIGURE 20–22

FIGURE 20–21

Answer questions 11 through 15 in reference to Figure 20–22.

11. Find R_T

12. Find X_{C_T}

13. Find Z

14. Find θ

15. If C_1 equals 0.1 μF, what is the frequency of the source voltage?

Answer questions 16 through 20 in reference to Figure 20–23.

16. Find I_R

17. Find I_C

18. Find I_T

19. Find θ

20. Find Z_T

FIGURE 20–23

CHALLENGE QUESTIONS

1. In your own words, describe two basic circuit operational differences between a series RC circuit and a series RL circuit.

2. In your own words, describe two basic circuit operational differences between a parallel RC circuit and a parallel RL circuit.

3. Are there logical comparisons between a series RC circuit and a parallel RL circuit? Explain.

4. Draw the circuit diagram of a phase "lead" RC circuit that will produce a 45-degree lead angle. The components you can use include a variable resistor whose range of R is from 0 to 10,000 ohms, and a 0.25 µF capacitor. Assume the frequency of operation is 120 Hz.

TEST YOURSELF

1. A two-component series RC circuit is operated at a frequency of 3,975 Hz. The circuit is comprised of a 0.04-µF capacitor and an undisclosed-value resistance. The circuit impedance is 1,250 Ω. The applied voltage is 25 V. Determine the following circuit parameters: R, X_C, I, V_R, V_C, and θ.

2. A two-branch parallel RC circuit has a total current of 5 mA. The circuit's capacitive reactance is 2 kΩ. The applied voltage is 5 V. Determine the following circuit parameters: R, I_C, I_R, Z, and θ.

PERFORMANCE PROJECTS CORRELATION CHART

Suggested performance projects that correlate with topics in this chapter are:

CHAPTER TOPIC	PERFORMANCE PROJECT	PROJECT NUMBER
Series RC Circuit Analysis	V, I, R, Z, and θ Relationships in a Series RC Circuit	56
Parallel RC Circuit Analysis	V, I, R, Z, and θ Relationships in a Parallel RC Circuit	57

After completing the above projects, perform the Summary checkout at the end of the performance projects section.

RLC Circuit
Analysis

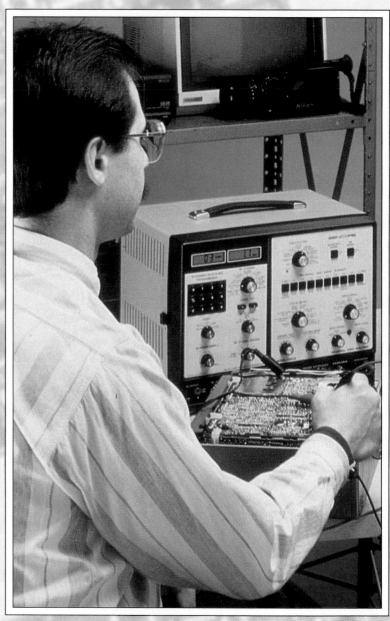

Courtesy of Sencore

21

KEY TERMS

Admittance *(Y)*

Apparent power *(S)*

Complex numbers

Imaginary numbers

Polar form notation

Power factor *(p.f.)*

Reactive power *(Q)*

Real numbers

Rectangular form notation

Susceptance *(B)*

True (or real) power *(P)*

Voltampere-reactive (VAR)

OUTLINE

Basic *RLC* circuit analysis and power in AC circuits

The series *RLC* circuit

The parallel *RLC* circuit

Power in AC circuits

Rectangular and polar vector analysis

Algebraic operations

Application of rectangular and polar analysis

CHAPTER PREVIEW

As a technician, you will work with many circuits that contain combinations of R, L, and C. You have already investigated the property of inductance in ac circuits, in pure inductive circuits, and in circuits with inductance and resistance. Likewise, you have seen how capacitance behaves in purely capacitive circuits and in circuits with both capacitance and resistance. Furthermore, you have learned to use the Pythagorean theorem and simple trigonometric functions to analyze these circuits. In this chapter, you will analyze circuits with various combinations of R, L, and C.

The outline of this chapter accommodates either programs that teach vector algebra and its manipulation of complex numbers or only the approaches learned to this point. The first section of the chapter will present RLC circuits and the basic analysis techniques you have been using and provide a discussion of power in ac circuits. The second section introduces rectangular and polar form vector analysis, sometimes called vector algebra. The last portion provides an opportunity to apply vector algebra to various configurations of RLC circuits.

OBJECTIVES

After studying this chapter, you should be able to:

* Solve RLC circuit problems using Pythagorean approach and trig functions

* Define and illustrate ac circuit parameters using both **rectangular** and **polar form notation**

* Define **real numbers** and **imaginary numbers**

* Define real power, **apparent power, power factor,** and **voltampere-reactive**

* Calculate values of real power, apparent power, and power factor, and draw the power triangle

* Analyze RLC circuits and state results in rectangular and polar forms

* Use the SIMPLER troubleshooting sequence to solve the Chapter Challenge problem

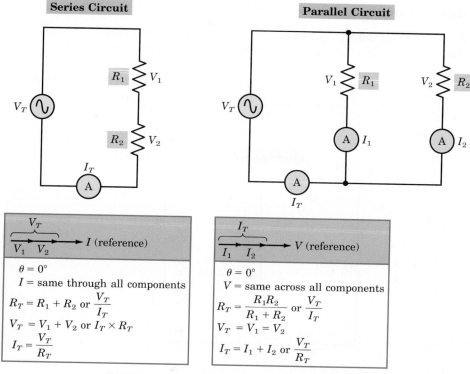

FIGURE 21–1   Purely resistive circuits

BASIC *RLC* CIRCUIT ANALYSIS AND POWER IN AC CIRCUITS

Review of Resistive Circuits

In studying circuits containing only resistances, you learned resistance in ac is treated the same as resistance in dc. Key facts about series and parallel resistive circuits are in Figure 21–1. Take a few minutes to review these facts.

Review of Inductive Circuits

As you know, inductance opposes a change in current. This feature causes the voltage to lead current by 90° in purely inductive circuits and by some angle between 0° and 90° in circuits with both L and R. Examine Figures 21–2 and 21–3 as you review inductive circuits. Remember in vector or

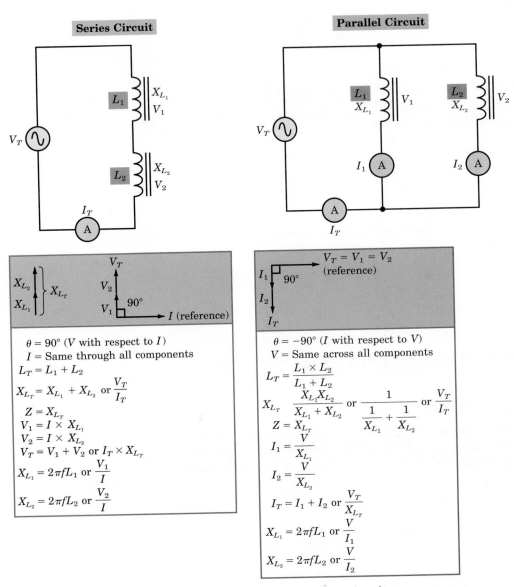

FIGURE 21–2   Purely inductive circuits

vector or phasor diagrams of circuit parameters V_L is plotted upward at 90° for *series* circuits with I as the reference vector. V_R is plotted at the 0° position. For the impedance diagram, X_L is plotted upward at 90°, and R is at the 0° position. For *parallel* circuits, I_L is plotted downward at −90°. When resistance exists, I_R is plotted at the 0° position with circuit V, which is the reference vector.

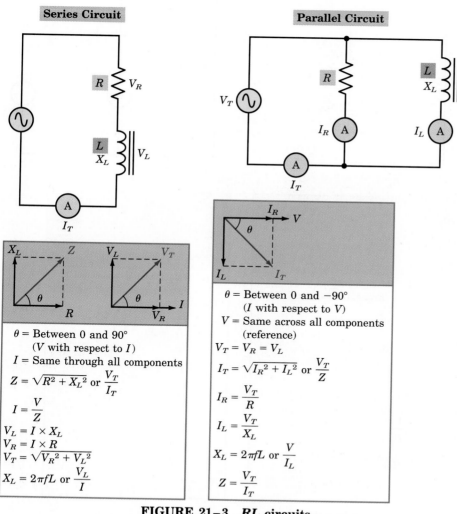

FIGURE 21–3 *RL* circuits

Review of Capacitive Circuits

Since it takes time for a potential difference to build up or change in value on a capacitor charging or discharging current action, capacitance opposes a voltage change. This feature causes current to lead voltage by 90° in purely capacitive circuits and by some angle less than 90° in circuits with both C and R.

Refer to Figures 21–4 and 21–5. Notice for series circuit V-I vector diagrams, V_C is plotted downward at $-90°$. For Z diagrams, X_C is plotted

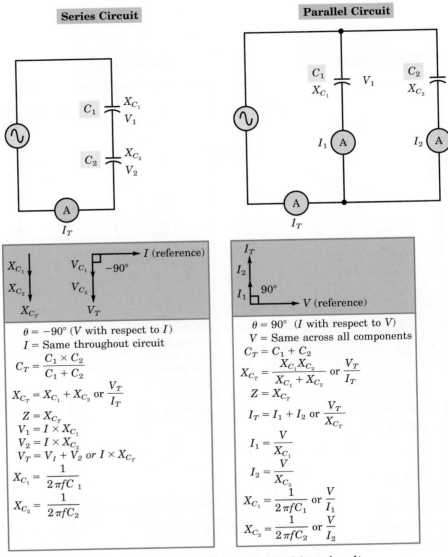

FIGURE 21–4   Purely capacitive circuits

downward at $-90°$. When resistance exists, V_R is plotted at the $0°$ position with the circuit I, which is the reference vector. X_C is plotted downward and R is plotted at $0°$ for series circuit impedance diagrams.

For parallel circuit *V-I* diagrams, I_C is plotted upward at $90°$, and I_R is plotted at the $0°$ position along with circuit V, which is the reference vector. Remember all of these facts as we move to the analysis of circuits containing all three components—R, L, and C.

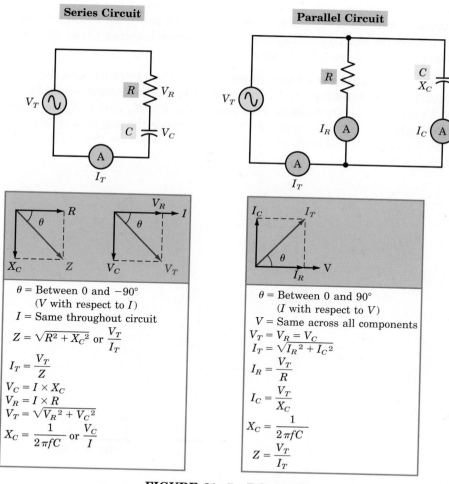

FIGURE 21–5   *RC* circuits

THE SERIES *RLC* CIRCUIT

A Simple Series *LC* Circuit Without Resistance

Let's look at what happens when we combine inductive reactance (X_L) and capacitive reactance (X_C) in a simple series circuit. Refer to Figure 21–6 as you study the following statements.

1. Since X_L is plotted upward and X_C is plotted downward, they *cancel* each other.

2. The result of combining them is that any remaining, or *net* reactance is the difference between the two reactances. That is, if X_L is greater than X_C, the net reactance is some value of X_L. If X_C is greater than X_L, the net circuit reactance is some value of X_C.

3. Apply the same rationale to voltages. Since the voltages are plotted 180° out-of-phase with each other and calculated by multiplying the common current (I) times each reactance, the resulting voltage of the two opposite-reactive voltages is the difference between them. In other words, V_{X_T} = largest V_X − smaller V_X.

4. Since the resulting reactance ohms are less than either reactance (due to the cancelling effect), the current is higher than it would be with either reactance alone. This means the $I \times X_L$ or $I \times X_C$ voltages are higher than the V applied! Obviously, since the opposite reactive voltages also cancel each other, the net voltage equals the applied voltage V_T.

FIGURE 21–6   Series *LC* circuit without resistance

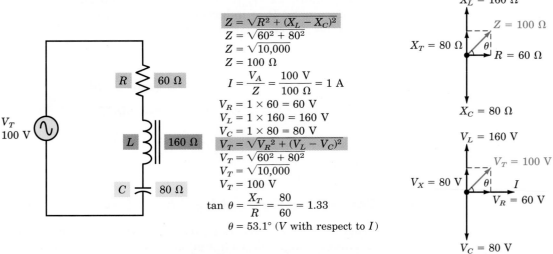

$$Z = \sqrt{R^2 + (X_L - X_C)^2}$$
$$Z = \sqrt{60^2 + 80^2}$$
$$Z = \sqrt{10,000}$$
$$Z = 100 \ \Omega$$
$$I = \frac{V_A}{Z} = \frac{100 \ V}{100 \ \Omega} = 1 \ A$$
$$V_R = 1 \times 60 = 60 \ V$$
$$V_L = 1 \times 160 = 160 \ V$$
$$V_C = 1 \times 80 = 80 \ V$$
$$V_T = \sqrt{V_R^2 + (V_L - V_C)^2}$$
$$V_T = \sqrt{60^2 + 80^2}$$
$$V_T = \sqrt{10,000}$$
$$V_T = 100 \ V$$
$$\tan \theta = \frac{X_T}{R} = \frac{80}{60} = 1.33$$
$$\theta = 53.1° \ (V \text{ with respect to } I)$$

FIGURE 21–7   Series *RLC* circuit

Series *RLC* Circuit Analysis

Refer to Figure 21–7 and note the *RLC* circuit analysis as you study the following procedures.

Computing Impedance by Pythagorean Approach

Using the concepts previously presented, the impedance must equal the vector sum of R and the remaining reactance. The formula is $Z = \sqrt{R^2 + (X_L - X_C)^2}$ when X_L is greater than X_C. If X_C is greater than X_L, then $Z = \sqrt{R^2 + (X_C - X_L)^2}$. In essence,

$$\boxed{\text{Formula 21–1} \quad Z = \sqrt{R^2 + X^2}}$$

 R value, $\boxed{x^2}$, $\boxed{+}$, X_{net}, $\boxed{x^2}$, $\boxed{=}$, $\boxed{\sqrt{x}}$

where net reactance (X) depends on the relative reactance values, that is

$$X = X_L - X_C$$

or

$$X = X_C - X_L$$

In Figure 21–7, $R = 60$ Ω, $X_L = 160$ Ω, and $X_C = 80$ Ω. The net reactance is 80 Ω of inductive reactance in series with 60 Ω of resistance. Using the Pythagorean formula to find Z:

$$Z = \sqrt{R^2 + X^2}$$

$$Z = \sqrt{60^2 + 80^2}$$

$$Z = \sqrt{3{,}600 + 6{,}400}$$

$$Z = \sqrt{10{,}000}$$

$$Z = 100\ \Omega$$

Computing Current by Ohm's Law

In series circuits, current is common throughout the circuit. Circuit current is found through Ohm's law where $I = V/Z$. Since V_A is given as 100 V, then

$$I = \frac{V}{Z} = \frac{100\text{ V}}{100\ \Omega} = 1\text{ A}$$

Computing Individual Voltage by Ohm's Law

The voltage across the resistor $= IR$, across the inductor $= IX_L$, and across the capacitor $= IX_C$.

$$V_R = 1\text{ A} \times 60\ \Omega = 60\text{ V}$$
$$V_L = 1\text{ A} \times 160\ \Omega = 160\text{ V}$$
$$V_C = 1\text{ A} \times 80\ \Omega = 80\text{ V}$$

Computing Total Voltage by Pythagorean Approach

Since the resistive and reactive voltages are out-of-phase, we must again use vector summation. As you know, total voltage is found by the formula $V_T = \sqrt{V_R^2 + (V_L - V_C)^2}$, if inductor voltage is greater than capacitor voltage. If V_C is greater than V_L, the formula is $V_T = \sqrt{V_R^2 + (V_C - V_L)^2}$. Simplifying, the formula becomes

Formula 21–2	$V_T = \sqrt{V_R^2 + V_X^2}$

V_R value, $\boxed{x^2}$, $\boxed{+}$, V_X value, $\boxed{x^2}$, $\boxed{=}$, $\boxed{\sqrt{x}}$

where V_X is *net reactive voltage*. Thus, in Figure 21–7:

$$V_T = \sqrt{60^2 + 80^2}$$

$$V_T = \sqrt{3{,}600 + 6{,}400}$$

$$V_T = \sqrt{10{,}000}$$

$$V_T = 100\text{ V}$$

Computing Phase Angle from Impedance and Voltage Data

Using impedance data, the formula is:

Formula 21–3	$\text{tangent of } \theta = \dfrac{X \text{ (net reactance)}}{R}$

X(net reactance) value, $\boxed{\div}$, R value, $\boxed{=}$

$$\tan \theta = \frac{80}{60}$$
$$\tan \theta = 1.33$$

The angle whose tangent = 1.33 is 53.1°. Since the net reactance is inductive reactance (plotted upward), the angle is +53.1°. NOTE: If the net reactance were capacitive reactance (plotted downward) of the same value, the phase angle would be −53.1° rather than +53.1°.

Using voltage data, the formula is:

Formula 21–4	$\text{tangent of } \theta = \dfrac{V_X}{V_R}$

$V_{X(\text{net})}$ value, $\boxed{\div}$, V_R value, $\boxed{=}$

$$\tan \theta = \frac{80}{60}$$
$$\tan \theta = 1.33$$

The angle whose tangent = 1.33 is 53.1°.

Since the net reactive voltage is inductive voltage, which leads current (the reference vector), the angle is +53.1°. NOTE: If the net reactive voltage were V_C (plotted downward) of the same value, the phase angle would be −53.1° rather than +53.1°.

PRACTICAL NOTES

With more than one resistor or one reactance of each kind in the circuit, combine resistances to one total R value; inductive reactances to one total X_L value; and capacitive reactances to one total X_C value. Then solve the problem with the standard procedures to solve series *RLC* circuits.

EXAMPLE

Refer to the series RLC circuit in Figure 21–8 and calculate Z, V_T, V_{L_1}, V_{C_1}, V_{R_1}, and θ.

In our example circuit, $R_1 = 20\ \Omega$ and $R_2 = 10\ \Omega$, so circuit $R = 30\ \Omega$.
$X_{L_1} = 50\ \Omega$ and $X_{L_2} = 40\ \Omega$, so circuit $X_L = 90\ \Omega$.
$X_{C_1} = 80\ \Omega$ and $X_{C_2} = 40\ \Omega$, so circuit $X_C = 120\ \Omega$.
Net reactance = $120\ \Omega\ (X_C) - 90\ \Omega\ (X_L) = 30\ \Omega$ of capacitive reactance.
$Z = \sqrt{R^2 + X^2} = \sqrt{30^2 + 30^2} = \sqrt{900 + 900} = \sqrt{1,800} = 42.42\ \Omega$
I is given as 1 ampere.
$V_R = IR = 1 \times 30 = 30\ \text{V}$
$V_L = IX_L = 1 \times 90 = 90\ \text{V}$
$V_C = IX_C = 1 \times 120 = 120\ \text{V}$

$R_T = 30\ \Omega$
$X_{L_T} = 90\ \Omega$
$X_{C_T} = 120\ \Omega$
X_C "NET" $= 30\ \Omega$
$Z = \sqrt{R^2 + X^2} = 42.42\ \Omega$
$I = 1\ \text{A}$
$V_T = \sqrt{V_R^2 + V_X^2} = 42.42\ \text{V}$
$\tan \theta = \dfrac{V_X}{V_R} = \dfrac{30}{30} = 1$
$\theta = -45°\ (V\ \text{with respect to}\ I)$

FIGURE 21–8

$$V_T = \sqrt{V_R{}^2 + V_X{}^2} = \sqrt{30^2 + 30^2} = \sqrt{900 + 900} = \sqrt{1,800} = 42.42 \text{ V}$$
$$V_{L_1} = IX_{L_1} = 1 \text{ A} \times 50 \text{ } \Omega = 50 \text{ V}$$
$$V_{C_1} = IX_{C_1} = 1 \text{ A} \times 80 \text{ } \Omega = 80 \text{ V}$$
$$V_{R_1} = IR_1 = 1 \text{ A} \times 20 \text{ } \Omega = 20 \text{ V}$$

$$\tan \theta = \frac{X}{R}, \text{ or } \frac{V_X}{V_R}$$

$$\tan \theta = \frac{30}{30} = 1$$

The angle whose tangent = 1 is 45°.

Now that you have seen how this is done, try the following practice problem.

PRACTICE PROBLEMS I

Refer to Figure 21–9 and calculate R, X_{C_T}, X_L, Z, V_T, V_{C_2}, V_L, V_R, and θ.

✔ **IN-PROCESS LEARNING CHECK I**

Fill in the blanks as appropriate.

1. When both capacitive and inductive reactance exist in the same circuit, the net reactance is _____.

2. When plotting reactances for a series circuit containing L and C, the X_L is plotted _____ from the reference, and the X_C is plotted _____ from the reference.

3. In a series circuit with both capacitive and inductive reactance, the capacitive voltage is plotted _____ degrees out of phase with the inductive voltage.

4. If inductive reactance is greater than capacitive reactance in a series circuit, the formula to find the circuit impedance is _____.

5. In a series *RLC* circuit, current is found by _____.

6. True or False. Total applied voltage in a series *RLC* circuit is found by the Pythagorean approach or by adding the voltage drops around the circuit.

7. If the frequency applied to a series *RLC* circuit decreases while the voltage applied remains the same, the component(s) whose voltage will increase is/are the _____.

FIND
$R =$
$X_{C_T} =$
$X_L =$
$Z =$
$V_T =$
$V_{C_2} =$
$V_L =$
$V_R =$
$\theta =$

FIGURE 21–9

THE PARALLEL *RLC* CIRCUIT

A Simple *LC* Parallel Circuit

Refer to Figure 21–10 as you study the following discussion on combining inductive and capacitive reactances in parallel.

1. Since I_L is plotted downward and I_C is plotted upward, obviously the reactive branch currents *cancel* each other.

2. Combining the reactive branch currents produces the resulting reactive current that is the difference between the opposite-reactance branch currents. That is, the resulting reactive current (I_X) is the larger reactive branch current minus the smaller reactive branch current.

3. In a parallel circuit, the voltage is the same across all branches, that is, V_T.

Parallel *RLC* Circuit Analysis

As was just stated, the *opposite-direction phasors* cancelling effect is also applicable to parallel circuits. But this cancellation effect is different between series and parallel circuits because of the circuit parameters involved. In series circuits, the current is common and the reactive voltage opposite-direction phasors cancel. However, in parallel circuits, the voltage

**FIGURE 21–10   Parallel *LC* circuit
without resistance**

$$\text{Tan } \theta = \frac{I_X}{I_R} = \frac{1}{2} = 0.5$$

$$\theta = 26.56° \ (I \text{ with respect to } V)$$

FIGURE 21–11   Parallel *RLC* circuit

is common and the reactive branch currents are opposite-direction phasors. Look at Figure 21–11 as you study the following discussion.

1. I_T is the vector sum of the circuit resistive and reactive branch currents.

2. The resulting reactive branch current value used to calculate total current is found by subtracting the smaller reactive branch current value from the larger reactive branch current value. This value (I_X) is then used to calculate I_T.

3. The general formula for I_T is $I_T = \sqrt{I_R^2 + (I_L - I_C)^2}$, where I_L is greater than I_C. If I_C is greater than I_L, the formula becomes $I_T = \sqrt{I_R^2 + (I_C - I_L)^2}$. Using the *net reactive current*, the formula becomes

$$\boxed{\text{Formula 21–5} \quad I_T = \sqrt{I_R^2 + I_X^2}}$$

I_R value, $\boxed{x^2}$, $\boxed{+}$, $I_{X(\text{net})}$, $\boxed{x^2}$, $\boxed{=}$, $\boxed{\sqrt{x}}$

Let's apply this knowledge and Ohm's law to solve a problem.

EXAMPLE

Refer again to the circuit in Figure 21–11 and find I_R, I_L, I_C, I_X, I_T, Z, and θ.

Use Ohm's law to solve for each branch current:

$$I_R = \frac{V_T}{R} = \frac{60}{30} = 2 \text{ A}$$

$$I_L = \frac{V_T}{X_L} = \frac{60}{60} = 1 \text{ A}$$

$$I_C = \frac{V_T}{X_C} = \frac{60}{30} = 2\text{ A}$$

Subtract the smaller reactive current from the larger to solve for I_X:

$$I_X = I_C - I_L = 2 - 1 = 1\text{ A}$$

Use the Pythagorean approach to solve for I_T:

$$I_T = \sqrt{I_R{}^2 + I_X{}^2} = \sqrt{2^2 + 1^2} = \sqrt{4 + 1} = \sqrt{5} = 2.24\text{ A}$$

Use Ohm's law to solve for Z:

$$Z = \frac{V_T}{I_T} = \text{approximately } 26.8\ \Omega$$

Use the tangent function to solve for the phase angle:

$$\text{Tan } \theta = \frac{I_X}{I_R} = \frac{1}{2} = 0.5$$

The angle whose tangent = 0.5 is 26.56°. Since the net reactive current is capacitive, current leads the reference circuit voltage and the phase angle (I_T with respect to V_T) is +26.56°.

Try solving the following practice problem.

PRACTICE PROBLEMS II

Refer to Figure 21–12 and solve for I_R, I_L, I_C, I_X, I_T, Z, and θ.

✓
IN-PROCESS LEARNING CHECK II

1. In parallel *RLC* circuits, capacitive and inductive branch currents are _____ out-of-phase with each other.

2. The formula to find total current in a parallel *RLC* circuit when I_C is greater than I_L is _____.

3. True or False. The quickest way to find the circuit impedance of a parallel *RLC* circuit is to use an impedance diagram.

4. If the frequency applied to a parallel *RLC* circuit increases while the applied voltage remains the same, the branch current(s) that will increase is/are the _____ branch(es).

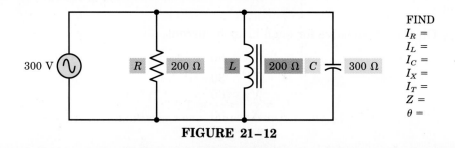

FIND
I_R =
I_L =
I_C =
I_X =
I_T =
Z =
θ =

FIGURE 21–12

POWER IN AC CIRCUITS

Background

In dc circuits, you know how to find the power expended by a circuit through the resistive power dissipation. That is,

$$I^2R, \frac{V^2}{R}, \text{ and } V \times I$$

In ac circuits with both resistance and reactance, the apparent power supplied by the source is not the real or true power dissipated in the form of heat. Obviously, this is because the reactances take power from the circuit during one portion of the cycle and return it to the circuit during another portion of the cycle. As you know, circuits with reactances cause the voltage and current to be out-of-phase with each other. Therefore, the product of the **out-of-phase** circuit v and i does not indicate the actual circuit power dissipation.

Power in a Purely Resistive Circuit

Look at Figure 21–13. The power is dissipated during both half cycles. During the positive alternation, the power graph is represented by the products of a series of instantaneous values of i and v along the waveform. During the negative alternation, the power graph represents the products of the series of instantaneous values of negative i and negative v along the waveform. The product of a negative times a negative, of course, comes out positive. In other words, the resistor dissipates power during both alterna-

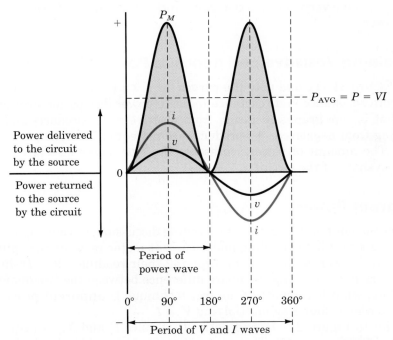

FIGURE 21–13 Instantaneous power graphs in a purely resistive circuit

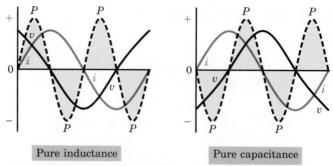

FIGURE 21–14   Instantaneous power graphs in purely reactive circuits

tions. Over the whole cycle, an average power dissipation equal to the product of current and voltage results. It can be mathematically shown that the average power dissipated equals the effective (rms) voltage value times the effective (rms) current value.

Power in Purely Reactive Circuits

Phase differences between voltage and current for purely inductive or capacitive circuits cause the power graphs, Figure 21–14. Because reactances take power from the circuit for one-quarter cycle and return it the next quarter cycle, you see a power graph of twice the source frequency. Notice, too, the average power over the complete cycle is zero, since the reactive instantaneous v times i products are positive half the time and negative half the time. This fact implies that purely inductive or capacitive reactances do not dissipate power. In reality, inductors (due to their wire resistance) have some resistive component, which dissipates some power in the coil. Good capacitors have very low effective resistance and typically dissipate almost zero power.

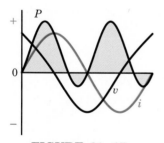

**FIGURE 21–15
Instantaneous power
graphs—R and X
combined**

Combining Resistive and Reactive Power Features

View Figure 21–15 and note the curves for $v \times i$ products in an ac circuit with both resistance and reactance indicate power dissipation over one cycle. That is, the graphs resulting from positive $v \times i$ products and those resulting from negative $v \times i$ products are not equal. There is power dissipated. The amount of power represents the *true power* dissipated by the resistive share of the circuit.

Apparent Power

To someone not familiar with our preceding discussion, it would appear that by taking a voltmeter reading and a current meter reading for a given ac circuit, the power is found by multiplying their readings ($V \times I$). But the meters are *not* indicating the phase difference between the two quantities. Therefore, what is computed from this technique is **apparent power.** Apparent power is found by multiplying $V \times I$.

Refer to Figure 21–16 and notice that since R and X_L are equal, the circuit phase angle is 45°. The apparent power is found by the product of V

$p.f. = \cos \theta = \cos 45° = 0.707$

$p.f. = \dfrac{R}{Z}$ or $\dfrac{V_R}{V_T}$

$S = VI = 141.5$ VA (apparent power)
$P = VI \cos \theta = I^2R = 100$ W (true power)
$Q = VI \sin \theta = 100$ VAR (reactive power)

**FIGURE 21–16 Power and power factor in a
series ac circuit**

applied times I. In this case, apparent power = 141.5 volts × 1 ampere = 141.5 VA. NOTE: Apparent power is expressed as **volt-amperes,** not watts! Also, note the abbreviation for apparent power is **S.**

With these thoughts, let's find how phase angle is used to find the real or true power supplied to and dissipated by the circuit.

True Power

True power dissipation is determined by finding the amount of power dissipated by the resistance in the circuit. If we assume a perfect inductor in Figure 21–16, the power dissipated by the resistor equals the true power dissipation. Because the I_R and V_R are in-phase with each other, the formula $V_R \times I_R$ is used. Therefore, true power = 100 V × 1 A = 100 W.

The I^2R Method

The preference is to say that true power equals I^2R, where R represents all circuit resistive quantities. This also includes resistor components and resistance values in reactive components. These R values are easily measured. It is not as easy (for example) to separate reactive and resistive voltage components in an inductor by measuring V_L.

The Power Factor Method

The power factor method considers the phase angle between current and voltage, created by the reactance. The value of the circuit **power factor (p.f.)** reflects how resistive or reactive the circuit is acting. A power factor of one indicates a purely resistive circuit. A power factor of zero represents a purely reactive circuit. Obviously, a power factor between zero and one indicates the circuit has both resistance and reactance, and true power is less than apparent power. Power factor is equal to the cosine of the phase angle.

| Formula 21–6 | Power factor $(p.f.)$ = $\cos\theta$ |

 Angle value, $\boxed{\cos}$

Then for our sample circuit in Figure 21–16:

| Formula 21–7 | Apparent power (S) = $V \times I$ |

 V value, $\boxed{\times}$, I value, $\boxed{=}$

$S = 141.5 \times 1 = 141.5$ volt-amperes

| Formula 21–8 | True power (P) = I^2R |

 I value, $\boxed{x^2}$, $\boxed{\times}$, R value, $\boxed{=}$

$P = 1^2 \times 100 = 100$ watts OR

| Formula 21–9 | P = Apparent power $\times$ power factor |

 Apparent power value, $\boxed{\times}$, power factor value, $\boxed{=}$

$P = 141.5 \times \cos\theta$ (where $\cos 45° = 0.707$)
$P = 141.5 \times 0.707 = 100$ watts

POWER FACTOR FORMULAS.   Power factor really expresses the ratio of true power to apparent power (which is the same ratio expressed by the $\cos\theta$, Formula 21–6).

| Formula 21–10 | $p.f.$ = $\dfrac{\text{True power } (P)}{\text{Apparent power } (S)}$ |

 True power value, $\boxed{\div}$, apparent power value, $\boxed{=}$

$$\frac{VI \times \cos\theta}{VI} \text{ OR } \frac{R}{Z} \text{ (Series)} = \frac{I_R}{I_T} \text{ (Parallel)}$$

EXAMPLE

Recall in *series circuits*

$$\cos\theta = \frac{R}{Z}\left(or\ \frac{V_R}{V_T}\right)$$

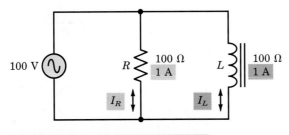

$$p.f. = \cos \theta = \frac{I_R}{I_T} \text{ where } I_T = \sqrt{I_R^2 + I_L^2}$$

$$p.f. = \cos \theta = \cos 45° = 0.707$$

**FIGURE 21–17   Power factor in a parallel
ac circuit**

For Figure 21–16, our series circuit example:

$$Z = \frac{V_T}{I_T} = \frac{141.5 \text{ V}}{1 \text{ A}} = 141.5 \text{ } \Omega$$

$$\cos \theta = \frac{R}{Z} = \frac{100}{141.5} = 0.707. \text{ Therefore, } p.f. = 0.707$$

(Incidentally, the angle whose cos = 0.707 is 45°, as you know.)

In *parallel circuits* the *power factor* = cos θ = I_R/I_T. Refer to Figure 21–17 and notice the calculations for power factor using the resistive and total current values. In this case, the voltage applied is 100 volts. Thus:

$$I_R = \frac{100}{100} = 1 \text{ A}$$

$$I_L = \frac{100}{100} = 1 \text{ A}$$

$$I_T = \sqrt{I_R^2 + I_L^2} = \sqrt{1^2 + 1^2} = \sqrt{2} = 1.414$$

$$p.f. = \frac{I_R}{I_T} = \frac{1}{1.414} = 0.707$$

Again, in this case $\theta = 45°$.

THE POWER TRIANGLE.   The relationship between true and apparent power is sometimes represented visually by a *power triangle* diagram.

Figure 21–18a is a power triangle for a series *RL* circuit where $R = X_L$. Apparent power (S) in volt-amperes is plotted on a line at an angle equal to θ from the *x*-axis. Its length, representing magnitude or value, is $V \times I$. Notice the apparent power is the result of two components:

1. The **true power *(P)*,** sometimes called *resistive power,* is plotted on the *x*-axis (horizontal) and has a magnitude of $VI \times \cos \theta$ watts.

2. The **reactive power *(Q)*,** often called **voltampere-reactive (VAR),** is plotted on the *y*-axis (vertical) and has a magnitude of $VI \times \sin \theta$.

Figure 21–18b is a power triangle for a circuit with resistance and capacitive reactance. From your knowledge of the Pythagorean theorem and

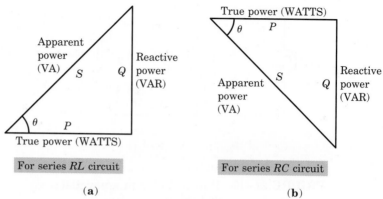

FIGURE 21–18   The power triangle

right-triangle analysis, you can see apparent power is computed from the formula:

| Formula 21–11 | $S = \sqrt{P^2 + Q^2}$ |

 True power value, $\boxed{x^2}$, $\boxed{+}$, net reactive power value, $\boxed{x^2}$, $\boxed{=}$, $\boxed{\sqrt{x}}$

where S = apparent power (measured in volt-amperes, VA)
P = true power (measured in watts, W)
Q = net reactive power (measured in volt-amperes reactive, VAR)

(NOTE: Power triangles are often used to analyze ac power distribution circuits.)

PRACTICAL NOTES

This formula is applicable to both series and parallel circuits. To summarize:

1. Knowing only the value of voltage or current does not indicate the amount of power actually being dissipated by the circuit.

2. A power triangle (different from V-I vector diagram) is used to diagram the various types of power in the ac circuit. Because the power waveform is a different frequency sinusoidal representation compared to the V and I quantities, we cannot plot the power values on the same vector diagram with V and I. Thus, the right-triangle (Pythagorean) approach displays the various types of power in the ac circuit.

3. The vertical element in the power triangle is the reactive power. The horizontal element in the triangle is the true or resistive power. Therefore, the hypotenuse of the triangle is the apparent power, which is the result of reactive and resistive power.

IN-PROCESS LEARNING CHECK III

1. Power dissipated by a perfect capacitor over one whole cycle of input is _____.
2. The formula for apparent power is _____, and the unit expressing apparent power is _____.
3. True power in *RLC* circuits is that power dissipated by the _____.
4. The formula for power factor is _____.
5. If the power factor is one, the circuit is purely _____.
6. The lower the power factor, the more _____ the circuit is acting.
7. In the power triangle, true power is shown on the _____ axis, reactive power (VAR) is shown on the _____ axis, and apparent power is the _____ of the right triangle.

RECTANGULAR AND POLAR VECTOR ANALYSIS

Overview of Rectangular and Polar Forms

In studying dc circuits, you learned that adding, subtracting, multiplying, and dividing electrical quantities was all that was needed to analyze circuits. When you studied ac circuits, you learned that out-of-phase quantities need a different treatment. You learned several approaches—graphic analysis (plotting and measuring vectors or phasors and angles); the Pythagorean theorem (to find resulting magnitudes or values); and simple trigonometry (to find related angles). Also, you learned trigonometry can be used to find magnitude information.

In this section, you will learn about algebraic vector analysis that involves both rectangular and polar forms to analyze vector quantities. You will learn how the *j* operator is used in complex numbers to distinguish the vertical vector from other components. Actually, these techniques are only a small extension of what you already know and use!

Phasors

You know that phasors represent sinusoidal ac quantities when they are the *same* frequency. Of course, they must be the same frequency so that when using phasor drawings the quantities compared maintain the same phase differential throughout the ac cycle. The normal direction of rotation of the constant-magnitude rotating radius vector (phasor) is counterclockwise from the 0° position (reference point). The relative position of the phasor shows its relationship to the reference (0°) position and to other phasors in the diagram. The phasor length indicates the magnitude, or relative value of the quantity represented, Figure 21–19. In ac, this is the effective (rms) value of *V, I,* and so forth.

Length of phasor denotes magnitude or value.
Angle of phasor denotes relative direction.

FIGURE 21–19 Phasor illustration

Rectangular Notation

In **rectangular form notation,** the magnitude and direction of a vector relates to both the x-axis (horizontal) and the y-axis (vertical) of the rectangular coordinate system, Figure 21–20. One way to find the location and magnitude of a given vector on this coordinate system is by **complex numbers.** A complex number locates a positional point with respect to two axes.

A complex number has two parts—**real** and **imaginary.** The first part, real term or number, indicates the horizontal component on the x-axis. The second element, imaginary term or number, indicates the vertical component on the y-axis. For example, if it is stated that a specific vector $= 4 + j3$, this means the point at the end of that vector is located by projecting upward from 4 on the horizontal axis until a point even with 3 on the $+j$ (vertical) axis is reached. (Incidentally, in this case, the length of this vector from the origin is 5, and its angle from the reference plane is 36.9°.)

Notice in Figure 21–20 the horizontal axis is also called the x-axis, the real axis, or the resistance axis. The vertical axis is also called the imaginary axis, or the reactance axis. The j operator designates values on that

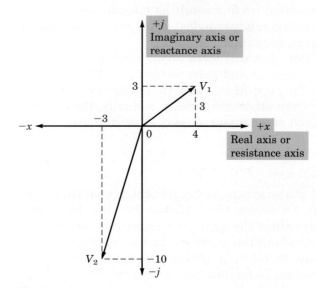

$V_1 = 4 + j3$
$V_2 = -3 - j10$

**FIGURE 21–20 Illustration of rectangular
notation form**

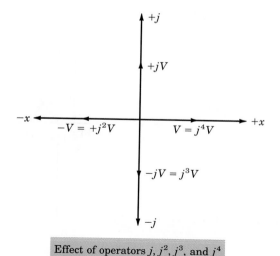

Effect of operators j, j^2, j^3, and j^4

FIGURE 21–21   The j operator

axis. The $+j$ indicates an upward direction, while $-j$ signifies a downward direction. When a number in a formula is preceded by the letter j, the number is a vertical (or y-axis) component.

Since the j operator is so important in the rectangular form of vector algebra, let's define the j operator more specifically:

$$j = \sqrt{-1}$$
$$j^2 = (\sqrt{-1})(\sqrt{-1}) = -1$$
$$j^3 = (j^2)(j) = (-1)(j) = -j$$
$$j^4 = (j^2)(j^2) = (-1)(-1) = +1$$

Look at Figure 21–21 to relate these equations to the rectangular coordinate format. Notice the various powers of the j operator rotate a phasor as follows:

- $+j$ is a 90° operator, rotating a phasor 90° in the counterclockwise (CCW) direction, or +90 degrees.

- $-j$ rotates a phasor 90° in the clockwise (CW) direction, or −90°.

- j^2 rotates the phasor 180°.

- j^3 rotates the phasor to the +270° position, or −90°.

- j^4 is +1, rotating the vector back to the starting point (0°).

EXAMPLE

If the value of V_T for a series *RL* circuit (with a perfect inductor) is $30 + j40$, it might be illustrated as in Figure 21–22. Observe V_T is 50 volts and θ is 53.1°. Suppose that V_T, expressed in rectangular form, is $4 + j3$. What is the value of V_T and the angle? This should be familiar to you.

Answer: 5 volts at an angle of 36.9°.

FIGURE 21–22   Example of rectangular notation

In our previous discussions, you saw how this was done graphically by projecting from 4 (horizontal) and $+j3$ (vertical) points on the coordinate system. The resultant vector and angle could then be measured. You have also previously learned that the resultant of these two projections can be found by vectorial addition; that is, by using the Pythagorean theorem to find the length of the resulting vector (the hypotenuse of the right triangle). The angle can be found by using the tangent function. The scientific calculator offers still another method, which is the preferred method. We will introduce you to this method after you are familiarized with basic concepts of polar notation.

Polar Notation

Refer to Figure 21–23 to see how vectors are represented on polar coordinates.

Whereas the rectangular form expresses vectors by the horizontal and vertical axes, the **polar form notation** expresses the length and angle of the resulting vector to define a vector. Previously, we stated one value of V_T is $30 + j40$ in rectangular notation. This same V_T is expressed in polar form as $50 \angle 53.1°$ V. (The resulting vector of the other two vectors described in the rectangular form.)

The advantage for using polar form notation is that the values are the same as those measured by instruments such as voltmeters, ammeters, and so on. That is, in our example above, a voltmeter would actually read 50 volts for V_T.

Converting From Rectangular to Polar Form

To convert a complex number in rectangular form to polar form simply find the vector resulting from the real term, j term, and angle.

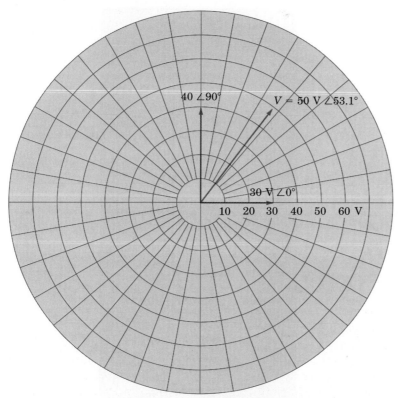

FIGURE 21–23   Illustration of vectors represented on polar coordinate system

EXAMPLE

In a series *RL* circuit, what is the polar notation for a value expressed in rectangular notation as 60 V $+j$70 V?
Use phasor addition to find the magnitude:

$$V_T = \sqrt{V_R{}^2 + V_L{}^2} = \sqrt{60^2 + 70^2} = \sqrt{8{,}500} = 92.2 \text{ V}$$

Use the tangent function to solve for the angle:

$$\tan \theta = \frac{V_L}{V_R} = \frac{70}{60} = 1.166$$

The angle whose tangent = 1.166 is 49.3°
Polar Form = 92.2 $\angle$49.3° V

You get the idea! To convert, solve for the magnitude by vectorially adding the real and the *j* terms; then solve for the angle, using the arctangent, as appropriate. The resulting magnitude and angle are then combined to express the polar notation for the vector.

**FIGURE 21–24 Sample
scientific calculator with *P-R* and
R-P conversion modes (Courtesy
of Texas Instruments)**

Using the Calculator to Convert from Rectangular to Polar Form

In Figure 21–24, you see one example of a scientific calculator. Let's see the steps you can use to perform the conversion of values expressed in rectangular form to find the equivalent in polar form.

Keys of particular interest will be the $\boxed{\text{X} \gtrless \text{Y}}$ key, and the $\boxed{\text{R} \triangleright \text{P}}$ key. Let's use the values given in our previous example and perform the conversion using the calculator.

EXAMPLE

1. Enter the real term value, in this case 60.

2. Press the $\boxed{\text{X} \gtrless \text{Y}}$ key. (This enters the real term portion.)

3. Enter the imaginary term value, in this case 70.

4. Press the $\boxed{\text{3rd}}$ key, then the $\boxed{\text{R} \triangleright \text{P}}$ key (rectangular to polar key). By pressing the $\boxed{\text{3rd}}$ and $\boxed{\text{R} \triangleright \text{P}}$ key in sequence, you have invoked the rectangular-to-polar conversion mode of the calculator.

5. Read the number in the calculator display area (92.19). This number represents the vector magnitude for the polar notation. Observe that this number agrees with what we found using the Pythagorean formula.

6. Press the $\boxed{\text{X} \triangleleft \text{Y}}$ key again to get the angle information. The readout says 49.3, indicating the angle is 49.3 degrees.

7. Combining the results of steps 5 and 6, you find that the polar notation indicates that 92.19 $\angle$ 49.3 degrees is the equivalent of the original 60 + *j*70 rectangular notation form.

NOTE: If the imaginary number is a −*j*, then enter the number, followed by the $\boxed{+/-}$ key, prior to pressing the $\boxed{\text{3rd}}$ key (in step 4).

Verify the following samples of rectangular and equivalent polar form notations.

Rectangular		*Polar*
3 + *j*4	=	5∠53.1°
4 − *j*3	=	5∠−36.9°
60 + *j*70	=	92.2∠49.3°

Now you try a couple conversions on your own.

PRACTICE PROBLEMS III

1. Convert 20 + *j*30 to polar form.
2. Convert 30 − *j*20 to polar form.

Converting From Polar to Rectangular Form

Later in the chapter, you will see the rectangular form is most useful to *add and subtract* vector values and the polar form is useful to *multiply and divide* vector values. For this reason, it is convenient to know how to convert from rectangular to polar form and from polar to rectangular form, as appropriate.

Recall that trig is used to solve for the sides of a right triangle, if the angle and one side are known. Let's put this fact to use.

Since the polar form expresses the hypotenuse of the right triangle plus the angle, we can solve for the other two sides of the triangle by using the appropriate trig functions. Obviously, knowing the sides allows us to express the real and *j* terms appropriately.

Refer to Figure 21–25 and note the horizontal component; i.e., the resistive or real term (*R* in series circuit impedance diagrams), V_R in series circuits, and I_R in parallel circuits is found by:

$$R = Z \cos \theta$$
$$V_R = V_T \cos \theta$$
$$I_R = I_T \cos \theta$$

Series circuit

$X = Z \sin \theta$

$R = Z \cos \theta$

$V_X = V_T \sin \theta$

$V_R = V_T \cos \theta$

Parallel circuit

$I_X = I_T \sin \theta$

$I_R = I_T \cos \theta$

**FIGURE 21–25   Converting from polar to
rectangular form**

Recall

$$\cos \theta = \frac{\text{adj}}{\text{hyp}}; \text{ therefore, adj} = \text{hyp} \times \cos \theta$$

In like manner, use trig to solve for the j (imaginary) term, (i.e., X in series impedance diagrams, V_X in series circuits, and I_X in parallel circuits). Referring to Figure 21–25 again, the reactive term(s) are found by:

$$X = Z \sin \theta$$
$$V_X = V_T \sin \theta$$
$$I_X = I_T \sin \theta$$

Recall

$$\sin \theta = \frac{\text{opp}}{\text{hyp}}; \text{ therefore, opp} = \text{hyp} \times \sin \theta$$

Once the real and j terms are known, we can express the answer in rectangular form.

Using the Calculator to Convert from Polar to Rectangular Form

You can use the trigonometry we have been discussing to solve for these values or, as you might guess, you can use the scientific caluclator to do the job. At this point, let's give you the calculator steps used in solving a sample problem. Let's assume that the polar form notation is 50 $\angle 37°$. To find the rectangular notation equivalent, use the following steps:

EXAMPLE

1. Enter 50, then press the $\boxed{\text{X} \gtrless \text{Y}}$ key.

2. Enter 37.

3. Press the $\boxed{\text{2nd}}$ key, then the $\boxed{\text{P} \triangleright \text{R}}$ key. The display shows the value of the real term in the rectangular expression, 39.93.

4. Press the $\boxed{\text{X} \triangleq \text{Y}}$ key. The display shows the value of the imaginary (j) component of the rectangular expression, $+j30.09$.

5. Combine the results of steps 3 and 4 to get the answer; that is, $50 \angle 37$ degrees in polar form $= 39.93 + j30.09$ degrees in rectangular form.

Both the trig-related formulas and the related calculator sequences follow for your convenience.

Formula 21–12	$Z \angle \theta^\circ = (Z \cos \theta) + (j\, Z \sin \theta)$

for a series circuit impedance diagram

 Z value, $\boxed{\text{X} \triangleq \text{Y}}$, angle value (θ), $\boxed{\text{2nd}}$, $\boxed{\text{P} \triangleright \text{R}}$, $\boxed{\text{X} \triangleq \text{Y}}$

Answer is combination of last two readouts; that is, $\boxed{\text{P} \triangleright \text{R}}$ gives the real term and $\boxed{\text{X} \triangleq \text{Y}}$ gives the imaginary (j) term.

Formula 21–13	$V_T \angle \theta^\circ = V_T \cos \theta + j\, V_T \sin \theta$

for a series circuit V-I vector diagram

V_T value, $\boxed{\text{X} \triangleq \text{Y}}$, angle value (θ), $\boxed{\text{2nd}}$, $\boxed{\text{P} \triangleright \text{R}}$, $\boxed{\text{X} \triangleq \text{Y}}$

Answer is combination of last two readouts; that is, $\boxed{\text{P} \triangleright \text{R}}$ gives the real term and $\boxed{\text{X} \triangleq \text{Y}}$ gives the imaginary (j) term.

Formula 21–14	$I_T \angle \theta^\circ = I_T \cos \theta + j\, I_T \sin \theta$

for a parallel circuit V-I vector diagram

I_T value, $\boxed{\text{X} \triangleq \text{Y}}$, angle value (θ), $\boxed{\text{2nd}}$, $\boxed{\text{P} \triangleright \text{R}}$, $\boxed{\text{X} \triangleq \text{Y}}$

Answer is combination of last two readouts; that is, $\boxed{\text{P} \triangleright \text{R}}$ gives the real term and $\boxed{\text{X} \triangleq \text{Y}}$ gives the imaginary (j) term.

EXAMPLES

Convert $45 \angle 30^\circ$ to rectangular form (trig method).

$$45 \angle 30^\circ = 45\,(\cos 30^\circ) + j45\,(\sin 30^\circ)$$
$$45 \angle 30^\circ = 45 \times 0.866 + j45 \times 0.5$$
$$45 \angle 30^\circ = 38.97 \ + j22.5$$

Convert $45 \angle 30^\circ$ to rectangular form (calculator method).

45, $\boxed{\text{X} \gtrless \text{Y}}$, 30, $\boxed{2^{\text{nd}}}$, $\boxed{\text{P} \triangleright \text{R}}$, (gives R value). Then press $\boxed{\text{X} \gtrless \text{Y}}$ again (gives reactive value). Verify the results by drawing a sketch to see if the results look logical.

PRACTICE PROBLEMS IV

1. Convert 66 $\angle 60°$ to rectangular form.
2. Convert 75 $\angle 48°$ to rectangular form.

Now that you are familiar with rectangular and polar notation and their relationships, we will show how these expressions are used mathematically.

ALGEBRAIC OPERATIONS

Addition (Rectangular form)

Basic rule: Add in-phase (resistive) terms and add out-of-phase (reactive) j terms.

EXAMPLE

Find the sum of V_1 and V_2.

$$\text{Example 1: } V_1 = 15 + j20 \qquad \text{Example 2: } V_1 = 20 + j20$$
$$\underline{V_2 = 20 - j10} \qquad\qquad \underline{V_2 = 30 + j10}$$
$$V_1 + V_2 = 35 + j10 \qquad\qquad V_1 + V_2 = 50 + j30$$

Subtraction (Rectangular form)

Basic rule: Change the sign of the subtrahend and add.

EXAMPLE

Subtract branch current I_2 from branch current I_1, where

$$I_1 = 5 - j6 \text{ A}$$
$$I_2 = 4 + j7 \text{ A}$$

Change the sign of branch current I_2 and add:

$$5 - j6$$
$$\underline{-4 - j7}$$
$$I_1 - I_2 = 1 - j13$$

EXAMPLE

When given total current and one branch current (I_1), find the other branch current (I_2).

$$I_T = I_1 + I_2; \text{ therefore, } I_2 = I_T - I_1.$$

where

$$I_T = 13 - j5.5 \text{ mA}$$
$$I_1 = 6 - j4.0 \text{ mA}$$

Change the sign of branch current 1, and add:

$$\begin{aligned} 13 &- j5.5 \\ -6 &+ j4.0 \\ \hline I_2 = 7 &- j1.5 \end{aligned}$$

PRACTICAL NOTES

If addition or subtraction is needed to analyze a problem and the terms are given in polar form, convert them to rectangular form because you cannot use polar form to add or subtract.

On the other hand, even though multiplication and division are possible in rectangular form, polar form is easier for multiplying and dividing. If you know the rectangular form, you will first convert it to the polar form to multiply or divide. For example, $V_T = I_T \times Z_T$.

Multiplication (Polar form)

Basic rule: Multiply the two magnitudes and add the angles algebraically.

EXAMPLE

If a current of 5 $\angle 30°$ A flows through a Z of 20 $\angle -15°$ Ω, what is the voltage?

$$V = IZ$$
$$V = (5 \angle 30°) \times (20 \angle -15°)$$
$$V = 100 \angle 15° \text{ V}$$

Division (Polar form)

Basic rule: Divide one magnitude by the other and subtract the angles algebraically.

EXAMPLE

Find the current in a circuit if the voltage = 35 $\angle 60°$ and the impedance = 140 $\angle -20°$.

$$I = \frac{V}{Z}$$

$$I = \frac{35 \angle 60°}{140 \angle -20°}$$

$$I = 0.25 \angle 80° \text{ A}$$

Raising to Powers (Polar form)

Basic rule: Raise the magnitude to the power indicated and multiply the angle by that power.

EXAMPLE

If $I = 3 \angle 20°$, then $I^2 = 9 \angle 40°$.

Taking a Root (Polar form)

Basic rule: Take the root of the magnitude and divide the angle by the root.

EXAMPLE

If $I^2 = 100 \angle 60°$, then $I = 10 \angle 30°$.

PRACTICE PROBLEMS V

1. Add $20 + j5$ and $35 - j6$
2. Add $34.9 - j10$ and $15.4 + j12$
3. Subtract $20 + j5$ from $35 - j6$
4. Subtract $22.9 - j7$ from $37.8 - j13$
5. Multiply $5 \angle 20° \times 25 \angle -15°$
6. Multiply $24.5 \angle 31° \times 41 \angle 23°$
7. Divide $5 \angle 20°$ by $25 \angle -15°$
8. Divide $33 \angle 34°$ by $13 \angle 45°$
9. Find I^2 if $I = 12 \angle 25°$
10. Find I if $I^2 = 49 \angle 66°$
11. Find the result of: $\dfrac{5 \angle 50° + (30 + j40)}{2.6 \angle -15°}$
12. Find the result of: $\dfrac{10 \angle 53.1° + (25 - j25)}{4.8 \angle 36°}$

APPLICATION OF RECTANGULAR AND POLAR ANALYSIS

Sample Series *RLC* Circuit Analysis and Applying Vector Analysis

Refer to Figure 21–26 as you study the following discussion on finding various circuit parameters. Note if addition and subtraction are needed, the calculations use rectangular notation; if multiplication or division is involved, the calculations use polar notation.

1. Find Z_T:

Polar Form:

Rectangular Form: $20 + j10$
Z_T = vector sum of impedances
$Z_T = \sqrt{R_T{}^2 + X_T{}^2}$
$Z_T = \sqrt{20^2 + 10^2}$
$Z_T = \sqrt{500}$

$Z = 22.36 \angle 26.56° \ \Omega$ $Z_T = 22.36 \ \Omega$

2. Find I_T (Assume I_T is at $0°$ since this is a series circuit.):

Polar Form: Rectangular Form:

$I_T = \dfrac{V_T}{Z_T}$

$I_T = \dfrac{60.00 \angle 26.56}{22.36 \angle 26.56}$

$I_T = 2.68 \angle 0° \ \text{A}$ $I_T = 2.68 + j0 \ \text{A}$

3. Find individual voltages:
   a. $V_{R_1} = I \times R_1 = 2.68 \angle 0° \times 10 \angle 0° = 26.8 \angle 0° \ \text{V}$
   b. $V_{R_2} = I \times R_2 = 2.68 \angle 0° \times 10 \angle 0° = 26.8 \angle 0° \ \text{V}$
   c. $V_L = I \times X_L = 2.68 \angle 0° \times 20 \angle 90° = 53.6 \angle 90° \ \text{V}$
   d. $V_C = I \times X_C = 2.68 \angle 0° \times 10 \angle -90° = 26.8 \angle -90° \ \text{V}$

FIGURE 21–26   Sample *RLC* circuit analysis

4. Verify total voltage via sum of individual drops. That is, convert each V from polar to rectangular form, then add:

Polar Form:
$V_{R_1} = 26.8 \angle 0°$
$V_{R_2} = 26.8 \angle 0°$
$V_L = 53.6 \angle 90°$
$V_C = 26.8 \angle -90°$

Rectangular Form:
a. $26.8 + j0$
b. $26.8 + j0$
c. $0.0 + j53.6$
d. $0.0 - j26.8$

Total Voltage = $53.6 + j26.8$ or $60 \angle 26.56°$
V_T = approximately $60 \angle 26.56°$ V (polar) or $53.6 + j26.8$ (rectangular)

Sample Parallel *RLC* Circuit Analysis

Remember for parallel circuits in dc, it is easier to use the conductance of branches or add branch currents, than to combine reciprocal resistances. Recall conductance is the reciprocal of resistance. That is, $G = 1/R$ siemens.

In ac circuits, we have to deal with pure resistance *and* reactance *and* the combination of resistance and reactance (impedance). The comparable term for conductance of a pure reactance is called **susceptance,** where susceptance $= 1/\pm X$. The symbol for susceptance is **B.** The reciprocal of impedance is called **admittance,** which is equal to $1/Z$. Thus:

• For pure resistance, conductance $(G) = 1/R$

• For pure reactance, susceptance $(B) = 1/\pm X$

• For combined R and X admittance $(Y) = 1/Z$

The phase angle designation for B or Y is opposite that expressed for X or Z, since it is a reciprocal relationship. This means reactance for an inductive branch is $+jX_L$, and inductive susceptance is $-jB_L$. Also, capacitive reactance is $-jX_C$ and capacitive susceptance is jB_C.

Formula 21–15	Inductive susceptance: $B_L = \dfrac{1}{X_L} = \dfrac{1}{2\pi fL}$

6.28, $\boxed{\times}$, frequency value, $\boxed{\times}$, inductance value, $\boxed{=}$, $\boxed{1/x}$

Formula 21–16	Capacitive susceptance: $B_C = \dfrac{1}{X_C} = 2\pi fC$

6.28, $\boxed{\times}$, frequency value, $\boxed{\times}$, capacitance value, $\boxed{=}$

Notice the inversion of the X_L and X_C formulas. Important admittance relationships are:

$$Y = G \pm jB = G + j(B_C - B_L)$$

$$\boxed{\text{Formula } 21\text{–}17 \quad Y = \sqrt{G^2 + B^2}}$$

Value of conductance *(G)*, $\boxed{x^2}$, $\boxed{+}$, value of susceptance *(B)*,

$\boxed{x^2}$, $\boxed{=}$, $\boxed{\sqrt{x}}$

$$\theta = \text{angle whose tangent} = \frac{B}{G}$$

$$\text{NOTE: } Y = \frac{I}{V}$$

$$I = V \times Y$$

$$V = \frac{I}{Y}$$

EXAMPLE

Refer to Figure 21–27. Use the admittance approach and find the total current of a parallel circuit operating at 60 Hz with the following parameters: $R = 500\ \Omega$; $L = 1$ H; $C = 4\ \mu\text{F}$; and $V = 100$ V.

$$G = \frac{1}{R} = \frac{1}{500} = 0.002\ \text{S}$$

$$B_L = \frac{1}{2\pi fL} = \frac{1}{6.28 \times 60 \times 1} = 0.00265\ \text{S}$$

$$B_C = 2\pi fC = 6.28 \times 60 \times (4 \times 10^{-6}) = 0.00151\ \text{S}$$

$$Y = G + j(B_C - B_L) = 0.002 + j(0.00151 - 0.00265)$$

$$Y = 0.002 - j0.00114 = 0.0023\angle-29.7°\ \text{S}$$

$$I_T = VY = 100\ \angle0° \times 0.0023\ \angle-29.7° = 0.23\ \angle-29.7°\ \text{A}$$

You can calculate the parameters using rectangular forms to add and subtract and polar forms to multiply and divide. When using these notations, it is a good idea to make appropriate conversions so each significant parameter is expressed in both forms. In this way, you use the form most applicable to the calculation you are performing.

$$G = \frac{1}{R} = \frac{1}{500} = 0.002\ S$$

$$-jB_L = \frac{1}{2\pi fL} = \frac{1}{6.28 \times 60 \times 1} = -j0.00265\ S$$

$$jB_C = 2\pi fC = 6.28 \times 60 \times (4 \times 10^{-6}) = j0.00151\ S$$

$$Y = G + j(B_C - B_L) = 0.002 + j(0.00151 - 0.0026\ S)$$

$$Y = 0.002 - j0.00114 = 0.0023\ \angle-29.7°\ S$$

$$I_T = VY = 100\ \angle0° \times 0.0023\ \angle-29.7°$$

$$I_T = 0.23\ \angle-29.7°\ A$$

FIGURE 21–27

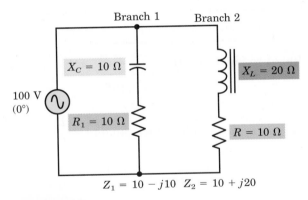

$$Z_T = \frac{Z_1 Z_2}{Z_1 + Z_2}$$

FIGURE 21–28   Sample combination circuit calculations

Sample Combination Circuit Calculations

Refer to Figure 21–28 and note the following calculations.

1. Find Z_T. Since in a parallel circuit

$$Z_T = \frac{Z_1 Z_2}{Z_1 + Z_2}$$

we need to know both the rectangular and polar forms of each branch impedance. These forms are a good starting point.

$$Z_1 = 10 - j10 \text{ (rectangular)} \quad 14.14 \angle -45° \text{ (polar)}$$
$$Z_2 = 10 + j20 \text{ (rectangular)} \quad 22.36 \angle 63.4° \text{ (polar)}$$
$$Z_T = \frac{Z_1 \times Z_2}{Z_1 + Z_2} = \frac{14.14 \angle -45° \times 22.36 \angle 63.4°}{10 - j10 + 10 + j20} = \frac{316.17 \angle 18.4°}{20 + j10}$$
$$Z_T = \frac{316.17 \angle 18.4°}{22.36 \angle 26.56°} = 14.13 \angle -8.16° \ \Omega$$

2. Assume 100 volts applied voltage at 0° and find the current.

$$\text{Branch 1 current} = \frac{V}{Z_1}$$
$$\text{Branch 1 current} = \frac{100 \angle 0°}{14.14 \angle -45°} = 7.07 \angle 45° \text{ A}$$
$$\text{Branch 2 current} = \frac{V}{Z_2}$$
$$\text{Branch 2 current} = \frac{100 \angle 0°}{22.36 \angle 63.4°} = 4.47 \angle -63.4° \text{ A}$$

3. To find I_T, convert each branch current to rectangular form and add the branch currents.

$I_1 = 7.07 \angle 45°$ (polar) $5 + j5$ (rectangular)
$I_2 = \underline{4.47 \angle -63.4°}$ (polar) $\underline{2 - j4}$ (rectangular)
$I_T = 7.07 \angle 8.13°$ A (polar) $7 + j1$ A (rectangular)

4. Individual component voltages are calculated using $I \times R$, $I \times X_L$, and $I \times X_C$, as appropriate. You can practice rectangular and polar notation and appropriate conversion techniques by doing this. If you do this, remember to sketch vector diagram(s), as appropriate, to check the logic of your results.

PRACTICAL NOTES

The calculations and techniques introduced in this chapter will be used by technicians with varying frequency, depending upon their specific job functions. Regardless of your job function, having this knowlege provides you with a stronger foundation in ac analysis. Some specific examples of where these calculations and techniques are used include radio and TV transmission systems and power transmission systems. Certainly, many other areas of endeavor for technicians can and will take advantage of the knowledge gained in this chapter.

SUMMARY

▶ Vectors in opposite directions cancel each other. If one has greater magnitude than the other, there is a net resulting vector equal to the larger value minus the smaller value.

▶ Power in ac circuits with both resistive and reactive components is described in three ways:

1. Apparent power *(S)*—applied voltage times the circuit current, expressed in volt-amperes (VA) and plotted as the hypotenuse of the power triangle.

2. Reactive power *(Q)* or voltampere-reactive (VAR)—equal to $VI \times \sin \theta$, and plotted on the vertical axis.

3. True power *(P)* in watts—equal to $VI \cos \theta$, and plotted on the horizontal axis.

Because of the right-triangle relationship: $S = \sqrt{P^2 + Q^2}$.

▶ Vector quantities are analyzed in several ways—graphic means, algebraic expressions based on rectangular coordinates, and expressions based on polar coordinates.

▶ Separating a vector quantity into its components is easily accomplished by simple trigonometric functions. These functions are sine, cosine, and tangent.

▶ Complex numbers express vector quantities relative to the rectangular notation system. The complex number has a term that defines the horizontal (real or resistive) component and a j term that expresses the reactive component or out-of-phase element. The general term for impedance is $R \pm jX$, for applied voltage is $V_R \pm V_X$, and for total current in a parallel circuit is $I_R \pm jI_X$.

▶ Vectors expressed algebraically in rectangular coordinates are added, subtracted, multiplied, and divided. However, this form of notation is most easily used to add and subtract.

To add, like terms are collected (i.e., resistive terms collected and j terms collected), and added algebraically. To subtract, change the sign of the subtrahend and add algebraically.

▶ Vectors expressed in polar form *cannot* be added or subtracted algebraically. However, this form is easily used to multiply, divide, and find square roots. To multiply polar expressions, multiply magnitudes and add the angles algebraically. To divide polar expressions, divide the magnitudes and subtract the angles algebraically. To raise a polar expression to a power, raise the magnitude by the given power and multiply the angle by the power. To take a root, take the root of the magnitude value and divide the angle by the root value.

▶ Because polar form facilitates multiplication and division and rectangular form expedites addition and subtraction, it is a good idea to convert from one form to the other, as appropriate.

▶ When two branches are in parallel, with each having resistive and reactive components, the resultant Z is found by the formula

$$Z = \frac{Z_1 Z_2}{(Z_1 + Z_2)}$$

▶ It is convenient to use the polar form for the numerator (for multiplying) and the rectangular form for the denominator (for adding). Then, convert the denominator to the polar form to allow division into the numerator to get the final answer.

▶ The ease that current flows through a resistance is called conductance, where $G = 1/R$. A comparable term for reactance is called susceptance, where $B = 1/\pm X$. When resistance and reactance are combined in a circuit, the term for ease of current flow is admittance (Y). Admittance is the reciprocal of impedance, where $Y = 1/Z$. The unit of measure for all these terms is siemens (S).

▶ Some important relationships between admittance, susceptance, and conductance are $Y = G \pm jB$, $Y = \sqrt{G^2 + B^2}$, and $\theta = B/G$.

FORMULAS AND CALCULATOR SEQUENCES SUMMARY

FORMULA 21–1 $Z = \sqrt{R^2 + X^2}$ (Series RLC circuits)

R value, $\boxed{x^2}$, $\boxed{+}$, X_{net}, $\boxed{x^2}$, $\boxed{=}$, $\boxed{\sqrt{x}}$

FORMULA 21–2 $V_T = \sqrt{V_R^2 + V_X^2}$ (Series RLC circuits)

V_R value, $\boxed{x^2}$, $\boxed{+}$, V_X value, $\boxed{x^2}$, $\boxed{=}$, $\boxed{\sqrt{x}}$

FORMULA 21–3   tangent of $\theta = \dfrac{X(\text{net reactance})}{R}$ (Series RLC circuits)

$X(\text{net reactance})$ value, $\boxed{\div}$, R value, $\boxed{=}$

FORMULA 21–4   tangent of $\theta = \dfrac{V_X}{V_R}$ (Series RLC circuits)

$V_{X(\text{net})}$ value, $\boxed{\div}$, V_R value, $\boxed{=}$

FORMULA 21–5 $I_T = \sqrt{I_R^2 + I_X^2}$ (Parallel RLC circuits)

I_R value, $\boxed{x^2}$, $\boxed{+}$, $I_{X(\text{net})}$, $\boxed{x^2}$, $\boxed{=}$, $\boxed{\sqrt{x}}$

FORMULA 21–6   Power factor *(p.f.)* = cos θ

Angle value, $\boxed{\cos}$

FORMULA 21–7   Apparent power *(S)* = $V \times I$

V value, $\boxed{\times}$, I value, $\boxed{=}$

FORMULA 21–8   True power *(P)* = I^2R

I value, $\boxed{x^2}$, $\boxed{\times}$, R value, $\boxed{=}$

FORMULA 21–9 P = Apparent power $\times$ power factor

Apparent power value, $\boxed{\times}$, power factor value, $\boxed{=}$

FORMULA 21–10 $p.f. = \dfrac{\text{True power } (P)}{\text{Apparent power } (S)}$

True power value, $\boxed{\div}$, apparent power value, $\boxed{=}$

NOTE: In series circuits $p.f.$ = cos θ = R/Z. In parallel circuits $p.f.$ = cos θ = I_R/I_T.

FORMULA 21–11 $S = \sqrt{P^2 + Q^2}$

True power value, $\boxed{x^2}$, $\boxed{+}$, net reactive power value, $\boxed{x^2}$, $\boxed{=}$, $\boxed{\sqrt{x}}$

FORMULA 21–12   Z $\angle\theta°$ = (Z cos θ) + (jZ sin θ) (For series circuits)

USING THE SCIENTIFIC CALCULATOR CONVERSION TECHNIQUE:

Z value, $\boxed{X \gtrless Y}$, angle value (θ), $\boxed{2^{\text{nd}}}$, $\boxed{P \triangleright R}$, $\boxed{X \gtrless Y}$

Answer is combination of last two readouts; that is, $\boxed{P \triangleright R}$ gives the real term and $\boxed{X \gtrless Y}$ gives the imaginary *(j)* term.

FORMULA 21–13 $V_T \angle\theta°$ = V_T cos θ + jV_T sin θ (For series circuits)

V_T value, $\boxed{X \gtrless Y}$, angle value (θ), $\boxed{2^{\text{nd}}}$, $\boxed{P \triangleright R}$, $\boxed{X \gtrless Y}$

Answer is combination of last two readouts; that is, $\boxed{P \triangleright R}$ gives the real term and $\boxed{X \gtrless Y}$ gives the imaginary *(j)* term.

FORMULA 21–14 $I_T \angle\theta°$ = I_T cos θ + jI_T sin θ (For parallel circuits)

I_T value, $\boxed{X \gtrless Y}$, angle value (θ), $\boxed{2^{\text{nd}}}$, $\boxed{P \triangleright R}$, $\boxed{X \gtrless Y}$

Answer is combination of last two readouts; that is, $\boxed{P \triangleright R}$ gives the real term and $\boxed{X \gtrless Y}$ gives the imaginary *(j)* term.

FORMULA 21–15 Inductive susceptance $B_L = \dfrac{1}{X_L} = \dfrac{1}{2\pi fL}$

6.28, $\boxed{\times}$, frequency value, $\boxed{\times}$, inductance value, $\boxed{=}$, $\boxed{1/x}$

FORMULA 21–16 Capacitive susceptance $B_C = \dfrac{1}{X_C} = 2\pi fC$

6.28, $\boxed{\times}$, frequency value, $\boxed{\times}$, capacitance value, $\boxed{=}$

FORMULA 21–17 $Y = \sqrt{G^2 + B^2}$

Value of conductance (G), $\boxed{x^2}$, $\boxed{+}$, value of susceptance (B) $\boxed{x^2}$, $\boxed{=}$, $\boxed{\sqrt{x}}$

REVIEW QUESTIONS

(Be sure to make appropriate vector diagram sketches to help clarify your thinking.)

1. Refer to Figure 21–29a. Find Z, I, V_C, V_L, V_R, and θ using the Pythagorean theorem, Ohm's law, and appropriate trig functions. Draw impedance and V-I vector diagrams, labeling all parts, as appropriate.

2. Refer to Figure 21–29b. Find Z, I_C, I_L, I_R, I_T, and θ using the Pythagorean theorem, Ohm's law, and appropriate trig functions. Draw an appropriate V-I vector diagram and label all parts, as appropriate.

3. Calculate true power, apparent power, and reactive power for the circuit of Question 1. What is the power factor?

4. Draw a power triangle for the circuit in Question 1.

5. A series circuit has a 20-Ω resistance and a 40-Ω inductive reactance. How do you state the circuit Z in rectangular form? If the inductive reactance is changed to an equal value of capacitive reactance, how do you state Z?

6. A series RLC circuit is composed of 10-Ω resistance, 20-Ω inductive reactance, and 30-Ω capacitive reactance. How is the circuit Z stated in rectangular form?

(a)

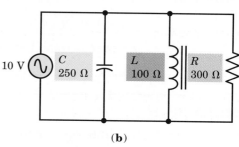

(b)

FIGURE 21–29

7. A parallel circuit with a 20-V source has a resistive branch of 20 Ω and a capacitive reactance branch of 20 Ω. What is the total

current stated in polar form? See Figure 21–30.

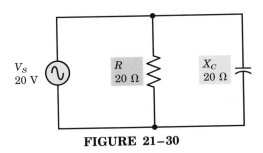

FIGURE 21–30

8. Convert $30 + j40$ to polar form.

9. Convert $14.14 \angle 45°$ to rectangular form.

10. A series *RLC* circuit has a 10-V, 1-kHz source, an *R* of 8 kΩ, an X_L of 6 kΩ, and an X_C of 4 kΩ. What is the total circuit current? (Express in polar form.) What is the true power?

11. What is the total impedance of a series *RLC* circuit that has a 20-kHz ac source, $C = 100$ pF, $L = 0.159$ H, and $R = 15$ kΩ? (Express in polar form and in rectangular form.)

12. A series *RLC* circuit has 100 kΩ of resistance, a capacitor with 0.0005 μF of capacity, and an inductor with 1.5 H of inductance. What is the value of voltage drop across the inductor, if the applied ac voltage is 100 V at a frequency of 10 kHz? (Express in polar form.)

13. If the frequency of the applied voltage increases, what happens to V_C?

14. What is the true power dissipated by a series *RLC* circuit with 100 V applied, if $R = 30$ Ω, $X_L = 40$ Ω, and $X_C = 10$ Ω?

15. A parallel *RC* circuit has a capacitor with 2 Ω of X_C and a resistor with 4 Ω of resistance. What is the impedance? (Express in polar form.)

16. A parallel *RC* circuit has a power factor of 0.5 and $I_T = 5$ A, $f = 1$ kHz, and $R = 60$ Ω. What is the value of *C*?

17. What is the total impedance of a parallel *RLC* circuit comprised of $R = 6$ Ω, $X_L = 12$ Ω, and $X_C = 24$ Ω?

CHALLENGE QUESTIONS

Referring to the answers you obtain for selected Review Questions, use the appropriate calculator key sequences and solve for the following.

1. Refer to Review Question 6. State the *Z* in polar form.

2. Refer to Review Question 7. State the total current in rectangular form.

3. Refer to Review Question 10. State the total current in rectangular form.

4. Refer to Review Question 15. State the impedance in rectangular form.

5. In your own words, define what is meant by the terms "real number" and "imaginary number."

6. In your own words, define what is meant by the term "complex number."

7. Explain why using measurements of a circuit's voltage with a voltmeter and its current with a current meter does not provide the necessary information to find the actual power dissipated by the circuit.

8. Explain how you can make measurements to determine the actual power dissipated by an ac circuit containing inductors and/or capacitors.

TEST YOURSELF

1. A parallel *RLC* circuit has 400-Ω resistance, 500-Ω inductive reactance branch, and 800-Ω capacitive reactance branch. What is the circuit admittance value?

2. What are the Z and θ of the circuit in Question 1? (Assume a voltage applied of 800 V.)

3. What value of X_C produces a $\theta = 0°$?

4. For the circuit in Question 1, what happens to the following parameters if frequency doubles? (Answer with "I" for increase, "D" for decrease, and "RTS" for remain the same.)
   a. I_R will
   b. I_C will
   c. I_L will
   d. I_T will
   e. Z_T will
   f. θ will

CHAPTER TROUBLESHOOTING CHALLENGE

Following the SIMPLER troubleshooting sequence, find the problem with this circuit. As you follow the sequence, record your responses to each step on a separate sheet of paper.

SYMPTOMS Gather, verify, and analyze symptom information. (Look at the "Starting Point Information.")

IDENTIFY Identify initial suspect area for the location of the trouble.

MAKE Make a decision about "What type of test to make" and "Where to make it." (To simulate making a decision about each test you want to make, select the desired test from the "TEST" column listing below.)

PERFORM Perform the test. (Look up the result of the test in Appendix C. Use the number in the "RESULTS" column to find the test result in the Appendix.)

LOCATE Locate and define a new "narrower" area in which to continue troubleshooting.

EXAMINE Examine available information and again determine "what type of test" and "where."

REPEAT Repeat the preceding analysis and testing steps until the trouble is found. What would you do to restore this circuit to normal operation? When you have solved the problem, compare your results with those shown on the following pages.

CHALLENGE CIRCUIT 11

STARTING POINT INFORMATION

1. Circuit diagram
2. $V_{R_3} = 0.057$ V

TEST	Results in Appendix C
$V_{\text{A-B}}$	(78)
$V_{\text{B-C}}$	(38)
$V_{\text{C-D}}$	(96)
$R_{\text{A-B}}$	(104)
$R_{\text{B-C}}$	(63)
$R_{\text{C-D}}$	(32)

CIRCUIT

Meter lead

Meter lead

R_3
100 Ω
D

L_1
11 H
400 Ω

L_2
11 H
400 Ω

C

275 Hz
7.8 V

C_1
0.05 μF
B

C_2
0.05 μF

R_1
10 kΩ

R_2
10 kΩ

A

CHALLENGE CIRCUIT 11

Symptoms the voltmeter measures approximately 0.057 V. If the circuit were operating as prescribed by the diagram, the voltmeter should measure about 0.112 V. This implies circuit current is lower than it normally would be. Thus, circuit impedance is higher than normal.

Identify initial suspect area: Between points C and D, the total inductive reactance of the parallel inductors should be about 9.5 kΩ. Between points B and C, the net capacitive reactance of the parallel capacitors should be about 5.8 kΩ. The net resistance between points A and B should be about 5 kΩ. If the circuit were operating correctly, the total impedance would be about 6.2 kΩ. However, it's acting like the circuit has about 14 kΩ total impedance.

Make test decision: The upward shift in Z could be caused by any R, C, or L opening or any C or L shorting. We'll first look at the resistors. Let's look at $V_{\text{A-B}}$.

Perform 1st Test: Test $V_{\text{A-B}}$. $V_{\text{A-B}}$ is about 3.11 V, again indicating lower than normal circuit current. We know the Rs are not shorted or the voltage would read zero volts. Also, since we know the circuit current is about 0.57 mA, then 0.57 mA times the normal equivalent R of approximately 5 kΩ comes close to that value. The resistors appear to be acting normally. Locate new suspect area: The Rs have been eliminated. Thus, the reactive components are the new suspect area.

Examine available data.

Repeat analysis and testing: Let's look at the inductive branches, isolating as appropriate.

Check $V_{\text{C-D}}$. $V_{\text{C-D}}$ is about 11 V. An open inductor could cause this situation so let's check that out.

Check $R_{\text{C-D}}$. $R_{\text{C-D}}$ is about 400 Ω. This should be about 200 Ω with the parallel circuit situation.

Check R_{L_1}. R_{L_1} is about 400 Ω, which is normal.

Check R_{L_2}. R_{L_2} is infinite ohms, which is an open coil.

Change L_2 and check the circuit operation. It's normal.

1st Test

2nd Test

3rd Test

4th Test

5th Test

6th Test

Series and Parallel Resonance

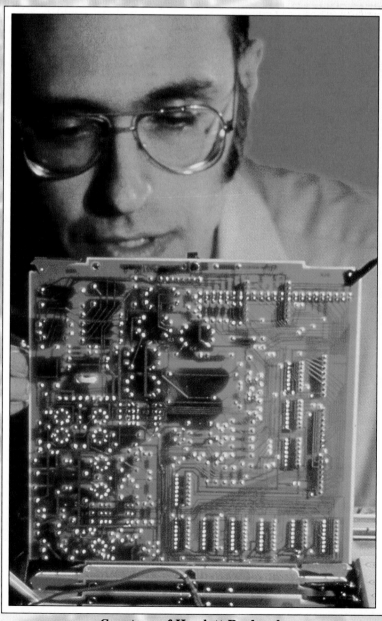

Courtesy of Hewlett-Packard

22

KEY TERMS

Bandpass

Bandpass filter

Bandstop filter

Bandwidth

High-pass filter

L/C ratio

Low-pass filter

Q factor

Resonance

Selectivity

Skirts

OUTLINE

X_L, X_C, and frequency

Series resonance characteristics

The resonant frequency formula

Some resonance curves

Q and resonant rise of voltage

Parallel resonance characteristics

Parallel resonance formulas

Effect of a coupled load on the tuned circuit

Q and the resonant rise of impedance

Selectivity, bandwidth, and bandpass

Measurements related to resonant circuits

Special notations about single components

Filter applications of nonresonant and resonant RLC circuits

CHAPTER PREVIEW

Series and parallel resonant circuits provide valuable capabilities in numerous electronic circuits and systems. When you select a radio or TV station, you are using the unique capability of a resonant circuit to select a narrow group of frequencies from a wide spectrum of frequencies that impact the receiver's antenna system. The transmitter that is transmitting the selected signals also has a number of "tuned" resonant circuits that assure only the desired frequencies are being transmitted and that spurious signals are not being transmitted. Both the transmitting and receiving antenna systems are further examples of special tuned or resonant circuits.

In addition to the valuable tuning application, resonance allows a related application called *filtering*. Filtering can involve either passing desired signals or voltages, or blocking undesired ones. Tuning and filtering are the most frequent applications for series and parallel resonance.

In this chapter, you will study the key characteristics of both series and parallel resonant circuits. Your knowledge of these circuits will be frequently utilized throughout your career.

OBJECTIVES

After studying this chapter, you should be able to:

- List the key characteristics of series and parallel resonant circuits
- Calculate the resonant frequency of circuits
- Calculate **L** or **C** values needed for **resonance** at a given f_r
- Calculate **Q** factor for series and parallel resonant circuits
- Determine **bandwidth** and **bandpass** of resonant circuits
- Draw circuit diagrams for three types of **filters**
- Use the SIMPLER troubleshooting sequence to solve the Chapter Challenge problems

X_L, X_C, AND FREQUENCY

Previously you learned that as frequency increases, X_L increases proportionately. The formula for inductive reactance expresses this by $X_L = 2\pi f L$. You also learned that as frequency increases, X_C decreases in inverse proportion, expressed in the formula

$$X_C = \frac{1}{2\pi f C}$$

Conversely, as frequency decreases, X_L decreases and X_C increases. In other words, X_L and X_C change in opposite directions as frequency changes.

To review, refer to Figure 22–1 and note the relationships between X_L, X_C, and frequency. Notice X_C is not illustrated at frequencies approaching

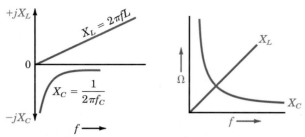

FIGURE 22–1   Relationshsips between X_L, X_C, and frequency

zero hertz. This is because X_C is approaching infinite ohms at these frequencies, thus is difficult to illustrate on such a graph.

SERIES RESONANCE CHARACTERISTICS

The implication of these statements about the relationships of X_L, X_C, and frequency is that for a given set of L and C values, there is a specific frequency where the *absolute* magnitudes of X_L and X_C are equal. Look at Figure 22–2 and notice that for this series RLC circuit, there is one frequency where this occurs. In a series circuit containing L and C, the condition when the magnitude of X_L equals the magnitude of X_C is called **resonance** and the frequency where this occurs is the *resonant frequency*. At resonance, some interesting observations can be made about circuit electrical parameters.

Refer to Figure 22–3 as you study the following list of features.

The characteristics of a series RLC circuit at resonance are:

1. $X_L = X_C$ (In this case, 100 Ω = 100 Ω.)

2. Z = minimum and equals R in circuit. (In this case, 10 Ω.)

3. I = maximum and equals V/R. (In this case, 50 mV/10 Ω = 5 mA.)

4. V_C and V_L are maximum. ($I \times X_C$ = 500 mV; $I \times X_L$ = 500 mV.)

5. $\theta = 0°$ (Since circuit is acting purely resistive.)

FIGURE 22–2 $X_L = X_C$ at resonant frequency (f_r)

FIGURE 22–3   **Sample series circuit parameters at resonance**

Let's examine these statements.

1. When X_L equals X_C, the reactances cancel, leaving zero Ω of net reactance in the circuit. The remaining opposition to current, under these conditions, is any resistance in the circuit. Generally, most of this resistance is the resistance in the inductor windings (unless we have purposely added an additional discrete resistor component). The coil resistance is shown in Figure 22–3 as r_s, indicating series resistance.

2. Because the reactances cancel each other, the Z is minimum and equals the R in the circuit.

3. When Z is minimum ($Z = R$), I is maximum. That is, $I = V/R$.

4. V_C and V_L are maximum since I is limited only by the R value and is therefore maximum. The voltage across the capacitor equals $I \times X_C$, and the voltage across the inductor equals $I \times X_L$. As noted in our sample circuit, the voltage across the coil and the capacitor is 10 times V applied. There is a *resonant rise* in voltage across the reactive components. We will examine this fact again later in the chapter.

5. Again, because the reactances cancel each other, the circuit acts purely resistive; therefore, phase angle (θ) is zero.

✓ IN-PROCESS LEARNING CHECK I

Fill in the blanks as appropriate.

1. In circuits containing both inductance and capacitance, when the source frequency increases, inductive reactance _____ and capacitive reactance _____.

2. In a series RLC (or LC) circuit, when X_L equals X_C, the circuit is said to be at _____.

3. In a series RLC (or LC) circuit, when X_L equals X_C, the circuit impedance is _____; the circuit current is _____; and the phase angle between applied voltage and current is _____ degrees.

4. In a series RLC (or LC) circuit, when X_L equals X_C, the voltage across the inductor or the capacitor is _____.

Since you now know some of the series resonance effects, you may wonder how the frequency of resonance is determined.

Let's deviate from examining circuit parameters to look at the basic formulas that allow you to determine the resonant frequency for a circuit with given component values, or to find the appropriate component values for a desired resonant frequency.

THE RESONANT FREQUENCY FORMULA

The Basic Formula

In our previous discussions, a fixed value of L and C were assumed, while f was changed until X_L equaled X_C. With fixed L and C values, there is only one possible resonant frequency. In many practical applications, either the C or L value or both can be varied to achieve resonance at various frequencies. An example is tuning in different radio stations on a receiver. Also, different **L/C ratios** for any given frequency can be obtained. You can see that a technician should know how to determine the resonant frequency of any given set of LC parameters. Let's look briefly at the derivation and formula for frequency of resonance. Starting with resonance $X_L = X_C$, where L is in henrys, C is in farads, and f is in Hz, the derivation is $X_L = X_C$.

$$2\pi f L = \frac{1}{2\pi f C}$$

$$2\pi f_r L = \frac{1}{2\pi f_r C}$$

$$2\pi L (f_r)^2 = \frac{1}{2\pi C}$$

$$f_r^2 = \frac{1}{(2\pi)^2 LC}$$

Formula 22–1	$f_r = \dfrac{1}{2\pi\sqrt{LC}}$ OR $\dfrac{0.159}{\sqrt{LC}}$

 Inductance value, $\boxed{\times}$, capacitance value, $\boxed{=}$, $\boxed{\sqrt{x}}$, $\boxed{\times}$, 6.28, $\boxed{=}$, $\boxed{1/x}$

EXAMPLE

What is the resonant frequency of a series LC circuit having a 200-μH inductor and a 400-pF capacitor?

$$f_r = \frac{1}{2\pi\sqrt{LC}}$$

$$f_r = \frac{1}{6.28 \times \sqrt{200 \times 10^{-6} \times 400 \times 10^{-12}}}$$

$$f_r = \frac{1}{6.28 \times \sqrt{80{,}000 \times 10^{-18}}}$$

$$f_r = \frac{1}{1{,}776 \times 10^{-9}} = 562.98 \text{ kHz}$$

200, $\boxed{\text{EE}}$, $\boxed{+/-}$, 6, $\boxed{\times}$, 400, $\boxed{\text{EE}}$, $\boxed{+/-}$, 12, $\boxed{=}$, $\boxed{\sqrt{x}}$, $\boxed{\times}$, 6.28, $\boxed{=}$, $\boxed{1/x}$

Now you try one!

PRACTICE PROBLEMS I

What is the resonant frequency of a series LC circuit containing an inductor with an inductance of 250 μH and a capacitor with a capacitance of 500 pF?

	L (μH)	C (pF)	f_r (MHz)
	100	100	1.59
	100	200	1.13
	400	400	0.398
(a)	400	10	2.52

$$\left(\text{Using } f_r = \frac{1}{2\pi\sqrt{LC}} \text{ OR } \frac{0.159}{\sqrt{LC}}\right)$$

Different LC products cause different f_rs

	L (μH)	C (pF)	f_r (MHz)
	500	51	1
	400	63.75	1
	300	85	1
(b)	200	127.5	1

$$\left(\text{Using } f_r = \frac{1}{2\pi\sqrt{LC}} \text{ OR } \frac{0.159}{\sqrt{LC}}\right)$$

Different values having same LC product yield same f_r

FIGURE 22–4   (a) Values of L and C determine f_r and (b) Different combinations of L and C can produce same f_r.

VARIATIONS OF THE f_R FORMULA. As you can see from the basic resonant frequency formula, the L and C values determine the frequency of resonance. That is, the larger the LC product is, the lower is the f_r, and vice versa, Figure 22–4a. Note also in Figure 22–4b how various combinations of L and C can be resonant at one particular frequency (in this case, 1 MHz).

It is frequently useful to find the L value needed with an existing C value or the C value needed with an existing L value to make the circuit resonant at some specified frequency. By rearranging the resonant frequency formula (i.e., squaring both sides to delete the square root portion), we can find L or C, if the other two variables are known. When this is done, the following useful formulas are derived:

$$L = \frac{1}{(2\pi)^2 f_r^2 C} = \frac{1}{4\pi^2 f_r^2 C} =$$

Formula 22–2	$L = \dfrac{0.02533}{f_r^2 C}$

0.02533, $\boxed{\div}$, $\boxed{(}$, resonant frequency value, $\boxed{x^2}$, $\boxed{\times}$, capacitance value, $\boxed{)}$, $\boxed{=}$

$$C = \frac{1}{(2\pi)^2 f_r^2 L} = \frac{1}{4\pi^2 f_r^2 L} =$$

Formula 22–3	$C = \dfrac{0.02533}{f_r^2 L}$

0.02533, $\boxed{\div}$, $\boxed{(}$, resonant frequency value, $\boxed{x^2}$, $\boxed{\times}$, inductance value, $\boxed{)}$, $\boxed{=}$

NOTE: These formulas are for your information—You need not memorize them!

OTHER USEFUL VARIATIONS. When dealing in radio frequency circuits, it is sometimes inconvenient to manage henrys, farads, and hertz, as in the previous formulas. It is often more convenient to deal with microhenrys (μH), picofarads (pF), and megahertz (MHz). The following formulas are based on these convenient units.

$$f^2 = \frac{25,330}{LC} \text{ or}$$

Formula 22–4	$L = \dfrac{25,330}{f^2 C}$

where L is in µH, C is in pF, f is in MHz or

 25,330, $\boxed{\div}$, $\boxed{(}$, resonant frequency value in MHz, $\boxed{x^2}$, $\boxed{\times}$,

capacitance value in pF, $\boxed{)}$, $\boxed{=}$

Formula 22–5	$C = \dfrac{25{,}330}{f^2 L}$

 25,330, $\boxed{\div}$, $\boxed{(}$, resonant frequency value in MHz, $\boxed{x^2}$, $\boxed{\times}$,

inductance value in µH, $\boxed{)}$, $\boxed{=}$

NOTE: Again, these formulas are for your information, but you may find occasion to use them.

EXAMPLE

A circuit has an L value of 100 µH. What C value will make the circuit resonant frequency equal 1 MHz?

$$\text{Using } C = \frac{0.02533}{f_r^2 L}$$
$$\text{where } L \text{ is in H, } C \text{ in F, } f \text{ in Hz.}$$
$$C = \frac{0.02533}{(1 \times 10^6)^2 \times 100 \times 10^{-6}}$$
$$C = \frac{0.02533}{1 \times 10^{12} \times 100 \times 10^{-6}}$$
$$C = \frac{25{,}330 \times 10^{-12}}{100} = 253.3 \text{ pF}$$

$$\text{Using } C = \frac{25{,}330}{f^2 L}$$
$$\text{where } L \text{ is in µH, } C \text{ in pF, and } f \text{ in MHz.}$$
$$C = \frac{25{,}330}{f^2 L} = \frac{25{,}330}{100} = 253.3 \text{ pF}$$

 25,330, $\boxed{\div}$, $\boxed{(}$, 1, $\boxed{x^2}$, $\boxed{\times}$, 100, $\boxed{)}$, $\boxed{=}$

Incidentally, you can also use the X_C equals X_L approach. Since you know the frequency and the L, you can calculate X_L using the $X_L = 2\pi f L$ formula. Then, knowing that X_C is that same value at resonance, you use the X_C formula to solve for C. That is

$$C = \frac{1}{2\pi f X_C}$$

See Figure 22–5.

Now you apply the appropriate formula(s) to solve for an unknown L when C and f_r are known.

Using basic units: $L = H$, $C = F$, and $f_r = Hz$
$L = \dfrac{0.02533}{f_r^2 C}$ and $C = \dfrac{0.02533}{f_r^2 L}$
Using convenient units: $L = \mu H$, $C = pF$, and $f_r = MHz$
$L = \dfrac{25{,}330}{f_r^2 C}$ and $C = \dfrac{25{,}330}{f_r^2 L}$
Using the $X_L = X_C$ approach:
If you know X_C, then $L = \dfrac{X_C}{2\pi f_r}$
If you know X_L, then $C = \dfrac{1}{2\pi f_r X_L}$

FIGURE 22–5   Some useful variations derived from the f_r and reactance formulas

PRACTICE PROBLEMS II

What L value is needed with a 100-pF capacitor to make the circuit resonant frequency equal 5 MHz?

SOME REASONANCE CURVES

Figure 22–6 shows the resonance curves (graphic plots) of current, impedance, and phase angle versus frequency for a series RLC circuit.

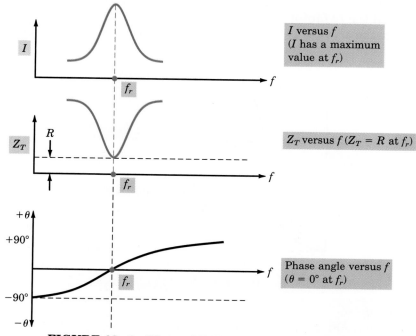

FIGURE 22–6   Plots of I, Z, and θ versus frequency

Plot of Current *(I)* Versus Frequency *(f)*

When frequency is considerably below the circuit resonant frequency, the current is low compared to current at resonance. This is because $Z = \sqrt{r^2 + (X_C - X_L)^2}$; therefore, circuit impedance *(Z)* is much higher than it is at resonance. For series *RLC* circuits below resonance, the net reactance is capacitive reactance. The same pattern is true above resonance, except the net reactance for a series *RLC* circuit is X_L. (NOTE: For series resonant circuits, plotting current rising to maximum value at resonance is the most significant aspect of the response curves.)

Plot of Impedance *(Z)* Versus Frequency

As you would expect, impedance is lowest at the resonant frequency, since the reactances cancel each other at resonance. The impedance graph in Figure 22–6 shows that *Z* equals *R* (resistance) at resonance. Thus, for any circuit, when *Z* is low, *I* is high and when *Z* is high, *I* is low.

Plot of Phase Angle Versus Frequency

At resonance, the circuit acts resistively and the phase angle is 0°. (This is because the reactances have cancelled.) Below resonance the circuit acts capacitively, since the net reactance is capacitive reactance. Above resonance, the circuit acts inductively since the net reactance is X_L.

Q AND RESONANT RISE OF VOLTAGE

Q and the Response Characteristic

The shape of a given circuit's resonance response relates to an important factor called the **Q factor,** or figure of merit, Figure 22–7. The higher the Q factor is, the sharper is the response characteristic. This sharp response

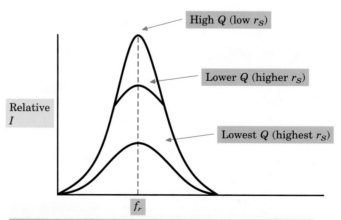

NOTE: r_S value does not change f_r but does affect response.

$$Q = \frac{X_L}{r_S}$$

FIGURE 22–7   Response curves versus circuit *Q*

curve is sometimes described as having steep **skirts,** or as indicating a circuit with high **selectivity** (i.e., one that is very frequency selective). The lower the Q factor is, the broader, or flatter the response curve and the less selective the circuit is.

Q Equals X_L/R or X_C/R

The Q factor of a resonant circuit relates to two parameters. The first parameter is associated with the ratio of reactance to resistance at the resonant frequency. In formula form, $Q = X_L/r_s$ or X_C/r_s at resonance, where r_s is the series resistance in the circuit, which often is the coil resistance. In other words, the Q of the coil (X_L over R) is frequently the same as the Q of the resonant circuit. Obviously, the lower the series circuit resistance is, the higher the Q is and the sharper the resonant circuit response. The higher the r_s is, the lower the Q is and the flatter the response. Note that since I is the same throughout the circuit, the ratio

$$\frac{I \times X_L}{I \times R}, \text{ or } \frac{V_L}{V_R}$$

is also the same as the Q factor. Likewise, the ratio of reactive power to true power in the circuit also equals the Q factor.

Q Related to *L/C* Ratio

A second related factor that affects the Q is the ratio of L to C; that is L/C. Since both L and C values affect the frequency of resonance, a *given* resonant frequency can be achieved by various LC values, refer again to Figure 22–4b. However, different combinations of LC values have different reactances at resonance. For this reason, it is best to have a *high L/C* ratio to achieve a high X_L at resonance; hence, a high X_L/R or Q factor, Figure 22–8.

Resonant Rise of Voltage

Refer to Figure 22–9 and note the voltages around the series circuit. The term *magnification factor* describes the fact that voltage across each of the

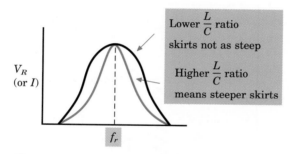

(Same I since r_S assumed to be constant)

FIGURE 22–8   Q related to *L/C* ratios

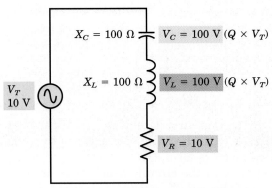

FIGURE 22–9   Magnification factor

reactive components at resonance is Q times V applied. This is because of the resonant rise of I at resonance, which is due to the reactances cancelling each other at the resonant frequency. Since I is limited only by R at that frequency, and since each reactive voltage drop is I times X with X being very high, the reactive voltage can be very high. In fact, at resonance, the reactive voltages are Q times V applied; thus, voltage has been magnified (Q times) across the reactive components. In other words, the higher the Q is, the greater the magnification is. As stated earlier, the ratio of reactive voltage to applied voltage at resonance equals the Q. If we measure the voltage across one reactive component and call it V_{out} and call V applied V_{in}, then

$$Q = \frac{V_{\text{out}}}{V_{\text{in}}}$$

PRACTICE PROBLEMS III

1. What is the Q of the circuit in Figure 22–10?
2. What is the V_C value at resonance, if the applied voltage in Figure 22–10 is 12 volts?
3. If r_s doubles in this circuit, what is the Q value?
4. If r_s is a smaller value than shown in Figure 22–10, are the response curve skirts more or less steep?

FIGURE 22–10

PARALLEL RESONANCE CHARACTERISTICS

A parallel resonant circuit is a frequently used circuit in receivers, transmitters, and frequency measuring devices. As you might expect, many parallel resonant circuit features are opposite to those of a series resonant circuit. Look at Figure 22–11 as you study the following discussion.

Highlighting the basic features of a high-Q parallel resonant circuit:

1. Z = maximum (and is purely resistive)

2. Line current = minimum

3. Current inside *tank* = maximum

4. $X_L = X_C$ (ideal components)

5. $\theta = 0°$ (and *p.f.* = 1)

Let's examine these statements further.

1. Z is maximum at resonance because the inductive and capacitive branch currents are 180° out-of-phase, thus cancel each other (assuming ideal L and C components). This means the vector sum of the branch currents (or I line) is very small. If the inductor and capacitor are perfect, the Z is infinite. Since practical inductors and capacitors have resistances and losses, there is an effective R involved, Figure 22–11. (For your information, $Z_T = L/C_{r_s}$.)

2. Since Z is maximum at resonance, obviously, the line current (or total current supplied by the source) is minimum.

3. Within the LC (tank) circuit, the circulating current can be very high. In fact, the current inside the tank circuit equals $Q \times I$ line and the total circuit impedance is $Q \times X_L$.

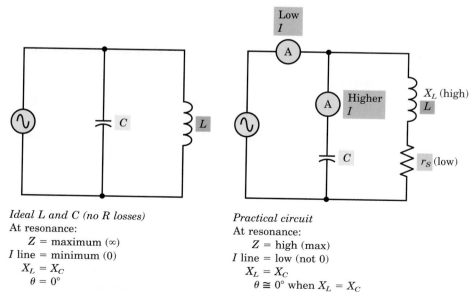

Ideal L and C (no R losses)
At resonance:
   Z = maximum (∞)
I line = minimum (0)
   $X_L = X_C$
    $\theta = 0°$

Practical circuit
At resonance:
   Z = high (max)
I line = low (not 0)
   $X_L = X_C$
    $\theta \cong 0°$ when $X_L = X_C$

FIGURE 22–11   Parallel resonant circuit characteristics

4. When $X_L = X_C$, the preceding characteristics are exhibited. In practical parallel resonant circuits having less-than-perfect components, some r_s value associated with the coil exists. This means when $X_L = X_C$, the Z_L of the inductive branch is slightly greater than the X_C of the capacitive branch. In low-Q circuits, this means the frequency needs to be slightly lower to allow more I_L, so the inductive branch current equals the capacitive branch current. For circuits having a Q above 10, however, this effect is negligible.

5. At resonance (for high-Q circuits), circuit current and applied voltage are virtually in-phase.

PRACTICAL NOTES

For low-Q circuits, the condition of unity power factor occurs at a slightly different frequency than when $X_L = X_C$. The frequency of $p.f. = 1$ is sometimes termed antiresonance, or the antiresonant frequency.

IN-PROCESS LEARNING CHECK II

1. In a high-Q parallel resonant circuit, Z is _____.
2. In a high-Q parallel circuit using ideal components, X_L _____ equal X_C at resonance.
3. The power factor of an ideal parallel resonant circuit is _____.
4. Phase angle in an ideal parallel resonant circuit is _____.
5. In a parallel resonant circuit, the current circulating in the tank circuit is _____ than the line current.

PARALLEL RESONANCE FORMULAS
For Q Greater than 10

When the circuit Q is high (when the r_s is low, or the equivalent R_p is high), the resonant frequency is determined by the LC component values as it is in series resonant circuits. With Qs greater than 10, the resonant frequency is calculated with the same formula used for series resonance. That is:

$$f_r = \frac{1}{2\pi\sqrt{LC}}$$

For Q Less than 10

When the circuit Q is low (i.e., when r_s is high or effective R_p is low, or when a reflected impedance from a load coupled to the tuned circuit is low causing an effective low value of equivalent R_p), the R, L, and C values enter in the formula for parallel resonance. That is, the resistance has a significant effect on the frequency of resonance so it should be considered. The following formula (only for your information) illustrates that resistance affects resonance in parallel LC circuits.

Formula 22-6	$f_r = \dfrac{1}{2\pi\sqrt{LC}} \times \sqrt{1 - \dfrac{C{r_s}^2}{L}}$

NOTE: Formula 22-6 is given as information only to show that resistance has effect on resonant frequency in low-Q, parallel resonant circuits. No computations using this formula will be required in this text. In most cases, the Q of parallel tuned circuits is high enough so that the effect of the resistance on the resonant frequency can be generally neglected.

To check your grasp of parallel resonant circuits thus far, try to solve the following problems.

PRACTICE PROBLEMS IV

1. Calculate the resonant frequency of a parallel LC circuit with the following parameters: $L = 200$ μH having a resistance of 10 Ω and $C = 20$ pF.

2. What is the Q of the circuit in Question 1?

3. What is the impedance at resonance of the circuit in Question 1?

4. Is the resonant frequency of the circuit in Question 1 lower or higher if $r_s = 1$ kΩ?

EFFECT OF A COUPLED LOAD ON THE TUNED CIRCUIT

Whenever a parallel resonant circuit delivers energy to a load (i.e., a load is coupled to the tuned circuit), the Q of the tank circuit is affected, Figure 22-12. If the power dissipated by the coupled load is greater than 10 times the power lost in the tank circuit reactive components, the tank impedance is high compared to the load impedance. Thus, for all practical purposes, the impedance of the parallel combination of the tank and load virtually equals the load resistance value. When a tank is heavily loaded by such a resistive load impedance, the Q of the system equals R over X, or $Q = R/X$, where R is the parallel load resistance and X is the reactance. (NOTE: This is the inverse of the X_L/R formula used for Q in series resonance and high-Q parallel resonant circuits). For example, if a resistive load of 4,000 ohms is connected across a parallel resonant circuit where X_C (or X_L) = 400 ohms,

$$Q = \frac{R}{X_L} = \frac{4,000}{400} = 10$$

The lower the reflected R value, the lower the Q. If load is "heavy" (i.e., much energy absorbed by load) $Z_T \approx Z$ of load and $Q = \dfrac{R_p}{X_L}$.

FIGURE 22–12   Effect of coupling a load to a parallel resonant circuit

Q AND THE RESONANT RISE OF IMPEDANCE

The preceding discussions show us that the current characteristic is the most meaningful response factor in series resonant circuits, where the impedance characteristic is the most significant response factor in parallel resonant circuits. The response characteristics of a parallel tuned circuit are also related to the Q factor. In Figure 22–13 you can see there is a resonant rise of impedance as the resonant frequency is approached. As we previously stated, the impedance at resonance is $Q \times X_L$. You can also see from this illustration that the higher the Q is, the sharper the response characteristics, and the lower the Q is, the flatter and broader the response.

SELECTIVITY, BANDWIDTH, AND BANDPASS

Selectivity

The sharper the response curve of a resonant circuit, the more selective it is. This means that only frequencies close to resonance appear near the max-

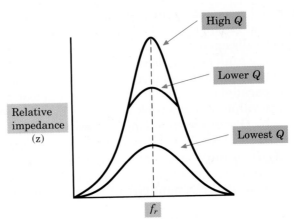

FIGURE 22–13   Resonant rise of Z for a parallel resonant circuit

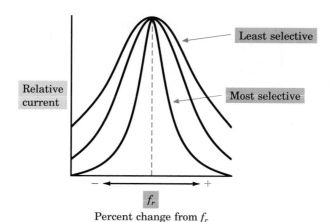

FIGURE 22–14   **Examples of various
selectivities**

imum response portion of the resonant circuit response. This indicates that
in series resonant circuits, the current value *falls off* rapidly from its maximum value at frequencies slightly below or above the resonant frequency.
In parallel resonant circuits, the impedance *falls off* rapidly from the maximum value obtained at resonance. Conversely, resonant circuits that have
less selectivity allow a wider band of frequencies close to their maximum
response levels. Selectivity, used to describe a resonant circuit, refers to the
capability of the circuit to differentiate between frequencies. Refer to Figure
22–14 for examples of various selectivities versus response characteristics.

Bandwidth and Bandpass

You have looked at various response curves and have noticed that none have
perfect vertical skirts. Therefore, to define **bandwidth,** some value or level
on the response curve is chosen that indicates all frequencies below that
level are rejected and frequencies above the chosen value are selected. The
accepted level is 70.7% of maximum response level.

For series resonant circuits, we consider those two points on the response curve where current is 70.7% of maximum. For parallel resonant
circuits, the points where Z is 70.7% of maximum are used.

Total bandwidth *(BW)* is calculated as:

$$\boxed{\text{Formula 22–7} \quad \Delta f = f_2 - f_1}$$

where: $\Delta f = BW$
f_2 = higher frequency 0.707 point
f_1 = lower frequency 0.707 point on response curve

half-power point (higher frequency), $\boxed{-}$, half-power point

(lower frequency), $\boxed{=}$

Also, $f_1 = f_r - \dfrac{\Delta f}{2}$, and $f_2 = f_r + \dfrac{\Delta f}{2}$.

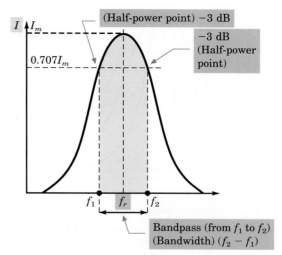

FIGURE 22–15   Bandwidth and bandpass

Notice that if $P = I^2R$ and at f_1 and f_2 the current level is 0.707 of maximum, then at those points on the response curve, $P = (0.707 \, I_{max})^2R = 0.5 \, P_{max}$. For this reason, these points (f_1 and f_2) are often called *half-power points*, Figure 22– 15.

PRACTICAL NOTES

The response curves are not perfectly symmetrical above and below resonance. For that reason, the f_1 and f_2 expressions are approximations. They are close to true for very high Q circuits.

Bandwidth Related to Q

You have learned the Q of a resonant circuit is the primary factor determining the response curve shape. Thus, BW (Δf) and Q are related. The formula showing this relationship is:

$$\text{Formula 22–8} \quad \Delta f = \frac{f_r}{Q}$$

 resonant frequency value, $\boxed{\div}$, Q value, $\boxed{=}$

Rearranging this formula, we can also say that:

$$\text{Formula 22–9} \quad Q = \frac{f_r}{\Delta f}$$

 resonant frequency value, $\boxed{\div}$, bandwidth value, $\boxed{=}$

Bandpass

As has been discussed, the bandwidth is the width of the band of frequencies where a given resonant circuit responds within specified criteria. For example, a given resonant circuit responds over a 20–kHz bandwidth with levels at or above 70.7% of maximum response level.

On the other hand, **bandpass** is the specific frequencies at the upper and lower limits of the bandwidth. For example, if the resonant frequency is 2,000 kHz, and the circuit has a bandwidth of 20 kHz, the bandpass is from 1,990 kHz to 2,010 kHz, (assuming a symmetrical response). This indicates the specific band of frequencies passed is from 1,990 to 2,010 kHz, and frequencies outside this limit cause a response that falls below the 70.7% level, or below the half-power points (sometimes expressed as the −3 dB points, as you will learn later). Again, Figure 22–15 illustrates bandwidth and bandpass.

EXAMPLE

1. What is the total bandwidth of a resonant circuit if f_r is 1,000 kHz and f_1 is 990 kHz?

   Since one half-power point is 10 kHz below the resonant frequency, it is assumed that if the response curve is symmetrical, the other half-power point is 10 kHz above the resonant frequency. This means the total BW = 20 kHz. That is, f_1 = 990 kHz and f_2 = 1,010 kHz.

2. What is the bandwidth of a resonant circuit if resonant frequency is 5 kHz and Q is 25?

$$\Delta f = \frac{f_r}{Q} = \frac{5,000}{20} = 250 \text{ Hz}$$

3. What is the bandpass of the circuit described in Question 2?

Bandpass = from 5,000 − 125 Hz to 5,000 + 125 Hz = 4,875 to 5,125 Hz.

PRACTICE PROBLEMS V

1. In a resonant circuit, f_r is 2,500 kHz and Δf is 12 kHz. What are the f_1 and f_2 values?

2. In a resonant circuit, Q is 50 and bandwidth is 10 kHz. What is the resonant frequency and the bandpass?

3. What is the f_1 value in a circuit whose resonant frequency is 3,200 kHz and whose bandwidth is 2 kHz?

MEASUREMENTS RELATED TO RESONANT CIRCUITS

Because resonant circuits are so prevalent in electronics, it is worthwhile to know some of their practical measurements. These measurements include measuring resonant frequency; determining Q and magnification factor; determining maximum current (for series resonant circuits) and maximum impedance (for parallel resonant circuits); determining bandwidth and bandpass; measuring the *tuning range* of a tunable resonant circuit (with variable C or L); and determining the tuning ratio. Let's briefly look at each of these measurements.

1. Measuring resonant frequency

   There are several ways to determine resonant frequency. Of course, if you are trying to measure at what frequency a tunable signal source is operating (what frequency it is tuned to), a digital frequency counter coupled to the output circuit might be used for direct frequency measurement, Figure 22–16. One way to get an approximate resonant frequency measurement of a "stand-alone" LC circuit is to use a metering device that reacts to circuit conditions. One type of metering device is called a tunable "dip" meter. This type device is used with "power-off" conditions to determine the resonant frequency for a given LC circuit. When these tunable devices are near (usually inductively coupled) to LC circuits, as the tuned circuit in the dip meter is tuned near to the resonant frequency of the LC circuit, the LC circuit absorbs energy from the dip meter circuit and causes the meter reading to "dip" downward. The frequency at which the dip occurred is then read from the calibrated dial on the dip meter. Other means may include field strength meters or oscilloscopes to monitor frequency outputs of signal sources. It is imperative that the technician be aware of the possibility of "detuning" the circuit under test, and use the proper test probes and/or coupling methods with the scope, or other test equipment to minimize detuning effects.

   Keep in mind the fact that there are various means for checking resonant frequencies of circuits. As you continue in your studies, you will probably have opportunity to get some practical experience in us-

FIGURE 22–16   A device used to determine f_r. (Courtesy of Hewlett-Packard Co.)

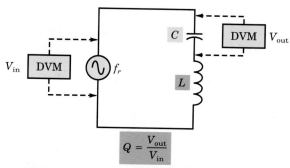

FIGURE 22–17   Determining Q of series resonant circuit

ing some of these devices, and you will learn the proper techniques for making these measurements.

2. Determining Q and magnification factor
   In a series resonant circuit, using a high input impedance DVM, or an oscilloscope allows the measurement of voltage across one reactive component (generally, the capacitor). When the circuit is tuned to resonance, the voltage is maximum. This voltage is then compared to the voltage source to determine the Q and the voltage magnification factor. That is,

$$Q = \frac{V_{out}}{V_{in}}$$

See Figure 22–17.

3. Determining maximum current *(I)* for series and maximum impedance *(Z)* for parallel resonant circuits
   In a series resonant circuit, a current meter or a voltmeter reading across a series resistor is used to determine maximum current. Read the current meter directly, or use Ohm's law where

$$\frac{V_R}{R} = I$$

to determine current when the circuit is tuned for maximum current or resonance, Figure 22–18.

   In a parallel resonant circuit, use a variable R in series with the source and adjust it so that V_{TANK} across the tuned circuit and V_R across the variable R are equal (at resonance). Remove the R from the circuit and measure its value. Its value is the same as the Z of the tuned circuit at resonance since the V drops were equal. Since we can calculate X_L from the resonant frequency formula and we know that

$$Q = \frac{Z_T}{X_L}$$

we can determine the Q, Figure 22–19.

Variable
f

C

L

Low Ω
R

A

Current
meter

Use a current meter, *or*, insert low R value, measure V_R,
and then calculate I. $\left(I = \dfrac{V_R}{R} \right)$

**FIGURE 22–18   Determining maximum I for
a series resonant circuit**

Adjust R so that $V_R = V_{\text{TANK}}$ at resonance
(then measure R . . . $R = Z_{\text{TANK}}$)

f_r

R

V

V_R

V V_{TANK}

**FIGURE 22–19   Determining maximum Z for a
parallel resonant circuit**

4. Determining bandwidth and bandpass
    For a series resonant circuit, tune the circuit through resonance and
    note the frequencies where current is 70.7% of maximum value. The
    difference between the lower frequency and the higher frequency is the
    bandwidth. The actual frequencies at those two points represent the
    bandpass.
        For a parallel resonant circuit, you can use the Z determining sys-
    tem and find the frequencies where Z is 70.7% of maximum value.
    Again, the difference between those frequencies is the bandwidth, and
    the precise lower-to-upper frequencies are the bandpass, Figure 22–20.

5. Measuring the tuning range
    If the tuned circuit has a variable C or L, or if both are variable, there
    are limits to the range of resonant frequencies to which the circuit can
    be tuned.

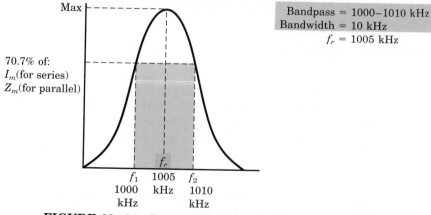

FIGURE 22–20   Determining bandwidth and bandpass

The various resonant frequency measurement devices previously mentioned can check the actual resonant frequencies that are capable of being tuned by the circuit. Simply measure the frequency at both ends of the tuning extremities, and you will know the circuit tuning range.

6. Determining the tuning ratio
From the resonant frequency formula, you can reason that the frequency change caused by a change in C or L is inversely proportional to the square root of the *change* in C or L. That is,

$$f_r = \frac{1}{2\pi\sqrt{LC}}$$

therefore, f_r is proportional to $1/\sqrt{LC}$.

If it is desired to tune through a range of frequencies from 1,000 to 6,000 kHz, this means this tuning ratio of 6:1 (highest to lowest frequency) requires a change of C or L in the ratio of 36:1. If C is the variable component and its minimum C value is 10 pF, it has to have a maximum C value of 36 × 10, or 360 pF to tune the desired range.

⚠ SAFETY HINTS

When working around equipment with resonant circuits, particularly in transmitter circuits, be aware that it is possible to get bad *rf burns* by touching or getting too near a live circuit or component. In addition, these circuits often have high dc voltages, as well as high radio frequency ac energy—If touched while the circuit is live, injury or death can result!

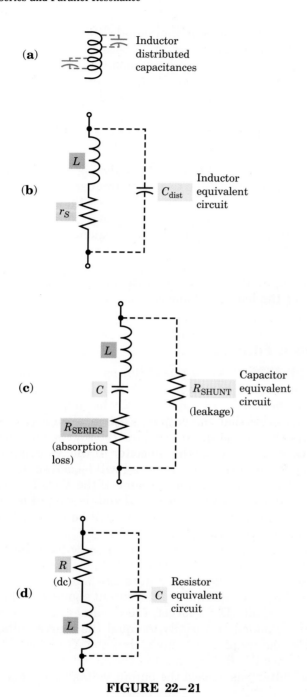

(a) Inductor distributed capacitances

(b) Inductor equivalent circuit

(c) Capacitor equivalent circuit

(d) Resistor equivalent circuit

FIGURE 22–21

networks and their common applications. Two general filter classifications that will be discussed are nonresonant filters and resonant filters.

Another way filters are categorized is by the frequency range where they are used. For example, power supply filters typically operate at frequencies of 60 Hz, 120 Hz, 400 Hz, and so forth. Audio filtering networks obviously operate in audio frequency environments. *Rf* filters are used at frequencies above the audio range in receivers, transmitters, and other *rf* systems.

Names of filter networks or individual filter components often reflect what they do; for example, low-pass, high-pass, bandpass, bandstop, smoothing, decoupling, and so on. Another common way to refer to filter networks is the way the components are *laid out;* for example, L-shaped, pi-shaped, T-shaped, and so on. With these few classifications in mind, let's look at some filter networks.

A General Principle

Remember this general principle while analyzing any type filter: A device or circuit that has low opposition to a given frequency or band of frequencies and is in series with the output or load passes that frequency or band of frequencies along to the output or load. A device or circuit that has a low opposition to a given frequency or band of frequencies and is in parallel (or shunt) with the output or load bypasses or prevents that frequency or band of frequencies at the output or load. The converse of these two statements is also true. High oppositions in series with the signal path prevent signals from arriving at the load. High oppositions in parallel allow signals at the output or load.

Nonresonant Filters

LOW-PASS FILTERS.   A **low-pass filter** passes along low frequency components of a given signal or waveform, while impeding the passage of higher frequency components. For example, look at the simple RC filter network in Figure 22–22. Notice that the output is being taken across the capacitor. The voltage distribution of the signal from the source will depend upon the relative values of R and X_C of the capacitor. If the X_C is high compared to the R, then a greater portion of the signal will be across the capacitor, and the network "output" signal will be greater. If the X_C is low with respect to the R, the resistor drops most of the signal and the output is very low. Since the capacitor is frequency-sensitive, its X_C will be high at low frequencies and low at high frequencies.

In Figure 22–23a, the source frequency is dc (or 0 hertz). The X_C is virtually infinite and almost all the input voltage appears across the capacitor or is present at the output. In Figure 22–23b, the source frequency is such that R and X_C are equal; thus, the output signal and the voltage drop across the R are equal. (Don't forget, these voltages are out-of-phase; they would have to be added vectorially to equal the source voltage value). In Figure 22–23c, the frequency is high, thus the X_C is low, and the output voltage is low since the R drops most of the signal voltage. As you can see, this "low-pass" filter has caused the output voltage to be maximum at low

**FIGURE 22–22   Example of a
low-pass filter**

FIGURE 22–23   Example of simple RC low-pass filter operation

frequencies and minimum at high frequencies. Thus, it is called a low-pass filter! The frequency response of a low-pass filter is shown in Figure 22–24. Various circuit and component configurations, other than the simple RC circuit previously shown, that provide low-pass filtering are shown in Figure 22–25.

FIGURE 22–24   Frequency response characteristics of a low-pass filter

Simple inductor filter

(a)

"L-type" filter

(b)

"Pi-type" filter

(c)

"T-type" filter

(d)

FIGURE 22–25   Examples of circuit configurations for low-pass filters

HIGH-PASS FILTERS. A **high-pass filter** passes the high frequency components of a signal (or group of signals) to the output or load, while greatly attenuating or preventing, the low frequency components from being passed to the output or load.

Figure 22–26a illustrates a simple RC, high-pass filter. We can use the same concepts to explain this filter as we did in explaining the simple RC low-pass filter. That is, the relative values of R and X_C will determine the

(a)

(b)

(c)

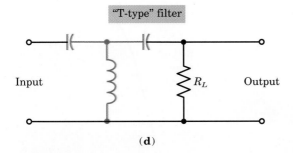

(d)

**FIGURE 22–26   Examples of different configurations
for high-pass filters**

amount of the input signal that appears at the output. The difference between the two types of filters, is that in this case we take the output across
the R, rather than the C. This means that the higher the X_C compared to the
R value, the lower the output will be, and vice versa.

Let's assume the same values for R and C in Figure 22–26a as was used
in our low-pass filter examples in Figure 22–23. If $R = 100 \ \Omega$ and the value
of $C = 1 \ \mu\text{F}$, the following conditions will exist.

1. With a 10-volt dc source as input, the output will be zero. This is because the capacitor (once charged) blocks dc. This "ultra-low" frequency (dc) is not passed to the output.

2. At a frequency of 1,592 hertz, R and X_C are equal, and the output will be 7.07 volts with a 10-volt ac source as the input.

3. At a frequency of 15,920 hertz, $R = 100$ ohms and $X_C =$ about 10 ohms; thus, the output will be about 9 volts.

Since this filter passes along the higher frequencies to the output, while passing very little signal at low frequencies, it is called a high-pass filter!

Examples of their applications include use as tone controls in radios and use in the antenna input circuits to receivers to prevent unwanted interference. Figure 22–26 shows several examples of configurations of R, L, and C components used as high-pass filters. Typically, the capacitors, which have a high opposition to low frequencies, are in series with the output or load. The inductors, which have a low opposition to low frequencies, are in parallel, or shunt with the output, thus effectively bypassing the low frequencies around the load. The result is the high frequency components of the input signal are passed to the load; and the low frequency components are greatly attenuated or prevented from reaching the load. Thus, the filter is called a high-pass filter.

Figure 22–27 shows the typical response characteristic of a high-pass filter.

NOTE: Low-pass and high-pass filter networks can be combined to pass a desired band of frequencies. Those frequencies passed along to the output or load are the frequencies that are not stopped by either filter network. This is one application of *bandpass filtering*. Typically however, this application is often fulfilled by resonant circuits.

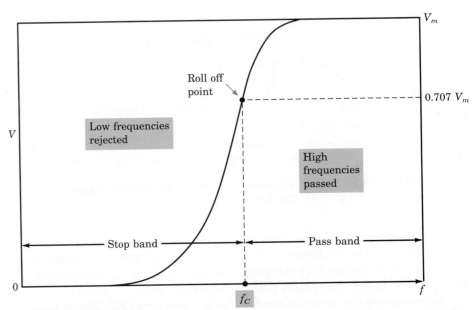

FIGURE 22–27   Typical response characteristic of a high-pass filter

PRACTICAL NOTES

As you know, points falling below 70.7% or the half-power points of a filter response are out of the band of frequencies that the filter will pass. In low-pass filters, Figures 22–24 and 22–25, the high frequencies are rejected. In high-pass filters, Figures 22–26 and 22–27, the lower frequencies are rejected. The frequency where rejection begins is sometimes called the *cutoff frequency* (f_c). For a simple RC network, Figure 22–26a, $X_C = R$ at this cutoff frequency. The formula to find the cutoff frequency for this RC network is:

| Formula 22–10 | $f_c = \dfrac{1}{2\pi RC}$ |

6.28, $\boxed{\times}$, resistor value, $\boxed{\times}$ capacitance value, $\boxed{=}$, $\boxed{1/x}$

To find the cutoff frequency where $X_L = R$ for an RL network similar to the one shown in Figure 22–24a, the formula is:

| Formula 22–11 | $f_c = \dfrac{1}{2\pi\left(\dfrac{L}{R}\right)}$ |

6.28, $\boxed{\times}$, $\boxed{(}$, inductance value, $\boxed{\div}$, resistance value, $\boxed{)}$,
$\boxed{=}$, $\boxed{1/x}$

Resonant Filters

THE BANDPASS FILTER.   Refer to Figure 22–28 and notice a combination of series and parallel resonant circuits are used to pass a desired band of frequencies. The desired output band of frequencies finds little opposition from the series resonant circuits in series with the output path and very high opposition from the parallel resonant circuits. Therefore, the desired band of frequencies appears at the output or load. This tunable filter is tuned with variable Cs or Ls to pass one desired band of frequencies while rejecting all other frequencies. An example of a tunable **bandpass filter** is

FIGURE 22–28   Resonant-type bandpass filter

the single tunable parallel resonant circuit that tunes different radio stations.

THE BANDSTOP FILTER. By placing the series resonant circuit(s) in shunt with the signal path and one or more parallel resonant circuit(s) in series with the signal path, a **bandstop filter** is developed, Figure 22–29. Also, note the frequency response characteristic.

The previous discussions have only introduced you to filters. We have not attempted to investigate filter design formulas (e.g., for *constant-k* or *m-derived filters*), or the wide spectrum of modern filters, such as YIG (yittrium-iron-garnet) filters, multichannel analog filters, and computer-controlled filters.

Perhaps your interest in studying filters was increased by this simple overview. Filter networks have numerous practical uses, from simple power supply filtering to signal processing in such high-technology areas as speech processing and medical instrumentation. As you continue in your studies, you will have the opportunity to study some of the basic applications of filters, coupling networks, and bypassing networks shown in Figure 22–30. The principles discussed in this chapter should help you in the more detailed studies.

FIGURE 22–29   Resonant-type bandstop filter and response characteristic

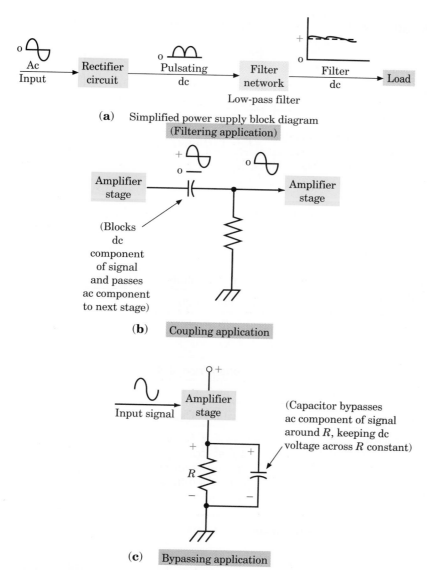

(a)   Simplified power supply block diagram
(Filtering application)

(b)   Coupling application

(c)   Bypassing application

FIGURE 22–30   Sample applications of filters

SUMMARY

▶ X_L and X_C change in opposite directions for a given frequency change. In other words, as f increases, X_L increases and X_C decreases. Also, as f decreases, X_L decreases and X_C increases.

▶ For any LC combination, there is one frequency where X_L and X_C are equal: the resonant frequency or the frequency of resonance.

▶ At resonance, series RLC circuits exhibit the characteristics of: $X_L = X_C$, maximum current, minimum Z (and $Z = R$), θ of 0°, and voltage across each reactance of Q times V applied.

▶ At resonance, parallel RLC circuits exhibit the characteristics of: $X_L = X_C$, minimum current, maximum Z ($Z = Q \times X_L$), $\theta = 0°$, and current within the tank circuit = Q times I line.

▶ See Figure 22–31 for comparison of series and parallel resonant circuits.

	SERIES RESONANCE	PARALLEL RESONANCE
SIMILARITIES	$X_L = X_C$	$X_L = X_C$
	$f_r = \dfrac{1}{2\pi\sqrt{LC}}$	$f_r = \dfrac{1}{2\pi\sqrt{LC}}$
	$\theta = 0°$	$\theta = 0°$ (High Q circuits)
	$p.f. = 1$	$p.f. = 1$ (High Q circuits)
	$f = \dfrac{f_r}{Q}$	$f = \dfrac{f_r}{Q}$
DIFFERENCES	$f_r = \dfrac{1}{2\pi\sqrt{LC}}$ no matter circuit R value	f_r can be affected by R
	$I = $ maximum	I line $=$ minimum
	$Z = $ minimum $(=R)$	$Z = $ maximum
	$Q = \dfrac{X_L}{r_s}$	$Q = \dfrac{R}{X_L}$ (low Q circuits)
	Resonant rise of I	Resonant rise of Z
	Resonant rise of reactive V $(Q_X V_A)$	$(Z = Q \times X_L)$
	Above $f_r = $ Inductive	Above $f_r = $ Capacitive
	Below $f_r = $ Capacitive	Below $f_r = $ Inductive $p.f. = 1$ called "antiresonance" . . . can be a different f than maximum Z point.

FIGURE 22–31   Similarities and differences between series and parallel resonance

▶ The general formula to find the resonant frequency for series LC circuits and for high-Q parallel LC circuits is:

$$f_r = \frac{1}{2\pi\sqrt{LC}}$$

▶ The Q of a resonant circuit is determined by the circuit resistance and the L/C ratio. The more the circuit resistive losses are, the lower the Q is. The higher the L/C ratio is, the higher the Q is for a given f_r.

▶ The higher the Q of a resonant circuit is, the greater the magnification factor is for voltage (series resonant circuit) and impedance (parallel resonant circuit).

▶ The higher the Q of a resonant circuit is, the sharper is its response curve and the more selective it is to a band of frequencies. High Q means narrow bandwidth and narrow bandpass.

▶ Bandwidth is often defined by the frequencies where the response curve for the circuit is above the 70.7% of the maximum level. This is sometimes called the half-power or −3 dB points on the response curve.

▶ The relationship of resonant frequency, Q, and bandwidth (BW) is shown by:

$$BW \ (\Delta f) = \frac{f_r}{Q}, \text{ or } Q = \frac{f_r}{\Delta f}$$

▶ *R, L,* and *C* components can be configured to produce filtering effects. Low-pass filters allow low frequencies to pass to the output, while attenuating high frequencies. High-pass filters pass high frequencies to the output while attenuating the low frequencies.

▶ Generally, to allow a certain frequency spectrum to pass to the output, a low impedance circuit or component is placed in series with the signal path, and/or a high impedance circuit or component is placed in shunt with the signal path.

▶ To prevent a certain frequency spectrum at the output of the filter, a high impedance circuit or component is placed in series with the signal path, and/or a low impedance circuit or component is placed in shunt with the signal path.

▶ Filters are categorized in various ways. Some classifications are physical layout shape (L, pi, T); usage (bypass, decouple, bandpass, bandstop, smoothing); component types (*LC, RC,* active, passive); number of sections they have (single-section, two-section); and the frequency spectrum in which they are used (power supply, audio, rf).

▶ Multiple-section filters generally provide more filtering than single-section filters.

FORMULAS AND CALCULATOR SEQUENCES SUMMARY

FORMULA 22–1   Resonant frequency: $f_r = \dfrac{1}{2\pi\sqrt{LC}}$

Inductance value, $\boxed{\times}$, capacitance value, $\boxed{=}$, $\boxed{\sqrt{x}}$, $\boxed{\times}$, 6.28, $\boxed{=}$, $\boxed{1/x}$

FORMULA 22–2   Inductance value for given f_r and C: $L = \dfrac{0.02533}{f_r^2 C}$

0.02533, $\boxed{\div}$, $\boxed{(}$, resonant frequency value, $\boxed{x^2}$, $\boxed{\times}$, capacitance value, $\boxed{)}$, $\boxed{=}$

FORMULA 22–3   Capacitance value for given f_r and L: $C = \dfrac{0.2533}{f_r^2 L}$

0.02533, $\boxed{\div}$, $\boxed{(}$, resonant frequency value, $\boxed{x^2}$, $\boxed{\times}$, inductance value, $\boxed{)}$, $\boxed{=}$

FORMULA 22–4   Inductance value (in μH) for given f_r and C: $L = \dfrac{25{,}330}{f^2 C}$

25,330, $\boxed{\div}$, $\boxed{(}$, resonant frequency value in MHz, $\boxed{x^2}$, $\boxed{\times}$, capacitance value in pF, $\boxed{)}$, $\boxed{=}$

FORMULA 22–5   Capacitance value (in pF) for given f_r and L: $C = \dfrac{25{,}330}{f^2 L}$

25,330, $\boxed{\div}$, $\boxed{(}$, resonant frequency value in MHz, $\boxed{x^2}$, $\boxed{\times}$, inductance value in μH, $\boxed{)}$, $\boxed{=}$

FORMULA 22–6 $f_r = \dfrac{1}{2\pi\sqrt{LC}} \times \sqrt{1 - \dfrac{Cr_s^{\,2}}{L}}$

FORMULA 22–7   Total bandwidth *(BW)*: $\Delta f = f_2 - f_1$

half-power point (higher frequency), $\boxed{-}$, half-power point (lower frequency), $\boxed{=}$

FORMULA 22–8   Bandwidth *(BW)* related to *Q*: $\Delta f = \dfrac{f_r}{Q}$

resonant frequency value, $\boxed{\div}$, *Q* value, $\boxed{=}$

FORMULA 22–9   *Q* value: $Q = \dfrac{f_r}{\Delta f}$

resonant frequency value, $\boxed{\div}$, bandwidth value, $\boxed{=}$

FORMULA 22–10   Cutoff frequency *(RC* filter): $f_c = \dfrac{1}{2\pi RC}$

6.28, $\boxed{\times}$, resistor value, $\boxed{\times}$ capacitance value, $\boxed{=}$, $\boxed{1/x}$

FORMULA 22–11   Cutoff frequency *(RL* filter): $f_c = \dfrac{1}{2\pi\left(\dfrac{L}{R}\right)}$

6.28, $\boxed{\times}$, $\boxed{(}$, inductance value, $\boxed{\div}$, resistance value, $\boxed{)}$, $\boxed{=}$, $\boxed{1/x}$

REVIEW QUESTIONS

1. State the condition that must exist if a series *RLC* circuit is at resonance.

2. Disregarding the generator or source impedance, a series *LC* circuit is comprised of a 100-pF capacitor and a 1-μH inductor having 10 Ω of resistance.
   a. What is the impedance at resonance?
   b. What is the resonant frequency?
   c. What is the *Q?*
   d. What is the bandwidth?
   e. What is the bandpass?
   f. Below the resonant frequency, does the circuit act inductively or capacitively?

3. The voltage applied to the circuit in Question 2 is 20 volts.
   a. What is the voltage across the capacitor at resonance?
   b. What is the current at resonance?
   c. What is the true power at resonance?
   d. What is the net reactive power?

4. A parallel *LC* circuit has a capacitive branch with a 30-pF capacitor and an inductive branch with an inductor having 28 μH and a r_s of 30 Ω.
   a. What is the f_r?

b. What is the Q? (Assume a Q greater than 10.)
c. What is the bandwidth?
d. What is the bandpass?
e. At resonance, what is the Z?
f. Below resonance, does the circuit act inductively or capacitively?

5. If the voltage applied to the circuit in Question 4 is 20 volts, what is the line current? What is the current through the capacitor branch?

6. Draw a pi-type LC filter that can filter a power supply output. Is this a low-pass or a high-pass filter?

7. Draw a circuit showing a series resonant and a parallel resonant circuit connected so the combination passes a band of frequencies close to the resonant frequencies of the circuits.

8. A series RLC circuit consists of a 100-Ω resistor, a 75-H inductor, and a 300-pF capacitor. What is the resonant frequency?

9. A series RLC circuit has a resonant frequency of 40 kHz and the Q is 20. At what frequency will the phase angle be 45°?

10. A series RLC circuit is resonant at 150 kHz, and has $R = 100\ \Omega$ and $C = 25$ pF. What is the inductance value?

11. A series RLC circuit is resonant at 21 kHz and has $R = 40\ \Omega$ and inductance $= 12$ H. What is the C value?

12. Given a series RLC circuit that is resonant at 1.5 MHz, and knowing that C has a reactance of 500 Ω, what is the R value for the circuit to have a bandwidth of 10 kHz?

13. A series RLC circuit is comprised of a 200-Ω resistor, a 36-mH inductor, and a capacitor. What is the voltage across the R at resonance if V applied is 100 V?

14. What is the resonant frequency of a parallel LC combination having a 200-μH inductor and a 0.0002-μF capacitor?

15. If the frequency of the source in Question 14 is adjusted to a new frequency above the resonance frequency, will I_T lead or lag V_A?

16. A 220-μH inductor has a distributed capacitance of 25 pF and an effective resistance of 80 Ω. What impedance will it exhibit at resonant frequency?

17. If a parallel LC circuit is acting inductively at a frequency of 150 kHz, to tune the circuit to resonance requires which of the following?
a. Increasing X_C
b. Decreasing X_L
c. Decreasing L
d. Increasing C

CHALLENGE QUESTIONS

Perform the required research to determine the answers to the following questions.

1. Is the tuned circuit at the input of a receiver typically a series or parallel LC circuit?

2. What is meant by a "double-tuned" transformer, as used between certain stages of radio or TV receivers?

3. Would the circuits described in Questions 1 and 2 be considered resonant or nonresonant filters?

4. Match the frequency response curves, shown in the right column, to the types of filters, listed in the left column. Place the letter identifying each response curve in the blank identifying the filter or circuit type that would have that type response.

_____ Parallel resonant
_____ Low-pass
_____ Bandpass
_____ High-pass
_____ Bandstop
_____ Series resonant

(a)

(b)

(c)

(d)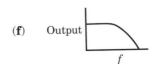

(e)

(f)

5. In your own words, define the meaning of the terms selectivity, bandwidth, and bandpass.

TEST YOURSELF

Answer the following questions with "I" for increase, "D" for decrease, and "RTS" for remain the same.

1. For a series resonant RLC circuit, if frequency is slightly increased:
   a. X_L will _____   d. Z will _____
   b. X_C will _____   e. I will _____
   c. R will _____   f. θ will _____

g. V_C will _____

h. V_L will _____

i. V_R will _____

j. V_T will _____

k. BW will _____

l. Q will _____

m. I_C will _____

n. I_L will _____

o. I_T will _____

2. For a series resonant *RLC* circuit, if frequency is slightly decreased:

a. X_L will _____

b. X_C will _____

c. R will _____

d. Z will _____

e. I will _____

f. θ will _____

g. V_C will _____

h. V_L will _____

i. V_R will _____

j. V_T will _____

k. BW will _____

l. Q will _____

m. I_C will _____

n. I_L will _____

o. I_T will _____

3. For a parallel resonant *RLC* circuit, if frequency is slightly increased:

a. X_L will_____

b. X_C will_____

c. R will_____

d. Z will_____

e. I will_____

f. θ will_____

g. V_C will_____

h. V_L will_____

i. V_R will_____

j. V_T will_____

k. BW will_____

l. Q will_____

m. I_C will_____

n. I_L will_____

o. I_T will_____

4. For a parallel resonant *RLC* circuit, if frequency is decreased:

a. X_L will _____

b. X_C will _____

c. R will _____

d. Z will _____

e. I will _____

f. θ will _____

g. V_C will _____

h. V_L will _____

i. V_R will _____

j. V_T will _____

k. BW will _____

l. Q will _____

m. I_C will _____

n. I_L will _____

o. I_T will _____

5. Research the terms *m*-derived and constant-*k* filters and write the definition of each.

CHAPTER TROUBLESHOOTING CHALLENGE

Following the SIMPLER troubleshooting sequence, find the problem with this circuit. As you follow the sequence, record your responses to each step on a separate sheet of paper.

CHALLENGE CIRCUIT 12

SYMPTOMS Gather, verify, and analyze symptom information. (Look at the "Starting Point Information.")

IDENTIFY Identify initial suspect area for the location of the trouble.

MAKE Make a decision about "What type of test to make" and "Where to make it." (To simulate making a decision about each test you want to make, select the desired test from the "TEST" column listing below.)

PERFORM Perform the test. (Look up the result of the test in Appendix C. Use the number in the "RESULTS" column to find the test result in the Appendix.)

LOCATE Locate and define a new "narrower" area in which to continue troubleshooting.

EXAMINE Examine available information and again determine "what type of test" and "where."

REPEAT Repeat the preceding analysis and testing steps until the trouble is found. What would you do to restore this circuit to normal operation? When you have solved the problem, compare your results with those shown on the following pages.

STARTING POINT INFORMATION

1. Circuit diagram
2. V_R is lower than it should be when C is adjusted so that the LC resonance is below the 318 kHz source signal frequency.

TEST	Results in Appendix C

V_R (when C set at the position for
318 kHz LC resonance) . . . .(55)
V_R (when C set at a position for LC
resonance above 318 kHz) . . . .(85)
V_R (when C set at a position for LC
resonance below 318 kHz) . . . .(30)

R_C (C set at a 318 kHz position). . . . .(58)
R_C (C set above 318 kHz position). . . . .(100)
R_C (C set below 318 kHz position). . . . .(72)
NOTE: R_C is resistance measured from stator-to-rotor plates across the capacitor.

Source lead   Scope lead

Scope lead

Symptoms When the variable C is adjusted so the LC circuit is resonant at about 318 kHz source signal (i.e., C is about 100 pF), the scope shows maximum deflection due to maximum I_R drop at series resonance (Figure 1). When the variable C is adjusted so the LC circuit resonance is slightly higher in frequency (plates more unmeshed), the V_R (and scope deflection) slightly decrease as would be expected (Figure 2). When the capacitor is adjusted so the LC resonance is slightly lower (plates more meshed), the V_R and scope deflection drop drastically (Figure 3). This indicates some drastic change in the circuit. The off-resonance impedance must have increased at an abnormal rate for this circuit.

Identify initial suspect area: Since the only item being varied is the capacitance (because of our tuning efforts), the variable tuning capacitor is the initial suspect area.

Make test decision: We can check the tuning capacitor's operation with an ohmmeter to make sure that it isn't shorting at some point in its tuning range. The ohmmeter check is an easy check to find any possible shorts.

Perform 1st Test: Perform a resistance check throughout the tuning range. As the plates approach being more fully meshed a short between the stator and rotor plates becomes evident. This appears to be our problem. (NOTE: No Figure shown for this test.)

Locate new suspect area: There is no new area because the variable capacitor has proven to be the likely culprit.

Examine available data.

2.5 mH
L

318 kHz

R
10 Ω

C

50–275 pF

CHALLENGE CIRCUIT 12

Repeat analysis and testing: Let's visually inspect the capacitor to see if we can see some shorted plates.

Inspect the capacitor with strong light. The result is you see two plates that are rubbing as the plates begin to more fully mesh. The problem is shorting plates on the tuning capacitor. (NOTE: No Figure shown for this test.)

Bend the plates (Figure 4) to eliminate the rubbing. Then, check with the ohmmeter again, while turning the rotor plates through their range. The result is everything is normal (Figure 5).

Symptoms (2)   Figure 2

Symptoms (3)   Figure 3

3rd Test (1)   Figure 4

3rd Test (2)   Figure 5

CHAPTER TROUBLESHOOTING CHALLENGE

Following the SIMPLER troubleshooting sequence, find the problem with this circuit. As you follow the sequence, record your responses to each step on a separate sheet of paper.

CHALLENGE CIRCUIT 13

SYMPTOMS Gather, verify, and analyze symptom information. (Look at the "Starting Point Information.")

IDENTIFY Identify initial suspect area for the location of the trouble.

MAKE Make a decision about "What type of test to make" and "Where to make it." (To simulate making a decision about each test you want to make, select the desired test from the "TEST" column listing below.)

PERFORM Perform the test. (Look up the result of the test in Appendix C. Use the number in the "RESULTS" column to find the test result in the Appendix.)

LOCATE Locate and define a new "narrower" area in which to continue troubleshooting.

EXAMINE Examine available information and again determine "what type of test" and "where."

REPEAT Repeat the preceding analysis and testing steps until the trouble is found. What would you do to restore this circuit to normal operation? When you have solved the problem, compare your results with those shown on the following pages.

STARTING POINT INFORMATION

1. Circuit diagram
2. Using the scope to monitor when V across the 1-kΩ R is minimum (indicating that the parallel LC impedance is maximum), we find that the range of frequencies tuned from maximum C value to minimum C value is slightly higher in frequency than should be expected from the rated values shown on the diagram.

TEST	Results in Appendix C
$R_{inductor}$	(87)
$R_{capacitor}$	(19)
C of capacitor when "fully meshed"	(60)
C of capacitor when "fully unmeshed"	(107)
L of inductor	(41)

CIRCUIT

13

Source lead

Source lead

Scope lead   Scope lead

Symptoms One would normally expect the tuning range of the *LC* circuit to be from about 190 kHz to 450 kHz as the *C* is varied between 50 and 275 μμF. It is actually tuning from about 220 kHz (Figure 1) to 580 kHz (Figure 2).

Identify initial suspect area: Since the frequencies are higher than expected, it looks like the *LC* product must be less than the rated values. Either the capacitor has decreased by losing plates (which is highly unlikely) or the inductor has decreased in value for some reason. We can't be sure at this point. The inductor is the initial suspect. However, we'll keep both the *C* and *L* in the initial suspect area.

Make test decision: What could make the inductor have less inductance than anticipated? Possibly, shorted turns which would make *L* decrease. Since the rated dc resistance of this coil is about 7 Ω, an ohmmeter check (with the inductor isolated from the rest of the circuit) might reveal if enough turns are shorted to make a difference in coil resistance.

Perform 1st Test: With *L* isolated and source removed, measure its resistance (Figure 3). *R* measures about 5 Ω. Even this little resistance change could be significant, since we're dealing with such a small rated resistance in the first place. A number of turns could be shorted.

Locate new suspect area: The hot zone is still the inductor. However, we need to make another inductor check so we can exclude the capacitor.

r_S
7 Ω

50–275 pF *C*

L
2.5 mH

R 1 kΩ

CHALLENGE CIRCUIT 13

Examine available data.

Repeat analysis and testing: Since the coil looks suspicious, let's substitute a known good coil of like ratings and see how the circuit operates.

Substitute an inductor of the same ratings. Then check the circuit tuning characteristics. The result is the component substitution worked. The circuit tunes with the expected range of frequencies, 190 kHz (Figure 4) to 450 kHz (Figure 5). The trouble was shorted turns on the inductor.

Symptoms (2)
Tuned above resonance

Figure 2

1st Test
Tuned below resonance

Figure 3

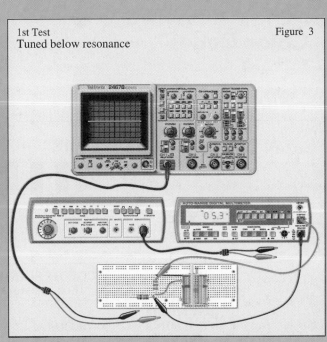

2nd Test (1)

Figure 4

2nd Test (2)
Checking operation
through tuning range

Figure 5

A. RESISTOR COLOR CODE

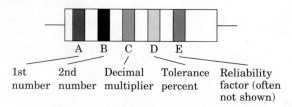

| | A | B | C | D | E |
| | 1st number | 2nd number | Decimal multiplier | Tolerance percent | Reliability factor (often not shown) |

COLOR SIGNIFICANCE CHART
(For resistors and capacitors)

COLOR	NUMBER COLOR REPRESENTS	DECIMAL MULTIPLIER	TOLERANCE PERCENT	VOLTAGE RATING	% CHANGE PER 1,000 HRS OPER.
Black	0	1	—	—	—
Brown	1	10	1*	100*	1%
Red	2	100	2*	200*	0.1%
Orange	3	1,000	3*	300*	0.01%
Yellow	4	10,000	4*	400*	0.001%
Green	5	100,000	5*	500*	—
Blue	6	1,000,000	6*	600*	—
Violet	7	10,000,000	7*	700*	—
Gray	8	100,000,000	8*	800*	—
White	9	1,000,000,000	9*	900*	—
Gold	—	0.1	5	1,000*	—
Silver	—	0.01	10	2,000*	—
No Color	—	—	20	500*	—

(*Applicable to capacitors only)

B. CAPACITOR COLOR CODES

5th and 6th band = *V* rating (multiply numbers by 100)

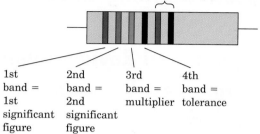

1st
band =
1st
significant
figure

2nd
band =
2nd
significant
figure

3rd
band =
multiplier

4th
band =
tolerance

See Color Significance Chart for meaning of colors

Band color coding system for tubular ceramic capacitors

1st
significant
figure

2nd
significant
figure

Multiplier

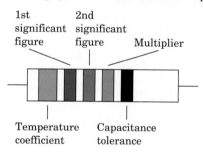

Temperature
coefficient

Capacitance
tolerance

C. SEMICONDUCTOR DIODE COLOR CODE(S)

(NOTE: Prefix "1N . . . " is understood.)

Double-width band = cathode

(Example: 1N400B)

(Example: 1N222)

1st
figure
yellow

2nd
figure
black

3rd
figure
black

Suffix
letter
red

1st
figure
red

2nd
figure
red

3rd
figure
red

(Bands grouped near cathode end)

Suffix colors
Brown = A
Red   = B
Orange = C
Yellow = D
Green   = E
Blue   = F

APPENDIX B

ANSWERS TO In-Process Learning Checks and Practice Problems
ANSWERS TO Odd-Numbered Review Questions, Challenge Questions, and Test Yourself

CHAPTER 1

✓ IN-PROCESS LEARNING CHECKS

I

1. Matter is anything that has **weight** and occupies **space.**

2. Three physical states of matter are **solid, liquid,** and **gas.**

3. Three chemical states of matter are **elements, compounds,** and **mixtures.**

4. The smallest particle that a compound can be divided into and retain its physical properties is the **molecule.**

5. The smallest particle that an element can be divided into but retain its physical properties is the **atom.**

6. The three parts of an atom that interest electronics students are the **electron, proton,** and **neutron.**

7. The atomic particle having a negative charge is the **electron.**

8. The atomic particle having a positive charge is the **proton.**

9. The atomic particle having a neutral charge is the **neutron.**

10. The **protons** and **neutrons** are found in the atom's nucleus.

11. The particle that orbits the nucleus of the atom is the **electron.**

II

1. An electrical system (circuit) consists of a source, a way to transport the electrical energy, and a **load.**

2. Static electricity is usually associated with **non-conductor-type** materials.

3. The basic electrical law is that **unlike** charges attract each other and **like** charges repel each other.

4. A positive sign or a negative sign often shows electrical **polarity.**

5. If two quantities are *directly* related, as one increases the other will **increase.**

6. If two quantities are *inversely* related, as one increases the other will **decrease.**

7. The unit of charge is the **coulomb.**

8. The unit of current is the **ampere.**

9. An ampere is an electron flow of one **coulomb** per second.

10. If two points have different electrical charge levels, there is a difference of **potential** between them.

11. The volt is the unit of **electromotive force,** or **potential** difference.

REVIEW QUESTIONS

1. Matter is anything that has weight and occupies space whether it is a solid, liquid, or gas.

3. Three chemical states of matter are elements, compounds, and mixtures. Copper is an element, sugar is a compound, and sand and gold dust are a mixture.

5. A compound is a form of matter that can be chemically divided into simpler substances and that has two or more types of atoms. Examples are water (hydrogen and oxygen) and sugar (carbon, hydrogen, and oxygen).

7. I.D. of charges on particles. See page 10.

9. Free electrons are electrons in the atomic structure that can be easily moved or removed from their original atom. These are outer-ring or valence electrons in conductor materials.

11. Chemical, mechanical, light, and heat.

13. Unlike charges attract and like charges repel each other.

15. Force of attraction or repulsion will be one-ninth the original force.

17. Unit of measure of electromotive force (emf) is the volt.

19. a. Unit of measure for current is the ampere.
    b. 4 amperes of current flow

21. A closed circuit has an unbroken path for current flow. An open circuit has discontinuity or a broken path; therefore, the circuit does not provide a continuous path for current.

TEST YOURSELF

1. Drawing:

1st shell = 2
2nd shell = 8
3rd shell = 18
4th shell = 2
Total electrons = 30

3. Zinc

5. 30

7. Circuit diagram:

Source / Load / Conductor wires

CHAPTER 2

PRACTICE PROBLEMS

I
The circular-mil cross-sectional area = $2.5^2 = 6.25$ CM.

II
The resistance of 300 feet of copper wire having a cross-sectional area of 2,048 CM = $1.523 \, \Omega$.

III
1. 500 feet of #12 wire has a resistance of approximately $0.81 \, \Omega$.

2. The wire gage of wire having a cross-sectional area of 10,380 CM is #10 wire.

3. #10 wire has an approximate resistance of $1.018 \, \Omega$ per thousand feet of length.

IV
1. 39 kΩ, 10% tolerance.
2. 18 kΩ, 5%, and 0.1%.

V
1. It is a TV receiver system.
2. The signal flows from left-to-right, as illustrated.
3. The receiver circuitry receives signals from the TV antenna.
4. Two types of outputs are picture information and audio sound information.

VI
1. There are 3 kinds of components (resistor, switch, battery).
2. The highest value resistor is 100 kΩ.
3. A SPST switch is used.
4. The voltage source is a battery.

VII
1. There are 5 resistors.
2. Resistor R_3. The lowest R value resistor is 1 kΩ.
3. There are 2 switches in the circuit.
4. The circuit applied voltage is 100 V.
5. S_2 is opened, and S_1 is closed.

✔ IN-PROCESS LEARNING CHECKS

I
1. Charge is represented by the letter **Q.** The unit of measure is the **coulomb,** and the abbreviation is **C.**
2. The unit of potential difference is the **volt.** The symbol is **V.**
3. The abbreviation for current is **I.** The unit of measure for current is the **ampere,** and the symbol is **A.**
4. The abbreviation for resistance is **R.** The unit of measure for resistance is the **ohm,** and the symbol **the Greek letter omega, Ω.**
5. Conductance is the **ease** with which current flows through a component or circuit. The abbreviation is **G.** The unit of measure for conductance is the **siemens,** and the symbol is **S.**
6. 0.0000022 amperes = 2.2 microamperes (μA). Could be expressed as 2.2×10^{-6}A.
7. One-thousandth is represented by the metric prefix **milli.** When expressed as a power of ten, it is **10^{-3}.**
8. 10,000 ohms might be expressed as **10 kΩ** (kilohms). Expressed as a whole number times a power of ten, **10×10^3 ohms.**

II
1. The greater the resistivity of a given conductor, the **higher** its resistance will be per given length.

2. The circular-mil area of a conductor is equal to the square of its **diameter** in **mils (thousandths of an inch).**

3. The smaller the diameter of a conductor, the **higher** its resistance will be per given length.

4. The higher the temperature in which a conductor must operate, the **higher** will be its resistance.

5. Wire tables show the AWG wire **size,** the cross-sectional area of wires in **circular-mils,** the resistance per given length of wire, and the operating **temperature** at which these parameters hold true.

III

1. Negative lead connected to **Point B;** positive lead connected to **Point A.**

2. Negative lead connected to **Point B,** positive lead connected to **Point C.**

3. The first thing to do is to remove the source voltage from the circuit.

REVIEW QUESTIONS

1. 0.003 V

3. A micro unit is one-thousandth of a milli unit.

5. Schematic diagram of battery, two resistors, and a SPST switch:

7. Resistor

9. Battery

11. Carbon or composition resistors and wirewound resistors. Carbon/composition resistors are composed of carbon or graphite mixed with powdered insulating material. Wirewound resistors are constructed with special resistance wire wrapped around a ceramic or insulator core.

13. Value, power rating, tolerance, and physical size

15. a. Avoid parts of body coming in contact with devices, hot solder, rotating machinery, and so on.
    b. Don't wear jewelry that might touch circuits or get caught in devices.
    c. Wear safety glasses (when soldering or working with rotating machinery).
    d. Be sure power is off circuitry, as appropriate. Insulate your body from ground.
    e. Avoid parts of body touching electrical contacts or wires.

CHALLENGE QUESTIONS

1. The resistor color code is necessary because many resistors are too small to enable printing their values on them. Also, color code provides a "universal" method, readable by anyone in the world, regardless of their language.

3. Generally, reliability refers to the ability of the component to perform as it is expected to, for the length of time intended, under normal operating conditions. Resistor reliability ratings typically indicate the percentage of failure for 1,000 hours of operation.

TEST YOURSELF

1. Block diagram of a simple power distribution system:

3. a. Brown, black, red, silver
   b. Red, violet, orange
   c. Brown, black, green, silver
   d. Brown, black, black, gold

CHAPTER 3

PRACTICE PROBLEMS

I

1. $I = V/R = 150$ V$/10$ $\Omega = 15$ A

2. 7.5 A

3. a. Remain the same
   b. 300 V$/20$ $\Omega = 15$ A

II

1. 2.5×10^4

2. 1.2335×10^1

3. 1.0×10^{-1}

4. 1.5×10^{-3}

5. 23,500

6. 1,000

7. 10,000

8. 0.015

III

1. 2.34×10^{-3}
2. 1.255×10^1
3. 1.2455×10^3

IV

1. 1,456,000
2. 0.0333
3. 1.5
4. 1,000,000

V

1. P = energy/time = 25 J/2 sec = 12.5 W
2. P = energy/time = 1 J/100 sec = 0.01 W or 10 mW

VI

1. Energy = Power × Time = 250 W × 8 hrs = 2,000 Wh, or 2 kWh

VII

1. $P = V \times I$ = 150 V × 10 mA = 1,500 mW, or 1.5 W
2. a. I would decrease to half the original value.
   b. Power diss. = $V \times I$ = 150 × 5 mA = 750 mW, or 0.75 W

VIII

1. $V_A = I \times R$ = 1.5 mA × 4.7 kΩ = 7.05 V
2. $I = V/R$ = 100 V/27 kΩ = 3.704 mA
3. $R = V/I$ = 100 V/19.6 mA = 5.102 kΩ

IX

1. P = 40 W; I = 2 A
2. I = 1 mA; R = 100 kΩ
3. I = 12 mA; P = 1.44 W (or 1,440 mW)

✔ IN-PROCESS LEARNING CHECKS

I

1. I = **1.5 A**, or **1,500 mA**. ($I = V/R$; I = 15 V/10 Ω = 1.5 A)
2. V = **30 V**. ($V = I \times R$; V = 2 mA × 15 kΩ = 30 V)
3. R = **27 kΩ**. ($R = V/I$; R = 135 V/5 mA = 27 kΩ)
4. a. Current would decrease.
   b. Voltage applied would be higher.
   c. Circuit resistance must have halved.

II

1. negative, positive
2. positive, negative
3. electron
4. one, changes
5. expenditure, work, work, energy
6. joules, seconds
7. kilowatt hour

1. a. $I = V/R$; $V = I \times R$; $R = V/I$
   b. $P = V \times I$; $P = I^2R$; $P = V^2/R$
3. a. 5.1×10^3
   b. 47×10^3
   c. 0.000001×10^3
   d. 0.039×10^3
5. Power will be nine times the original power.
7. a. Work = expenditure of energy. Energy = ability to do work. Power = the rate of using energy or performing work.
   b. Work is measured in foot poundals, or foot pounds, or horsepower. Energy is measured in joules. Power (electrical) is measured in Watts.
9. $V = I \times R$ = 3 mA × 12 kΩ = 36 V
11. $R = V/I$ = 41 V/50 mA = 820 Ω
13. I = 3.162 A
15. $R = V/I$ = 100 V/8.5 mA = 11.765 kΩ

CHALLENGE QUESTIONS

1. Since 1 hp = 746 W, then 3.5 hp = 3.5 × 746 = 2,611 W.
3. If power has increased to 8 times original power while R has decreased to half value, the voltage must have doubled.

TEST YOURSELF

1. a. Schematic diagram.

   b. $P = I^2R$; $I^2 = P/R$; $I = \sqrt{P/R}$; $I = \sqrt{1,000/10} = \sqrt{100}$ = 10 A
   c. $V_A = I \times R$ = 10 A × 10 Ω = 100 V
   d. See diagram.

CHAPTER 4

PRACTICE PROBLEMS

I

1. Total resistance = 74 kΩ
2. Total resistance = 16 kΩ

II

1. R_1 = 47 Ω; R_T = 100 Ω
2. New R_T = 10.5 kΩ
3. R_2 = 3.9 kΩ

III

1. $V_A = 411$ V; $I_T = 3$ mA; $V_{R_2} = 81$ V; $V_{R_1} = 300$ V
2. $V_A = 342.5$ V; $I_T = 2.5$ mA; $V_{R_2} = 67.5$ V; $V_{R_1} = 250$ V

IV

1. $V_X = (R_X/R_T) \times V_T = (2.7$ kΩ$/17.4$ kΩ$) \times 50$ V $= 0.155 \times 50 = 7.75$ V (V_{R_1})
2. $(10$ kΩ$/17.4$ kΩ$) \times 50$ V $= 0.575 \times 50 = 28.75$ (V_{R_3})
3. $7.75 + 13.5 + 28.75 = 50$ V (Yes, they add up correctly.)

V

$V_2 = 30$ V; $V_A = 75$ V

VI

$V_T = 40$ V
$P_3 = 40$ mW
$R_1 = 5$ kΩ
$R_2 = 5$ kΩ
$R_3 = 10$ kΩ

VII

$R_T = 10$ kΩ $+ 47$ kΩ $+ 39$ kΩ $= 96$ kΩ
$I_T = V_A/R_T = 240$ V$/96$ kΩ $= 2.5$ mA
$V_{R_2} = I \times R_2 = 2.5$ mA $\times 47$ kΩ $= 117.5$ V

VIII

$R_T = 15$ kΩ $+ 2.7$ kΩ $+ 27$ kΩ $+ 1.5$ kΩ $= 46.2$ kΩ
$I_T = I_{R_1} = 10$ V$/15$ kΩ $= 0.67$ mA
$V_T = I_T \times R_T = 0.67$ mA $\times 46.2$ kΩ $= 30.95$ V

IX

$R_T = 10$ kΩ $+ 1$ kΩ $+ 10$ kΩ $+ 1$ kΩ $+ 3.9$ kΩ $= 25.9$ kΩ
$I_T = I_{R_1} = 5$ V$/10$ kΩ $= 0.5$ mA
$V_T = I_T \times R_T = 0.5$ mA $\times 25.9$ kΩ $= 12.95$ V

X

$P_1 = 90$ W; $P_2 = 90$ W; $P_3 = 90$ W; $P_T = 270$ W;
$I_T = 3$ A;
$V_1 = 30$ V; $V_2 = 30$ V; $V_3 = 30$ V; $R_3 = 33.3\%$ of R_T;
$P_3 = 33.3\%$ of P_T

XI

1. a. increase; b. decrease; c. decrease; d. increase; e. decrease
2. Yes

XII

1. With the 47-kΩ resistor shorted the voltage drops are:
   10-kΩ resistor drops 14.6 V
   27-kΩ resistor drops 39.4 V
   47-kΩ resistor drops 0 V
   100-kΩ resistor drops 146 V
2. They increase.

XIII

Givens: $R_1 = 20$ kΩ; $V_1 = \dfrac{2}{5} V_A$; $V_2 = \dfrac{1.5}{5} V_A$;
$V_A = 50$ V; $R_2 = 15$ kΩ; $R_3 = 15$ kΩ;
$V_2 = 15$ V; $V_3 = 15$ V; $P_T = 50$ mW; $P_1 = 20$ mW; $P_2 = 15$ mW; $P_3 = 15$ mW; $I = 1$ mA

XIV

1. $V_{R_1} = 4.5$ V
2. $V_{R_2} = 1.5$ V

XV

D to C = 94 V; C to B = 54 V; B to A = 40 V; D to B = 148 V

XVI

1. Voltage across $R_2 = 27.49$ V (or approximately 27.5 V).
2. Voltage at top of R_2 is $+27.5$ V with respect to point C.
3. Voltage across $R_1 = 9.995$ V (or approximately 10 V).
4. Voltage at point C with respect to point A = -37.5 V.

XVII

1. Minimum power rating should be 2 times (60 V × 30 mA), or $2 \times 1,800$ mW $= 3,600$ mW (3.6 W).
2. R dropping equals $\dfrac{110 \text{ V}}{30 \text{ mA}} = 3.66$ kΩ. Minimum power rating should be 2 times (110 V × 30 mA), or $2 \times 3,300$ mW $= 6,600$ mW (6.6 W).

✓ IN-PROCESS LEARNING CHECKS

I

1. The primary identifying characteristic of a series circuit is that the **current** is the same throughout the circuit.
2. The total resistance in a series circuit must be greater than any one **resistance** in the circuit.
3. In a series circuit, the highest value voltage is dropped by the **largest or highest** value resistance, and the lowest value voltage is dropped by the **smallest or lowest** value resistance.

4. *V* drop by the other resistor must be ***100 V.***

5. Total resistance ***increases.*** Total current ***decreases.*** Adjacent resistor's voltage drop ***decreases.***

6. Applied voltage is ***130 V.***

REVIEW QUESTIONS

1. $R_T = \dfrac{10\ V}{5\ A} = 2\ \Omega$; each resistor = 1 Ω

3. Diagram, labeling, and calculations for a series circuit.

a. $R_T = 50 + 40 + 30 + 20 = 140\ \Omega$
b. V applied = $I \times R$ = 2 A × 140 Ω = 280 V
c. $V_1 = 100$ V
   $V_2 = 80$ V
   $V_3 = 60$ V
   $V_4 = 40$ V
d. $P_T = I_T \times V_T$ = 2 A × 280 V = 560 W
e. $P_2 = 2$ A × 80 V = 160 W
   $P_4 = 2$ A × 40 V = 80 W
f. one-seventh V_A dropped by R_4
g. (1) Total resistance increases.
   (2) Total current decreases.
   (3) V_1 decreases; V_2 decreases; V_4 decreases.
   (4) P_T decreases since I decreases and V remains the same.
h. (1) Total resistance decreases.
   (2) Total current increases.
   (3) V_1 increases; V_3 increases; V_4 increases (due to increased I with Rs remaining the same)

5. Diagram of three sources to acquire 60 V, if sources are 100 V, 40 V, and 120 V.

7. Use power-off testing. This prevents damage to the power supply and other components (due to abnormally high currents) and allows measuring resistances with an ohmmeter to troubleshoot.

9. d.

11. Diagram of a three-resistor voltage divider.

13. 1.435 mA (approximately 1.44 mA)

15. $V_{R_3} = 6.74$ V

17. With R_3 shorted, voltmeter reading = 22.2 V

19. Percentage would stay the same, since R is the same percentage of R_T as before.

21. V applied = 125 V

23. R_1 is set at 3.81 kΩ
   a. $I_T = 4.2$ mA
   b. R_1 dissipates 16% of total power.
   c. $I_T = 4$ mA with R_1 set at middle of its range
   d. voltmeter would indicate 50 V.

25. a. $P_{R_2} = 423$ mW
   b. $P_T = 819$ mW
   c. $I_T = 3$ mA

CHALLENGE QUESTIONS

1. The difference between scientific and engineering notation is that scientific notation always uses a power of 10 that keeps the number between 1 and 10 times the power of 10. Engineering notation, on the other hand, uses numbers times a power of ten that has a multiple or submultiple of the exponent 3. This is because metric units are typically expressed in values which take advantage of this. For example, kilo = 1,000 (10^3), milli = 0.001 (10^{-3}), and so on.

3. a. Current is the same throughout all components in a series circuit.
   b. The largest value R drops the greatest voltage; the smallest value R drops the least voltage in a series circuit.
   c. The largest R in series dissipates the most power; the smallest value R in series dissipates the least power.

d. If a component in a series opens, current through all components drops to zero; however, applied voltage will appear across the opened component.
e. If a component in a series circuit shorts, current through all components will increase; however, voltage across the shorted component will decrease to zero.
f. Two means of gathering symptom information regarding a malfunctioning circuit are by observation (using the five senses), observing built-in system meters, monitoring output indicators, such as picture tubes or speakers, talking with someone who observed the problem, and so on.
g. The divide-and-conquer rule indicates that the first test in a series of components (where any one of them could be at fault) is to be made at, or just before, the middle of the questionable area. This eliminates half the number of components in the questionable area with just one check.

TEST YOURSELF

1. Circuit diagram where $V_3 = 3\ V_1$; $V_2 = 2\ V_1$; and $V_A = 60$ V.

a. $R_1 = 20$ kΩ
b. $V_1 = 10$ V
   $V_2 = 20$ V
   $V_3 = 30$ V
c. $I = 0.5$ mA; $P_T = 30$ mW

CHAPTER 5

PRACTICE PROBLEMS

I
$V_{R_1} = V_{R_2} = 2$ mA × 27 kΩ = 54 V
$V_A = V_{R_1} = V_{R_2} = 54$ V

II
$I_T = I_1 + I_2 + I_3 + I_4 + I_5$
The lowest R value branch passes the most current.
The highest R value branch passes the least current.

III
$V_2 = 50$ V; $V_1 = 50$ V; $V_A = 50$ V; $I_1 = 5$ A; $I_T = 6$ mA

IV
1. $R_1 = 15$ kΩ; $I_2 = 2.5$ mA; $I_3 = 2.5$ mA; $R_2 = 30$ kΩ
2. $R_1 = 27$ kΩ; $I_2 = 3.78$ mA; $I_3 = 2.23$ mA; $R_2 = 33$ kΩ; $R_T = 11.75$ kΩ.

V
1. $V_{R_1} = 30$ V; $V_T = 30$ V
2. $R_T = 1.67$ kΩ

VI
1. $R_T = 5.35$ kΩ
2. $R_T = 0.76$ kΩ (or 760 Ω)

VII
Total resistance = 25 Ω

VIII
1. $R_T = 12$ Ω
2. $R_T = 7.67$ kΩ

IX
$R_T = 27.15$ Ω

X
$R_2 = 100$ Ω

XI
$P_T = 2,250$ mW or 2.25 W

XII

XIII
1. $I_1 = 1.85$ mA (through 27-kΩ resistor)
   $I_2 = 1.07$ mA (through 47-kΩ resistor)
   $I_3 = 0.50$ mA (through 100-kΩ resistor)
2. $R_T = 1.54$ Ω
   $I_1 = 0.616$ A (through 10-Ω resistor)
   $I_2 = 1.1$ A (through 5.6-Ω resistor)
   $I_3 = 2.28$ A (through 2.7-Ω resistor)

XIV
1. $I_1 = 1.65$ mA
   $I_2 = 0.35$ mA
2. $I_1 = 16.4$ mA
   $I_2 = 3.6$ mA

REVIEW QUESTIONS
1. $R_T = 17.15$ kΩ; product-over-the-sum method
3. $I_2 = 30$ mA; Kirchhoff's current law

5. a. I_1 will RTS
      I_2 will D
      I_3 will RTS
      I_T will D
      P_T will D
      V_1 will RTS
      V_T will RTS

   b. I_1 will D
      I_2 will D
      I_3 will I
      I_T will I

7. I_1 = 12 mA
   I_2 = 6 mA (since I branch is inverse to R branch).

9. P_{R_1} = 0.5 A × 50 V = 25 W

11. a. M_1 = 2.13 mA
    b. M_2 = 37 mA
    c. M_3 = 39.13 mA
    d. M_4 = 10 mA
    e. M_5 = 49.13 mA
    f. P_{R_1} = 213 mW
    g. P_{R_2} = 3.7 W
    h. P_{R_3} = 1 W
    i. P_T = 4.91 W
    j. New P_T = 1.21 W

13. R_X = 31.59 kΩ

15. a. R_4
    b. P_{R_4} = 71.7 mW

17. V_T = 18 V

19. I_{R_2} = 5.45 mA

21. R_1 dissipates the most power.

23. I_T would be 34 mA if R_2 were changed to a 1.5-kΩ resistor value.

25. Power dissipated by R_3 would not change as long as V_T and R_3 were the same values.

CHALLENGE QUESTIONS

1. The purpose of performing "mental approximations" is to provide a rapid check as to whether you might have made a mistake in solving a problem. If your approximation comes out completely different from your other calculations, chances are you have a simple mathematical error that needs correcting.

3. a. 4.864864865 03
   b. 4.864864865 03

TEST YOURSELF

1. Circuit diagram of a three-branch parallel circuit.

CHAPTER 6

PRACTICE PROBLEMS

I

1. The only component carrying total current is R_7.
2. The only single components in parallel with each other are R_2 and R_3.
3. The only single components that pass the same current are R_1, R_4, and R_5.

II

1. R_T for Figure 6–11 = 12.1 kΩ
2. R_T for Figure 6–12 = 7.63 kΩ

III

1. I_T = 2 A; I_1 = 1.33 A; I_2 = 0.67 A; I_3 = 0.33 A.
2. I_T = 6.17 mA; I_1 = 2.36 mA; I_2 = 1.38 mA; I_3 = 2.43 mA

IV

1. *voltage across:* R_1 = 30 V; R_2 = 100 V; R_3 = 25 V; R_4 = 50 V; R_5 = 50 V; R_6 = 25 V.
   current through: R_1 = 3 A; R_2 = 0.5 A; R_3 = 2.5 A; R_4 = 0.5 A; R_5 = 2 A; and R_6 = 2.5 A.

2. *voltage across:* R_1 = 50 V; R_2 = 10. 7 V; R_3 = 8.9 V; R_4 = 15.2 V; R_5 = 4.45 V; R_6 = 0.72 V; R_7 = 1.94 V; R_8 = 2.0 V; R_9 = 2.0 V.
   current through: R_1 = 5 mA; R_2 = 2.28 mA; R_3 = 2.28 mA; R_4 = 2.72 mA; R_5 = 2.0 mA; R_6 = 0.72 mA; R_7 = 0.72 mA; R_8 = 0.36 mA; R_9 = 0.36 mA.

V

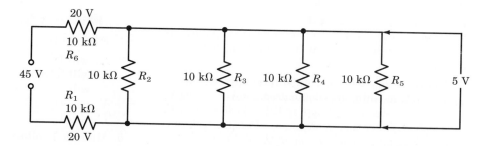

VI

1. Circuit diagram:

2. Load #1 = 1 kΩ
   Load #2 = 1.5 kΩ
   Load #3 = 2.5 kΩ
   R_1 divider resistor = 304 Ω
   R_2 divider resistor = 95 Ω
   R_3 divider resistor = 50 Ω

3. See labeling of diagram in answer 1.

VII

1. V_{R_L} = 150 V; P_{R_L} = 300 mW
2. R_L = 75 kΩ
3. R_2 = 50 kΩ
4. The change, as specified, means R_2 has decreased.
5. The change, as specified, means R_1 has increased.
6. V_{R_2} equals 166.66 V.
7. If R_1 opens, V_{R_L} drops to zero volts.
   If R_2 shorts, V_{R_L} drops to zero volts.

✔ IN-PROCESS LEARNING CHECKS

I

For Figure 6–13a: R_T = 10 Ω
For Figure 6–13b: R_T = 120 Ω

II

I_1 = 0.4 mA; I_T = 1 mA; V_T = 85 V; R_T = 85 kΩ; P_3 = 36 mW; and the 100-kΩ resistor (R_3) is dissipating the most power.

III

1. Figure 6–25: The defective component is R_5. It has drastically increased in value from 3.3 kΩ to 17.3 kΩ.

2. Figure 6–26: (NOTE: Total circuit current *should be* 2 mA, meaning R_3 should drop 2 mA × 47 kΩ V, or 94 V. Instead it is dropping 121 − 17.6 = 103.4 V, which is higher than it should be.) Possible bad components could include R_1 or R_2 decreased in value, or R_3 increased in value.

3. Figure 6–27: If R_2 increases in value, V_{R_1} increases and V_{R_3} decreases. (This is due to total circuit resistance increasing, which decreases the total current through R_3, resulting in R_3 dropping less voltage. Kirchhoff's voltage law would indicate, therefore, that the voltage across parallel resistors R_1 and R_2 has to increase so the loop voltages equal V applied.)

ANSWERS TO SPECIAL "THINKING EXERCISE" (PAGE 223)

If R_2 opens:
V_T remains the same.
V_1 decreases due to lower total current through R_1.
V_2 increases because V_T stays the same, and V_1 and V_4 decrease due to lower total current through them.
V_3 increases for the same reason V_2 increases.
V_4 decreases due to lower I_T caused by higher R_T.
I_T decreases due to higher R_T with same applied voltage.
I through R_1 decreases due to higher circuit total R.
I through R_2 decreases to zero due to its opening.
I through R_3 increases due to higher V across it.
I through R_4 decreases due to lower I_T through it.
P_T decreases due to lower I_T with same V_T.
P_3 increases due to higher I through same R value.
P_4 decreases due to lower current through same R value.

REVIEW QUESTIONS

1. R_T = 22 Ω
3. R_T will D
   R_4 will RTS
   V_1 will I
   V_3 will D
   P_T will I
5. Minimum voltage is 13 V.
   Maximum voltage is 18 V.

7. $R_4 = 6$ kΩ (so that branches voltage equals 80 V). $I_2 = 10$ mA (since $I_T = 20$ mA and I_1 can be calculated as 10 mA, due to V across the 8-kΩ resistor having to be 80 V. That is, the 100 V applied minus the 20-V drop across the 1-kΩ resistor that has I_T passing through it).

9. Light gets dimmer, due to the additional current that passes through the resistor in series with the source. This drops the voltage across the remaining sections of the circuit.

11.

a. $R_T = 169$ kΩ
b. $I_T = 2$ mA
c. $V_{R_1} = 14$ V
d. $V_{R_2} = 14$ V
e. $V_{R_3} = 14$ V
f. $V_{R_4} = 14$ V
g. $V_{R_5} = 62$ V
h. $V_{R_6} = 62$ V
i. $V_{R_7} = 138$ V
j. $V_{R_8} = 200$ V

13. Prior to connecting voltmeter, $V_{R_1} = 5$ V.

15. Approximately 4 V.

17. Yes, connecting the meter changes R_T and I_T. A higher I_T will pass through R_1, causing its voltage to increase from 5 V to 6 V. Less voltage will be dropped across R_2 than was present before the meter was connected.

19. Total current with the meter across one resistor = 60 mA.

21. If the meter were connected from the bottom of R_2 to the top of R_1, the voltages across the two resistors would not be affected.

23. $R_3 = 1$ kΩ; $R_2 = 495$ Ω; $R_1 = 490$ Ω.

25. R_1 would have to be decreased in value.

CHALLENGE QUESTIONS

1. A series-parallel circuit has portions that are connected in series and portions that are connected in parallel. The series portions must be analyzed using series circuit principles, and the parallel portions must be analyzed using parallel circuit principles. The results of the analysis of each type circuit portion must be combined with the results from other connected portions in order to solve the total circuit.

3. Series portions of series-parallel circuits are identified as those in which the same (common) current passes through the components, and by the fact that the components are in tandem.

5. First, set the calculator to the engineering mode. Next, put in one resistor's value, then press the 1/x key, then the plus (+) key. Put in the second resistor's value, press the 1/x key, then the + key. Input the third resistor's value, press the 1/x key, press the equals (=) key, then press the 1/x key one more time for the answer.

7. Connecting a voltmeter will change circuit conditions by virtue of the meter demanding some current (even though small), which also introduces extra current through components in series with the metered portion of the circuit, thus changing voltage drops around the circuit, and causing a higher total current from the source.

9. If R_2 opened:
   a. Voltage across load connected to R_1 would increase.
   b. The voltage across R_1 would decrease (since R_T has increased and I_T had decreased).

TEST YOURSELF

1. $R_1 = 0.5$ kΩ
   $R_2 = 0.5$ kΩ
   $R_3 = 0.75$ kΩ
   $R_4 = 6$ kΩ
   (NOTE: Current through $R_4 = 10$ mA; through $R_3 = 20$ mA; through $R_2 = 30$ mA; through $R_1 = 60$ mA. R_1 drops 30 V with 60 mA through it; therefore = 0.5 kΩ. R_2 drops 15 V with 30 mA through it; therefore = 0.5 kΩ; R_3 drops 15 V with 20 mA through it; therefore = 0.75 kΩ; and R_4 drops 60 V with 10 mA through it; therefore = 6 kΩ.)

CHAPTER 7

PRACTICE PROBLEMS

I

$I = 4$ A
$V_L = 16$ V
$P_T = 80$ W
$P_L = 64$ W
Efficiency = 80%

II

1. I through R_1 = 0 mA
   V across R_1 with respect to point A = 0 V
   I through R_2 = 7.5 mA
   V across R_2 with respect to point A = +75 V
   I through R_3 = 7.5 mA
   V across R_3 with respect to point A = −75 V

2. I through R_1 = 10 mA
   V across R_1 with respect to point A = +100 V
   I through R_2 = 12.5 mA
   V across R_2 with respect to point A = −125 V
   I through R_3 = 2.5 mA
   V across R_3 with respect to point A = +25 V

III

1. If R_L = 175 Ω, I_L = 0.25 A and V_L = 43.75 V (Figure 7–10)

2. If I_L = 125 mA, then R_L = 375 Ω (Figure 7–10)

3. I_L = 1.16 mA; V_{R_L} = 0.55 V

4. In Figure 7–13, R_{TH} = 4 kΩ and V_{TH} = 50 V.
   If R_L = 16 kΩ, then V_L = 40 V and I_L = 2.5 mA.

IV

If R_L is 60 Ω, I_L = 0.588 A and V_L = 35.28 V.

V

1. If R_L = 100 Ω in Figure 7–18a: I_L = 0.4 A; V_L = 40 V.

2. If R_L changes from 25 Ω to 50 Ω, I_L decreases (from 1 A to 0.66 A), and V_L increases (from 25 V to 33.5 V).

✔ IN-PROCESS LEARNING CHECKS

I

1. For maximum power transfer to occur between source and load, the load resistance should be **equal to** the source resistance.

2. The higher the efficiency of power transfer from source to load, the **greater** the percentage of total power is dissipated by the load.

3. Maximum power transfer occurs at **50%** efficiency.

4. If the load resistance is less than the source resistance, efficiency is **smaller** *than* the efficiency at maximum power transfer.

5. To analyze a circuit having two sources, the superposition theorem indicates that Ohm's law **can** be used.

6. What is the key method in using the superposition theorem to analyze a circuit with more than one source? **Noting the direction of current flow and polarity of voltage drops.**

7. Using the superposition theorem, if the sources are considered "voltage" sources, are these sources considered shorted or opened during the analysis process? **Shorted.**

8. When using the superposition theorem when determining the final result of your analysis, the calculated parameters are combined, or superimposed, **algebraically.**

REVIEW QUESTIONS

1. Matching the impedance of an antenna system to the impedance of a transmitter output stage, and matching the output impedance of an audio amplifier to the impedance of a speaker.

3. V_{R_1} = 20 V
   V_{R_2} = 20 V

5. V_{TH} = 40.5 V
   R_{TH} = 14.85 kΩ
   I_L = 0.96 mA
   V_L = 26.1 V

7.

I_N 2.72 mA R_N 14.85 kΩ R_L 27 kΩ

9. To "Thevenize" a circuit:
   1) open or remove R_L.
   2) determine open circuit V at points where R_L is to be connected. (This value is V_{TH}.)
   3) determine resistance looking toward source from R_L connection points, assuming the source is shorted.
   4) draw the Thevenin equivalent circuit with V_{TH} as source and R_{TH} in series with source.
   5) calculate I_L and V_L for given value(s) of R_L using two-resistor series circuit analysis techniques.

11. The maximum power that can be delivered to the load in Figure 7–22 is 1.25 W.

13. 1 ohm

15. R_N = 3.67 Ω; I_N = 10 A

17. R_N = 15 Ω; $I_N = \dfrac{V_{TH}}{R_{TH}} = \dfrac{25\ V}{15\ Ω}$ = 1.67 A

19. Thevenin's theorem

CHALLENGE QUESTIONS

1. It is important for a voltage source to have as low an internal resistance (r_{int}) as possible in order that source output voltage changes will be minimized as current demands for the connected circuit change. Ideally; a voltage source would have zero internal resistance; thus it would not have any internal voltage drop for any current level.

3. a. The reason for carefully observing the direction of current flow when using the superposition theorem is to provide proper assignment of polarities enabling proper algebraic addition of the parameter results when combining the circuit actions of each source considered independently.
   b. Components through which currents created by the multiple sources are aiding or opposing.

TEST YOURSELF

1. R_{TH} is the fixed-value, Thevenin equivalent-circuit series resistor, which, for calculation purposes, is considered to be in series with R_L. Finding this value enables using a simple two-resistor series circuit analysis to calculate circuit parameters for various values of R_L. R_{TH} is determined by looking back into the network at the points where R_L will be connected, and considering the voltage source to be shorted (or zero ohms).

3. R_N is the fixed-value, Norton equivalent-circuit parallel resistor considered to be in parallel with the "constant current" source and R_L. R_N (like R_{TH}) is also found by looking back into the network at the points R_L is to be connected, and considering the voltage source as being replaced by its internal R (R_{int}), or zero ohms, in this case. Knowing the R_N value (along with the equivalent circuit I_N value) enables reducing circuit analysis to a simple two-resistor, parallel circuit, current-division approach.

5. a. Certain semiconductor circuits, select power supply circuits, et al.
   b. Superposition

CHAPTER 8

PRACTICE PROBLEMS

I

1. $I_A = 0.55$ A
2. $I_B = 0.25$ A
3. $V_1 = 9.6$ V
4. $V_2 = 8.25$ V
5. $V_3 = 12.1$ V
6. $V_4 = 8.25$ V
7. $V_5 = 11.75$ V
8. $V_{A-B} =$ close to 0 V

Considered from Points A and B with respect to the opposite ends of resistors R_2 and R_4, respectively: Point A is +8.25 V with respect to the common connection points of R_2 and R_4. Point B is +8.25 V with respect to the common connection points of R_2 and R_4. Thus, Point A and Point B are at the same potential. Consid-

ered from Points A and B with respect to the opposite ends of R_3 and R_5, respectively: Point A is −12.1 V with respect to the common connection points of R_3 and R_5. Point B is −11.75 V with respect to the common connection points of R_3 and R_5. Therefore, Point A is −0.6 V with respect to Point B, or Point B is +0.6 V with respect to point A. (NOTE: Variance is due to the fact that the math was not carried out to a large number of decimal places.)

II

1. $I_A = 0.714$ A
2. $I_B = 0.143$ A
3. $V_1 = 7.14$ V; current through $R_1 = 0.714$ A
4. $V_2 = 12.85$ V; current through $R_2 = 0.857$ A
5. $V_3 = 2.145$ V; current through $R_3 = 0.143$ A

III

1. $V_{R_1} = 18.5$ V
2. $V_{R_2} = 41.5$ V
3. $V_{R_3} = 6.5$ V
4. Current through $R_1 = 0.925$ A
5. Current through $R_2 = 4.15$ A
6. Current through $R_3 = 3.25$ A

Solution

$$I_1 + I_2 = I_3$$

Loop with 60-V source:

$$\left(\frac{V_{R_1}}{R_1}\right) + \left(\frac{V_{R_3}}{R_3}\right) = \frac{V_{R_2}}{R_2}$$

$$V_{R_1} + V_{R_2} = 60$$

$$V_{R_1} = 60 - V_{R_2}$$

Loop with 48-V source:

$$V_{R_3} + V_{R_2} = 48$$

$$V_{R_3} = 48 - V_{R_2}$$

Substituting these terms into $I_1 + I_2 = I_3$:

$$\left(60 - \frac{V_{R_2}}{20}\right) + \left(48 - \frac{V_{R_2}}{2}\right) = \frac{V_{R_2}}{10}$$

Multiplying each term by 20 to move and/or remove the denominators:

$$(60 - V_{R_2}) + 10(48 - V_{R_2}) = 2\,V_{R_2}$$

$$60 - V_{R_2} + 480 - 10\,V_{R_2} = 2\,V_{R_2}$$

$$540 = 13\,V_{R_2}$$

$V_{R_2} = 41.5$ V

$I_{R_2} = \dfrac{V_{R_2}}{10} = \dfrac{41.5 \text{ V}}{10 \ \Omega} = 4.15$ A

$V_{R_3} = 48 - V_{R_2} = 48 - 41.5 = 6.5$ V

$I_{R_3} = \dfrac{V_{R_3}}{2} = \dfrac{6.5 \text{ V}}{2 \ \Omega} = 3.25$ A

$V_{R_1} = 60 - 41.5 = 18.5$ V

$I_{R_1} = \dfrac{18.5}{20} = 0.925$ A

Verifying:

$V_{R_1} + V_{R_2} = V_{S_1}; 18.5 \text{ V} + 41.5 \text{ V} = 60$ V

$V_{R_2} + V_{R_3} = V_{S_2}; 41.5 \text{ V} + 6.5 \text{ V} = 48$ V

IV

1. $R_A = 31.82 \ \Omega; R_B = 70 \ \Omega; R_C = 46.67 \ \Omega$
2. $R_1 = 22 \ \Omega; R_2 = 10 \ \Omega; R_3 = 15 \ \Omega$
3. Yes

V

$(R_1 = 7 \ \Omega; R_2 = 3.5 \ \Omega; R_3 = 2.33 \ \Omega)$
$R_T = (R_D + R_2)\|(R_E + R_3) + R_1$
$R_T = 13.44 \ \Omega$
$I_T = V_T/R_T = 84 \text{ V}/13.44 \ \Omega = 6.25$ A

 IN-PROCESS LEARNING CHECKS

I

1. Path
2. Mesh
3. + (positive)
4. +, −

II

1. Current
2. Major node
3. Reference node
4. Current

III

1. (a) T (tee) (b) π (pi)
2. AC (or electrical)
3. Current
4. is not

REVIEW QUESTIONS

1. $I_A = 0.8$ A
3. $V_{R_1} = 8$ V

5. $V_{R_3} = 4$ V
7. $I_{R_2} = 1.2$ A
9. I_1 (current through R_1) = 2.13 A
11. I_3 (current through R_2) = 2.34 A
13. $V_{R_2} = 11$ V
15. $R_A = 45 \ \Omega$
17. $R_C = 30 \ \Omega$

CHALLENGE QUESTIONS

1. A mesh current differs from an actual circuit current in that the mesh current is not considered to split and branch. It is considered to flow in only one loop.
3. Delta and wye circuit connections may be found in industry in such applications as transformer, motor, and generator circuits.

TEST YOURSELF

1. *Mesh*—An unbroken or closed loop. A set of branches forming a closed path for current in a network. *Loop*—A complete current path, or electrical circuit. *Node*—A junction or current branch point in an electrical circuit.
3. Typically one equation is written for each mesh; therefore, for a three-mesh circuit there would be three equations.

CHAPTER 9

PRACTICE PROBLEMS

I

1. $A = 1.5 \text{ V} - \dfrac{1.25 \text{ V}}{0.1 \text{ A}} = \dfrac{0.25 \text{ V}}{0.1 \text{ A}} = 2.5 \ \Omega$
2. Since we defined useful life voltage between 1.4 V and 1.6 V, this cell is *bad*.

II

1. It will deliver 7.5 amperes for 20 hours $\left(\dfrac{150 \text{ ampere-hours}}{20 \text{ time in hours}}\right)$.
2. C rating would be 60/12.5, meaning it could deliver 12.5 amperes for 4.8 hours.

 IN-PROCESS LEARNING CHECKS

I

1. The carbon-zinc type dry cell is composed of a negative electrode made of **zinc** and a positive electrode made of **carbon.** The electrolyte solution in this

type cell is **ammonium chloride,** called **sal ammoniac.**

2. The process which causes hydrogen bubbles to form on the positive electrode, and which is detrimental to the cell's operation, is called **polarization.**

3. Manganese dioxide is often added to the cell to prevent the process mentioned in question 2. This agent is sometimes known as a **depolarizer** agent.

4. Another problem that can occur in dry cells is that of little batteries being formed due to impurities suspended in the cell's electrolyte near the zinc electrode. This type activity is known as **local action.** To reduce this undesired activity, mercury is put on the zinc by a process called **amalgamation.**

5. The shelf life of dry cells is **hurt** by storing the cells in hot conditions.

6. Nominal cell terminal voltage of dry cells is about **1.5 V.**

REVIEW QUESTIONS

1. Positive ion is an atom that has lost electron(s).

3. Amalgamation is the process of coating the inside of the zinc electrode with mercury in a carbon-zinc cell in order to minimize local action.

5. Manganese dioxide

7. Mercury cell

9. 100 ampere-hours

TEST YOURSELF

1. The source's positive terminal is connected to the battery's positive terminal and the source's negative terminal is connected to the battery's negative terminal. This reverses the chemical action in the battery and recharges it.

3. a. Ni-cad cells are in the range of 110–1,200 mA per hr.
   b. Alkaline cells are in the range of 150–750 mA (for 10 hrs).

5. a. AAA (carbon-zinc): V = 1.5 V; Current = not given in catalog; mA/hr = not given in catalog
   b. AA (carbon-zinc): V = 1.5 V; Current = not given in catalog; mA/hr = not given in catalog
   c. A (carbon-zinc): not a standard type
   d. C (carbon-zinc): V = 1.5 V; Current = not given in catalog; mA/hr = not given in catalog
   e. D (carbon-zinc): V = 1.5 V; Current = not given in catalog; mA/hr = not given in catalog

7. Current-delivering capability primarily determined by size of active area of the electrodes.

9. Some applications of lithium cells are in computers, cameras, watches, and calculators.

CHAPTER 10

PRACTICE PROBLEMS
I

II
Figure 10–13a = Move away from each other.
Figure 10–13b = Move toward one another.
Figure 10–13c = Move away from each other.

III
Figure 10–17a = N at right end, S at left end; Figure 10–17b = N at left end, S at right end; and Figure 10–17c = N at left end, S at right end.

IN-PROCESS LEARNING CHECKS
I

1. Small magnets that are suspended and free to move align in a **North** and **South** direction.

2. Materials that lose their magnetism after the magnetizing force is removed are called **temporary magnets.** Materials that retain their magnetism after the magnetizing force is removed are called **permanent magnets.**

3. A wire carrying dc current **does** establish a magnetic field. Once a magnetic field is established, if the current level doesn't change, the field is **stationary.**

4. A law of magnetism is like poles **repel** each other and unlike poles **attract** each other.

5. A maxwell represents **one** line of force.

6. A weber represents 10^8 lines of force.

7. Lines of force are **continuous.**

8. Lines of force related to magnets exit the **North** pole of a magnet and enter the **South** pole.

9. Nonmagnetic materials do **not** stop the flow of magnetic flux lines through themselves.

10. Yes, magnetism **can** be induced from one magnetic object to another object, if it is a magnetic material.

II
1. For a given coil dimension and core material, what two factors primarily affect the strength of an electromagnet? **Number of turns** and **amount of current.**

2. The left-hand rule for determining the polarity of electromagnetics states that when the fingers of your left hand point **in the same direction** as the current passing through the coil, the thumb points toward the **North** pole of the electromagnet.

3. Adjacent current-carrying conductors that are carrying current in the same direction tend to **attract each other.**

4. If you grasp a current-carrying conductor so that the thumb of your left hand is in the direction of current through the conductor, the fingers "curled around the conductor" **will indicate** the direction of the magnetic field around the conductor.

5. When representing an end view of a current conductor pictorially, it is common to show current coming out of the paper via a **dot.**

III

1. A *B-H* curve is also known as a **magnetization** curve.

2. *B* stands for flux **density.**

3. *H* stands for magnetizing **intensity.**

4. The point where increasing current through a coil causes no further significant increase in flux density is called **saturation.**

5. The larger the area inside a "hysteresis loop" the **larger** the magnetic losses represented.

REVIEW QUESTIONS

1. *Magnetism* is the property causing certain materials to develop attraction or repulsion when the molecular structure has certain alignments. Also, magnetism causes induced voltage or current into conductors when there is relative motion between the magnet and the conductor. (Magnetism is the property associated with the setting up of a magnetic field around a current-carrying conductor, called electromagnetism). *Flux* = Magnetic lines of force. *Field* = Area of magnetic influence of a magnet. *Pole* = Either the N-seeking or S-seeking end of a magnet. *Magnetic Polarity* = Identification of the N-seeking or S-seeking ends of a magnet; relative direction of a magnet's flux.

3. Like poles repel and unlike poles attract.

5. North and South poles are indicated.

7. Permeability

9. Using SI units and terms:
Flux—Weber (lines/10^8)
Flux Density—Webers/sq meter (Teslas or T)

11. Permeability—μ (mu) Teslas/(AT/meter)
Relative Permeability—$μ_r$ (flux w/core material/flux w/vacuum core)

13. Magnetomotive force is the force by which a magnetic field is produced and relates to the ampere-turns involved in an electromagnet. Magnetic field intensity adds the dimension of ampere-turns per a given length of the electromagnetic coil. In the SI system, field intensity is AT/meter.

15. Reluctance will triple.

17. Flux density will increase.

19. A magnetic shield effectively "shunts" the magnetic flux around the object being shielded due to the shield's providing an easy path for flux through itself.

CHALLENGE QUESTIONS

1. Magnetic polarity is a way of indicating magnetic differences between two physical points or ends of a magnet or electromagnet. Typically, we speak of magnetic poles (indicated by magnetic polarity) as being North-seeking or South-seeking poles. Also, an arbitrary standard has been established that indicates that magnetic flux lines exit the North pole of a magnet, and enter into the South pole of the magnet.

TEST YOURSELF

1. Hysteresis loop showing saturation, residual magnetism, and coercive force areas is shown.

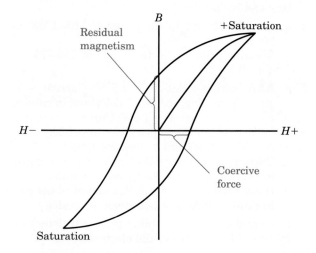

3. a. $\dfrac{10^8}{10^6} = 100$ lines of flux

   b. Teslas = webers per square meter. Since there are 100 lines in 0.005 m², there are 1/0.005 × 100 lines in a square meter, or lines = 200 × 100 = 20,000 number of webers per square meter (teslas) = 20,000/100,000,000 = 0.0002 T (or 0.2 mT or 200 μT)

CHAPTER 11

PRACTICE PROBLEMS

I

$$R_S = \frac{V_m}{I_S}$$
$$R_S = \frac{100 \text{ mV}}{99 \text{ mA}}$$
$$R_S = 1.01 \ \Omega$$

II

$$R_{Mult} = \frac{V \text{ (range)}}{I \text{ (fs)}} - R_M$$
$$R_{Mult} = \frac{250}{50 \times 10^{-6}} - 2,000$$
$$R_{Mult} = 5,000,000 - 2,000$$
$$R_{Mult} = 4,998,000 \text{ or } 4.998 \text{ M}\Omega$$

III

1. Ohms-per-volt rating of a 20-microampere movement is $\dfrac{1}{fs} = \dfrac{1}{20 \times 10^{-6}} = 50,000$ ohms per volt.

2. Full-scale current rating of a meter rated at 2,000-ohms per volt is 1/2,000 = 0.5 mA movement.

✔ IN-PROCESS LEARNING CHECK

I

1. Three essential elements in an analog meter movement include a **moving** element, an **indicating** means, and some form of **damping.**

2. The purpose of the permanent magnet in a d'Arsonval meter movement is to establish a fixed **magnetic field.**

3. The movable element in a moving coil meter moves due to the interaction of the fixed **magnetic field** and the magnetic field of the movable coil when current passes through the coil.

4. Electrical conduction of current in a movable meter movement is accomplished through the **springs,** which also provides a desired **mechanical** force against which the movable coil must act.

5. The element in a moving-coil movement producing electrical damping is the **coil frame.**

REVIEW QUESTIONS

1. Full-scale current rating = 2 mA
$$R \text{ meter} = \frac{200 \text{ mV}}{2 \text{ mA}} = 100 \ \Omega$$

3. A shorting-type switch is where the movable contact contacts the next contact position before leaving the previous contact position. This switch is frequently used in multiple-range current meters to assure the meter is never unshunted and therefore has some measure of protection from overcurrent.

5. Circuit for measuring specified currents is shown.

I through middle branch

7. Ohms-per-volt sensitivity =
$$\frac{1}{(fs)} = \frac{1}{40 \ \mu\text{A}} = \frac{25 \text{ k}\Omega}{\text{V}}$$
$$R_M + R_{Mult} = 500 \times 25 \text{ k}\Omega = 12,500 \text{ k}\Omega, \text{ or } 12.5 \text{ M}\Omega$$

9. The resulting metering circuit resistance for the conditions stated would be 50 Ω. (Shunt R = meter R since both branches are to carry 1 mA for a total of 2 mA. Meter resistance = V_M/I_M or 100 mV/1 mA = 100 Ω.)

11. A voltmeter should have a high resistance to prevent appreciably changing circuit conditions when it is connected in parallel with the portion of the circuit where the measurement is being made.

13. Midscale = 1,500 Ω
Quarter scale = 4,500 Ω

15. With voltmeter checks, it is not necessary to break the circuit and insert the meter in series, as is required for direct current meter measurements.

17. Damping is the designed prevention of excess mechanical pointer oscillation above and below an indicated meter reading in a meter movement.

19. It loads the circuit under test much less, thus giving a truer picture of the circuit operation; it is easier to read without error.

TEST YOURSELF

1. 25 V
3. 50 V

CHAPTER 12

PRACTICE PROBLEMS

I

1. The sine of 35° = 0.5735.
2. Cycle 1 ends and cycle 2 begins at point G.
3. Maximum rate of change for the waveform shown occurs at points J, N, R, and T.

II

1.

2.

III

1. 10 V
2. 9 V
3. 90%
4. No
5. Yes
6. a. 120.8 V
   b. 171 V
   c. 120.8 V
   d. 171 V
   e. 108.9 V

IV

1. a. V_1 = 50 V
   b. V_2 = 50 V
   c. I_p = 7.07 mA
   d. P_p = 999.7 mW
   e. P_1 = 250 mW
2. a. I_1 = 10 mA
   b. I_2 = 10 mA
   c. I_p = 28.28 mA
   d. $V_{p\text{-}p}$ = 282.8 V
3. a. 66.7 V
   b. 33.3 V

c. 33.3 V
d. 444.89 mW

V

The duty cycle = Pulse width/Time for one period times 100. Duty cycle = (1 μs/15 μs) × 100 = 6.67% duty cycle

VI

1. V_{avg} = Baseline value + (duty cycle × p-p amplitude)
   V_{avg} = 0 + (pulse width/period × p-p amplitude)
   V_{avg} = 0 + [(2 ms/30 ms) × 6 V] = 0 + 0.066 × 6 = 0.396 V
2. −2.604 V

✔ IN-PROCESS LEARNING CHECKS

I

1. The basic difference is that dc (direct current) is in one direction and of one polarity. Alternating current (ac) periodically changes direction and polarity.
2. The reference, or zero-degree position is horizontally to the right when describing angular motion.
3. The y-axis is the vertical axis in the coordinate system.
4. The second quadrant in the coordinate system is that quadrant between 90 and 180 degrees.
5. A vector represents a given quantity's magnitude and direction with respect to location in space.
6. A phasor represents a rotating vector's relative position with respect to time.

II

1. 10-kHz
2. 0.0025 seconds, or 2.5 msec
3. 50 μsec
4. Amplitude
5. 45° and 135°; 225° and 315°
6. 1 and 2
7. 0.067 μsec
8. 40-Hz
9. As frequency increases, T *decreases.*
10. The longer a given signal's period, the *longer* the time for each alternation.

REVIEW QUESTIONS

1. Differences between ac and dc:
   a. dc is of one polarity and ac changes polarity periodically

b. dc is of one amplitude and ac is constantly changing amplitude

c. dc is unidirectional and ac is bidirectional.

3. $f = \dfrac{1}{t} = \dfrac{1}{0.2\ \mu sec} = 5\ MHz$

5. A 10 V rms, 1 MHz ac signal is illustrated as follows.

7. 400 Hz.

9. P-to-p voltage is approximately 63 volts.
($P = V^2/R$; 50 mW $= V^2/10$ kΩ; 500 $= V^2$; 22.36 $= V$ rms; rms × 2.828 = p-to-p V; 22.36 × 2.828 = 63.24 V)

11. If the rms value of an ac voltage doubles, the peak-to-peak value must **double.**

13. Frequency has been reduced to one-third the original frequency.

15. Time is one half the time of a period, where $T = 1/f = 1/500 = 0.002$ seconds. Therefore, one alternation equals 0.001 seconds, or 1 millisecond.

CHALLENGE QUESTIONS

1. The terms fundamental and harmonic, as applied to ac signals, can be described as follows: *Fundamental* refers to the "fundamental frequency" of the waveform or signal and is the lowest frequency or principle component (and usually, highest amplitude element) that makes up the composite waveform. The fundamental frequency is used as the reference frequency for the signal. *Harmonic* refers to a sinusoidal signal that is a multiple of the fundamental frequency. For example, a waveform with twice the frequency of the fundamental is called the "second harmonic."

3.

TEST YOURSELF

1. Labeled waveform diagram showing angles and amplitudes is shown. (NOTE: Cosine wave is leading the sine wave by 90°.)

a. See diagram.

b. Sine wave at 45° $= +7.07$ units
   Sine wave at 90° $= +10$ units

c. Cosine wave at 0° $= +10$ units
   Cosine wave at 45° $= +7.07$ units
   Cosine wave at 90° $= 0$ units

d. Cosine leading sine by 90°

3. Diagram of the three-resistor series-parallel circuit is shown.

a. $V_{R_1} = 20$ V
b. $I_{R_1} = 2$ mA
c. $P_{R_1} = 40$ mW
d. $I_{R_2} = 1$ mA
e. $V_{R_3} = 100$ V
f. $\theta = 0°$

CHAPTER 13

PRACTICE PROBLEMS

I

When the vertical frequency is four times the horizontal frequency, the waveform shows four cycles of the given waveform.

II

1. 3.5 V (rms: 5 V peak voltage)
2. a. 12 V (p-to-p) b. 4.24 V (rms)

III

1. +1 V dc
2. +0.8 V dc

IV

1. 30 V (p-to-p); 10.6 V (rms)
2. 1.76 V/div (rms)
3. Use the 10 position setting of the vertical volts/div control.

V

1. 1.5 V (p-to-p); 0.53 V (rms)
2. 3 V (p-to-p); 1.06 V (rms)

VI

72 degrees per major division

VII

1. a. Period (T) = 0.25 s
   b. Frequency = 4 Hz
2. a. Period (T) = 50 μs
   b. Frequency = 20 kHz

IN-PROCESS LEARNING CHECKS

I

1. The part of the oscilloscope producing the visual display is the *cathode-ray tube.*
2. The scope control that influences the brightness of the display is the *intensity* control.
3. A waveform is moved up or down on the screen by using the *vertical position* control.
4. A waveform is moved left or right on the screen by using the *horizontal position* control.
5. For a signal fed to the scope's vertical input, the controls that adjust the number of cycles seen on the screen are the horizontal *time/div.* and the horizontal *time variable* controls.
6. The control(s) that help keep the waveform from moving or jiggling on the display are associated with the *synchronization* circuitry.

REVIEW QUESTIONS

1. Name is cathode-ray tube and abbreviation is CRT.
3. To protect and add life to coating on CRT screen
5. vertical input terminal(s)
7. "Ext Trig" terminal (Note: Other names are also used)
9. vertical (or y-axis)

11. a. 30 V p-p
    b. T = 0.004 s
    c. horizontal time/div. and/or time variable
    d. 0.0004 div/s
    e. rms = 10.6 V; p = 15 V; p-p = 30 V

CHALLENGE QUESTIONS

1. List of control names should be based on the specific oscilloscope names students can observe on your equipment in your training setting:
   a. Control the trace focus _____.
   b. Control the trace intensity _____.
   c. Control(s) the vertical position _____.
   d. Control(s) the horizontal position _____.
   e. Vertical attenuator/amplifier controls _____
      _____.
   f. Horizontal sweep frequency controls _____
      _____.

TEST YOURSELF

1. Diagram B: vertical position and horizontal gain controls
   Diagram C: vertical position and horizontal gain controls
   Diagram D: vertical gain and horizontal position controls
   Diagram E: horizontal gain control
   Diagram F: vertical gain control
   Diagram G: horizontal sweep frequency controls (horizontal time/div. and time variable)
   Diagram H: vertical gain control
   Diagram I: vertical and horizontal gain controls
   Diagram J: horizontal sweep frequency controls (horizontal time/div. and time variable)
   Diagram K: horizontal gain control
   Diagram L: vertical position, vertical and horizontal gain controls
   Diagram M: vertical gain control

CHAPTER 14

PRACTICE PROBLEMS

I

Since 1 weber = 10^8 flux lines and cutting 1 Wb/sec induces 1 V, then cutting 2×10^8 flux lines (2 Wb) in 0.5 sec = 4 V

II

$$V_L = L \frac{di}{dt}$$

$$V_L = 5 \left(\frac{3}{1}\right) = 15 \text{ V}$$

III

10 mH + 6 mH + 100 µH = 16.1 mH total inductance

IV

1. Three equal value inductors (15 H) in parallel have a total equivalent inductance = 1/3 of one of them. L_T = 5 henrys.

2. L_T of 10 mH, 12 mH, 15 mH, and 20 mH inductors in parallel is 3.33 mH.

V

a. dc current through coil = 5 A

b. energy = $\dfrac{LI^2}{2}$ = $\dfrac{1(25)}{2}$ = 12.5 J

VI

Two seconds = approximately 2.4 TC.
After 2.4 TC, V_R is about 91% of V applied, or 9.1 V.
After 2.4 TC, V_L is about 9% of V applied, or 0.9 V.

VII

0.92 mA

 IN-PROCESS LEARNING CHECKS

I

1. All other factors remaining the same, when an inductor's number of turns increases four times, the inductance value ***increases*** by a factor of ***sixteen times.***

2. All other factors remaining the same, when an inductor's diameter triples, the inductance value ***increases*** by a factor of ***nine times.*** (NOTE: A is directly proportional to d^2).

3. All other factors remaining the same, when an inductor's length increases, the inductance value ***decreases.***

4. When the core of an air-core inductor is replaced by material having a permeability of ten, the inductance value ***increases*** by a factor of ***ten times.***

REVIEW QUESTIONS

1. Inductance (or self-inductance) is the property of a circuit that opposes a change in current flow.

3. a. number of turns on the coil
   b. length of the coil
   c. cross-sectional area of the coil
   d. relative permeability of the core material

5. Amount of emf induced in a circuit is related to the number of magnetic flux lines cut or linked per unit time.

7. Inductance will decrease.

9. Total inductance will increase.

11. Inductance will increase to 400 mH, (L varies as the square of the turns).
    Expressed in µH = 400,000 µH, or 0.4 H.

13. New inductance is 500 × 100 µH = 0.05 H.

15. $L_T = L_1 + L_2 + L_3$ = 0.1 H + 0.2 H + 0.001 H = 0.301 H, or 301 mH

17. 1 TC = L/R = 250 × 10^{-3}/100 = 0.0025 sec, or 2.5 ms. It will take 5 times 2.5 ms, or 12.5 ms for current to completely change from one level to another.

19. Tapping a coil at midpoint yields an inductance of one-fourth the value of the total inductance of the coil. In the case of the 0.4 H coil, L equals about 0.1 H at the 250 turn tap.

CHALLENGE QUESTIONS

1. V_L after 0.6 µs ≈ 6.76 V

TEST YOURSELF

1. Since the inductance change is proportional to the square of the turns
   $$\frac{L_1}{L_2} = \frac{N_1^2}{N_2^2}; \frac{4}{6} = \frac{1,000^2}{N_2^2}; N_2 = 1224.74 \text{ turns}$$

3. Because the voltage-per-turn is less than the varnish insulation breakdown voltage. (The voltage per turn in this case is approximately 2 V per turn.)

CHAPTER 15

PRACTICE PROBLEMS

I

1. 3 kΩ
2. 333.3 Ω

II

1. 6,280 Ω
2. 79.6 mH, or approximately 80 mH
3. 238.85 Hz, or approximately 239 Hz

III

1. 5 kΩ
2. X_L = 5,024 Ω, so $I \times X_L$ = 3 mA × 5,025 Ω = approximately 15 V
3. 40 mA

 IN-PROCESS LEARNING CHECKS

I

1. When current increases through an inductor, the cemf ***hinders*** the current increase.

2. When current decreases through an inductor, the cemf **hinders** the current decrease.

3. In a pure inductor, the **voltage** leads the **current** by 90 degrees.

4. What memory aid helps to remember the relationship described in Question 3? **"Eli"** the ice man.

5. The opposition that an inductor shows to ac is termed **inductive reactance.**

6. The opposition that an inductor shows to ac **increases** as inductance increases.

7. The opposition that an inductor shows to ac **decreases** as frequency decreases.

8. X_L is **directly** related to inductance value.

9. X_L is **directly** related to frequency.

REVIEW QUESTIONS

1. $L = \dfrac{X_L}{2\pi f} = \dfrac{376.8}{6.28 \times 100} = \dfrac{376.8}{628} = 0.6$ H

   $X_L = 4.5 \times 376.8\ \Omega = 1{,}695.6\ \Omega$

3. X_L is six times greater than the original value.

5. Circuit I will increase, since X_L will decrease and V is assumed to remain the same.

7. Q will decrease since $Q = X_L/R$. X_L will decrease while R stays the same.

9. Total inductance $= \dfrac{2 \times 6}{2 + 6} = \dfrac{12}{8} = 1.5$ mH

   Frequency $= \dfrac{X_L}{2\pi L} = 39.8$ kHz or ≈ 40 kHz

11. $Q = \dfrac{X_L}{R} = \dfrac{6.28 \times 2 \times 10^6 \times 10 \times 10^{-6}}{10}$

    $= \dfrac{125.6}{10} = 12.56$

    Q will increase, since $Q = X_L/R$ and X_L triples while R only doubles.

    Q is not an appropriate consideration for inductors in dc circuits. X_L is reactance to ac due to the changing current inducing back-emf.

13. $V_T = 45$ V
    $V_{L_1} = 15$ V
    $I_T = 0.75$ mA

15. ≈ 6 V

CHALLENGE QUESTIONS

1. Inductive reactance is directly related to the value of inductance because the induced cemf in an inductor

is directly related to the inductance value, and the inductive reactance is a direct result of cemf.

3. In most cases, a high Q coil is preferred since it will have less losses and dissipate less electrical energy in accomplishing its assigned task for the circuit.

TEST YOURSELF

1. Total inductance is 3 H
   (NOTE: L_1 computes to be 2 H via the $L = X_L/2\pi f$ formula.)

CHAPTER 16

PRACTICE PROBLEMS

I
5 pounds at an angle of 36.9° from horizontal.

II
$V_s = \sqrt{V_R^2 + V_L^2} = \sqrt{25 + 25} = \sqrt{50} = 7.07$ V

III

1. $\text{Cos } \theta = \dfrac{\text{adj}}{\text{hyp}} = \dfrac{5}{10}$ and $\cos^{-1} = 60°$

2. Tangent of 45° $= 1.0$

3. Sine of 30° $= 0.5$

IV
V applied $= 150$ V

V

1.

2. $\theta = 33.69°$ (Tan θ = opp/adj = 20/30 = 0.6666. Angle whose tangent = 0.6666 is 33.69°.)

3. Tangent function

4. Angle would have been greater (56.3°).

VI

1. $V_R = 35.35$ V (Cos θ = adj/hyp and cos 45° = 0.7071. Therefore, 0.7071 = adj/50 = 35.35.)

2. a. $V_R = 32.5$ V
   b. $V_L = 56.29$ V

3. a. Angle = 56.3°
   b. $V_T = 43.2$ V

VII

1.

2.

$X_L = 20\ \Omega$ ---- $Z = 25\ \Omega$

$\theta = 53.1°$

$R = 15\ \Omega$

3. $Z = \sqrt{15^2 + 20^2}$

$Z = \sqrt{625}$

$Z = 25\ \Omega$

4. $\operatorname{Tan} \theta = \dfrac{X_L}{R}$

$\operatorname{Tan} \theta = \dfrac{20}{15}$

$\operatorname{Tan} \theta = 1.33$

Angle $= 53.1°$

VIII

1. $I_T = \sqrt{I_R^2 + I_L^2}$

$I_T = \sqrt{5^2 + 3^2}$

$I_T = \sqrt{34}$

$I_T = 5.83\ \text{A}$

2. $Z = \dfrac{V_T}{I_T}$

$Z = \dfrac{300}{5.83}$

$Z = 51.45\ \Omega$

3. $\theta -30.9°$ (Tan θ = opp/adj = 3 A/5 A = 0.6. Angle whose tangent equals 0.6 is 30.9°.)

IX

$Z = 47 \times 100/(\sqrt{47^2 + 100^2}) = 4,700/110.5 = 42.5\ \Omega$

✔ IN-PROCESS LEARNING CHECKS

I

1. In a purely resistive ac circuit, the circuit voltage and current are **in**-phase.

2. In a purely inductive ac circuit, the current through the inductance and the voltage across the inductance are **90** degrees out-of-phase. In this case, the **voltage** leads the **current.**

3. A quantity expressing both magnitude and direction is a **vector** quantity.

4. The length of the vector expresses the **magnitude** of a quantity.

REVIEW QUESTIONS

1. a. $36\ \Omega$

   b. $Z = \sqrt{R^2 + X_L^2} = \sqrt{20^2 + 30^2} = \sqrt{1,300} =$ approximately $36\ \Omega$

   c. Cos $\theta = R/Z = 20/36 = 0.555$. Angle whose cosine equals 0.555 is 56.3°; thus, angle between R and $Z = 56.3°$.

3. Diagrams for the specified series RL circuit are shown.

Circuit diagram   V-I vector diagram   Impedance diagram

Since $\theta = 45°$, then R must $= X_L$, and V_R must $= V_L$.

$Z = \sqrt{R^2 + X_L^2}$

$Z = \sqrt{100 \times 10^6}$

$Z = 14.14 \times 10^3 = 14.14\ \text{k}\Omega$

5. a. R will RTS
   b. L will RTS
   c. X_L will I
   d. Z will I
   e. θ will D
   f. I_T will D
   g. V_T will RTS

7. a. R will I
   b. L will RTS
   c. X_L will RTS
   d. Z will I
   e. θ will D
   f. I_T will D
   g. V_T will RTS

9. a. R will I
   b. L will RTS
   c. X_L will RTS
   d. Z will I
   e. θ will I
   f. I_T will D
   g. V_T will RTS

11. a. Z = vector sum of 10 kΩ R and approximate 20 kΩ X_L = 22.36 kΩ
    b. $V_R = I \times R$ = (67 V/22.36 kΩ) × 10 kΩ = approximately 3 mA × 10 kΩ = 30 V
    c. θ = approximately 63.43°

13. a. I = 5 mA
    b. Z = 40 kΩ
    c. V_L = 105.4 V
    d. X_L = 21.08 kΩ
    e. $L = X_L/2\pi f$ = 56 H

CHALLENGE QUESTIONS

1. 57, sin (value in readout would then be 0.8386)

3. 34, cos (value in readout would then be 0.829)

5. Advantage is not having to know the hypotenuse value to calculate.

TEST YOURSELF

1. Diagrams and calculations for the specified series RL circuit are shown.

$X_L = 2\pi f L = 6.28 \times 1 \times 10^3 \times 10 = 62.8$ kΩ

$\text{Sin } \theta = \dfrac{X_L}{Z}; \sin 60° = \dfrac{62.8 \text{ kΩ}}{Z}; 0.866 = \dfrac{62.8 \text{ kΩ}}{Z};$

$0.866 \, Z = 62.8 \text{ kΩ}; Z = \dfrac{62.8 \text{ kΩ}}{0.866} = 72.517$ kΩ

$\text{Tan } \theta = \dfrac{\text{opp}}{\text{adj}} = \dfrac{X_L}{R}; \tan 60° = \dfrac{62.8 \text{ kΩ}}{R};$

$1.73 = \dfrac{62.8 \text{ kΩ}}{R}$

$1.73 \, R = 62.8 \text{ kΩ}; R = \dfrac{62.8 \text{ kΩ}}{1.73} = 36.3$ kΩ

$I = \dfrac{V_T}{Z}; I = \dfrac{72.5 \text{ V}}{72.5 \text{ kΩ}} = 1$ mA

$V_R = I \times R = 1 \text{ mA} \times 36.3 \text{ kΩ} = 36.3$ V
$V_L = I \times X_L = 1 \text{ mA} \times 62.8 \text{ kΩ} = 62.8$ V

3. The circuit diagram for the specified circuit is shown. I_T = approximately $\dfrac{140 \text{ V}}{14.14 \text{ kΩ}} \cong 10$ mA

$Z = \sqrt{R^2 + X_L^2}$
$X_L = 2\pi f L = 6.28 \times 200 \times 10^3 \times 4 \times 10^{-3} = 5.024$ kΩ (or 5 kΩ) each inductor
$R = 5 \text{ kΩ} + 5 \text{ kΩ} = 10$ kΩ
$Z = \sqrt{200 \times 10^6} = 14.14$ kΩ (or 14 kΩ)
$X_{L_T} \cong 10$ kΩ

CHAPTER 17

PRACTICE PROBLEMS

I

$M = k\sqrt{L_1 \times L_2}$
$M = 0.95\sqrt{10 \times 15}$
$M = 0.95\sqrt{150}$
$M = 0.95 \times 12.25 = 11.63$ H

II

$L_T = L_1 + L_2 - 2M$
$L_T = 10 \text{ H} + 10 \text{ H} - (2 \times 5 \text{ H})$
$L_T = 20 \text{ H} - 10 \text{ H} = 10$ H

III

1. $L_T = \dfrac{1}{\dfrac{1}{L_1 + M} + \dfrac{1}{L_2 + M}}$

$L_T = \dfrac{1}{\dfrac{1}{8 + 4} + \dfrac{1}{8 + 4}}$

$L_T = \dfrac{1}{\dfrac{1}{12} + \dfrac{1}{12}}$

$L_T = \dfrac{1}{\dfrac{1}{6}} = 1 \times \dfrac{6}{1} = 6$ H

2. $L_T = \cfrac{1}{\cfrac{1}{L_1 - M} + \cfrac{1}{L_2 - M}}$

$L_T = \cfrac{1}{\cfrac{1}{8 - 4} + \cfrac{1}{8 - 4}}$

$L_T = \cfrac{1}{\cfrac{1}{4} + \cfrac{1}{4}}$

$L_T = \cfrac{1}{\cfrac{1}{2}} = 1 \times \cfrac{2}{1} = 2\ H$

IV

1. $\dfrac{N_s}{N_P} = \dfrac{V_s}{V_P}$; s-p turns ratio $= \dfrac{300}{100}$ or 3:1

2. Turns ratio equals voltage ratio. Since the transformer steps-up the voltage six times, the secondary must have six times the turns of the primary. Therefore, the s-p turns ratio is 6:1.

3. Since the s-p turns ratio is 5:1, the secondary has five times the voltage of the primary, or $5 \times 50\ V = 250\ V$.

4. Voltage ratio with s-p turns ratio of 1:4 is 1:4 (secondary-to-primary). This is a step-down transformer, since the primary voltage is four times that of the secondary.

V

Since the impedance is transformed in relation to the square of the turns ratio, when the secondary has twice the number of turns of the primary, the impedance at the primary looks like one-fourth the impedance across the secondary, or $2,000/4 = 500\ \Omega$.

VI

1. The *primary-to-secondary* impedance ratio is related to the square of the turns relationship of primary-to-secondary. Since the primary has 5 times as many turns as the secondary, the primary impedance is 5^2 times greater than secondary. Therefore the p-s Z ratio is 25:1.

2. The impedance looking into the primary is $(16)^2 \times 4\ \Omega = 256 \times 4 = 1,024\ \Omega$.

✔ IN-PROCESS LEARNING CHECKS

I

1. Producing voltage via a changing magnetic field is called electromagnetic *induction.*

2. Inducing voltage in one circuit by varying current in another circuit is called *mutual inductance.*

3. The fractional amount of the total flux that links two circuits is called the *coefficient* of coupling, which is represented by the letter *k.* When 100% of the flux links the two circuits, the *coefficient* of coupling has a value of *1.*

4. The closer coils are, the *higher* the coupling factor produced. Compared to parallel coils, perpendicular coils have a *lower* degree of coupling.

REVIEW QUESTIONS

1. $\dfrac{V_s}{V_P} = \dfrac{I_P}{I_s}$

$\dfrac{240\ V}{120\ V} = \dfrac{I_P}{100\ mA}$

$I_P = 200\ mA$

3. $\dfrac{N_P}{N_s} = \dfrac{V_P}{V_s}; \dfrac{N_P}{N_s} = \dfrac{120\ V}{6\ V}; \dfrac{N_P}{N_s} = 20{:}1$

5. $M = k\sqrt{L_1 \times L_2} = 0.6\sqrt{2 \times 2} = 0.6 \times 2 = 1.2\ H$

7. $L_T = \cfrac{1}{\cfrac{1}{L_1 - M} + \cfrac{1}{L_2 - M}}$

where $M = k\sqrt{L_1 \times L_2} = 0.5\sqrt{10 \times 10} = 0.5 \times 10 = 5H$

$L_T = \cfrac{1}{\cfrac{1}{10 - 5} + \cfrac{1}{10 - 5}}$

$L_T = \dfrac{1}{0.4} = 2.5\ H$

9. $\dfrac{50}{500} = 0.1$ volts-per-turn (primary)

11. 2,000 turns

13. 1:4.75

15. 62.5 V

CHALLENGE QUESTIONS

1. A "flyback" transformer (sometimes called a "horizontal transformer") is typically used in the horizontal deflection circuits of televisions. It primarily provides high voltage to the picture tube or CRT. It may also provide low filament voltage for the high voltage rectifier. The scanning circuit, in which it operates, uses a frequency of 15,750 Hz.

TEST YOURSELF

1. $\dfrac{N_P}{N_s} = \sqrt{\dfrac{Z_P}{Z^s}}$

$$\frac{N_P}{200} = \sqrt{\frac{5,000}{16}}$$
$$N_P = 200 \times 17.67 = 3,535.5$$

3. If the impedance ratio (s-p) of an ideal transformer is 1:36, the turns ratio (s-p) is the square root of 1:36, or 1:6. This means the primary has six times as many turns as the secondary. Since the number of turns on the primary is 1,200, the $N_s = 1,200/6 = 200$ turns.

CHAPTER 18

PRACTICE PROBLEMS

I

1. $C = \dfrac{Q}{V} = \dfrac{100 \times 10^{-6}}{25} = 4 \times 10^{-6} = 4\ \mu F$

2. $Q = CV = 10 \times 10^{-6} \times 250 = 2,500\ \mu C$

3. $V = \dfrac{Q}{C} = \dfrac{50 \times 10^{-6}}{2 \times 10^{-6}} = 25\ V$

II

$$E = \frac{1}{2}CV^2 = \frac{10 \times 10^{-6} \times (100)^2}{2} = \frac{100,000 \times 10^{-6}}{2}$$
$$= 50,000\ \mu J\ \text{or}\ 0.05\ J$$

III

1. $C = \dfrac{2.25\ kA}{10^7{}_s} \times (n-1) = \dfrac{2.25 \times 10 \times 1}{10^7 \times 0.1}$

$C = \dfrac{22.5}{10^6} \times (3) = 22.5 \times 10^{-6} \times 3 = 67.5\ \mu F$

2. $C = \dfrac{8.85\ kA}{10^{12}{}_s} = \dfrac{8.85 \times 1 \times 0.2}{10^{12} \times 0.005} = \dfrac{1.77 \times 10^{-12}}{0.005}$
$= 354\ pF$

IV

1. $C_T = \dfrac{C_1 \times C_2}{C_1 + C_2} = \dfrac{5 \times 10^{-6} \times 20 \times 10^{-6}}{25 \times 10^{-6}} =$

$\dfrac{100 \times 10^{-12}}{25 \times 10^{-6}} = 4 \times 10^{-6}$, or $4\ \mu F$

2. Total capacitance of four 20-μF capacitors in series is equal to one of the equal value capacitors divided by 4. Thus, $20/4 = 5\ \mu F$, $C_T = C_1/C_N$.

3. V_{C_1} = four-fifths V_T, or 80 V. V_{C_2} = one-fifth V_T, or 20 V.

$(Q_T = C_T \times V_T; = 4\ \mu F \times 100\ V = 400\ \mu C)$

$\left(V_{C_1} = \dfrac{Q}{C_1}; = \dfrac{400\ \mu C}{5\ \mu F} = 80\ V\right)$

$\left(V_{C_2} = \dfrac{Q}{C_2}; = \dfrac{400\ \mu C}{20\ \mu F} = 20\ V\right)$

V

1. C_T = sum of individual capacitances; therefore = 60 μF.

2. $Q_{C_1} = C_1 \times V = 12\ \mu F \times 60\ V = 720\ \mu C$
$Q_{C_1} = Q_{C_2} = Q_{C_3} = Q_{C_4} = Q_{C_5}$
Q_T = sum of all the charges = $5 \times 720\ \mu C = 3,600\ \mu C$.

VI

1. One time constant for circuit = $0.001 \times 10^{-6} \times 10 \times 10^3$
1 TC = 0.01 milliseconds (or 10 microseconds)
Time allowed for charge = 20 μs, or 2 TC
After 2 TC, capacitor is charged to 86.5% of V applied.
In this case, V_C = 86.5% of 100 V, or 86.5 V, after 20 μs.

2. The voltage across the resistor 15 μs after the switch is closed will be equal to the percentage of V applied that V_R drops to after 1.5 TC.
(15 μs = 1.5 × the 10-μs TC). According to the TC chart, the resistor voltage is approximately 22% of V applied, or 22 V. Charging current = 2.2 mA.

3. V_C 30 μs after the switch is closed $\cong$ 95 V.

4. V_R 25 μs after the switch is closed $\cong$ 8.2 V.

✔ IN-PROCESS LEARNING CHECKS

I

1. A capacitor is an electrical component consisting of two conducting surfaces called **plates** that are separated by a nonconductor called the **dielectric.**

2. A capacitor is a device that stores electrical **charge** when voltage is applied.

3. Electrons **do not** travel through the capacitor dielectric.

4. During charging action, one capacitor plate collects **electrons,** making that plate **negative.** At the same time, the other plate is losing **electrons** to become **positive.**

5. Once the capacitor has charged to the voltage applied, it acts like an **open** circuit to dc. When the level of dc applied voltage increases, the capacitor **charges** to reach the new level. When the level of dc applied decreases, the capacitor **discharges** to reach the new level.

6. When voltage is first applied to a capacitor, **maximum** charge current occurs.

7. As the capacitor becomes charged, current **decreases** through the circuit in series with the capacitor.

REVIEW QUESTIONS

1. New capacitance is four times the original capacitance, since C is directly related to the area of the plates facing each other and inversely related to the dielectric thickness.

3. $C = \dfrac{Q}{V} = \dfrac{200 \text{ pC}}{50} = 4 \text{ pF}$

5. $C_T \text{ (pF)} = \dfrac{1}{\dfrac{1}{C_1} + \dfrac{1}{C_2} + \dfrac{1}{C_3}} = \dfrac{1}{\dfrac{1}{100} + \dfrac{1}{400} + \dfrac{1}{1,000}}$

$C_T = \dfrac{1}{0.01 + 0.0025 + 0.001} = \dfrac{1}{0.0135} = 74.07 \text{ pF}$

$C_T = 100 \text{ pF} + 400 \text{ pF} + 1,000 \text{ pF} = 1,500 \text{ pF}$

7. $Q_{C_1} = C_1 \times V_1 = 10 \ \mu\text{F} \times 100 \text{ V} = 1,000 \ \mu\text{C}$
$Q_{C_2} = C_2 \times V_2 = 20 \ \mu\text{F} \times 100 \text{ V} = 2,000 \ \mu\text{C}$
$Q_{C_3} = C_3 \times V_3 = 50 \ \mu\text{F} \times 100 \text{ V} = 5,000 \ \mu\text{C}$
$C_T = C_1 + C_2 + C_3 = 80 \ \mu\text{F}$
$Q_T = Q_1 + Q_2 + Q_3 = 8,000 \ \mu\text{C}$

9. $v_R = V \text{ applied} \times \epsilon^{-t/\tau} = 200 \times 2.718^{-1.2} = 60.23 \text{ V}$
$v_C = V_A - V_R$, or approximately $200 - 60 = 140 \text{ V}$

CHALLENGE QUESTIONS

1. Students should reflect the ranges and operational procedures for measuring capacitance with a DMM in terms of the equipment or vendor data available to them.

3. Students should reflect the ranges and operational procedures for measuring capacitance with an LCR meter in terms of the equipment or vendor data available to them.

TEST YOURSELF

1. Total capacitance of circuit shown is calculated using knowledge that capacitors in series add like resistors in parallel, and capacitors in parallel add like resistors in series. Thus, total capacitance of the two branches (points A to B) = 20 μF. This 20 μF is in series with C_6's 10 μF and C_1's 10 μF. Simplifying further, the two 10 μFs in series are equivalent to 5 μF. This value is in series with the 20 μF of the two branches.

$$\frac{20 \times 5}{20 + 5} = 4 \ \mu\text{F total capacitance}$$

3. $Q_{C_6} = 320 \ \mu\text{C}$

5. Voltage applied is 3,000 V. (Since the voltage across C_1 and across C_6 are each double that across points A and B, so 1,200 V + 1,200 V + 600 V = 3,000 V.) (NOTE: This cannot be actually done, since the voltage ratings of C_1 and C_6 were described as being

300 V each. Therefore, it would not be possible to have 1,200 V across each without their breaking down. But, for purposes of achieving 600 V across points A and B, it takes 3,000 V applied, as shown.)

7. Charge on first capacitor ($Q = C \times V$) = 15 μF × 150 V = 2,250 μC. Charge on second capacitor = 15 μF × 300 V = 4,500 μC. Total charge to be distributed on total capacitance is 6,750 μC. Since capacitors are connected in parallel, total C = 30 μF. $V = Q/C$ = 6,750 μC/30 μF = 225 V.

CHAPTER 19

PRACTICE PROBLEMS

I

1. Yes, the X_C formula verifies the values of X_C shown in Figure 19–3.

2. $X_C = 1/(2\pi fC) = 1/(6.28 \times 1,000 \times 10^{-6}) = 159 \ \Omega$

3. X_C remains the same if f is doubled and C is halved simultaneously.

II

1. $f = 1/(2\pi C X_C) = 1/(6.28 \times 500 \times 10^{-12} \times 1,000) = 318 \text{ kHz}$

2. $C = 1/(2\pi f X_C) = 1/(6.28 \times 2,000 \times 3,180) = 0.025 \ \mu\text{F}$

III

$X_{C_1} = 1/(2\pi fC) = 1/(6.28 \times 1,000 \times 2 \times 10^{-6}) = 79.5 \ \Omega$
$X_{C_2} = 1/(2\pi fC) = 1/(6.28 \times 1,000 \times 4 \times 10^{-6}) = 39.75 \ \Omega$
$X_{C_T} = X_{C_1} + X_{C_2} + X_{C_3} + X_{C_4} = 79.5 + 39.75 + 200 + 300 = 619.25 \ \Omega$

IV

1. $X_{C_T} = 1/(2\pi f C_T) = 1/(6.28 \times 1,000 \times 6 \times 10^{-6}) = 26.5 \ \Omega$

2. $X_{C_T} = 1/(2\pi f C_T) = 1/(6.28 \times 500 \times 25 \times 10^{-6}) = 12.7 \ \Omega$

3. $X_{C_T} = 1,600/4 = 400 \ \Omega$

V

1. $X_{C_1} = 1,000 \ \Omega$; $X_{C_2} = 1,000 \ \Omega$; $X_{C_3} = 2,000 \ \Omega$; $X_{C_T} = 4,000 \ \Omega$; $C_2 = 0.795 \ \mu\text{F}$; $C_3 = 0.3975 \ \mu\text{F}$; $V_2 = 25 \text{ V}$; $I_T = 25 \text{ mA}$

2. $C_1 = 0.5 \ \mu\text{F}$; $X_{C_1} = 1,000 \ \Omega$; $X_{C_2} = 1,000 \ \Omega$; $X_{C_3} = 500 \ \Omega$; $I_2 = 2 \text{ mA}$; $I_T = 8 \text{ mA}$

IN-PROCESS LEARNING CHECKS

I

1. For a capacitor to develop a potential difference between its plates, there must be a **charging** current.

2. For the potential difference between capacitor plates to decrease (once it is charged), there must be a **discharging** current.

3. When ac is applied to capacitor plates, the capacitor will alternately **charge** and **discharge.**

4. The value of instantaneous capacitor current directly relates to the value of **capacitance** and the rate of change of **voltage.**

5. The amount of charging or discharging current is maximum when the rate of change of voltage is **maximum.** For sine-wave voltage, this occurs when the sine wave is near **zero** points.

6. Capacitor current **leads** the voltage across the capacitor by **90** degrees. The expression that helps you remember this relationship is Eli the **ice** man.

REVIEW QUESTIONS

1. $X_C = \dfrac{1}{6.28fC} = \dfrac{1}{7.85 \times 10^{-6}} = 127.38 \text{ k}\Omega$

3. X_C decreases to one-third its original value, since

X_C is inverse to frequency $\left(X_C = \dfrac{1}{2\pi fC} \right)$

5. For the circuit shown:

$C_T = \dfrac{C_1 \times C_2}{C_1 + C_2} = \dfrac{0.636 \times 1.59}{0.636 + 1.59} = 0.454 \text{ µF}$

$X_{C_T} = X_{C_1} + X_{C_2} = 25 \text{ k}\Omega + 10 \text{ k}\Omega = 35 \text{ k}\Omega$

$V_A = 70 \text{ V (sum of } V_{C_1} \text{ and } V_{C_2})$

$C_1 = \dfrac{1}{6.28 \times 10 \text{ Hz} \times X_C}$; where $X_C = \dfrac{V}{I} =$

$\dfrac{50 \text{ V}}{2 \text{ mA}} = 25 \text{ k}\Omega$. Thus, $C_1 =$

$\dfrac{1}{6.28 \times 10 \times 25 \text{ k}\Omega} = \dfrac{1}{1570 \times 10^3} = 0.636 \text{ µF}.$

$C_2 = \dfrac{1}{6.28 \times 10 \text{ Hz} \times X_C}$, where $X_C = \dfrac{V}{I} = \dfrac{20 \text{ V}}{2 \text{ mA}} =$

$10 \text{ k}\Omega$. Thus, $C_2 = \dfrac{1}{6.28 \times 10 \times 10 \text{ k}\Omega} =$

$\dfrac{1}{628 \times 10^3} = 1.59 \text{ µF}$

7. a. C_1 will RTS
   b. C_2 will RTS
   c. C_T will RTS
   d. X_{C_1} will D
   e. X_{C_2} will D
   f. X_{C_T} will D
   g. I will I
   h. V_{C_1} will RTS
   i. V_{C_2} will RTS
   j. V_A will RTS

9. Total current increases to 12.56 mA, if $C_2 = C_1$ and all other factors remain the same.

11. X_C of $C_1 = 1,000 \text{ }\Omega$

13. $C_T = 0.063 \text{ µF}$

15. increase

CHALLENGE QUESTIONS

1. A capacitor can be said to be an "open circuit" with respect to dc because the insulating dielectric material between the capacitor's plates breaks the dc current path. Only very small "leakage" current can pass through the insulating material.

TEST YOURSELF

1. Capacitive voltage divider with the parameters specified is shown.

159 V 2 kHz

C_1 2 µF; X_{C_1} = 39.75 Ω; V_{C_1} = 39.75 V

C_2 1 µF; X_{C_2} = 79.5 Ω; V_{C_2} = 79.5 V

C_3 2 µF; X_{C_3} = 39.75 Ω; V_{C_3} = 39.75 V

I = 1 A

CHAPTER 20

PRACTICE PROBLEMS

I

1. $V_T = \sqrt{V_R^2 + V_C^2} = \sqrt{80^2 + 60^2} = \sqrt{10,000}$
   $V_T = 100 \text{ V}$

$V_C = 60 \text{ V}$ θ $V_R = 80 \text{ V}$ $V_T = 100 \text{ V}$

2. $Z = \sqrt{R^2 + X_C^2} = \sqrt{160^2 + 120^2} = \sqrt{40,000}$
   $Z = 200 \text{ }\Omega$

$X_C = 120 \text{ }\Omega$ θ $R = 160 \text{ }\Omega$ $Z = 200 \text{ }\Omega$

53 V

60 V θ

V_T 80.06 V

II

1. $V_T = 80 \text{ V}; \theta = 48.5°$

$R = 26.5 \text{ }\Omega$

$X_C = 30 \text{ }\Omega$ θ 48.5°

$Z = 40 \text{ }\Omega$

2. $\theta = 48.54°$

III

Angle of output voltage "lead" equals arctan X_C/R.
With R set at 500 Ω:
$X_C = 1/(2\pi fC) = 1/(6.28 \times 400 \times 0.5 \times 10^{-6}) = 796\ \Omega$
$R = 500\ \Omega$
$X_C/R = 796/500 = 1.59$
arctan 1.59 = angle of $57.8°$
Angle of lead = $57.8°$

IV

Lag angle = $90°$ minus circuit phase angle
Phase angle = arctan X_C/R
$X_C = 1/(2\pi fC) = 1/(6.28 \times 400 \times 0.5 \times 10^{-6}) = 796\ \Omega$
$R = 750\ \Omega$
$X_C/R = 796/750 = 1.06$
arctan 1.06 = angle of $46.7°$
Angle of lag = $90 - 46.7°$; Angle of lag = $43.3°$

V

1. $I_T = 22.36$ mA
2. $\theta = 63.43°$
3. $Z = 7.11$ kΩ

$I_C = 20$ mA $I_T = 22$ mA
$\theta = 63°$
V_T $I_R = 10$ mA

✔ IN-PROCESS LEARNING CHECKS

I

1. In series RC circuits, the **current** is the reference vector, shown at $0°$.
2. In the $V\text{-}I$ vector diagram of a series RC circuit, the circuit voltage is shown **lagging** the circuit current.
3. In series RC circuits, an impedance diagram **can** be drawn.
4. Using the Pythagorean approach, the formula to find V_T in a series RC circuit is
   $V_T = \sqrt{V_R^2 + V_C^2}$.
5. Using the Pythagorean approach, the formula to find Z in a series RC circuit is $Z = \sqrt{R^2 + X_C^2}$.
6. The trig function used to solve for phase angle when you do not know the hypotenuse value is the **tangent** function.

II

1. In parallel RC circuits, **voltage** is the reference vector in $V\text{-}I$ vector diagram.

2. In parallel RC circuits, current through the resistor(s) is plotted **out-of-phase** with the circuit current vector.
3. In parallel RC circuits, current through capacitor branch(es) is plotted at **+90 degrees** with respect to circuit applied voltage.
4. The Pythagorean formula to find circuit current in parallel RC circuits is $I_T = \sqrt{I_R^2 + I_C^2}$.
5. **No.**
6. **Yes.**
7. **No.**

REVIEW QUESTIONS

1. $Z = \sqrt{R^2 + X_C^2} = \sqrt{64 \times 10^6 + 225 \times 10^6} =$
   $\sqrt{289 \times 10^6} = 17$ kΩ

3. $C = \dfrac{1}{2\pi fX_C} = \dfrac{1}{6.28 \times 1,060 \times 15\text{ k}\Omega} = \dfrac{1}{99.8 \times 10^6}$
   $\times 10^6 = 0.01\ \mu$F

5. a. R will RTS
   b. X_C will D
   c. Z will D
   d. I_T will I
   e. θ will D
   f. V_C will D
   g. V_R will I
   h. V_T will RTS

7. $R = \dfrac{V_R}{I} = \dfrac{150\text{ V}}{10\text{ mA}} = 15$ kΩ

9. $f = \dfrac{1}{2\pi CX_C} = \dfrac{1}{6.28 \times 0.02 \times 10^{-6} \times 30 \times 10^3} =$
   $\dfrac{1}{3.77 \times 10^{-3}} = 265$ Hz

11. $R_T = 20$ kΩ + 18 kΩ + 30 kΩ = 68 kΩ
13. $Z = \sqrt{R^2 + X_C^2} = \sqrt{(68 \times 10^3)^2 + (52 \times 10^3)^2} =$
    $\sqrt{7,328 \times 10^6} = 85.6$ kΩ

15. $f = \dfrac{1}{2\pi CX_C} = \dfrac{1}{6.28 \times 0.1 \times 10^{-6} \times 30 \times 10^3} =$
    $\dfrac{1}{0.0188} = 53$ Hz

17. $I_C = \dfrac{320\text{ V}}{16\text{ k}\Omega} = 20$ mA

19. Tan $\theta = \dfrac{I_C}{I_R} = \dfrac{20}{32} = 0.625$. Angle whose tangent
    equals 0.625 is $32°$. $\theta = 32°$

CHALLENGE QUESTIONS

1. Two basic circuit operational differences between series RC circuits and series RL circuits are:

   a. Circuit current leads circuit voltage in series RC circuits and lags circuit voltage in series RL circuits.

   b. As frequency of operation is increased, or C is increased in series RC circuits, the phase angle between circuit current and circuit voltage decreases. For a series RL circuit, as frequency increases, or L is increased, the circuit phase angle increases.

3. Yes, there are similar circuit effects when comparing a series RC circuit and a parallel RL circuit. For example, as frequency of operation is increased, a *series RC circuit* will cause phase angle to decrease. The same is true for a *parallel RL circuit;* that is, an increase in f will cause the phase angle to decrease.

TEST YOURSELF

1. $R = 750 \ \Omega$
   $X_C = 1,000 \ \Omega$
   $I = 20$ mA
   $V_R = 15$ V
   $V_C = 20$ V
   $\theta = -53.1°$

   $$X_C = \frac{1}{6.28 \times 3,975 \times 0.04 \times 10^{-6}} = 1,000 \ \Omega$$

   $$\text{Tan } \theta = \frac{\text{opp}}{\text{adj}} = \frac{-X_C}{R} = \frac{-1,000}{750} = -1.33$$

   $\theta =$ the angle whose tan $= -1.33$
   $\theta = -53.1°$ (V_T lags reference vector for series circuits, I_T.)
   $R = Z \cos \theta = 1,250 \times \cos 53.1° = 1,250 \times 0.6 = 750 \ \Omega$

CHAPTER 21

PRACTICE PROBLEMS

I

$R = 200 \ \Omega$
$X_{C_T} = 500 \ \Omega$
$X_L = 250 \ \Omega$
$Z = \sqrt{R^2 + X^2} = \sqrt{200^2 + 250^2} = \sqrt{102,500} = 320.15 \ \Omega$
$V_T = I \times Z = 0.25 \times 320.15 = 80.04$ V
$V_{C_2} = I \times X_{C_2} = 0.25 \times 250 = 62.5$ V
$V_L = I \times X_L = 0.25 \times 250 = 62.5$ V
$V_R = I \times R = 0.25 \times 200 = 50$ V
$$\theta = \frac{\text{angle whose tangent equals } X(\text{net})}{R}$$

Tan $\theta = 250/200 = 1.25$; therefore, $\theta = -51.34°$ (V with respect to I)

II

$I_R = \dfrac{V}{R} = \dfrac{300}{200} = 1.5$ A
$I_L = \dfrac{V}{X_L} = \dfrac{300}{200} = 1.5$ A
$I_C = \dfrac{V}{X_C} = \dfrac{300}{300} = 1$ A
$I_X = I_L - I_C = 1.5 - 1 = 0.5$ A
$I_T \sqrt{I_R^2 + I_X^2} = \sqrt{1.5^2 + 0.5^2} = \sqrt{2.5} = 1.58$ A
$Z = \dfrac{V}{I} = \dfrac{300}{1.58} = 189.8 \ \Omega$
$\theta =$ angle whose tan equals $I_X/I_R = 0.5/1.5 = 0.333$
Therefore, $\theta = -18.43°$ (I with respect to V)

III

1. $20 + j30$ in rectangular form converts to $36.05 \angle 56.31°$ in polar form.

2. $30 - j20$ in rectangular form converts to $36.05 \angle -33.7°$ in polar form.

IV

1. $66 \angle 60°$ in polar form converts to $33 + j57.15$ in rectangular form.

2. $75 \angle 48°$ converts to $50.2 + j55.7$ in rectangular form.

V

1. $20 + j5$
   $\underline{35 - j6}$
   $55 - j1$

2. $34.9 - j10$
   $\underline{15.4 + j12}$
   $50.3 + j2$

3. $35 - j \, 6$ $35 - j6$
   $\underline{20 + j5 \text{ (change sign)}}$ $\underline{-20 - j5}$
   $15 - j11$

4. $37.8 - j13$ $37.8 - j13$
   $\underline{22.9 - j7 \text{ (change sign)}}$ $\underline{-22.9 + j7}$
   $14.9 - j6$

5. $5 \angle 20° \times 25 \angle -15° = 125 \angle 5°$

6. $24.5 \angle 31° \times 41 \angle 23° = 1,004.5 \angle 54°$

7. $\dfrac{5 \angle 20°}{25 \angle -15°} = 0.2 \angle 35°$

8. $\dfrac{33 \angle 34°}{13 \angle 45°} = 2.54 \angle -11°$

9. $I = 12 \angle 25°$, then $I^2 = 144 \angle 50°$

10. If $I^2 = 49 \angle 66°$, then $I = 7 \angle 33°$

11. $\dfrac{5 \angle 50° + (30 + j40)}{2.6 \angle -15°} =$

$\dfrac{(3.2 + j3.8) + (30 + j40)}{2.6 \angle -15°} = \dfrac{33.2 + j43.8}{2.6 \angle -15°}$

$= \dfrac{54.9 \ \angle 52.8°}{2.6 \angle -15°} = 21.1 \ \angle 67.8°$

12. $\dfrac{10 \angle 53.1° + (25 - j25)}{4.8 \angle 36°} = \dfrac{31 - j17}{4.8 \angle 36°} = \dfrac{35 \angle -28.7°}{4.8 \angle 36°}$

$= 7.3 \ \angle -64.7°$

✔ IN-PROCESS LEARNING CHECKS

I

1. When both capacitive and inductive reactance exist in the same circuit, the net reactance is **the difference between their reactances.**

2. When plotting reactances for a series circuit containing L and C, the X_L is plotted **upward** from the reference, and the X_C is plotted **downward** from the reference.

3. In a series circuit with both capacitive and inductive reactance, the capacitive voltage is plotted **180** degrees out-of-phase with the inductive voltage.

4. If inductive reactance is greater than capacitive reactance in a series circuit, the formula to find the circuit impedance is $Z = \sqrt{R^2 + (X_L - X_C)^2}$.

5. In a series RLC circuit, current is found by $\dfrac{V}{Z}$.

6. **False.**

7. If the frequency applied to a series RLC circuit decreases while the voltage applied remains the same, the component(s) whose voltage will increase is/are the **capacitor(s).**

II

1. In parallel RLC circuits, capacitive and inductive branch currents are **180 degrees** out-of-phase with each other.

2. The formula to find total current in a parallel RLC circuit when I_C is greater than I_L is

$I_T = \sqrt{I_R^2 + (I_C - I_L)^2}$.

3. **False.**

4. If the frequency applied to a parallel RLC circuit increases while the applied voltage remains the same, the branch current(s) that will increase is/are the **capacitive** branch(es).

III

1. Power dissipated by a perfect capacitor over one whole cycle of input is **zero.**

2. The formula for apparent power is $S = V \times I,$ and the unit expressing apparent power is **volt-amperes.**

3. True power in RLC circuits is dissipated by the **resistances.**

4. The formula for power factor is **p.f. = cos θ.**

5. If the power factor is one, the circuit is purely **resistive.**

6. The lower the power factor, the more **reactive** the circuit is acting.

7. In the power triangle, true power is on the **horizontal** axis, reactive power (VAR) is on the **vertical** axis, and apparent power is the **hypotenuse** of the right triangle.

REVIEW QUESTIONS

1. $Z = \sqrt{R^2 + X_T^2} = \sqrt{300^2 + 150^2} = \sqrt{112,500} = 335.4 \ \Omega$

$I = \dfrac{V_T}{Z} = \dfrac{10 \ V}{0.3354 \ k\Omega} = 29.8 \ mA$

$V_C = I \times X_C = 29.8 \ mA \times 250 \ \Omega = 7.45 \ V$
$V_L = I \times X_L = 29.8 \ mA \times 100 \ \Omega = 2.98 \ V$
$V_R = I \times R = 29.8 \ mA \times 300 \ \Omega = 8.94 \ V$

$\text{Cos } \theta = \dfrac{\text{adj}}{\text{hyp}} = \dfrac{R}{Z} = \dfrac{300}{335.4} = 0.89$

$\theta = $ angle whose cos $= 0.89 = 26.56°$ (I_T will lead V_T.)

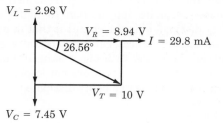

3. True power $= I^2R = 0.0298^2 \times 300 = 0.266 \ W$
Apparent power $= V \times I =$
$10 \ V \times 0.0298 \ mA = 0.298 \ VA$
Reactive power $= V \times I \times \sin \theta = 10 \ V \times 0.0298 \times 0.447 = 0.133 \ VAR$

Power factor $= \dfrac{\text{true power}}{\text{apparent power}} = \dfrac{0.266}{0.298} = 0.89$

(or p.f. $= \text{Cos } \theta = \cos 26.56° = 0.89$)

5. $Z = 20 + j40$

   If the circuit were an RC circuit, it would be stated as

   $Z = 20 - j40$

7. $I_R = \dfrac{20}{20} = 1$ A

   $I_C = \dfrac{20}{20} = 1$ A

   $I_T = 1.414$ A at an angle of 45°

   $I_T = 1.414 \angle 45°$

9. Converting $14.14 \angle 45°$ to rectangular form: $10 + j10$. That is, $R = Z \times \cos\theta = 14.14 \times \cos 45° = 10$, since the angle is 45°; R and X must be equal; thus $Z = R + jX_L$

   $14.14 = 10 + j10$

11. $Z = 61.43 \angle -75.96°$ kΩ (polar form)

    $Z = 14,900 - j59,600$ (rectangular form)

13. V_C will decrease.

15. $Z = 1.79 \angle -63.4°$ Ω

17. $Z = 5.82 \angle 14°$ Ω

CHALLENGE QUESTIONS

1. $10 - j10 = 14.14 \angle -45°$ in polar form.

3. $1.21 \angle -14° = 1.17 - j0.29$

5. A "real number" in math can be any rational or irrational number. In complex numbers related to the coordinate system, the real term in the complex number is plotted on the horizontal (or x) axis of the coordinate system. Imaginary numbers are complex numbers; they involve the square roots of negative numbers. The imaginary unit, or j operator, represents the square root of -1. For our purposes, the j operator is the vertical element used in ac analysis.

7. The reason simply measuring voltage with a voltmeter and the current with a current meter does not provide adequate information to find the actual power dissipated by the circuit is that in ac circuits containing reactances, currents and voltages are not generally in-phase with one another. Thus, taking the product of values that are not in-phase will not yield the actual power being dissipated.

TEST YOURSELF

1. $G = \dfrac{1}{400} = 0.00250$

   $-jB_L = \dfrac{1}{500} = 0.00200$

   $jB_C = \dfrac{1}{800} = 0.00125$

$Y = G + j(B_C - B_L)$

$Y = 0.0025 + (0.00125 - 0.002)$

$Y = 0.0025 - 0.00075 = 0.00175$ siemens

3. 500 Ω

CHAPTER 22

PRACTICE PROBLEMS

I

$f_r = \dfrac{1}{(2\pi\sqrt{LC})} = \dfrac{1}{(6.28 \times \sqrt{250 \times 10^{-6} \times 500 \times 10^{-12}})}$

$= 450.4$ kHz

II

$L = \dfrac{25,330}{f^2 C} = 10.13$ μH

III

1. $Q = \dfrac{X_L}{R} = \dfrac{1,000}{50} = 20$

2. $V_C = I \times X_C$

   $I = \dfrac{V_T}{Z} = \dfrac{12}{50} = 0.24$ A

   $V_C = 0.24$ A $\times 1,000$ $\Omega = 240$ V

3. If r_s doubles, $Q = \dfrac{1,000}{100} = 10$.

4. If r_s is made smaller, the skirts are steeper.

IV

1. $f_r = \dfrac{1}{(2\pi\sqrt{LC})} = 2.517$ MHz

2. $Q = \dfrac{X_L}{R} = 316.13$

3. $Z = Q \times X_L =$ approximately 999 kΩ

4. If r_s is 1 kΩ; the f_r is lower.

V

1. $f_1 = 2,494$ kHz and $f_2 = 2,506$ kHz

2. $f_r = 500$ kHz and bandpass = from 495 kHz to 505 kHz

3. $f_1 = 3,200$ kHz $- \dfrac{2 \text{ kHz}}{2} = 3,200$ kHz $- 1$ kHz $= 3,199$ kHz

✓ IN-PROCESS LEARNING CHECKS

I

1. In circuits containing both inductance and capacitance, when the source frequency increases, inductive reactance **increases** and capacitive reactance **decreases.**

2. In a series RLC (or LC) circuit, when X_L equals X_C, the circuit is said to be at **resonance.**

3. In a series RLC (or LC) circuit, when X_L equals X_C, the circuit impedance is **minimum;** the circuit current is **maximum;** and the phase angle between applied voltage and current is **zero** degrees.

4. In a series RLC (or LC) circuit, when X_L equals X_C, the voltage across the inductor or the capacitor is **maximum.**

II

1. In a high-Q parallel resonant circuit, Z is **maximum.**

2. In a high-Q parallel circuit using ideal components, X_L **does** equal X_C at resonance.

3. The power factor of an ideal parallel resonant circuit is **1.**

4. Phase angle in an ideal parallel resonant circuit is **0 degrees.**

5. In a parallel resonant circuit, the current circulating in the tank circuit is **higher** than the line current.

REVIEW QUESTIONS

1. $X_L = X_C$

3. a. $V_C = Q \times V_A = 9.98 \times 20$; $V_C = 199.6$ V

   b. I at resonance $= \dfrac{V_A}{R} = \dfrac{20 \text{ V}}{10 \text{ }\Omega} = 2$ A

   c. True power $= I^2 R = 2^2 \times 10 \text{ }\Omega = 40$ W

   d. Net reactive power is 0 VAR.

5. I line $= \dfrac{V}{Z} = \dfrac{20 \text{ V}}{31.018 \text{ k}\Omega} = 0.6447$ mA

   $I_C = Q \times I$ line $= 32.155 \times 0.6447$ mA $= 20.73$ mA

7. Circuit is shown.

Resonant bandpass filter

Input   Output

9. Phase angle will be 45° at a frequency of 39 kHz. (Start solution with $\Delta f = f_r/Q$, 45° phase angle occurs at a half power point.)

11. $C = \dfrac{0.02533}{f_{r^2}L} = 4.79$ pF

13. At resonance, voltage across $R = V$ applied or 100 V.

15. I_T will lead V_A due to higher capacitive branch current and lower inductive branch current. I_T would be capacitive.

17. To tune to resonance, increase C (choice d). (Since it was originally acting inductively, this indicates current through inductive branch was greater than current through capacitive branch. By increasing C, X_C decreases and I_C increases. Changed by the proper amount, resonance is reached.)

CHALLENGE QUESTIONS

1. The tuned circuit in a receiver input stage is typically a parallel LC circuit.

3. They would be considered resonant filters.

5. *Selectivity*—the ability to select desired signal frequencies, while rejecting undesired signal frequencies. Referring to resonant circuits, selectivity indicates the sharpness of the response curve. A narrow and more pointed response curve results in a higher degree of frequency selectiveness, or selectivity. *Bandwidth*—the frequency limits (*total difference in frequency* between upper and lower frequencies) over which a specified response level is shown by a frequency-sensitive circuit, such as a resonant circuit. The cutoff level is defined at 70.7% of peak response, sometimes called the half-power points on the response curve. *Bandpass*—the band of frequencies that passes through a filter or circuit with minimal attenuation or degradation; often *designated by specific frequencies*. When specified, they are designated indicating that the lower frequency is at the lower frequency "half-power point" of the response curve and the upper frequency is designated as the frequency that is at the upper "half-power point" on the response curve. Bandpass is often expressed as *the number of hertz between these two frequency response limits.*

TEST YOURSELF

1. a. X_L will I
   b. X_C will D
   c. R will RTS
   d. Z will I
   e. I will D
   f. θ will I
   g. V_C will D
   h. V_L will I
   i. V_R will D
   j. V_T will RTS
   k. BW will RTS
   l. Q will RTS (close to the same)

  m. I_C will D
  n. I_L will D
  o. I_T will D

3. a. X_L will I
  b. X_C will D
  c. R will RTS
  d. Z will D
  e. I will I
  f. θ will I
  g. V_C will RTS
  h. V_L will RTS
  i. V_R will RTS
  j. V_T will RTS
  k. BW will RTS
  l. Q will RTS
  m. I_C will I
  n. I_L will D
  o. I_T will I

5. The m-derived filter is a modified version of a constant-k filter. Its characteristics are such that the "m" factor relates to the ratio of the filter cutoff frequency to the frequency of infinite attenuation. In general, m ranges in values between 0 and 1. (Typically around 0.6.) The constant-k filter is one where the product of the impedances in the filter "arms" (typically a capacitive element and an inductive element) are constant at any frequency. The "k" represents whatever this "constant" factor is, (e.g., $Z_1 \times Z_2 = k^2$). The purpose for such filter design parameters as the constant-k and m-derived is to have filters whose input and output impedances are relatively constant over the frequency range of expected operation, or the passband. That is, to maintain their impedance match between output (load) and input (source) connected to the filter network.

Chapter Challenge Test Results

Find the number listed next to the test you chose and record the result.

1. 1.4 V
2. signal normal
3. 2 kΩ
4. 2 V
5. 1.5 V
6. high
7. 1.5 V
8. 0 V
9. 0 V
10. 3 V
11. 0 V
12. 7 V
13. ∞ Ω
14. 0 Ω
15. ≈0 V
16. 1 kΩ
17. 400 Ω
18. 1.5 V
19. ∞ Ω
20. 1.5 V
21.
22. 10 V
23. no signal
24. circuit operates normally
25. 10 kΩ
26. slightly high
27. 0 V
28. 0 V
29. 0 Ω
30. greatly below max value

31. 5 V
32. 400 Ω
33. ∞ Ω
34. 6.4 V
35. noticeably high
36. 4 kΩ
37. 10 kΩ
38. 3.3 V
39. 14 V
40. 0 V
41. ≈1.5 mH
42. 3.0 V
43. 7 V
44. 1.5 V
45. high
46. 0 Ω
47. 0 V
48. 2.3 V
49. 14 V
50. 0.5 V
51. 1.5 V
52. 10 V
53. (14 VAC)
54. 10 kΩ
55. max value
56. 10 V
57. signal normal
58. ∞ Ω
59. 1.5 kΩ
60. 275 pF

61. 14 V
62. 1 kΩ
63. ∞ Ω
64. 1.5 V
65. ≈6 VDC
66. 0 V
67. 7.5 V
68. 10 kΩ
69. low
70. signal normal
71. no change in operation
72. 0 Ω
73. 1.5 V
74. (120 VAC)
75. 10 V
76. slightly low
77. 5.9 V
78. 3.1 V
79. 7.5 V
80. 7 V
81. 2 V
82. 0 V
83. 10 V
84. 0 Ω
85. slightly below max value
86. 1.5 V
87. 5 Ω
88. 14 V
89. 2.3 V

90. ∞ Ω
91. 0 V
92. normal
93. 10 kΩ
94. ≈5 V
95. 10 V
96. ≈11 V
97. 0 V
98. low
99. 1.5 V
100. ∞ Ω
101. 0 V
102. 0 V
103. 0 V
104. ≈5 kΩ
105. 0 V
106. no signal
107. 50 pF
108. ∞ Ω
109. 12 kΩ
110. 0 V
111. no change in operation
112. 1.5 V
113. signal normal
114. 297 Ω
115. 0 V
116. ≈5 V
117. within normal range

Copper-Wire Table

Wire Size A.W.G. (B&S)	Diam. in Mils[1]	Circular Mil Area	Turns per Linear Inch (25.4 mm)[2]			Cont.-duty current[3] single wire in open air	Cont.-duty current[3] wires or cables in conduits or bundles	Feet per Pound (0.45 kg) Bare	Ohms per 1000 ft. 25°C	Current Carrying Capacity[4] at 700 C.M. per Amp.	Diam. in mm.	Nearest British S.W.G. No.
			Enamel	S.C.E.	D.C.C.							
1	289.3	83690	—	—	—	—	—	3.947	0.1264	119.6	7.348	1
2	257.6	66370	—	—	—	—	—	4.977	0.1593	94.8	6.544	3
3	229.4	52640	—	—	—	—	—	6.276	0.2009	75.2	5.827	4
4	204.3	41740	—	—	—	—	—	7.914	0.2533	59.6	5.189	5
5	181.9	33100	—	—	—	—	—	9.980	0.3195	47.3	4.621	7
6	162.0	26250	—	—	—	—	—	12.58	0.4028	37.5	4.115	8
7	144.3	20820	—	—	—	—	—	15.87	0.5080	29.7	3.665	9
8	128.5	16510	7.6	—	7.1	73	46	20.01	0.6405	23.6	3.264	10
9	114.4	13090	8.6	—	7.8	—	—	25.23	0.8077	18.7	2.906	11
10	101.9	10380	9.6	9.1	8.9	55	33	31.82	1.018	14.8	2.588	12
11	90.7	8234	10.7	—	9.8	—	—	40.12	1.284	11.8	2.305	13
12	80.8	6530	12.0	11.3	10.9	41	23	50.59	1.619	9.33	2.053	14
13	72.0	5178	13.5	—	12.8	—	—	63.80	2.042	7.40	1.828	15
14	64.1	4107	15.0	14.0	13.8	32	17	80.44	2.575	5.87	1.628	16
15	57.1	3257	16.8	—	14.7	—	—	101.4	3.247	4.65	1.450	17
16	50.8	2583	18.9	17.3	16.4	22	13	127.9	4.094	3.69	1.291	18
17	45.3	2048	21.2	—	18.1	—	—	161.3	5.163	2.93	1.150	18
18	40.3	1624	23.6	21.2	19.8	16	10	203.4	6.510	2.32	1.024	19
19	35.9	1288	26.4	—	21.8	—	—	256.5	8.210	1.84	0.912	20
20	32.0	1022	29.4	25.8	23.8	11	7.5	323.4	10.35	1.46	0.812	21
21	28.5	810	33.1	—	26.0	—	—	407.8	13.05	1.16	0.723	22
22	25.3	642	37.0	31.3	30.0	—	5	514.2	16.46	0.918	0.644	23
23	22.6	510	41.3	—	37.6	—	—	648.4	20.76	0.728	0.573	24
24	20.1	404	46.3	37.6	35.6	—	—	817.7	26.17	0.577	0.511	25
25	17.9	320	51.7	—	38.6	—	—	1031	33.00	0.458	0.455	26
26	15.9	254	58.0	46.1	41.8	—	—	1300	41.62	0.363	0.405	27
27	14.2	202	64.9	—	45.0	—	—	1639	52.48	0.288	0.361	29
28	12.6	160	72.7	54.6	48.5	—	—	2067	66.17	0.228	0.321	30
29	11.3	127	81.6	—	51.8	—	—	2607	83.44	0.181	0.286	31
30	10.0	101	90.5	64.1	55.5	—	—	3287	105.2	0.144	0.255	33
31	8.9	80	101	—	59.2	—	—	4145	132.7	0.114	0.227	34
32	8.0	63	113	74.1	61.6	—	—	5227	167.3	0.090	0.202	36
33	7.1	50	127	—	66.3	—	—	6591	211.0	0.072	0.180	37
34	6.3	40	143	86.2	70.0	—	—	8310	266.0	0.057	0.160	38
35	5.6	32	158	—	73.5	—	—	10480	335	0.045	0.143	38–39
36	5.0	25	175	103.1	77.0	—	—	13210	423	0.036	0.127	39–40
37	4.5	20	198	—	80.3	—	—	16660	533	0.028	0.113	41
38	4.0	16	224	116.3	83.6	—	—	21010	673	0.022	0.101	42
39	3.5	12	248	—	86.6	—	—	26500	848	0.018	0.090	43
40	3.1	10	282	131.6	89.7	—	—	33410	1070	0.014	0.080	44

[1] A mil is 0.001 inch. A circular mil is a square mil $\times \frac{\pi}{4}$. The circular mil (c.m.) area of a wire is the square of the mil diameter.

[2] Figures given are approximate only; insulation thickness varies with manufacturer.

[3] Max. wire temp. of 212°F (100°C) and max. ambient temp. of 135°F (57°C).

[4] 700 circular mils per ampere is a satisfactory design figure for small transformers, but values from 500 to 1000 c.m. are commonly used.

Useful Conversion Factors

TO CONVERT	INTO	MULTIPLY BY
ampere-turns/cm	amp-turns/inch	2.54
ampere-turns/inch	amp-turns/cm	0.3937
ampere-turns/inch	amp-turns/meter	39.37
ampere-turns/meter	amp-turns/inch	0.0254
Centigrade	Farenheit	$(C° \times 9/5) + 32$
centimeters	feet	3.281×10^{-2}
centimeters	inches	0.3937
centimeters/sec	feet/sec	0.03281
circular mils	sq mils	0.7854
circumference	radians	6.283
circular mils	sq inches	7.854×10^{-7}
coulombs	faradays	1.036×10^{5}
degrees	radians	0.01745
dynes	joules/cm	10^{-7}
dynes	joules/meter (newtons)	10^{-5}
ergs	foot-pounds	7.367×10^{-6}
ergs	joules	10^{-7}
farads	microfarads	10^{6}
faradays	ampere-hours	26.8
faradays	coulombs	9.649×10^{4}
feet	centimeters	30.48
foot-pounds	ergs	1.356×10^{7}
gausses	lines/sq inch	6.452
gausses	webers/sq meter	10^{-4}
gilberts	ampere-turns	0.7958
grams	dynes	980.7
grams	pounds	2.205×10^{-3}
henries	millihenries	1,000
horsepower	foot-lbs/sec	550
horsepower	kilowatts	0.7457
horsepower	watts	745.7
inches	centimeters	2.54
inches	millimeters	25.4
joules	ergs	10^{7}
joules	watt-hrs	2.778×10^{-4}
kilolines	maxwells	1,000
kilowatts	foot-lbs/sec	737.6
kilowatts	horsepower	1.341
kilowatts	watts	1,000
kilowatt-hrs	joules	3.6×10^{6}
lines/sq inch	webers/sq meter	$1,550 \times 10^{-5}$
maxwells	kilolines	0.001
maxwells	webers	10^{-8}
microfarad	farads	10^{-6}
microns	meters	1×10^{-6}
millihenries	henries	0.001
mils	inches	0.001
nepers	decibels	8.686
ohms	megohms	10^{-6}
quadrants	degrees	90
quadrants	radians	1.571
radians	degrees	57.3
square inches	sq cms	6.452
temperature (°F)−32	temperature (°C)	5/9
watts	horsepower	1.341×10^{-3}
watts (Abs.)	joules/sec	1
webers	maxwells	10^{8}
webers/sq meter	webers/sq inch	6.452×10^{-4}
yards	meters	0.9144

APPENDIX F

Schematic Symbols

APPENDIX G
Glossary

ac: abbreviation for alternating current. The letters "ac" are also used as a prefix, or modifier to designate voltages, waveforms, and so forth that periodically alternate in polarity.

admittance (Y): the ability of a circuit with both resistance and reactance to pass ac current; the reciprocal of impedance.

alternation: one-half of a cycle. Alternations are often identified as the positive alternation or the negative alternation when dealing with ac quantities.

amalgamation: in carbon-zinc dry cell manufacturing, coating the inside of the zinc electrode with a thin layer of mercury to help prevent local action.

ampere (A): the basic unit of current flow. An ampere of current flow represents electron movement at a rate of one coulomb per second; that amount of current flowing through one ohm of resistance with one volt applied.

ampere-hour rating: a method of rating cells or batteries based on the amount of current they supply over a specified period with specified conditions.

analog operation: circuit operation where the outputs vary continuously over a range as a direct result of the changing input levels; also called linear operation.

apparent power (S): the product of V and I in an ac circuit, without regard to phase difference between V and I. The unit of measure is volt-amperes. In the power triangle it is the vector resulting from true and reactive power.

assumed voltage method: a method of finding total resistance in parallel circuits by assuming a circuit applied voltage value to find branch currents, which enables finding total current and total circuit resistance with less complex mathematics than other methods for certain situations.

AT: the abbreviation for ampere-turns or magnetomotive force.

atom: the basic building block of matter composed of different types of particles. Major atom particles are the electron, proton, and neutron.

attenuator: a controllable voltage divider network in oscilloscopes that helps adjust the amplitude of input signals to a useable size.

autotransformer: a special single-winding transformer where part of the winding is in both the primary and secondary circuit. These transformers can be made to step-up or step-down, depending on where on the winding the primary source and the secondary load are connected.

average value (for one-half cycle): in ac sinusoidal values, the average height of the curve above the zero axis, expressed as 0.637 times maximum value. (NOTE: The average height over an *entire* cycle is zero; however, the average height of one alternation is as defined above.)

B: the symbol representing flux density.

backoff ohmmeter scale: the reverse (right-to-left) scale characteristic of series-type ohmmeters. Zero ohms is indicated on the scale's right end and infinite ohms on the scale's left end. Also, this scale is a nonlinear scale. That is, calibration marks and values are not evenly spaced across the scale.

bandpass: the band of frequencies that passes through a filter with minimal attenuation or degradation; often designated by specific frequencies, where the lower frequency is the frequency at the low-frequency half-power point, and the upper frequency is the frequency at the upper half-power point on the resonance curve; also called pass-band.

bandpass filter: a filter that passes a selected band of frequencies with little loss but greatly attenuates frequencies either above or below the selected band.

bandstop filter: a filter that stops (greatly attenuates) a selected band of frequencies while allowing all other frequencies to pass with little attenuation; also called band-reject filter.

bandwidth: the frequency limits where a specified response level is shown by a frequency-sensitive circuit, such as a resonant circuit. The cutoff level is defined at 70.7% of peak response, sometimes called the half-power points on the response curve.

battery: a dc voltage source containing a combination of cells, connected to produce higher voltage or current than a single cell produces alone.

bilateral resistance: any resistance having equal resistance in either direction; that is, R is the same for current passing either way through the component.

bleeder: the resistive component assuring that the minimum current drawn from a power supply system is sufficient to achieve reasonable voltage regulation over the range of output currents demanded from the supply; serves to discharge the filter capacitors when the power supply is turned off, which adds safety for people working on electronic equipment fed by the power supply. This resistive component is also called bleeder resistor.

bleeder current: the fixed current through a resistive system that helps keep voltage constant under varying load conditions.

bleeder resistor: the resistor connected in parallel with power supply to draw a bleeder current.

block diagrams: graphic illustrations of systems or subsystems by means of blocks (or boxes) that contain information about the function of each block. These blocks are connected by lines that show direction of flow (signal, fluid, electrical energy, etc.) and how the blocks interact. Block diagrams can illustrate any kind of system: electrical, electronic, hydraulic, mechanical, and so forth. "Flow" generally moves from left to right through the diagram but is not a requirement.

bridge circuit: a special series-parallel circuit usually composed of two, two-resistor branches. Depending on the relative values of these resistors, the bridge output (between the center points of each branch) may be balanced or unbalanced.

bridging or shunting: connecting a component, device, or circuit in parallel with an existing component, device, or circuit.

capacitance: the ability to store electrical energy in the form of an electrostatic field.

capacitive reactance (X_C): the opposition a capacitor gives to ac or pulsating dc current. The symbol used to represent capacitive reactance is X_C.

capacitor: a device consisting of two or more conductors, separated by nonconductor(s). When charged, a capacitor stores electrical energy in the form of an electrostatic field and blocks dc current.

cell: (relates to voltage sources) a single (stand-alone) element or unit within a combination of similar units that convert chemical energy into electrical energy. The chemical action within a cell produces dc voltage at its output terminals or contacts.

cemf (counter-emf or back-emf): the voltage induced in an inductance that opposes a current change and is present due to current changes causing changing flux linkages with the conductor(s) making up the inductor.

charge (electrical): For our purposes, charge can be thought of as the electrical energy present where there is an accumulation of excess electrons, or a deficiency of electrons. For example, a capacitor stores electrical energy in the form of "charge" when one of its plates has more electrons than the other.

charging a capacitor: moving electrons to one plate of a capacitor and moving electrons off the other plate; resultant charge movement is called charging current.

circuit: a combination of elements or components that are connected to provide paths for current flow to perform some useful function.

coefficient of coupling (k): the amount or degree of coupling between two circuits. In inductively coupled circuits, it is expressed as the fractional amount of the total flux in one circuit that links the other circuit. If all the flux links, the coefficient of coupling is 1. The symbol for coefficient of coupling is k. When k is multiplied by 100, the percentage of total flux linking the circuits is expressed.

complex numbers: term that expresses a combination of a *real* and an *imaginary* term which must be added vectorially or with phasors. The real term is either a positive or negative number displayed on the horizontal axis of the rectangular coordinate system. The imaginary term, *j operator,* expresses a value on the vertical or *j*-axis. The value of the real term is expressed first. For example, $5 + j10$ means plus 5 units (to the right) on the horizontal axis and up 10 units on the vertical axis.

compound: a form of matter that can be chemically divided into simpler substances and that has two or more types of atoms.

conductance: the ease with which current flows through a component or circuit; the opposite of resistance.

conductor: a material that has many free electrons due to its atoms' outer ring having less than 4 electrons, which is less than the 8 needed for chemical and/or electrical stability.

constant current source: a current source whose output current is constant even with varying load resistances. Its output voltage varies to enable constant output current, as appropriate.

constant voltage source: a voltage source where output voltage (voltage applied to circuitry connected to source output terminals) remains virtually constant even under varying current demand (load) situations.

coordinate system: in ac circuit analysis, a system to indicate phase angles within four, 90° quadrants. The vertical component is the "Y" axis. The horizontal component is the "X" axis.

coordinates: the vertical and horizontal axes that describe the position and magnitude of a vector.

copper losses: the I^2R losses in the copper wire of the transformer windings. Since current passes through the wire and the wire has resistance to current flow, an I^2R loss is created in the form of heat, rather than this energy being a useful part of energy transfer from primary source to secondary load.

core: soft iron piece to help strengthen magnetic field; helps produce a uniform air gap between the permanent magnet's pole pieces and itself.

core losses: energy dissipated in the form of heat by the transformer core that subtracts from the energy transferred from primary to secondary; basically comprised of eddy-current losses (due to spurious induced currents in the core material creating an I^2R loss) and hysteresis losses, (due to energy used in constantly changing the polarity of the magnetic domains in the core material).

cosine: a trigonometric function of an angle of a right triangle; the ratio of the adjacent side to the hypotenuse.

cosine function: a trigonometric function used in ac circuit analysis; relationship of a specific angle of a right triangle to the side adjacent to that angle and the hypotenuse where $\cos \theta$ = adjacent/hypotenuse.

coulomb: the basic unit of charge; the amount of electrical charge represented by 6.25×10^{18} electrons.

covalent bonding: the pairing of adjacent atom electrons to create a more stable atomic structure.

CRT (cathode-ray tube): the oscilloscope component where electrons strike a luminescent screen material causing light to be emitted. Waveforms are displayed on an oscilloscope's CRT.

current (I): the progressive movement of electrons through a conductor. Current is measured in amperes.

current divider circuit: any parallel circuit divides total current through its branches in inverse relationship to its branch R values.

current leading voltage: sometimes used to describe circuits where the circuit current leads the circuit voltage, or the circuit voltage lags the circuit current.

current ratio: the relationship of the current values of secondary and primary, or vice versa. In an ideal transformer, the current ratio is the *inverse* of the voltage and turns ratios; $I_P/I_S = N_S/N_P = V_S/V_P$.

cycle: a complete series of values of a periodic quantity that recur over a period of time.

damping (electrical): the process of preventing an instrument's pointer from oscillating above and below

the measured value. Electrical damping is due to the magnetic field produced by induced current opposing the motion that induced it.

delta (Δ): When used in mathematical equations, a term meaning "a small change in." When used in conjunction with an electrical network configuration, it describes the general triangular appearance of the schematic diagram layout which depicts the electrical connections between components. Also, sometimes, "delta networks" are laid out in a fashion that they are called pi (π) networks.

delta network: An interconnection of components in a closed ring configuration that schematically can be represented as a triangular shape, similar to the Greek letter delta. A set of three branches connected so as to provide one closed mesh. Delta networks often appear in three-phase power systems, but also appear in other electrical and electronic circuits.

delta-wye conversion: the mathematical process of converting network component values, used in a delta network, to appropriate values to be used in a "Wye" network, and yet present the same impedance and load conditions to a source feeding such a network.

depolarizer: in a dry cell, the material to help prevent or lessen polarization. A common depolarizing material is manganese dioxide.

dielectric: nonconductive material separating capacitor plates.

dielectric constant: the relative permittivity of a given material compared to the permittivity of vacuum or air; the relative ability of a given material to store electrostatic energy (per given volume) to that of vacuum or air.

dielectric strength: the highest voltage a given dielectric material tolerates before electrically breaking down or rupturing; rated in volts per thousandths of an inch, V per mil.

digital operation: circuit operation where the output varies in discrete steps or in an on-off, high-low, 1-0 category of operation.

digital operation: circuit operation where the output varies in discrete steps or in an on-off, high-low, 1-0 category of operation.

discharging a capacitor: balancing (or neutralizing) the charges on capacitor plates by allowing excess electrons on the negative charged plate to move through a circuit to the positive charged plate. When opposite capacitor plates have the same charge, the voltage between them is zero and the capacitor is fully discharged.

dropping resistor: a resistor connected in series with a given load that drops the circuit applied voltage to the level required by the load when passing the rated load current through it. (Recall a *voltage drop* is the difference in voltage between two points caused by a loss of pressure (or loss of emf) as current flows through a component offering opposition to current flow.)

effective value (rms): the value of ac voltage or current that produces the same heating effect in a resistor as that produced by a dc voltage or current of equal value. In ac, this value is 0.707 times maximum value.

efficiency: percentage, or ratio, of useful output (such as power) delivered compared to amount of input required to deliver that output. For example, in electrical circuits, the amount of power input to a device, component stage, or system compared to the amount of power delivered to the load. Generic formula is Efficiency $(\%) = (P_{out}/P_{in}) \times 100$.

electrical charge: an excess or deficiency of electrons compared to the normal neutral condition of having an equal number of electrons and protons in each atom of the material.

electrode: generally, one of the electrical terminals that is a conducting path for electrons in or out of a cell, battery, or other current generating device.

electrolyte: a liquid, paste, or substance through which electrical conduction occurs due to chemical processes of the electrolyte and other materials immersed or exposed to it.

electromagnetism: the property of a conductor to produce a magnetic field when current is passing through it.

electron: the negatively charged particle in an atom orbiting the atom's nucleus.

electrostatic (electric) field: the region surrounding electrically charged bodies; area where an electrically charged particle experiences force.

element: a form of matter that cannot be chemically divided into simpler substances and that has only one type of atom.

energy: the ability to do work, or that which is expended in doing work. The basic unit of energy is the erg, where 980 ergs = 1 gram cm of work. A common unit in electricity/electronics is the joule, which equals 10^7 ergs.

ε: the Greek letter epsilon used to signify the mathematical value of 2.71828. This number, raised to various exponents that relate to time constants, can be used to determine electrical circuit parameter values in circuits where exponential changes occur.

equation: a mathematical statement of equality between two quantities or sets of quantities. Example, $A = B + C$.

equivalent resistance: the total resistance of two or more resistances in parallel.

Faraday's law: the law states the amount of induced voltage depends on the rate of cutting magnetic flux lines. In other words, greater number of flux lines cut per unit time produces greater induced voltage.

filter: in power supplies, the component or combination of components that reduce the ac component of the pulsating dc output of the rectifier prior to the output fed to a load.

flux: relating to magnetism; the magnetic lines of force.

flux density (B): the magnetic flux per unit cross-sectional area; unit of measure is the tesla (T) representing 1 Wb per square meter.

focus control: a scope control that adjusts the narrowness of the electron stream striking the screen to create a clear and sharp presentation.

force: in electrical charges, the amount of attraction or repulsion between charges. In electromotive force, the potential difference in volts.

frame/movable coil: the parts in a moving-coil meter movement that move in response to the interaction between the fixed magnetic field and the field of the current-carrying, movable coil.

free electrons: the electrons in the outer ring of conductor materials that are easily moved from their original atom. Sometimes termed valence electrons.

frequency (f): the number of cycles occurring in a unit of time. Generally, frequency is the number of cycles per second, and hertz (Hz) is cycles per second. In formulas, frequency is expressed as f, where $f = 1/T$.

full-scale current: the value of current that causes full-scale, or full-range deflection of the pointer or indicator device in a current meter movement; the maximum current that should pass through the moving-coil in a current meter movement.

gain controls: oscilloscope controls that adjust the horizontal and vertical size of the image displayed.

ground reference: a common reference point for electrical/electronic circuits; a common line or conducting surface in electrical circuits used as the point where electrical measurements are made; a common line or conductor where many components make connection to one side of a power source. Typically, ground reference is the chassis, or a common printed circuit "bus" (or line) called chassis ground. In power line circuits, may also be "earth ground."

H: the symbol representing magnetic field intensity.

henry: the basic unit of inductance, abbreviated "H". One henry of inductance is that amount of inductance having one volt induced cemf when the rate of current change is one ampere per second.

Hertz (Hz): the unit of frequency expressing a frequency of one cycle per second.

high-pass filter: a filter that passes frequencies *above* a specified cutoff frequency with little attenuation. Frequencies below the cutoff are greatly attenuated.

horizontal (X) amplifier: the scope amplifier circuitry that amplifies signals fed to the horizontal deflection circuitry.

I_m or I_{fs}: full-scale current value of the meter movement.

imaginary numbers: complex numbers; the imaginary part of a complex number that is a real number multiplied by the square root of minus one.

impedance: the total opposition an ac circuit displays to current at a given frequency; results from both resistance and reactance. Symbol representing impedance is Z and unit of measure is the ohm.

impedance ratio: the relationship of impedances for the transformer windings; directly related to the *square* of the turns ratio; $Z_P/Z_S = N_P{}^2/N_S{}^2$. Reflected impedance is the impedance reflected back to the primary from the secondary. This value depends on the square of the turns ratio and value of the load impedance connected across the secondary.

inductance (self-inductance): the property of a circuit to oppose a change in current flow.

induction: the ability to induce voltage in a conductor; ability to induce magnetic properties from one object to another by magnetically linking the objects.

inductive reactance: the opposition an inductance gives to ac or pulsating current.

in-line: typically means in series with the source.

instantaneous value: the value of a sinusoidal quantity at a specific moment, expressed with lower case letters, such as e or v for instantaneous voltage and i for instantaneous current.

insulator: a material in which the atoms' outermost ring electrons are tightly bound and not free to move from atom to atom, as in a conductor. Typically, these materials have close to the 8 outer shell electrons required for chemical/electrical stability.

intensity control: a scope control that adjusts the number of electrons striking the screen per unit time; thus controlling the brightness of the screen display.

ion: atoms that have gained or lost electrons and are no longer electrically balanced, or neutral atoms. Atoms that lose electrons become positive ions. Atoms that gain electrons become negative ions.

I_s: current through the shunt resistance.

isolation transformer: a 1:1 turns ratio transformer providing electrical isolation between circuits connected to its primary and secondary windings.

joule: the unit of energy. One joule of energy is the amount of energy moving one coulomb of charge between two points with a potential difference of one volt of emf; the work performed by a force of one newton acting through a distance of one meter equals one joule of energy. Also, 3.6×10^6 joules = 1 kilowatt hour.

Kirchhoff's current law: the value of current entering a point must equal the value of current leaving that same point in the circuit.

Kirchhoff's voltage law: the arithmetic sum of voltage drops around a closed loop equals V applied; the algebraic sum of voltages around the entire closed loop, including the source, equals zero.

L, L_T: L is the abbreviation representing inductance or inductor. L_T is the abbreviation for total inductance. L_1, L_2, and so on represent specific individual inductances.

L/C ratio: the ratio of inductance to capacitance in an LC resonant circuit. Generally, the higher this ratio is, the higher is the Q of the circuit and the more selective its response.

leading in phase: in ac circuits the electrical quantity that first reaches its maximum positive point is specified as leading the electrical quantity that reaches that point later.

left-hand rules: related to magnetism; when the thumb and fingers of the left hand determine the direction of magnetic fields of current-carrying conductors to find the polarity of an electromagnet's poles.

Lenz's law: the law states the direction of an induced voltage or current opposes the change causing it.

linear network: a circuit whose electrical behavior does not change with different voltage or current values.

lines: in magnetic circuits, lines represent flux.

lines of force (flux): imaginary lines representing direction and location of the magnetic influence of a magnetic field relative to a magnet.

Lissajous pattern(s): scope patterns on the CRT screen when sine-wave signals are simultaneously applied to the horizontal and vertical deflection systems. These patterns are useful to determine phase and frequency relationships between the two signals.

load: the amount of current or power drain required from the source by a component, device, or circuit.

loaded voltage divider: a network of resistors designed to create various voltage levels of output from one source voltage; where parallel loads, which demand current, are connected.

loading effect: changing the electrical parameters of an existing circuit when a load component, device, or circuit is connected. With voltmeters, the effect of the meter circuit's resistance when in parallel with the portion of the tested circuit; thus, causing a change in the circuit's operation and some change in the electrical quantity values throughout the circuit.

local action: in carbon-zinc cells, the formation of multiple little local cells created by zinc impurities that are suspended in electrolyte near the zinc electrode. Local action causes the cell to be eaten away and degenerates the cell's chemical capability, even when the cell is not in use.

long time-constant circuit: when the circuit conditions are such that the t/RC ratio = 0.1 (or less). This means the RC time required to charge or discharge the capacitor is long compared with the time allowed during each alternation or cycle.

loop: in electrical networks, a complete electrical circuit or current path.

low-pass filter: a filter that permits all frequencies *below* a specified cutoff frequency to pass with little attenuation. Frequencies above the cutoff frequency are attenuated greatly.

L/R time constant: the time required for an exponentially changing quantity to change by 63.2% of the total value of change to be achieved. For a circuit containing inductance, it is the time (in seconds) required for the current to acquire a level that is 63.2% of its final value. One time constant (in seconds) in a series RL or inductor circuit equals L/R, where L is in henrys and R is in ohms.

magnetic field: the field of influence or area where magnetic effects are observed surrounding magnets; a region of space where magnetic effects are observed.

magnetic field intensity (H): the amount of magnetomotive force per unit length. In the MKSA system, it is the ampere-turns per meter (AT/meter).

magnetic polarity: the relative direction of a magnet's flux with respect to its ends or poles; defined as the North-seeking pole and the South-seeking pole of a magnet.

magnetism: the property causing forces of attraction or repulsion in certain materials; property causing voltage to be induced in nearby conductors when there is relative motion between the magnetic object and the conductor.

magnetomotive force (mmf or F): the force causing magnetic lines of force through a medium, thus establishing a magnetic field; developed by current through a coil, where the amount of mmf relates to the amount of current and the number of turns on the coil ($F = AT$).

magnitude: the comparative size or amount of one quantity with respect to another quantity of the same type.

matter: anything that has weight and occupies space, whether it is a solid, liquid, or gas.

maximum power transfer theorem: a theorem that states when the resistance (or impedance) of the load is properly matched to the internal resistance of the power source, maximum power transfer can take place between the source and the load.

Maxwell (Mx): one line of force, or one flux line.

mesh: sometimes called loops. In electrical networks, a set of branches that form a complete electrical path. If any branch is omitted, the remainder of the circuit does not form a complete path.

meter shunt (current meter): the parallel resistance used with current meter movements to allow them to measure currents higher than their full-scale current rating; a percentage of the total current being measured is shunted, or bypassed around the meter by the meter shunt.

molecule: the smallest particle of a compound that resembles the compound substance itself.

moving-coil (d'Arsonval) movement: a type of meter movement. A coil is suspended within a permanent magnetic field and caused to move when current passes through it. Movement is due to the interaction of its magnetic field with the permanent magnet's field.

μ_r: the symbol for relative permeability.

multimeter: an instrument designed to measure several types of electrical quantities, for example, current, voltage, and resistance; commonly found in the form of the VOM and digital multimeter (DMM).

multiple-source circuit: a circuit containing more than one voltage or current source.

multiplier resistor(s) (voltmeter): the current-limiting series resistance(s) used with current meter movements, enabling them to be used as voltmeters that measure voltages much greater than the basic movement's voltage drop at full-scale current. The value of the multiplier resistance is that its resistance (plus the meter's resistance) limits current to the full-scale value at the desired voltage-range level.

mutual inductance (M or L_M): the property that enables a current change in one conductor or coil to induce a voltage in another conductor or coil, and vice versa. This induced voltage results from magnetic flux of one conductor or coil linking the other conductor or coil. The symbol for mutual inductance is M or L_M and the unit of measure is the henry.

network: a combination of electrically connected components.

neutron: a particle in the nucleus of the atom that displays no charge. Its mass is approximately equal to the proton.

node: a current junction or branching point in a circuit.

Norton's theorem: a theorem regarding circuit networks stating that the network can be replaced (and/or analyzed) by using an equivalent circuit consisting of a single "constant-current" source (called I_N) and a single shunt resistance (called R_N).

ohm (Ω): the basic unit of resistance; the amount of electrical resistance limiting the current to one ampere with one volt applied.

Ohm's law: a mathematical statement describing the relationships among current, voltage, and resistance in electrical circuits. Common equations are $V = I \times R$, $I = V/R$, and $R = V/I$.

ohms-per-volt rating: a voltmeter rating indicating how many ohms of resistance are needed to limit the current through the meter to its full-scale value when one volt is applied to the meter circuit.

open circuit: any break in the current path that is undesired, such as a broken wire or component, or designed, such as open switch contacts in a lighting circuit.

oscilloscope: a device that visually displays signal waveforms so time and amplitude parameters are easily determined.

parallel branch: a single current path within a circuit having two or more current paths where each of the paths are connected to the same voltage points; a single current path within a parallel circuit.

parallel circuit: a circuit with two or more paths for current flow where all components are connected between the same voltage points.

peak-to-peak value (p-p): the difference between the positive peak value and the negative peak value of a periodic waveform; computed for the sine wave as 2.828 times effective value.

peak value (V_p or maximum value, V_{max}): the maximum positive, negative value, or amplitude that a sinusoidal quantity reaches, sometimes referred to as peak value. It is calculated as 1.414 times effective value.

period (T): the time required for one complete cycle. In formulas, it is expressed as T, where $T = 1/f$.

permeability (mu or μ): the ability of a material to pass, conduct, or concentrate magnetic flux.

phase angle (θ): the difference in time or angular degrees between ac electrical quantities (or periodic functions) when the two quantities reach a certain point in their periodic function. This difference is expressed by time difference (the fraction of a period involved), or by angular difference (expressed as degrees or radians). (NOTE: A radian = 57.3°. $2\pi \times$ a radian = 360°.) An example is the difference in phase between circuit voltage and circuit current in ac circuits that contain reactances, such as inductive reactance.

phasor: a quantity that expresses position relative to time.

pointer: the element in a meter that provides the indicating function of how much movement has taken place by the movable coil.

polar form notation: expressing circuit parameters with respect to polar coordinates; expressing a point with respect to distance and angle (direction) from a fixed reference point on the polar axis. For example, $25 \angle 45°$.

polarity: in an electrical circuit, a means of designating differences in the electrical charge condition of two points, or a means of designating direction of current flow. In a magnetic system, a means of designating the magnetic differences between two points or locations.

polarization: relating to cells or batteries, the undesired formation of hydrogen bubbles around the positive electrode that degenerates the cell's operation.

pole pieces: special meter pieces used to help linearize the magnetic field in a moving-coil movement.

position control(s) (vertical and horizontal): scope controls that move the display vertically or horizontally; thus centering the display on the screen.

potential: in electrical circuits, the potential, or ability, to move electrons. In electromotive force, the difference in charge levels at two points. This potential difference is measured in volts.

power: the rate of doing work. In mechanical terms, a horsepower equals work done at a rate of 550 foot pounds per second. In terms of electrical power, it is electrical work at a rate of one joule per second.

power factor (p.f.): the ratio of true power to apparent power. It is equal to the cosine of the phase angle (θ).

powers of ten: working with small or large numbers can be simplified by converting the unit to a more manageable number times a power of ten. An example is 0.000006 ampere expressed as 6×10^{-6} amperes. Some of the frequently used powers of ten are 10^{-12} for pico (p) units, 10^{-6} for micro (μ) units, 10^{-3} for milli units, 10^3 for kilo units, and 10^6 for mega units.

product-over-the-sum method: a method of finding total resistance in parallel circuits by using two resistances at a time in the formula $R_T = R_1 \times R_2/ R_1 + R_2$.

proton: the positively charged particle in an atom located in the nucleus of the atom. Its weight is approximately 1,836 times that of the electron, and approximately the same weight as the neutron.

pulsating dc: a varying voltage that is one polarity, as opposed to reversing polarity as ac does.

Pythagorean theorem: a theorem expressing the mathematical relationship of the sides of a right triangle, where the square of the hypotenuse is equal to the sum of the squares of the other two sides. That is, $c^2 = a^2 + b^2$. Equation used in electronics is $c = \sqrt{a^2 + b^2}$.

Q (figure of merit): a number indicating the ratio of the amount of stored energy in the inductor magnetic field to the amount of energy dissipated. Found by using the ratio of the inductance's inductive reactance (X_L) to the resistance of the inductor (R). That is, $Q = X_L/R$.

Q factor: the relative quality or figure of merit of a given component or circuit; often related to the ratio of X_L/R of the inductor in the resonant circuit. Other circuit losses can also enter into the effective Q of the circuit, such as the loading that a load coupled to the resonant circuit causes and the effect of R introduced in the circuit. The less the losses are, the higher the Q is. The sharpness of the resonant circuit response curve is directly related to the Q of the circuit.

quadrants: segments of the coordinate system representing 90° angles, starting at a 0° reference plane.

range select (switch): the switch used with meter movements that connects or disconnects various shunt resistances values (for current or ohmmeter measurements), or multiplier resistances for the voltage measurement mode.

RC time constant: the time, in seconds, represented by the product of R in ohms and C in farads; time required for a capacitor to charge or discharge 63.2% of the change in voltage level applied. Five (5) RC time constants are needed for a capacitor to completely charge or discharge to a changed voltage level.

reactance: the opposition a capacitor or inductor gives to alternating current or pulsating direct current.

reactive power (Q): the vertical component of the ac power triangle that equals $VI \sin \theta$; expressed as VAR (voltampere-reactive).

real numbers: in math, any rational or irrational number.

rectangular form notation: expressing a set of parameters with the complex number system. For example, $R \pm jX$ indicates a combination of resistance and reactance. Thus, $5 + j10$ means 5 ohms of resistance in series with 10 ohms of inductive reactance.

relative permeability (μ_r): comparing a material's permeability with air or a vacuum; relative permeability is not a constant, since it varies for a given material depending on the degree of magnetic field intensity present; relative permeability
$$(\mu_r) = \frac{\text{flux density with given core}}{\text{flux density with vacuum core}}.$$

reluctance (R): the opposition of a given path to the flow of magnetic lines of force through it.

residual magnetism: the magnetism remaining in the core material of an electromagnet after the magnetizing force is removed.

resistance (R): in an electrical circuit, the opposition to electron movement or current flow. Its value is affected by many factors, including the dimensions and material of the conductors, the physical makeup and types of the components in the circuit, and temperature.

resistor: an electrical component that provides resistance in ohms and that controls or limits current or distributes or divides voltage.

resistor color code: a system to display the ohmic value and tolerance of a resistor by means of colored stripes or bands around the body of the resistor.

resonance: a special circuit condition existing when opposite types of reactances in a circuit are equal and have cancelling effects (i.e., $X_L = X_C$).

ripple: the ac component in a pulsating dc signal, such as the output of a power supply.

R_M: meter movement resistance value.

R_{Mult}: resistance value of the resistor used that enables a basic current meter to measure voltage by its current-limiting effect.

saturation: in magnetism, the condition when increasing the magnetizing force does not increase the flux density of a magnetic material.

scalar: a quantity expressing only quantity, such as a "real number."

scale: the face of a meter containing calibration marks that interpret the pointer's position in appropriate units of measure.

schematic diagrams: graphic illustrations of circuits or subcircuits and/or systems or subsystems. Schematics are more detailed than block diagrams and give specific information regarding how components are connected. Schematics show the types, values, and ratings of component parts, and in many cases show the electrical values of voltage, current, and so forth at various points throughout the circuit.

selectivity: referring to resonant circuits, selectivity indicates the sharpness of the response curve. A narrow and more pointed response curve results in a higher degree of frequency selectiveness or selectivity.

sensitivity rating: a means of rating meters, or meter movements to indicate how much current is required to cause full-scale deflection; how much voltage is dropped by the meter movement at full-scale current value. Also, voltmeters are rated by the ohms-per-volt required to limit current to full-scale value. (This is inversely related to the current sensitivity rating. That is, the reciprocal of the full-scale current value yields the ohms-per-volt rating.)

series circuit: a circuit in which components are connected in tandem or in a string providing only one path for current flow.

series-parallel circuit: a circuit comprised of a combination of both series-connected and parallel-connected components and/or subcircuits.

series-type ohmmeter circuit: the ohmmeter circuit where the meter, its zero-adjust resistance, and its internal voltage source are in series. When the test leads are connected across an unknown resistance to be measured, that resistance is placed in series with the meter circuit.

short circuit: generally, an undesired very low resistance path across two points in a circuit. It may be across one component, several components, or across the entire circuit. There are designed shorts in circuits, such as purposeful "jumpers" or closed switch contacts.

short time-constant circuit: describes a circuit where the frequency is such that the capacitor has at least ten times one RC time constant to charge or discharge. A short time-constant circuit is sometimes described as one where: $t/RC = 10$ (or greater).

siemens: the unit used to measure or quantify conductance; the reciprocal of the ohm ($G = 1/R$).

sine: a trigonometric function of an angle of a right triangle; the ratio of the opposite side to the hypotenuse.

sine function: a trigonometric function used in ac circuit analysis; expresses the relationship of a specified angle of a right triangle to the side opposite that angle and the hypotenuse where sin θ = opposite/hypotenuse.

sine wave: a wave or waveform expressed in terms of the sine function relative to time.

sinusoidal quantity: a quantity that varies as a function of the sine or cosine of an angle.

skirts: term typically used to describe the portions of a resonance response curve that descend down from the peak response point. The steepness of the skirts reflects the selectivity of the circuit.

source: in electrical circuits, the source is the device supplying the circuit applied voltage, or the electromotive force causing current to flow through the circuit. Sources range from simple flashlight cells to complex electronic circuits.

springs: the elements in a moving-coil meter movement that conduct current to the movable coil and provide a mechanical force against which the moving coil must operate.

steady-state condition: the condition where circuit values and conditions are stable or constant; opposite of transient conditions, such as switch-on or switch-off conditions, when values are changing from one state to another.

stops: pegs that limit the left and right movement of the meter's pointer and related moving elements.

subcircuit: a part or portion of a larger circuit.

superposition theorem: superimposing one set of values on another to create a set of values different from either set of the superimposed values alone; the algebraic sum of the superimposed sets of values.

susceptance (B): the relative ease that ac current passes through a pure reactance; the reciprocal of reactance; the imaginary component of admittance.

sweep circuits: circuits in an oscilloscope that cause the horizontal trace and the rapid retrace of the electron beam across the CRT.

sweep frequency control(s): scope controls that set the number of times the horizontal trace occurs per unit time. Frequently, this control is calibrated in time/div. to measure the time per division traced on the display. (NOTE: The formula $f = 1/T$ is then used to determine frequency.)

switches: devices for making or breaking continuity in a circuit.

synchronization: the process of bringing two entities into time coherence. In a scope, the circuitry and process that cause the beginning of the horizontal trace to be synchronized with the vertical input signal.

synchronization control(s): scope controls that control the timing of the horizontal sweep in synchronization with the signal(s) being tested.

tangent function: a trigonometric function used in ac circuit analysis; expresses the relationship of a specified angle of a right triangle to the side opposite and side adjacent to that angle where $\tan \theta = $ opposite/adjacent.

TC: the abbreviation often used to represent time constant.

tesla: a unit of flux density representing one weber per square meter in the MKSA system of units.

Thevenin's theorem: a theorem used to simplify analysis of circuit networks because the network can be replaced by a simplified equivalent circuit consisting of a single voltage source (called V_{TH}) and a single series resistance (called R_{TH}).

thin-film process: a process where thin layers of conductive or nonconductive materials are used to form integrated circuits.

transformers: devices that transfer energy from one circuit to another through electromagnetic (mutual) induction.

true (or real) power (P): the power expended in the ac circuit resistive parts that equals $VI \cos \theta$, or apparent power $\times \cos \theta$. The unit of measure is watts.

turns ratio: the number of turns on the separate windings of a transformer; expressed as the number of turns on the secondary versus the number of turns on the primary (N_S/N_P); frequently stated as the ratio of primary-to-secondary turns.

universal (Ayrton or ring) shunt: a method of connecting and selectively tapping resistances so various series and shunt resistance values are connected to the meter circuit, thus affording multiple current ranges. Circuitry assures that the meter is never without some shunt resistance value; thus, providing some measure of meter protection.

valence electrons: the electrons in the outer shell of an atom

vector: a quantity that illustrates both magnitude and direction. A vector's direction represents position relative to space.

vertical (Y) amplifier: the scope amplifier circuitry that amplifies signals fed to the vertical input jack(s).

vertical volts/div. control(s): scope controls that adjust the amplification or attenuation of the input signal fed to the vertical deflection system. Generally, this control has positions that provide calibrated vertical deflections in ranges of millivolts per division or volts per division; thus enabling measurement of waveform voltages.

volt (V): the basic unit of potential difference or electromotive force (emf); the amount of emf causing a current flow of one ampere through a resistance of one ohm.

voltage: the amount of emf available to move a current.

voltage divider action: the dividing of a specific source voltage into various levels of voltages by means of series resistors. The distribution of voltages at various points in the circuit is dependent on what proportion each resistor is of the total resistance in the circuit.

voltage rate of change (dV/dt): the amount of voltage level change per unit time. For example, a rate of change of one volt per second, 10 volts per μ second, and so forth.

voltage ratio: in an ideal transformer with 100% coupling ($k = 1$), the voltage ratio is the same as the turns ratio, since the voltage induced in each turn of the secondary and the primary windings is the same for a given rate of flux change; $V_S/V_P = N_S/N_P$.

voltampere-reactive (VAR): the unit of reactive power in contrast to real power in watts; 1 VAR is equal to one reactive volt-ampere.

watt: the basic unit of electrical power. The basic formula for electrical power is P (in watts) equals V (in volts) times I (in amperes).

weber: the unit of magnetic flux indicating 10^8 maxwells (flux lines).

Wheatstone bridge circuit: a special application of the bridge circuit commonly used to measure unknown resistances. It utilizes a sensitive current or voltage meter to determine when the bridge is balanced.

work: the expenditure of energy, where work = force $\times$ distance. The basic unit of work is the foot poundal. NOTE: At sea level, 32.16 foot poundals = one foot pound.

wye network: also, sometimes called a "T" network. A wye network consists of three branches whose connections all join at one end of each branch, and are separate electrical points at the other end of each branch. This electrical connection system can be schematically represented in the shape of a Y or a T, hence the names Wye and Tee are used in describing this electrical network.

wye-delta conversion: the mathematical process of converting network component values, used in a wye network, to appropriate values to be used in a delta network, and yet present the same impedance and load conditions to a source feeding such a network.

zero-adjust control: the adjustable series resistor in an ohmmeter used to regulate current through the meter movement so that with the test probes across zero ohms (or touching each other), there is full-scale current through the meter, and the pointer is located above the zero-ohms mark on the scale.

INDEX

Abbreviations, 37
Absolute permeability, 354
ac, 90, 92
    distinguished from dc, 423–26
    in-phase or out-of-phase
        quantities, 437–38
    purely resistive circuit, 442–43
Accelerating anodes, 460
ac voltage generation, 426–31
ac voltage measurement, 469–70
Admittance, 742
Air-core coils, 586, 602
Alkaline cell, 323
Alternation, 90, 92, 433
Amalgamation, 315
Ammeter, 56, 58
Ampere, 24, 37
Ampere-hour rating, 318–19
Amplitude, 522
Analog meters, 55
Angle of rotation, 430
Angular motion, 424–26
Angular velocity, 526
Apparent power, 724–25, 727–28
Applied voltage, 115–16
Assumed mesh current, 287
Atoms, 7, 9
    atomic number and weight,
        11–12
    ions, 14
    shells of, 13
    structural model of, 10
    valence electrons, 13–14
Attenuator test probes, 473
Audio frequency applications, 570
Audio transformers, 601
Autotransformer, 600
Average value, 440–41

$B, H,$ and permeability, 351,
    353–55
Back-emf, 492
Backoff ohmmeter scale, 400
Bandpass, 773, 776
Bandpass filter, 785–86
Bandstop filter, 786
Bandwidth, 771–72, 776
    and Q, 772
Battery, 135, 312, 325
    capacity ratings of, 318–19
    chemical action in, 312–14
    deep-cycle, 318

lead-acid, 317
    construction and operation of,
        320–21
nickel-cadmium, 318
nickel-metal-hydride, 318
nonrechargeable, 323–24
rechargeable, 317–18, 323
    charging methods, 321–23
    data on recharging, 320
    smart chargers for, 320
shelf-life of, 325
B-H curve, 356
Bilateral resistance, 257
Bleeder current, 231
Bleeder resistor, 231
Block diagrams, 61
Block level troubleshooting, 150
Breakdown voltage, 623
Bridge circuits, 238
    analysis of using conversion,
        303–304
    Wheatstone, 238–43

Capacitance
    factors affecting value of, 621–23
    farad as unit of, 619–20
    formulas of, 624–25
    stray or distributed, 644, 778
    total, in series and parallel,
        626–29, 712–13
Capacitive reactance
    concept of, 659
    current and voltage distribution
        through capacitive circuits,
        667
    definition of, 656
    in parallel, 664–66
    in series, 663–64
    V and I relationships
        in a purely capacitive ac
            circuit, 657-58
        in a purely resistive ac
            circuit, 657
    X_c
        calculation of, 660–63
        relationship to capacitance
            value, 659
        relationship to frequency of
            ac, 659–60
Capacitor, 614–15, 644
    ceramic, 637–38
    charging, 616–17

color coding, 641
discharging, 618
electrolytic, 638–39
energy stored in electrostatic
    field of, 621
fixed and variable, 636–37
mica, 637
paper and plastic, 637
parallel
    capacitance of, 630
    charge distribution related to,
        630–31
ratings for, 640
series capacitors voltage
    formula, 629
tantalum, 640
troubleshooting, 641–43
Carbon resistors, 49
Carbon-zinc cell, 314–17
Cathode-ray tube, 458–59
Cell, 135, 312, 325
    capacity ratings of, 318–19
    chemical action in, 312–14
    dry cell, 314–17
    nonrechargeable, 323–24
    rechargeable cell, 317–18, 323
        data on recharging, 320
        smart chargers, 320
    shelf-life of, 325
Ceramic capacitors, 637–38
Chassis ground, 137–38, 230
Circuit phase angle, 557
Clamp-on current probe, 405
Clamp-on measurement devices, 405
Coefficient of coupling, 585–86
Color coding
    capacitors, 641
    resistors, 50–51
Complex numbers, 730
    converting from polar to
        rectangular form, 735–38
    converting from rectangular to
        polar form, 732–35
Component equivalent circuits, 778
Component level troubleshooting,
    151
Compounds, 8
Conductance, 37
Conductors, 16
    functions of, 41
    resistance of, 42
        calculating, 44–45

resistivity of, 42–43
types of, 41–42
wires and their resistances,
    45–47
Constant current source, 273
Control electrode, 460
Conventional current, 89
Coordinate system, 425
Copper loss, 598–99
Core, 380
Core loss, 599
Core materials, 355–59
Cosine, 429, 552
Coulomb, 20
Counterbalance, 381
Counter-emf, 490, 492
Coupled load
    effect on tuned circuit, 769
Coupling circuit, 701
Current
    ac and dc, 90, 92
    ampere as unit of, 24
    analogy of current flow, 23–24
    definition of, 23
    formula relating current to
        coulombs of charge and
        time, 24–25
    measuring with a meter, 56, 58
    measuring with an oscilloscope,
        467–74
    in parallel circuit, 166–69
    at resonance, 764
    in series-parallel circuit, 213–16
Current-carrying conductor
    direction of field around, 340
    field around multiple-turn coil,
        342–43
    left-hand rule for, 340–41
    parallel, force between, 342
Current dividers
    formula for parallel circuit with
        any number of branches,
        189–90
    formula for two-branch parallel
        circuit, 190–92
Current leading voltage, 679
Current meters, 383–85, 390
    troubleshooting, 406, 408
Current meter shunts, 384
    multiple-range meter circuit,
        389
    resistance value calculations,
        385–89
Current ratio, 593–95
Cutoff frequency, 785
Cycle, 427

Damping, 382
D'Arsonval movement, 379–82
    advantages and disadvantages
        of, 382
    parts of, 380–81
    principle of operation, 381–82
dc, 90, 92
    distinguished from ac, 423–26
dc measuring instruments, 378,
    410–11
    current meters, 383–85, 390
        shunts, 385–89
    d'Arsonval movement, 379–82
    meter movement requirements,
        379
    multimeters, 401
        digital multimeter (DMM),
            402, 404
        related devices, 404–405
        VOM, 402
    ohmmeters, 398–401
    troubleshooting, 406–10
    voltmeters, 390–98
dc voltage measurement, 470
Deep-cycle battery, 318
Deflection plates, 460
Delta configuration, 297–98
Delta/wye conversion, 300–302
Depolarizer, 315
Diagrams
    block, 61
    schematic, 61, 64–65
    symbols used in, 62–63
Diamagnetic materials, 370
Dielectric, 614
    constants, 622–23
    material, 622
    strength, 623
    thickness, 622
Differentiated output, 697
Digital meters, 55
Digital multimeter (DMM), 402,
    404
Digital storage oscilloscope, 456,
    458
Dry cell, 314
    construction of, 315
    local action of, 315
    operating characteristics of,
        316–17
    polarization in, 315
    shelf life of, 315–16
Duty cycle, 446–47

Eddy current losses, 599
Edison cell, 323

Effective value, 440
Electrical circuits, 16, 29
    closed, 26
    open, 26–27
Electrical energy, 92
    formula for measuring
        consumption of, 93
    joule as unit of, 92, 499
Electrical potential
    definition of, 21
    means of producing, 22
    volt as unit of, 21–22
Electrical power, 92
Electrical shock, 67
Electrical system, 17
Electrodes, 312
Electrolyte, 312
Electrolytic capacitors, 638–39
    dry, 639
    replacing, 643
    wet, 639
Electromagnetic fields, 461
Electromagnetic induction, 359–67,
    584
Electromagnetism, 345–46. See also
    Magnetism
    current-carrying conductors
        direction of field around, 340
        left-hand rule for, 340–41
        parallel, force between, 342
    electromagnets
        factors influencing field
            strength of, 344
        left-hand rule for determining
            polarity of, 343
    field around a multiple-turn coil,
        342–43
    induction
        Faraday's law, 363
        generator action, 361–63
        generator effect in a motor,
            366–67
        Lenz's law, 364–65
        motor action, 359–60
        motor effect in a generator,
            365–66
Electromotive force (emf), 21, 490
Electron flow approach, 88–89
Electrons, 10–11
    energies that change balance of,
        15–16
    valence, 13–14
Electrostatic fields, 461, 615–16
Elements, 7
Energy, 92
Engineer, 5

Epsilon, 504, 634–35
    graph based on raised to
        negative powers, 505
    relationship of curves to circuit
        parameters, 505–506
    relationship of negative powers
        for and L/R time constant,
        506–508
Equivalent circuit resistance, 170

Farad, 619–20
Faraday's law, 363, 491–92
Ferrites, 370, 602
Fields of force, 21
Filtering, 756
Filters, 778
    frequency range of, 779
    nonresonant
        high-pass, 782–84
        low-pass, 780–81
    resonant
        bandpass, 785–86
        bandstop, 786
Flux density, 336, 353–54
Flux lines, 335–37
Flux linkage, 585–86
Focusing anode, 460
Foot pound, 92
Frequency, 435
    cutoff, 785
    measuring with oscilloscope,
        475–76
    range in filters, 779
    resonant, 757
    X_L, X_C and, 756–57
Full scale current, 384

Gain control, 464
Gas, 7
Generator
    motor effect in, 365–66
Generator action, 361–63
Ground reference, 230

Hall effect, 369
Harmonics, 445–46
Henry, 492, 584
Hertz, 435
High-pass filter, 782–84
High-Q circuit, 767–69
High-voltage probe, 405
Horizontal axis, 730
Horizontal sweep frequency
    controls, 462–63
Hot soldering irons, 68
Hysteresis loop, 356–59

Hysteresis loss, 599

I_M, 384–88
Imaginary numbers, 730
Impedance, 560
    in parallel RL circuit, 568
    at resonant frequency, 764
    in series RL circuit, 561–62
Impedance matching, 257
Impedance ratio, 595–98
In-circuit testing, 642
Inductance, 361–63
    characteristics of, 488–91
    definition of, 488
    epsilon for calculating circuit
        parameters, 504–509
    factors that determine, 493–95
    Henry as unit of, 492
    L/R time constant, 500–504
    in parallel circuit, 497–98,
        710–11
    in series circuit, 496, 710–11
    stray or distributed, 644, 778
Inductive coupling, 585
Inductive reactance, 520, 531–33
    concept of, 523–24
    in series and parallel, 529–31
    V and I relationships
        in a purely inductive ac
            circuit, 521–23
        in a purely resistive ac
            circuit, 521
    X_L
        formula for calculating,
            527–29
        and frequency of ac, 525–26
        and inductance value, 524–25
Inductors
    air-core, 510
    characteristics of, 488–91
    energy stored in magnetic field
        of, 498–99
    laminated iron-core, 510
    opposition to ac current in. See
        Inductive reactance
    powdered iron-core, 510
    Q rating of, 533–34
    self-resonance of, 778
    troubleshooting, 511
    variable ferrite-core, 510
    windings, 510
Industrial components and devices,
    348–50
Instantaneous values, 442–43
Instrument, 378
Insulator materials, 16

Integrated output, 697
International System (SI), 351
Ions, 14

Jewel bearings, 381
J operator, 730–31
Joule, 92, 499

Kilowatthours, 93
Kirchhoff's current law, 167–69
Kirchhoff's voltage law, 116–17

Laminations, 599
Law of electrical charges, 17–18
L circuit, 546
L/C ratio, 759
    and Q factor, 765
Lead-acid battery, 317
    construction and operation of,
        320–21
Leading voltage, 523
Leakage flux, 368
Left-hand rule
    for current-carrying conductors,
        340–41
    for determining polarity of
        electromagnets, 343
Lenz's law, 364–65, 492, 522
Linear network, 257
Lines of force, 335–37
Liquid, 7
Lissajous patterns, 474
Lithium cell, 324
Load, 230
Load current, 230
Loaded voltage dividers, 228–38
    three-element, 233–36
    two-element, 232–33
    variable capability of
        potentiometer, 236–37
Loading effects, 396–97, 411
Local action, 315
Long time-constant circuit, 697
Loop, 286
Loop/mesh analysis, 286–92
Low-pass filter, 780–81
Low-Q circuit, 768–69
L/R time constant, 500–504
    rapidly collapsing field and,
        509–10

Magnetic field, 335
Magnetic field intensity, 353
Magnetic linkage, 585
Magnetic polarity, 343
Magnetic shields, 368

Magnetic storage media, 347–48
Magnetism, 345–46. *See also*
  Electromagnetism
  applications of
    industrial components and
      devices, 348–50
    magnetic storage media,
      347–48
  classifications of materials, 370
  core material considerations,
    355
    B-H curve, 356
    hysteresis loop, 356–59
  Hall effect, 369
  history of, 335
  laws, rules, and terms of, 335–37
  lines of force, 337
  magnetic circuit, 351
  magnetic shields, 368
  permanent magnets, 338–39
  relationships among *B, H,* and
    permeability factors, 351,
    353–55
  shaping fields via pole pieces,
    368
  toroidal coil form, 367–68
  units, terms, symbols, and
    formulas, 351–52
Magnetomotive force, 353
Magnification factor, 765–66
Magnitude, 430
Matter
  chemical states of, 7–9
  composition of, 9
  definition of, 7
  physical states of, 7
Maximum power transfer theorem,
  257–62
  efficiency factor, 260–61
Mechanical work
  formula for, 92
Megahertz, 761
Mercury cell, 324–25
Mesh, 286
Meters, 54–60, 378
  basic movement requirements of,
    379
  current, 383–85, 390
    shunts, 385–89
Metric system, 38–40
Mica capacitors, 637
Microhenry, 761
Mixtures, 8
Molecules, 9
Motor
  generator effect in, 366–67

Motor action, 359–60
Moveable coil, 380
Moving coil movement, 379–82
Multimeters, 401
  digital multimeter, 402, 404
  probes and clamp-on devices,
    405
  VOM, 402
Multiple-range meter circuit, 389
Multiple-source circuit, 263–64
Multiplier resistor, 390–93
Murray loop, 242–43
Mutual inductance, 584
  between coils other than
    transformers
    parallel coils, 589–90
    relationship of *k* and *M,*
      590
    series coils, 588–89
  between windings, 586–88

Network analysis
  bridge circuit analysis using
    conversion, 303–304
  delta-wye conversion, 300–302
  loop/mesh analysis approach,
    286–92
  nodal analysis approach, 292–96
  wye-delta conversion, 298–300
Network theorems
  conversions between Norton and
    Thevenin, 277–79
  maximum power transfer,
    257–62
  Norton's, 273–77
  superposition, 263–66
  Thevenin's, 267–72
Neutrons, 10–11
Nickel-cadmium cell, 318, 325
Nickel-metal-hydride cell, 318
Nodal analysis, 292–96
Node, 292
Nonresonant filters, 780–84
Nonsinusoidal waveforms
  average value for, 447–48
  formation of, 444–45
  parameters for, 446
  pulse-width and duty cycle of,
    446–47
  types of, 444
Norton's theorem, 273–77
  conversions with Thevenin's
    theorem, 277–79

Ohm, 26, 37
Ohmmeter, 58, 398–401

troubleshooting with, 406,
    408–9, 511, 605, 642–43
Ohm's law, 74
  rearranging to find unknown
    quantity, 79–81
  relationship of current to
    resistance, 75–78
  relationship of current to
    voltage, 75
  solving problems using, 98–100
  versions of, 96–98
Ohms-per-volt rating, 394–96
1:1 turns-ratio isolation
    transformer, 601
Opens
  in capacitors, 641–42
  in inductors, 511
  in parallel circuit, 182–83
  in series circuit, 127–28
  in series-parallel circuit, 223–24
Oscilloscope, 477
  categories of, 456, 458
  cathode-ray tube, 458–59
  horizontal and vertical signals to
    view a waveform, 465–67
  horizontal section, 464
  horizontal sweep frequency
    control, 462–63
  intensity and focus controls, 460
  measuring frequency with,
    475–76
  measuring voltage and current
    with, 467–74
    ac voltage, 469–70
    attenuator test probes, 473
    calibration voltage approach,
      468–69
    dc voltage, 470
    volts-per-division control
      setting approach, 471–72
  for phase comparisons, 474–75
  position controls, 461–62
  synchronization controls, 464
  vertical section, 463
Out-of-circuit testing, 642
Out-of-phase circuit, 723

Paper capacitors, 637
Parallel capacitors
  capacitance of, 630
  charge distribution related to,
    630–31
Parallel circuit
  compared with series circuit,
    186–87
  current dividers

formula for any number of
    branches, 189–90
formula for two branches,
    190–92
current in, 166–69
definition of, 164
designing to specifications, 186,
    188
features of, 165
opens in, 182–83
power in, 180–82, 727
resistance in, calculating
    assumed voltage method,
        178–79
    conductance method, 173–74
    design formula for, 179–80
    Ohm's law method, 170–71
    product-over-the-sum
        method, 174–77
    reciprocal method, 171–73
shorts in, 184–85
sources in parallel, 189
total resistance in, 170–80
troubleshooting, 184, 186
voltage in, 166
Parallel coils, 589–90
Parallel RC circuit
    solving for impedance in, 692–93
    solving for I_T using Pythagorean
        theorem, 690–91
    solving for phase angle using
        trigonometry, 691
Parallel resonance
    bandwidth and bandpass in
        circuit, 776
    characteristics of, 767–68
    coupled load effect on tuned
        circuit, 769
    formulas for, 768–69
    maximum impedance (Z) for
        circuit, 775
    Q factor and rise of impedance,
        770
    selectivity, bandwidth, and
        bandpass, 770–73
    tuning range in circuit, 776
    tuning ratio and, 777
Parallel RLC circuit analysis
    opposite-direction phasors
        cancelling effect, 720–21
    simple LC circuit, 720
Parallel RL circuit analysis, 565–70
Paramagnetic materials, 370
Peak-to-peak value, 439
Period, 435
Periodic waves, 434

Permanent magnets, 338–39, 380
Permittivity, 622
Phase angle, 437, 522, 544, 568
    with circuit at resonance, 764
Phase measurement
    Lissajous pattern technique, 474
    overlaying technique for, 474–75
Phasors, 424, 438, 729
    opposite-direction, 720–21
Picofarad, 761
Plastic capacitors, 637
Pointer, 381, 399
Polar form notation, 732
    analysis applications, 741–45
    division in, 739–40
    multiplication in, 739
    raising to powers in, 740
    taking a root in, 740
Polarity, 18–20, 522
    of voltages, 89–90
Polarization, 315
Pole pieces, 380
    for shaping fields, 368
Position controls, 461-62
Potentiometer, 49-50, 231–32
Potentiometers
    varying voltage output, 236–37
    as volume controllers, 137
Powdered iron cores, 602
Power, 92
    in ac circuits
        apparent, 724–25
        I^2 method, 725
        power factor method, 725–27
        power triangle, 727–28
        purely reactive circuit, 724
        purely resistive circuit,
            723–24
        resistive/reactive
            combinations, 724
        true, 725
    formula relating to circuits, 94
    formula relating to electrical
        energy, 92
    in parallel circuit, 180–82
    in series circuit, 125–26
    in series-parallel circuit, 221–22
    variations on formula for,
        96–100
Power factor, 725–27
Power frequency applications, 570
Powers of ten
    application of metric prefixes,
        84–85
    basic rules for, 85–87
    and scientific notation, 81–84

Powerstats, 600
Power transformers, 599–601
Power triangle, 727–28
Precision resistors, 53–54
Primary cells and batteries, 323–24
Protons, 10–11
Pulse-width, 446
Purely capacitive ac circuit
    V and I relationships in, 657–58
Purely inductive ac circuit
    V and I relationships in, 521–23
Purely reactive ac circuits
    power in, 724
Purely resistive ac circuit, 442–43
    in parallel, 545–46
    power in, 723–24
    in series, 545
    V and I relationships in, 521,
        657
Pythagorean theorem, 550–51

Q factor
    determining in series resonant
        circuit, 775
    and L/C ratio, 765
    Q equals X_L/R or X_C/R, 765
    and resonant rise of impedance,
        770
    and resonant rise of voltage,
        765–66
    and response characteristic,
        764–65
Q rating, 533–34
Quadrants, 425

Radians of angular measure,
    428–31
Radio frequency, 570, 761
Range selector switch, 393–94
RC circuits in ac
    distinguished from RL circuits,
        695
    parallel RC analysis, 690–95
    series RC analysis, 679–89
    troubleshooting, 700
    waveshaping and timing
        applications, 696–97
R circuit, 545–46
RC "lag" network, 685–87
RC "lead" network, 685–86
RC time constant, 631–33
    exponential formulas to find
        voltage, 634–35
Real numbers, 730
Rectangular form notation, 730–32
    addition in, 738

analysis applications, 741-45
subtraction in, 738-39
Rectangular wave, 444
Reference node, 293
Reference points, 18-20
chassis ground as, 137-38, 230
and polarity and value of
voltage, 139-40
Relative permeability, 355
Reluctance, 354
Residual magnetism, 356
Resistance
of conductors, 42, 44-45
definition of, 26
measuring with a meter, 58
ohm as unit of, 26
in parallel circuits, 170-80, 709
in series circuits, 109-11, 709
Resistive power, 727
Resistivity, 42-44
Resistors
color coding of, 50-54
construction of, 49-50
forms of and symbols for, 48-49
labeled, 50, 54
types of, 47-48
uses of, 47
Resonance
definition of, 757
filter applications, 778-86
measurements related to circuits
at, 773-77
parallel circuit at, 767-73
series circuit at, 757-66
single components, 778
Resonance curves, 763-64
Resonant filters, 785-86
Resonant frequency, 757
formula for, 759-60
variations of, 761-63
measuring, 774-75
Resultant voltage, 134-35
RF transformers and coils, 602
R_L, 230
RLC circuit analysis
circuit review, 709-13
parallel RLC circuit, 720-22
power in ac circuits, 723-28
rectangular and polar vector
analysis, 729-45
series RLC circuit, 714-19
RL circuits in ac, 544
applications for, 570, 572
circuit analysis, 550-54
of parallel circuit, 565-70
of series circuit, 554-65

distinguished from RC circuits,
695
simple L circuit, 546
simple R circuit, 545-46
vectors for magnitude and
direction, 547-49
RL circuits in dc, 500-504
R_M, 384-88
R_{Mult}, 391
Rowland's law, 351-52
R_S, 386-88

Safety, 67-68
Sawtooth wave, 444, 465
Scale, 381
Schematic diagrams, 61, 64-65
Scientific calculator, 118-25, 545
converting from polar to
rectangular form using,
736-38
converting from rectangular to
polar form using, 734-35
using EE or EXP key, 121
using engineering mode, 121-23
using scientific notation mode,
124
Scientific notation, 81-84
Secondary cells and batteries,
317-18, 323
Selectivity, 765, 770-71
Self-inductance, 490, 492-93, 584
Self-resonance, 778
Semiconductors, 16
Sense dots, 587
Sensitivity, 383-84
Series capacitors
capacitance of, 627-28
voltage distribution related to,
628-29
voltage formula for, 629
Series circuit
compared with parallel circuit,
186-87
definition of, 109
opens in, 127-28
power in, 125-26
resistance in, 109-10
sample design problem, 133-34
shorts in, 128-30
total resistance formula, 110
Ohm's law approach, 111
troubleshooting, 131-32
block level, 150
component level, 151
SIMPLER sequence for,
144-49

voltage-divider action in, 136-41
voltage-dropping resistor for,
141-42
voltage in, 112
applied voltage, 115-16
individual component
voltages, 113-14
Kirchhoff's law of, 116-17
voltage divider rule, 114-15
voltage sources in, 134-36
Series coils, 588-89
Series-dropping resistor, 141-42
Series-parallel circuit
analyzing and identifying
portions of, 206-208
characteristics of, 205
current in, 213-16
definition of, 205
designing to specifications, 228
loaded voltage dividers, 228-38
three-element with multiple
loads, 233-36
two-element with load,
232-33
variation of with
potentiometer, 236-38
opens in, 223-24
power in, 221-22
shorts in, 225-26
total resistance in
using Ohm's law, 209
using "outside toward the
source" approach, 209-11
using reduce-and-redraw
approach, 211-12
troubleshooting, 224, 226
voltage in, 217-20
Wheatstone bridge circuit,
238-43
Series RC circuit, 679
positional information for
phasors, 680
solving for phase angle using
trigonometry, 682
solving for V_T in V-I vector
diagram, 681
solving for Z in the impedance
diagram, 681
solving magnitude values by
Pythagorean theorem,
680-81
solving the angle from
impedance diagram,
683-84
solving the angle from V-I vector
diagram, 683

Series resonance
    bandwidth and bandpass in
        circuit, 776
    characteristics of, 757–59
    maximum current (*I*) for circuit,
        775
    Q and magnification factor in
        circuit, 775
    Q factor and rise of voltage,
        764–66
    resonance curves, 763–64
    resonant frequency formula,
        759–63
    tuning range in circuit, 776–77
    tuning ratio and, 777
Series *RLC* circuit analysis
    current computed by Ohm's law,
        716
    impedance computed by
        Pythagorean theorem,
        715–16
    individual voltage computed by
        Ohm's law, 716
    phase angle computed from
        impedance and voltage
        data, 717–19
    simple circuit without
        resistance, 714
    total voltage computed by
        Pythagorean theorem, 716
Series *RL* circuit analysis, 563–65
    impedance analysis in, 560–62
    Pythagorean theorem for,
        555–56
    trigonometry for angle
        information in, 556–58
    trigonometry for voltages in,
        559–60
Series-type ohmmeter circuit,
    398–401
Shorts
    in capacitors, 641, 643
    in inductors, 511
    in parallel circuit, 184–85
    in series circuit, 128–30
    in series-parallel circuit, 225–26
Short time-constant circuit, 697
Shunt, 385–89
Siemens, 37
SIMPLER sequence, 144–49
Sine, 429, 552
Sine wave, 90, 429–33, 465, 522,
    659
    current and voltage values,
        439–41
    period and frequency in, 434–36

phase relationships, 437–38
    rate of change of voltage in,
        431–32
    waveform descriptors, 433–34
Skirts, 765
Smart chargers, 320
Sneak paths, 409–10
Solar cell, 324
Solid, 7
Spindle, 381
Springs, 381
Square wave, 444
    as calibrating voltage, 468–69
Static electricity
    coulomb as unit of charge, 20
    Coulomb's formula relating to
        electrical charges, 20
    definition of, 17
    and fields of force, 21
    and law of electrical charges,
        17–18
    polarity and reference points,
        18–20
Steady-state condition, 257
Stops, 381
Substitution box, 701
Superposition theorem, 263–66
Susceptance, 742
Sweep circuitry, 463
Sweep frequency controls, 463
Symbols
    basic electronic, 62–63
    for meters, 55
Synchronization controls, 464

Tangent, 429, 552
Tantalum capacitors, 640
Technician, 5
Technologist, 5
Thevenin's theorem, 267–72
    conversions with Norton's
        theorem, 277–79
Timing circuit, 701
Tolerance band colors, 51–52
Toroidal coil form, 367–68
Total capacitance, 626–29
Total circuit current, 166–67
Total circuit power, 126
Total current, 567
Total power, 180
Total resistance, 110–11
    assumed voltage method, 178–79
    conductance method, 173–74
    Ohm's law method, 170–71, 209
    "outside toward the source"
        method, 209–11

product-over-the-sum method,
    174–77
    reciprocal method, 171–73
    reduce-and-redraw method,
        211–12
Transformation ratio, 591
Transformers
    audio, 601
    definition of, 586
    energy losses of, 598–99
    power, 599–601
    primary windings, 587
    ratios of, 590–98
    rf and coils, 602
    secondary windings, 587
    troubleshooting, 602, 604–606
Transistor bias circuit, 137
Trigonometry, 552–54
    solving for voltages in series *RL*
        circuits using, 559–60
Troubleshooting
    capacitors, 641–43
    DC measuring instruments,
        406–10
    inductors, 511
    parallel circuit, 184, 186
    RC circuit, 700
    series circuit
        block level, 150
        component level, 151
        SIMPLER sequence for,
        144–49
    series-parallel circuit, 224, 226
    transformers, 602, 604–606
True power, 725, 727–28
Tuning range, 774, 776
Tuning ratio, 777
Turns ratio, 591

Units of measure, 37
Universal Time Constant Graph,
    502, 504

Valence electrons, 13–14
Variable controls, 463
Variacs, 600
Vectors, 424, 544, 547
    angular velocity of, 526
    plotting and measuring, 548–49
    rectangular and polar analysis
        of, 729–45
Vertical amplifier, 463
Vertical attenuator, 463
Vertical axis, 730
Vertical volts/div, 463
Volt, 21–22, 37

Voltage, 37
    leading, 523
    measuring with a meter, 55–56
    measuring with an oscilloscope,
        467–74
    in parallel circuit, 166
    polarity of, 89–90
    in series circuit, 112–17
    in series-parallel circuit, 217–21
Voltage-divider action, 136–41
Voltage divider rule, 114–15
Voltage dividers
    loaded, 228–38
Voltage ratio, 592–93
Voltaic cell, 312
Voltampere-reactive (VAR), 727
Voltmeter, 55–56, 398
    connecting to circuit being
        tested, 391–92
    loading effect, 396–97
    multiplier resistor purpose,
        390–91
    multiplier values
        approximating, 393
        calculating, 392–93

    range selector switch, 393–94
    ohms-per-volt rating, 394–96
    troubleshooting, 406–408
Volts-per-division control setting,
        471–72
VOM, 402

Watt, 92
Watthour, 93
Wattsecond, 93
Weber, 336
Wheatstone bridge circuit, 238–43
Windings
    open, 604
    primary, 587
    secondary, 587
    shorted, 604
Wire
    sizing by gage numbers, 45
    tables of, 45–47
Wire-wound resistors, 49
Work, 92
Wye configuration, 297–98
Wye/delta conversion, 298–300

X-axis, 730
X_C
    and capacitance value, 659
    combined with X_L in simple
        series circuit, 714
    and frequency of ac, 659–60
    methods for calculating, 660–63
    relationship with X_L and
        frequency, 756–57
X_L
    combined with X_C in simple
        series circuit, 714
    formula for calculating, 527–29
    relationship to frequency of ac,
        525–26
    relationship to inductance value,
        524–25
    relationship with X_C and
        frequency, 756–57

Y-axis, 730

Zero-adjust control, 399
Zinc-chloride cell, 324